Zhu · Zhong · Chen · Zhang Flow Around Bodies

Zhu You-lan   Zhong Xi-chang
Chen Bing-mu   Zhang Zuo-min

Difference Methods for Initial-Boundary-Value Problems and Flow Around Bodies

With 217 Figures and 40 Tables

Springer-Verlag
Berlin Heidelberg New York
London Paris Tokyo

Science Press
Beijing

Zhu You-lan Zhong Xi-chang
Chen Bing-mu Zhang Zuo-min

Computing Center
Chinese Academy of Sciences
Beijing
The People's Republic of China

Revised edition of the original Chinese edition published by Science Press Beijing 1980 as the fourth volume in the Series in Pure and Applied Mathematics.

Distribution rights throughout the world, excluding The People's Republic of China, granted to Springer-Verlag Berlin Heidelberg New York London Paris Tokyo

Mathematics Subject Classification (1980): 35-XX, 65-XX, 76-XX

ISBN 3-540-10887-4 Springer-Verlag Berlin Heidelberg New York
ISBN 0-387-10887-4 Springer-Verlag New York Berlin Heidelberg

Library of Congress Cataloging-in-Publication Data.
Ch'u pien chih wen t'i ch'a fen fang fa chi chiao liu. English. Difference methods for initial-boundary-value problems and flow around bodies / Zhu You-lan ... [et al.]. p. cm.
Translation of: Ch'u pien chih wen t'i ch'a fen fang fa chi chiao liu.
Half title: Initial-boundary-value problems and flow around bodies. "Revised edition of the original Chinese edition published by Science Press Beijing 1980 as the fourth volume in the Academia Sinica's series in pure and applied mathematics"–T.p.verso.
Bibliography: p. Includes index.
ISBN 0-387-10887-4 (U.S.)
1. Initial value problems. 2. Boundary value problems. 3. Difference equations. 4. Aerodynamics, Supersonic. I. Chu, Yu-lan. II. Title. III. Title: Initial-boundary-value problems and flow around bodies. IV. Series: Ch'un ts'ui shu hsüeh yü ying yung shu hsüeh chuan chu ; ti 4 hao. QA378.C46813 1988 515.3'5–dc19 88-20115 CIP

Printed in Germany

Typesetting: Science Press, Beijing, The People's Republic of China
Printing: Druckhaus Beltz, Hemsbach/Bergstr.; binding: J. Schäffer, Grünstadt
2141/3140-543210

Preface to the English Edition

Since the appearance of computers, numerical methods for the discontinuous solutions of quasi-linear hyperbolic systems of partial differential equations have been one of the most important research subjects in numerical analysis. The methods for this type of problems, generally speaking, can be classified into two categories. One category is the shock-capturing method, where the shocks are obtained with a uniform scheme, but are usually smeared. A typical representative of this category is the artificial viscosity method of von Neumann and Richtmyer. The other category consists of methods which give definite locations of the shocks and accurate flow fields, but are redundant to programming. The singularity-separating difference method developed by the authors belongs to the latter category. This method has a high accuracy and a solid theoretical basis, and has been successfully used for the solution of various problems. It was successful in the seventies, in computing supersonic flow around combined bodies and in the early eighties, in solving complicated unsteady flow, the Stefan problem, combustion problem and hyperbolic equations with a nonconvex equation of state. In this book we shall introduce our work of the seventies.

This monograph is divided into two parts. In Part I a numerical method for the initial-boundary-value problems of hyperbolic systems is discussed. In Part II the application of the method to computation of inviscid supersonic flow is described. The authors' work on the method of lines, which has been used to compute subsonic-transonic flow around blunt bodies and conical flow to provide the initial values of supersonic flow field computation, is also briefly described.

The prerequisites for reading this book are a knowledge of calculus and of numerical analysis and familiarity with the basic methods of mathematical physics. For the theoretical proofs, some functional analysis and matrix theory is required.

The authors wish to pay particular tribute to Professors Feng Kang and Zhuang Fenggan, who reviewed the Chinese manuscript of this book and gave a number of valuable suggestions during the course of this project. Sincere thanks are owed to Professor Huang Dun and Professor Translator-editor Sun Xianrou for reading the English manuscript. Gratitude is owed to Wang Ruquan, Li Yinfan, Wu Huamo,

Liu Xuezong, Fu Dexun and Cai Dayong for their comments and also to Bai Degin, Ma Dehui, Su Anjie and Wang Hui for their assistance in the difficult mathematical typing.

Also the authors are grateful to Wang Ruquan, Zhang Guanquan, Qin Bailiang and others who worked with us for a short period in the computation of flow around bodies.

Zhu Youlan
Zhong Xichang
Chen Bingmu
Zhang Zuomin

Beijing
June, 1987.

CONTENTS

Chapter 4 Inviscid Steady Flow

Chapter 5 Calculation of Supersonic Flow around Blunt Bodies

Chapter 6 Calculation of Supersonic Conical Flow

Chapter 7 Solution of Supersonic Regions of Flow around Combined Bodies

PART I

NUMERICAL METHODS

Chapter 1

Numerical Methods for Initial–Boundary–Value Problems for First Order Quasilinear Hyperbolic Systems in Two Independent Variables

Introduction

When discussing numerical methods for hyperbolic systems, it is usual to construct difference schemes and do theoretical analysis only for pure initial–value problems. However, most of the problems which exist in practice are initial–boundary–value problems. When applying the results from the pure initial–value problems (PIVP) to the initial–boundary–value problems (IBVP), difficulties are encountered since we usually do not know how to calculate the bounday points and how to ascertain whether an algorithm for boundary points is reasonable.

There have been several works written on initial–boundary–value problems[1–14], but they have been imperfect. Therefore, further development of numerical methods for initial–boundary–value problems is urgently needed. In this chapter, we shall discuss this problem thoroughly and systematically in the case of two independent variables. That is, we shall carefully describe a difference method for initial–boundary–value problems of the first order quasi–linear hyperbolic systems in two independent variables. Firstly, a way of constructing schemes for initial–boundary–value problems is given and several schemes are presented. Then, the stability of several classes of schemes for initial–boundary–value problems with variables coefficients is discussed. Finally, a method of solving difference equations——a block–double–sweep method for "segmental", incomplete linear algebraic systems——is described, and the stability of this direct method for the systems of difference equations with variable coefficients is discussed. In passing, three appendices which are extensions of the text are given.

The difference method described in this chapter has the following features.

(i) It is a singularity-separating difference method. That is, jump conditions are used at discontinuities, and the appearance of differences over discontinuities or "large" weak-discontinuities is not permitted in any difference equation. Therefore, the truncation error of this method is smaller, and fairly accurate results can be obtained by using a coarse net.

(ii) It adopts such a curvilinear coordinate system that all the discontinuities and "large" weak-discontinuities are coordinate lines. This makes singularity-separating very easy.

(iii) Difference schemes are constructed for initial-boundary-value problems, and theoretical analysis is done also for initial-boundary-value problems. Therefore, the schemes have a reliable theoretical foundation.

(iv) Explicit, implicit and mixed schemes are considered. Therefore, this method is suitable for many cases, and a stable, economical scheme can be chosen according to the features of the problems. Because of its universality, the accurate calculation of the intersection of discontinuities is possible if the structure of solutions near the point of intersection is clear. The accurate calculations of the reflection of discontinuities and of automatically formed discontinuities are also possible. As we know, it is not easy to solve these problems accurately.

(v) When implicit schemes are used, difference equations are solved by a direct method. And the soundness of this method has been pro ven, i.e., this method has been proven to be stable.

Comparing this method with other similar methods[C26, C72], we can see that it has the following features: this method is universal, i.e., it is suitable in all cases; it has a reliable theoretical foundation, i.e., in the case of variable coefficients, we have proved that the schemes for initial-boundary-value problems are stable with respect to initial values and boundary values in L_2, and that the direct method of solving difference systems is stable.

§ 1 Formulation of Problems

Consider the following initial-boundary-value problem for hyperbolic systems with two independent variables.

(i) In L regions $t\geqslant 0$, $x_{l-1}(t)\leqslant x\leqslant x_l(t)$ $(l=1, 2, \cdots, L)$, a first order quasi-linear hyperbolic system

$$\frac{\partial \widetilde{\boldsymbol{U}}}{\partial t}+\widetilde{A}(\widetilde{\boldsymbol{U}}, x, t)\frac{\partial \widetilde{\boldsymbol{U}}}{\partial x}=\widetilde{\boldsymbol{F}}(\widetilde{\boldsymbol{U}}, x, t) \tag{1.1}$$

is given.

(ii) At $t=0$, initial conditions

$$\begin{cases} \widetilde{\boldsymbol{U}}(x, 0)=\widetilde{\boldsymbol{D}}_l(x), & x_{l-1}(0)\leqslant x\leqslant x_l(0), \quad l=1, 2, \cdots, L, \\ x_l(0)=c_{0,l}, & x_l'(0)=c_{1,l}, \quad\quad\quad\quad\quad\quad l=0, 1, \cdots, L \end{cases} \tag{1.2}$$

are given.

(iii) On boundaries, boundary conditions and internal boundary conditions compatible with the differential equations (1.1)

$$\begin{cases} \boldsymbol{B}_0(\widetilde{\boldsymbol{U}}_{x_0^+}, x_0, x_0', t)=0, \\ \boldsymbol{B}_l(\widetilde{\boldsymbol{U}}_{x_l^-}, \widetilde{\boldsymbol{U}}_{x_l^+}, x_l, x_l', t)=0, \quad l=1, 2, \cdots, L-1, \\ \boldsymbol{B}_L(\widetilde{\boldsymbol{U}}_{x_L^-}, x_L, x_L', t)=0 \end{cases} \tag{1.3}$$

are given.

The shapes of boundaries $x_l(t)\,(l=0, 1, \cdots, L)$ and $\widetilde{\boldsymbol{U}}(x, t)$ in the region $t\geqslant 0,\ x_0(t)\leqslant x\leqslant x_L(t)$ need to be determined. Here, x, t are independent variables. $\widetilde{\boldsymbol{U}}$, $\widetilde{\boldsymbol{F}}$, $\widetilde{\boldsymbol{D}}_l$ are N-dimensional vectors. $c_{0,l}$, $c_{1,l}$ are scalars. $\widetilde{A}$ is an $N\times N$ matrix, and there are a real invertible matrix

$$\widetilde{\widetilde{G}}=\begin{pmatrix} \widetilde{\boldsymbol{G}}_1^* \\ \widetilde{\boldsymbol{G}}_2^* \\ \vdots \\ \widetilde{\boldsymbol{G}}_N^* \end{pmatrix}$$

and a real diagonal matrix

$$\widetilde{\Lambda}=\begin{pmatrix} \widetilde{\lambda}_1 & & 0 \\ & \widetilde{\lambda}_2 & \\ & & \ddots \\ 0 & & \widetilde{\lambda}_N \end{pmatrix},$$

such that

$$\widetilde{A}=\widetilde{\widetilde{G}}^{-1}\widetilde{\Lambda}\widetilde{\widetilde{G}}, \tag{1.4}$$

where $\widetilde{\boldsymbol{G}}_n^*$ is the transpose of a vector $\widetilde{\boldsymbol{G}}_n$. $\boldsymbol{B}_l\,(l=0, 1, \cdots, L)$ are γ_l-dimensional vectors, and their elements are nonlinear functions of the listed arguments. $\widetilde{\boldsymbol{U}}_{x_l^+}$ and $\widetilde{\boldsymbol{U}}_{x_l^-}$ denote the respective values of $\widetilde{\boldsymbol{U}}(x, t)$ on the upper and the lower sides of a boundary $x_l(t)$ (if $x_l(t)$ is a discontinuity, the values on the two sides are different).

$$x_l'=\frac{d}{dt}x_l(t).$$

Without loss of generality, suppose

$$\widetilde{\lambda}_1\geqslant\widetilde{\lambda}_2\geqslant\cdots\geqslant\widetilde{\lambda}_N, \tag{1.5}$$

and

$$\begin{cases} \widetilde{\lambda}_{1,x_l^+}\geqslant\cdots\geqslant\widetilde{\lambda}_{N-p_l,x_l^+}>x_l'\geqslant\widetilde{\lambda}_{N-p_l+1,x_l^+}\geqslant\cdots\geqslant\widetilde{\lambda}_{N,x_l^+}, & l=0, 1, \cdots, L-1, \\ \widetilde{\lambda}_{1,x_l^-}\geqslant\cdots\geqslant\widetilde{\lambda}_{q_l,x_l^-}\geqslant x_l'>\widetilde{\lambda}_{q_l+1,x_l^-}\geqslant\cdots\geqslant\widetilde{\lambda}_{N,x_l^-}, & l=1, 2, \cdots, L, \end{cases} \tag{1.6}$$

where the meaning of the subscript $x_l^\pm$ of $\lambda_{n,x_l^\pm}$ is the same as that of the subscript $x_l^\pm$ of $\boldsymbol{U}_{x_l^\pm}$. (In other cases the subscript $x_l^\pm$ also has the same meaning.)

The compatibility of the boundary and internal boundary conditions (1.3) with the differential equations (1.1) means that the following conditions are satisfied:

(1)[1]

$$\begin{cases} \gamma_0+p_0=N+1, \\ \gamma_l+p_l+q_l=2N+1, \quad l=1, 2, \cdots, L-1, \\ \gamma_L+q_L=N+1; \end{cases} \tag{1.7}$$

(2) if $t, x_0, x_1, \cdots, x_L, \boldsymbol{F}_0^+, \boldsymbol{F}_1^-, \boldsymbol{F}_1^+, \boldsymbol{F}_2^-, \cdots, \boldsymbol{F}_{L-1}^+, \boldsymbol{F}_L^-$ are properly given, the systems

$$\begin{cases} \boldsymbol{B}_0(\widetilde{\boldsymbol{U}}_{x_0^+}, x_0, x_0', t)=0, \\ \overline{\widetilde{G}}_{(0)}^+\widetilde{\boldsymbol{U}}_{x_0^+}=\boldsymbol{F}_0^+, \end{cases}$$
$$\begin{cases} \overline{\widetilde{G}}_{(l)}^-\widetilde{\boldsymbol{U}}_{x_l^-}=\boldsymbol{F}_l^-, \\ \boldsymbol{B}_l(\widetilde{\boldsymbol{U}}_{x_l^-}, \widetilde{\boldsymbol{U}}_{x_l^+}, x_l, x_l', t)=0, \\ \overline{\widetilde{G}}_{(l)}^+\widetilde{\boldsymbol{U}}_{x_l^+}=\boldsymbol{F}_l^+, \qquad l=1, 2, \cdots, L-1, \end{cases} \tag{1.8}$$
$$\begin{cases} \overline{\widetilde{G}}_{(L)}^-\widetilde{\boldsymbol{U}}_{x_L^-}=\boldsymbol{F}_L^-, \\ \boldsymbol{B}_L(\widetilde{\boldsymbol{U}}_{x_L^-}, x_L, x_L', t)=0 \end{cases}$$

always have reasonable solutions $\widetilde{\boldsymbol{U}}_{x_0^+}, x_0'$; $\widetilde{\boldsymbol{U}}_{x_l^-}, \widetilde{\boldsymbol{U}}_{x_l^+}, x_l'$ $(l=1, 2, \cdots, L-1)$; $\widetilde{\boldsymbol{U}}_{x_L^-}, x_L'$ respectively.

Here,

$$\overline{\widetilde{G}}_{(l)}^+=\left.\begin{pmatrix} \widetilde{\boldsymbol{G}}_{N-p_l+1}^* \\ \vdots \\ \widetilde{\boldsymbol{G}}_N^* \end{pmatrix}\right|_{x=x_l^+(t)},$$

$$\overline{\widetilde{G}}_{(l)}^-=\left.\begin{pmatrix} \widetilde{\boldsymbol{G}}_1^* \\ \vdots \\ \widetilde{\boldsymbol{G}}_{q_l}^* \end{pmatrix}\right|_{x=x_l^-(t)},$$

and $\boldsymbol{F}_l^+, \boldsymbol{F}_l^-$ are p_l, q_l-dimensional vectors respectively.

Moreover, we also suppose that the initial-value conditions (1.2) satisfy the boundary and internal boundary conditions (1.3), and that the lines $t=$constant are space-like.

The following is an explanation concerning the physical background of the above problem and the reliability of formulation.

In the inviscid steady supersonic flow around bodies, the following problem often appears (see Fig. 1.1). At $z=0$ (in order to follow the usual convention, we regard x, z as independent variables here), the flow parameters are given. For $z\geqslant 0$, the shape of the main shock and the flow parameters between the main shock and the body need to be determined. This field has the following features. There are several discontinuities (internal shocks, contact discontinuities) and weak-discontinuities (boundaries of expansion waves) whose locations are unknown. In those regions between discontinuities and weak-discontinuities, the flow parameters vary smoothly and the partial differential equations which

1) Condition (2) implies Condition (1). However, in order to emphasize it, Condition (1) is also listed.

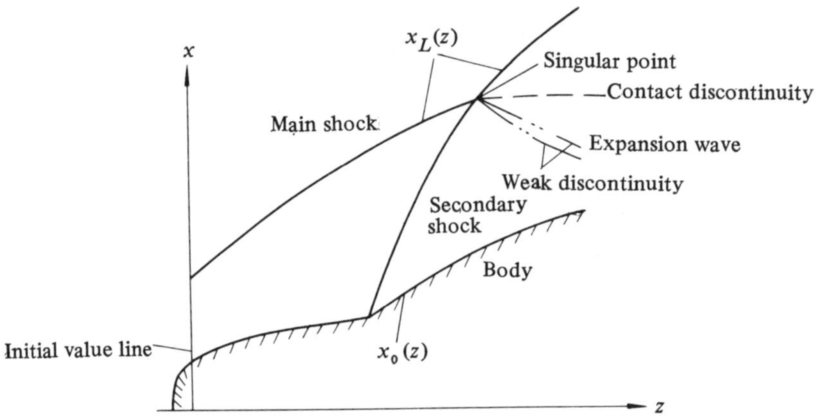

Figure 1.1 Inviscid steady supersonic flow around blunt bodies

describe the relations among the flow parameters are given. On the boundaries, some relations between flow parameters and the shapes of boundaries are given. On the internal boundaries, some relations between flow parameters on both sides of the boundaries and the shapes of boundaries are given. Therefore the problem we need to solve is the multi-boundary initial-boundary-value problem mentioned above. Moreover, there are many types of boundaries, i.e., p_l and q_l may take many values.

In unsteady flow, for example, in the blast wave, the situation is similar.

As for the reliability of the formulation, we want to point out the meaning of the compatibility conditions (1.7) and (1.8). The inequalities (1.6) indicate that on the upper sides of the boundaries x_l $(l=1, 2, \cdots, L-1)$, the backward characteristics corresponding to $\tilde{\lambda}_{N-p_l+1}, \cdots, \tilde{\lambda}_N$ can be drawn, and that on the lower sides, the backward characteristics corresponding to $\tilde{\lambda}_1, \cdots, \tilde{\lambda}_{q_l}$ can be drawn (see Fig. 1.2). Therefore, the differential equations and initial conditions should be able to determine those $\tilde{\boldsymbol{G}}_n^*\tilde{\boldsymbol{U}}$ on the above backward characteristics. That is, we can obtain (p_l+q_l) conditions. In $\boldsymbol{B}_l(l=1, 2, \cdots, L-1)$, there are $(2N+2)$ unknowns: $\tilde{\boldsymbol{U}}_{x_l^+}$, $\tilde{\boldsymbol{U}}_{x_l^-}$, x_l, x_l'. However, between x_l and x_l', there is a relation

$$x_l=\int x_l' dt. \tag{1.9}$$

Consequently, in order to determine all the unknowns on the boundary x_l, $(2N+1-p_l-q_l)$ more conditions are needed, i.e., $\gamma_l(l=1, 2, \cdots, L-1)$ should satisfy (1.7). The condition (1.8) means that not only the number

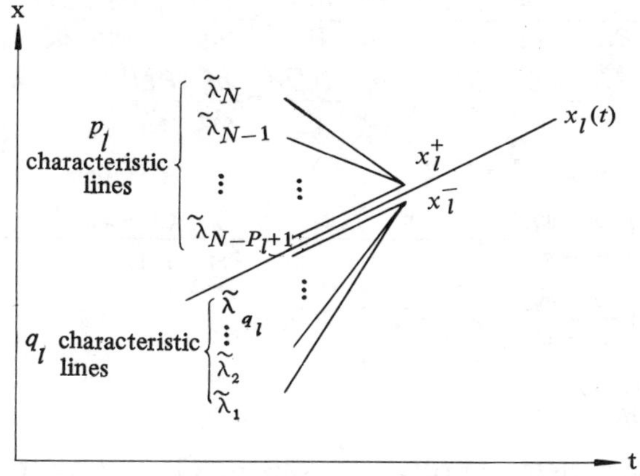

Figure 1.2 Distribution of characteristic curves on a boundary

of conditions should satisfy (1.7), but also those internal boundary conditions together with the differential equations and initial conditions should uniquely determine these unknowns on the boundary x_l. When $l=0$ or L, the situation is similar.

Finally, we point out the following fact. If the physical meaning of t is time, the line $t=$constant is automatically a space–like line. However, if the physical meaning of t is not time, in order to guarantee that $t=$ constant is a space–like line, one must choose independent variables carefully. (Of course, it is first required that the system (1.1) in the region of solution be hyperbolic.)

§ 2 Four Model Problems

For the purposes of solving a problem numerically, particularly when treating boundary points, it is desirable to have the problem formulated in a rectangular region. Therefore, we introduce a curvilinear coordinate system ξ, t:

$$\begin{cases} \xi = f_l\left(\dfrac{x - x_{l-1}(t)}{x_l(t) - x_{l-1}(t)}\right) + l - 1, & \text{if } x_{l-1}(t) \leqslant x \leqslant x_l(t), \\ t = t, & l = 1, 2, \cdots, L. \end{cases} \tag{2.1}$$

Here, $f_l(y)$ is a given, monotone, continuous function satisfying the conditions $f_l(0)=0$ and $f_l(1)=1$. Moreover, we want it to be a smooth function. $f_l(y)=y$ is the simplest form. In the coordinate system ξ, t, the region of solution is $t \geqslant 0$, $0 \leqslant \xi \leqslant L$, and the boundaries are $\xi = l (l=0,1,\cdots,$

L). That is, in this coordinate system, (1.1)—(1.3) becomes a problem in a rectangular region. Let $\widetilde{\boldsymbol{U}}(x, t)=\boldsymbol{U}(\xi, t)$. Clearly, there are the following relations among their derivatives

$$\begin{cases} \dfrac{\partial \widetilde{\boldsymbol{U}}}{\partial t}=\dfrac{\partial t}{\partial t}\dfrac{\partial \boldsymbol{U}}{\partial t}+\dfrac{\partial \xi}{\partial t}\dfrac{\partial \boldsymbol{U}}{\partial \xi}=\dfrac{\partial \boldsymbol{U}}{\partial t}+\dfrac{\partial \xi}{\partial t}\dfrac{\partial \boldsymbol{U}}{\partial \xi}, \\ \dfrac{\partial \widetilde{\boldsymbol{U}}}{\partial x}=\dfrac{\partial t}{\partial x}\dfrac{\partial \boldsymbol{U}}{\partial t}+\dfrac{\partial \xi}{\partial x}\dfrac{\partial \boldsymbol{U}}{\partial \xi}=\dfrac{\partial \xi}{\partial x}\dfrac{\partial \boldsymbol{U}}{\partial \xi}, \end{cases}$$

where, for $x_{l-1}\leqslant x\leqslant x_l$ $(l=1, 2, \cdots, L)$,

$$\begin{cases} \dfrac{\partial \xi}{\partial t}=f_l'\left(\dfrac{x-x_{l-1}}{x_l-x_{l-1}}\right)\dfrac{(-x_{l-1}')(x_l-x_{l-1})-(x-x_{l-1})(x_l'-x_{l-1}')}{(x_l-x_{l-1})^2}, \\ \dfrac{\partial \xi}{\partial x}=f_l'\left(\dfrac{x-x_{l-1}}{x_l-x_{l-1}}\right)\dfrac{1}{x_l-x_{l-1}}>0, \\ f_l'=\dfrac{df_l(y)}{dy}. \end{cases} \tag{2.2}$$

Therefore, in the new coordinate system, (1.1) may be rewritten as follows:

$$\frac{\partial \boldsymbol{U}}{\partial t}+A\frac{\partial \boldsymbol{U}}{\partial \xi}=\boldsymbol{F},$$
$$t\geqslant 0, \quad l-1\leqslant \xi\leqslant l, \quad l=1, 2, \cdots, L,$$

where

$$A(\boldsymbol{U}, \boldsymbol{X}, \boldsymbol{X}'\ \xi, t)=\frac{\partial \xi}{\partial t}E+\frac{\partial \xi}{\partial x}\widetilde{A},$$
$$\boldsymbol{F}(\boldsymbol{U}, \boldsymbol{X}, \xi, t)=\widetilde{\boldsymbol{F}}(\widetilde{\boldsymbol{U}}, x, t).$$

Here, E is an $N\times N$ unit matrix, and $\boldsymbol{X}$ is the transpose of the vector $(x_0, x_1, \cdots, x_L)$. Hence the problem (1.1)—(1.3) in § 1 can be rewritten in the following form.

(i) In L regions, $t\geqslant 0$, $l-1\leqslant\xi\leqslant l$ $(l=1, 2, \cdots, L)$, a system

$$\frac{\partial \boldsymbol{U}}{\partial t}+A(\boldsymbol{U}, \boldsymbol{X}, \boldsymbol{X}', \xi, t)\frac{\partial \boldsymbol{U}}{\partial \xi}=\boldsymbol{F}(\boldsymbol{U}, \boldsymbol{X}, \xi, t) \tag{2.3}$$

is given.

(ii) At $t=0$, initial conditions

$$\begin{cases} \boldsymbol{U}(\xi, 0)=\boldsymbol{D}_l(\xi), \quad l-1\leqslant\xi\leqslant l, \quad l=1, 2, \cdots, L, \\ \boldsymbol{X}(0)=\boldsymbol{C}_0, \ \boldsymbol{X}'(0)=\boldsymbol{C}_1 \end{cases} \tag{2.4}$$

are given.

(iii) On boundaries, boundary conditions and internal boundary conditions

$$\begin{cases} \boldsymbol{B}_0(\boldsymbol{U}_{0^+}, x_0, x_0', t)=0, \\ \boldsymbol{B}_l(\boldsymbol{U}_{l^-}, \boldsymbol{U}_{l^+}, x_l, x_l', t)=0, \quad l=1, 2, \cdots, L-1, \\ \boldsymbol{B}_L(\boldsymbol{U}_{L^-}, x_L, x_L', t)=0 \end{cases} \tag{2.5}$$

are given.

$\boldsymbol{X}(t)$ and $\boldsymbol{U}(\xi, t)$ in the region $t \geqslant 0, 0 \leqslant \xi \leqslant L$ need to be determined. In the above expressions, $\boldsymbol{X}' \equiv \frac{d\boldsymbol{X}}{dt}$; $\boldsymbol{C}_0$ and $\boldsymbol{C}_1$ are $(L+1)$–dimensional vectors whose components are $c_{0,l}$ and $c_{1,l}$ respectively;

$$\boldsymbol{D}_l(\xi) = \widetilde{\boldsymbol{D}}_l(x(\xi, t))\,|_{t=0};$$

and $\boldsymbol{U}_{l^+}$ and $\boldsymbol{U}_{l^-}$ denote the values of $\boldsymbol{U}$ on the upper and the lower sides of the line $\xi = l$ respectively. (From now on, the subscripts l^+ and l^- have the same meaning.)

Let

$$\Lambda(\boldsymbol{U}, \boldsymbol{X}, \boldsymbol{X}', \xi, t) = \frac{\partial \xi}{\partial t} E + \frac{\partial \xi}{\partial x} \widetilde{\Lambda}.$$

We then have

$$A = \overline{G}^{-1} \Lambda \overline{G}, \tag{2.6}$$

where

$$\overline{G}(\boldsymbol{U}, \boldsymbol{X}, \xi, t) = \overline{\widetilde{G}}(\widetilde{\boldsymbol{U}}, x, t).$$

Moreover, $\frac{\partial \xi}{\partial x} > 0$. Therefore, (1.5) can be rewritten a

$$\lambda_1 \geqslant \lambda_2 \geqslant \cdots \geqslant \lambda_N, \tag{2.7}$$

where λ_n is the n–th diagonal element of Λ.

Besides, we have

$$\left.\frac{\frac{\partial \xi}{\partial t}}{\frac{\partial \xi}{\partial x}}\right|_{x = x_l^*} = -x_l', \quad l = 0, 1, \cdots, L,$$

and

$$\Lambda|_{\xi = l^\pm} = \left[\frac{\partial \xi}{\partial x}\left(\widetilde{\Lambda} + \frac{\frac{\partial \xi}{\partial t}}{\frac{\partial \xi}{\partial x}} E\right)\right]\Bigg|_{x = x_l^*} = \left(\frac{\partial \xi}{\partial x}\right)_{x_l^*} (\widetilde{\Lambda}_{x_l^*} - x_l' E).$$

Thus from (1.6) we can derive

$$\begin{cases} \lambda_{1, l^+} \geqslant \cdots \geqslant \lambda_{N - p_l, l^+} > 0 \geqslant \lambda_{N - p_l + 1, l^+} \geqslant \cdots \geqslant \lambda_{N, l^+}, \\ \qquad l = 0, 1, \cdots, L-1, \\ \lambda_{1, l^-} \geqslant \cdots \geqslant \lambda_{q_l, l^-} \geqslant 0 > \lambda_{q_l + 1, l^-} \geqslant \cdots \geqslant \lambda_{N, l^-}, \\ \qquad l = 1, 2, \cdots, L. \end{cases} \tag{2.8}$$

As in § 1, it is still required that $\boldsymbol{B}_l$ satisfy the compatibility conditions (1.7) and (1.8). According to the notation here, (1.8) might be described as follows:

When t, x_0, x_1, $\cdots$, x_L, $\boldsymbol{F}_0^+$, $\boldsymbol{F}_1^-$, $\boldsymbol{F}_1^+$, $\boldsymbol{F}_2^-$, $\cdots$, $\boldsymbol{F}_{L-1}^+$, $\boldsymbol{F}_L^-$ are properly given, the systems

$$\begin{cases} \boldsymbol{B}_0(\boldsymbol{U}_{0^+}, x_0, x_0', t) = 0, \\ \overline{G}^+_{(0)} \boldsymbol{U}_{0^+} = \boldsymbol{F}^+_0, \end{cases}$$

$$\begin{cases} \overline{G}^-_{(l)} \boldsymbol{U}_{l^-} = \boldsymbol{F}^-_l, \\ \boldsymbol{B}_l(\boldsymbol{U}_{l^-}, \boldsymbol{U}_{l^+}, x_l, x_l', t) = 0, \\ \overline{G}^+_{(l)} \boldsymbol{U}_{l^+} = \boldsymbol{F}^+_l, \quad l = 1, 2, \cdots, L-1, \end{cases} \tag{2.9}$$

$$\begin{cases} \overline{G}^-_{(L)} \boldsymbol{U}_{L^-} = \boldsymbol{F}^-_L, \\ \boldsymbol{B}_L(\boldsymbol{U}_{L^-}, x_L, x_L', t) = 0 \end{cases}$$

have reasonable solutions $\boldsymbol{U}_{0^+}, x_0'$; $\boldsymbol{U}_{l^-}, \boldsymbol{U}_{l^+}, x_l'$ $(l=1, 2, \cdots, L-1)$; $\boldsymbol{U}_{L^-}, x_L'$ respectively, where

$$\overline{G}^+_{(l)} = \begin{pmatrix} \boldsymbol{G}^*_{N-q_l+1} \\ \vdots \\ \boldsymbol{G}^*_N \end{pmatrix}_{\xi=l^+}, \quad \overline{G}^-_{(l)} = \begin{pmatrix} \boldsymbol{G}^*_1 \\ \vdots \\ \boldsymbol{G}^*_{q_l} \end{pmatrix}_{\xi=l^-},$$

$\boldsymbol{G}^*_n$ being the n-th row of $\overline{G}$.

We can see from the above calculation that, by using coordinate transformation, a problem with "moving boundaries" can become a problem with "fixed boundaries". However, the coefficients of the partial differential equations change slightly. In the present case, the coefficients depend not only on the independent variables and the unknown vector $\boldsymbol{U}$, but also on the unknowns $\boldsymbol{X}$ and $\boldsymbol{X}'$. A change of this kind does not give rise to any difficulty in the numerical calculation. But the new problem has "fixed boundaries", which is very convenient in the accurate treatment of boundaries.

In the following, we shall discuss numerical methods only for the initial-boundary-value problem (2.3)—(2.5) ($\boldsymbol{B}_l$ satisfy the conditions (1.7) and (2.9)). Of course, we again suppose that the lines $t=$constant are space-like, and that the initial conditions (2.4) satisfy the boundary conditions (2.5).

In solving general initial-boundary-value problems, one encounters the following difficulties. The first is how to get a system of difference equations such that the sum of the number of difference equations and the number of boundary conditions is equal to the number of the unknowns. The second is to prove that the scheme satisfies certain minimal theoretical requirements, for example, that it is stable and convergent in the linear case.

In what follows, we explain these problems in the case $L=1$. In the region $t \geqslant 0$, $0 \leqslant \xi \leqslant 1$, we take a net: $\xi_m = m\Delta\xi_1$ $(m=0, 1, \cdots, M_1)$, $t_k = k\Delta t$ $(k=0, 1, \cdots)$, where $\Delta\xi_1 = \dfrac{1}{M_1}$. Thus, at each t-level, the unknowns are $\boldsymbol{U}(\xi_m, t_k)$ $(m=0, 1, \cdots, M_1)$, x_0, x_1, x_0', x_1', and the number of unknowns is $(M_1+1)N+4$. We have $\gamma_0+\gamma_1=2N+2-p_0-q_1$ boundary conditions and two relations (1.9) for $l=0$ and 1 already, so we need $M_1N-N+p_0+q_1$ equations. If $p_0+q_1=N$, the M_1N difference equations needed may be obtained by directly discretizing (2.3). In the paper [4], which deals with the initial-boundary-value problems with $p_0+q_1=N$, the difference equations needed are obtained in this way. In the case $p_0+q_1\neq N$, there is a problem if we try to get the equations needed in the above way. The problem is that the number of difference equations is always equal to an integral multiple of N, so it is impossible for the number of equations to be equal to the number of unknowns. Id the solution, N difference equations approximating (2.3) are used to determine the unknowns at every interior point, and special treatment is applied for boundary points. For example, in order to calculate the boundary points, (2.3) is rewritten as

$$G_n^*\frac{\partial U}{\partial t}+\lambda_n G_n^*\frac{\partial U}{\partial \xi}=f_n,\quad n=1,\ 2,\ \cdots,\ N, \tag{2.10}$$

and at $\xi=0$ the difference equations approximating the equations in (2.10) corresponding to $n=N-p_0+1,\ \cdots,\ N$ are used; at $\xi=1$ the equations corresponding to $n=1,\ 2,\cdots,\ q_1$ are used. In this way, the $(M_1-1)N+p_0+q_1$ equations needed can be obtained. This procedure has been adopted in our early work and by some others. Its advantage is that it always makes the number of equations equal to the number of unknowns. However, this method does have certain problems. For example, the accuracies at the interior points and at the boundary points might be different. Also, the main problem is that because the schemes for the interior points and for the boundary points are different in many cases, then generally speaking, it does not seem to be sufficient for the stability of initial-boundary-value schemes[1)] that the von Neumann condition is satisfied everywhere. Intuitively, this is easily understood. In fact, when we use the von Neumann condition to discuss stability, we implicitly suppose that the same algorithm is adopted near this point. The algorithm near the boundaries has a "sudden change", so an analysis of this kind loses its effectiveness. In order to guarantee that initial-boundary-value schemes are stable, one must consider how to match the schemes for boundary points with those for interior points in constructing schemes. It may be true that if the same scheme is adopted both at the boundary points and at the interior points, then the stability of the initial-boundary-value schemes can be more easily guaranteed. Therefore, it is very necessary to present a new way of constructing schemes. This way should guarantee that the number of equations is always equal

1) We call a scheme for initial-boundary-value problems an initial-boundary-value scheme.

to the number of unknowns and that the difference schemes for the boundary points and the interior points are the same.

Obviously, only the uncentered schemes can be used at the boundary points. For explicit schemes approximating (2.3), in order to guarantee that the schemes are stable for all λ_n, they should be centered. Thus, if at the interior points the difference equations are deduced from (2.3), it is more difficult to adopt the same scheme for the interior and the boundary points. In order that this may be easily realized, the difference equations at interior points are also deduced from (2.10).

The reason why the above procedure always makes the number of equations equal to the number of unknowns is that the numbers of difference equations are different for different λ_n according to the values $\lambda_n(0, t)$ and $\lambda_n(1, t)$. In the following we shall explain this fact. Suppose $p_0+q_1 \geqslant N$. Obviously, establishing difference equations for (2.3) is equivalent to establishing equations for (2.10), $n=1, 2, \cdots, N$. Thus, when $m=1, \cdots, M_1-1$, for every λ_n, one difference equation is established in the above treatment. Therefore, in the above procedure, if $\lambda_n(0, t)>0$ and $\lambda_n(1, t)\geqslant 0$ (when $n=1, \cdots, N-p_0$, λ_n has these features), M_1 difference equations ($m=1, 2, \cdots, M_1$) are established; if $\lambda_n(0, t)\leqslant 0$ and $\lambda_n(1, t)\geqslant 0$ ($n=N-p_0+1, \cdots, q_1$), M_1+1 equations ($m=0, 1, \cdots, M_1$) are established; if $\lambda_n(0, t)\leqslant 0$ and $\lambda_n(1, t)<0$ ($n=q_1+1, \cdots, N$), M_1 equations are established. That is, the total number of difference equations is

$$(N-p_0)M_1+(q_1-N+p_0)(M_1+1)+M_1(N-q_1)=M_1N-N+p_0+q_1.$$

For $p_0+q_1<N$, the situation is similar. When $n=1, \cdots, q_1, N-p_0+1, \cdots, N$, M_1 equations are established for every λ_n. If $\lambda_n(0, t)>0$ and $\lambda_n(1, t)<0$ ($n=q_1+1, \cdots, N-p_0$), M_1-1 equations ($m=1, 2, \cdots, M_1-1$) are established. Therefore, the total number of difference equations is also

$$(p_0+q_1)M_1+(N-p_0-q_1)(M_1-1)=M_1N-N+p_0+q_1.$$

Consequently, we present the following way of constructing difference schemes. Let Δt bean increment of t, and let $\Delta\xi_l=\dfrac{1}{M_l}$ be increments of ξ in the interval $[l-1, l]$, $l=1, 2, \cdots, L-1$, where M_l are integers. That is, the lines of the net are

$$t=0, \Delta t, 2\Delta t, \cdots, \quad \xi=0, \Delta\xi_1, 2\Delta\xi_1, \cdots, 1,$$
$$1+\Delta\xi_2, \cdots, 2, 2+\Delta\xi_3, \cdots, L.$$

In the l-th interval $l-1\leqslant\xi\leqslant l$ ($l=1, 2, \cdots, L$), the difference equations corresponding to λ_n, which approximate the equation (2.10), are established as follows.

(i) If $\lambda_{n,l-1^+}>0$ and $\lambda_{n,l^-}\geqslant 0$, M_l difference equations "in this interval" are established by using an identical scheme at the points $\frac{\xi-l+1}{\Delta\xi_l}=1$, $2, \cdots, M_l$. A difference equation "in this interval" means that only the values of functions in this interval appear in the equation.

(ii) If $\lambda_{n,l-1^+}\leqslant 0$ and $\lambda_{n,l^-}<0$, M_l difference equations in this interval are established by an identical scheme at the points $\frac{\xi-l+1}{\Delta\xi_l}=0, 1, \cdots, M_l-1$.

(iii) If $\lambda_{n,l-1^+}\leqslant 0$ and $\lambda_{n,l^-}\geqslant 0$, M_l+1 difference equations in this interval are established by an identical scheme at the points $\frac{\xi-l+1}{\Delta\xi_l}=0, 1, \cdots, M_l$.

(iv) If $\lambda_{n,l-1^+}>0$ and $\lambda_{n,l^-}<0$, M_l-1 equations in this interval are established by an identical scheme at the points $\frac{\xi-l+1}{\Delta\xi_l}=1, 2, \cdots, M_l-1$.

Then, the sum of the number of difference equations and the number of boundary conditions is always equal to the number of unknowns. In fact, the total number of unknowns here is

$$\sum_{l=1}^{L} N(M_l+1)+2(L+1).$$

In the l-th interval, we have $(M_l-1)N+p_{l-1}+q_l$ equations, so that the number of difference equations is

$$\sum_{l=1}^{L}(M_l-1)N+\sum_{l=1}^{L}(p_{l-1}+q_l).$$

According to (1.7), the number of boundary conditions is

$$\sum_{l=0}^{L}\gamma_l=2LN-\sum_{l=1}^{L}(p_{l-1}+q_l)+L+1.$$

Moreover, from (1.9) we can also obtain $L+1$ relations. Therefore, the total number of conditions is

$$\sum_{l=1}^{L}(M_l+1)N+2(L+1).$$

That is, it is always true that the two numbers are equal. Of course, this fact alone does not ensure that this method is sound. In § 4, for several classes of schemes obtained by this method or by a slightly modified method, we shall prove the following result: if the von Neumann condition (3.2) and the condition (3.3), which almost guarantees that the systems of difference equations are well-conditioned, are satisfied everywhere, then these classes of initial-boundary-value schemes are stable. Therefore, this method is probably sound.

Consequently, the task of constructing difference schemes for general initial-boundary-value problems can be reduced to that of constructing schemes for the following four model problems:

$$A_1:\quad\begin{cases}\dfrac{\partial u}{\partial t}+\lambda(u,\xi,t)\dfrac{\partial u}{\partial \xi}=f(u,\xi,t), & t\geqslant 0,\ 0\leqslant\xi\leqslant 1,\\ \lambda(u(0,t),0,t)>0, & \lambda(u(1,t),1,t)\geqslant 0,\\ u(\xi,0)=d(\xi), & 0\leqslant\xi\leqslant 1,\\ u(0,t)=e_0(t), & t\geqslant 0;\end{cases}\tag{2.11}$$

$$A_2:\quad\begin{cases}\dfrac{\partial u}{\partial t}+\lambda(u,\xi,t)\dfrac{\partial u}{\partial \xi}=f(u,\xi,t), & t\geqslant 0,\ 0\leqslant\xi\leqslant 1,\\ \lambda(u(0,t),0,t)\leqslant 0, & \lambda(u(1,t),1,t)<0,\\ u(\xi,0)=d(\xi), & 0\leqslant\xi\leqslant 1,\\ u(1,t)=e_1(t), & t\geqslant 0;\end{cases}$$

$$B:\quad\begin{cases}\dfrac{\partial u}{\partial t}+\lambda(u,\xi,t)\dfrac{\partial u}{\partial \xi}=f(u,\xi,t), & t\geqslant 0,\ 0\leqslant\xi\leqslant 1,\\ \lambda(u(0,t),0,t)\leqslant 0, & \lambda(u(1,t),1,t)\geqslant 0,\\ u(\xi,0)=d(\xi), & 0\leqslant\xi\leqslant 1;\end{cases}$$

$$C:\quad\begin{cases}\dfrac{\partial u}{\partial t}+\lambda(u,\xi,t)\dfrac{\partial u}{\partial \xi}=f(u,\xi,t), & t\geqslant 0,\ 0\leqslant\xi\leqslant 1,\\ \lambda(u(0,t),0,t)>0, & \lambda(u(1,t),1,t)<0,\\ u(\xi,0)=d(\xi), & 0\leqslant\xi\leqslant 1,\\ u(0,t)=e_0(t), & t\geqslant 0,\\ u(1,t)=e_1(t), & t\geqslant 0.\end{cases}$$

In § 3 we shall discuss how to construct schemes for these model problems which satisfy the condition: An identical algorithm is adopted for the interior and boundary points; and both the von Neumann condition (3.2) and the "well-conditioned" condition (3.3) are satisfied at every point.

We must point out that among the four model problems, A_1, A_2, B are more basic. In fact, a model C can be divided into a model A_1 and a model A_2. This fact can be explained as follows: taking a characteristic $\frac{d}{dt}\xi_1(t)=\lambda$ in the region of the solution as a boundary (suppose $\xi_1(0)=\xi_1\in(0,1)$), we can rewrite a model problem C as

$$\begin{cases}\dfrac{\partial u}{\partial t}+\lambda(u,\xi,t)\dfrac{\partial u}{\partial \xi}=f(u,\xi,t), & t\geqslant 0,\ 0\leqslant\xi\leqslant\xi_1(t),\\ \lambda(u(0,t),0,t)>0,\\ \dfrac{d\xi_1}{dt}=\lambda(u(\xi_1,t),\xi_1,t), & \xi_1(0)=\xi_1,\\ u(\xi,0)=d(\xi), & 0\leqslant\xi\leqslant\xi_1(t),\\ u(0,t)=e_0(t), & t\geqslant 0,\end{cases}$$

and

$$\begin{cases} \dfrac{\partial u}{\partial t}+\lambda(u, \xi, t)\dfrac{\partial u}{\partial \xi}=f(u, \xi, t), \quad t\geqslant 0, \ \xi_1(t)\leqslant\xi\leqslant 1, \\ \dfrac{d\xi_1}{dt}=\lambda(u(\xi_1, t), \xi_1, t), \quad \xi_1(0)=\bar{\xi}_1, \\ \lambda(u(1, t), 1, t)<0, \\ u(\xi, 0)=d(\xi), \quad \xi_1(t)\leqslant\xi\leqslant 1, \\ u(1, t)=e_1(t), \quad t\geqslant 0. \end{cases}$$

By applying a transformation of coordinates, the two problems above can be easily transformed into a model A_1 and a model A_2. However, for the models A_1, A_2 and B, the situation is different. When one takes an arbitrary line as an internal boundary, at least one of the new initial-boundary-value problems belongs to the set $\{A_1, A_2, B\}$. (When we take a characteristic as an internal boundary, the models A_1, A_2, B are divided into A_1 and B, A_2 and B, B and B respectively.)

For the model problems A_1, A_2, B and C, we have $N=1$, $p_0=0$, 1, 1, 0, and $q_1=1$, 0, 1, 0 respectively. Thus, the models A_1 and A_2 satisfy the condition $p_0+q_1=N$, the model B satisfies $p_0+q_1>N$, and the model C satisfies $p_0+q_1<N$. From the discussion above, we can see that for the models, a problem with $p_0+q_1<N$ can be divided into several problems with $p_0+q_1\geqslant N$. However, it is impossible to divide a problem with $p_0+q_1\geqslant N$ into several problems with $p_0+q_1<N$. This fact is also true for the general initial-boundary-value problems. For example, a problem with $p_{l-1}+q_l<N$ can be divided into several problems with $p_{l-1}+q_l\geqslant N$. In fact, if at points

$$\xi=\bar{\xi}_1, \bar{\xi}_2, \cdots, \bar{\xi}_{N-p_l-1-q_l} \quad (1>\bar{\xi}_1>\bar{\xi}_2>\cdots>\bar{\xi}_{N-p_l-1-q_l}>0),$$

we take as internal boundaries some characteristics corresponding to $\lambda_{q_l+1}, \lambda_{q_l+2}, \cdots, \lambda_{N-p_l-1}$ respectively, then a problem with $p_{l-1}+q_l<N$ is divided into $N-p_{l-1}-q_l+1$ problems with $\tilde{p}_{i-1}+\tilde{q}_i\geqslant N$. However, it is impossible to divide a problem with $p_{l-1}+q_l\geqslant N$ into several problems with $\tilde{p}_{i-1}+\tilde{q}_i<N$ by adding some internal boundaries. The reason is the following: for every internal boundary we have $\tilde{p}_i+\tilde{q}_i\geqslant N$, which means $p_{l-1}+q_l+\sum_{i=1}^{k}(\tilde{p}_i+\tilde{q}_i)\geqslant(k+1)N$, k being the number of new internal boundaries, and p_{l-1}, q_l, $\tilde{p}_i$, $\tilde{q}_i$ all are non-negative integers, so it is impossible to divide the integers into $k+1$ pairs with the sum being less than N. Therefore, in our discussions, we shall pay a lot of attention to the problems with $p_{l-1}+q_l\geqslant N$.

We also want to point out that if $\lambda=0$ is not allowed, then the model problems B and C cannot occur, i.e., p_0+q_1 is always equal to 1. Therefore, the existence of the points of $\lambda=0$ leads to occurrence of the problems with $p_0+q_1\neq 1$, and to a need to discuss initial-boundary-value problems which have one equation and no boundary condition, or one equation and two boundary conditions. This case is similar to and has a close relationship to singularities in ordinary differential equations. Consider the equation

$$-\lambda(\xi)\frac{du}{d\xi}=u.$$

If $\lambda(\xi)\neq 0$ in the interval $[0, 1]$, then u is determined in $[0, 1]$ when a condition is given at either end of the interval $[0, 1]$. However, if a saddle point $(\lambda(\xi_0)=0, \frac{d}{d\xi}\lambda(\xi_0)>0)$ appears, no boundary condition is needed; and if a knot $(\lambda(\xi_0)=0, \frac{d}{d\xi}\lambda(\xi_0)<0)$ appears in $(0, 1)$, one boundary condition is needed at each end. Obviously, the case with a saddle point is similar to the model B, and the case with a knot in $(0, 1)$ is similar to the model C. The reason for this similarity is the following: (2.11) may be approximated by

$$-\lambda^k\frac{du^{k+1}}{d\xi}=\frac{u^{k+1}-u^k}{\Delta t}-f^k,$$

where the superscript k means that $t=k\Delta t$. Thus, if u^k is given, we should solve a boundary problem for ordinary differential equations. For the model B, if there is only one point of $\lambda=0$, then $\frac{d\lambda}{d\xi}\geqslant 0$ at this point. For the model C, if there is only one point of $\lambda=0$ (this point must belong to $(0, 1)$), then $\frac{d\lambda}{d\xi}\leqslant 0$ at this point. Therefore, if there is only one point of $\lambda=0$ and $\frac{d\lambda}{d\xi}\neq 0$ at this point, then the ordinary differential equation corresponding to the model B has a saddle point, and the equation corresponding to the model C has a knot in the interval $(0, 1)$.

For the problems with several zero-points of λ and for the problems with some points where $\lambda=\frac{d\lambda}{d\xi}=0$ (suppose that the orders of zero-points are integers), the situation is similar. For an ordinary differential equation corresponding to the model B, the number of saddle points in $[0, 1]$ is one more than the number of knots in $(0, 1)$. For an equation corresponding to the model C, the number of saddle points in $[0, 1]$ is one less than the number of knots in $(0, 1)$. Here, if at a point,

$$\lambda=\frac{d\lambda}{d\xi}=\cdots=\frac{d^{i-1}\lambda}{d\xi^{i-1}}=0, \quad \frac{d^i\lambda}{d\xi^i}\neq 0,$$

we regard it as an i-multiple zero-point where several saddle points and knots are "stuck" together. If i is an even number, the singularity is regarded as a point with $\frac{i}{2}$ saddle points and $\frac{i}{2}$ knots. If i is odd and

$\frac{d^i\lambda}{d\xi^i}>0$, it is regarded as a point with $\frac{i+1}{2}$ saddle points and $\frac{i-1}{2}$ knots. If i is odd and $\frac{d^i\lambda}{d\xi^i}<0$, it is considered as a point with $\frac{i-1}{2}$ saddle points and $\frac{i+1}{2}$ knots.

When there exist zero-points of λ, some new problems of formulation of initial-boundary-value problems and numerical methods for them arise. Thus, we do need a special designation for the place where $\lambda=0$. If one of λ_n is equal to zero at a line $f_1(\xi, t)=0$, we say that (2.3) is degenerate at the line, and call the line a degenerate line of (2.3). Given this definition, a general initial-boundary-value problem is one in which the system of differential equations may be degenerate at some lines and which has several boundaries. In this chapter, we shall discuss numerical methods for the initial-boundary-value problems with both the features.

Finally, we shall compare the singularity-separating difference method with the "through" difference method, the method of characteristics, and the shock-fitting method without the transformation of coordinates, showing its virtues and defects.

In this method only the quantities on the same side of a discontinuity appear in every difference equation. The jump conditions are the only equations in which the quantities on the two sides of a discontinuity appear. Therefore, the truncation error of this difference method is much less than that of the "through" difference method, and accurate results may be obtained by using fewer net-points. Moreover, when solving a hyperbolic system, the following situation may be met. In one part of the region of solution, functions vary rapidly (in the regions near singularities, derivatives are close to infinity); in other parts, functions vary slowly. In order to save computation time, it is better to adopt a nonequidistant net, that is, in the region where functions vary rapidly, the fine mesh sizes are taken, and in the other regions, the coarse mesh sizes are taken. For the hyperbolic systems, the boundaries between these regions are usually characteristics (these characteristics usually belong to weak discontinuities). Therefore, a reasonable nonequidistant net can be obtained by taking the characteristics as internal boundaries and adopting different mesh sizes in different regions. In this way, not only will a reasonable arrangement of net points automatically be obtained, but also the differences over weak-discontinuities will not appear in any difference equation and the truncation error will be less (when there are some differences over "large" weak-discontinuities in difference equations, the truncation error is quite large). This means that at the weak-discontinuity points, the "through" method is also not as good as the method here. In brief, when compared to the "through" difference method, it has the virtue that it can give more accurate results by using fewer net points. Furthermore, the singularity-separating difference method for initial-boundary-value problems has quite a complete theoretical foundation, so it might be believed that the computed results are reasonable. However, when the "through" method is adopted, the treatment of boundary points is sometimes carelessly carried out, so one cannot be completely sure that the computed results are reasonable. The disadvantage of this method is that the code is more complicated than that of the "through" method. The reason for this is that in the case of the singularity-separating method, many boundaries must be treated.

Now, we shall compare this method with the natural method of characteristics and with the net-characteristics method. The net of this method is more flexible than that of the natural method of characteristics and it can be more easily generalized to the case with three independent variables than the natural method of characteristics. Nevertheless, the main advantage is that it has more flexibility than the two methods of characteristics when choosing difference schemes. Because of this fact, the method here may suit various needs, and by using it, interaction and reflection of discontinuities, and automatically formed discontinuities can be accurately calculated. These problems have always been difficult to solve completely.

In the method discussed in this book, the lines of discontinuities are always lines of coordinates, so the algorithm for boundary points is accurate and simple and the complete theoretical analysis of methods can be more easily performed. This is its advantage as compared with the shock-fitting method without the transformation of coordinates.

§ 3 Some Difference Schemes

In § 2, we pointed out that the construction of difference schemes for initial-boundary-value problems can be reduced to that for four model problems. We shall now discuss how to construct "continuous" difference schemes for model problems. This is only in connection with the types of models, so we omit the subscript n and suppose $0\leqslant\xi\leqslant1$ in this section. In the following, a quantity f_m^k with a superscript k and a subscript m will denote the value of f at $t=t_k\equiv k\Delta t$ and $\xi=\xi_m\equiv m\Delta\xi$. Moreover, we shall use the following symbols:

$$\Delta\xi=\frac{1}{M},\quad \sigma=\lambda\frac{\Delta t}{\Delta\xi},\quad \Delta_+\boldsymbol{U}_m=\boldsymbol{U}_{m+1}-\boldsymbol{U}_m,$$

$$\Delta_-\boldsymbol{U}_m=\boldsymbol{U}_m-\boldsymbol{U}_{m-1},\quad \Delta_0\boldsymbol{U}_m=\frac{1}{2}(\boldsymbol{U}_{m+1}-\boldsymbol{U}_{m-1}),$$

$$(\sigma G^* \Delta_\sigma U)_m = \begin{cases} \sigma_m G_m^* \Delta_+ U_m, & \sigma_m \leqslant 0, \\ \sigma_m G_m^* \Delta_- U_m, & \sigma_m \geqslant 0, \end{cases}$$

$$\Delta_{+2} U_m = \frac{1}{2}(U_{m+2} - U_m), \quad \Delta_{-2} U_m = \frac{1}{2}(U_m - U_{m-2}),$$

$$(\sigma G^* \Delta_{\sigma 2} U)_m = \begin{cases} \sigma_m G_m^* \Delta_{+2} U_m, & \sigma_m \leqslant 0, \\ \sigma_m G_m^* \Delta_{-2} U_m, & \sigma_m \geqslant 0, \end{cases}$$

$$\mu f_m = \frac{1}{2}(f_{m+\frac{1}{2}} + f_{m-\frac{1}{2}}).$$

Some similar symbols will be adopted for u in (2.11). $\mathcal{G}(\lambda)$ denotes a set of integers, whose definition is as follows: it is the set

$1, 2, \cdots, M$, if λ corresponds to the model A_1,

$0, 1, \cdots, M-1$, if λ corresponds to the model A_2,

$0, 1, \cdots, M$, if λ corresponds to the model B,

or $1, 2, \cdots, M-1$, if λ corresponds to the model C.

$[m_1, m_2]$ denotes the set $m_1, m_1+1, \cdots, m_2$ if $m_1 \leqslant m_2$; it is an empty set if $m_1 > m_2$, m_1 and m_2 being nonnegative integers.

In this section we shall present several schemes which satisfy the following three conditions when they are applied to the four models.

Condition (i) For any $m \in \mathcal{G}(\lambda)$, we can obtain a difference equation in which only the values at points in the interval $[0, 1]$ appear, and the scheme satisfies the von Neumann condition everywhere. That is, for the homogeneous equation

$$\frac{\partial u}{\partial t} + \lambda \frac{\partial u}{\partial \xi} = 0,$$

we can obtain the following form of difference equations

$$\sum_{h=H_1(m)}^{H_2(m)} r_{h,m}^k u_{m+h}^{k+1} = \sum_{h=H_1(m)}^{H_2(m)} s_{h,m}^k u_{m+h}^k, \tag{3.1}$$

$$m \in \mathcal{G}(\lambda), \quad H_1(m) \leqslant 0, \quad H_2(m) \geqslant 0,$$

$$0 \leqslant m + H_1(m) \leqslant m + H_2(m) \leqslant M,$$

where

$$r_{h,m}^k \equiv r_{h,m}^k(\Delta\xi, \Delta t) = r_h(\xi_m, t_k, \Delta\xi, \Delta t),$$

$$s_{h,m}^k \equiv s_{h,m}^k(\Delta\xi, \Delta t) = s_h(\xi_m, t_k, \Delta\xi, \Delta t),$$

and we suppose

$$\sum_{h=H_1}^{H_2} r_{h,m}^k = \sum_{h=H_1}^{H_2} s_{h,m}^k = 1.$$

Moreover, for any $m \in \mathcal{G}(\lambda)$, the von Neumann condition

$$r_m^{k*}(\theta) r_m^k(\theta) - s_m^{k*}(\theta) s_m^k(\theta) \geqslant 0 \tag{3.2}$$

holds, where

$$r_m^k(\theta)=\sum_{h=H_1}^{H_2} r_{h,m}^k(0,\ 0)e^{ih\theta},$$

$$s_m^k(\theta)=\sum_{h=H_1}^{H_2} s_{h,m}^k(0,\ 0)e^{ih\theta},$$

and $r_m^{k*}(\theta)$ and $s_m^{k*}(\theta)$ are the conjugate complex numbers of $r_m^k(\theta)$ and $s_m^k(\theta)$.

Condition (ii) The "well-conditioned" condition

$$r_m^{k*}(\theta)r_m^k(\theta)\geqslant c_0>0,\quad m\in\mathscr{G}(\lambda) \tag{3.3}$$

holds everywhere, where c_0 is a constant. (The condition almost guarantees that the difference systems of pure-initial problems are well-conditioned, so it is called the "well-conditioned" condition. For explicit schemes, it always holds.)

Condition (iii) The scheme is Lipschitz continuous, i.e., $r_{h,m}^k$ and $s_{h,m}^k$ satisfy the Lipschitz condition with respect to both t and ξ

$$\begin{aligned}&|r_{h,m+1}^k-r_{h,m}^k|\leqslant\tilde{c}_0\Delta\xi,\quad |r_{h,m}^{k+1}-r_{h,m}^k|\leqslant\tilde{c}_0\Delta t,\\ &|s_{h,m+1}^k-s_{h,m}^k|\leqslant\tilde{c}_0\Delta\xi,\quad |s_{h,m}^{k+1}-s_{h,m}^k|\leqslant\tilde{c}_0\Delta t.\end{aligned} \tag{3.4a}$$

Moreover, they also satisfy

$$\begin{aligned}&|r_{h,m}^k(\Delta\xi,\ \Delta t)-r_{h,m}^k(0,\ 0)|\leqslant\tilde{c}_0(\Delta\xi+\Delta t),\\ &|s_{h,m}^k(\Delta\xi,\ \Delta t)-s_{h,m}^k(0,\ 0)|\leqslant\tilde{c}_0(\Delta\xi+\Delta t).\end{aligned} \tag{3.4b}$$

In the following, sometimes we make condition (iii) stronger. For example, we require that $\dfrac{\partial r_h}{\partial\xi}$ and $\dfrac{\partial s_h}{\partial\xi}$ be continuous, and that r_h and s_h be section-wise analytic in closed intervals. Sometimes, we make condition (iii) weaker. For example, we only require that the scheme be section-wise Lipschitz continuous, i.e., after [0, 1] is divided into several subintervals, (3.4a) is required to hold in each of these subintervals. Of course, in this case, the quantities in each difference equation must belong to an identical subinterval.

In the next section, for several classes of initial-boundary-value schemes, we shall prove that the schemes are stable if these conditions hold. Therefore, it is reasonable to take these conditions as an criterion we shall follow in constructing schemes for initial-boundary-value problems.

The following are four schemes which satisfy the three conditions above.

1. *First Order Uncentered Four-point Schemes*

If the equation (2.10) is approximated by

$$\begin{aligned}&\boldsymbol{G}_m^{*k}\boldsymbol{U}_m^{k+1}+\alpha(\sigma^k\boldsymbol{G}^{*k}\Delta_\sigma\boldsymbol{U}^{k+1})_m\\ &\quad=\boldsymbol{G}_m^{*k}\boldsymbol{U}_m^k-(1-\alpha)(\sigma\boldsymbol{G}^*\Delta_\sigma\boldsymbol{U})_m^k+\Delta tf_m^k,\end{aligned} \tag{3.5}$$

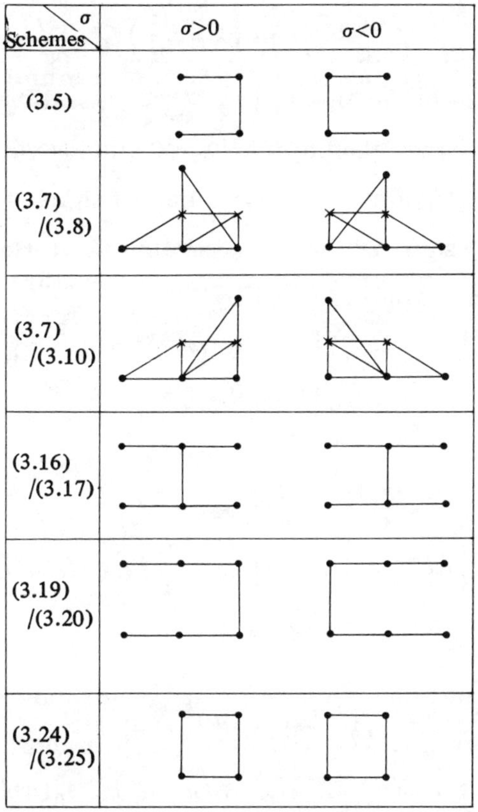

Figure 3.1 Graphic representation of several schemes

then for all the models A_1, A_2, B and C, we can obtain the difference equations desired, and the scheme is Lipschitz continuous if λ satisfies the Lipschitz condition. Moreover, when the condition

$$|\sigma| \begin{cases} \leqslant \dfrac{1}{1-2\alpha}, & \text{if} -\infty<\alpha<\dfrac{1}{2}, \\ <\infty, & \text{if } \dfrac{1}{2}\leqslant\alpha<\infty \end{cases} \tag{3.6}$$

is fulfilled, the conditions (3.2) and (3.3) hold. Therefore, this is a good scheme for initial–boundary–value problems. When $\alpha=0$, it is the C–I–R scheme[16]. Keller and Thomée have presented a similar scheme. Unfortunately, they are all schemes of the first order.

2. *Two-step Second Order Explicit Schemes*

At the first step, (2.10) is approximated by

$$G_m^{*k} U_m^{k+\alpha} = G_m^{*k} U_m^k - \alpha(\sigma G^* \Delta_\sigma U)_m^k + \alpha \Delta t f_m^k. \tag{3.7}$$

For the second step, (2.10) is first rewritten as

$$G^* \left(\frac{\partial U}{\partial t} \mp \frac{\Delta\xi}{\Delta t} \frac{\partial U}{\partial \xi} \right) + \left(\lambda \pm \frac{\Delta\xi}{\Delta t} \right) G^* \frac{\partial U}{\partial \xi} = f.$$

Here the sign is selected in this way: if $\lambda_{m+\frac{1}{2}}^{k+\alpha} > 0$ and $\lambda_{m-\frac{1}{2}}^{k+\alpha} > 0$, then at the point m the upper sign is chosen; if $\lambda_{m+\frac{1}{2}}^{k+\alpha} < 0$ and $\lambda_{m-\frac{1}{2}}^{k+\alpha} < 0$, then the lower sign is chosen; if $\lambda_{m+\frac{1}{2}}^{k+\alpha}$ and $\lambda_{m-\frac{1}{2}}^{k+\alpha}$ have different signs (in this case $\lambda_{m+\frac{1}{2}}^{k+\alpha} \approx \lambda_{m-\frac{1}{2}}^{k+\alpha} \approx 0$), then the sign is chosen according to certain requirements. Then, the equation is approximated by

$$\begin{aligned} \mu G_{m\pm\frac{1}{2}}^{*k+\alpha} U_m^{k+1} = {} & \mu G_{m\pm\frac{1}{2}}^{*k+\alpha} U_{m\pm 1}^k - (\mu \sigma_{m\pm\frac{1}{2}}^{k+\alpha} \pm 1) \mu G_{m\pm\frac{1}{2}}^{*k+\alpha} \Delta_\pm U_m^{k+\alpha} \\ & + \Delta t \mu f_{m\pm\frac{1}{2}}^{k+\alpha} \equiv S_{1,a} U_m^k; \end{aligned} \tag{3.8a}$$

or

$$\begin{aligned} \mu G_{m\pm\frac{1}{2}}^{*k+\alpha} U_m^{k+1} = {} & \mu G_{m\pm\frac{1}{2}}^{*k+\alpha} U_{m\pm 1}^k - (\mu \sigma_{m\pm\frac{1}{2}}^{k+\alpha} \pm 1) \\ & \times \left[\mu G_{m\pm\frac{1}{2}}^{*k+\alpha} \Delta_\pm U_m^{k+\alpha} + \left(\alpha - \frac{1}{2} \right) \Delta_\mp (\sigma G^* \Delta_\sigma U)_{m\pm 1}^k \right] \\ & + \Delta t \mu f_{m\pm\frac{1}{2}}^{k+\alpha} \equiv S_{1,b} U_m^k; \end{aligned} \tag{3.8b}$$

or

$$\begin{aligned} \mu G_{m\pm\frac{1}{2}}^{*k+\alpha} U_m^{k+1} = {} & \mu G_{m\pm\frac{1}{2}}^{*k+\alpha} U_{m\pm 1}^k - (\mu \sigma_{m\pm\frac{1}{2}}^{k+\alpha} \pm 1) \\ & \times \Big[\mu G_{m\pm\frac{1}{2}}^{*k+\alpha} \Delta_\pm U_m^{k+\alpha} + \left(\alpha - \frac{1}{2} \right) (2\Delta_\mp (\sigma G^* \Delta_\sigma U)_{m\pm 1}^k \\ & \mp \Delta_\mp (G^* \Delta_\mp U)_{m\pm 1}^k) \Big] + \Delta t \mu f_{m\pm\frac{1}{2}}^{k+\alpha} \equiv S_{1,c} U_m^k. \end{aligned} \tag{3.8c}$$

Here, the purpose of adding the second order terms in (3.8b) and (3.8c) is to extend or change the range in the α, σ–plane where the condition (3.2) holds. Clearly, when $\sigma_m^{k+\alpha} = 0$, the three expressions can be rewritten as

$$\mu G_{m\pm\frac{1}{2}}^{*k+\alpha} U_m^{k+1} = \mu G_{m\pm\frac{1}{2}}^{*k+\alpha} U_m^k + \Delta t \mu f_{m\pm\frac{1}{2}}^{k+\alpha} + O(\Delta\xi) U_m^k,$$

where $O(\Delta\xi)$ is an operator in which the moduli of coefficients are less than $k\Delta\xi$, k being a constant. Therefore, it does not make any essential difference that we take the simpler formula

$$G_m^{*k+\alpha} U_m^{k+1} = G_m^{*k+\alpha} U_m^k + \Delta t f_m^{k+\alpha}$$

instead of (3.8a)—(3.8c) when $\sigma_m^{k+\alpha} = 0$. In the above schemes, a difference equation at a point m involves the values at the points $m-1$, m, $m+1$ (but if $\lambda \approx 0$, an equation may involve the values at four points), and it can be applied to the model C. Moreover, the conditions

$$\frac{1}{2}\leqslant\alpha,\quad 0\leqslant|\sigma|\leqslant\frac{1}{2\alpha},\tag{3.9a}$$

$$|\sigma|\leqslant 1,\tag{3.9b}$$

$$\frac{1}{2}\leqslant\alpha\leqslant 1,\quad |\sigma|\leqslant 1\tag{3.9c}$$

guarantee respectively that the schemes (3.8a), (3.8b) and (3.8c) satisfy the von Neumann condition (3.2). The Richtmyer[21] scheme, the MacCormack[23] scheme and the schemes here with $\alpha=\frac{1}{2}$ are the variants of the Lax–Wendroff[18] scheme.

However, in general, these schemes cannot be applied to the models A_1, A_2 and B, unless $\lambda\geqslant 0$ at $\xi=0$ and $\lambda\leqslant 0$ at $\xi=1$. In fact, if $\lambda\neq 0$ at a boundary point and we need a difference equation at the boundary point, then only the uncentered schemes can be used in order to guarantee that only the quantities at points in the interval $0\leqslant\xi\leqslant 1$ appear in the difference equation. Thus, a single scheme is not enough to solve initial–boundary–value problems. We need to construct some other schemes. The first step is the same. However, for the second step, (2.2) is first rewritten as

$$G^*\left(\frac{\partial U}{\partial t}\pm\frac{\Delta\xi}{\Delta t}\,\frac{\partial U}{\partial\xi}\right)+\left(\lambda\mp\frac{\Delta\xi}{\Delta t}\right)G^*\,\frac{\partial U}{\partial\xi}=f.$$

Here, if $\lambda>0$, then the upper signs are chosen; if $\lambda<0$, then the lower signs are chosen; if $\lambda\approx 0$, then the signs are chosen according to certain requirements. Then the equation is approximated by

$$\begin{aligned}\mu G^{*k+\alpha}_{m\mp\frac{1}{2}}U^{k+1}_m=&\,\mu G^{*k+\alpha}_{m\mp\frac{1}{2}}U^k_{m\mp1}-(\mu\sigma^{k+\alpha}_{m\mp\frac{1}{2}}\mp1)\mu G^{*k+\alpha}_{m\mp\frac{1}{2}}\Delta_\mp U^{k+\alpha}_m\\&+\Delta t\mu f^{k+\alpha}_{m\mp\frac{1}{2}}\equiv S_{2,a}U^k_m;\end{aligned}\tag{3.10a}$$

or

$$\begin{aligned}\mu G^{*k+\alpha}_{m\mp\frac{1}{2}}U^{k+1}_m=&\,\mu G^{*k+\alpha}_{m\mp\frac{1}{2}}U^k_{m\mp1}-(\mu\sigma^{k+\alpha}_{m\mp\frac{1}{2}}\mp1)\\&\times\left[\mu G^{*k+\alpha}_{m\mp\frac{1}{2}}\Delta_\mp U^{k+\alpha}_m+\left(\alpha-\frac{1}{2}\right)\Delta_\mp(\sigma G^*\Delta_\sigma U)^k_m\right]\\&+\Delta t\mu f^{k+\alpha}_{m\mp\frac{1}{2}}\equiv S_{2,b}U^k_m;\end{aligned}\tag{3.10b}$$

or

$$\begin{aligned}\mu G^{*k+\alpha}_{m\mp\frac{1}{2}}U^{k+1}_m=&\,\mu G^{*k+\alpha}_{m\mp\frac{1}{2}}U^k_{m\mp1}-(\mu\sigma^{k+\alpha}_{m\mp\frac{1}{2}}\mp1)\\&\times[\mu G^{*k+\alpha}_{m\mp\frac{1}{2}}\Delta_\mp U^{k+\alpha}_m+(2\alpha-1)\Delta_\mp(\sigma G^*\Delta_\sigma U)^k_m]\\&\pm\left(\alpha-\frac{1}{2}\right)\Delta_\mp(\sigma G^*\Delta_\sigma U)^k_m+\Delta t\mu f^{k+\alpha}_{m\mp\frac{1}{2}}\equiv S_{2,c}U^k_m.\end{aligned}\tag{3.10c}$$

The purpose of adding the second order terms in (3.10b) and (3.10c) is also to extend or change the range in the α, σ–plane where the condition (3.2) holds. In these schemes, if $\lambda\geqslant\varepsilon>0$, the difference equations at a point m involve the points m, $m-1$ and $m-2$; if $\lambda\leqslant -s<0$, they involve the points $m, m+1, m+2$; if $\lambda\approx 0$, they can be rewritten as

$$\mu G^{*k+\alpha}_{m\mp\frac{1}{2}} U^{k+1}_m = \mu G^{*k+\alpha}_{m\mp\frac{1}{2}} U^k_m + \Delta t\mu f^{k+\alpha}_{m\mp\frac{1}{2}} + O(\Delta\xi)U^k_m.$$

Therefore, as λ changes its sign, different schemes will be adopted, but condition (iii) still holds if λ is smooth. As we have done for (3.8a—c), we modify (3.10a—c) for simplicity in the following way: when $\sigma^{k+\alpha}_m=0$, instead of (3.10a—c), we take

$$G^{*k+\alpha}_m U^{k+1}_m = G^{*k+\alpha}_m U^k_m + \Delta t f^{k+\alpha}_m.$$

Obviously, when these schemes are applied to the model B, the first part of condition (i) is satisfied, i.e., only the values at the points in the interval [0, 1] are involved in difference equations. Moreover, the conditions

$$\begin{cases} -\frac{1}{2}\leqslant \alpha<\frac{1}{2}, & 0\leqslant|\sigma|\leqslant 1, \\ \alpha=\frac{1}{2}, & 0\leqslant|\sigma|\leqslant 2; \end{cases} \tag{3.11a}$$

$$0\leqslant|\sigma|\leqslant 2; \tag{3.11b}$$

$$\frac{1}{2}\leqslant\alpha\leqslant 1,\quad 0\leqslant|\sigma|\leqslant 2 \tag{3.11c}$$

guarantee respectively that the schemes (3.10a), (3.10b) and (3.10c) satisfy the von Neumann condition. Therefore, they are applicable to the model B. The schemes (3.10b) and (3.10c) are not applicable to the models A_1, A_2 and C. This is because the values at the point $m=-1$ (or $M+1$) appear in the difference equations at the point $m=1$ (or $M-1$), i.e. because the first part of condition (i) is not satisfied. The scheme (3.10a) is also not applicable to A_1, A_2 and C, because the schemes actually adopted at the points $m=1$ and $m=2$ (or $m=M-1$ and $M-2$) are different, and condition (iii) is not fulfilled.

However, taking the scheme (3.7)/(3.8) ((3.7)/(3.8) means that (3.7) is used at the first step and (3.8j) is used at the second step, j=a, b or c), and (3.7)/(3.10) as bases, we can construct schemes applicable to every case in the following way. (i) If $\lambda\leqslant 0$ at the point $\xi=0$, and if the maximal region where $\lambda\leqslant 0$ and which includes the point $\xi=0$ is $0\leqslant\xi\leqslant\xi_2$, then we take

$$m_1=\min\left\{E\left(\frac{\xi_1}{\Delta\xi}\right), M-1\right\},\quad m_0=0,$$

where $E\left(\frac{\xi_1}{\Delta\xi}\right)$ is the integer part of $\frac{\xi_1}{\Delta\xi}$; if $\lambda>0$ at the point $\xi=0$, then let $[m_0, m_1]$ be an empty set and $m_1=0$. (ii) If $\lambda\geqslant 0$ at the point $\xi=1$, and if the maximal region where $\lambda \geqslant 0$ and which includes the point $\xi = 1$ is $\xi_2 \leqslant \xi \leqslant 1$, then we take

$$m_2 = \max\left\{M - E\left(\frac{1-\xi_2}{\Delta\xi}\right), 1\right\}, \quad m_3 = M;$$

if $\lambda < 0$ at $\xi = 1$, then let $m_2 = M$ and $[m_2, m_3]$ be an empty set. After having m_0, m_1, m_2, m_3, at the second step, we calculate U_m^{k+1} by using

$$\begin{cases} (\beta_{1,m}\mu G^{*k+\alpha}_{m\pm\frac{1}{2}} + (1-\beta_{1,m})\mu G^{*k+\alpha}_{m\mp\frac{1}{2}}) U_m^{k+1} \\ \quad = \beta_{1,m} S_{1,j} U_m^k + (1-\beta_{1,m}) S_{2,j} U_m^k, \quad m \in [m_0, m_1], \\ \mu G^{*k+\alpha}_{m\pm\frac{1}{2}} U_m^{k+1} = S_{1,j} U_m^k, \quad m \in [m_1+1, m_2-1], \\ ((1-\beta_{2,m})\mu G^{*k+\alpha}_{m\pm\frac{1}{2}} + \beta_{2,m}\mu G^{*k+\alpha}_{m\mp\frac{1}{2}}) U_m^{k+1} \\ \quad = (1-\beta_{2,m}) S_{1,j} U_m^k + \beta_{2,m} S_{2,j} U_m^k, \quad m \in [m_2, m_3]. \end{cases} \tag{3.12}$$

$$(j = a, \text{ b or c.})$$

Here, we require that

$$\beta_{1,m}\big|_{m=m_0} = 0, \quad \beta_{1,m}\big|_{m=m_1} = 1,$$
$$\beta_{2,m}\big|_{m=m_2} = 0, \quad \beta_{2,m}\big|_{m=m_3} = 1,$$

that $\beta_{1,m}$ and $\beta_{2,m}$ be smooth functions of m, and that $0 \leqslant \beta_{1,m} \leqslant 1$, $0 \leqslant \beta_{2,m} \leqslant 1$. For example, let

$$\beta_{1,m} = \frac{m - m_0}{m_1 - m_0}, \quad \beta_{2,m} = \frac{m - m_2}{m_3 - m_2}, \tag{3.13}$$

where if the denominator and numerator are equal to zero, let $\beta_{1,m} = 0$, $\beta_{2,m} = 1$. In this case, if λ satisfies the Lipschitz condition with respect to t and ξ, then condition (iii) holds. Moreover, if the schemes (3.7)/(3.8) and (3.7)/(3.10) satisfy the condition (3.2), then the scheme (3.7)/(3.12) also satisfies (3.2).

If we want $\beta_{1,m}$ and $\beta_{2,m}$ to be smoother and only the scheme (3.7)/(3.10) to be used near the boundaries, then let

$$\begin{cases} \beta_{1,m} = \begin{cases} 0, & m_0 \leqslant m \leqslant m_0', \\ \left(\dfrac{m - m_0'}{m_1 - m_0'}\right)^2 \left(3 - 2\dfrac{m - m_0'}{m_1 - m_0'}\right), & m_0' < m \leqslant m_1, \end{cases} \\ \beta_{2,m} = \begin{cases} \left(\dfrac{m - m_2}{m_3' - m_2}\right)^2 \left(3 - 2\dfrac{m - m_2}{m_3' - m_2}\right), & m_2 \leqslant m < m_3', \\ 1, & m_3' \leqslant m \leqslant m_3, \end{cases} \end{cases} \tag{3.14}$$

where

$$m_0' = m_0 + E\left(\frac{m_1 - m_0}{2}\right), \quad m_3' = m_3 - E\left(\frac{m_3 - m_2}{2}\right).$$

Clearly,

$$\begin{cases} \left.\dfrac{\partial \beta_{1,m}}{\partial m}\right|_{m=m_0'} = \left.\dfrac{\partial \beta_{1,m}}{\partial m}\right|_{m=m_1} = 0, \\ \left.\dfrac{\partial \beta_{2,m}}{\partial m}\right|_{m=m_2} = \left.\dfrac{\partial \beta_{2,m}}{\partial m}\right|_{m=m_3'} = 0. \end{cases}$$

Therefore, when the scheme (3.7)/(3.12) with (3.14) is applied to the models, if λ and $\frac{\partial\lambda}{\partial\xi}$ are continuous and λ is section–wise analytic, then the coefficients of the schemes have the same properties as λ does.

It is easily seen that because the scheme (3.8) is gradually connected to (3.10) in the scheme (3.12), not only can this scheme be applied to every model, but also the coefficients of this scheme are smooth functions. Moreover, the conditions

$$\alpha=\frac{1}{2},\quad |\sigma|\leqslant 1, \tag{3.15a}$$

$$|\sigma|\leqslant 1, \tag{3.15b}$$

$$\frac{1}{2}\leqslant\alpha\leqslant 1,\quad |\sigma|\leqslant 1 \tag{3.15c}$$

guarantee respectively that the schemes (3.7)/(3.12j), j=a, b, c, satisfy the von Neumann condition.

Finally, we should point out that these schemes have second order accuracy when $\alpha=\frac{1}{2}$. Consider (3.7)/(3.10), and suppose $\lambda>0$. It is clear that (3.7) is a first order scheme. That is, at the levels $k+\frac{1}{2}$, every quantity has an error of $O(\Delta t^2)$. Therefore

$$\mu G^{*k+\frac{1}{2}}_{m-\frac{1}{2}}\,\frac{U^{k+1}_m-U^k_{m-1}}{\Delta t}=\left[G^*\left(\frac{\partial U}{\partial t}+\frac{\Delta\xi}{\Delta t}\,\frac{\partial U}{\partial\xi}\right)\right]^{k+\frac{1}{2}}_{m-\frac{1}{2}}+O(\Delta t^2),$$

$$\mu G^{*k+\frac{1}{2}}_{m-\frac{1}{2}}\,\frac{U^{k+\frac{1}{2}}_m-U^{k+\frac{1}{2}}_{m-1}}{\Delta\xi}=\left[G^*\frac{\partial U}{\partial\xi}\right]^{k+\frac{1}{2}}_{m-\frac{1}{2}}+O(\Delta t^2)$$

$$+\frac{1}{\Delta\xi}(O_m(\Delta t^2)-O_{m-1}(\Delta t^2)),$$

where $[f]^{k+\frac{1}{2}}_{m-\frac{1}{2}}$ denotes the exact value of f at $t=\left(k+\frac{1}{2}\right)\Delta t$, $\xi=\xi_{m-\frac{1}{2}}$, an O_m, O_{m-1} are the errors of $\mu G^{*k+\frac{1}{2}}_{m-\frac{1}{2}}U^{k+\frac{1}{2}}_m$, $\mu G^{*k+\frac{1}{2}}_{m-\frac{1}{2}}U^{k+\frac{1}{2}}_{m-1}$ respectively. Obviously, whether this scheme is a second order one depends on whether $O_m(\Delta t^2)-O_{m-1}(\Delta t^2)$ is a quantity of $O(\Delta t^3)$. If the algorithms for every point are the same, it is evidently true. However, the algorithms for interior points and boundary points are different because the boundary conditions are used in the algorithm for boundary points. Thus, the forms of truncation errors of U are different. Therefore, in order to prove the scheme to be a second order one, some calculations are necessary. $G^{*k}_m U^{k+\frac{1}{2}}_m$ and $G^{*k}_{m-1}U^{k+\frac{1}{2}}_{m-1}$ are obtained by using an identical method, so

$$G_m^{*k}U_m^{k+\frac{1}{2}} - G_m^{*k}[U]_m^{k+\frac{1}{2}} - (G_{m-1}^{*k}U_{m-1}^{k+\frac{1}{2}} - G_{m-1}^{*k}[U]_{m-1}^{k+\frac{1}{2}}) = O(\Delta t^3).$$

Therefore, the only thing we need to do is to prove that the difference between

$$O_m(\Delta t^2) - O_{m-1}(\Delta t^2)$$

and

$$G_m^{*k}U_m^{k+\frac{1}{2}} - G_m^{*k}[U]_m^{k+\frac{1}{2}} - G_{m-1}^{*k}U_{m-1}^{k+\frac{1}{2}} + G_{m-1}^{*k}[U]_{m-1}^{k+\frac{1}{2}}$$

is $O(\Delta t^3)$. In fact, we have

$$\begin{aligned}
&G_m^{*k}U_m^{k+\frac{1}{2}} - G_m^{*k}[U]_m^{k+\frac{1}{2}} - G_{m-1}^{*k}U_{m-1}^{k+\frac{1}{2}} + G_{m-1}^{*k}[U]_{m-1}^{k+\frac{1}{2}} \\
&\quad - \mu G^{*k+\frac{1}{2}}_{m-\frac{1}{2}}U_m^{k+\frac{1}{2}} + [G^*]^{k+\frac{1}{2}}_{m-\frac{1}{2}}[U]_m^{k+\frac{1}{2}} + \mu G^{*k+\frac{1}{2}}_{m-\frac{1}{2}}U_{m-1}^{k+\frac{1}{2}} - [G^*]^{k+\frac{1}{2}}_{m-\frac{1}{2}}[U]_{m-1}^{k+\frac{1}{2}} \\
&= \left(G_m^{*k} - \mu G^{*k+\frac{1}{2}}_{m-\frac{1}{2}}\right)\left(U_m^{k+\frac{1}{2}} - [U]_m^{k+\frac{1}{2}}\right) \\
&\quad - \left(G_{m-1}^{*k} - \mu G^{*k+\frac{1}{2}}_{m-\frac{1}{2}}\right)\left(U_{m-1}^{k+\frac{1}{2}} - [U]_{m-1}^{k+\frac{1}{2}}\right) \\
&\quad - \left(\mu G^{*k+\frac{1}{2}}_{m-\frac{1}{2}} - [G^*]^{k+\frac{1}{2}}_{m-\frac{1}{2}}\right)\left([U]_m^{k+\frac{1}{2}} - [U]_{m-1}^{k+\frac{1}{2}}\right) = O(\Delta t^3),
\end{aligned}$$

so the conclusion is true. Here, we suppose that $\Delta t/\Delta\xi$ is bounded, and that U and G^* satisfy the Lipschitz condition with respect to ξ and with respect to U, t, ξ respectively.

In the case $\lambda<0$, and for the schemes (3.7)/(3.8) and (3.7)/(3.12), the conclusion can be proved in the same way. Therefore, if $\alpha=\frac{1}{2}$, these two-step schemes for initial-boundary-value problems have second order accuracy.

In passing, we may note the following fact: the scheme (3.7) can be rewritten in the form of the scheme of characteristics with the linear interpolation

$$G_m^{*k}U_m^{k+\alpha} = G_m^{*k}((1-\alpha|\sigma|)U_m^k + \alpha|\sigma|U_{m\mp1}^k) + \alpha\Delta t f_m^k,$$

where the negative sign in "$\mp$" is chosen if $\sigma>0$ and the positive sign is chosen if $\sigma<0$; and then the schemes (3.7)/(3.8b) and (3.7)/(3.10b) are close to the scheme of characteristics with the centered and uncentered quadratic interpolations respectively. In fact, it can be easily checked that for the homogeneous equation with a constant coefficient

$$\frac{\partial u}{\partial t} + \lambda\frac{\partial u}{\partial \xi} = 0,$$

they are the same schemes as the method of characteristics using centered quadratic interpolation

$$u_m^{k+1} = \frac{1}{2}(1+\sigma)\sigma u_{m-1}^k + (1+\sigma)(1-\sigma)u_m^k - \frac{1}{2}(1-\sigma)\sigma u_{m+1}^k$$

and the method of characteristics using uncentered quadratic interpolation

$$u_m^{k+1} = -\frac{1}{2}(1-|\sigma|)|\sigma|u_{m\mp2}^k + |\sigma|\ (2-|\sigma|)u_{m\mp1}^k + \frac{1}{2}(2-|\sigma|)(1-|\sigma|)u_m^k$$

respectively, where the negative sign in "$\mp$" is selected if $\sigma>0$ and the positive sign is chosen if $\sigma<0$.

3. *Second Order Implicit Schemes*

We have already discussed second order explicit schemes. However, if $|\lambda|$ is very large and the functions do not vary rapidly with respect to t, then the implicit schemes are more suitable than the explicit schemes. The reason for this is the following: if an explicit scheme is adopted, then step Δt must be very small; however, from the viewpoint of the truncation error, a very small Δt is not necessary; and thus, the implicit scheme should be adopted in order to be able to use a larger Δt. In the following, we shall give a universal scheme based on two six-point schemes. The two basic schemes are a centered and an uncentered six-point schemes.

(i) *A Centered Six-point Scheme*

To begin with the difference equation is

$$G_m^{*k} U_m^{k+\alpha} + \alpha\sigma_m^k G_m^{*k} \Delta_0 U_m^{k+\alpha} = G_m^{*k} U_m^k + \alpha \Delta t f_m^k, \tag{3.16}$$

which, in what follows, will then be abbreviated to:

$$R_3 U_m^{k+\alpha} = S_3 U_m^k.$$

In the second step, the difference equation is

$$\begin{aligned} &G_m^{*k+\alpha} U_m^{k+1} + \alpha\sigma_m^{k+\alpha} G_m^{*k+\alpha} \Delta_0 U_m^{k+1} \\ &= G_m^{*k+\alpha} U_m^k - (1-\alpha)\sigma_m^{k+\alpha} G_m^{*k+\alpha} \Delta_0 U_m^k + \Delta t f_m^{k+\alpha}, \end{aligned} \tag{3.17}$$

which, in what follows, will be abbreviated to:

$$R_4 U_m^{k+1} = S_4 U_m^k.$$

In this scheme, the equations at m involve the values at the points $m-1$, m, $m+1$. When

$$\alpha \geqslant \frac{1}{2}, \tag{3.18}$$

the von Neumann condition is satisfied. The condition (3.3) is always satisfied for this scheme. Thus, it is applicable to the model C. However, generally, it is not applicable to the models A_1, A_2 and B. Crank and Nicolson[15] have presented a similar scheme for parabolic equations. The scheme here is for hyperbolic equations and a two-step scheme.

(ii) *An Uncentered Six-point Scheme*

As was the case in the second scheme, the centered scheme is not satisfactory for the initial-boundary-value problems. The following is an uncentered implicit scheme which will be combined with (3.16)/

(3.17) in the same way as (3.7)/(3.10) has been combined with (3.7)/(3.8).

For the first step, the approximate equation is

$$\begin{aligned}G^{*k}_m U^{k+\alpha}_m + \alpha[2(\sigma^k G^{*k}\Delta_\sigma U^{k+\alpha})_m - (\sigma^k G^{*k}\Delta_{\sigma 2} U^{k+\alpha})_m]\\ = G^{*k}_m U^k_m + \alpha\Delta t f^k_m,\end{aligned} \tag{3.19}$$

which, in what follows, will be abbreviated to:

$$R_5 U^{k+\alpha}_m = S_5 U^k_m.$$

In the second step, the difference equation is

$$\begin{aligned}G^{*k+\alpha}_m U^{k+1}_m + \alpha[2(\sigma^{k+\alpha} G^{*k+\alpha}\Delta_\sigma U^{k+1})_m - (\sigma^{k+\alpha} G^{*k+\alpha}\Delta_{\sigma 2} U^{k+1})_m]\\ = G^{*k+\alpha}_m U^k_m - (1-\alpha)[2(\sigma^{k+\alpha} G^{*k+\alpha}\Delta_\sigma U^k)_m\\ - (\sigma^{k+\alpha} G^{*k+\alpha}\Delta_{\sigma 2} U^k)_m] + \Delta t f^{k+\alpha}_m,\end{aligned} \tag{3.20}$$

which is abbreviated to:

$$R_6 U^{k+1}_m = S_6 U^k_m.$$

In this scheme, if $\lambda > 0$, the equations at the point m involve the values at the points $m-2$, $m-1$ and m; if $\lambda < 0$, they involve the values at the points m, $m+1$ and $m+2$. If $\lambda \approx 0$, the equations can be rewritten as

$$G^{*k+\alpha}_m U^{k+1}_m + O(\Delta\xi) U^{k+1}_m = G^{*k+\alpha}_m U^k_m + O(\Delta\xi) U^k_m + \Delta t f^{k+\alpha}_m,$$

where $O(\Delta\xi)$ denotes an operator in which the moduli of coefficients are less than $k\Delta\xi$, k being a constant. Therefore, as the sign of λ changes, condition (iii) still holds even though the equations used change from (3.16)/(3.17) to (3.19)/(3.20) or from (3.19)/(3.20) to (3.16)/(3.17). It is easy to prove that when

$$\alpha \geqslant \frac{1}{2}, \tag{3.21}$$

the von Neumann condition is satisfied and that when $\alpha \geqslant 0$, the condition (3.3) holds.

Therefore, for the model B in which $\lambda \leqslant 0$ near the boundary $\xi = 0$ and $\lambda \geqslant 0$ near the boundary $\xi = 1$, this scheme can be applied. Moreover, conditions (i)—(iii) hold if (3.21) is valid and if λ satisfies the Lipschitz condition with respect to ξ and t. It is not applicable in the other cases.

Taking (3.16)/(3.17) and (3.19)/(3.20) as basic schemes, we can construct a universal scheme, as has been done for (3.7)/(3.12). For the first step, the difference equations are

$$\begin{cases}\beta_{1,m} R_3 U^{k+\alpha}_m + (1-\beta_{1,m}) R_5 U^{k+\alpha}_m\\ \quad = \beta_{1,m} S_3 U^k_m + (1-\beta_{1,m}) S_5 U^k_m, & m \in [m_0, m_1],\\ R_3 U^{k+\alpha}_m = S_3 U^k_m, & m \in [m_1+1, m_2-1],\\ (1-\beta_{2,m}) R_3 U^{k+\alpha}_m + \beta_{2,m} R_5 U^{k+\alpha}_m\\ \quad = (1-\beta_{2,m}) S_3 U^k_m + \beta_{2,m} S_5 U^k_m, & m \in [m_2, m_3].\end{cases} \tag{3.22}$$

In the second step, the equations are

$$\begin{cases} \beta_{1,m}R_4U_m^{k+1}+(1-\beta_{1,m})R_6U_m^{k+1} \\ \quad =\beta_{1,m}S_4U_m^k+(1-\beta_{1,m})S_6U_m^k, & m\in[m_0,\ m_1], \\ R_4U_m^{k+1}=S_4U_m^k, & m\in[m_1+1,\ m_2-1], \\ (1-\beta_{2,m})R_4U_m^{k+1}+\beta_{2,m}R_6U_m^{k+1} \\ \quad =(1-\beta_{2,m})S_4U_m^k+\beta_{2,m}S_6U_m^k, & m\in[m_2,\ m_3]. \end{cases} \tag{3.23}$$

The two-step scheme is a second order one if $\alpha=\dfrac{1}{2}$. Of course, in the first step, the differential equations can also be approximated by

$$G_m^{*k}U_m^{k+\alpha}+\alpha(\sigma^kG^{*k}\Delta_\sigma U^{k+\alpha})_m=G_m^{*k}U_m^k+\alpha\Delta tf_m^k, \tag{3.5$'$}$$

which is the scheme (3.5) with $\alpha=1$. Clearly, (3.5)′/(3.23) is also a second order scheme if $\alpha=\dfrac{1}{2}$.

4. "*Section-wise*" *Mixed Second Order Schemes*

The number of operations for implicit schemes is greater than that for explicit schemes. In order to decrease the number of operations, implicit and explicit schemes can be combined in the following way: if λ is small, explicit schemes are adopted; if λ is large, implicit schemes are adopted. For example, one can combine (3.7)/(3.12) and (3.22)/(3.23). However, the implicit schemes now treat only the regions with $|\lambda|>\varepsilon$, where ε is a positive constant, so we can choose a simpler implicit scheme to match (3.7)/(3.12).

We shall consider the following four-point implicit scheme which also is a two-step one and similar to those schemes in the references [2], [4] and [19]. In the first step, the difference equation is

$$\frac{1}{2}\mu G^{*k}_{m\mp\frac{1}{2}}(U_m^{k+\alpha}+U_{m\mp1}^{k+\alpha})+\alpha\mu\sigma^k_{m\mp\frac{1}{2}}\mu G^{*k}_{m\mp\frac{1}{2}}\Delta_\mp U_m^{k+\alpha}$$
$$=\frac{1}{2}\mu G^{*k}_{m\mp\frac{1}{2}}(U_m^k+U_{m\mp1}^k)+\alpha\Delta t\mu f^k_{m\mp\frac{1}{2}}. \tag{3.24}$$

For the second step, the formula is

$$\frac{1}{2}\mu G^{*k+\alpha}_{m\mp\frac{1}{2}}(U_m^{k+1}+U_{m\mp1}^{k+1})+\alpha\mu\sigma^{k+\alpha}_{m\mp\frac{1}{2}}\mu G^{*k+\alpha}_{m\mp\frac{1}{2}}\Delta_\mp U_m^{k+1}$$
$$=\frac{1}{2}\mu G^{*k+\alpha}_{m\mp\frac{1}{2}}(U_m^k+U_{m\mp1}^k)$$
$$-(1-\alpha)\mu\sigma^{k+\alpha}_{m\mp\frac{1}{2}}\mu G^{*k+\alpha}_{m\mp\frac{1}{2}}\Delta_\mp U_m^k+\Delta t\mu f^{k+\alpha}_{m\mp\frac{1}{2}}, \tag{3.25}$$

where the upper sign is selected if $\lambda>0$, and the lower sign is chosen if $\lambda<0$ Clearly, it is a second order scheme if $\alpha=\dfrac{1}{2}$. If

$$\alpha\geqslant\frac{1}{2}, \tag{3.26}$$

the von Neumann condition holds. If

$$0<C_1' \leqslant \min\{|\sigma|, |\alpha|\}, \quad C_1' \text{ being a constant,} \tag{3.27}$$

is valid, the condition (3.3) holds. Moreover, if $\lambda>0$ and $m\in[1, M]$ or if $\lambda<0$ and $m\in[0, M-1]$, then the difference equations involve only the values at the points in the interval $[0, 1]$. Therefore, it is applicable to the models A_1 and A_2 in which $|\lambda|$ is always greater than ε.

(3.7)/(3.12) and (3.24)/(3.25) can be combined as follows. Let a constant $\varepsilon\in(0, \Delta\xi/\Delta t]$ and let $\lambda(\xi)$ be a continuous function. Obviously, the interval $[0, 1]$ can be divided into several subintervals such that $|\lambda|\geqslant\varepsilon$ or $|\lambda|\leqslant\varepsilon$ in each subinterval. Therefore, any model problem can be divided into several submodel problems in which $|\lambda|-\varepsilon$ does not change its sign. The types of submodels contain the models A_1 and A_2 with $|\lambda|\geqslant\varepsilon$ and the models A_1, A_2, B and C with $|\lambda|\leqslant\varepsilon$. On the other hand, (3.7)/(3.12) is applicable to any model with $|\lambda|\leqslant\varepsilon \leqslant \frac{\Delta\xi}{\Delta t}$, and (3.24)/(3.25) is applicable to the models A_1 and A_2 with $|\lambda|\geqslant\varepsilon >0$. Therefore, the combined scheme (3.7)/(3.12)—(3.24)/(3.25) is a universal scheme for initial–boundary–value problems and it satisfies conditions (i), (ii) and the "weakened" condition (iii).

This scheme can be used for all cases, including all the models and including the case where λ is very large, and its algorithm is simpler than (3.22)/(3.23). Therefore, much attention is paid to this scheme in this book. In Sections 5 and 6, we shall present a method for solution of those difference systems obtained by this scheme, and make a theoretical analysis of the method. In Chapter 2 a generalized form of this scheme in three independent variables will be presented. In Chapter 7 the generalized scheme will be applied to the calculation of complicated problems.

In passing, we would like to point out that it seems to be difficult to construct a third order explicit scheme which satisfies conditions (i)—(iii) and which can be used in practice. The reason for this is the following. If an uncentered explicit scheme

$$u_m^{k+1}=s_0u_m^k+s_{\pm1}u_{m\pm1}^k+s_{\pm2}u_{m\pm2}^k+s_{\pm3}u_{m\pm3}^k$$

approximates the equation $\dfrac{\partial u}{\partial t}+\lambda\dfrac{\partial u}{\partial\xi}=0$, and is a third order one, then it should take this form

$$u_m^{k+1}=\frac{(\sigma\mp1)(\sigma\mp2)(\sigma\mp3)}{\mp6}u_m^k+\frac{\sigma(\sigma\mp2)(\sigma\mp3)}{\pm2}u_{m\mp1}^k$$

$$+\frac{\sigma(\sigma\mp1)(\sigma\mp3)}{\mp2}u_{m\mp2}^k+\frac{\sigma(\sigma\mp1)(\sigma\mp2)}{\pm6}u_{m\mp3}^k,$$

where the upper sign is chosen if $\lambda>0$ and the lower sign is taken if

$\lambda<0$. Thus, it can be proved that in this case, (3.2) is equivalent to

$$\sigma=0, \text{ or } 1\leqslant|\sigma|\leqslant 2, \text{ or } |\sigma|=3.$$

Therefore, it is possible that the scheme is stable for initial-boundary-value problems only if

$$1\leqslant|\sigma|\leqslant 2.$$

Clearly, this condition is too strong from the viewpoint of its application.

It seems appropriate to say something about a scheme which we have used, which does not satisfy conditions (i)—(iii) of this section, and which seems to be unstable. Further, we shall present our opinion concerning "adoptable ranges" for an unstable scheme.

The scheme is the high order method of lines (see the appendix of Chapter 5).

The method is as follows: by using the Lagrange interpolation polynomial, $\dfrac{\partial \boldsymbol{U}_m}{\partial \xi}$ can be expressed as $\sum\limits_{j=0}^{M} c_{m,j}\boldsymbol{U}_j$, where $\boldsymbol{U}_j=\boldsymbol{U}(\xi_j, t)$, and $c_{m,j}$ are certain constants. Therefore, from (2.10), we obtain a system of ordinary differential equations

$$\boldsymbol{G}_m^*\frac{d\boldsymbol{U}_m}{dt}=-\lambda_m\boldsymbol{G}_m^*\sum_{j=0}^{M}c_{m,j}\boldsymbol{U}_j+\boldsymbol{f}_m,\quad m\in\mathscr{G}(\lambda).$$

After we have such a system for every λ, the original problem becomes an initial value problem for ordinary differential equations which should be simultaneously solved with certain nonlinear equations (boundary conditions). Moreover, if the nonlinear equations are also changed into ordinary differential equations, then the problem becomes aninitial problem for ordinary differential equations completely. This system can be solved by using the usual methods of solving ordinary differential equations, for example, by the Runge–Kutta method.

We have obtained some results by using this scheme. If the exact solution is smooth, this method can give good results by using a small M; if the exact solution is not smooth, the computed results will have large oscillations.

The stability of this scheme has not been carefully discussed. From the computed results, and for other consideration, it is doubtful that this scheme is stable if M is large. However, if M is small and if the true solution is smooth, actual computation shows us that it is a successful method: the amount of computation is smaller and the results have quite high accuracy. Therefore, we should consider whether or not an unstable scheme can be used under certain conditions, and our opinion is that it can. The reason for this is as follows: Suppose that we have a scheme

$$\boldsymbol{X}^{k+1}=A\boldsymbol{X}^k,$$

and that the eigenvalues of A are $|\tilde{\lambda}_0|\geqslant|\tilde{\lambda}_1|\geqslant\cdots\geqslant|\tilde{\lambda}_M|$. Usually, some of them are less than or equal to 1 in absolute values, the others are greater than 1 in absolute values. Also suppose that $|\lambda_m|>1$ for $m\leqslant m^*$ and $|\lambda_m|\leqslant 1$ for $m>m^*$, and that the $M+1$ independent eigenvectors are $\boldsymbol{Y}_m$. Thus, $\boldsymbol{X}^0$ can be written as

$$\boldsymbol{X}^0=\sum_{m=0}^{M} x_m^0 \boldsymbol{Y}_m,$$

and $\boldsymbol{X}^k$ has the expression

$$\boldsymbol{X}^k=\sum_{m=0}^{M} x_m^0 \tilde{\lambda}_m^k \boldsymbol{Y}_m.$$

Therefore, if $|x_m^0|$ is very small when $m\leqslant m^*$, and if k is not large, then the instability of this scheme does not have a serious influence on computation, and the computed results can be accepted. Of course, we assume that the word of computer contains a great number of bits, so the rounding errors are not important in computation. Moreover, if a scheme is unstable, usually only the $\tilde{\lambda}_m$ corresponding to the components with high frequency are greater than 1 in absolute values. Therefore, instability of this kind is dangerous only when the initial values are not smooth. If the initial values are smooth, the number of steps is not large and the rounding errors can be neglected, then the computed results may be accepted even though the scheme is unstable. The method of lines with a small value for M probably belongs to this case. Perhaps, this analysis can also be applied when the method of lines is used for mixed-type differential equations (see Chapter 3).

We should also discuss difference schemes for (1.9) and the application of the boundary conditions (2.5). (1.9) can be approximated by

$$x_i^{k+\alpha}=x_i^k+x_i'^k\alpha\Delta t, \tag{3.28}$$

$$x_i^{k+1}=x_i^k+x_i'^{k+\alpha}\Delta t. \tag{3.29}$$

This scheme is also a second order scheme if $\alpha=\dfrac{1}{2}$. It can be incorporated with the preceding two-step difference schemes for partial differential equations in solving (2.3)—(2.5).

There are two different ways in which boundary conditions can be used. In order to obtain all the quantities desired, the first way is to solve the boundary conditions and the difference equations obtained from (2.10) and (1.9) simultaneously. In this case, if explicit schemes are adopted for partial differential equations, then systems of nonlinear equations need to be solved at boundary points. If implicit schemes are adopted, then a system which is composed of a large number of linear equations and a few nonlinear equations needs to be solved. The other way is as follows. Differentiating the boundary condition (2.5) with respect to t, and thus we obtain an "incomplete" system of ordinary differential equations

$$\begin{cases} \dfrac{\partial \boldsymbol{B}_0}{\partial \boldsymbol{U}_{0^+}} \dfrac{\partial \boldsymbol{U}_{0^+}}{\partial t} + \dfrac{\partial \boldsymbol{B}_0}{\partial x_0} \dfrac{d x_0}{dt} + \dfrac{\partial \boldsymbol{B}_0}{\partial x'_0} \dfrac{d x'_0}{dt} + \dfrac{\partial \boldsymbol{B}_0}{\partial t} = 0, \\ \dfrac{\partial \boldsymbol{B}_l}{\partial \boldsymbol{U}_{l^-}} \dfrac{\partial \boldsymbol{U}_{l^-}}{\partial t} + \dfrac{\partial \boldsymbol{B}_l}{\partial \boldsymbol{U}_{l^+}} \dfrac{\partial \boldsymbol{U}_{l^+}}{\partial t} + \dfrac{\partial \boldsymbol{B}_l}{\partial x_l} \dfrac{d x_l}{dt} + \dfrac{\partial \boldsymbol{B}_l}{\partial x'_l} \dfrac{d x'_l}{dt} + \dfrac{\partial \boldsymbol{B}_l}{\partial t} = 0, \\ \qquad l = 1, 2, \cdots, L-1, \\ \dfrac{\partial \boldsymbol{B}_L}{\partial \boldsymbol{U}_{L^-}} \dfrac{\partial \boldsymbol{U}_{L^-}}{\partial t} + \dfrac{\partial \boldsymbol{B}_L}{\partial x_L} \dfrac{d x_L}{dt} + \dfrac{\partial \boldsymbol{B}_L}{\partial x'_L} \dfrac{d x'_L}{dt} + \dfrac{\partial \boldsymbol{B}_L}{\partial t} = 0, \end{cases} \tag{3.30}$$

which can be rewritten as

$$\frac{\partial \boldsymbol{B}}{\partial \boldsymbol{X}} \frac{\partial \boldsymbol{X}}{\partial t} + \frac{\partial \boldsymbol{B}}{\partial t} = 0,$$

where

$$\boldsymbol{B}^* = (\boldsymbol{B}_0^*, \boldsymbol{B}_1^*, \cdots, \boldsymbol{B}_L^*),$$
$$\boldsymbol{X}^* = (\boldsymbol{U}_{0^+}^*, x_0, x'_0, \boldsymbol{U}_{1^-}^*, \cdots, x'_L).$$

Then, (3.30) is approximated by

$$\left(\frac{\partial \boldsymbol{B}}{\partial \boldsymbol{X}}\right)^k \boldsymbol{X}^{k+\alpha} = \left(\frac{\partial \boldsymbol{B}}{\partial \boldsymbol{X}}\right)^k \boldsymbol{X}^k - \left(\frac{\partial \boldsymbol{B}}{\partial t}\right)^k \alpha \Delta t, \tag{3.31}$$

$$\left(\frac{\partial \boldsymbol{B}}{\partial \boldsymbol{X}}\right)^{k+\alpha} \boldsymbol{X}^{k+1} = \left(\frac{\partial \boldsymbol{B}}{\partial \boldsymbol{X}}\right)^{k+\alpha} \boldsymbol{X}^k - \left(\frac{\partial \boldsymbol{B}}{\partial t}\right)^{k+\alpha} \Delta t. \tag{3.32}$$

(3.31)/(3.32), like (3.28)/(3.29), has second order accuracy if $\alpha = \dfrac{1}{2}$, and it can be incorporated with the preceding two-step schemes for partial differential equations in solving (2.3)—(2.5). When using the second method, the systems of nonlinear equations are not encountered and the amount of computation is smaller. The first method is adopted in Chapter 7, and the second method is applied in the appendix of Chapter 5.

Finally, we shall show how conditions such as (3.6) are derived. (Please read Lemma 4.3 of Section 4 before reading the following part. because certain symbols and conclusions there will be used here.)

For the linear homogeneous equation

$$\frac{\partial u}{\partial t} + \lambda(\xi, t) \frac{\partial u}{\partial \xi} = 0, \tag{3.33}$$

the scheme (3.5) has the following form,

$$u_m^{k+1} + \alpha(\sigma^k \Delta_\sigma u^{k+1})_m = u_m^k - (1-\alpha)(\sigma \Delta_\sigma u)_m^k,$$

or

$$(1+\alpha|\sigma|_m^k) u_m^{k+1} - \alpha|\sigma|_m^k u_{m\mp1}^{k+1} = (1-(1-\alpha)|\sigma|_m^k) u_m^k + (1-\alpha)|\sigma|_m^k u_{m\mp1}^k,$$

where the subscript $m-1$ in $u_{m\mp1}$ is chosen if $\sigma > 0$ and the subscript $m+1$ is chosen if $\sigma < 0$. Therefore, the expressions for $r(\theta)$ and $s(\theta)$ are

$$r(\theta) = (1+\alpha|\sigma|) - \alpha|\sigma| e^{\mp i\theta},$$
$$s(\theta) = (1-(1-\alpha)|\sigma|) + (1-\alpha)|\sigma| e^{\mp i\theta},$$

and we know from Lemma 4.3 that

$$r^*(\theta)r(\theta)=1+4(1+a|\sigma|)a|\sigma|\sin^2\frac{\theta}{2},$$

$$s^*(\theta)s(\theta)=1-4(1-(1-a)|\sigma|)(1-a)|\sigma|\sin^2\frac{\theta}{2},$$

and that the condition (3.2) is equivalent to

$$a_1=4[(1+a|\sigma|)a|\sigma|+(1-(1-a)|\sigma|)(1-a)|\sigma|]=4|\sigma|(1+(2a-1)|\sigma|)\geqslant 0,$$

which can be rewritten as

$$1+(2a-1)|\sigma|\geqslant 0,$$

or

$$\begin{cases} |\sigma|\leqslant\dfrac{1}{1-2a}, & \text{if } a<\dfrac{1}{2},\\ |\sigma|<\infty, & \text{if } a\geqslant\dfrac{1}{2}. \end{cases}$$

This means that for the scheme (3.5), the condition (3.6) is equivalent to (3.2). (3.3) can be rewritten as

$$1+4(1+a|\sigma|)a|\sigma|\sin^2\frac{\theta}{2}\geqslant c_0>0.$$

Clearly, if the inequality holds for $\theta=0$ and π, then it holds for all θ. Obviously, it holds if $\theta=0$, and if $\theta=\pi$, it becomes

$$1+4(1+a|\sigma|)a|\sigma|=(1+2a|\sigma|)^2\geqslant c_0>0.$$

Thus, it holds if $1+2a|\sigma|$ is not close to zero. The condition (3.2) requires $1+(2a-1)|\sigma|\geqslant 0$. Therefore, if (3.2) holds, i.e., if (3.6) holds, then $1+2a|\sigma|$ cannot be close to zero. That is, when (3.2) holds, (3.3) also holds.

For (3.33), the schemes (3.7) and (3.8a) have the following forms

$$u_m^{k+a}=u_m^k-a(\sigma\Delta_\sigma u)_m^k,$$

$$u_m^{k+1}=u_{m\pm1}^k-(\mu\sigma_{m\pm\frac{1}{2}}^{k+a}\pm1)\Delta_\pm u_m^{k+a},$$

or

$$u_m^{k+a}=(1-a|\sigma|_m^k)u_m^k+a|\sigma|_m^k u_{m\mp1}^k,$$

$$u_m^{k+1}=u_{m\pm1}^k+(|\mu\sigma_{m\pm\frac{1}{2}}^{k+a}|+1)u_m^{k+a}-(|\mu\sigma_{m\pm\frac{1}{2}}^{k+a}|+1)u_{m\pm1}^{k+a}.$$

Thus

$$r(\theta)=1,$$

$$\begin{aligned} s(\theta)&=e^{\pm i\theta}+(|\sigma|+1)((1-a|\sigma|)+a|\sigma|e^{\mp i\theta})\\ &\quad-(|\sigma|+1)e^{\pm i\theta}((1-a|\sigma|)+a|\sigma|e^{\mp i\theta})\\ &=[1-(|\sigma|+1)(1-a|\sigma|)]e^{\pm i\theta}+(|\sigma|+1)(1-2a|\sigma|)\\ &\quad+(|\sigma|+1)a|\sigma|e^{\mp i\theta}, \end{aligned}$$

We know from Lemma 4.3 that in this case, (3.2) is equivalent to

$$\begin{aligned} a_1&=4(s'_{-1}s'_0+s'_0s'_1+4s'_{-1}s'_1)=4(1-s'_0)s'_0+16s'_{-1}s'_1\\ &=4(1-(|\sigma|+1)(1-2a|\sigma|))(|\sigma|+1)(1-2a|\sigma|)\\ &\quad+16(1-(|\sigma|+1)(1-a|\sigma|))(|\sigma|+1)a|\sigma|\\ &=4(|\sigma|+1)|\sigma|[(-1+2a|\sigma|+2a)(1-2a|\sigma|)\\ &\quad+4a|\sigma|(-1+a|\sigma|+a)]=4(|\sigma|+1)|\sigma|(2a-1)\geqslant 0, \end{aligned}$$

$$a_1+a_2=4(1-s'_0)s'_0\geqslant 0,$$

i.e., (3.2) is equivalent to

$$\begin{cases} (|\sigma|+1)|\sigma|(2a-1)\geqslant 0,\\ 0\leqslant s'_0=(|\sigma|+1)(1-2a|\sigma|)=1-|\sigma|^2-(2a-1)|\sigma|(|\sigma|+1)\leqslant 1, \end{cases}$$

which can be rewritten as

$$\alpha\geqslant\frac{1}{2},\quad |\sigma|\leqslant\frac{1}{2\alpha}.$$

Therefore, for the scheme (3.7)/(3.8a), (3.2) is equivalent to (3.9a). (In the expressions above, s'_h denotes $s_h(0, 0)$. In what follows, the same symbol will be adopted.)

(3.8b) and (3.8c) are obtained by adding a term of the second order in (3.8a), and the term of the second order can be expressed in the form $\beta(u^k_{m-1}-2u^k_m+u^k_{m+1})$ if $\Delta x=\Delta t=0$. Therefore, if s'_h still denotes the coefficients of (3.8a), then for (3.7)/(3.8b) and (3.7)/(3.8c), (3.2) is equivalent to

$$\begin{cases} a_1=4(1-(s'_0-2\beta))(s'_0-2\beta)+16(s'_{-1}+\beta)(s'_1+\beta) \\ \quad =4(1-s'_0)s'_0+16s'_{-1}s'_1+8\beta\geqslant 0, \\ 0\leqslant s'_0-2\beta\leqslant 1. \end{cases}$$

For (3.8b), $\beta=-(|\sigma|+1)|\sigma|\left(\alpha-\frac{1}{2}\right)$. Hence, (3.2) is equivalent to

$$\begin{cases} a_1=4(|\sigma|+1)|\sigma|(2\alpha-1)-4(|\sigma|+1)|\sigma|(2\alpha-1)=0, \\ 0\leqslant s'_0-2\beta=1-|\sigma|^2\leqslant 1, \end{cases}$$

i.e.,

$$|\sigma|\leqslant 1.$$

Therefore, for (3.7)/(3.8b), (3.2) is equivalent to (3.9b). For (3.8c), $\beta=-(|\sigma|+1)\left(\alpha-\frac{1}{2}\right)(2|\sigma|-1)$. Hence, (3.2) is equivalent to

$$\begin{cases} a_1=4(|\sigma|+1)(2\alpha|\sigma|-|\sigma|-4\alpha|\sigma|+2|\sigma|+2\alpha-1) \\ \quad =4(1-|\sigma|^2)(2\alpha-1)\geqslant 0, \\ 0\leqslant s'_0-2\beta=2(1-\alpha)(1-|\sigma|^2)=1-|\sigma|^2-(2\alpha-1)(1-|\sigma|^2)\leqslant 1, \end{cases}$$

i.e.,

$$\begin{cases} 0\leqslant(1-|\sigma|^2)(2\alpha-1), \\ 0\leqslant(1-|\sigma|^2)(1-\alpha), \end{cases}$$

which can be rewritten as

$$\frac{1}{2}\leqslant\alpha\leqslant 1,\quad |\sigma|\leqslant 1.$$

Therefore, for (3.7)/(3.8c), (3.2) is equivalent to (3.9c).

For (3.33), the scheme (3.10a) has the following form

$$u^{k+1}_m=u^k_{\mp 1}-(\mu\sigma^{k+\alpha}_{m\mp\frac{1}{2}}\mp 1)\Delta_{\mp}u^{k+\alpha}_m=u^k_{m\mp 1}-(|\mu\sigma^{k+\alpha}_{m\mp\frac{1}{2}}|-1)(u^{k+\alpha}_m-u^{k+\alpha}_{m\mp 1}).$$

Thus, for (3.7)/(3.10a), we have

$$\begin{aligned} r(\theta)&=1, \\ s(\theta)&=e^{\mp i\theta}-(|\sigma|-1)((1-\alpha|\sigma|)+\alpha|\sigma|e^{\mp i\theta}) \\ &\quad+(|\sigma|-1)e^{\mp i\theta}((1-\alpha|\sigma|)+\alpha|\sigma|e^{\mp i\theta}) \\ &=-(|\sigma|-1)(1-\alpha|\sigma|)+(1+(|\sigma|-1)(1-2\alpha|\sigma|))e^{\mp i\theta} \\ &\quad+(|\sigma|-1)\alpha|\sigma|e^{\mp i2\theta}. \end{aligned}$$

We know from Lemma 4.3 that for (3.7)/(3.10a), (3.2) is equivalent to

$$\begin{cases} a_1=4(1-s'_{\mp 1})s'_{\mp 1}+16s'_0s'_{\mp 2} \\ \quad =-4(|\sigma|-1)(1-2\alpha|\sigma|)(1+(|\sigma|-1)(1-2\alpha|\sigma|)) \\ \qquad -16(|\sigma|-1)^2\alpha|\sigma|(1-\alpha|\sigma|) \\ \quad =4(1-|\sigma|)|\sigma|[(1+2\alpha-2\alpha|\sigma|)(1-2\alpha|\sigma|)+4(|\sigma|-1)\alpha(1-\alpha|\sigma|)] \\ \quad =4(1-|\sigma|)|\sigma|(1-2\alpha)\geqslant 0, \\ a_1+a_2=4(1-s'_{\mp 1})s'_{\mp 1}\geqslant 0, \end{cases}$$

i.e.,

$$\begin{cases} |\sigma|(1-|\sigma|)(1-2a)\geqslant 0, \\ 0\leqslant s'_{\mp 1}=1-(1-|\sigma|)(1-2a|\sigma|)=(1+2a)|\sigma|-2a|\sigma|^2 \\ \quad =1-(1-|\sigma|)^2-|\sigma|(1-|\sigma|)(1-2a)\leqslant 1, \end{cases}$$

which can be rewritten as

$$\begin{cases} |\sigma|(1-|\sigma|)(1-2a)\geqslant 0, \\ (1+2a)|\sigma|-2a|\sigma|^2\geqslant 0. \end{cases}$$

The first inequality of the two inequalities is equivalent to

$$|\sigma|=0, \quad \text{or} \quad \begin{cases} a\leqslant\frac{1}{2}, \\ |\sigma|\leqslant 1, \end{cases} \quad \text{or} \quad \begin{cases} a\geqslant\frac{1}{2}, \\ |\sigma|\geqslant 1. \end{cases}$$

The second inequality is equivalent to

$$|\sigma|=0, \quad \text{or} \quad \begin{cases} a\geqslant 0, \\ |\sigma|\leqslant 1+\frac{1}{2a}, \end{cases} \quad \text{or} \quad \begin{cases} a\leqslant 0, \\ |\sigma|\geqslant 1+\frac{1}{2a}. \end{cases}$$

Therefore, for (3.7)/(3.10a), (3.2) is equivalent to

$$|\sigma|=0;$$

or

$$\begin{cases} \frac{1}{2}\leqslant a, \\ 1\leqslant|\sigma|\leqslant 1+\frac{1}{2a}; \end{cases}$$

or

$$\begin{cases} -\frac{1}{2}\leqslant a\leqslant\frac{1}{2}, \\ |\sigma|\leqslant 1; \end{cases}$$

or

$$\begin{cases} a\leqslant -\frac{1}{2}, \\ 1+\frac{1}{2a}\leqslant|\sigma|\leqslant 1. \end{cases}$$

The useful regions are

$$-\frac{1}{2}\leqslant a<\frac{1}{2}, \quad |\sigma|\leqslant 1,$$

and

$$a=\frac{1}{2}, \quad |\sigma|\leqslant 2.$$

That is, for (3.7)/(3.10a), if (3.11a) holds, (3.2) is satisfied.

(3.10b) and (3.10c) are obtained by adding a term of the second order in (3.10a). Thus, for (3.10b) and (3.10c), a_1 is equal to a_1 of (3.10a) plus 8β (for (3.8b) and (3.8c) the situation is similar). For (3.10b),

$$\beta=-(|\sigma|-1)\left(a-\frac{1}{2}\right)|\sigma|.$$

Thus, for (3.7)/(3.10b), (3.2) is equivalent to

$$\begin{cases} a_1=4(1-|\sigma|)|\sigma|(1-2a)-4(|\sigma|-1)(2a-1)|\sigma|=0, \\ 0\leqslant 1+(|\sigma|-1)(1-2a|\sigma|)+(|\sigma|-1)(2a-1)|\sigma| \\ \quad =|\sigma|(2-|\sigma|)=1-(1-|\sigma|)^2\leqslant 1, \end{cases}$$

i.e., it is equivalent to (3.11b)

$$0\leqslant|\sigma|\leqslant 2.$$

For (3.10c),

$$\beta=-(|\sigma|-1)(2\alpha-1)|\sigma|+\left(\alpha-\frac{1}{2}\right)|\sigma|.$$

Thus, for (3.7)/(3.10c), (3.2) is equivalen t to

$$\begin{cases} a_1=4(1-|\sigma|)|\sigma|(1-2\alpha)+8(1-|\sigma|)(2\alpha-1)|\sigma|+4(2\alpha-1)|\sigma| \\ \quad =4(2\alpha-1)|\sigma|(2-|\sigma|)\geqslant 0, \\ 0\leqslant 1+(|\sigma|-1)(1-2\alpha|\sigma|)+2(|\sigma|-1)(2\alpha-1)|\sigma|-(2\alpha-1)|\sigma| \\ \quad =1+(|\sigma|-1)(1-2\alpha|\sigma|+(2\alpha-1)|\sigma|)-(2\alpha-1)|\sigma|(2-|\sigma|) \\ \quad =1-(1-|\sigma|)^2-(2\alpha-1)|\sigma|(2-|\sigma|) \\ \quad =2(1-\alpha)|\sigma|(2-|\sigma|)\leqslant 1, \end{cases}$$

i.e.,

$$\begin{cases}(2\alpha-1)|\sigma|(2-|\sigma|)\geqslant 0, \\ (1-\alpha)|\sigma|(2-|\sigma|)\geqslant 0.\end{cases}$$

The last two inequalities can be rewritten as

$$\frac{1}{2}\leqslant\alpha\leqslant 1, \quad 0\leqslant|\sigma|\leqslant 2.$$

That is, for (3.7)/(3.10c), (3.2) is equivalent to (3.11c).

For (3.7)/(3.12j) (j=a, b, c),

$$s(\theta)=\beta s^{\mathrm{I}}(\theta)+(1-\beta)s^{\mathrm{II}}(\theta), \quad 0\leqslant\beta\leqslant 1,$$

where $s^{\mathrm{I}}(\theta)$ denotes $s(\theta)$ of (3.7)/(3.8j) and $s^{\mathrm{II}}(\theta)$ denotes $s(\theta)$ of (3.7)/(3.10j). Thus

$$|s(\theta)|\leqslant\beta|s^{\mathrm{I}}(\theta)|+(1-\beta)|s^{\mathrm{II}}(\theta)|\leqslant\max\{s^{\mathrm{I}}(\theta), s^{\mathrm{II}}(\theta)\}.$$

Moreover, for explicit schemes, (3.2) is equivalent to $|s(\theta)|\leqslant 1$. Therefore, (3.7)/(3.12j) will be stable if α and σ belong to the region where both (3.7)/(3.8j) and (3.7)/(3.10j) are stable. In α, $|\sigma|$-plane, the region where both (3.9j) and (3.11j) hold is that specified by (3.15j), so (3.7)/(3.12j) is stable if (3.15j) holds. (Of course, the region where (3.7)/(3.12j) is stable is probably larger than that specified by (3.15j).)

For (3.33), the first steps of (3.16)/(3.17) and (3.19)/(3.20) are not needed, so (3.18) and (3.21) are derived from (3.17) and (3.20) respectively.

For (3.33), the scheme (3.17) has the following form

$$u_m^{k+1}+\alpha\sigma_m^{k+\alpha}\Delta_0 u_m^{k+1}=u_m-(1-\alpha)\sigma_m^{k+\alpha}\Delta_0 u_m^k,$$

and the expressions of $r(\theta)$ and $s(\theta)$ are

$$r(\theta)=\frac{\alpha\sigma}{2}e^{i\theta}+1-\frac{\alpha\sigma}{2}e^{-i\theta},$$

$$s(\theta)=-\frac{(1-\alpha)\sigma}{2}e^{i\theta}+1+\frac{(1-\alpha)\sigma}{2}e^{-i\theta}.$$

We know from Lemma 4.3 that

$$r^*(\theta)r(\theta)=1-[4(1-r_0')r_0'+16r_{-1}'r_1']\sin^2\frac{\theta}{2}+16r_{-1}'r_1'\sin^4\frac{\theta}{2}$$

$$=1+4\alpha^2\sigma^2\sin^2\frac{\theta}{2}-4\alpha^2\sigma^2\sin^4\frac{\theta}{2},$$

$$s^*(\theta)s(\theta)=1-[4(1-s_0')s_0'+16s_{-1}'s_1']\sin^2\frac{\theta}{2}+16s_{-1}'s_1'\sin^4\frac{\theta}{2}$$

$$=1+4(1-\alpha)^2\sigma^2\sin^2\frac{\theta}{2}-4(1-\alpha)^2\sigma^2\sin^4\frac{\theta}{2}.$$

and that (3.2) is equivalent to

$$\begin{cases} a_1=4\alpha^2\sigma^2-4(1-\alpha)^2\sigma^2=4(2\alpha-1)\sigma^2\geqslant 0, \\ a_1+a_2=0, \end{cases}$$

so it is easily derived that (3.2) is equivalent to (3.18)

$$\alpha\geqslant\frac{1}{2}.$$

Moreover, (3.3) can be written in the form

$$r^*(\theta)r(\theta)=1+4\alpha^2\sigma^2\sin^2\frac{\theta}{2}\left(1-\sin^2\frac{\theta}{2}\right)\geqslant c_0>0,$$

so it holds automatically and we can take c_0 as unity.

For (3.33), the scheme (3.20) has the following form

$$\begin{aligned} &u_m^{k+1}+\alpha[2(\sigma^{k+\alpha}\Delta_\sigma u^{k+1})_m-(\sigma^{k+\alpha}\Delta_{\sigma 2}u^{k+1})_m] \\ &\quad=u_m^k-(1-\alpha)[2(\sigma^{k+\alpha}\Delta_\sigma u^k)_m-(\sigma^{k+\alpha}\Delta_{\sigma 2}u^k)_m], \end{aligned}$$

which can be rewritten as

$$\begin{aligned} &u^{k+1}+\alpha\left[2|\sigma_m^{k+\alpha}|(u_m^{k+1}-u_{m\mp1}^{k+1})-|\sigma_m^{k+\alpha}|\frac{1}{2}(u_m^{k+1}-u_{m\mp2}^{k+1})\right] \\ &\quad=u_m^k-(1-\alpha)\left[2|\sigma_m^{k+\alpha}|(u_m^k-u_{m\mp1}^k)-|\sigma_m^{k+\alpha}|\frac{1}{2}(u_m^k-u_{m\mp2}^k)\right], \end{aligned}$$

so $r(\theta)$ and $s(\theta)$ are

$$r(\theta)=\left(1+\frac{3\alpha}{2}|\sigma|\right)-2\alpha|\sigma|e^{\mp i\theta}+\frac{\alpha|\sigma|}{2}e^{\mp i2\theta},$$

$$s(\theta)=\left(1-\frac{3(1-\alpha)}{2}|\sigma|\right)+2(1-\alpha)|\sigma|e^{\mp i\theta}-\frac{(1-\alpha)|\sigma|}{2}e^{\mp i2\theta}.$$

We know from Lemma 4.3 that

$$\begin{aligned} r^*(\theta)r(\theta)&=1-\left[-4(1+2\alpha|\sigma|)2\alpha|\sigma|+16\left(1+\frac{3\alpha}{2}|\sigma|\right)\frac{\alpha|\sigma|}{2}\right] \\ &\quad\times\sin^2\frac{\theta}{2}+16\left(1+\frac{3\alpha}{2}|\sigma|\right)\frac{\alpha|\sigma|}{2}\sin^4\frac{\theta}{2} \\ &=1+4\alpha^2|\sigma|^2\sin^2\frac{\theta}{2}+4(2\alpha|\sigma|+3\alpha^2|\sigma|^2)\sin^4\frac{\theta}{2}, \end{aligned}$$

$$\begin{aligned} s^*(\theta)s(\theta)&=1+4(1-\alpha)^2|\sigma|^2\sin^2\frac{\theta}{2}+4(-2(1-\alpha)|\sigma| \\ &\quad+3(1-\alpha)^2|\sigma|^2)\sin^4\frac{\theta}{2}, \end{aligned}$$

and that (3.2) is equivalent to

$$\begin{cases} a_1=4\alpha^2|\sigma|^2-4(1-\alpha)^2|\sigma|^2=4(2\alpha-1)|\sigma|^2\geqslant 0, \\ a_1+a_2=4\alpha^2|\sigma|^2-4(1-\alpha)^2|\sigma|^2+4(2\alpha|\sigma|+3\alpha^2|\sigma|^2+2(1-\alpha)|\sigma| \\ \qquad -3(1-\alpha)^2|\sigma|^2)=8|\sigma|+16(2\alpha-1)|\sigma|\geqslant 0. \end{cases}$$

Therefore, in this case, (3.2) is equivalent to (3.21)

$$\alpha\geqslant\frac{1}{2}.$$

Now, (3.3) has the following form

$$1+4\alpha^2|\sigma|^2\sin^2\frac{\theta}{2}+4(2\alpha|\sigma|+3\alpha^2|\sigma|^2)\sin^4\frac{\theta}{2}\geqslant c_0>0,$$

which always holds if $\alpha>0$. That is, for (3.20), (3.2) is equivalent to (3.21) and (3.3) always holds.

Finally, we consider (3.24)/(3.25). As in the case of (3.16)/(3.17), the first step of (3.24)/(3.25) is not needed for the linear equation (3.33). For (3.33), (3.25) has the following form

$$\frac{1}{2}(u_m^{k+1}+u_{m\mp1}^{k+1})+\alpha\mu\sigma_{m\mp\frac{1}{2}}^{k+\alpha}\Delta_\mp u_m^{k+1}=\frac{1}{2}(u_m^k+u_{m\mp1}^k)-(1-\alpha)\,\mu\sigma_{m\mp\frac{1}{2}}^{k+\alpha}\Delta_\mp u_m^k,$$

which can be rewritten as

$$\frac{1}{2}(u_m^{k+1}+u_{m\mp1}^{k+1})+\alpha\left|\mu\sigma_{m\mp\frac{1}{2}}^{k+\alpha}\right|(u_m^{k+1}-u_{m\mp1}^{k+1})$$
$$=\frac{1}{2}(u_m^k+u_{m\mp1}^k)-(1-\alpha)\left|\mu\sigma_{m\mp\frac{1}{2}}^{k+\alpha}\right|(u_m^k-u_{m\mp1}^k),$$

and $r(\theta)$ and $s(\theta)$ are

$$\begin{cases} r(\theta)=\left(\frac{1}{2}+\alpha|\sigma|\right)+\left(\frac{1}{2}-\alpha|\sigma|\right)e^{\mp i\theta}, \\ s(\theta)=\left(\frac{1}{2}-(1-\alpha)|\sigma|\right)+\left(\frac{1}{2}+(1-\alpha)|\sigma|\right)e^{\mp i\theta}. \end{cases}$$

Further, we know from Lemma 4.3 that

$$r^*(\theta)r(\theta)=1-4(1-r_0')r_0'\sin^2\frac{\theta}{2}=1-(1-4\alpha^2|\sigma|^2)\sin^2\frac{\theta}{2},$$

$$s^*(\theta)s(\theta)=1-(1-4(1-\alpha)^2|\sigma|^2)\sin^2\frac{\theta}{2},$$

and that (3.2) is equivalent to

$$a_1=(1-4(1-\alpha)^2|\sigma|^2)-(1-4\alpha^2|\sigma|^2)=4(2\alpha-1)|\sigma|^2\geqslant 0.$$

Therefore, it is derived that (3.2) is equivalent to (3.26)

$$\alpha\geqslant\frac{1}{2}.$$

Now, (3.3) has the form

$$1+(4\alpha^2|\sigma|^2-1)\sin^2\frac{\theta}{2}\geqslant c_0>0,$$

which is equivalent to

$$4\alpha^2|\sigma|^2\geqslant c_0,$$

so (3.3) holds if (3.27) is valid.

§ 4 The Stability of Difference Schemes for Initial–Boundary–Value Problems and the "Condition" of Systems of Difference Equations

1. *Definitions and Symbols*

In this section we shall discuss the following linear initial–boundary–value problem

$$\begin{cases} \frac{\partial \boldsymbol{U}}{\partial t}+A(x,t)\frac{\partial \boldsymbol{U}}{\partial x}=C(x,t)\boldsymbol{U}+\boldsymbol{F}(x,t), & t\geqslant 0,\ 0\leqslant x\leqslant 1, \\ \boldsymbol{U}(x,0)=\boldsymbol{D}(x), & 0\leqslant x\leqslant 1, \\ B_0(t)\boldsymbol{U}(0,t)=\boldsymbol{E}_0(t), & t\geqslant 0, \\ B_1(t)\boldsymbol{U}(1,t)=\boldsymbol{E}_1(t), & t\geqslant 0, \end{cases}$$

where A, C, B_0 and B_1 are $N\times N$, $N\times N$, $(N-p)\times N$ and $(N-q)\times N$-matrices respectively; $\boldsymbol{U}$, $\boldsymbol{F}$, $\boldsymbol{E}_0$, $\boldsymbol{E}_1$ and $\boldsymbol{D}$ are N, N, $(N-p)$, $(N-q)$ and N-dimensional vectors respectively. Here, we require again that the boundary conditions satisfy (1.7), (1.8), i.e., that they be compatible with the differential equations. In the present case, (1.8) means that $(B_0\bar{G}^{-1}(0, t))_{1\sim(N-p)}$ and $(B_1\bar{G}^{-1}(1, t))_{(q+1)\sim N}$ are invertible, where $(B_j\bar{G}^{-1})_{n_1\sim n_2}$ denotes the matrix which is composed of the n_1-th to the n_2-th columns of $B_j\bar{G}^{-1}$. Obviously, we might further postulate that $(B_0\bar{G}^{-1}(0, t))_{1\sim(N-p)}$ and $(B_1\bar{G}^{-1}(1, t))_{(q+1)\sim N}$ are unit matrices of orders $N-p$ and $N-q$ respectively. In order to make the solutions smooth, we also assume that the initial values satisfy the boundary conditions.

The differential equations above can be rewritten as

$$\frac{\partial \bar{G}\boldsymbol{U}}{\partial t}+\Lambda\frac{\partial \bar{G}\boldsymbol{U}}{\partial x}=\left(\frac{\partial \bar{G}}{\partial t}+\Lambda\frac{\partial \bar{G}}{\partial x}+\bar{G}C\right)\boldsymbol{U}+\bar{G}\boldsymbol{F},$$

so we need to consider only equations of the form $\dfrac{\partial \boldsymbol{U}}{\partial t}+\Lambda\dfrac{\partial \boldsymbol{U}}{\partial x}=C\boldsymbol{U}+\boldsymbol{F}$. Furthermore, the lower order term $C\boldsymbol{U}$ does not usually influence the "condition" of systems of difference equations and the stability of the schemes. For simplicity, we suppose that $C=0$. Moreover, we discuss only the stability with respect to initial values and boundary values, so we also let $\boldsymbol{F}=0$. That is, we shall discuss the following initial-boundary-value problem

$$\begin{cases}\dfrac{\partial \boldsymbol{U}}{\partial t}+\Lambda(x, t)\dfrac{\partial \boldsymbol{U}}{\partial x}=0, \quad t\geqslant 0,\ 0\leqslant x\leqslant 1, & (4.1)\\ \boldsymbol{U}(x, 0)=\boldsymbol{D}(x), \quad 0\leqslant x\leqslant 1, & (4.2)\\ B_0(t)\boldsymbol{U}(0, t)=\boldsymbol{E}_0(t), \quad t\geqslant 0, & (4.3)\\ B_1(t)\boldsymbol{U}(1, t)=\boldsymbol{E}_1(t), \quad t\geqslant 0. & (4.4)\end{cases}$$

Here, $(B_0)_{1\sim(N-p)}$ and $(B_1)_{(q+1)\sim N}$ are unit matrices; Λ satisfies (2.8) when $x=0$ and $x=1$; and the initial values (4.2) satisfy (4.3) and (4.4). Also, we assume that Λ, B_0, B_1, $\boldsymbol{E}_0$, $\boldsymbol{E}_1$ and $\boldsymbol{D}(x)$ are sufficiently smooth functions of the independent variables.

From the boundary conditions (4.3) and (4.4), we can easily deduce the following inequalities

$$\begin{cases}\displaystyle\sum_{n\in\mathscr{G}_0(\Lambda_0)}|u_{n,0}|^2\leqslant c_1\sum_{n\in\bar{\mathscr{G}}_0(\Lambda_0)}|u_{n,0}|^2+c_1\|\boldsymbol{E}_0\|^2,\\ \displaystyle\sum_{n\in\mathscr{G}_1(\Lambda_1)}|u_{n,M}|^2\leqslant c_1\sum_{n\in\bar{\mathscr{G}}_1(\Lambda_1)}|u_{n,M}|^2+c_1\|\boldsymbol{E}_1\|^2,\end{cases} \tag{4.5}''$$

where $u_{n,m}=u_n(m\Delta x, t)$, u_n being the n-th component of vector $\boldsymbol{U}(x, t)$; c_1 is a positive constant; the definitions of the sets $\mathscr{G}_0(\Lambda_0)$ and $\mathscr{G}_1(\Lambda_1)$ are as follows: $\mathscr{G}_0(\Lambda_0)$ is composed of all n in $[1, N]$ with $\lambda_n(0, t)>0$, i.e., it is the set $[1, N-p]$; $\mathscr{G}_1(\Lambda_1)$ is composed of all n in $[1, N]$ with $\lambda_n(1, t)<0$, i.e., it is the set $[q+1, N]$. $\bar{\mathscr{G}}_j(\Lambda_j)$ is the set of the elements in the set $[1, N]$ which do not belong to $\mathscr{G}_j(\Lambda_j)$, $j=0, 1$; and

$$\|E_0\|^2=\sum_{n=1}^{N-p}|e_{0,n}|^2, \quad \|E_1\|^2=\sum_{n=1}^{N-q}|e_{1,n}|^2,$$

$e_{0,n}$ and $e_{1,n}$ being the n-th elements of the respective vectors E_0 and E_1.

In the following discussion of stability, we assume that the inequalities

$$\begin{cases}\sum_{n\in\mathscr{G}_0(\Lambda_0)}|u_{n,0}|^2\leqslant c_1\sum_{n\in\overline{\mathscr{G}}_0(\Lambda_0)}|\lambda_n(0, t)|\,|u_{n,0}|^2+c_1\|E_0\|^2,\\ \sum_{n\in\mathscr{G}_1(\Lambda_1)}|u_{n,M}|^2\leqslant c_1\sum_{n\in\overline{\mathscr{G}}_1(\Lambda_1)}|\lambda_n(1, t)|\,|u_{n,M}|^2+c_1\|E_1\|^2\end{cases} \tag{4.5}$$

hold. They are slightly stronger than (4.5)″ and are fulfilled in the following two cases.

(1) If the elements of B_j satisfy the condition

$$|b_{j,i,n}|^2\leqslant c_1'|\lambda_n(j, t)|, \tag{4.6}$$

where $b_{j,i,n}$ is the element of B_j located in the i-th row and in the n-th column, and c_1' is a constant, then (4.5) holds. In fact, in the present case,

$$u_{i,0}=-\sum_{n\in\overline{\mathscr{G}}_0(\Lambda_0)}b_{0,i,n}u_{n,0}+e_{0,i}, \quad \text{for } i\in\mathscr{G}_0(\Lambda_0),$$

so

$$\begin{aligned}|u_{i,0}|^2&\leqslant(N+1)\Big[\sum_{n\in\overline{\mathscr{G}}_0(\Lambda_0)}|b_{0,i,n}|^2|u_{n,0}|^2+|e_{0,i}|^2\Big]\\ &\leqslant(N+1)\Big[c_1'\sum_{n\in\overline{\mathscr{G}}_0(\Lambda_0)}|\lambda_n(0, t)|\,|u_{n,0}|^2+|e_{0,i}|^2\Big],\end{aligned}$$

which can yield the first inequality of (4.5) by letting

$$c_1=(N+1)\max\{Nc_1', 1\}.$$

In the same way, the second inequality can be obtained. In these calculations, the relation

$$\left(\sum_{n=1}^{N+1}a_n\right)^2\leqslant(N+1)\sum_{n=1}^{N+1}a_n^2$$

is used.

For example, if $|\lambda_n|\geqslant\varepsilon>0$ on boundaries, ε being a positive constant, and if $|b_{j,i,n}|^2$ is bounded by $c_1'\varepsilon$, then (4.6) is valid.

(2) If for any n,

$$\begin{cases}|\lambda_n(j, t)|\geqslant\varepsilon>0 \quad \text{or } \lambda_n(j, t)\equiv0,\\ j=0, 1, \quad t\geqslant0;\end{cases} \tag{4.7}$$

and if for these eigenvalues, which are always equal to zero on boundaries,

$$\lambda_{N-p+1}(0, t),\ \lambda_{N-p+2}(0, t),\ \cdots,\ \lambda_q(1, t),\ \lambda_{q-1}(1, t),\ \cdots,$$

the corresponding difference equations at the boundary points are

$$\begin{cases} u_n(0, (k+1)\Delta t) = u_n(0, k\Delta t), & n = N-p+1, N-p+2, \cdots, \\ u_n(1, (k+1)\Delta t) = u_n(1, k\Delta t), & n = q, q-1, \cdots, \end{cases} \tag{4.8}$$

then (4.3) and (4.4) can be rewritten in this form

$$\begin{cases} (B_0 - B_0\widetilde{E}_0)\boldsymbol{U}(0, t) = \boldsymbol{E}_0(t) - B_0\widetilde{E}_0\boldsymbol{U}(0, t) \\ \quad = \boldsymbol{E}_0(t) - B_0\widetilde{E}_0\boldsymbol{D}(0), \\ (B_1 - B_1\widetilde{E}_1)\boldsymbol{U}(1, t) = \boldsymbol{E}_1(t) - B_1\widetilde{E}_1\boldsymbol{D}(1). \end{cases}$$

Here $\widetilde{E}_0$ and $\widetilde{E}_1$ are $N\times N$–diagonal matrices, and the $(N-p+1)$–th, $(N-p+2)$–th, $\cdots$ diagonal elements of $\widetilde{E}_0$ and the q–th, $(q-1)$–th, $\cdots$ diagonal elements of $\widetilde{E}_1$ are equal to one and all other elements are equal to zero, i.e., the $(N-p+1)$–th, $(N-p+2)$–th, $\cdots$ columns of $B_0 - B_0\widetilde{E}_0$ and the q–th, $(q-1)$–th, $\cdots$ columns of $B_1 - B_1\widetilde{E}_1$ are zero–vectors. Therefore, $B_j - B_j\widetilde{E}_j$, $j=0, 1$, satisfy the condition (4.6), and we can obtain the inequalities

$$\begin{cases} \sum\limits_{n\in\mathscr{G}_0(A_0)} |u_{n,0}|^2 \leqslant c_1 \sum\limits_{n\in\overline{\mathscr{G}}_0(A_0)} |\lambda_n(0, t)|\,|u_{n,0}|^2 \\ \quad + c_1\|\boldsymbol{E}_0 - B_0\widetilde{E}_0\boldsymbol{D}(0)\|^2, \\ \sum\limits_{n\in\mathscr{G}_1(A_1)} |u_{n,M}|^2 \leqslant c_1 \sum\limits_{n\in\overline{\mathscr{G}}_1(A_1)} |\lambda_n(1, t)|\,|u_{n,M}|^2 \\ \quad + c_1\|\boldsymbol{E}_1 - B_1\widetilde{E}_1\boldsymbol{D}(1)\|^2, \end{cases} \tag{4.5$'$}$$

which are similar to (4.5).

Therefore, (4.5) or (4.5)$'$ holds if the following condition is satisfied:

Condition (iv) The coefficient matrices B_0 and B_1 of the boundary conditions satisfy (4.6), or the eigenvalues λ_n satisfy (4.7) at boundaries and the schemes adopted at $m=0$, M, degenerate into (4.8) when $\sigma=0$.

In the following, we will put restriction (iv) on the boundary conditions and on the schemes for boundary points. (4.6) and (4.7) are valid insome physical problems. However, there is perhaps the case in which (4.7) is not valid (for example, one of λ has may be equal to zero some where at the boundaries but not always equal to zero), and also (4.6) is not satisfied. Probably, this case will not create difficulties in the discussion of stability of initial–boundary–value difference schemes, but the discussion will become more tedious and the definition of stability will need to be slightly changed. It is not a necessary restriction that the schemes adopted at $m=0$, M degenerate into (4.8) when $\sigma=0$, although we put this condition on the schemes. However, for those schemes we discuss in this chapter, i.e. for (3.5), (3.7)/(3.12), (3.22)/(3.23), and the mixed scheme (3.7)/(3.12)—(3.24)/(3.25), relations (4.8) hold. Therefore, the main point of (iv)is to require that (4.6) or (4.7) hold.

Regarding the "condition" of the equations and the stability of difference schemes, we define: if for any solution of the difference equations approximating the problem (4.1)—(4.4)

$$\begin{cases} \sum_{h=H_1}^{H_2} r_{h,n,m}^k u_{n,m+h}^{k+1} = \sum_{h=H_1}^{H_2} s_{h,n,m}^k u_{n,m+h}^k, \\ \quad m \in \mathscr{G}(\lambda_n), \quad n=1, 2, \cdots, N, \\ B_0^{k+1} \boldsymbol{U}(0, t_{k+1}) = \boldsymbol{E}_0^{k+1}, \\ B_1^{k+1} \boldsymbol{U}(1, t_{k+1}) = \boldsymbol{E}_1^{k+1}, \\ \quad k=0, 1, \cdots, \\ \boldsymbol{U}(m\Delta x, 0) = \boldsymbol{D}(m\Delta x), \quad m=0, 1, \cdots, M, \end{cases} \tag{4.9}$$

there exists a positive constant α_1 independent of M such that the inequality

$$\|\boldsymbol{U}(x, t_{k+1})\|_x^2 \leqslant \alpha_1 \left[\sum_{n=1}^{N} \sum_{m \in \mathscr{G}(\lambda_n)} |f_{n,m}^{k+1}|^2 + \|\boldsymbol{E}_0^{k+1}\|^2 + \|\boldsymbol{E}_1^{k+1}\|^2 \right] \Delta x$$

$$(k=0, 1, \cdots) \tag{4.10}$$

holds, then we say that the system (4.9) is uniformly well-conditioned with respect to M in L_2[4]. Here,

$$u_{n,m}^k = u_n(m\Delta x, \ k\Delta t), \quad B_j^k = B_j(k\Delta t), \quad \boldsymbol{E}_j^k = \boldsymbol{E}_j(k\Delta t), \quad j=0, \ 1,$$

$$\|\boldsymbol{U}(x, t_k)\|_x^2 = \sum_{n=1}^{N} \|u_n^k\|_x^2, \qquad \|u_n^k\|_x^2 = \sum_{m=0}^{M} |u_{n,m}^k|^2 \Delta x,$$

$$f_{n,m}^{k+1} = \sum_{h=H_1}^{H_2} s_{h,n,m}^k u_{n,m+h}^k, \quad t_k = k\Delta t.$$

If for any solution of the difference scheme (4.9), the inequality

$$\|\boldsymbol{U}(x, t)\|_x^2 \leqslant \alpha_2 [e^{\alpha_3 t} \|\boldsymbol{U}(x, 0)\|_x^2 + \|e^{\frac{1}{2}\alpha_3(t-t')} \boldsymbol{E}_0(t')\|_{t',t}^2$$

$$+ \|e^{\frac{1}{2}\alpha_3(t-t')} \boldsymbol{E}_1(t')\|_{t',t}^2] \tag{4.11}$$

is fulfilled, then we say that the difference scheme (4.9) is stable with respect to initial and boundary values in the sense of the L_2 norm, where α_2 and α_3 are positive constants, t is equal to an integer times Δt, and

$$\|e^{\frac{1}{2}\alpha_3(t-t')} \boldsymbol{E}_j(t')\|_{t',t}^2 = \Delta t \sum_{k=0}^{\frac{t}{\Delta t}} e^{\alpha_3(t-k\Delta t)} \|\boldsymbol{E}_j(k\Delta t)\|^2.$$

Here, we want to point out that B_j and $\boldsymbol{E}_j$ in (4.9) may be the same as in (4.3), (4.4), or they may be different from those in (4.3), (4.4) if the boundary conditions are rewritten as they have been rewritten in the above case (2).

The following symbols are introduced:

(1)

$$\begin{cases} \delta_0(\lambda_{n,0}) = \begin{cases} 1, & \text{if} \quad \lambda_{n,0} > 0, \quad \text{i.e.,} \quad n \in \mathscr{G}_0(\Lambda_0), \\ 0, & \text{if} \quad \lambda_{n,0} \leqslant 0, \quad \text{i.e.,} \quad n \in \overline{\mathscr{G}_0}(\Lambda_0), \end{cases} \\ \delta_1(\lambda_{n,M}) = \begin{cases} 1, & \text{if} \quad \lambda_{n,M} < 0, \quad \text{i.e.,} \quad n \in \mathscr{G}_1(\Lambda_1), \\ 0, & \text{if} \quad \lambda_{n,M} \geqslant 0, \quad \text{i.e.,} \quad n \in \overline{\mathscr{G}_1}(\Lambda_1). \end{cases} \end{cases} \tag{4.12}$$

According to these symbols, $\mathscr{G}(\lambda_n)$ in § 3 can be rewritten as

$$\mathscr{G}(\lambda_n)=[\delta_0(\lambda_{n,0}),\ M-\delta_1(\lambda_{n,M})].$$

(2)

$$\begin{cases} T_n^k = \sum\limits_{m\in\mathscr{G}(\lambda_n)} \left(\sum\limits_{h=H_1}^{H_2} r_{h,n,m}^{k-1} u_{n,\ m+h}^k\right)^2 \Delta x, \\ T^k = \sum\limits_{n=1}^{N} T_n^k + \|B_0^k U(0,\ t_k)\|^2 \Delta x + \|B_1^k U(1,\ t_k)\|^2 \Delta x. \end{cases} \tag{4.13}$$

With these symbols, the definition (4.10) may be rewritten as

$$\|U(x,\ t_{k+1})\|_x^2 \leqslant \alpha_1 T^{k+1} \tag{4.10$'$}$$

For simplicity, we shall use the superscripts and subscripts in the following way. If no confusion results when a superscript or a subscript is omitted, then we shall omit it. If there is a common superscript or subscript for all the quantities in an expression, then we omit the superscrpit or subscript for each quantity, and add brackets and put the superscript or subscript outside the brackets. For example, (4.13) may be rewritten as

$$\begin{cases} T_n^k = \sum\limits_{m\in\mathscr{G}(\lambda_n)} \left(\sum\limits_{h=H_1}^{H_2} r_{h,m}^{k-1}\, u_{m+h}^k\right)_n^2 \Delta x, \\ T^k = \sum\limits_{n=1}^{N} T_n^k + \|B_0^k U(0,\ t_k)\|^2 \Delta x + \|B_1^k U(1,\ t_k)\|^2 \Delta x. \end{cases}$$

(3) When λ_n corresponds to the model A_1, i.e., $n\in\mathscr{G}_0(\Lambda_0)\cap\overline{\mathscr{G}}_1(\Lambda_1)$, we define

$$\tilde{u}_{n,m}=c_3^{m/M}u_{n,m}, \tag{4.14a}$$

where c_3 is any given positive number. When λ_n corresponds to A_2, B and C, i.e., $n\in\overline{\mathscr{G}}_0(\Lambda_0)\cap\mathscr{G}_1(\Lambda_1)$, $\overline{\mathscr{G}}_0(\Lambda_0)\cap\overline{\mathscr{G}}_1(\Lambda_1)$ and $\mathscr{G}_0(\Lambda_0)\cap\mathscr{G}_1(\Lambda_1)$, we define respectively

$$\tilde{u}_{n,m}=c_3^{(1-m/M)}u_{n,m}, \tag{4.14b}$$

$$\tilde{u}_{n,m}=c_3 u_{n,m}, \tag{4.14c}$$

and

$$\tilde{u}_{n,m}=u_{n,m}. \tag{4.14d}$$

We also define

$$\|\tilde{u}_n\|_x^2=\sum_{m=0}^{M}|\tilde{u}_{n,m}|^2\Delta x, \tag{4.14e}$$

$$\|\tilde{U}\|_x^2=\sum_{n=1}^{N}\|\tilde{u}_n\|_x^2. \tag{4.14f}$$

Obviously, between $\|\tilde{U}\|_x^2$ and $\|U\|_x^2$, there is the following relation

$$\begin{aligned} \min\{1,c_3^2\}\|U\|_x^2 &= \min_{n,m}\left\{\left(\frac{\tilde{u}}{u}\right)_{n,m}^2\right\}\|U\|_x^2 \leqslant \|\tilde{U}\|_x^2 \\ &\leqslant \max_{n,m}\left\{\left(\frac{\tilde{u}}{u}\right)_{n,m}^2\right\}\|U\|_x^2 = \max\{1,\ c_3^2\}\|U\|_x^2. \end{aligned} \tag{4.15}$$

(4) Furthermore, we define

$$\begin{cases} \widetilde{T}_n^k = \sum\limits_{m\in\mathscr{G}(\lambda_n)} \left(\dfrac{\widetilde{u}_m^k}{u_m^k} \sum\limits_{h=H_1}^{H_2} r_{h,m}^{k-1}\, u_{m+h}^k\right)_n^2 \Delta x \\ \quad = \sum\limits_{m\in\mathscr{G}(\lambda_n)} \left(\sum\limits_{h=H_1}^{H_2} \widetilde{r}_{h,m}^{k-1}\, \widetilde{u}_{m+h}^k\right)_n^2 \Delta x, \\ \widetilde{T}^k = \sum\limits_{n=1}^{N} \widetilde{T}_n^k + c_2^2 \| B_0^k U(0,\ t_k)\|^2 \Delta x + c_2^2 \| B_1^k U(1,\ t_k)\|^2 \Delta x, \end{cases} \tag{4.16}$$

where c_2 is a given positive constant;

$$\widetilde{r}_{h,m}^{k-1} = \frac{\widetilde{u}_m^k}{u_m^k}\ \frac{u_{n+h}^k}{\widetilde{u}_{m+h}^k}\ r_{h,m}^{k-1} = \left(\ +O\left(\frac{c_3}{M}\right)\right) r_{h,m}^{k-1};$$

and $O\left(\dfrac{c_3}{M}\right)$ denotes a quantity whose modulus is less than $c\dfrac{c_3}{M}$, c being a positive constant. Evidently, between T and $\widetilde{T}$, there is the following relationship

$$\min\{1,\ c_2^2,\ c_3^2\} T \leqslant \widetilde{T} \leqslant \max\{1,\ c_2^2,\ c_3^2\} T. \tag{4.17}$$

From the inequalities (4.15) and (4.17), we know that (4.10)′ is equivalent to

$$\|\widetilde{U}(x,\ t_{k+1})\|_x^2 \leqslant \widetilde{\alpha}_1 \widetilde{T}^{k+1}, \tag{4.10$''$}$$

i. e., whether a system is well-conditioned may be reduced to whether one can find c_2, c_3 and $\widetilde{\alpha}_1$ such that (4.10)″ holds.

2. *Basic Theorems and Basic Lemmas*

There are the following results on the "condition" of systems and the stability of schemes.

Theorem 4.1 If for any bounded c_3 and sufficiently small Δx, the inequality

$$\begin{aligned} \widetilde{T}_n + c_4 \Delta x \|\widetilde{u}_n\|_x^2 \geqslant & -c_5 |\widetilde{u}_{n,0}|^2 \Delta x \delta_0(\lambda_{n,0}) \\ & - c_5 |u_{n,M}|^2 \Delta x \delta_1(\lambda_{n,M}) + c_6 \sum_{m\in\mathscr{G}(\lambda_n)} |\widetilde{u}_{n,m}|^2 \Delta x \end{aligned} \tag{4.18}$$

is satisfied when a scheme is applied to the homogeneous model problems A_1, A_2, B and C, where c_5 and c_6 are positive constants independent of c_3, and c_4 is a positive constant less than $k_1+k_2c_3$, k_1 and k_2 being constants, then the system of difference equations (4.9) is always assured to be well-conditioned when the scheme is applied to the initial-boundary-value problem (4.1)—(4.4).

Proof As has been pointed out, whether the system (4.9) is well-conditioned is reduced to whether (4.10)″ holds. It is easy to deduce (4.10)″ from (4.18) and (4.5)″. In fact, summing up (4.18) over n, we get

$$\begin{aligned} \sum_{n=1}^{N} \widetilde{T}_n + c_4 \Delta x \sum_{n=1}^{N} \|\widetilde{u}_n\|_x^2 \geqslant & -c_5 \sum_{n\in\mathscr{G}_0(\Lambda_0)} |\widetilde{u}_{n,0}|^2 \Delta x \\ & - c_5 \sum_{n\in\mathscr{G}_1(\Lambda_1)} |\widetilde{u}_{n,M}|^2 \Delta x + c_6 \sum_{n=1}^{N} \sum_{m\in\mathscr{G}(\lambda_n)} |\widetilde{u}_{n,m}|^2 \Delta x, \end{aligned}$$

and by (4.14a—d), (4.5)″ can be rewritten as

$$\begin{cases}\|\boldsymbol{E}_0\|^2\geqslant\dfrac{1}{c_1}\sum\limits_{n\in\mathscr{G}_0(\Lambda_0)}|\tilde{u}_{n,0}|^2-\dfrac{1}{c_3^2}\sum\limits_{n\in\mathscr{G}_0(\Lambda_0)}|\tilde{u}_{n,0}|^2,\\ \|\boldsymbol{E}_1\|^2\geqslant\dfrac{1}{c_1}\sum\limits_{n\in\mathscr{G}_1(\Lambda_1)}|\tilde{u}_{n,M}|^2-\dfrac{1}{c_3^2}\sum\limits_{n\in\mathscr{G}_1(\Lambda_1)}|\tilde{u}_{n,M}|^2.\end{cases}$$

Therefore, noting the second and the third expressions of (4.9), we can obtain the following inequality

$$\begin{aligned}&\sum_{n=1}^{N}\tilde{T}_n+c_2^2\|B_0\boldsymbol{U}(0,t)\|^2\Delta x+c_2^2\|B_1\boldsymbol{U}(1,t)\|^2\Delta x+c_4\Delta x\|\tilde{\boldsymbol{U}}\|_x^2\\ &\geqslant\left(\frac{c_2^2}{c_1}-c_5\right)\sum_{n\in\mathscr{G}_0(\Lambda_0)}|\tilde{u}_{n,0}|^2\Delta x-\frac{c_2^2}{c_3^2}\sum_{n\in\bar{\mathscr{G}}_0(\Lambda_0)}|\tilde{u}_{n,0}|^2\Delta x\\ &\quad+\left(\frac{c_2^2}{c_1}-c_5\right)\sum_{n\in\mathscr{G}_1(\Lambda_1)}|\tilde{u}_{n,M}|^2\Delta x-\frac{c_2^2}{c_3^2}\sum_{n\in\bar{\mathscr{G}}_1(\Lambda_1)}|\tilde{u}_{n,M}|^2\Delta x\\ &\quad+c_6\sum_{n=1}^{N}\sum_{m\in\mathscr{G}(\lambda_n)}|\tilde{u}_{n,m}|^2\Delta x,\end{aligned}$$

from which, the inequality

$$\tilde{T}\geqslant\frac{c_6}{4}\|\tilde{\boldsymbol{U}}\|_x^2,$$

i.e., (4.10)″ can be deduced by letting $\dfrac{c_2^2}{c_1}-c_5\geqslant\dfrac{c_6}{2}$, $\dfrac{c_2^2}{c_3^2}\leqslant\dfrac{c_6}{2}$, $c_4\Delta x\leqslant\dfrac{c_6}{4}$. That is, the conclusion is proved.

Theorem 4.2 If for any bounded c_3 and sufficiently small Δx, (4.18) and the inequality

$$\begin{aligned}\tilde{T}_n^{k+1}-\tilde{T}_n^k\leqslant&[c_7\delta_0(\lambda_{n,0})-c_8|\sigma_{n,0}|(1-\delta_0(\lambda_{n,0}))]|\tilde{u}_{n,0}^k|^2\Delta x\\ &+[c_7\delta_1(\lambda_{n,M})-c_8|\sigma_{n,M}|(1-\delta_1(\lambda_{n,M}))]|\tilde{u}_{n,M}^k|^2\Delta x\\ &+c_4\Delta x\|\tilde{u}_n^k\|_x^2\end{aligned}\tag{4.19}$$

are satisfied when a scheme is applied to the homogeneous model problems A_1, A_2, B and C, where c_4 is a positive constant less than $k_1+k_2c_3$, k_1 and k_2 being constants, and c_7, c_8 are positive constants independent of c_3, then the scheme (4.9) can be assured to be stable with respect to initial and boundary values when it is applied to the initial-boundary-value problem (4.1)—(4.4) with (4.5).

Proof Summing up (4.19) over n, we get

$$\begin{aligned}\sum_{n=1}^{N}\tilde{T}_n^{k+1}-\sum_{n=1}^{N}\tilde{T}_n^k\leqslant&c_7\sum_{n\in\mathscr{G}_0(\Lambda_0)}|\tilde{u}_{n,0}^k|^2\Delta x\\ &-c_8\sum_{n\in\mathscr{G}_0(\Lambda_0)}|\sigma_{n,0}||\tilde{u}_{n,0}^k|^2\Delta x+c_7\sum_{n\in\mathscr{G}_1(\Lambda_1)}|\tilde{u}_{n,M}^k|^2\Delta x\\ &-c_8\sum_{n\in\bar{\mathscr{G}}_1(\Lambda_1)}|\sigma_{n,M}||\tilde{u}_{n,M}^k|^2\Delta x+c_4\Delta x\|\tilde{\boldsymbol{U}}^k\|_x^2.\end{aligned}$$

Further, noting the second and the third expressions of (4.9), (4.14a—d), (4.16) and (4.5), and taking $c_3^2\geqslant\max\left\{\dfrac{c_1c_7}{c_8}\dfrac{\Delta x}{\Delta t},\ \dfrac{2c_2^2}{c_6}\right\}$, we can derive the inequality

$$\begin{aligned}\widetilde{T}^{k+1}-\widetilde{T}^{k}\leqslant & c_7\sum_{n\in\mathscr{G}_0(\Lambda_0)}|u_{n,0}^k|^2\Delta x-\frac{\Delta t}{\Delta x}c_8c_3^2\sum_{n\in\overline{\mathscr{G}}_0(\Lambda_0)}|\lambda_{n,0}|\,|u_{n,0}^k|^2\Delta x\\ &+c_7\sum_{n\in\mathscr{G}_1(\Lambda_1)}|u_{n,M}^k|^2\Delta x-\frac{\Delta t}{\Delta x}c_8c_3^2\sum_{n\in\overline{\mathscr{G}}_1(\Lambda_1)}|\lambda_{n,M}|\,|u_{n,M}^k|^2\Delta x\\ &+c_2^2\Delta x[\|\boldsymbol{E}_0^{k+1}\|^2-\|\boldsymbol{E}_0^k\|^2+\|\boldsymbol{E}_1^{k+1}\|^2-\|\boldsymbol{E}_1^k\|^2]+c_4\Delta x\|\widetilde{\boldsymbol{U}}^k\|_x^2\\ \leqslant & c_1c_7\Delta x[\|\boldsymbol{E}_0^k\|^2+\|\boldsymbol{E}_1^k\|^2]+c_2^2\Delta x[\|\boldsymbol{E}_0^{k+1}\|^2-\|\boldsymbol{E}_0^k\|^2\\ &+\|\boldsymbol{E}_1^{k+1}\|^2-\|\boldsymbol{E}_1^k\|^2]+c_4\Delta x\|\widetilde{\boldsymbol{U}}^k\|_x^2.\end{aligned}$$

It follows from the proof of Theorem 4.1 that (4.10)″ is valid. Thus, there exists a constant α_3 such that the above inequality can be rewritten as

$$\begin{aligned}\widetilde{T}^{k+1}\leqslant &(1+\alpha_3\Delta t)\widetilde{T}^k+\sum_{j=0}^{1}\{c_1c_7\|\boldsymbol{E}_j^k\|^2+c_2^2(\|\boldsymbol{E}_j^{k+1}\|^2-\|\boldsymbol{E}_j^k\|^2)\}\Delta x\\ \leqslant &(1+\alpha_3\Delta t)^k\widetilde{T}^1+\sum_{j=0}^{1}\Big\{c_1c_7\sum_{i=0}^{k-1}(1+\alpha_3\Delta t)^i\|\boldsymbol{E}_j^{k-i}\|^2\\ &+c_2^2\sum_{i=0}^{k-1}(1+\alpha_3\Delta t)^i(\|\boldsymbol{E}_j^{k-i+1}\|^2-\|\boldsymbol{E}_j^{k-i}\|^2)\Big\}\Delta x\\ =&(1+\alpha_3\Delta t)^k\widetilde{T}^1+\sum_{j=0}^{1}\Big\{c_1c_7\sum_{i=0}^{k-1}(1+\alpha_3\Delta t)^i\|\boldsymbol{E}_j^{k-i}\|^2\\ &+c_2^2\Big[\|\boldsymbol{E}_j^{k+1}\|^2+\sum_{i=0}^{k-2}(1+\alpha_3\Delta t)^i\alpha_3\Delta t\|\boldsymbol{E}_j^{k-i}\|^2\\ &-(1+\alpha_3\Delta t)^{k-1}\|\boldsymbol{E}_j^1\|^2\Big]\Big\}\Delta x\\ \leqslant &(1+\alpha_3\Delta t)^k\widetilde{T}^1+\max\{c_1c_7,\ c_2^2\}\Delta x\sum_{j=0}^{1}\sum_{i=1}^{k+1}(1+\alpha_3\Delta t)^{k+1-i}\|\boldsymbol{E}_j^i\|^2\\ \leqslant & e^{\alpha_3 t_k}\widetilde{T}^1+c_9\sum_{j=0}^{1}\|e^{\frac{1}{2}\alpha_3(t_{k+1}-t')}\boldsymbol{E}_j(t')\|_{t',t_{k+1}}^2,\end{aligned}$$

where $c_9=\dfrac{\max\{c_1c_7,\ c_2^2\}}{\Delta t}\Delta x$. In deriving the last inequality the following fact is used: if

$$Y^{(k+1)}\leqslant aY^{(k)}+b^{(k)}\quad(k=0,\ 1,\ \cdots),$$

then

$$\begin{aligned}Y^{(k+1)}&\leqslant aY^{(k)}+b^{(k)}\leqslant a(aY^{(k-1)}+b^{(k-1)})+b^{(k)}\\ &\leqslant\cdots\leqslant a^{k+1}Y^{(0)}+\sum_{i=0}^{k}a^ib^{(k-i)}.\end{aligned}$$

Furthermore, because the conditions of Theorem 4.1 are satisfied here and because of some other reasons, we can find two positive constants k_3 and k_4 such that

$$k_3\|\boldsymbol{U}(x,\ k\Delta t)\|_x^2\leqslant\widetilde{T}^k\ \text{ and }\ \widetilde{T}^1\leqslant k_4\|\boldsymbol{U}(x,\ 0)\|_x^2.$$

From these, letting $\alpha_2=\max\left\{\dfrac{k_4}{k_3},\ \dfrac{c_9}{k_3}\right\}$, we can easily obtain the following inequality

$$\|U(x,t)\|_x^2 \leqslant \alpha_2\left[e^{\alpha_3 t}\|U(x,0)\|_x^2+\sum_{j=0}^{1}\|e^{\frac{1}{2}\alpha_3(t-t')}E_j(t')\|_{t',t}^2\right].$$

The conclusion is thus proved.

We know from Theorems 4.1 and 4.2 that if one can prove (4.18) and (4.19) to be valid when the adopted scheme is applied to the model problems, then it can be said with certainty that the system of difference equations is well-conditioned and the difference scheme is stable when the scheme is applied to the initial-boundary-value problem (4.1)—(4.4) with (4.5).

In what follows, we shall show whether the inequalities (4.18) and (4.19) are valid or not is closely connected with whether the conditions (3.3) and (3.2) are valid or not. To do this, we first introduce the following two lemmas.

Lemma 4.1 If

$$\sum_l (-1)^{H_3(l)} d_l^*(\theta) d_l(\theta) \equiv 0,$$

then the matrix

$$Q = \sum_l (-1)^{H_3(l)} D_l^* D_l$$

is a "pseudo-null" matrix, i.e., the sums of the elements on every diagonal line of the matrix are all equal to zero, where $H_3(l)$ is equal to either 0 or 1;

$$d_l(\theta) = \sum_{h=H_1}^{H_2} d_{l,h} e^{ih\theta}; \quad D_l = (d_{l,H_1}, d_{l,H_1+1}, \cdots, d_{l,H_2});$$

and the symbol "$*$" represents conjugate transpose for vectors and conjugation for scalar quantities.

Proof Because of

$$\sum_l (-1)^{H_3(l)} \left(\sum_{h=H_1}^{H_2} d_{l,h} e^{ih\theta}\right)^* \left(\sum_{h=H_1}^{H_2} d_{l,h} e^{ih\theta}\right)$$

$$= \sum_l (-1)^{H_3(l)} \left(\sum_{h=0}^{H-1} d_{l,H_1+h} e^{ih\theta}\right)^* \left(\sum_{h=0}^{H-1} d_{l,H_1+h} e^{ih\theta}\right)$$

$$= \sum_l (-1)^{H_3(l)} \sum_{h=-(H-1)}^{H-1} \left(\sum_j d_{l,j}^* d_{l,j+h}\right) e^{ih\theta}$$

$$= \sum_{h=-(H-1)}^{H-1} \left(\sum_l (-1)^{H_3(l)} \sum_j d_{l,j}^* d_{l,j+h}\right) e^{ih\theta} \equiv 0,$$

for any h, we have

$$\sum_l (-1)^{H_3(l)} \sum_j d_{l,j}^* d_{l,j+h} = 0,$$

where $H = H_2 - H_1 + 1$, and j in summation formula runs over all the values satisfying both $H_1 \leqslant j+h \leqslant H_2$ and $H_1 \leqslant j \leqslant H_2$.

On the other hand

$$Q=\sum_l(-1)^{H_3(l)}D_l^*D_l$$

$$=\sum_l(-1)^{H_3(l)}\begin{pmatrix} d^*_{l,H_1}d_{l,H_1}, & d^*_{l,H_1}d_{l,H_1+1}, & \cdots, & d^*_{l,H_1}d_{l,H_2} \\ d^*_{l,H_1+1}d_{l,H_1}, & d^*_{l,H_1+1}d_{l,H_1+1}, & \cdots, & d^*_{l,H_1+1}d_{l,H_2} \\ \cdots & \cdots & \cdots & \cdots \\ d^*_{l,H_2}d_{l,H_1}, & d^*_{l,H_2}d_{l,H_1+1}, & \cdots, & d^*_{l,H_2}d_{l,H_2} \end{pmatrix},$$

and the sum of elements on the h-th diagonal line is $\sum_l(-1)^{H_3(l)}\cdot\sum_j d^*_{l,j}d_{l,j+h}$, where the h-th diagonal line denotes the main diagonal line if $h=0$, the h-th upper-diagonal line if $h>0$ and the $|h|$-th lower-diagonal line if $h<0$. Therefore, the conclusion is valid.

Lemma 4.2 If $d_1^*(\theta)d_1(\theta)-d_2^*(\theta)d_2(\theta)\geqslant 0$, then the matrix $D_1^*D_1-D_2^*D_2$ can be represented by a sum of one nonnegative definite matrix Z and one pseudo-null matrix Q.

Proof Because of $d_1^*(\theta)d_1(\theta)-d_2^*(\theta)d_2(\theta)\geqslant 0$, one can find a set of $c_l(\theta)$ such that $d_1^*(\theta)d_1(\theta)-d_2^*(\theta)d_2(\theta)\equiv\sum_l c_l^*(\theta)c_l(\theta)$. (The Fejér–Riesz Theorem tells us that in the present case, there exists $c(\theta)$ such that

$$d_1^*(\theta)d_1(\theta)-d_2^*(\theta)d_2(\theta)=c^*(\theta)c(\theta).$$

Sometimes, it can also be represented by a sum of several nonnegative terms. Thus, the conclusion above is true.) Hence, we know from Lemma 4.1 that

$$Q=D_1^*D_1-D_2^*D_2-\sum_l C_l^*C_l$$

is a pseudo-null matrix. Obviously, $\sum_l C_l^*C_l$ is a nonnegative definite matrix. Therefore, the conclusion of this Lemma is true.

From Lemma 4.2, it can be easily seen that the conditions (3.2) and (3.3) are closely related to (4.18) and (4.19) respectively.

In fact, the difference equations in (4.9) may be rewritten as

$$\sum_{h=H_1}^{H_2}\tilde r^k_{h,n,m}\tilde u^{k+1}_{n,m+h}=\sum_{h=H_1}^{H_2}\tilde s^k_{h,n,m}\tilde u^k_{n,m+h}, \tag{4.20}$$

where

$$\frac{\tilde r^k_{h,n,m}}{r^k_{h,n,m}}=\frac{\tilde s^k_{h,n,m}}{s^k_{h,n,m}}=\frac{\tilde u^k_{n,m}}{u^k_{n,m}}\,\frac{u^{k+1}_{n,m+h}}{\tilde u^{k+1}_{n,m+h}}=\frac{\tilde u^k_{n,m}}{u^k_{n,m}}\,\frac{u^k_{n,m+h}}{\tilde u^k_{n,m+h}}.$$

Moreover, from

$$\left|\frac{\tilde u^k_{n,m}}{u^k_{n,m}}\,\frac{u^k_{n,m+h}}{\tilde u^k_{n,m+h}}-1\right|=O(\Delta x),$$

and (3.4b), we can obtain

$$\begin{cases}|\tilde r^k_{h,n,m}(\Delta x,\ \Delta t)-r^k_{h,n,m}(0,\ 0)|\leqslant O(\Delta x),\\ |\tilde s^k_{h,n,m}(\Delta x,\ \Delta t)-s^k_{h,n,m}(0,\ 0)|\leqslant O(\Delta x).\end{cases}$$

Thus, we have

$$\begin{cases} \tilde{T}_n^{k+1} = \sum\limits_{m\in \mathscr{G}(\lambda_n)} \left(\sum\limits_{h=H_1}^{H_2} \tilde{r}_{h,n,m}^k \tilde{u}_{n,m+h}^{k+1} \right)^2 \Delta x \\ \quad = \sum\limits_{m\in \mathscr{G}(\lambda_n)} \left(\sum\limits_{h=H_1}^{H_2} \tilde{s}_{h,n,m}^k \tilde{u}_{n,m+h}^k \right)^2 \Delta x \\ \quad = \sum\limits_{m\in \mathscr{G}(\lambda_n)} (S^*S\tilde{U}, \tilde{U})_{n,m,H_1,H}^k \Delta x + O(\Delta x) \|\tilde{u}_n^k\|_x^2, \\ \tilde{T}_n^k = \sum\limits_{m\in \mathscr{G}(\lambda_n)} \left(\sum\limits_{h=H_1}^{H_2} \tilde{r}_{h,n,m}^{k-1} \tilde{u}_{n,m+h}^k \right)^2 \Delta x \\ \quad = \sum\limits_{m\in \mathscr{G}(\lambda_n)} R^*R\tilde{U}, \tilde{U})_{n,m,H_1,H}^k \Delta x + O(\Delta x) \|\tilde{u}_n^k\|_x^2. \end{cases} \tag{4.21}$$

When deducing (4.21), we use the condition (3.4a), assume $\Delta t/\Delta x$ to be bounded, and adopt the following symbols:

$$\begin{cases} R_{n,m}^k = (r_{H_1,n,m}^k,\ r_{H_1+1,n,m}^k,\ \cdots,\ r_{H_2,n,m}^k) \big|_{\Delta x=\Delta t=0}, \\ S_{n,m}^k = (s_{H_1,n,m}^k,\ s_{H_1+1,n,m}^k,\ \cdots,\ s_{H_2,n,m}^k) \big|_{\Delta x=\Delta t=0}, \\ (A\tilde{U}, \tilde{U})_{n,m,H_1,H}^k = \sum\limits_{i=1}^{H} \sum\limits_{j=1}^{H} (a_{i,j} \tilde{u}_{m+H_1-1+i} \tilde{u}_{m+H_1-1+j})_n^k, \\ a_{i,j} \text{ being the element located on row } i \text{ and column } j \\ \text{of } A,\ H = H_2 - H_1 + 1. \end{cases} \tag{4.22}$$

From Lemma 4.2, it follows that if the conditions (3.3) and (3.2) are satisfied, then there exist certain nonnegative definite matrices $\tilde{Z}_H$, Z_H and certain pseudo-null matrices $\tilde{Q}_H$, Q_H, such that

$$(R^*R\tilde{U}, \tilde{U})_m - c_0 |\tilde{u}_m|^2 = (\tilde{Z}_H \tilde{U}, \tilde{U})_m + (\tilde{Q}_H \tilde{U}, \tilde{U})_m, \tag{4.23}$$

$$((R^*R - S^*S)\tilde{U}, \tilde{U})_m = (Z_H \tilde{U}, \tilde{U})_m + (Q_H \tilde{U}, \tilde{U})_m, \tag{4.24}$$

where the subscript H of $\tilde{Z}$, Z, $\tilde{Q}$, Q shows that the matrices are of $H\times H$ order. Therefore, we have

$$\tilde{T}_n \geqslant [c_0 \sum_{m\in \mathscr{G}(\lambda_n)} |\tilde{u}_{n,m}|^2 + \sum_{m\in \mathscr{G}(\lambda_n)} (\tilde{Q}_H \tilde{U}, \tilde{U})_m] \Delta x - O(\Delta x) \|\tilde{u}_n\|_x^2 \tag{4.25}$$

and

$$\tilde{T}_n^{k+1} - \tilde{T}_n^k \leqslant - \sum_{m\in \mathscr{G}(\lambda_n)} (Q_H \tilde{U}, \tilde{U})_m \Delta x + O(\Delta x) \|\tilde{u}_n\|_x^2. \tag{4.26}$$

From the property of pseudo-null matrices, we know that if every element of $\tilde{Q}_H$ and Q_H satisfies the Lipschitz condition as x varies, then in the above inequality,

$$\sum_{m\in \mathscr{G}(\lambda_n)} (\tilde{Q}_H \tilde{U}, \tilde{U})_m \quad \text{and} \quad - \sum_{m\in \mathscr{G}(\lambda_n)} (Q_H \tilde{U}, \tilde{U})_m$$

may be replaced by the sums of

$$O(\Delta x) \sum_{m=0}^{M} |\tilde{u}_{n,m}|^2$$

and certain quadratic forms of $\tilde{u}_{n,m}$ at points near the boundaries. Therefore, (4.25) and (4.26) are similar to (4.18) and (4.19) respectively.

In the following, using the results of Theorems 4.1, 4.2 and Lemmas 4.1, 4.2, we shall discuss the "condition" of systems of difference equations and the stability of the horizontal two-point schemes, the horizontal three-point explicit schemes, a class of horizontal four-point explicit schemes and the section-wise mixed schemes. We shall show the following results: When the schemes (3.5), (3.7)/(3.8), (3.7)/(3.10), (3.7)/(3.12), (3.25) and the scheme obtained by combining (3.7)/(3.12) and (3.25) are applied to the initial-boundary-value problem (4.1)—(4.4) with (4.6) or (4.7), if conditions (i), (ii) and (iii) of § 3 are valid, then the systems of difference equations are well-conditioned and the difference schemes are stable. Here, the horizontal H-point scheme (4.9) is a scheme in which $H_2(m)-H_1(m)+1\leqslant H$ for any m. If among all the coefficients $r^k_{h,m}$, only $r^k_{0,m}\neq 0$, we call it an explicit scheme. Moreover, we can assume $r^k_{0,m}=1$ without loss of generality. If among all the coefficients $r^k_{h,m}$, more than one coefficient is not equal to zero, it is called an implicit scheme. The section-wise mixed scheme is a scheme in which the algorithms might be different for different eigenvalues λ_n and different subintervals. In the mixed schemes, only the values at the points which belong to the same subinterval appear in each difference equation and the number of subintervals is bounded. For example, the fourth scheme in § 3 belongs to this type of scheme.

Before deducing further results by using Theorems 4.1—4.2, we must point out the following facts:

Lemma 4.3 For the horizontal H-point schemes (3.1), $r^*(\theta)r(\theta)$ and $s^*(\theta)s(\theta)$ can be expressed in the following forms

$$r^*(\theta)r(\theta)=1+\sum_{j=1}^{H-1}a_{r,j}\sin^{2j}\frac{\theta}{2}, \tag{4.27a}$$

$$s^*(\theta)s(\theta)=1+\sum_{j=1}^{H-1}a_{s,j}\sin^{2j}\frac{\theta}{2}, \tag{4.27b}$$

and the condition (3.2) can be rewritten as

$$\sum_{j=1}^{H-1}a_j\sin^{2j}\frac{\theta}{2}\geqslant 0 \quad (a_j=a_{r,j}-a_{s,j}). \tag{4.27c}$$

Furthermore, for the horizontal two-point, three-point and four-point schemes, (4.27c) is respectively equivalent to

(1)

$$a_1\geqslant 0, \tag{4.28a}$$

(2)

$$\begin{cases} a_1\geqslant 0, \\ a_1+a_2\geqslant 0, \end{cases} \tag{4.28b}$$

(3)

$$\begin{cases} a_1\geqslant 0, \\ a_1+a_2+a_3\geqslant 0, \\ 2a_1+a_2+2[a_1(a_1+a_2+a_3)]^{\frac{1}{2}}\geqslant 0. \end{cases} \tag{4.28c}$$

In (4.27a—c), the expressions of $a_{r,j}$ and $a_{s,j}$ are

$$\begin{cases} a_{r,1} = -4\left[\sum_{h=H_1}^{H_2-1} r'_h r'_{h+1} + 4\sum_{h=H_1}^{H_2-2} r'_h r'_{h+2} + 9\sum_{h=H_1}^{H_2-3} r'_h r'_{h+3}\right], \\ a_{r,2} = 16\left[\sum_{h=H_1}^{H_2-2} r'_h r'_{h+2} + 6\sum_{h=H_1}^{H_2-3} r'_h r'_{h+3}\right], \\ a_{r,3} = -64\sum_{h=H_1}^{H_2-3} r'_h r'_{h+3}, \\ a_{s,1} = -4\left[\sum_{h=H_1}^{H_2-1} s'_h s'_{h+1} + 4\sum_{h=H_1}^{H_2-2} s'_h s'_{h+2} + 9\sum_{h=H_1}^{H_2-3} s'_h s'_{h+3}\right], \\ a_{s,2} = 16\left[\sum_{h=H_1}^{H_2-2} s'_h s'_{h+2} + 6\sum_{h=H_1}^{H_2-3} s'_h s'_{h+3}\right], \\ a_{s,3} = -64\sum_{h=H_1}^{H_2-3} s'_h s'_{h+3}, \end{cases} \tag{4.29}$$

where $s'_h = s_h(\Delta x, \Delta t)\,|_{\Delta x=\Delta t=0}$, $r'_h = r_h(\Delta x, \Delta t)\,|_{\Delta x=\Delta t=0}$, and the schemes with $H<4$ are regarded as four-point schemes in which some r_h and s_h are always equal to zero.

Proof Because

$$\begin{aligned} r^*(\theta)r(\theta) &= \left(\sum_{h=H_1}^{H_2} r'_h e^{-ih\theta}\right)\left(\sum_{h=H_1}^{H_2} r'_h e^{ih\theta}\right) \\ &= \sum_{h_1=H_1}^{H_2}\sum_{h=H_1}^{H_2} r'_h r'_{h_1} e^{i(h_1-h)\theta} = \sum_{j=-(H-1)}^{H-1}\ \sum_{H_1<h,h+j<H_2} r'_h r'_{h+j} e^{ij\theta} \\ &= \sum_{h=H_1}^{H_2} r'^2_h + 2\sum_{j=1}^{H-1}\ \sum_{H_1<h<h+j<H_2} r'_h r'_{h+j}\cos j\theta, \end{aligned}$$

and for any positive integer j,

$$\cos j\theta = \sum_{l=0}^{j} d_{j,l}\sin^{2l}\frac{\theta}{2},$$

we get

$$r^*(\theta)r(\theta) = \sum_{j=0}^{H-1} a_{r,j}\sin^{2j}\frac{\theta}{2}.$$

In a similar way, we can also get

$$s^*(\theta)s(\theta) = \sum_{j=0}^{H-1} a_{s,j}\sin^{2j}\frac{\theta}{2}.$$

Moreover,

$$a_{r,0} = \sum_{h=H_1}^{H_2} r'^2_h + 2\sum_{j=1}^{H-1}\ \sum_{H_1<h<h+j<H_2} r'_h r'_{h+j} = \left(\sum_{h=H_1}^{H_2} r'_h\right)^2 = 1,$$

$$a_{s,0} = \left(\sum_{h=H_1}^{H_2} s'_h\right)^2 = \left(\sum_{h=H_1}^{H_2} r'_h\right)^2 = a_{r,0}.$$

Therefore, let $a_j = a_{r,j} - a_{s,j}$, the von Neumann condition can be expressed in the following form

$$r^*(\theta)r(\theta) - s^*(\theta)s(\theta) = \sum_{j=1}^{H-1} a_j\sin^{2j}\frac{\theta}{2} \geqslant 0.$$

In the case $H=4$,

$$r^*(\theta)r(\theta)-s^*(\theta)s(\theta)=a_1\sin^2\frac{\theta}{2}+a_2\sin^4\frac{\theta}{2}+a_3\sin^6\frac{\theta}{2}$$

$$=\sin^2\frac{\theta}{2}\Big(a_1\cos^4\frac{\theta}{2}+(2a_1+a_2)\cos^2\frac{\theta}{2}\sin^2\frac{\theta}{2}$$

$$+(a_1+a_2+a_3)\sin^4\frac{\theta}{2}\Big)$$

$$=\sin^2\frac{\theta}{2}\Big\{[2a_1+a_2+2(a_1(a_1+a_2+a_3))^{\frac{1}{2}}]\sin^2\frac{\theta}{2}\cos^2\frac{\theta}{2}$$

$$+\left[a_1^{\frac{1}{2}}\cos^2\frac{\theta}{2}-(a_1+a_2+a_3)^{\frac{1}{2}}\sin^2\frac{\theta}{2}\right]^2\Big\},$$

so we know that the condition (3.2) is valid if $a_1\geqslant 0$, $a_1+a_2+a_3\geqslant 0$, and $2a_1+a_2+2(a_1(a_1+a_2+a_3))^{\frac{1}{2}}\geqslant 0$. Moreover, letting $\sin^2\frac{\theta}{2}\approx 0$, $\cos^2\frac{\theta}{2}=0$, and $a_1^{\frac{1}{2}}\cos^2\frac{\theta}{2}-(a_1+a_2+a_3)^{\frac{1}{2}}\sin^2\frac{\theta}{2}=0$, we can also see that these conditions are necessary. That is, in this case, the von Neumann condition is equivalent to (4.28c).

In the case $H=3$, $r'_h r'_{h+3}=s'_h s'_{h+3}=0$, so $a_3=0$. Hence, it can be easily seen that (4.28b) is equivalent to (4.28c). That is, in this case, (4.28b) is equivalent to (3.2). In the case $H=2$, we have $a_2=a_3=0$, so (4.28a) is equivalent to (4.28c), and also to (3.2).

As for (4.29), by using

$$\cos\theta=1-2\sin^2\frac{\theta}{2},$$

$$\cos 2\theta=1-8\sin^2\frac{\theta}{2}+8\sin^4\frac{\theta}{2},$$

$$\cos 3\theta=1-18\sin^2\frac{\theta}{2}+48\sin^4\frac{\theta}{2}-32\sin^6\frac{\theta}{2},$$

it can be easily obtained. Lemma 4.3 is completed.

According to this lemma and certain conclusions obtained in the process of proving Lemma 4.2, we can obtain the following conclusion: for the schemes with $H=2$, 3, 4, we can construct matrices Z_H and Y_H such that the matrix $Q_H=-Y_H-Z_H$ is a pseudo-null matrix. Defining

(1)
$$Z_2=B_2^*B_2, \tag{4.30a}$$

where

$$B_2=\left(\frac{a_1}{4}\right)^{\frac{1}{2}}(1,\ -1);$$

(2)
$$Z_3=B_3^*B_3+C_3^*C_3, \tag{4.30b}$$

where

$$B_3=\left(\frac{a_1}{16}\right)^{\frac{1}{2}}(1,\ 0,\ -1),$$

$$C_3=\left(\frac{a_1+a_2}{16}\right)^{\frac{1}{2}}(1,\ -2,\ 1);$$

(3) $$Z_4=B_4^*B_4+C_4^*C_4, \tag{4.30c}$$

where

$$B_4=\left(\frac{2a_1+a_2+2(a_1(a_1+a_2+a_3))^{\frac{1}{2}}}{64}\right)^{\frac{1}{2}}(1,\ -1,\ -1,\ 1),$$

$$C_4=\left(\frac{a_1}{64}\right)^{\frac{1}{2}}(1,\ 1,\ -1,\ -1)+\left(\frac{a_1+a_2+a_3}{64}\right)^{\frac{1}{2}}$$
$$\times(1,\ -3,\ 3,\ -1)$$
$$=\frac{1}{8}\big(a_1^{\frac{1}{2}}+(a_1+a_2+a_3)^{\frac{1}{2}},\ a_1^{\frac{1}{2}}-3(a_1+a_2+a_3)^{\frac{1}{2}},$$
$$-a_1^{\frac{1}{2}}+3(a_1+a_2+a_3)^{\frac{1}{2}},\ -a_1^{\frac{1}{2}}-(a_1+a_2+a_3)^{\frac{1}{2}}\big),$$

and

(4) $$Y_H=S_H^*S_H-R_H^*R_H, \tag{4.30d}$$

where the subscript H means that the quantities correspond to the H-point schemes. We can prove the conclusion straightforwardly.

In fact, by using

$$\sin^2\frac{\theta}{2}=\frac{1}{4}(1-e^{-i\theta})(1-e^{i\theta}),$$

$$\cos^2\frac{\theta}{2}=\frac{1}{4}(1+e^{-i\theta})(1+e^{i\theta}),$$

we get

(1) $$a_1\sin^2\frac{\theta}{2}=\left[\left(\frac{a_1}{4}\right)^{\frac{1}{2}}(1-e^{i\theta})\right]^*\left[\left(\frac{a_1}{4}\right)^{\frac{1}{2}}(1-e^{i\theta})\right];$$

(2) $$a_1\sin^2\frac{\theta}{2}+a_2\sin^4\frac{\theta}{2}$$
$$=a_1\sin^2\frac{\theta}{2}\cos^2\frac{\theta}{2}+(a_1+a_2)\sin^4\frac{\theta}{2}$$
$$=\left[\left(\frac{a_1}{16}\right)^{\frac{1}{2}}(1-e^{i2\theta})\right]^*\left[\left(\frac{a_1}{16}\right)^{\frac{1}{2}}(1-e^{i2\theta})\right]$$
$$+\left[\left(\frac{a_1+a_2}{16}\right)^{\frac{1}{2}}(1-2e^{i\theta}+e^{i2\theta})\right]^*$$
$$\times\left[\left(\frac{a_1+a_2}{16}\right)^{\frac{1}{2}}(1-2e^{i\theta}+e^{i2\theta})\right];$$

(3) $$a_1\sin^2\frac{\theta}{2}+a_2\sin^4\frac{\theta}{2}+a_3\sin^6\frac{\theta}{2}$$
$$=(2a_1+a_2+2(a_1(a_1+a_2+a_3))^{\frac{1}{2}})\sin^4\frac{\theta}{2}\cos^2\frac{\theta}{2}$$

$$+\left(a_1^{\frac{1}{2}}\cos^2\frac{\theta}{2}-(a_1+a_2+a_3)^{\frac{1}{2}}\sin^2\theta\right)^2\sin^2\frac{\theta}{2}$$

$$=\frac{2a_1+a_2+2(a_1(a_1+a_2+a_3))^{\frac{1}{2}}}{64}\,|(1-e^{i\theta})^2(1+e^{i\theta})|^2$$

$$+\frac{1}{64}\,|a_1^{\frac{1}{2}}(1+2e^{i\theta}+e^{i2\theta})$$

$$+(a_1+a_2+a_3)^{\frac{1}{2}}(1-2e^{i\theta}+e^{i2\theta})\,|^2\,|1-e^{i\theta}|^2$$

$$=\frac{2a_1+a_2+2(a_1(a_1+a_2+a_3))^{\frac{1}{2}}}{64}\,|1-e^{i\theta}-e^{i2\theta}+e^{i3\theta}|^2$$

$$+\frac{1}{64}\,|a_1^{\frac{1}{2}}(1+e^{i\theta}-e^{i2\theta}-e^{i3\theta})$$

$$+(a_1+a_2+a_3)^{\frac{1}{2}}(1-3e^{i\theta}+3e^{i2\theta}-e^{i3\theta})\,|^2,$$

and furthermore, (4.30a—c) follow.

Moreover, the matrices Z_3 and Z_4 have the following properties.

(1) If $a_2=O(\Delta x)$, then Z_3 can be rewritten as

$$Z_3=\tilde{Z}_{3,1}+\tilde{Z}_{3,2}+O(\Delta x), \tag{4.31a}$$

where $\tilde{Z}_{3,1}$ and $\tilde{Z}_{3,2}$ are nonnegative definite matrices of the form

$$\tilde{Z}_{3,1}=\begin{pmatrix} \multicolumn{2}{c}{Z_{3,1}} & 0 \\ & & 0 \\ 0 & 0 & 0 \end{pmatrix},\quad \tilde{Z}_{3,2}=\begin{pmatrix} 0 & 0 & 0 \\ 0 & \multicolumn{2}{c}{Z_{3,2}} \\ 0 & & \end{pmatrix},$$

the relation $Z_{3,1}+Z_{3,2}=Z_2$ between $Z_{3,1}$ and $Z_{3,2}$ being valid.

(2) If $a_3=O(\Delta x)$, then Z_4 can be rewritten as

$$Z_4=\tilde{Z}_{4,1}+\tilde{Z}_{4,2}+O(\Delta x), \tag{4.31b}$$

where $\tilde{Z}_{4,1}$ and $\tilde{Z}_{4,2}$ are nonnegative definite matrices of the form

$$\tilde{Z}_{4,1}=\begin{pmatrix} & & & 0 \\ & Z_{4,1} & & 0 \\ & & & 0 \\ 0 & 0 & 0 & 0 \end{pmatrix},\quad \tilde{Z}_{4,2}=\begin{pmatrix} 0 & 0 & 0 & 0 \\ 0 & & & \\ 0 & & Z_{4,2} & \\ 0 & & & \end{pmatrix},$$

the relation $Z_{4,1}+Z_{4,2}=Z_3+O(\Delta x)$ between $Z_{4,1}$ and $Z_{4,2}$ being valid.

(3) If $a_2=O(\Delta x)$ and $a_3=O(\Delta x)$, then Z_4 can be rewritten as

$$Z_4=\tilde{Z}_{4,3}+\tilde{Z}_{4,4}+O(\Delta x), \tag{4.31c}$$

where $\tilde{Z}_{4,3}$ and $\tilde{Z}_{4,4}$ are nonnegative definite matrices of the form

$$\tilde{Z}_{4,3}=\begin{pmatrix} \multicolumn{2}{c}{Z_{4,3}} & 0 & 0 \\ & & 0 & 0 \\ 0 & 0 & 0 & 0 \\ 0 & 0 & 0 & 0 \end{pmatrix},\quad \tilde{Z}_{4,4}=\begin{pmatrix} 0 & 0 & 0 & 0 \\ 0 & 0 & 0 & 0 \\ 0 & 0 & \multicolumn{2}{c}{Z_{4,4}} \\ 0 & 0 & & \end{pmatrix},$$

the relation $Z_{4,3}+Z_{4,4}=Z_2$ between $Z_{4,3}$ and $Z_{4,4}$ being valid.

The relations (4.31a—c) can be deduced also in a straightforward way. In fact, (4.30b) can be rewritten as

$$Z_3=\frac{1}{16}\begin{pmatrix} 2a_1+a_2 & -2a_1-2a_2 & a_2 \\ -2a_1-2a_2 & 4a_1+4a_2 & -2a_1-2a_2 \\ a_2 & -2a_1-2a_2 & 2a_1+a_2 \end{pmatrix}.$$

Thus, if $a_2=O(\Delta x)$, then

$$Z_3=\frac{1}{16}\begin{pmatrix} 2a_1 & -2a_1 & 0 \\ -2a_1 & 4a_1 & -2a_1 \\ 0 & -2a_1 & 2a_1 \end{pmatrix}+O(\Delta x)$$

$$=\frac{1}{8}\begin{pmatrix} a_1 & -a_1 & 0 \\ -a_1 & a_1 & 0 \\ 0 & 0 & 0 \end{pmatrix}+\frac{1}{8}\begin{pmatrix} 0 & 0 & 0 \\ 0 & a_1 & -a_1 \\ 0 & -a_1 & a_1 \end{pmatrix}+O(\Delta x).$$

Letting $Z_{3,1}=\frac{1}{8}\begin{pmatrix} a_1 & -a_1 \\ -a_1 & a_1 \end{pmatrix}=Z_{3,2}$, we can write it in the form of (4.31a). Moreover,

$$Z_{3,1}+Z_{3,2}=\frac{1}{4}\begin{pmatrix} a_1 & -a_1 \\ -a_1 & a_1 \end{pmatrix}=Z_2.$$

Therefore, the property (1) is true.

(4.30c) can be rewritten as

$$Z_4=\frac{1}{64}\begin{pmatrix} 2a_1+2\bar{a}_3+4\bar{a}_2 & -4\bar{a}_2-4\bar{a}_3 & -2a_1+2\bar{a}_3 & 0 \\ -4\bar{a}_2-4\bar{a}_3 & 8\bar{a}_3 & 4\bar{a}_2-4\bar{a}_3 & 0 \\ -2a_1+2\bar{a}_3 & 4\bar{a}_2-4\bar{a}_3 & 2a_1+2\bar{a}_3-4\bar{a}_2 & 0 \\ 0 & 0 & 0 & 0 \end{pmatrix}$$

$$+\frac{1}{64}\begin{pmatrix} 0 & 0 & 0 & 0 \\ 0 & 2a_1+2\bar{a}_3-4\bar{a}_2 & 4\bar{a}_2-4\bar{a}_3 & -2a_1+2\bar{a}_3 \\ 0 & 4\bar{a}_2-4\bar{a}_3 & 8\bar{a}_3 & -4\bar{a}_2-4\bar{a}_3 \\ 0 & -2a_1+2\bar{a}_3 & -4\bar{a}_2-4\bar{a}_3 & 2a_1+2\bar{a}_3+4\bar{a}_2 \end{pmatrix}$$

$$+\frac{1}{64}\begin{pmatrix} -a_3 & a_3 & a_3 & -a_3 \\ a_3 & -a_3 & -a_3 & a_3 \\ a_3 & -a_3 & -a_3 & a_3 \\ -a_3 & a_3 & a_3 & -a_3 \end{pmatrix},$$

where $\bar{a}_3=a_1+a_2+a_3$, $\bar{a}_2=[a_1(a_1+a_2+a_3)]^{\frac{1}{2}}$. Furthermore, the first and the second matrices of the above expression are respectively

$$\begin{pmatrix} & & & 0 \\ & D_1^*D_1 & & 0 \\ & & & 0 \\ 0 & 0 & 0 & 0 \end{pmatrix}, \quad \begin{pmatrix} 0 & 0 & 0 & 0 \\ 0 & & & \\ 0 & & D_2^*D_2 & \\ 0 & & & \end{pmatrix},$$

where

$$D_1=\frac{\sqrt{2}}{8}(a_1^{\frac{1}{2}}+\bar{a}_3^{\frac{1}{2}},\ -2\bar{a}_3^{\frac{1}{2}},\ -a_1^{\frac{1}{2}}+\bar{a}_3^{\frac{1}{2}}),$$

$$D_2=\frac{\sqrt{2}}{8}(-a_1^{\frac{1}{2}}+\bar{a}_3^{\frac{1}{2}},\ -2\,\bar{a}_3^{\frac{1}{2}},\ a_1^{\frac{1}{2}}+\bar{a}_3^{\frac{1}{2}}).$$

Thus, if $a_3=O(\Delta x)$, by letting $Z_{4,1}=D_1^*D_1$ and $Z_{4,2}=D_2^*D_2$, the above expression of Z_4 can be then rewritten as (4.31b). Moreover, if $a_3=O(\Delta x)$, obviously $D_1^*D_1+D_2^*D_2=Z_3+O(\Delta x)$. Therefore, the property (2) is true.

Z_4 can also be rewritten as

$$Z_4=\frac{1}{64}\left(\begin{array}{cccc} 8a_1 & -8a_1 & 0 & 0 \\ -8a_1 & 8a_1 & 0 & 0 \\ 0 & 0 & 8a_1 & -8a_1 \\ 0 & 0 & -8a_1 & 8a_1 \end{array}\right.$$

$$+\left(\begin{array}{cc} 2a_2+a_3+4(\bar{a}_2-a_1) & -4(\bar{a}_2-a_1)-4a_2-3a_3 \\ -4(\bar{a}_2-a_1)-4a_2-3a_3 & -4(\bar{a}_2-a_1)+10a_2+9a_3 \\ 2a_2+3a_3 & 8(\bar{a}_2-a_1)-8a_2-9a_3 \\ -a_3 & 2a_2+3a_3 \end{array}\right.$$

$$\left.\begin{array}{cc} 2a_2+3a_3 & -a_3 \\ 8(\bar{a}_2-a_1)-8a_2-9a_3 & 2a_2+3a_3 \\ -4(\bar{a}_2-a_1)+10a_2+9a_3 & -4(\bar{a}_2-a_1)-4a_2-3a_3 \\ -4(\bar{a}_2-a_1)-4a_2-3a_3 & 2a_1+a_3+4(\bar{a}_2-a_1) \end{array}\right)\right).$$

If $a_2=O(\Delta x)$ and $a_3=O(\Delta x)$, every element of the second matrix is a quantity of $O(\Delta x)$. Thus, letting

$$Z_{4,3}=Z_{4,4}=\frac{1}{8}\begin{pmatrix} a_1 & -a_1 \\ -a_1 & a_1 \end{pmatrix},$$

we can get (4.31c).

3. *The Stability of the Horizontal Two-point Schemes and the "Condition" of Their Difference Systems*

On the horizontal two-point schemes, we give Lemma 4.4 and Theorem 4.3.

Lemma 4.4 Consider the horizontal two-point schemes approximating (4.1). If the conditions (3.3) and (3.2) are satisfied, then for the schemes with $H_1=0$, the inequality $r_0'\geqslant\frac{1}{2}+\varepsilon$ must hold when $\sigma<0$; $r_0'\leqslant\frac{1}{2}-\varepsilon$ must hold when $\sigma>0$; and for the schemes with $H_1=-1$, the inequality $r'_0 \geqslant \frac{1}{2} + \varepsilon$ must hold when $\sigma > 0$; $r'_0 \leqslant \frac{1}{2} - \varepsilon$ must hold when $\sigma < 0$, ε being a positive constant greater than $\frac{1}{2}\sqrt{c_0}$.

Proof The general form of the horizontal two-point schemes is

$$r_0 u_m^{k+1} + r_{\pm1} u_{m\pm1}^{k+1} = s_0 u_m^k + s_{\pm1} u_{m\pm1}^k,$$

where "+" and "−" in "±" correspond to $H_1 = 0$ and -1 respectively (in the following, the same rule will be adopted).

The expression can be rewritten as

$$r_0 \left(u + \frac{\partial u}{\partial t} \Delta t\right)_m^k + r_{\pm1} \left(u + \frac{\partial u}{\partial t} \quad \pm \frac{\partial u}{\partial x} \Delta x\right)_m^k$$
$$= s_0 u_m^k + s_{\pm1} \left(u \pm \frac{\partial u}{\partial x} \Delta x\right)_m^k + O(\Delta x^2).$$

According to (4.1), there exists the relation

$$\frac{\partial u}{\partial t} = -\lambda \frac{\partial u}{\partial x}.$$

Thus, from the expression above, we can deduce

$$\begin{cases} r'_0 + r'_{\pm1} = s'_0 + s'_{\pm1}, \\ r'_0(-\sigma) + r'_{\pm1}(-\sigma \pm 1) = \pm s'_{\pm1}. \end{cases}$$

In § 3 we suppose $r'_0 + r'_{\pm1} = 1$, so they can be further rewritten as

$$\begin{cases} r'_{\pm1} = 1 - r'_0, \\ s'_0 = r'_0 \pm \sigma, \\ s'_{\pm1} = 1 - (r'_0 \pm \sigma). \end{cases}$$

According to Lemma 4.3, the condition (3.2) in this case can be rewritten as

$$\frac{a_1}{4} = s'_0 s'_{\pm1} - r'_0 r'_{\pm1} = (r'_0 \pm \sigma) - (r'_0 \pm \sigma)^2 - r'_0(1 - r'_0)$$
$$= \pm\sigma(1 - 2r'_0) - \sigma^2 \geqslant 0,$$

and (3.3) can be expressed as

$$1 - 4r'_0 r'_{\pm1} \sin^2 \frac{\theta}{2} = 1 - 4(r'_0 - r'^2_0) \sin^2 \frac{\theta}{2} \geqslant c_0 > 0 \quad (c_0 \leqslant 1),$$

which is equivalent to

$$r'_0 \leqslant \frac{1}{2} - \frac{\sqrt{c_0}}{2} \quad \text{or} \quad r'_0 \geqslant \frac{1}{2} + \frac{\sqrt{c_0}}{2}.$$

Therefore, for the schemes with $H_1 = 0$, if $\sigma < 0$, and if (3.2) and (3.3) are fulfilled, then $r'_0 \geqslant \frac{1}{2} + \varepsilon$, where $\varepsilon \geqslant \frac{1}{2}\sqrt{c_0}$; if $\sigma > 0$, then $r'_0 \leqslant \frac{1}{2} - \varepsilon$; for the schemes with $H_1 = -1$, if $\sigma > 0$ and if (3.2) and (3.3) are fulfilled, then $r'_0 \geqslant \frac{1}{2} + \varepsilon$; if $\sigma < 0$, then $r'_0 \leqslant \frac{1}{2} - \varepsilon$.

From Theorems 4.1, 4.2 and Lemmas 4.1—4.4, the following results can be easily obtained.

Theorem 4.3 If a horizontal two-point scheme (4.9) approximating the initial-boundary-value problem (4.1)—(4.4) satisfies conditions (i)—(iii) in § 3, and the boundary conditions and the schemes at the boundaries satisfy condition (iv), then the system of difference equations (4.9) is well-conditioned and the scheme (4.9) is stable.

Proof To begin with, we prove that (4.18) is valid when the scheme is applied to the model problem A_1.

As a preparation of the whole proof, we prove that in this case, $H_1(1)=H_1(M)=-1$ and all the coefficients $r'_{0,m}\geqslant\frac{1}{2}+\varepsilon$. In fact, condition (i) is satisfied, so $H_1(M)=-1$ is evident. We now prove $H_1(1)=-1$. If this were not so, $H_1(1)$ should be equal to 0. Obviously, $\sigma>0$ when $m=1$. Thus, according to Lemma 4.4, $r'_{0,1}\leqslant\frac{1}{2}-\varepsilon$ in this case. As pointed out above, $H_1(M)=-1$. Therefore, there must be a place where H_1 changes. We suppose $H_1(m_1)=0$ and $H_1(m_1+1)=-1$. According to condition (iii), we have $r_{1,m_1}=O(\Delta x)$ and $r_{-1,m_1+1}=O(\Delta x)$, so $r'_{0,m_1}=1-O(\Delta x)$. This would contradict $r'_{0,1}\leqslant\frac{1}{2}-\varepsilon$. This is due to the following reason. In this situation, (3.3) is reduced to $r'_0\leqslant\frac{1}{2}-\varepsilon$ or $r'_0\geqslant\frac{1}{2}+\varepsilon$. Hence, if (3.3) is satisfied everywhere and if r'_0 satisfies the Lipschitz condition with respect to x, then either all r'_0 are greater than or equal to $\frac{1}{2}+\varepsilon$, or all r'_0 are less than or equal to $\frac{1}{2}-\varepsilon$. Therefore, $H_1(1)\neq 0$, i.e., $H_1(1)=-1$. This is the result desired. In addition, according to Lemma 4.4 and the above proof, we can also show that any r'_0 is greater than or equal to $\frac{1}{2}+\varepsilon$.

We now prove that if H_1 is equal to -1 everywhere, then (4.18) is valid. For the horizontal two-point schemes, (3.3) can be rewritten as

$$r^*(\theta)r(\theta)=1-4r'_0(1-r'_0)\sin^2\frac{\theta}{2}\geqslant c_0>0,$$

$$1-4r'_0(1-r'_0)\sin^2\frac{\theta}{2}-c'_0>0,$$

where

$$0<c'_0=\frac{c_0}{2}\leqslant\frac{1}{2}.$$

Therefore, letting

$$1-4r'_0(1-r'_0)\sin^2\frac{\theta}{2}-c'_0=(ae^{-i\theta}+b)^*(ae^{-i\theta}+b),$$

we have

$$r^*(\theta)r(\theta)=c'_0+(ae^{-i\theta}+b)^*(ae^{-i\theta}+b),$$

where the real numbers a and b are not unique. In the case $H_1=-1$ and $r'\geqslant\frac{1}{2}+\varepsilon,\ \varepsilon\geqslant\frac{1}{2}\sqrt{c_0}$, we choose

$$\begin{cases} a=\dfrac{2r'_0(1-r'_0)}{\sqrt{1-c'_0}+\sqrt{(2r'_0-1)^2-c'_0}}=\dfrac{\sqrt{1-c'_0}-\sqrt{(2r'_0-1)^2-c'_0}}{2}, \\ b=\dfrac{\sqrt{1-c'_0}+\sqrt{(2r'_0-1)^2-c'_0}}{2}. \end{cases}$$

Thus, according to Lemma 4.1, we can construct the following pseudo-null matrix corresponding to $r(\theta)$:

$$\tilde{Q}_2=R^*R-\begin{pmatrix}0 & 0\\ 0 & c'_0\end{pmatrix}-\begin{pmatrix}a^2 & ab\\ ab & b^2\end{pmatrix}=\begin{pmatrix}r'^2_{-1}-a^2 & 0\\ 0 & r'^2_0-c'_0-b^2\end{pmatrix}.$$

Moreover, according to $(2r'_0-1)^2-c'_0\geqslant\frac{c_0}{2}>0$, we know that every element of $\tilde{Q}_2$ satisfies the Lipschitz condition with respect to x if condition (iii) is valid. Therefore, we have the following inequality:

$$\begin{aligned}\sum_{m=1}(R^*R\tilde{U},\tilde{U})_m\Delta x &\geqslant c'_0\sum_{m=1}^{M}|\tilde{u}_m|^2\Delta x+\sum_{m=1}^{M}(\tilde{Q}_2\tilde{U},\tilde{U})_m\Delta x\\ &\geqslant c'_0\sum_{m=1}^{M}|\tilde{u}_m|^2\Delta x+(r'^2_{-1}-a^2)_1|\tilde{u}_0|^2\Delta x\\ &\quad+(r'^2_0-c'_0-b^2)_M|\tilde{u}_M|^2\Delta x-O(\Delta x)\sum_{m=0}^{M}|\tilde{u}_m|^2\Delta x\\ &\geqslant c_6\sum_{m=1}^{M}|\tilde{u}_m|^2\Delta x-c_5|\tilde{u}_0|^2\Delta x-O(\Delta x)\sum_{m=0}^{M}|\tilde{u}_m|^2\Delta x,\end{aligned}$$

where

$$c_5>|r'^2_{-1,1}-a_1^2|,\quad c_6=\min\{c'_0,\ (r'^2_0-b^2)_M\}.$$

In addition,

$$\begin{cases}\dfrac{\partial b^2}{\partial c'_0}=2b\,\dfrac{1}{4}\left[\dfrac{-1}{\sqrt{1-c'_0}}+\dfrac{-1}{\sqrt{(2r'_0-1)^2-c'_0}}\right]\\ \qquad=\dfrac{-b^2}{\sqrt{1-c'_0}\sqrt{(2r'_0-1)^2-c'_0}}<0,\\ (r'^2_0-b^2)\,|_{c'_0=0}=0,\end{cases}$$

and $c'_0>0$, so $(r'^2_0-b^2)_M>0$, i.e., $c_6>0$. Therefore, owing to $\delta_0(\lambda_{n,0})=1$ and $\delta_1(\lambda_{n,M})=0$ for the model A_1, the above inequality can be rewritten as (4.18):

$$\begin{aligned}\tilde{T}_n+c_4\Delta x\|\tilde{u}_n\|_x^2\geqslant &-c_5|\tilde{u}_{n,0}|^2\Delta x\delta_0(\lambda_{n,0})-c_5|\tilde{u}_{n,M}|^2\Delta x\delta_1(\lambda_{n,M})\\ &+c_6\sum_{m\in\mathscr{G}(\lambda_n)}|\tilde{u}_{n,m}|^2\Delta x.\end{aligned}$$

As for the case where H_1 is not always equal to -1, (4.18) can be obtained in a similar way. In order to prove (4.18), besides $H_1(1)=$

$H_1(M)=-1$, the following two facts need to be used.

(1) For $H_1=0$, we may construct a pseudo-null matrix of the following form

$$\tilde{\tilde{Q}}_2=R^*R-\begin{pmatrix} c'_0 & 0 \\ 0 & 0 \end{pmatrix}-\begin{pmatrix} b^2 & ab \\ ab & a^2 \end{pmatrix}=\begin{pmatrix} r_0'^2-c'_0-b^2 & 0 \\ 0 & r_1'^2-a^2 \end{pmatrix}.$$

(2) When H_1 changes its value, there must be $r_1=O(\Delta x)$, $r_{-1}=O(\Delta x)$, and $r'_0=1+O(\Delta x)$. Thus $r_0'^2-c'_0-b^2=O(\Delta x)$, $(1-r'_0)^2-a^2=O(\Delta x)$, i.e., every element of $\tilde{Q}_2$ and $\tilde{\tilde{Q}}_2$ is a quantity of $O(\Delta x)$.

It is easy to prove that (4.18) is valid for the model A_2. Let $x'=1-x$ and $\lambda'=-\lambda$, the equation $\frac{\partial u}{\partial t}+\lambda\frac{\partial u}{\partial x}=0$ can be rewritten as $\frac{\partial u}{\partial t}+\lambda'\frac{\partial u}{\partial x'}=0$. Hence, a model problem A_2 can be easily rewritten as a problem A_1. Therefore, from the fact that (4.18) is valid for the model A_1, it certainly follows that (4.18) is valid for the model A_2.

We now prove that (4.18) is satisfied for the model B. According to condition (i), we have $H_1(0)=0$ and $H_1(M)=-1$. Therefore, in the region $m\approx 0$, the model B is similar to the model A_2, and in the region $m\approx M$, the model B is similar to the model A_1. Moreover, the treatment for the interior points is the same for all the models. Therefore, the above method may be used to prove the conclusion for the model B. Of course, because H_1 must change at some points in this case, the proof will be longer.

Lastly, we discuss the model C. We first prove that in this situation, $H_1(1)=-1$ and $H_1(M-1)=0$. Obviously, $\sigma>0$ at $m=1$. Hence, if $H_1(1)=0$, then $r'_{0,1}\leqslant\frac{1}{2}-\varepsilon$. As pointed out at the beginning, this means that $r'_{0,m}\leqslant\frac{1}{2}-\varepsilon$ everywhere. Therefore, according to the fact that $\sigma<0$ at $m=M-1$, we can definately say that $H_1(M-1)=-1$. In this case, there must be points where H_1 changes its values and $r'_0=1+O(\Delta x)\geqslant\frac{1}{2}+\varepsilon$. This contradicts $r'_{0,m}\leqslant\frac{1}{2}-\varepsilon$. Consequently, $H_1(1)=0$ is impossible. By the same argument, we can prove $H_1(M-1)=-1$ to be impossible. Therefore, $H_1(1)=-1$ and $H_1(M-1)=0$, i.e., at $m\approx 1$, the model C is similar to the model A_1 and at $m\approx M-1$, the model C is similar to the model A_2. Thus, the preceding method may be used to prove that (4.18) is valid when the scheme is applied to the model C.

We have now proved that (4.18) is satisfied for all the models. In the following, we shall prove that (4.19) is valid for all the models, We shall pay most attention to the model problem A_1 with $H_1\equiv -1$. As having been pointed out for the horizontal two-point schemes, we can find a pseudo-null matrix

$$Q_2=-Y_2-Z_2,$$

where the expressions of Y_2 and the nonnegative definite matrix Z_2 are (4.30d) and (4.30a) respectively. We know from these that the expression of Q_2 for $H_1=-1$ is

$$Q_2=\begin{pmatrix}(1-r'_0)^2 & r'_0(1-r'_0)\\ r'_0(1-r'_0) & r'^2_0\end{pmatrix}-\begin{pmatrix}(1-s'_0)^2 & s'_0(1-s'_0)\\ s'_0(1-s'_0) & s'^2_0\end{pmatrix}$$

$$-(s'_0(1-s'_0)-r'_0(1-r'_0))\begin{pmatrix}1 & -1\\ -1 & 1\end{pmatrix}$$

$$=\begin{pmatrix}s'_0-r'_0 & 0\\ 0 & r'_0-s'_0\end{pmatrix}.$$

Therefore, we have

$$\sum_{m=1}^{M}(Y_2\tilde{U},\ \tilde{U})_m\Delta x\leqslant-\sum_{m=1}^{M}(Q_2\tilde{U},\ \tilde{U})_m\Delta x$$

$$\leqslant(r'_{0,0}-s'_{0,0})\,|\tilde{u}_0|^2\Delta x$$

$$-(r'_{0,M}-s'_{0,M})\,|\tilde{u}_M|^2\Delta x+O(\Delta x)\sum_{m=0}^{M}|\tilde{u}_m|^2\Delta x.$$

We have shown in the proof of Lemma 4.4 that $r'_0-s'_0=\sigma$ for the scheme with $H_1=-1$. Therefore, noticing (4.21), we can rewrite the inequality as

$$\tilde{T}^{k+1}_n-\tilde{T}^k_n\leqslant\sigma_{n,0}|\tilde{u}^k_{n,0}|^2\Delta x-\sigma_{n,M}|\tilde{u}^k_{n,M}|^2\Delta x+O(\Delta x)\,\|\tilde{u}^k_n\|^2_x.$$

In addition, $\delta_0(\lambda_{n,0})=1$, $\delta_1(\lambda_{n,M})=0$ and $\sigma_{n,M}\geqslant0$ for the model A_1, so the inequality can be further rewritten as (4.19):

$$\tilde{T}^{k+1}_n-\tilde{T}^k_n\leqslant[c_7\delta_0(\lambda_{n,0})-|\sigma_{n,0}|(1-\delta_0(\lambda_{n,0}))]\,|\tilde{u}^k_{n,0}|^2\Delta x$$

$$+[c_7\delta_1(\lambda_{n,M})-|\sigma_{n,M}|(1-\delta_1(\lambda_{n,M}))]\,|\tilde{u}^k_{n,M}|^2\Delta x+c_4\Delta x\|\tilde{u}^k_n\|^2_x.$$

If H_1 is not always equal to -1, (4.19) can be proved in a similar way. The only differences are as follows. When H_1 is always equal to -1, the fact that Q_2 is a pseudo-null matrix guarantees that the coefficients of $|\tilde{u}_{n,m}|^2$ on the right hand sides of these inequalities, $m\in[1,M-1]$, are quantities of $O(\Delta x)$. When H_1 is not always equal to -1, for the points where H_1 does not change its value, the same result is still obtained from the fact that Q_2 is a pseudo-null matrix. However, for the points where H_1 changes its value, this result cannot be obtained from that fact. It is obtained, instead, from $s'_0=1+O(\Delta x)$ and $r'_0=1+O(\Delta x)$, i.e., from the fact that the elements of Q_2 are quantities of $O(\Delta x)$.

As has been the case in proving (4.18), the above method proving (4.19) for model A_1 can also be used for the models A_2, B and C.

The inequalities (4.18) and (4.19) have been proved, and condition (iv) guarantees that (4.5) or (4.5)′ is satisfied. Thus, according to Theorems 4.1 and 4.2, the system of difference equations (4.9) is well–

conditioned and the difference scheme (4.9) is stable.

From this theorem, the following corollaries can be easily obtained.

Corollary 1 When the scheme (3.5) is applied to the initial–boundary–value problem (4.1)—(4.4) with (4.7) or (4.6), if the condition (3.6) is valid and if every element of the matrix Λ satisfies the Lipschitz condition with respect to x, t:

$$\begin{cases} |\lambda_n(x+\Delta x,\ t)-\lambda_n(x,\ t)|\leqslant c_1\Delta x, \\ |\lambda_n(x,\ t+\Delta t)-\lambda_n(x,\ t)|\leqslant c_1\Delta t, \end{cases} \tag{4.32}$$

then the difference system is well–conditioned and the difference scheme is stable.

Corollary 2 When the scheme (3.25) (for linear equations, the first step (3.24) is not required) is applied to the initial–boundary–value problem (4.1)—(4.4) with $|\lambda|\geqslant\varepsilon>0$, if every element of matrix Λ satisfies the conditions (4.32), then the system of difference equations is well–conditioned and the difference scheme is stable.

Babenko et al.[4] and Thomée[2] have discussed the stability of (3.25) for initial–boundary–value problems. However, they discussed only the stability with respect to initial values. Moreover, they have proved in [4] only that the scheme (3.25) satisfies a necessary condition of stability——the Godunov–Pyabenkii spectrum criterion[3]. (Actually, a scheme with three independent variables is discussed in [4]. However, if one particular independent variable is dropped, it becomes (3.25). Therefore, the result in [4] includes this conclusion.)

Now let us prove Corollary 1. When (3.6) is valid, (3.2) and (3.3) can be derived. Therefore, conditions (i) and (ii) of § 3 are satisfied. When the conditions (4.32) are valid, it is easily known from the expression of the scheme (3.5) that condition (iii) of § 3 is satisfied. Obviously, this scheme can be written as $u_m^{k+1}=u_m^k$ when $\sigma=0$, i.e., condition (iv) of this section is valid. All the conditions of Theorem 4.3 are satisfied, so this corollary is true.

It is also easy to prove Corollary 2. In fact, when (3.26) is valid, (3.2) is satisfied; when $|\lambda|\geqslant\varepsilon>0$, (3.27) can be derived from (3.26); (3.4a—b) can be obtained from (4.32). Thus, conditions (i)—(iii) are satisfied. In addition, (4.6) is satisfied when $|\lambda|\geqslant\varepsilon>0$. That is, all the conditions of Theorem 4.3 are valid. Therefore, Corollary 2 is true.

4. *The Stability of the Horizontal Three–point Explicit Schemes*

In the following, we shall give two Lemmas and Theorem 4.4 about the stability of the horizontal three–point explicit schemes.

Lemma 4.5 For a horizontal three–point explicit scheme with $H_1=-2$ or 0 approximating (4.1), if the von Neumann condition holds, and if $\sigma=0$, then

$$s'_h=\begin{cases}1, & h=0,\\ 0, & h\neq 0.\end{cases}$$

Proof First of all, we consider the case $H_1=-2$, i.e.,

$$u_m^{k+1}=s_{-2}u_{m-2}^k+s_{-1}u_{m-1}^k+s_0u_m^k.$$

Evidently, from the relation, we can obtain

$$\left(u+\frac{\partial u}{\partial t}\Delta t\right)_m^k=s_{-2}\left(u-\frac{\partial u}{\partial x}2\Delta x\right)_m^k+s_{-1}\left(u-\frac{\partial u}{\partial x}\Delta x\right)_m^k$$
$$+s_0u_m^k+O(\Delta t^2).$$

Moreover, the scheme approximates $\frac{\partial u}{\partial t}+\lambda\frac{\partial u}{\partial x}=0$. Hence, we have

$$\begin{cases}s'_{-2}+s'_{-1}+s'_0=1,\\ -2s'_{-2}-s'_{-1}=-\sigma,\end{cases}$$

or

$$\begin{cases}s'_0=1-\frac{1}{2}(\sigma+s'_{-1}),\\ s'_{-2}=\frac{1}{2}(\sigma-s'_{-1}).\end{cases}\tag{4.33}$$

Therefore, by using Lemma 4.3, the von Neumann condition can be rewritten as

$$\begin{cases}a_1=4[s'_{-2}s'_{-1}+s'_{-1}s'_0+4s'_{-2}s'_0]=4s'_{-1}(1-s'_{-1})+8(\sigma-s'_{-1})\left(1-\frac{1}{2}(\sigma+s'_{-1})\right)\\ \quad=4s'_{-1}-4s'^2_{-1}+8(\sigma-s_{-1}')-4(\sigma^2-s'^2_{-1})\\ \quad=8\sigma-4\sigma^2-4s'_{-1}\geqslant 0,\\ a_1+a_2=4[s'_{-2}s'_{-1}+s'_{-1}s'_0]=4s'_{-1}(1-s'_{-1})\geqslant 0.\end{cases}\tag{4.34}$$

The second inequality is equivalent to $0\leqslant s'_{-1}\leqslant 1$, and the first inequality becomes $-s'_{-1}\geqslant 0$ when $\sigma=0$. Hence

$$s'_{-1}=0,$$

and

$$s'_{-2}=\frac{1}{2}(\sigma-s'_{-1})=0,\quad s'_0=1-\frac{1}{2}(\sigma+s'_{-1})=1,$$

i.e., the conclusion is proved.

In the case $H_1=0$, the conclusion can be obtained in the same way. Therefore, the proof is completed.

In passing, we point out that in the case $H_1=-1$, there is a different conclusion. In that case, the conclusion is that if the von Neumann condition is satisfied and if $\sigma=0$, then

$$0\leqslant s'_{-1}\leqslant\frac{1}{2},\quad 0\leqslant s'_1\leqslant\frac{1}{2},\quad s'_0=1-(s'_{-1}+s'_1).$$

Lemma 4.6 For a horizontal three-point explicit scheme approximating (4.1), if the von Neumann condition is satisfied, then H_1 is equal to -2 or -1 for $\sigma>0$, while H_1 is equal to -1 or 0 for $\sigma<0$.

Proof The proof is simple. We need to prove only that if $\sigma>0$ and $H_1=0$, or if $\sigma<0$ and $H_1=-2$, then the von Neumann condition must not be satisfied.

Consider the case $H_1=-2$. In the process of proving Lemma 4.5, we have pointed out that in this case, the von Neumann condition is equivalent to

$$\begin{cases} a_1=8\sigma-4\sigma^2-4s'_{-1}\geqslant 0, \\ a_1+a_2=4s'_{-1}(1-s'_{-1})\geqslant 0. \end{cases}$$

Moreover, the second inequality is equivalent to $0\leqslant s'_{-1}\leqslant 1$. Therefore, when $\sigma<0$, the first inequality contradicts the second inequality. This means that the von Neumann condition must not be satisfied. In the case $H_1=0$ and $\sigma>0$, the conclusion can be obtained in a similar way.

The following Theorem 4.4 follows from the lemmas and the preceding theorems.

Theorem 4.4 For a horizontal three-point explicit scheme (4.9) approximating the initial-boundary-value problem (4.1)—(4.4), if conditions (i) and (iii) of § 3 and condition (iv) of this section are satisfied, then the scheme is stable.

Proof For an explicit scheme, (4.18) is automatically satisfied. Therefore, according to Theorem 4.2, the only thing we need to do is to prove that (4.19) holds for the models A_1, A_2, B and C. We now prove that (4.19) holds when the scheme is applied to the model A_1. To do this, let us first investigate the terms in (4.19) related to the region near $m=M$. Obviously, $H_1=-2$ for $m=M$. We suppose that $H_1=-2$ for

$$m\in[M-M_1-1,\ M].$$

According to Lemma 4.3 and its corollary, we can construct the following pseudo-null matrix

$$Q_3=-Y_3-Z_3, \tag{4.35}$$

and have the following inequality

$$\begin{aligned}\sum_{m=M-M_1}^{M}(Y_3\widetilde{U},\widetilde{U})_m &= -\sum_{m=M-M_1}^{M}(Z_3\widetilde{U},\widetilde{U})_m-\sum_{m=M-M_1}^{M}(Q_3\widetilde{U},\widetilde{U})_m \\ &\leqslant -\sum_{m=M-M_1}^{M}(Q_3\widetilde{U},\widetilde{U})_m.\end{aligned} \tag{4.36}$$

In the following, we shall investigate the sum on the right-hand side of the inequality (4.36). For this purpose,

$$-\sum_{m=M-M_1}^{M}(Q_3\widetilde{U},\widetilde{U})_m$$

is rewritten as

$$\sum_{h=-2}^{2}\sum_{m=M-M_1-2}^{M} b_{m,m+h}\tilde{u}_m\tilde{u}_{m+h},$$

where $b_{m,m+h}=b_{m+h,m}$ and $b_{m,m+h}=0$ for $m+h>M$. Because the sum of all the elements on every diagonal of Q_3 is equal to zero, and because every element of Q_3 satisfies the Lipschitz condition with respect to x under the conditions of this theorem, $b_{m,m}$, $b_{m,m+1}$, $b_{m,m+2}$, $b_{m+1,m}$ and $b_{m+2,m}$ are quantities of $O(\Delta x)$ for $m\in[M-M_1,\ M-2]$. Moreover, $b_{M,M}$ etc. have the following expression:

$$\begin{pmatrix} b_{M-1,M-1} & b_{M-1,M} \\ b_{M,M-1} & b_{M,M} \end{pmatrix}$$

$$=\begin{pmatrix} \frac{1}{4}(a_1+a_2)+\frac{1}{16}(2a_1+a_2)+s_0'^2+s_{-1}'^2-1 & -\frac{1}{8}(a_1+a_2)+s_0's_{-1}' \\ -\frac{1}{8}(a_1+a_2)+s_0's_{-1}' & \frac{1}{16}(2a_1+a_2)+s_0'^2-1 \end{pmatrix}_M$$

$$+O(\Delta x),$$

where the subscript M of this matrix means that each element of this matrix is a quantity at $m=M$. Furthermore, according to (4.33) and (4.34), we have the following three equalities:

(1) $$\frac{1}{4}(a_1+a_2)+\frac{1}{16}(2a_1+a_2)+s_0'^2+s_{-1}'^2-1$$

$$=s_{-1}'(1-s_{-1}')+\frac{1}{4}[2s_{-1}'(1-s_{-1}')+(\sigma-s_{-1}')(2-\sigma-s_{-1}')]$$

$$+\frac{1}{4}(2-\sigma-s_{-1}')^2+s_{-1}'^2-1$$

$$=\frac{1}{2}(1-s_{-1}')(s_{-1}'-2)+\frac{1}{2}(2-\sigma-s_{-1}')(1-s_{-1}')=\frac{\sigma}{2}(s_{-1}'-1);$$

(2) $$\frac{1}{16}(2a_1+a_2)+s_0'^2-1$$

$$=\frac{1}{4}[2s_{-1}'(1-s_{-1}')+(\sigma-s_{-1}')(2-\sigma-s_{-1}')]$$

$$+\frac{1}{4}(2-\sigma-s_{-1}')^2-1$$

$$=\frac{1}{2}(1-s_{-1}')(s_{-1}'-2)-s_{-1}'+\frac{1}{2}(2-\sigma-s_{-1}')(1-s_{-1}')$$

$$=\frac{\sigma}{2}(s_{-1}'-1)-s_{-1}';$$

(3) $$\frac{-1}{8}(a_1+a_2)+s_0's_{-1}'$$

$$=\frac{-1}{2}s_{-1}'(1-s_{-1}')+\frac{1}{2}(2-\sigma-s_{-1}')s_{-1}'=\frac{1}{2}s_{-1}'(1-\sigma).$$

Therefore, the matrix above can be rewritten as

$$\begin{pmatrix} b_{M-1,M-1} & b_{M-1,M} \\ b_{M,M-1} & b_{M,M} \end{pmatrix}$$

$$=\begin{pmatrix} \frac{\sigma}{2}(s'_{-1}-1) & \frac{s'_{-1}}{2}(1-\sigma) \\ \frac{s'_{-1}}{2}(1-\sigma) & \frac{\sigma}{2}(s'_{-1}-1)-s'_{-1} \end{pmatrix}_M + O(\Delta x)$$

$$=\begin{pmatrix} \frac{\sigma}{2}(s'_{-1}-1) & \frac{s'_{-1}}{2}(1-\sigma) \\ \frac{s'_{-1}}{2}(1-\sigma) & \frac{(\sigma-1)^2 s'^2_{-1}}{2\sigma(s'_{-1}-1)} \end{pmatrix}_M$$

$$+\begin{pmatrix} 0 & 0 \\ 0 & \frac{\sigma(s'_{-1}-1)}{2}-s'_{-1}-\frac{(\sigma-1)^2 s'^2_{-1}}{2\sigma(s'_{-1}-1)} \end{pmatrix}_M + O(\Delta x). \quad (4.37)$$

In proving Lemma 4.5, we have pointed out that $0\leqslant s'_{-1}\leqslant 1$ must hold in this case. Evidently, $\sigma\geqslant 0$ for $m=M$. Thus in the last expression of (4.37), the first diagonal element of the first matrix

$$\frac{\sigma(s'_{-1}-1)}{2}$$

is less than or equal to zero. Moreover, its determinant is equal to zero. Therefore, it is a nonpositive definite matrix.

From

$$\begin{cases} 0\leqslant s'_{-1}\leqslant 1, \\ a_1=8\sigma-4\sigma^2-4s'_{-1}\geqslant 0, \\ \sigma\geqslant 0, \end{cases}$$

we have

$$-2s'_{-1}\sigma+s'_{-1}\sigma^2+s'^2_{-1}\leqslant 0,$$

$$(\sigma-s'_{-1})^2\leqslant(1-s'_{-1})\sigma^2,$$

$$\frac{(\sigma-s'_{-1})^2}{2\sigma(1-s'_{-1})}\leqslant\frac{\sigma}{2},$$

so, for the non-zero element of the second matrix in the last expression of (4.37), there exists the following relation:

$$\frac{\sigma(s'_{-1}-1)}{2}-s'_{-1}-\frac{(\sigma-1)^2 s'^2_{-1}}{2\sigma(s'_{-1}-1)}$$

$$=\frac{\sigma^2(s'_{-1}-1)^2-2s'_{-1}\sigma(s'_{-1}-1)-(\sigma-1)^2 s'^2_{-1}}{2\sigma(s'_{-1}-1)}$$

$$=\frac{\sigma^2 s'^2_{-1}-2\sigma^2 s'_{-1}+\sigma^2-2\sigma s'^2_{-1}+2s'_{-1}\sigma-\sigma^2 s'^2_{-1}+2\sigma s'^2_{-1}-s'^2_{-1}}{2\sigma(s'_{-1}-1)}$$

$$=\frac{2s'_{-1}\sigma(1-\sigma)+\sigma^2-s'^2_{-1}}{2\sigma(s'_{-1}-1)}=-\sigma+\frac{(\sigma-s'_{-1})^2}{2\sigma(1-s'_{-1})}\leqslant-\frac{\sigma}{2}.$$

Hence we have

$$(\tilde{u}_{M-1}\ \tilde{u}_M)\begin{pmatrix} b_{M-1,M-1} & b_{M-1,M} \\ b_{M,M-1} & b_{M,M}\end{pmatrix}\begin{pmatrix}\tilde{u}_{M-1}\\ \tilde{u}_M\end{pmatrix}$$

$$\leqslant -\frac{\sigma_M}{2}|\tilde{u}_M|^2+O(\Delta x)[|\tilde{u}_{M-1}|^2+|\tilde{u}_M|^2]. \tag{4.38}$$

Let us now investigate the terms related to the region near $m=1$. According to Lemma 4.6, if $\sigma>0$ and if the von Neumann condition holds, we must have $H_1=-2$ or -1. Moreover, according to condition (i), it is not possible that $H_1(1)=-2$. Therefore, $H_1(1)=-1$. Suppose $H_1(m)=-1$ for $m\in[1, M_2+1]$. As we have done in the region near $m=M$, we construct the matrix (4.35) and get

$$\sum_{m=1}^{M_2}(Y_3\tilde{U},\tilde{U})_m\leqslant -\sum_{m=1}^{M_2}(Q_3\tilde{U},\tilde{U})_m. \tag{4.39}$$

In the following, we rewrite

$$-\sum_{m=1}^{M_2}(Q_3\tilde{U},\tilde{U})_m$$

as

$$\sum_{h=-2}^{2}\sum_{m=0}^{M_2+1}b_{m,m+h}\tilde{u}_m\tilde{u}_{m+h},$$

where $b_{m,m+h}=b_{m+h,m}$. We can see from the property of Q_3 that $b_{m,m-2}$, $b_{m,m-1}$, $b_{m,m}$, $b_{m-1,m}$, $b_{m-2,m}$ are quantities of $O(\Delta x)$, that $b_{m,m+h}=0$ for $m+h<0$ and that

$$\begin{pmatrix} b_{0,0} & b_{0,1}\\ b_{1,0} & b_{1,1}\end{pmatrix}$$

$$=\begin{pmatrix} \frac{1}{16}(2a_1+a_2)+s_{-1}'^2 & -\frac{1}{8}(a_1+a_2)+s_{-1}'s_0' \\ -\frac{1}{8}(a_1+a_2)+s_{-1}'s_0' & \frac{1}{4}(a_1+a_2)+\frac{1}{16}(2a_1+a_2)+s_{-1}'^2+s_0'^2-1\end{pmatrix}_0$$

$$+O(\Delta x).$$

For a horizontal three-point explicit scheme with $H_1=-1$ approximating (4.1), there are the relations

$$\begin{cases} s_{-1}'+s_0'+s_1'=1,\\ -s_{-1}'+s_1'=-\sigma,\end{cases}$$

i.e.,

$$\begin{cases} s_1'=\frac{1}{2}(1-s_0'-\sigma),\\ s_{-1}'=\frac{1}{2}(1-s_0'+\sigma).\end{cases}$$

Thus we have

$$\begin{cases} a_1=4(s_{-1}'s_0'+s_0's_1'+4s_{-1}'s_1')=4[s_0'(1-s_0')+(1-s_0')^2-\sigma^2]\\ \quad=4(1-s_0'-\sigma^2),\\ a_1+a_2=4(s_{-1}'s_0'+s_0's_1')=4s_0'(1-s_0'),\end{cases}$$

and

$$\begin{cases} \frac{1}{16}(2a_1+a_2)+s'^2_{-1} \\ \quad =\frac{1}{4}(1-s'_0-\sigma^2)+\frac{1}{4}s'_0(1-s'_0)+\frac{1}{4}(1-s'_0+\sigma)^2 \\ \quad =\frac{1}{4}(1-s'_0)(1+s'_0+1-s'_0+2\sigma) \\ \quad =\frac{1}{2}(1-s'_0)(\sigma+1), \\ -\frac{1}{8}(a_1+a_2)+s'_{-1}s'_0=-\frac{1}{2}s'_0(1-s'_0)+\frac{1}{2}s'_0(1-s'_0+\sigma)=\frac{s'_0\sigma}{2}, \\ \frac{1}{4}(a_1+a_2)+\frac{1}{16}(2a_1+a_2)+s'^2_{-1}+s'^2_0-1 \\ \quad =s'_0(1-s'_0)+\frac{1}{2}(1-s'_0)(\sigma+1)+s'^2_0-1 \\ \quad =\frac{1}{2}(1-s'_0)(\sigma-1). \end{cases}$$

Therefore, the above 2×2-matrix can further be rewritten as

$$\begin{pmatrix} b_{0,0} & b_{0,1} \\ b_{1,0} & b_{1,1} \end{pmatrix}=\frac{1}{2}\begin{pmatrix} (1-s'_0)(\sigma+1) & s'_0\sigma \\ s'_0\sigma & (1-s'_0)(\sigma-1) \end{pmatrix}_0+O(\Delta x)$$

$$=\frac{1}{2}\begin{pmatrix} \frac{\sigma^2 s'^2_0}{(1-s'_0)(\sigma-1)} & \sigma s'_0 \\ \sigma s'_0 & (1-s'_0)(\sigma-1) \end{pmatrix}_0$$

$$+\frac{1}{2}\begin{pmatrix} (1-s'_0)(\sigma+1)-\frac{\sigma^2 s'^2_0}{(1-s'_0)(\sigma-1)} & 0 \\ 0 & 0 \end{pmatrix}_0+O(\Delta x). \tag{4.40}$$

Because the von Neumann condition is now equivalent to

$$\begin{cases} \frac{1}{4}a_1=1-s'_0-\sigma^2\geqslant 0, \\ \frac{1}{4}(a_1+a_2)=s'_0(1-s'_0)\geqslant 0, \end{cases}$$

we can derive

$$0\leqslant s'_0\leqslant 1,\quad \sigma^2\leqslant 1-s'_0\leqslant 1,$$

and

$$0\leqslant s'_0-s'^2_0-s'_0\sigma^2=(1-s'_0)(1-\sigma^2)-(1-s'_0)^2(1-\sigma^2)-\sigma^2 s'^2_0,$$

$$\frac{(1-s'_0)^2(1-\sigma^2)+\sigma^2 s'^2_0}{2(1-s'_0)(1-\sigma)}\leqslant\frac{1}{2}(1+\sigma),$$

from which the following inequalities follow:

$$\left\{\begin{aligned}&\frac{\sigma^2 s_0'^2}{2(1-s_0')(\sigma-1)}\leqslant 0,\\&\frac{1}{2}\left[(1-s_0')(\sigma+1)-\frac{\sigma^2 s_0'^2}{(1-s_0')(\sigma-1)}\right]\\&\quad=\frac{(1-s_0')^2(1-\sigma^2)+\sigma^2 s_0'^2}{2(1-s_0')(1-\sigma)}\leqslant\frac{1}{2}(1+\sigma).\end{aligned}\right. \tag{4.41}$$

Therefore, noticing that the determinant of the first matrix in the last expression of (4.40) is equal to zero, we have

$$(\tilde{u}_0\ \tilde{u}_1)\begin{pmatrix}b_{0,0} & b_{0,1}\\ b_{1,0} & b_{1,1}\end{pmatrix}\begin{pmatrix}\tilde{u}_0\\ \tilde{u}_1\end{pmatrix}$$

$$\leqslant\frac{1}{2}(1+\sigma_0)|\tilde{u}_0|^2+O(\Delta x)(|\tilde{u}_0|^2+|\tilde{u}_1|^2). \tag{4.42}$$

We now investigate the related inequalities in $[1, M]$. As pointed out, $H_1(1)=-1$ and $H_1(M)=-2$. Hence, there must exist places where H_1 changes its value. To begin with, we consider the case with only one place where H_1 changes its value, that is, we suppose $M_2+2=M-M_1-1$. Because H_1 changes its value as m changes from M_2+1 to M_2+2, we must have $s_{-2}=O(\Delta x)$ and $s_1=O(\Delta x)$ for $m\approx M_2$ according to condition (iii), and $a_2=-16s_{-2}'s_0'$ or $-16s_{-1}'s_1'$ is also a quantity of $O(\Delta x)$. Therefore, (4.31a) holds, i.e., Z_3 is almost equal to the sum of the nonnegative definite matrices $\tilde{Z}_{3,1}$ and $\tilde{Z}_{3,2}$. In this case, by introducing the matrices $W_{3,1}=-Y_3-\tilde{Z}_{3,1}$ and $W_{3,2}=-Y_3-\tilde{Z}_{3,2}$ at $m=M_2+1$ and $M-M_1-1$ respectively, and introducing the pseudo-null matrices Q_3 at the other points, the inequality (4.19) can be obtained. This is because (1) we can immediately get the following inequality:

$$\begin{aligned}\sum_{m=1}^{M}(Y_3\boldsymbol{U},\boldsymbol{U})_m\leqslant&-\sum_{m=1}^{M_2}(Q_3\boldsymbol{U},\boldsymbol{U})_{m,-1,3}-(W_{3,1}\boldsymbol{U},\boldsymbol{U})_{M_2+1,-1,3}\\&-(W_{3,2}\boldsymbol{U},\boldsymbol{U})_{M-M_1-1,-2,3}-\sum_{m=M-M_1}^{M}(Q_3\boldsymbol{U},\boldsymbol{U})_{m,-2,3},\end{aligned} \tag{4.43}$$

by using the fact that Z_3, $\tilde{Z}_{3,1}$ and $\tilde{Z}_{3,2}$ are nonnegative definite matrices; and (2) the inequality (4.19) can be easily obtained from (4.43). In fact, according to (4.31a), when m is less than or equal to, and close to M_2, we have

$$\begin{aligned}-(Q_3\boldsymbol{U},\boldsymbol{U})_{m,-1,3}=&(Y_3\boldsymbol{U},\boldsymbol{U})_{m,-1,3}+(\tilde{Z}_{3,1}\boldsymbol{U},\boldsymbol{U})_{m,-1,3}\\&+(\tilde{Z}_{3,2}\boldsymbol{U},\boldsymbol{U})_{m,-1,3}+(O(\Delta x)\boldsymbol{U},\boldsymbol{U})_{m,-1,3}\\=&(\tilde{Y}_3\boldsymbol{U},\boldsymbol{U})_{m,-1,2}+(Z_{3,1}\boldsymbol{U},\boldsymbol{U})_{m,-1,2}\\&+(Z_{3,2}\boldsymbol{U},\boldsymbol{U})_{m+1,-1,2}+(O(\Delta x)\boldsymbol{U},\boldsymbol{U})_{m,-1,3};\end{aligned}$$

when m is greater than or equal to and close to $M-M_1$, there is the following relation

$$-(Q_3\widetilde{U},\widetilde{U})_{m,-2,3}=(Y_3\widetilde{U},\widetilde{U})_{m,-2,3}+(\widetilde{Z}_{3,1}\widetilde{U},\widetilde{U})_{m,-2,3}$$
$$+(\widetilde{Z}_{3,2}\widetilde{U},\widetilde{U})_{m,-2,3}+(O(\Delta x)\widetilde{U},\widetilde{U})_{m,-2,3}$$
$$=(\widetilde{Y}_3\widetilde{U},\widetilde{U})_{m,-1,2}+(Z_{3,1}\widetilde{U},\widetilde{U})_{m-1,-1,2}$$
$$+(Z_{3,2}\widetilde{U},\widetilde{U})_{m,-1,2}+(O(\Delta x)\widetilde{U},\widetilde{U})_{m,-2,3};$$

and when, $m=M_2+1,\ M-M_1-1$, we have

$$-(W_{3,1}\widetilde{U},\widetilde{U})_{M_2+1,-1,3}=(Y_3\widetilde{U},\widetilde{U})_{M_2+1,-1,3}+(\widetilde{Z}_{3,1}\widetilde{U},\widetilde{U})_{M_2+1,-1,3}$$
$$=(\widetilde{Y}_3\widetilde{U},\widetilde{U})_{M_2+1,-1,2}+(Z_{3,1}\widetilde{U},\widetilde{U})_{M_2+1,-1,2}$$
$$+(O(\Delta x)\widetilde{U},\widetilde{U})_{M_2+1,-1,3};$$

$$-(W_{3,2}\widetilde{U},\widetilde{U})_{M-M_1-1,-2,3}=(Y_3\widetilde{U},\widetilde{U})_{M-M_1-1,-2,3}$$
$$+(\widetilde{Z}_{3,2}\widetilde{U},\widetilde{U})_{M-M_1-1,-2,3}$$
$$=(\widetilde{Y}_3\widetilde{U},\widetilde{U})_{M-M_1-1,-1,2}$$
$$+(Z_{3,2}\widetilde{U},\widetilde{U})_{M-M_1-1,-1,2}$$
$$+(O(\Delta x)\widetilde{U},\widetilde{U})_{M-M_1-1,-2,3};$$

where the 2×2-matrix $\widetilde{Y}_{3,m}$ is composed of the elements in $Y_{3,m}$ corresponding to $\tilde{u}_{m-1}$ and $\tilde{u}_m$. From these relations and the relation $Z_{3,1}+Z_{3,2}=Z_2$, we have

$$-\sum_{m=M_2-k}^{M_2}(Q_3\widetilde{U},\widetilde{U})_{m,-1,3}-(W_{3,1}\widetilde{U},\widetilde{U})_{M_2+1,-1,3}$$
$$-(W_{3,2}\widetilde{U},\widetilde{U})_{M-M_1-1,-2,3}-\sum_{m=M-M_1}^{M-M_1+k}(Q_3\widetilde{U},\widetilde{U})_{m,-2,3}$$
$$=\sum_{m=M_2-k}^{M-M_1+k}(\widetilde{Y}_3\widetilde{U},\widetilde{U})_{m,-1,2}+(Z_{3,1}\widetilde{U},\widetilde{U})_{M_2-k,-1,2}$$
$$+\sum_{m=M_2-k+1}^{M-M_1+k-1}(Z_2\widetilde{U},\widetilde{U})_{m,-1,2}+(Z_{3,2}\widetilde{U},\widetilde{U})_{M-M_1+k,-1,2}$$
$$+\sum_{m=M_2-k}^{M_2+1}(O(\Delta x)\widetilde{U},\widetilde{U})_{m,-1,3}+\sum_{m=M-M_1-1}^{M-M_1+k}(O(\Delta x)\widetilde{U},\widetilde{U})_{m,-2,3}$$
$$=-\sum_{m=M_2-k+1}^{M-M_1+k-1}((-\widetilde{Y}_3-Z_2)\widetilde{U},\widetilde{U})_{m,-1,2}+(\widetilde{Y}_3\widetilde{U},\widetilde{U})_{M_2-k,-1,2}$$
$$+(Z_{3,1}\widetilde{U},\widetilde{U})_{M_2-k,-1,2}+(\widetilde{Y}_3\widetilde{U},\widetilde{U})_{M-M_1+k,-1,2}$$
$$+(Z_{3,2}\widetilde{U},\widetilde{U})_{M-M_1+k,-1,2}+\sum_{m=M_2-k}^{M_2+1}(O(\Delta x)\widetilde{U},\widetilde{U})_{m,-1,3}$$
$$+\sum_{m=M-M_1-1}^{M-M_1+k}(O(\Delta x)\widetilde{U},\widetilde{U})_{m,-2,3}.$$

Moreover, in the case $H_1=-2$, by using (4.29), $s'_{-2}=O(\Delta x)$ and $s'_{-2}+s'_{-1}+s'_0=1$, we can see that there is the following equality:

$$\frac{a_1}{4}=s'_{-2}s'_{-1}+s'_{-1}s'_0+4s'_{-2}s'_0=s'_{-1}s'_0+O(\Delta x)$$
$$=\frac{1}{2}[(1-s'_{-2})^2-(s'^2_{-1}+s'^2_0)]+O(\Delta x)=\frac{1}{2}(1-s'^2_{-1}-s'^2_0)+O(\Delta x),$$

and that the matrix

$$-\widetilde{Y}_3-Z_2=-\begin{pmatrix} s'^2_{-1} & s'_{-1}s'_0 \\ s'_{-1}s'_0 & s'^2_0-1 \end{pmatrix}-\frac{a_1}{4}\begin{pmatrix} 1 & -1 \\ -1 & 1 \end{pmatrix}.$$

$$=-\begin{pmatrix} s'^2_{-1}+\frac{a_1}{4} & s'_{-1}s'_0-\frac{a_1}{4} \\ s'_{-1}s'_0-\frac{a_1}{4} & s'^2_0-1+\frac{a_1}{4} \end{pmatrix}$$

is almost a pseudo-null matrix, i.e., the sum of elements on every diagonal of the matrix is equal to $O(\Delta x)$. In the case $H_1=-1$, there exists a similar result. Therefore, we can see that the coefficients of the quadratic form on the right-hand side of (4.43) are quantities of $O(\Delta x)$ not only for $m\in[2, M_2-1]$ and $[M-M_1, M-2]$ but also for $m\approx M_2$. Consequently, using the inequalitis (4.38) and (4.42), we can rewrite (4.43) as

$$\sum_{m=1}^{M}(Y_3\widetilde{U},\widetilde{U})_m=\sum_{m=1}^{M}(S^*S\widetilde{U},\widetilde{U})_m-\sum_{m=1}^{M}(R^*R\widetilde{U},\widetilde{U})_m$$
$$\leqslant\frac{1}{2}(1+\sigma_0)|\widetilde{u}_0|^2-\frac{\sigma_M}{2}|\widetilde{u}_M|^2+O(\Delta x)\sum_{m=0}^{M}|\widetilde{u}_m|^2.$$

Noticing the relation (4.21) and the fact that for the model problem A_1,

$$\delta_0(\lambda_{n,0})=1,\quad \delta_1(\lambda_{n,M})=0,\quad \sigma_{n,M}\geqslant 0,$$

we can further rewrite the above inequality as (4.19):

$$\begin{aligned}\widetilde{T}^{k+1}_n-\widetilde{T}^k_n\leqslant&[c_7\delta_0(\lambda_{n,0})-c_8|\sigma_{n,0}|(1-\delta_0(\lambda_{n,0}))]|\widetilde{u}^k_{n,0}|^2\Delta x\\&+[c_7\delta_1(\lambda_{n,M})-c_8|\sigma_{n,M}|(1-\delta_1(\lambda_{n,M}))]|\widetilde{u}^k_{n,M}|^2\Delta x\\&+c_4\Delta x\|\widetilde{u}^k_n\|^2_x,\end{aligned}$$

where $c_7\geqslant\frac{1}{2}(1+\sigma_0)$ and $c_8=\frac{1}{2}$.

We have discussed the case in which $\delta H_1(m)\equiv H_1(m)-H_1(m-1)$ is equal to zero everywhere, except one place where $H_1(m)=-2$ and $H_1(m-1)=-1$. However, the conclusion is also valid in the case with several places where $\delta H_1(m)\neq 0$. In that case, δH_1 might be equal to -2, 1 or 2, besides 0 or -1. Moreover, for $\delta H_1=-1$, besides the case in which $H_1(m)=-2$ and $H_1(m-1)=-1$, one of the other possibilities is $H_1(m)=-1$ and $H_1(m-1)=0$. Therefore, the situation is complicated, and the corresponding expressions become longer. In the following, we shall not give concrete expressions, instead we shall just point out the key points of the proof. If $\delta H_1(m)=1$, instead of Q_3, we construct

$$W_{3,3}=-Y_3-Z_3-\widetilde{Z}_{3,3},$$
$$W_{3,4}=-Y_3-Z_3-\widetilde{Z}_{3,4},$$

respectively at the points $m-1$ and m, where

$$\tilde{Z}_{3,3}=\begin{pmatrix} 0 & 0 & 0 \\ 0 & & \\ 0 & & Z_{3,1} \end{pmatrix}, \quad \tilde{Z}_{3,4}=\begin{pmatrix} Z_{3,2} & & 0 \\ & & 0 \\ 0 & 0 & 0 \end{pmatrix}.$$

If $\delta H_1(m)=-1$, instead of Q_3, we construct $W_{3,1}$ and $W_{3,2}$ at the points $m-1$ and m respectively. If $\delta H_1(m)=\pm 2$, the scheme is almost a horizontal one-point scheme, so every element of Q_3 is a quantity of $O(\Delta x)$. In this case, we still construct Q_3 because Q_3 and $-Y_3$ have the same effect. In this way, we can always obtain the desired inequality (4.19).

We have proved that (4.19) is valid for the model A_1. If we wish to prove that (4.19) is valid for the other models, we need only to notice the following facts.

The model A_2 is almost the same as the model A_1, so it is easy to show that the inequality (4.19) is valid for the model A_2. The model B is similar to the model A_2 when $m\approx 0$, and similar to the model A_1 when $m\approx M$. The model C is similar to A_1 when $m\approx 0$, and similar to A_2 when $m\approx M$. Thus, for the models B and C, certain inequalities similar to (4.38) and (4.42) can be obtained near the boundary points. In the above discussion on the interior points, we have considered all the possibilities of $\delta H_1(m)$. Hence, the result can be used for any model. Therefore, it is easy to show that (4.19) is valid for the models B and C.

(4.19) is satisfied; (4.5) or (4.5)$'$ is fulfilled according to condition (iv); and (4.18) is always valid for an explicit scheme, so the scheme is stable according to Theorem 4.2.

Lastly, we give two explanatory notes.

(1) If a scheme "contains" more than three but a limited number of points in the m-direction, and if the scheme has the following form

$$u_m^{k+1}=\sum_{h=H_1}^{H_1+2} s_{h,m}^k u_{m+h}^k+\sum_h O_h(\Delta x)u_{m+h}^k,$$

then the above proof is also valid, i.e., the scheme remains stable if it satisfies the conditions of this theorem.

(2) In condition (iv), it is required that the scheme be reduced to $u_m^{k+1}=u_m^k$ when $m=0$ and M if $\sigma=0$. For a horizontal three-point explicit scheme, this requirement is almost satisfied if condition (i) holds. In fact, according to condition (i), we have $H_1=0$ for $m=0$. Hence, by using Lemma 4.5, we can obtain the following conclusion: if condition (i) which contains the von Neumann condition holds, then we must have $s_0'=1, s_1'=s_2'=0$ if $\sigma=0$, i.e., $u_0^{k+1}=u_0^k+\sum_{h=0}^{2} O_h(\Delta x)u_h^k$. Therefore, in that case, the relation $u_0^{k+1}=u_0^k$ almost holds. When $m=M$ the situation is similar.

From Theorem 4.4 we can obtain the following corollary.

Corollary When the schemes (3.7)/(3.8) and (3.7)/(3.10) are used for the initial-boundary-value problem (4.1)—(4.4) with (4.7) or (4.6), if:

(1) only u_m^k with $m \in [0, M]$ appear in the system of difference equations (4.9);

(2) the scheme satisfies the von Neumann condition at every point(for example, (3.9j) is satisfied if the scheme (3.7)/(3.8j) is adopted, or(3.11j) is satisfied if the scheme (3.7)/(3.10j) is adopted, j=a, b, or c);

(3) the matrix A satisfies the Lipschitz condition (4.32) with respect to x and t;

then, the scheme is stable.

It is easy to prove this corollary. (1) and (2) guarantee that condition (i) holds. (3) guarantees that condition (iii) holds. Moreover, condition (iv) is satisfied because we assume (4.7) or (4.6) to hold and because the scheme (3.7)/(3.10j) adopted for both $m=0$ and M is reduced to $u_m^{k+1}=u_m^k$ if $\sigma=0$. Therefore, this corollary is deduced from Theorem 4.4.

5. *The Stability of a Certain Class of Horizontal Four-point Explicit Schemes*

Firstly we shall discuss second order schemes.

Lemma 4.7 For a horizontal four-point second-order explicit scheme appoximating (4.1), we must have $a_1=0$, and the von Neumann condition (3.2) can be reduced to

$$\begin{cases} a_2 \geqslant 0, \\ a_2+a_3 \geqslant 0. \end{cases} \tag{4.28c)'}$$

Proof According to Lemma 4.3, the condition (3.2) is equivalent to (4.28c). Hence, (3.2) must be equivalent to (4.28c)′ if $a_1=0$. Therefore, the only thing we need to do is to prove $a_1=0$. In the following, we shall do some straightforward calculations. Consider the case $H_1=-2$. In this case, the scheme can be written as

$$u_m^{k+1}=s_{-2}u_{m-2}^k+s_{-1}u_{m-1}^k+s_0u_m^k+s_1u_{m+1}^k,$$

and there is the following relation

$$\begin{aligned}
&\left(u+\frac{\partial u}{\partial t}\Delta t+\frac{1}{2}\frac{\partial^2 u}{\partial t^2}\Delta t^2\right)_m^k \\
&\quad = s_{-2}\left(u-\frac{\partial u}{\partial x}2\Delta x+\frac{1}{2}\frac{\partial^2 u}{\partial x^2}4\Delta x^2\right)_m^k \\
&\qquad +s_{-1}\left(u-\frac{\partial u}{\partial x}\Delta x+\frac{1}{2}\frac{\partial^2 u}{\partial x^2}\Delta x^2\right)_m^k+s_0u_m^k \\
&\qquad +s_1\left(u+\frac{\partial u}{\partial x}\Delta x+\frac{1}{2}\frac{\partial^2 u}{\partial x^2}\Delta x^2\right)_m^k+O(\Delta t^3).
\end{aligned}$$

On the other hand, it follows from (4.1) that

$$\begin{cases} \dfrac{\partial u}{\partial t}=-\lambda\dfrac{\partial u}{\partial x}, \\ \dfrac{\partial^2 u}{\partial t\,\partial x}=-\lambda\dfrac{\partial^2 u}{\partial x^2}-\dfrac{\partial \lambda}{\partial x}\dfrac{\partial u}{\partial x}, \\ \dfrac{\partial^2 u}{\partial t^2}=-\lambda\dfrac{\partial^2 u}{\partial t\,\partial x}-\dfrac{\partial \lambda}{\partial t}\dfrac{\partial u}{\partial x} \\ \qquad =\lambda^2\dfrac{\partial^2 u}{\partial x^2}+\left(\lambda\dfrac{\partial \lambda}{\partial x}-\dfrac{\partial \lambda}{\partial t}\right)\dfrac{\partial u}{\partial x}, \end{cases}$$

Therefore, there are the relations

$$\begin{cases} s'_{-2}+s'_{-1}+s'_0+s'_1=1, \\ -2s'_{-2}-s'_{-1}+s'_1=-\sigma, \\ 4s'_{-2}+s'_{-1}+s'_1=\sigma^2, \end{cases}$$

i.e.,

$$\begin{cases} s'_{-1}=\bar{s}_{-1}-3s'_{-2}, & \bar{s}_{-1}=\dfrac{\sigma(\sigma+1)}{2}, \\ s'_0=\bar{s}_0+3s'_{-2}, & \bar{s}_0=-(\sigma-1)(\sigma+1), \\ s'_1=\bar{s}_1-s'_{-2}, & \bar{s}_1=\dfrac{(\sigma-1)\sigma}{2}. \end{cases}$$

From the expressions for $\bar{s}_i$, we can obtain

$$\begin{aligned} \frac{1}{4}a_1 &= \bar{s}_{-1}\bar{s}_0+\bar{s}_0\bar{s}_1+4\bar{s}_{-1}\bar{s}_1 \\ &\quad +s'_{-2}(\bar{s}_{-1}+3\bar{s}_{-1}-3\bar{s}_0-\bar{s}_0+3\bar{s}_1+4\bar{s}_0-4\bar{s}_{-1}-12\bar{s}_1+9\bar{s}_1) \\ &\quad +s'^2_{-2}(-3-9-3+12+12-9) \\ &= \bar{s}_{-1}\bar{s}_0+\bar{s}_0\bar{s}_1+4\bar{s}_{-1}\bar{s}_1 \\ &= \sigma(\sigma-1)(\sigma+1)\left(-\frac{1}{2}-\frac{\sigma}{2}-\frac{\sigma}{2}+\frac{1}{2}+\sigma\right)=0. \end{aligned}$$

For $H_1=-3,\ -1,\ 0$, the situation is similar. For example, in the case $H_1=-3$, we have

$$\begin{cases} s'_0=\bar{s}_0-s'_{-3}, & \bar{s}_0=\dfrac{1}{2}(\sigma-1)(\sigma-2), \\ s'_{-1}=\bar{s}_{-1}+3s'_{-3}, & \bar{s}_{-1}=-\sigma(\sigma-2), \\ s'_{-2}=\bar{s}_{-2}-3s'_{-3}, & \bar{s}_{-2}=\dfrac{1}{2}\sigma(\sigma-1), \end{cases}$$

and

$$\frac{1}{4}a_1=\bar{s}_{-2}\bar{s}_{-1}+\bar{s}_{-1}\bar{s}_0+4\bar{s}_{-2}\bar{s}_0=0.$$

Therefore, this lemma is proved.

Theorem 4.5 A horizontal four-point second order explicit scheme which approximates the initial-boundary-value problem (4.1)—(4.4) is stable, if conditions (i), (iii) of § 3 and condition (iv) of this section hold and if the following condition holds:

Condition (v) For every n, if m belongs to the boundaries of the set

$\mathscr{G}_m(\lambda_n)$, i.e., if m is equal to the maximum or minimum of the set $\mathscr{G}_m(\lambda_n)$, the actually adopted scheme is a horizontal three-point explicit one.

Proof As we did in proving Theorem 4.4, first of all, we shall prove that the inequality (4.19) is valid for the model A_1.

According to condition (i), $H_1(M)=-3$. Moreover, according to conditions (i) and (v) and Lemma 4.6, the scheme at $m=1$ must be a horizontal three-point scheme with $H_1(1)=-1$. In the following, we shall consider it as a horizontal four-point scheme with $H_1(1)=-1$. When $m=M-1$, we have $H_1(M-1)=-3$ or -2 because of condition (i). When $m=2$, we have $H_1(2)=-2$ or -1 or 0 because of condition (i). However, $H_1(2)=0$ is impossible. In fact, the scheme adopted at $m=1$ must be a horizontal three-point scheme with $H_1(1)=-1$, so the von Neumann condition requires

$$a_1+a_2=4(s'_{-1}s'_0+s'_0s'_1)\geqslant 0.$$

On the other hand, there are the following relations

$$\begin{cases} s'_{-1}+s'_0+s'_1=1, \\ -s'_{-1}+s'_1=-\sigma. \end{cases}$$

Hence the inequality above can be rewritten as $s'_0(1-s'_0)\geqslant 0$, i.e., $0\leqslant s'_0\leqslant 1$, and there is the following relation

$$s'_{-1}=\frac{1+\sigma-s'_0}{2}\geqslant\frac{\sigma}{2}.$$

Furthermore, from $s'_{-1,1}\geqslant\frac{\sigma_1}{2}$, we can obtain $s'_{-1,2}\geqslant\frac{\sigma_1}{2}-O(\Delta x)$ because of condition (iii). Obviously, $\sigma_1\geqslant s>0$, so $s'_{-1,2}>0$ if Δx is sufficiently small. Therefore, $H_1(2)\leqslant -1$, i.e., $H_1(2)$ is equal to -2 or -1[1].

In the following the case $H_1(2)=-2=H_1(M-1)$ will be considered, and we assume further $H_1(m)=-2$ for $m\in[2, M-1]$. In this case, we construct

$$\begin{cases} W_{4,5}=-Y_4-\tilde{Z}_{4,5}, \quad \tilde{Z}_{4,5}=\begin{pmatrix} 0 & 0 & 0 & 0 \\ 0 & & & \\ 0 & & Z_{4,1} & \\ 0 & & & \end{pmatrix}, \\[2ex] W_{4,6}=-Y_4-\tilde{Z}_{4,6}, \quad \tilde{Z}_{4,6}=\begin{pmatrix} & & & 0 \\ & Z_{4,2} & & 0 \\ & & & 0 \\ 0 & 0 & 0 & 0 \end{pmatrix} \end{cases}$$

at $m=M$ and 1 respectively, and

$$Q_4=-Y_4-Z_4$$

1) This conclusion can be obtained from other angles.

at $m\in[2, M-1]$, where Y_4 stands for $-R^*R+S^*S$ with $H=4$. Because $\tilde{Z}_{4,5}$, $\tilde{Z}_{4,6}$, and Z_4 are nonnegative definite matrices, we have

$$\sum_{m=1}^{M}(Y_4\tilde{U}, \tilde{U})_m \leqslant -(W_{4,6}\tilde{U}, \tilde{U})_{1,-1,4} - \sum_{m=2}^{M-1}(Q_4\tilde{U}, \tilde{U})_{m,-2,4} - (W_{4,5}\tilde{U}, \tilde{U})_{M,-3,4}. \qquad (4.44)$$

In the following, we shall rewrite the quadratic form on the right-hand side of this inequality as

$$\sum_{h=-3}^{3}\sum_{m=0}^{M} b_{m,m+h}\tilde{u}_m\tilde{u}_{m+h}$$

and investigate it, where $b_{m,m+h}=b_{m+h,m}$, and where $b_{m,m+h}=0$ if $m+h<0$ or $m+h>M$. According to Lemma 4.7, we have $a_1=0$ for the second-order schemes under consideration. Hence, every element of Q_4 satisfies the Lipschitz condition, and $b_{m,m+h}=O(\Delta x)$ if $m\in[4, M-4]$ and $h\in[-3, 3]$.

If $m\approx M$, we must have $s_1=O(\Delta x)$ and $a_3=O(\Delta x)$. Hence, by using (4.31b), we have

$$\begin{aligned}-(Q_4\tilde{U}, \tilde{U})_{m,-2,4} &= (Y_4\tilde{U}, \tilde{U})_{m,-2,4}+(Z_4\tilde{U}, \tilde{U})_{m,-2,4}\\ &= (\tilde{Y}_4\tilde{U}, \tilde{U})_{m,-2,3}+(Z_{4,1}\tilde{U}, \tilde{U})_{m,-2,3}\\ &\quad+(Z_{4,2}\tilde{U}, \tilde{U})_{m+1,-2,3}+(O(\Delta x)\tilde{U}, \tilde{U})_{m,-2,4},\end{aligned}$$

where $\tilde{Y}_{4,m}$ is a 3×3-matrix, and is composed of the elements of $Y_{4,m}$ corresponding to $\tilde{u}_{m-2}$, $\tilde{u}_{m-1}$ and $\tilde{u}_m$. Moreover, there is the following relation

$$-(W_{4,5}\tilde{U}, \tilde{U})_{M,-3,4}=(\tilde{Y}_4\tilde{U}, \tilde{U})_{M,-2,3}+(Z_{4,1}\tilde{U}, \tilde{U})_{M,-2,3}.$$

Therefore, we have

$$\begin{aligned}&-\sum_{m=M-k}^{M-1}(Q_4\tilde{U}, \tilde{U})_{m,-2,4}-(W_{4,5}\tilde{U}, \tilde{U})_{M,-3,4}\\ &=\sum_{m=M-k}^{M}(\tilde{Y}_4\tilde{U}, \tilde{U})_{m,-2,3}+(Z_{4,1}\tilde{U}, \tilde{U})_{M-k,-2,3}\\ &\quad+\sum_{m=M-k+1}^{M}(Z_3\tilde{U}, \tilde{U})_{m,-2,3}+\sum_{m=M-k}^{M-1}(O(\Delta x)\tilde{U}, \tilde{U})_{m,-2,4}\\ &=-\sum_{m=M-k+1}^{M}((-\tilde{Y}_4-Z_3)\tilde{U}, \tilde{U})_{m,-2,3}+(\tilde{Y}_4\tilde{U}, \tilde{U})_{M-k,-2,3}\\ &\quad+(Z_{4,1}\tilde{U}, \tilde{U})_{M-k,-2,3}+\sum_{m=M-k}^{M-1}(O(\Delta x)\tilde{U}, \tilde{U})_{m,-2,4}.\end{aligned}$$

On the other hand, according to (4.29) and because of $s'_1=O(\Delta x)$ and $s'_{-2}+s'_{-1}+s'_0+s'_1=1$, we have

$$\frac{1}{16}a_2=-(s'_{-2}s'_0+s'_{-1}s'_1)-6s'_{-2}s'_1=-s'_{-2}s'_0+O(\Delta x),$$

$$-\frac{4}{16}(a_1+a_2)=-\frac{1}{4}[4(s'_{-2}s'_{-1}+s'_{-1}s'_0+s'_0s'_1)-60s'_{-2}s'_1]$$

$$=-(s'_{-2}s'_{-1}+s'_{-1}s'_0)+O(\Delta x),$$

$$\frac{8a_1+6a_2}{16}=\frac{1}{8}[16(s'_{-2}s'_{-1}+s'_{-1}s'_0+s'_0s'_1)$$

$$+16(s'_{-2}s'_0+s'_{-1}s'_1)-144s'_{-2}s'_1]$$

$$=2(s'_{-2}s'_{-1}+s'_{-1}s'_0+s'_{-2}s'_0)+O(\Delta x)$$

$$=(1-s'_1)^2-(s'^2_{-2}+s'^2_{-1}+s'^2_0)+O(\Delta x)$$

$$=1-(s'^2_{-2}+s'^2_{-1}+s'^2_0)+O(\Delta x).$$

Therefore, the sum of elements on each diagonal of the matrix

$$-\widetilde{Y}_4-Z_3=-\begin{pmatrix} s'^2_{-2} & s'_{-2}s'_{-1} & s'_{-2}s'_0 \\ s'_{-1}s'_{-2} & s'^2_{-1} & s'_{-1}s'_0 \\ s'_0s'_{-2} & s'_0s'_{-1} & s'^2_0-1 \end{pmatrix}$$

$$-\frac{1}{16}\begin{pmatrix} 2a_1+a_2 & -2a_1-2a_2 & a_2 \\ -2a_1-2a_2 & 4a_1+4a_2 & -2a_1-2a_2 \\ a_2 & -2a_1-2a_2 & 2a_1+a_2 \end{pmatrix}$$

is a quantity of $O(\Delta x)$ and every element of this matrix satisfies the Lipschitz condition. That is, this matrix is similar to Q_3 of a horizontal three-point scheme. Hence, the results which we have derived at $m\approx M$ in proving Theorem 4.4 can also be obtained in the case here. When $m\approx 1$, the situation is similar. Therefore, we have the following result similar to that for horizontal three-point schemes:

$$\sum_{m=1}^{M}(Y_4\widetilde{U},\widetilde{U})_m\leqslant\frac{1}{2}(1+\sigma_0)|\widetilde{u}_0|^2$$

$$-\frac{\sigma_M}{2}|\widetilde{u}_M|^2+O(\Delta x)\sum_{m=0}^{M}|\widetilde{u}_m|^2, \tag{4.45}$$

which can be rewritten as

$$\widetilde{T}^{k+1}_n-\widetilde{T}^k_n\leqslant\frac{1}{2}(1+\sigma_{n,0})|\widetilde{u}^k_{n,0}|^2\Delta x$$

$$-\frac{\sigma_{n,M}}{2}|\widetilde{u}^k_{n,M}|^2\Delta x+O(\Delta x)\|\widetilde{u}^k_n\|^2_x.$$

Noticing $\delta_0(\lambda_{n,0})=1$, $\delta_1(\lambda_{n,M})=0$ and $\sigma_{n,M}\geqslant 0$ for the model A_1, and letting $c_7\geqslant\frac{1}{2}(1+\sigma_{n,0})$ and $c_8=\frac{1}{2}$, we can obtain the inequality (4.19) from the inequality above. Thus the conclusion is proved.

We have proved the conclusion in the case where δH_1 is equal to zero at every interior point. When $\delta H_1\neq 0$ at some points, the conclusion can still be proved if the following modifications are made.

(1) If $\delta H_1(m)=-1$, then, instead of Q_4,

$$\begin{cases} W_{4,1} = -Y_4 - \widetilde{Z}_{4,1} \\ W_{4,2} = -Y_4 - \widetilde{Z}_{4,2} \end{cases}$$

are taken at the points $m-1$ and m respectively.

(2) If $\delta H_1(m)=1$, instead of Q_4,

$$\begin{cases} W_{4,7} = -Y_4 - Z_4 - \widetilde{Z}_{4,5}, \\ W_{4,8} = -Y_4 - Z_4 - \widetilde{Z}_{4,6} \end{cases}$$

are taken at the points $m-1$ and m respectively.

(3) If $\delta H_1(m) = -2$, instead of Q_4,

$$W_{4,3} = -Y_4 - \widetilde{Z}_{4,3}$$

is taken at the points $m-2$ and $m-1$, and

$$W_{4,4} = -Y_4 - \widetilde{Z}_{4,4}$$

is taken at the points m and $m+1$.

(4) If $\delta H_1(m)=2$, instead of Q_4,

$$W_{4,9} = -Y_4 - Z_4 - \widetilde{Z}_{4,7}, \quad \widetilde{Z}_{4,7} = \begin{pmatrix} 0 & 0 & 0 & 0 \\ 0 & 0 & 0 & 0 \\ 0 & 0 & \multicolumn{2}{c}{\multirow{2}{*}{$Z_{4,3}$}} \\ 0 & 0 & & \end{pmatrix}$$

is taken at the points $m-2$ and $m-1$, and

$$W_{4,10} = -Y_4 - Z_4 - \widetilde{Z}_{4,8}, \quad \widetilde{Z}_{4,8} = \begin{pmatrix} \multicolumn{2}{c}{\multirow{2}{*}{$Z_{4,4}$}} & 0 & 0 \\ & & 0 & 0 \\ 0 & 0 & 0 & 0 \\ 0 & 0 & 0 & 0 \end{pmatrix}$$

is taken at the points m and $m+1$ respectively.

(5) If $\delta H_1(m) = \pm 3$, the scheme is almost a horizontal one-point scheme. Hence, Q_4 and Y_4 are quantities of $O(\Delta x)$, and the above conclusion can be proved by taking Q_4 or $-Y_4$ instead of Q_4.

The reasons why the conclusion is valid are (i) there are the following relations

$$Z_{4,1} + Z_{4,2} = Z_3 + O(\Delta x),$$
$$Z_{4,3} + Z_{4,4} = Z_2 + O(\Delta x);$$

(ii) when $\delta H_1 = \pm 1 (\pm 2)$, all the elements of Y_4 other than a matrix of order three (order two) are quantities of $O(\Delta x)$, $\cdots$, and so on.

We have proved this theorem in the case $H_1(2) = H_1(M-1) = -2$. Howevre, $H_1(2) = -1$ or $H_1(M-1) = -3$ is possible. In that case, only the following small change is needed. If $H_1(2) = -1$, instead of $W_{4,6}$,

$$W_{4,8} = -Y_4 - Z_4 - \widetilde{Z}_{4,6}$$

is taken at $m=1$. If $H_1(M-1) = -3$, instead of $W_{4,5}$,

$$W_{4,7} = -Y_4 - Z_4 - \widetilde{Z}_{4,5}$$

is taken at $m=M$. In this way, (4.45) and (4.19) can also be obtained.

Therefore, in all possible cases, we have proved that (4.19) is valid for the model A_1. In proving Theorem 4.4, the following has been done: By using the result that (4.19) is valid for the model A_1, we have proved that (4.19) is also valid for the other models. In this case, the same thing can be done. That is, we can prove that (4.19) is satisfied for all the models by using the result for the model A_1. Therefore, according to Theorem 4.2, the scheme is stable.

Also, the two explanations given at the end of the proof of Theorem 4.4 are still true.

From this theorem, we get the following corollary.

Corollary When the scheme (3.7)/(3.12) with $\alpha=\dfrac{1}{2}$ and with $\beta_{1,m}$ and $\beta_{2,m}$ in (3.13) is used for the initial-boundary-value problem (4.1)—(4.4) with (4.7) or (4.6), if the von Neumann condition is satisfied, for example, $|\sigma_n|\leqslant 1$ everywhere, and if every element of the matrix A satisfies (4.32), then it is stable.

It is easy to prove this corollary. The von Neumann condition is satisfied, for example, $|\sigma_n|\leqslant 1$, so condition (i) is fulfilled. The condition (4.32) and the fact that $\beta_{1,m}$ and $\beta_{2,m}$ are in the form (3.13) guarantee that condition (iii) is fulfilled. It is easy to see that conditions (iv) and (v) are satisfied in this case. Moreover, if $\lambda\neq 0$, it is always a horizontal four-point scheme. Of course, if $\lambda\approx 0$, it might not be a horizontal four-point scheme. However, it should almost be a horizontal one-point scheme. Therefore, according to the explanation (1) in the proof of Theorem 4.4, this situation does not influence the validity of the proof. That is, we know from the conclusion of Theorem 4.5 that the scheme is stable.

This theorem is of great practical value. When an initial-boundary-value problem is solved and we are interested in first order schemes, we can realize aim by using horizontal two-point schemes. Therefore, generally speaking, the horizontal multi-point schemes are considered only when we are interested in high-order schemes. It is not always possible for a horizontal three-point explicit scheme with second-order accuracy to satisfy conditions (i) and (iii). For example, for the model A_1 with $\lambda=\text{constant}>0$, we must have $H_1(1)=-1$ for $m=1$ and $H_1(M)=-2$ for $m=M$. When a horizontal three-point scheme with second order accuracy is used for an equation with a constant coefficient, s_h are uniquely determined if H_1 is given. Hence, when $\delta H_1\neq 0$, condition (iii) is usually not fulfilled. Therefore, if we want to find a scheme which is of a second order accuracy, satisfies conditions (i) and (iii), and is applicable to every problem, we have to consider the class of horizontal four-point explicit schemes. The horizontal four-point explicit scheme is that with a minimal number of points among the "universal" schemes satisfying conditions (i) and (iii). Therefore, if we want to use a second order explicit scheme, we will naturally choose a horizontal four-point explicit scheme. Therefore this class of schemes and Theorem 4.5 have great practical value.

Now let us discuss the first order schemes. We shall give a lemma before proving Theorem 4.6.

Lemma 4.8 If $f(x)\geqslant 0$ and f is a section-wise analytic function (the number of sections is limited) in $[0, 1]$, and if f_x is a continuous function, then $\sqrt{f}$ satisfies the Lipschitz condition in $[\varepsilon, 1-\varepsilon]$, $0<\varepsilon<\frac{1}{2}$, that is, there exists a constant c such that the following inequality

$$|\sqrt{f(x'')}-\sqrt{f(x')}|\leqslant c(x''-x') \quad (\varepsilon\leqslant x'\leqslant x''\leqslant 1-\varepsilon)$$

holds.

Proof First of all, we shall prove the following fact: there exists a constant c such that this inequality

$$\left|\frac{1}{2}f_x\Big/\sqrt{f}\right|\leqslant c$$

holds for all x which belong to $[\varepsilon, 1-\varepsilon]$ and at which $f(x)\neq 0$. If it were not true, i.e., if there were no such constant c, we could find a sequence $x_0, x_1, \cdots, x_n, \cdots$, which converges to $\bar{x}\in[\varepsilon, 1-\varepsilon]$, such that

$$\left|\frac{1}{2}f_x(x_i)\Big/\sqrt{f(x_i)}\right|\to\infty.$$

Obviously, we have $f(\bar{x})=0$. Because $f(x)\geqslant 0$, $\bar{x}$ is an interior point in $[0, 1]$ and f_x is a continuous function, this relation $f_x(\bar{x})=0$ follows. Without loss of generality, we suppose that $x_0, x_1, \cdots, x_n, \cdots$, and $\bar{x}$ belong to the interval $[a, b]$ in which f is analytic. Hence, we have

$$f_x(x_i)=f_{xx}(\tilde{x}_i)(x_i-\bar{x}) \quad (x_i\lesseqgtr\tilde{x}_i\lesseqgtr\bar{x}),$$

$$f(x_i)=\frac{1}{2}f_{xx}(\tilde{\tilde{x}}_i)(x_i-\bar{x})^2 \quad (x_i\lesseqgtr\tilde{\tilde{x}}_i\lesseqgtr\bar{x}),$$

and

$$\lim_{x_i\to\bar{x}}\left|\frac{\frac{1}{2}f_x(x_i)}{\sqrt{f(x_i)}}\right|=\lim_{x_i\to\bar{x}}\frac{\frac{1}{2}|f_{xx}(\tilde{x}_i)|}{\sqrt{\frac{1}{2}f_{xx}(\tilde{\tilde{x}}_i)}}=\sqrt{\frac{1}{2}f_{xx}(\bar{x})}\,{}^{1)}.$$

This result contradicts the relation $\left|\frac{1}{2}f_x(x_i)/\sqrt{f(x_i)}\right|\to\infty$. Therefore, the conclusion is proved.

We now prove that for any positive constant ε_1, the following

---

1) If $f_{xx}(\bar{x})\neq 0$, the last equality is evidently true. If $f_{xx}(\bar{x})=0$, one can prove that the limit value is equal to zero. Thus the last equality is also true.

inequality

$$|\sqrt{f(x'')+\varepsilon_1}-\sqrt{f(x')+\varepsilon_1}|\leqslant c(x''-x'),$$
$$(\varepsilon\leqslant x'\leqslant x''\leqslant 1-\varepsilon) \tag{4.46}$$

is true. In fact, the only thing we need to do is to prove $\left|\frac{d\sqrt{f(x)+\varepsilon_1}}{dx}\right|$ $\leqslant c$ in $[\varepsilon, 1-\varepsilon]$. If f is equal to zero everywhere in an adjacent region of x, $\left|\frac{d\sqrt{f(x)+\varepsilon_1}}{dx}\right|\leqslant c$ is evidently true. We now suppose that there are non-zero points of f in any adjacent region of x. In this case, we can find such a sequence $x_0, x_1, \cdots, x_n, \cdots$, that it converges to x and for any x_i, $f(x_i)\neq 0$. On the other hand, $\frac{d\sqrt{f+\varepsilon_1}}{dx}=\frac{1}{2}f_x/\sqrt{f+\varepsilon_1}$ is a continuous function. Hence,

$$\left|\frac{d\sqrt{f(x)+\varepsilon_1}}{dx}\right|=\lim_{x_i\to x}\left|\frac{d\sqrt{f(x_i)+\varepsilon_1}}{dx}\right|$$
$$=\lim_{x_i\to x}\left|\frac{\frac{1}{2}f_x(x_i)}{\sqrt{f(x_i)+\varepsilon_1}}\right|\leqslant\lim_{x_i\to x}\left|\frac{\frac{1}{2}f_x(x_i)}{\sqrt{f(x_i)}}\right|\leqslant c.$$

This is to say, the inequality

$$\left|\frac{d\sqrt{f(x)+\varepsilon_1}}{dx}\right|\leqslant c$$

is true in $[\varepsilon, 1-\varepsilon]$.

Because ε_1 is arbitrary, letting $\varepsilon_1\to 0$, we obtain Lemma 4.8.

By using the above lemma and several preceding results, the following result can be easily obtained.

Theorem 4.6 Consider a horizontal four-point explicit scheme (4.9) approximating the initial-boundary-value problem (4.1)—(4.4). The scheme is stable if condition (i) of § 3 and condition (iv) of this section hold, and if the two following conditions hold:

Condition (iii)′ (3.4a) and (3.4b) hold; $\frac{\partial r_h}{\partial x}$ and $\frac{\partial s_h}{\partial x}$ are continuous functions; and r_h and s_h are section-wise analytic functions.

Condition (v)′ The actually adopted scheme in $[0, \varepsilon_0]$ and $[1-\varepsilon_1, 1]$ is a horizontal three-point scheme, where ε_0 and ε_1 are positive constants.

Proof The proof of this theorem is almost the same as that for Theorem 4.5, there being only a small difference between them for the following reason. According to (4.30c), $a_1=0$ for a second order scheme. Hence every element of the matrices appearing in the proof can be obtained from the coefficients of schemes through several additions, subtractions and multiplications. Therefore, if the coefficients of schemes satisfy the Lipschitz condition in [0, 1], then the elements of matrices must also satisfy the Lipschitz condition in $[0, 1]$. In this case, $\sqrt{a_1(a_1+a_2+a_3)}$ appears in the matrices Z_4, $Z_{4,1}$ and $Z_{4,2}$. Thus, we cannot obtain this conclusion only from the condition that the coefficients of schemes satisfy the Lipschitz condition. However, instead of (iii), we now assume that condition (iii)′,which is stronger than (iii), holds, i.e., we now assume that the derivatives of the coefficients are continuous, and that the coefficients are section-wise analytic. Moreover, according to Lemma 4.3, $a_1(a_1+a_2+a_3)$ is nonnegative if condition (i) containing the von Neumann condition is satisfied. Hence, by using Lemma 4.8, we can see that $\sqrt{a_1(a_1+a_2+a_3)}$ satisfies the Lipschitz condition in $[\varepsilon, 1-\varepsilon]$, $\varepsilon>0$. This means that all elements of the matrices appearing in the proof satisfy the Lipschitz condition in $[\varepsilon, 1-\varepsilon]$. Because of $\varepsilon\neq 0$, the process of proving Theorem 4.5 cannot be completely applied to the case here. This is why we assume that condition (v)′ holds, instead of condition (v) which is weaker than (v)′.

Consider the model A_1, and let $H_1(1)=-1$, $H_1(M)=-3$, and $H_1=-2$ for all the other points. In this case, (4.19) can be proved in the following way. As we have done in proving Theorem 4.5, we still introduce $W_{4,5}$ if $m=M$, $W_{4,6}$ if $m=1$, and Q_4 if $m\Delta x\in[\varepsilon_0, 1-\varepsilon_1]$. However, we take

$$Q'_{4,m}=-Y_{4,m}-\widetilde{Z}_{4,1,m-1}-\widetilde{Z}_{4,2,m},$$

instead of $Q_{4,m}$ if $m\Delta x\in(\Delta x, \varepsilon_0)$, and

$$Q''_{4,m}=-Y_{4,m}-\widetilde{Z}_{4,1,m}-\widetilde{Z}_{4,2,m+1},$$

if $m\Delta x\in(1-\varepsilon_1, 1)$. Therefore, we have the following inequality

$$\begin{aligned}\sum_{m=1}^{M}(Y_4\widetilde{U}, \widetilde{U})_m\leqslant &-(W_{4,6}\widetilde{U}, \widetilde{U})_{1,-1,4}\\ &-\sum_{m\Delta x\in(\Delta x,\varepsilon_0)}(Q'_4\widetilde{U}, \widetilde{U})_{m,-2,4}-\sum_{m\Delta x\in[\varepsilon_0,1-\varepsilon_1]}(Q_4\widetilde{U}, \widetilde{U})_{m,-2,4}\\ &-\sum_{m\Delta x\in(1-\varepsilon_1,1)}(Q''_4\widetilde{U}, \widetilde{U})_{m,-2,4}-(W_{4,5}\widetilde{U}, \widetilde{U})_{1,-3,4}.\end{aligned}\tag{4.47}$$

We rewrite the quadratic form on the right-hand side of the above inequality as

$$\sum_{h=-3}^{3}\sum_{m=0}^{M}b_{m,m+h}\widetilde{u}_m\widetilde{u}_{m+h},$$

where $b_{m,m+h}=b_{m+h,m}$, and $b_{m,m+h}=0$ if $m+h<0$ or $m+h>M$, and investigate the quadratic form. Let $0<\varepsilon<\min\{\varepsilon_0, \varepsilon_1\}$. We have pointed out that $\sqrt{a_1(a_1+a_2+a_3)}$ satisfies the Lipschitz condition in $[\varepsilon, 1-\varepsilon]$. Hence, every element of $Q_{4,m}-Q'_{4,m}$ for $m\Delta x\in(\Delta x, \varepsilon_0)\cap[\varepsilon, 1-\varepsilon]$, and every element of $Q_{4,m}-Q''_{4,m}$ for $m\Delta x\in(1-\varepsilon_1, 1)\cap[\varepsilon, 1-\varepsilon]$ are quantities of $O(\Delta x)$. Therefore, from the fact that Q_4 is a pseudo-null matrix, and the fact that every element of Q_4 satisfies the Lipschitz condition, we can see that $b_{m,m+h}=O(\Delta x)$ for $m\Delta x\in[\varepsilon, 1-\varepsilon]$. In $[0, \varepsilon]$ and $[1-\varepsilon, 1]$, the adopted scheme is a horizontal three-point scheme. Furthermore, the concrete forms of $W_{4,6}$, Q_4', Q_4'' and $W_{4,5}$ guarantee that the coefficients of the quadratic form on the right-hand side of (4.47) for $m\Delta x\in[\Delta x, \varepsilon]$ and $[1-\varepsilon, 1]$ are the same as those of (4.43). Therefore, (4.19) can be obtained from (4.47) in the same way as it can be obtained from (4.43).

The conclusion for other cases can be proved in the same way.

Corollary When the scheme (3.7)/(3.12) with $\beta_{1,m}$ and $\beta_{2,m}$ in (3.14) is applied to the initial-boundary-value problem (4.1)—(4.4) with (4.7) or (4.6), if (1) the von Neumann condition is satisfied everywhere, for example, (3.15j) corresponding to (3.7)/(3.12j) holds everywhere, j=a, b, or c; and (2) λ_n satisfies (4.32), $\dfrac{\partial\lambda_n}{\partial x}$ is a continuous function in [0, 1] and λ_n is a section-wise analytic function, then the scheme is stable.

The process of proving this corollary is similar to that of proving the corollary of Theorem 4.5. The differences are as follows. In the case of proving the corollary of Theorem 4.5, $\beta_{1,m}$ and $\beta_{2,m}$ have the form (3.13), and the inequality (4.32) holds, so conditions (iii) and (v) hold. In the case here, $\beta_{1,m}$ and $\beta_{2,m}$ have the form (3.14), the inequality (4.32) holds, $\dfrac{\partial\lambda_n}{\partial x}$ is a continuous function and λ_n is a section-wise analytic function, so conditions (iii)′ and (v)′ hold.

6. *The Stability of Section-wise Mixed Schemes and the "Condition" of Their Difference Systems*

Finally, we shall consider section-wise mixed schemes.

Theorem 4.7 Let the scheme (4.9) approximate the initial-boundary-value problem (4.1)—(4.4), and let it be a section-wise mixed scheme, and suppose the inequality (4.5) is satisfied. If the inequalities (4.18) and (4.19) hold when the scheme adopted in any section for any n is applied to the corresponding model problem; and if the number of sections is finite and $|\lambda|\geqslant\varepsilon>0$ holds on the borders of any two sections, ε being a constant, then the scheme is stable and the system of equations is well-conditioned.

Proof Firstly, we shall prove the following fact. For a fixed n, if (4.19) is satisfied both in the interval $[0, x_1]$ and in the interval $[x_1, 1]$, where $x_1=M'\Delta x$, then (4.19) is also valid in the interval [0, 1]. We denote $\tilde{u}_{n,m}$, c_3, $\tilde{T}_n$ etc. by $\bar{u}_{n,m}$, $\bar{c}_3$, $\bar{T}_n$ etc. in the first interval and they are denoted by $\bar{\bar{u}}_{n,m}$, $\bar{\bar{c}}_3$, $\bar{\bar{T}}_n$ etc. in the second interval. Hence the inequalities (4.19) in the two intervals are

$$\begin{cases} \overline{T}_n^{k+1} - \overline{T}_n^k \leqslant [c_7\delta_0(\lambda_{n,0}) - c_8|\sigma_{n,0}|(1-\delta_0(\lambda_{n,0}))]\,|\overline{u}_{n,0}^k|^2\Delta x \\ \quad + [c_7\delta_1(\lambda_{n,M'}) - c_8|\sigma_{n,M'}|(1-\delta_1(\lambda_{n,M'}))]\,|\overline{u}_{n,M'}^k|^2\Delta x \\ \quad + c_4\|\overline{u}_n^k\|_x^2\Delta x; \\ \overline{\overline{T}}_n^{k+1} - \overline{\overline{T}}_n^k \leqslant [c_7\delta_0(\lambda_{n,M'}) - c_8|\sigma_{n,M'}|(1-\delta_0(\lambda_{n,M'}))]\,|\overline{\overline{u}}_{n,M'}^k|^2\Delta x \\ \quad + [c_7\delta_1(\lambda_{n,M}) - c_8|\sigma_{n,M}|(1-\delta_1(\lambda_{n,M}))]\,|\overline{\overline{u}}_{n,M}^k|^2\Delta x + c_4\|\overline{\overline{u}}_n^k\|_x^2\Delta x. \end{cases}$$

To begin with, we shall consider the case $\lambda_{n,M'} \geqslant \varepsilon > 0$, and let

$$\widetilde{T}_n = \overline{T}_n + \overline{\overline{T}}_n,$$

$$\widetilde{u}_{n,m} = \begin{cases} \overline{u}_{n,m}, & m \leqslant M', \\ \overline{\overline{u}}_{n,m}, & m > M', \end{cases}$$

$$\|\widetilde{u}_n^k\|_x^2 = \|\overline{u}_n^k\|_x^2 + \|\overline{\overline{u}}_n^k\|_x^2 - |\overline{\overline{u}}_{n,M'}^k|^2\Delta x.$$

Obviously, if we can prove

$$\begin{aligned} &[c_7\delta_1(\lambda_{n,M'}) - c_8|\sigma_{n,M'}|(1-\delta_1(\lambda_{n,M'}))]\,|\overline{u}_{n,M'}^k|^2 \\ &\quad + [c_7\delta_0(\lambda_{n,M'}) - c_8|\sigma_{n,M'}|(1-\delta_0(\lambda_{n,M'}))]\,|\overline{\overline{u}}_{n,M'}^k|^2 + c_4\Delta x\,|\overline{\overline{u}}_{n,M'}^k|^2 \\ &= -c_8\sigma_{n,M'}|\overline{u}_{n,M'}^k|^2 + (c_7 + c_4\Delta x)\,|\overline{\overline{u}}_{n,M'}^k|^2 \leqslant 0, \end{aligned} \tag{4.48}$$

then we can immediately obtain the inequality (4.19) in the interval $[0, 1]$:

$$\begin{aligned} \widetilde{T}_n^{k+1} - \widetilde{T}_n^k \leqslant\ & [c_7\delta_0(\lambda_{n,0}) - c_8|\sigma_{n,0}|(1-\delta_0(\lambda_{n,0}))]\,|\widetilde{u}_{n,0}^k|^2\Delta x \\ & + [c_7\delta_1(\lambda_{n,M}) - c_8|\sigma_{n,M}|(1-\delta_1(\lambda_{n,M}))]\,|\widetilde{u}_{n,M}^k|^2\Delta x \\ & + c_4\|\widetilde{u}_n^k\|_x^2\Delta x. \end{aligned}$$

It is then easy to make (4.48) hold. According to (4.14a—d), we have $\overline{u}_{n,M'}^k = \overline{c}_3 u_{n,M'}^k$ in this case. Hence if we let $\overline{c}_3$ be sufficiently large, then (4.48) will hold. Therefore, the conclusion holds when $\lambda_{n,M'} \geqslant \varepsilon$. In the case $\lambda_{n,M'} \leqslant -\varepsilon$, the proof is similar.

In the case with two intervals, we have proved that if (4.19) holds in every interval, then (4.19) also holds in $[0, 1]$. In the case with more than two intervals, this conclusion still holds. This can be proved by using the inductive method.

For the inequality (4.18), there is a similar conclusion. In fact, if $\lambda_{n,M'} > \varepsilon$, then we have

$$\overline{T}_n + c_4\Delta x\|\overline{u}_n\|_x^2 \geqslant -c_5|\overline{u}_{n,0}|^2\Delta x\delta_0(\lambda_{n,0}) + c_6\sum_{m=\delta_0(\lambda_{n,0})}^{M'}|\overline{u}_{n,m}|^2\Delta x,$$

$$\overline{\overline{T}}_n + c_4\Delta x\|\overline{\overline{u}}_n\|_x^2 \geqslant -c_5|\overline{\overline{u}}_{n,M'}|^2\Delta x - c_5|\overline{\overline{u}}_{n,M}|^2\Delta x\delta_1(\lambda_{n,M}) + c_6\sum_{m=M'+1}^{M-\delta_1(\lambda_{n,M})}|\overline{\overline{u}}_{n,m}|^2\Delta x$$

in the two intervals, from which we can obtain

$$\begin{aligned}\overline{T}_n+\overline{\overline{T}}_n+c_4\Delta x[\|\overline{u}_n\|_x^2+\|\overline{\overline{u}}_n\|_x^2-|\overline{\overline{u}}_{n,M'}|^2\Delta x]\\ \geqslant -c_5|\overline{u}_{n,0}|^2\Delta x\delta_0(\lambda_{n,0})-c_5|\overline{\overline{u}}_{n,M}|^2\Delta x\delta_1(\lambda_{n,M})\\ +\frac{c_6}{2}\Big[\sum_{m=\delta_0(\lambda_{n,0})}^{M'}|\overline{u}_{n,m}|^2\Delta x+\sum_{m=M'+1}^{M-\delta_1(\lambda_{n,M})}|\overline{\overline{u}}_{n,M}|^2\Delta x\Big]\\ +\Big[\frac{c_6}{2}|\overline{u}_{n,M'}|^2-(c_5+c_4\Delta x)|\overline{\overline{u}}_{n,M'}|^2\Big]\Delta x.\end{aligned}$$

If we let $\bar{c}_3$ be sufficiently large, not only does (4.48) hold, but also

$$\frac{c_6}{2}|\overline{u}_{n,M'}|^2-(c_5+c_4\Delta x)|\overline{\overline{u}}_{n,M'}|^2\geqslant 0$$

holds. Hence the inequality above can be rewritten as (4.18):

$$\begin{aligned}\widetilde{T}_n+c_4\Delta x\|\widetilde{u}_n\|_x^2\geqslant -c_5|\widetilde{u}_{n,0}|^2\Delta x\delta_0(\lambda_{n,0})\\ -c_5|\widetilde{u}_{n,M}|^2\Delta x\delta_1(\lambda_{n,M})+\frac{c_6}{2}\sum_{m\in\mathscr{G}_m(\lambda_n)}|\widetilde{u}_{n,m}|^2\Delta x.\end{aligned}$$

In the case $\lambda_{n,M'}<-\varepsilon$ and the case with more than two subintervals, the same result can be obtained in a similar way.

Because (4.18) and (4.19) hold, we know from Theorems 4.1 and 4.2 that this theorem is true.

The corollary below follows from the preceding theorems.

Corollary When a scheme combining (3.7)/(3.12j) and (3.25) is applied to the initial-boundary-value problem (4.1)—(4.4) with (4.7) or (4.6), if

(1) every element of Λ satisfies (4.32), $\frac{\partial\lambda_n}{\partial x}$, $n=1, 2, \cdots, N$, are continuous, and λ_n are section-wise analytic in closed intervals, the number of closed intervals being finite. (when $\alpha=\frac{1}{2}$, every element of Λ is just required to satisfy (4.32).)

(2) $|\lambda_n|\leqslant\varepsilon\leqslant\frac{\Delta x}{\Delta t}$ when (3.7)/(3.12j) is used; and $|\lambda_n|\geqslant\varepsilon>0$ when (3.25) is used; moreover, the number of sections is finite;

(3) (3.14) is adopted as $\beta_{1,m}$ and $\beta_{2,m}$ for (3.7)/(3.12j) (if $\alpha=\frac{1}{2}$, (3.13) can also be adopted as $\beta_{1,m}$ and $\beta_{2,m}$); and α satisfies the following conditions:

$$\begin{cases}\alpha=\frac{1}{2}, \text{ if (3.7)/(3.12a) and (3.25) are adopted;}\\ \alpha\geqslant\frac{1}{2}, \text{ if (3.7)/(3.12b) and (3.25) are adopted;}\\ \frac{1}{2}\leqslant\alpha\leqslant 1, \text{ if (3.7)/(3.12c) and (3.25) are adopted,}\end{cases}$$

then the scheme is unconditionally stable, i.e., it is stable for any bounded $\frac{\Delta t}{\Delta x}$, and the system of difference equations is well-conditioned.

It is easy to obtain this corollary from Theorem 4.7, the proof of the other preceding theorems, (3.15j), (3.26), ⋯, and so on.

§ 5 Solution of Systems of Difference Equations

If an implicit scheme is adopted, a simultaneous system involving the unknowns at a large number of points needs to be solved. Hence, a method for solving the difference system should be given. There are two different types of methods for solving the system: the iteration method and the direct method. If an iteration method is adopted and the number of iterations is finite, then the scheme is essentially an explicit scheme, but no longer the original one. If a direct method is adopted, then there are two possibilities. If the boundary conditions are non-linear and if they are used according to the first way given in § 3, then a system which is composed of a large number of linear equations and a few nonlinear equations needs to be solved; if the boundary conditions are used according to the second way given in § 3, then a system of linear equations needs to be solved. The latter case is a special example of the former one, so the basic idea of methods for solving systems in the former case and the methods themselves can be used in the latter case. Therefore we should pay most attention to the former case. Because of the particularity of the latter case, a few other methods with smaller amount of computation can be found. Therefore we shall give a brief discussion about the latter case as well.

For the systems which are composed of a large number of linear equations and a few non-linear equations and whose matrices of coefficients are multi-diagonal, we have presented a direct method called the block-double-sweep method for "incomplete" systems of linear algebraic equations. In Appendix 2, we shall describe this method for the general case. In this section, we describe a special block-double-sweep method for "incomplete" systems of linear algebraic equations, which has been used for practical computation.

We pointed out in § 2 that the case $p_{l-1}+q_l<N$ can be "transferred" into the case $p_{l-1}+q_l\geqslant N$, and that the model problems appearing in the case $p_{l-1}+q_l\geqslant N$ are merely A_1, A_2 and B. Furthermore, λ of the model B is less than zero at one end and greater than zero at the other end. Hence, usually $|\lambda|\leqslant\varepsilon$, ε being a positive constant and not very large. λ in model A_1 is greater than zero at one end and greater than or equal to zero at the other end. Therefore usually $\lambda\geqslant-\varepsilon$. λ of the model A_2 is less than zero at one end and less than or equal to zero at the other end. Therefore usually $\lambda\leqslant\varepsilon$. Consequently, if we take $\Delta t\leqslant\dfrac{\Delta\xi}{\varepsilon}$ and use the schemes (3.7)/(3.12) and (3.24)/(3.25) in the combined

way given in § 3, i.e., (3.7)/(3.12) is used for $|\lambda| \leqslant \varepsilon$ and (3.24)/(3.25) is used for $|\lambda| > \varepsilon$, then the situation will be as follows: for the model B, (3.7)/(3.12) is adopted everywhere; for the model A_1, (3.7)/(3.12) is adopted if $\lambda \leqslant \varepsilon$ and (3.24)/(3.25) is adopted if $\lambda > \varepsilon$; for the model A_2, (3.7)/(3.12) is adopted if $\lambda \geqslant -\varepsilon$ and (3.24)/(3.25) is adopted if $\lambda < -\varepsilon$.

Consequently, in some practical problems, we need a method for the systems of difference equations which are obtained in the following way: the implicit scheme (3.24)/(3.25) is adopted if λ corresponds to the model A_1 and is greater than ε or if λ corresponds to the model A_2 and is less than $-\varepsilon$; an explicit scheme is adopted in the other cases. Moreover, the initial-boundary-value problem has the relation $p_{l-1}+q_l \geqslant N$. In this section, we only discuss a method for such a system of difference equations.

Generally speaking, the method can be divided into three steps. The first step is the procedure of elimination (the direct sweep); the second step is to calculate the quantities at the boundary points by using an iteration and the third step is the calculation of the unknowns (the inverse sweep). In the following, we describe the procedures in detail. According to (2.8), if $p_{l-1}+q_l \geqslant N$, the equations (2.10) in the interval $l-1 \leqslant \xi \leqslant l$ can be divided into three groups: the equations with $n=1$ through $N-p_{l-1}$ are called the first group, every equation of which corresponds to a model problem A_1; the equations with $n=N-p_{l-1}+1$ through q_l are called the second group, every equation of which corresponds to a model problem B; the equations with $n=q_l+1$ through N are called the third group, every equation of which corresponds to a model problem A_2. We have assumed that for the model problems A_1 and A_2, the mixed scheme (3.7)/(3.12)—(3.24)/(3.25) is adopted, and that for the model problem B, the scheme (3.7)/(3.12) is adopted. Therefore either when the quantities at the level $t=(k+\alpha)\Delta t$ are calculated from the quantities at the level $t=k\Delta t$, or when the quantities at the level $t=(k+1)\Delta t$ are calculated from the levels $t=k\Delta t$ and $t=(k+\alpha)\Delta t$, the system has the following form:

$$\begin{cases} \overline{G}_1^{(m+1)}\boldsymbol{U}_{m+1}+\Omega_1^{(m+1)}\overline{G}_1^{(m+1)}\boldsymbol{U}_m=\boldsymbol{F}_1^{(m+1)}, \quad m=0, 1, \cdots, M-1, & (5.1\text{a}) \\ \overline{G}_2^{(m)}\boldsymbol{U}_m=\boldsymbol{F}_2^{(m)}, \quad m=0, 1, \cdots, M, & (5.1\text{b}) \\ \Omega_3^{(m)}\overline{G}_3^{(m)}\boldsymbol{U}_{m+1}+\overline{G}_3^{(m)}\boldsymbol{U}_m=\boldsymbol{F}_3^{(m)}, \quad m=0, 1, \cdots, M-1, & (5.1\text{c}) \end{cases}$$

where (5.1a, b, c) are the difference equations corresponding to the first, second, third groups of equations respectively (in (5.1a, b, c) the subscript l is omitted). Hence in (5.1a, b, c), every row of $\overline{G}_1$, $\overline{G}_2$ or $\overline{G}_3$ is one of $\boldsymbol{G}_n^*$; $\overline{G}_1$, $\overline{G}_2$, $\overline{G}_3$ are respectively composed of those $\boldsymbol{G}_n^*$ with $n=1$ through $N-p_{l-1}$, $N-p_{l-1}+1$ through q_l, q_l+1 through N, and they are $s_1\times N$, $s_2\times N$, $s_3\times N$-matrices respectively, where $s_1=N-p_{l-1}$, $s_2=p_{l-1}+$

q_l-N, $s_3=N-q_l$; $\Omega_1^{(m)}$, $\Omega_3^{(m)}$ are $s_1\times s_1$, $s_3\times s_3$-diagonal matrices respectively, whose diagonal elements are equal to zero if the corresponding equation is approximated by (3.7)/(3.12), or $\left(\frac{1}{2}-\alpha|\sigma_n|\right)\Big/\left(\frac{1}{2}+\alpha|\sigma_n|\right)$ if the corresponding equation is approximated by (3.24)/(3.25), and whose diagonal elements are all less than $1-\varepsilon$ in norm, ε being a positive constant; $\boldsymbol{U}_m$, $\boldsymbol{F}_1^{(m+1)}$, $\boldsymbol{F}_2^{(m)}$, $\boldsymbol{F}_3^{(m)}$ are N, s_1, s_2, s_3-dimensional vectors respectively.

The procedure of elimination is the procedure in which $\boldsymbol{U}_1$, $\boldsymbol{U}_2$, $\cdots$, $\boldsymbol{U}_{M-1}$ are eliminated from the system (5.1), and $s_1+2s_2+s_3$ equations involving only $\boldsymbol{U}_0$ and $\boldsymbol{U}_M$ are obtained in every interval. In (5.1b), the equations with $m=0$, M involve only $\boldsymbol{U}_0$, $\boldsymbol{U}_M$ respectively. Therefore, s_1+s_3 more equations involving only $\boldsymbol{U}_0$ and $\boldsymbol{U}_M$ need to be derived. The details of the procedure are as follows. Let

$$\mu_0^{(1)}=\overline{G}_1^{(1)},\quad \mu_1^{(1)}=\Omega_1^{(1)}\overline{G}_1^{(1)},\quad \boldsymbol{\mu}^{(1)}=\boldsymbol{F}_1^{(1)},$$
$$\nu_0^{(1)}=\Omega_3^{(0)}\overline{\boldsymbol{G}}_3^{(0)},\quad \nu_1^{(1)}=\overline{G}_3^{(0)},\quad \boldsymbol{\nu}^{(1)}=\boldsymbol{F}_3^{(0)}.$$

Then the equations (5.1a) and (5.1c) with $m=0$ can be rewritten as

$$\mu_0^{(1)}\boldsymbol{U}_1+\mu_1^{(1)}\boldsymbol{U}_0=\boldsymbol{\mu}^{(1)},$$
$$\nu_0^{(1)}\boldsymbol{U}_1+\nu_1^{(1)}\boldsymbol{U}_0=\boldsymbol{\nu}^{(1)}.$$

Having $\mu_0^{(1)}$, $\mu_1^{(1)}$, $\boldsymbol{\mu}^{(1)}$, $\nu_0^{(1)}$, $\nu_1^{(1)}$, $\boldsymbol{\nu}^{(1)}$, we can carry out the following computation beginning with $m=1$ for the system

$$\begin{cases}\overline{G}_1^{(m+1)}\boldsymbol{U}_{m+1}+\Omega_1^{(m+1)}\overline{G}_1^{(m+1)}\boldsymbol{U}_m=\boldsymbol{F}_1^{(m+1)}, & (5.2\text{a})\\ \overline{G}_2^{(m)}\boldsymbol{U}_m=\boldsymbol{F}_2^{(m)}, & (5.2\text{b})\\ \Omega_3^{(m)}\overline{\boldsymbol{G}}_3^{(m)}\boldsymbol{U}_{m+1}+\overline{G}_3^{(m)}\boldsymbol{U}_m=\boldsymbol{F}_3^{(m)}, & (5.2\text{c})\\ \mu_0^{(m)}\boldsymbol{U}_m+\mu_1^{(m)}\boldsymbol{U}_0=\boldsymbol{\mu}^{(m)}, & (5.2\text{d})\\ \nu_0^{(m)}\boldsymbol{U}_m+\nu_1^{(m)}\boldsymbol{U}_0=\boldsymbol{\nu}^{(m)}. & (5.2\text{e})\end{cases}$$

From (5.2b—d),

$$\boldsymbol{U}_m=\overline{\overline{G}}^{(m)-1}\boldsymbol{F}^{(m)}-\overline{\overline{G}}^{(m)-1}H^{(m)}\boldsymbol{U}_{m+1}-\overline{\overline{G}}^{(m)-1}\tilde{\mu}_1^{(m)}\boldsymbol{U}_0 \tag{5.3}$$

is obtained by using the pivoting–elimination method. At the same time, $\boldsymbol{U}_m$ is eliminated from (5.2a) and (5.2e), and the following relations

$$\left.\begin{aligned}\mu_0^{(m+1)}\boldsymbol{U}_{m+1}+\mu_1^{(m+1)}\boldsymbol{U}_0=\boldsymbol{\mu}^{(m+1)},\\ \nu_0^{(m+1)}\boldsymbol{U}_{m+1}+\nu_1^{(m+1)}\boldsymbol{U}_0=\boldsymbol{\nu}^{(m+1)}\end{aligned}\right. \tag{5.4$'$}$$

are obtained. After that, for the two equations and the corresponding equations in (5.1a—c), the above calculation is repeated until $m=M-1$, i.e., until we obtain

$$\begin{cases}\mu_0^{(M)}\boldsymbol{U}_M+\mu_1^{(M)}\boldsymbol{U}_0=\boldsymbol{\mu}^{(M)},\\ \nu_0^{(M)}\boldsymbol{U}_M+\nu_1^{(M)}\boldsymbol{U}_0=\boldsymbol{\nu}^{(M)}.\end{cases} \tag{5.4}$$

In the above relations, the following symbols

$$\bar{\bar{G}}^{(m)}=\begin{pmatrix}\mu_0^{(m)}\\ \bar{G}_2^{(m)}\\ \bar{G}_3^{(m)}\end{pmatrix},\quad H^{(m)}=\begin{pmatrix}O\\ O\\ \Omega_3^{(m)}\bar{G}_3^{(m)}\end{pmatrix},$$
$$\tilde{\mu}_1^{(m)}=\begin{pmatrix}\mu_1^{(m)}\\ O\\ O\end{pmatrix},\quad \boldsymbol{F}^{(m)}=\begin{pmatrix}\boldsymbol{\mu}^{(m)}\\ \boldsymbol{F}_2^{(m)}\\ \boldsymbol{F}_3^{(m)}\end{pmatrix} \tag{5.5}$$

are used. Moreover, for $\mu_0^{(m+1)}, \cdots, \boldsymbol{\nu}^{(m+1)}$, there are the following expressions

$$\left\{\begin{array}{l}\mu_0^{(m+1)}=\bar{G}_1^{(m+1)}-\Omega_1^{(m+1)}\bar{G}_1^{(m+1)}\bar{\bar{G}}^{(m)^{-1}}H^{(m)},\\ \mu_1^{(m+1)}=-\Omega_1^{(m+1)}\bar{G}_1^{(m+1)}\bar{\bar{G}}^{(m)^{-1}}\tilde{\mu}_1^{(m)},\\ \boldsymbol{\mu}^{(m+1)}=\boldsymbol{F}_1^{(m+1)}-\Omega_1^{(m+1)}\bar{G}_1^{(m+1)}\bar{\bar{G}}^{(m)^{-1}}\boldsymbol{F}^{(m)},\\ \nu_0^{(m+1)}=-\nu_0^{(m)}\bar{\bar{G}}^{(m)^{-1}}H^{(m)},\\ \nu_1^{(m+1)}=\nu_1^{(m)}-\nu_0^{(m)}\bar{\bar{G}}^{(m)^{-1}}\tilde{\mu}_1^{(m)},\\ \boldsymbol{\nu}^{(m+1)}=\boldsymbol{\nu}^{(m)}-\nu_0^{(m)}\bar{\bar{G}}^{(m)^{-1}}\boldsymbol{F}^{(m)},\end{array}\right. \tag{5.6}$$

where $\bar{\bar{G}}^{(m)}$, $H^{(m)}$, $\tilde{\mu}_1^{(m)}$ are $N\times N$-matrices, $\boldsymbol{F}^{(m)}$ is an N-dimensional vector, the respective dimensions of $\mu_0^{(m+1)}, \cdots, \boldsymbol{\nu}^{(m+1)}$ are the same as those of $\mu_0^{(1)}, \cdots, \boldsymbol{\nu}^{(1)}$, and O stands for a null-matrix with a proper order. We want to point out here that if the number of rows of μ_1 is small, the following technique should be adopted in order to decrease the number of operations and save the storage locations. Before (5.3) is obtained from (5.2b—d) and $\boldsymbol{U}_m$ is eliminated from (5.2a) and (5.2e), the equations (5.2) are rewritten as

$$\begin{pmatrix}O\\ O\\ \Omega_3^{(m)}\bar{G}_3^{(m)}\\ \bar{G}_1^{(m+1)}\\ O\end{pmatrix}\boldsymbol{U}_{m+1}+\begin{pmatrix}\mu_0^{(m)}\\ \bar{G}_2^{(m)}\\ \bar{G}_3^{(m)}\\ \Omega_1^{(m+1)}\bar{G}_1^{(m+1)}\\ \nu_0^{(m)}\end{pmatrix}\boldsymbol{U}_m+\begin{pmatrix}E_{s_1}\\ O\\ O\\ O\\ O\end{pmatrix}\mu_1^{(m)}\boldsymbol{U}_0$$
$$+\begin{pmatrix}O\\ O\\ O\\ O\\ E_{s_3}\end{pmatrix}\nu_1^{(m)}\boldsymbol{U}_0=\begin{pmatrix}\boldsymbol{\mu}^{(m)}\\ \boldsymbol{F}_2^{(m)}\\ \boldsymbol{F}_3^{(m)}\\ \boldsymbol{F}_1^{(m)}\\ \boldsymbol{\nu}^{(m)}\end{pmatrix},$$

where E_{s_1}, E_{s_3} are $s_1\times s_1$, $s_3\times s_3$-unit matrices respectively, and O stands for a certain null-matrix. Then the computation is carried out with the coefficients of $\boldsymbol{U}_{m+1}$, $\boldsymbol{U}_m$, $\mu_1^{(m)}\boldsymbol{U}_0$, $\nu_1^{(m)}\boldsymbol{U}_0$ and the terms on the right hand side. Because the pivot element of elimination method is selected among the coefficients of the matrix $\bar{\bar{G}}^{(m)}$ and because the corresponding coefficients of $\nu_1^{(m)}\boldsymbol{U}_0$ are equal to zero, the coefficients of $\nu_1^{(m)}\boldsymbol{U}_0$ do not change in the procedure of elimination. Therefore, the computation may be carried out only with $2N+s_1+1$ columns, i.e., with the coefficients of $\boldsymbol{U}_{m+1}$, $\boldsymbol{U}_m$, $\mu_1^{(m)}\boldsymbol{U}_0$ and the terms on the right hand side. The number of operations in this case is less than the number in the case where the computation is carried out with $3N+1$ columns, i.e., with the coefficients of $\boldsymbol{U}_{m+1}$, $\boldsymbol{U}_m$, $\boldsymbol{U}_0$ and the terms on the right hand side. Of course, we do not get $\overline{\overline{G}}^{(m)-1}\tilde{\mu}_1^{(m)}$, $\mu_1^{(m+1)}$, $\nu_1^{(m+1)}$ now. Instead, we obtain

$$\overline{\overline{G}}^{(m)-1}\begin{pmatrix} E_{s_1} \\ O \\ O \end{pmatrix}, \qquad -\Omega_1^{(m+1)}\overline{G}_1^{(m+1)}\overline{\overline{G}}^{(m)-1}\begin{pmatrix} F_{s_1} \\ O \\ O \end{pmatrix},$$

$$-\nu_0^{(n)}\overline{\overline{G}}^{(m)-1}\begin{pmatrix} E_{s_1} \\ O \\ O \end{pmatrix},$$

as the matrix of coefficients of $\mu_1^{(m)}\boldsymbol{U}_0$. Thus, if we want to get $\overline{\overline{G}}^{(m)-1}\tilde{\mu}_1^{(m)}$, $\mu_1^{(m+1)}$, $\nu_1^{(m+1)}$, we have to carry out the following calculation:

$$\begin{cases} \overline{\overline{G}}^{(m)-1}\tilde{\mu}_1^{(m)} = \overline{\overline{G}}^{(m)-1}\begin{pmatrix} E_{s_1} \\ O \\ O \end{pmatrix}\mu_1^{(m)}, \\ \mu_1^{(m+1)} = -\Omega_1^{(m+1)}\overline{G}_1^{(m+1)}\overline{\overline{G}}^{(m)-1}\begin{pmatrix} E_{s_1} \\ O \\ O \end{pmatrix}\mu_1^{(m)}, \\ \nu^{(m+1)} = \nu_1^{(m)} - \nu_0^{(m)}\overline{\overline{G}}^{(m)-1}\begin{pmatrix} E_{s_1} \\ O \\ O \end{pmatrix}\mu_1^{(m)}. \end{cases}$$

However, it is not necessary to calculate $\overline{\overline{G}}^{(m)-1}\tilde{\mu}_1^{(m)}$. If $s_1<\frac{N}{2}$, instead of $\overline{\overline{G}}^{(m)-1}\tilde{\mu}_1^{(m)}$, we may keep $\overline{\overline{G}}^{(m)-1}\begin{pmatrix} E_{s_1} \\ O \\ O \end{pmatrix}$ and $\mu_1^{(m)}$, which benefits the decrease of the storage locations and the number of operations in the procedure of calculation of the unknowns. Therefore, if s_1 is small, it is reasonable to compute the coefficients of $\mu_1^{(m)}\boldsymbol{U}_0$ instead of those of $\boldsymbol{U}_0$ and to keep $\overline{\overline{G}}^{(m)-1}\begin{pmatrix} E_{s_1} \\ O \\ O \end{pmatrix}$ and $\mu_1^{(m)}$. If s_3 is small, there exists a similar problem for the treatment of the coefficients of $\boldsymbol{U}_{m+1}$ in (5.3). Besides that, if one wants to decrease the memory capacity, then one may eliminate several rows of the matrices of coefficients of $\boldsymbol{U}_{m+1}$ and $\boldsymbol{U}_0$ in (5.3) by using the method in Appendix 2 which causes the number of non-zero coefficients in (5.3) to decrease to N^2+N.

We point out that for this method, the procedure of elimination may start from $m=1$ or $m=M-1$, or even from a middle point. When a specific problem needs to be solved, a method should be chosen such that the amount of computation and memory capacity are small, and such that it is easy to make a program. For example, if $\Omega_3^{(0)}$ has s_4 null-diagonal-elements, then $\nu_0^{(1)}$ must have s_4 null-rows. This fact makes $\nu_0^{(m+1)}$ have s_4 null-rows, and makes s_4 rows of $\nu_1^{(m+1)}$ and s_4 elements of $\boldsymbol{\nu}^{(m+1)}$ constant. Hence we have s_4 equations in the system of equations

$$\nu_0^{(m)}\boldsymbol{U}_m+\nu_1^{(m)}\boldsymbol{U}_0=\boldsymbol{\nu}^{(m)}$$

which may not enter the calculation in the procedure of elimination. That is, if the procedure of elimination starts from $m=1$, the amount of computation and the memory capacity are small. Therefore, the procedure of elimination in this case should start from $m=1$. Obviously, if several diagonal elements of $\Omega_1^{(M)}$ are equal to zero, then the procedure of elimination should start from $m=M-1$.

Also, we want to point out that the number of rows of μ_0 or ν_0 may be equal to zero in some special cases. In these cases, there is no corresponding equation.

The second step is the calculation of quantities at the boundary points. After (5.4) is obtained at every interval, using these relations, the equations (5.1b) with $m_l=0$, M_l $(l=1, 2, \cdots, L)$, and the boundary conditions (2.5) in which x_l have been determined according to (3.28) or (3.29), we can compute $\boldsymbol{U}_{l,0}$, $\boldsymbol{U}_{l,M_l}$ $(l=1, 2, \cdots, L)$ and x_l' $(l=0, 1,\cdots, L)$ by the following iteration:

$$\begin{cases}\boldsymbol{B}_0(\boldsymbol{U}_{1,0}^{(k+1)},\ x_0,\ x_0'^{(k+1)},\ t)=0,\\ \overline{G}_2(1)^{(0)}\boldsymbol{U}_{1,0}^{(k+1)}=\boldsymbol{F}_2(1)^{(0)},\\ \nu_1(1)^{(M_1)}\boldsymbol{U}_{1,0}^{(k+1)}=\boldsymbol{\nu}(1)^{(M_1)}-\nu_0(1)^{(M_1)}\boldsymbol{U}_{1,M_1}^{(k)},\end{cases}\tag{5.7a}$$

$$\begin{cases}\mu_0(l)^{(M_l)}\boldsymbol{U}_{l,M_l}^{(k+1)}=\boldsymbol{\mu}(l)^{(M_l)}-\mu_1(l)^{(M_l)}\boldsymbol{U}_{l,0}^{(k)},\\ \overline{G}_2(l)^{(M_l)}\boldsymbol{U}_{l,M_l}^{(k+1)}=\boldsymbol{F}_2(l)^{(M_l)},\\ \boldsymbol{B}_l(\boldsymbol{U}_{l,M_l}^{(k+1)},\ \boldsymbol{U}_{l+1,0}^{(k+1)},\ x_l,\ x_l'^{(k+1)},\ t)=0,\\ \overline{G}_2(l+1)^{(0)}\boldsymbol{U}_{l+1,0}^{(k+1)}=\boldsymbol{F}_2(l+1)^{(0)},\\ \nu_1(l+1)^{(M_{l+1})}\boldsymbol{U}_{l+1,0}^{(k+1)}=\boldsymbol{\nu}(l+1)^{(M_{l+1})}-\nu_0(l+1)^{(M_{l+1})}\boldsymbol{U}_{l+1,M_{l+1}}^{(k)},\\ \qquad l=1,\ 2,\ \cdots,\ L-1,\end{cases}\tag{5.7b}$$

$$\begin{cases}\mu_0(L)^{(M_L)}\boldsymbol{U}_{L,M_L}^{(k+1)}=\boldsymbol{\mu}(L)^{(M_L)}-\mu_1(L)^{(M_L)}\boldsymbol{U}_{L,0}^{(k)},\\ \overline{G}_2(L)^{(M_L)}\boldsymbol{U}_{L,M_L}^{(k+1)}=\boldsymbol{F}_2(L)^{(M_L)},\\ \boldsymbol{B}_L(\boldsymbol{U}_{L,M_L}^{(k+1)},\ x_L,\ x_L'^{(k+1)},\ t)=0,\\ \qquad k=0,\ 1,\ \cdots,\end{cases}\tag{5.7c}$$

where l in $\overline{G}_2(l)^{(m_l)}$, $\boldsymbol{F}_2(l)^{(m_l)}$, $\nu_0(l)^{(M_l)}$, $\cdots$, $\boldsymbol{\mu}(l)^{(M_l)}$ denotes that they are

$\overline{G}_2^{(m)}$, $F_2^{(m)}$, $\nu_0^{(M)}$, $\cdots$, $\mu^{(M)}$ in the l-th interval respectively; $U_{l,m_l}^{(k+1)}$ denotes the value of U at the point $\xi = m_l \dfrac{1}{M_l} + (l-1)$ in the l-th interval $l-1 \leqslant \xi \leqslant l$, $k+1$ being the index of iteration and $U_{l+1,0} \neq U_{l,M_l}$ being possible because of the discontinuity of the solution. The iteration (5.7) is carried out in the following way: after $U_{l,0}^{(k)}$, $U_{l,M_l}^{(k)}$ $(l=1, 2, \cdots, L)$ are given, $U_{l,0}^{(k+1)}$, $U_{l,M_l}^{(k+1)}$ $(l=1, 2, \cdots, L)$ and $x_l'^{(k+1)}$ $(l=0, 1, \cdots, L)$ can be obtained from (5.7a—c); taking $U_{l,0}^{(k+1)}$, $U_{l,M_l}^{(k+1)}$ as the new given values, the next approximate values $U_{l,0}^{(k+2)}$, $U_{l,M_l}^{(k+2)}$ and $x_l'^{(k+2)}$ can be obtained. The procedure continues until the differences between two adjacent approximate values are sufficiently small. At the beginning, the corresponding values at the last t-level can be taken as $U_{l,0}^{(0)}$ and $U_{l,M_l}^{(0)}$.

So far, we have not given a complete algorithm for the solution of the unknowns at the boundary points. The iteration (5.7a—c) only simplifies the solution of a nonlinear system involving all the unknowns at the boundary points into the solution of $L+1$ nonlinear systems every one of which involves only the unknowns at one boundary point, i.e., the iteration (5.7a—c) simplifies the original problem into the solution of (5.7a), (5.7b) $(l=1, 2, \cdots, L-1)$ and (5.7c) (suppose that the values on the right hand side are given). We think that a method for the system involving only the unknowns at one boundary point should be constructed according to the special features of the system. Moreover, there are usually some effective methods for many problems in practice. Therefore, we shall not discuss this problem in this chapter. We shall do this problem for certain boundary conditions in the problem of flow around bodies in Chapter 7.

The third step is to calculate the values at interior points. After the values of U at boundary points are obtained, by using (5.3)

$$U_m = \overline{\overline{G}}^{(m)-1} F^{(m)} - \overline{\overline{G}}^{(m)-1} H^{(m)} U_{m+1} - \overline{\overline{G}}^{(m)-1} \tilde{\mu}_1^{(m)} U_0,$$

U_{M-1}, U_{M-2}, $\cdots$, U_1 can be obtained successively. (If the procedure of elimination starts from $m=M-1$, the order of calculation is U_1, U_2 $\cdots$, U_{M-1}.) This is the procedure of calculation of the unknowns.

It should be pointed out that the procedure of elimination is not necessary in those intervals with $s_1=0$ (in those intervals with $s_3=0$, the situation is similar). We suppose that $s_1(1)=0$ ($s_1(1)$ stands for s_1 with $l=1$), and that the procedure of elimination is not carried out in the interval with $l=1$. In this case, because of the relation $s_1(1)=0$ which makes the dimension of $\mu_0(1)^{(M_1)}$ equal to zero, the unknowns at all the boundary points except the point $\xi=0$ can be obtained by solving (5.7b) and (5.7c) after the procedures of elimination in the other intervals are completed. After that, the procedure of calculation of the unknowns can be carried out in all the other intervals. The unknowns in the first interval can be obtained directly from U_{M_1} and from the equations in that interval

$$\begin{cases} \overline{G}_2^{(m)} U_m = F_2^{(m)}, \\ \Omega_3^{(m)} \overline{G}_3^{(m)} U_{m+1} + \overline{G}_3^{(m)} U_m = F_3^{(m)}, \end{cases} \quad m = M_1-1,\ M_1-2,\ \cdots,\ 0 \tag{5.8}$$

in the following way: when U_{m+1} is known, U_m can be computed from (5.8) by using the elimination method, $m = M_1-1, \cdots, 0$. After U_0 is obtained, x_0' may be determined through $B_0(U_{1,0}, x_0, x_0', t) = 0$. In this process of calculation, the procedure of elimination in the first interval is cancelled, i.e., only the procedure of calculation of the unknowns is needed. Hence the number of operations is decreased. In fact, if the calculation is carried out in the original manner, then before U_{M_1-1}, $U_{M_1-2}, \cdots, U_1$ are calculated, the relation (5.3)

$$U_m = \overline{\overline{G}}^{(m)^{-1}} F^{(m)} - \overline{\overline{G}}^{(m)^{-1}} H^{(m)} U_{m+1}$$

is obtained from (5.8) by using the elimination. Now U_m is computed directly from (5.8) when U_{m+1} has been obtained. Clearly, the number of operations of the latter way is less than that of the former.

In the case $s_3(L) = 0$ in the L-th interval or in the case $s_1(1) = 0$ and $s_3(L) = 0$, the procedures of elimination in the corresponding intervals can be cancelled in a similar way. However, if $s_1 = 0$ or $s_3 = 0$ at several boundary points, the situation is slightly complicated. In order to cancel the procedure of elimination in all the intervals with $s_1 = 0$ or $s_3 = 0$, we present a method called the "group-dividing" method. In addition to a decrease in the number of operations, this method can also decrease the memory capacity of the coefficients which are obtained in the procedure of elimination and which are used in the procedure of calculation of the unknowns.

In order to make the description of this method clear, let us introduce several technical terms. The upper boundary of a group of intervals B_u is a boundary with $s_3 = 0$ or a boundary on which the boundary conditions contain in essence only the quantities on its lower side. The lower boundary of a group of intervals B_l is a boundary with $s_1 = 0$ or a boundary on which the boundary conditions contain in essence only the quantities on its upper side. B_n is a boundary which is neither B_u nor B_l, i.e., which is not a boundary of a group of intervals. Clearly, the upper boundary of the L-th interval and the lower boundary of the first interval are B_u and B_l. If the quantities on the upper side (or on the lower side) of a boundary have been obtained, then the boundary must be B_u (or B_l). A set of intervals is called a group of intervals, if the uppermost boundary of the intervals is B_u, the lowest boundary is B_l, and the middle boundaries are B_n. Obviously, for every group of intervals, the system can be independently solved. In that case, a boundary with $s_1 = 0$ or $s_3 = 0$

must be the uppermost or lowest boundary. Therefore, the system for each group may be solved by using the method mentioned above in which the procedure of elimination in the intervals with $s_1=0$ or $s_3=0$ is cancelled. Now, the group-dividing method may be described as follows: To begin with, we find out all the groups of intervals among all the intervals, and carry out the computation. Now, because all the quantities in these groups of intervals have been obtained, B_u and B_l of every group become B_l and B_u respectively if the quantities on the other side of the boundary are unknown. Then we find out all the groups of intervals among all the remaining intervals and carry out the computation again. This process continues until all the quantities are obtained. Because the number of intervals is L and because there is always at least one group of intervals before the computation is completed, all the quantities can be obtained after at most L such computations.

In the following, we shall illustrate the group-dividing method with two examples.

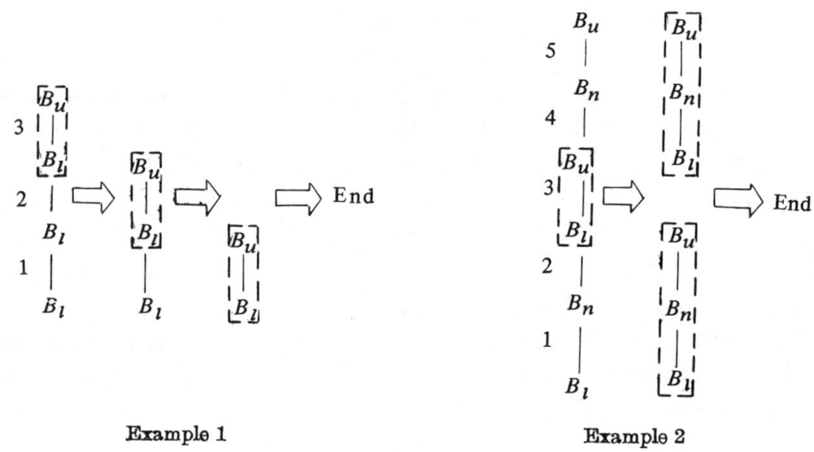

Figure 5.1

In the first example, there are four boundaries. From the top to the bottom, they are B_u, B_l, B_l and B_l. From the above description, we know that the third interval constitutes a group of intervals, and that the quantities in the interval can be first obtained. After the quantities in the third interval are obtained, the boundary 2 becomes B_u from B_l. Hence the second interval constitutes a group and the quantities in this interval can be independently calculated. Finally, the first interval constitutes a group, and the quantities there can be calculated. In this example, every group contains only one interval. Therefore, all the quantities are obtained after 3 cycles.

In the second example, there are 6 boundaries. Besides B_u and B_l, there is also B_n. At the beginning, the interval 3 constitutes a group of intervals, and the quantities in the interval 3 can be calculated. After that, the boundaries 2 and 3 become respectively B_u and B_l from B_l and B_u, and the intervals 1 and 2, 4 and 5 constitute two groups of intervals. The quantities in the intervals 4 and 5, and in the intervals 1 and 2 can be obtained.

In the following, we shall discuss the physical foundation of the group-dividing method. We have pointed out that if a boundary is B_u, then there are two possibilities. One is $s_3=N-q=0$, i.e., $q=N$. In this case, the quantities on the lower side of the boundary can be determined by certain difference equations in the interval below it. The other is that the quantities on the upper side of the boundary are known. In this case, besides the boundary conditions, only certain difference equations in the interval below it are needed in order to determine the quantities on the lower side of the boundary. In order to determine the quantities on the upper side of a boundary B_l, we only need the boundary conditions and certain difference equations in the interval above it. Therefore the quantities in the region between B_u and B_l can be independently determined.

Then, why can the procedure of elimination in the intervals with $s_1(l)=0$ be cancelled? It is because of the following fact. We use an explicit scheme for the second group of equations, i.e., for $n=N-p_{l-1}+1(=1),\cdots,q_l$, so we always determine the quantities at the point $m=M_l$ before determining the quantities at the interior points. After that, because each equation in the second group of difference equations contains only the quantities U_m, and because each equation in the third group of difference equations contains only the quantities U_{m+1} and U_m, the quantities at the point m can be calculated in succession by using the quantities at the point $m+1$, $m=M_l-1$, M_l-2, $\cdots$, 0. In the case $s_3(l)=0$, the situation is similar. That is, the fact that the procedure of elimination can be cancelled in certain cases is associated with the adopted scheme. The procedure of elimination cannot be cancelled for all schemes. However, the group-dividing method above is applicable to any scheme of § 3.

We shall now discuss the difference between the double-sweep method here and the usual double-sweep method[4] on the concrete process of calculation and on the range of application. In the method here, the procedure of elimination starts with eliminating U_1 or U_{M-1}. In the usual method, the procedure of elimination starts with eliminating U_0 or U_M. Consequently, the usual method requires that the boundary conditions be linear at least at one end of the interval. For the method here, the boundary conditions can be nonlinear, and they can even be nonlinear internal boundary conditions. Hence the method here has a wider range of application. The difference of the physical meaning is as follows: In the procedure of elimination of the method here, the derived relations are the constraint conditions which the difference equations in the discussed interval impose on U_0 and U_M. However, they are not related to the boundary conditions. These constraint conditions then reflect only the property of the differential equations, and can be used for various boundary conditions. In the procedure of elimination of the usual method, the derived relations are the constraint conditions which the difference equations in the discussed interval and the boundary conditions at one end of the interval impose on the quantities at the other end of the interval. The derived conditions are applicable only in the special case.

If the boundary conditions are linear, then the system which should be solved is linear, and it is better to use the usual double-sweep method because of its smaller amount of computation. If the boundary conditions are nonlinear, but are used in the second way given in § 3, i.e., if the boundary conditions are differentiated with respect to t, and the derived ordinary differential equations are approximated by (3.31)/(3.32), then the situation is similar. The usual double-sweep method for the case here can be described as follows. Suppose the boundary conditions are

$$\begin{cases} \mu_3(0)U_{1,0}=x'_0\mu_4(0)+\mu_5(0), & (5.9\text{a}) \\ \mu_2(l)U_{l,M_l}+\mu_3(l)U_{l+1,0}=x'_l\mu_4(l)+\mu_5(l), \quad l=1, 2, \cdots, L-1, & (5.9\text{b}) \\ \mu_2(L)U_{L,M_L}=x'_L\mu_4(L)+\mu_5(L), & (5.9\text{c}) \end{cases}$$

where $\mu_2(l)$—$\mu_5(l)$ stand for μ_2—μ_5 on the l-th boundary. In each interval, the equations remain to be (5.1a—c). In this case, the computation can be carried out in the following manner.

(1) By using the elimination method, the following relations with $l=1$

$$x'_{l-1}=\mu_6(l-1)U_{l,0},+\mu'_s(l-1) \qquad (5.10)$$

$$\mu_0(l)^{(0)}U_{l,0}=\mu(l)^{(0)} \qquad (5.11)$$

are obtained from (5.9a).

(2) For $l=1, 2, \cdots, L-1$, the process of eliminating the unknowns from (5.1a—c) in the l-th interval and (5.11) is repeated as follows.

Starting from $m=0$, for the system

$$\begin{cases} \overline{G}_1^{(m+1)}U_{m+1}+\Omega_1^{(m+1)}\overline{G}_1^{(m+1)}U_m=F_1^{(m+1)}, & (5.12\text{a}) \\ \overline{G}_2^{(m)}U_m=F_2^{(m)}, & (5.12\text{b}) \\ \Omega_3^{(m)}\overline{G}_3^{(m)}U_{m+1}+\overline{G}_3^{(m)}U_m=F_3^{(m)}, & (5.12\text{c}) \\ \mu_0^{(m)}U_m=\mu^{(m)}, & (5.12\text{d}) \end{cases}$$

the following computation is carried out by using the elimination method: from (5.12b—d) we obtain the relations

$$U_m = \overline{\overline{G}}^{(m)^{-1}} F^{(m)} - \overline{\overline{G}}^{(m)^{-1}} H^{(m)} U_{m+1}, \tag{5.13}$$

and at the same time,

$$\mu_0^{(m+1)} U_{m+1} = \mu^{(m+1)} \tag{5.14$'$}$$

is obtained from (5.12c). The definition of $\overline{\overline{G}}^{(m)}$, $H^{(m)}$, $F^{(m)}$ and the recurrence expressions of $\mu_0^{(m+1)}$, $\mu^{(m+1)}$ in the above equations are formally the same as those in (5.5) and (5.6). However, because $\mu_0^{(1)}$, $\mu^{(1)}$ here are usually different from $\mu_0^{(1)}$, $\mu^{(1)}$ there, $\mu_0^{(m+1)}$, $\mu^{(m+1)}$ etc. in the two processes of calculation are different. After (5.14)′ is obtained, the above computation for a higher m is repeated until

$$\mu_0^{(M)} U_M = \mu^{(M)} \tag{5.14}$$

is obtained. Then, the following equations

$$U_{l,M_l} = \mu_7(l) U_{l+1,0} + \mu_8(l), \tag{5.15}$$

$$x_l' = \mu_6(l) U_{l+1,0} + \mu_5'(l), \tag{5.10}$$

$$\mu_0(l+1)^{(0)} U_{l+1,0} = \mu(l+1)^{(0)} \tag{5.11}$$

are obtained from (5.14) and (5.9b) by using the elimination method.

(3) After (5.11) corresponding to the L–th interval is derived, we repeat the first part of (2), and (5.14) corresponding to the L–th interval can then be obtained. Furthermore, U_{L,M_L} and x_L' can be determined from (5.14) and (5.9c).

(4) From U_{L,M_L}, U at all other points can be obtained by using (5.13) and (5.15). x_l' $(l = L-1, \cdots, 0)$ can be determined from (5.10).

All the foregoing descriptions about the group–dividing method, about the method of cancelling the procedure of elimination in the intervals with $s_1 = 0$ or $s_3 = 0$, about the directions of the direct–sweep and the inverse–sweep, and so on, apply to the situation here. We shall not discuss these problems in detail for this case.

Lastly, we want to point out the following fact which will be proved in the next section. The procedure of elimination and the procedure of calculation of the unknowns in this section are stable, and the iteration (5.7a—c) is convergent under quite weak conditions. Therefore, the method given here is sound.

§6 The Stability of the Procedure of Elimination and the Procedure of Calculation of the Unknowns, and the Convergence of Iteration

In this section, we shall discuss certain theoretical problems about the method for the solution of difference systems described in § 5. For simplicity, we consider only the case with one interval. It can be proved by using the same method that these results also hold in the case with several intervals.

Concerning the stability of a general procedure of calculation, we give the following definition. Suppose we have a procedure of calculation as follows:

$$X_i^{(m+1)}=f_i(X_1^{(m)}, X_2^{(m)}, \cdots, X_I^{(m)}), \quad i=1, 2, \cdots, I, \quad m=0, 1, 2, \cdots.$$

If for the variations $\delta X_i^{(0)}$ of $X_i^{(0)}$ ($i=1, 2, \cdots, I$) and the variations $\delta X_i^{(m)}$ of $X_i^{(m)}$, there exist the following relations

$$\|\delta X_i^{(m)}\| \leqslant c[\|\delta X_1^{(0)}\|+\cdots+\|\delta X_I^{(0)}\|], \quad i=1, 2, \cdots, I, \quad m=0, 1, 2, \cdots,$$

where c is a constant independent of m, then the procedure of calculation is stable.

Before analysing stability, we give a lemma.

Lemma 6.1 Suppose S is an $I\times I$–nonnegative matrix whose eigenvalues are not greater than one in norm, and suppose its elementary divisors corresponding to the eigenvalues with a unit norm are linear. Moreover, let $Z^{(m)}$ ($m=0, 1, \cdots$) be a sequence of nonnegative vectors. If the relation

$$Z^{(m+1)} \leqslant SZ^{(m)}$$

holds, then there exists a nonnegative matrix C' such that the relation

$$Z^{(m)} \leqslant C'Z^{(0)}$$

holds, where these inequalities between vectors denote that the same inequalities between all corresponding elements hold. Therefore, if for a sequence of the vector

$$Z^{(m)*}=(\|\delta X_1\|, \|\delta X_2\|, \cdots, \|\delta X_I\|)^{(m)}, \quad m=0, 1, 2, \cdots,$$

the relation $Z^{(m+1)} \leqslant SZ^{(m)}$ holds, then the procedure of calculation is stable.

Proof The proof of this lemma is simple. Because the elements of S and Z are nonnegative real numbers, from

$$Z^{(1)} \leqslant SZ^{(0)}$$

we derive

$$Z^{(2)} \leqslant SZ^{(1)} \leqslant S^2Z^{(0)},$$

and furthermore,

$$Z^{(m)} \leqslant SZ^{(m-1)} \leqslant \cdots \leqslant S^mZ^{(0)}.$$

If the Jordan canonical matrix of S is Λ_s, and if

$$S=\tilde{S}^{-1}\Lambda_s\tilde{S},$$

then

$$S^m=\tilde{S}^{-1}\Lambda_s^m\tilde{S}.$$

Because every eigenvalue of S is not greater than one in norm, and because its elementary divisors corresponding to the eigenvalues with a unit norm are linear, every element of Λ_s^m is less in norm than a positive constant independent of m. Thus, every element of S^m is also less than a positive constant independent of m. This means that the conclusion of this lemma holds.

We can obtain the following theorem from Lemma 6.1.

Theorem 6.1 If

(1) in the region $t\geqslant 0$, $0\leqslant\xi\leqslant 1$, the square of the determinant of $\overline{G}$, i.e., the Gram determinant $\Gamma(\boldsymbol{G}_1, \boldsymbol{G}_2, \cdots, \boldsymbol{G}_N)$ has an upper bound (for example, equal to one), and has a lower bound greater than zero, i.e.,

$$0<\tilde{c}_1\leqslant\Gamma(\boldsymbol{G}_1, \boldsymbol{G}_2, \cdots, \boldsymbol{G}_N)\leqslant 1; \qquad (6.1a)$$

(2) in the interval $0\leqslant\xi\leqslant 1$, $\boldsymbol{G}_n$ satisfies the Lipschitz condition with respect to ξ, i.e.,

$$\|\boldsymbol{G}_{n,m+1}-\boldsymbol{G}_{n,m}\|\leqslant\tilde{c}_2\Delta\xi, \quad n=1, 2, \cdots, N, \qquad (6.1b)$$

where $\boldsymbol{G}_{n,m}$ stands for the value of $\boldsymbol{G}_n$ at $\xi=m\Delta\xi$;

(3) in the region $t\geqslant 0, 0\leqslant\xi\leqslant 1, \lambda_1, \lambda_2, \cdots, \lambda_N$ are uniformly bounded;

(4) $\Delta\xi$ is sufficiently small, and $\dfrac{\Delta t}{\Delta\xi}$ satisfies the inequality

$$0\leqslant\frac{\Delta t}{\Delta\xi}\leqslant\tilde{c}_3,$$

$\tilde{c}_3$ being a positive constant;

then both the procedure of elimination i.e., the procedure of calculating $\boldsymbol{\mu}_0^{(m+1)}, \cdots, \boldsymbol{\nu}^{(m+1)}$ $(m=1, 2, \cdots, M-1)$ by (5.6), and the procedure of calculation of the unknowns, i.e., the procedure of calculating $\boldsymbol{U}_{M-1}$, $\boldsymbol{U}_{M-2}, \cdots, \boldsymbol{U}_1$ by (5.3), are stable.

If besides the conditions (1)—(4), the problem satisfies the following condition:

(5) the equations

$$\begin{cases}\boldsymbol{B}_0(\boldsymbol{U}_0, x_0')=0,\\ \overline{G}_{2,0}\boldsymbol{U}_0=\boldsymbol{F}_2^{(0)}, & t\geqslant 0, \ \xi=0,\\ \overline{G}_{3,0}\boldsymbol{U}_0=\boldsymbol{F}_3^{(0)},\end{cases} \qquad (6.1c)$$

$$\begin{cases}\overline{G}_{1,M}\boldsymbol{U}_M=\boldsymbol{F}_1^{(M)},\\ \overline{G}_{2,M}\boldsymbol{U}_M=\boldsymbol{F}_2^{(M)}, & t\geqslant 0, \ \xi=1,\\ \boldsymbol{B}_1(\boldsymbol{U}_M, x_1')=0\end{cases} \qquad (6.1d)$$

always have solution; every element of the Jacobi determinants of the expressions on the left hand sides of (6.1c) with respect to $\boldsymbol{U}_0$, x_0', and (6.1d) with respect to $\boldsymbol{U}_M$, x_1', is a continuous bounded function, and the squares of the determinants are greater than a positive constant $\tilde{c}_4$, where $\overline{G}_{j,m}$ $(j=1, 2, 3)$ with a subscript m stand for their values at $\xi=m\Delta\xi$, and $\boldsymbol{F}_2^{(0)}$, $\boldsymbol{F}_3^{(0)}$, $\boldsymbol{F}_1^{(M)}$, $\boldsymbol{F}_2^{(M)}$ are any given vectors, then the iteration (5.7) is convergent.

Proof Before proving this theorem, let us point out the following facts. The conditions (3) and (4) guarantee that every element of the diagonal matrices Ω_1, Ω_3 of the equations (5.1a), (5.1c) is less than $1-\varepsilon$ in norm, ε being a constant $\in(0, 1)$. According to the conditions (1) and (2), $\Gamma(\boldsymbol{G}_1, \boldsymbol{G}_2, \cdots, \boldsymbol{G}_N)$ has a lower bound greater than zero, and every element of $\boldsymbol{G}_1, \boldsymbol{G}_2, \cdots, \boldsymbol{G}_N$ is bounded. Therefore, we can always find two constants c_0, c_1 and a small positive number ε_1 such that if $\overline{\overline{G}}$ is sufficiently close to $\overline{G}$ at a point ξ, i.e., if

$$\|\overline{\overline{G}}-\overline{G}\|\leqslant\varepsilon_1, \tag{6.2}$$

then $\overline{\overline{G}}$ must be an invertible matrix and

$$\|\overline{\overline{G}}^{-1}\|\leqslant c_0, \quad \|\overline{\overline{G}}\|\leqslant c_1. \tag{6.3}$$

If the condition (2) holds and $\Delta\xi$ is sufficiently small, then the following conditions

$$\begin{cases} \|\overline{G}_1^{(m+1)}-\overline{G}_1^{(m)}\|\leqslant\delta_1=\dfrac{\varepsilon}{(1-\varepsilon)c_0c_1}\dfrac{\varepsilon_1}{2}, \\ \|\overline{G}_3^{(m)}-\overline{G}_3^{(m+1)}\|\leqslant\delta_1, \\ \|\overline{G}^{(m)}-(\overline{G})_m\|\leqslant\dfrac{\varepsilon_1}{2}, \quad m=1, 2, \cdots, M-1 \end{cases} \tag{6.4}$$

hold, where $(\overline{G})_m$ stands for $\overline{G}$ at $\xi=m\Delta\xi$, and

$$\overline{G}^{(m)}=\begin{pmatrix}\overline{G}_1^{(m)}\\ \overline{G}_2^{(m)}\\ \overline{G}_3^{(m)}\end{pmatrix}.$$

Also, we want to point out the following facts. In the same way as we do for a square matrix, we can define a norm for a general matrix as follows:

$$\|A\|=\max\frac{\|A\boldsymbol{X}\|}{\|\boldsymbol{X}\|}.$$

Moreover, a series of operational rules such as $\|AB\|\leqslant\|A\|\,\|B\|$ still hold. In addition, without loss of generality, we shall suppose c_0, $c_1\geqslant1$, $\varepsilon_1\leqslant\varepsilon<1$.

By using the facts listed above, we now prove that if the conditions (1)—(4) hold, then $\overline{\overline{G}}^{(m)}$ is always invertible, and

$$\|\overline{\overline{G}}^{(m)}\|\leqslant c_1, \quad \|\overline{\overline{G}}^{(m)-1}\|\leqslant c_0. \tag{6.5}$$

Moreover, there are the following results for $\mu_0^{(m)}, \cdots, \boldsymbol{\nu}^{(m)}$:

$$\begin{cases} \mu_0^{(m)}=\overline{G}_1^{(m)}-\omega_1^{(m)}\overline{\overline{G}}^{(m)}, \quad \|\omega_1^{(m)}\|\leqslant\dfrac{\varepsilon_1}{2c_1}\leqslant\dfrac{\varepsilon_1}{2}, & \text{(6.6a)} \\ \mu_1^{(m)}=\omega_4^{(m)}\overline{G}_1^{(1)}, \quad \|\omega_4^{(m)}\|\leqslant\left(1-\dfrac{\varepsilon}{2}\right)^{m-1}(1-\varepsilon), & \text{(6.6b)} \end{cases}$$

$$\left\{\begin{array}{ll}\|\mu^m\|\leqslant\dfrac{1-\left(1-\dfrac{\varepsilon}{2}\right)^m}{\dfrac{\varepsilon}{2}}c_2\leqslant\dfrac{2}{\varepsilon}c_2, & (6.6c)\\ \nu_0^{(m)}=\omega_5^{(m)}\overline{G}_3^{(m-1)},\quad \|\omega_5^{(m)}\|\leqslant\left(1-\dfrac{\varepsilon}{2}\right)^{m-1}(1-\varepsilon), & (6.6d)\\ \nu_1^{(m)}=\overline{G}_3^{(0)}+\omega^{(m)}\overline{G}_1^{(1)}, & \\ \|\omega_6^{(m)}\|\leqslant\begin{cases}0, & m=1\\ (1-\varepsilon)^2c_0\delta_1\displaystyle\sum_{l=1}^{m-2}\left(1-\dfrac{\varepsilon}{2}\right)^{2l}, & m\geqslant 2\end{cases}\leqslant\dfrac{\varepsilon_1(1-\varepsilon)}{2\left(1-\dfrac{\varepsilon}{4}\right)c_1}, & (6.6e)\\ \|\nu^{(m)}\|\leqslant\dfrac{1-\left(1-\dfrac{\varepsilon}{2}\right)^m}{\dfrac{\varepsilon}{2}}c_3\leqslant\dfrac{2}{\varepsilon}c_3,\quad m=1,2,\cdots,M, & (6.6f)\end{array}\right.$$

where

$$c_2=\max_{m=1,2,\cdots,M}\|F_1^{(m)}\|+\frac{\varepsilon_1}{2c_1}\Big(\max_{m=1,2,\cdots,M-1}\|F_2^{(m)}\|+\max_{m=0,1,\cdots,M-1}\|F_3^{(m)}\|\Big),$$

$$c_3=\frac{2}{\varepsilon}c_2+\max_{m=1,2,\cdots,M-1}\|F_2^{(m)}\|+\max_{m=0,1,\cdots,M-1}\|F_3^{(m)}\|.$$

Now, we proceed to prove these results with induction. For $m=1$, noticing the definitions of $\mu_0^{(1)}$, $\cdots$, $\nu^{(1)}$, $\overline{\overline{G}}^{(1)}$, the estimates (6.2)—(6.4), and the definitions of c_2, c_3, we see that (6.5) and (6.6) hold. Suppose the conclusion is true for m. We need to prove that it is also true for $m+1$.

Because $\overline{\overline{G}}^{(m)}$ is invertible, we can let $\overline{G}_1^{(m+1)}=\overline{G}_1^{(m)}+\omega_2^{(m)}\overline{\overline{G}}^{(m)}$, where the $s_1\times N$-matrix $\omega_2^{(m)}$ is equal to $(\overline{G}_1^{(m+1)}-\overline{G}_1^{(m)})\overline{\overline{G}}^{(m)-1}$. By using (6.4) and (6.5), we can obtain

$$\|\omega_2^{(m)}\|\leqslant\|\overline{G}_1^{(m+1)}-\overline{G}_1^{(m)}\|\,\|\overline{\overline{G}}^{(m)-1}\|\leqslant\frac{\varepsilon}{(1-\varepsilon)c_1}\,\frac{\varepsilon_1}{2}.$$

Furthermore, from (6.6a) and the definition (5.5) of $\overline{\overline{G}}^{(m)}$, we have

$$\overline{G}_1^{(m+1)}=\mu_0^{(m)}+(\omega_1^{(m)}+\omega_2^{(m)})\overline{\overline{G}}^{(m)}=(\widetilde{E}_1+\omega_1^{(m)}+\omega_2^{(m)})\overline{\overline{G}}^{(m)},$$

i.e.

$$\overline{G}_1^{(m+1)}\overline{\overline{G}}^{(m)-1}=\widetilde{E}_1+\omega_1^{(m)}+\omega_2^{(m)},\tag{6.7}$$

where

$$\widetilde{E}_1=(E_{s_1},0)$$

is an $s_1\times N$-matrix. Hence we can derive from (5.6) and (5.5)

$$\begin{aligned}\mu_0^{(m+1)}&=\overline{G}_1^{(m+1)}-\Omega_1^{(m+1)}\overline{G}_1^{(m+1)}\overline{\overline{G}}^{(m)-1}H^{(m)}\\&=\overline{G}_1^{(m+1)}-\Omega_1^{(m+1)}(\widetilde{E}_1+\omega_1^{(m)}+\omega_2^{(m)})_{s_5\sim N}\Omega_3^{(m)}\overline{G}_3^{(m)}\\&=\overline{G}_1^{(m+1)}-\Omega_1^{(m+1)}(\omega_1^{(m)}+\omega_2^{(m)})_{s_5\sim N}\Omega_3\overline{G}_3^{(m)},\end{aligned}\tag{6.8}$$

where $(A)_{s_5\sim N}$ stands for the matrix which is composed of the s_5-th column through the N-th column of A, $s_5=N-s_3+1$. (We shall use similar symbols hereafter.) Moreover, because of

$$\|(\omega_1^{(m)}+\omega_2^{(m)})_{s_1\sim N}\|\leqslant\|\omega_1^{(m)}+\omega_2^{(m)}\|\leqslant\frac{\varepsilon_1}{2c_1}+\frac{\varepsilon}{(1-\varepsilon)c_1}\,\frac{\varepsilon_1}{2}=\frac{\varepsilon_1}{2c_1(1-\varepsilon)},$$

$$\|\overline{G}_3^{(m)}\|\leqslant\|\overline{\overline{G}}^{(m)}\|\leqslant c_1,$$

and noticing that $\Omega_1^{(m+1)}$, $\Omega_3^{(m)}$ are less than $1-\varepsilon$ in norm, we have

$$\|\Omega_1^{(m+1)}\overline{G}_1^{(m+1)}\overline{\overline{G}}^{(m)^{-1}}H^{(m)}\|\leqslant(1-\varepsilon)^2\frac{\varepsilon_1}{2c_1(1-\varepsilon)}\,c_1=\frac{1-\varepsilon}{2}\,\varepsilon_1,$$

and further

$$\begin{aligned}\|\overline{\overline{G}}^{(m+1)}-(\overline{G})_{m+1}\|&\leqslant\|\overline{\overline{G}}^{(m+1)}-\overline{G}^{(m+1)}\|+\|\overline{G}^{(m+1)}-(\overline{G})_{m+1}\|\\&=\|\mu_0^{(m+1)}-\overline{G}_1^{(m+1)}\|+\|\overline{G}^{(m+1)}-(\overline{G})_{m+1}\|\\&\leqslant\frac{1-\varepsilon}{2}\,\varepsilon_1+\frac{\varepsilon_1}{2}<\varepsilon_1.\end{aligned}$$

(In the process of deriving the relations above, we assume that if there is the relation

$$\boldsymbol{X}_N^*=(\boldsymbol{X}_{s_1}^*,\underbrace{0,\cdots,0}_{N-s_1})$$

between an N-dimensional vector $\boldsymbol{X}_N$ and an s_1-dimensional vector $\boldsymbol{X}_{s_1}$, then there is the relation $\|\boldsymbol{X}_N\|=\|\boldsymbol{X}_{s_1}\|$. This is true for the common norms: the L_1-norm, L_2-norm and L_∞-norm.) Therefore, we know from the foregoing results that $\overline{\overline{G}}^{(m+1)}$ is invertible, and that

$$\|\overline{\overline{G}}^{(m+1)^{-1}}\|\leqslant c_0,\quad\|\overline{\overline{G}}^{(m+1)}\|\leqslant c_1,$$

i.e., the estimate (6.5) is obtained.

We can now let

$$\overline{G}_3^{(m)}=\overline{G}_3^{(m+1)}+\omega_3^{(m)}\overline{\overline{G}}^{(m+1)}=(\widetilde{E}_3+\omega_3^{(m)})\overline{\overline{G}}^{(m+1)},\tag{6.9}$$

where $\widetilde{E}_3$, $\omega_3^{(m)}$ are $s_3\times N$-matrices,

$$\widetilde{E}_3=(0,\ E_{s_3}),\quad\omega_3^{(m)}=(\overline{G}_3^{(m)}-\overline{G}_3^{(m+1)})\overline{\overline{G}}^{(m+1)^{-1}},$$

and

$$\|\widetilde{E}_3+\omega_3^{(m)}\|\leqslant1+\frac{\varepsilon}{(1-\varepsilon)c_1}\,\frac{\varepsilon_1}{2}<\frac{1-\dfrac{\varepsilon}{2}}{1-\varepsilon}.\tag{6.10}$$

Hence, after letting

$$\omega_1^{(m+1)}=\Omega_1^{(m+1)}(\omega_1^{(m)}+\omega_2^{(m)})_{s_1\sim N}\Omega_3^{(m)}(\widetilde{E}_3+\omega_3^{(m)}),$$

(6.8) can be rewritten as

$$\mu_0^{(m+1)}=\overline{G}_1^{(m+1)}-\omega_1^{(m+1)}\overline{\overline{G}}^{(m+1)},$$

where $\omega_1^{(m+1)}$ satisfies

$$\|\omega_1^{(m+1)}\|\leqslant\frac{\varepsilon_1(1-\varepsilon)}{2c_1}\,\frac{1-\dfrac{\varepsilon}{2}}{1-\varepsilon}<\frac{\varepsilon_1}{2c_1}.$$

Therefore, the conclusion (6.6a) on $\mu_0^{(m+1)}$ holds. Moreover, from (5.6), (5.5), (6.7), (6.6b) etc., we have

$$\begin{aligned}\mu_1^{(m+1)} &= -\Omega_1^{(m+1)}\overline{G}_1^{(m+1)}\overline{\overline{G}}^{(m)-1}\tilde{\mu}_1^{(m)} \\ &= -\Omega_1^{(m+1)}(\widetilde{E}_1+\omega_1^{(m)}+\omega_2^{(m)})_{1\sim s_1}\mu_1^{(m)} \\ &= -\Omega_1^{(m+1)}(\widetilde{E}_1+\omega_1^{(m)}+\omega_2^{(m)})_{1\sim s_1}\omega_4^{(m)}\overline{G}_1^{(1)},\end{aligned} \tag{6.11}$$

$$\begin{aligned}\|\Omega_1^{(m+1)}(\widetilde{E}_1+\omega_1^{(m)}+\omega_2^{(m)})_{1\sim s_1}\| &\leqslant (1-\varepsilon)\left(1+\frac{\varepsilon_1}{2(1-\varepsilon)c_1}\right) \\ &\leqslant 1-\frac{\varepsilon}{2}.\end{aligned}$$

Hence, if let

$$\omega_4^{(m+1)} = -\Omega_1^{(m+1)}(\widetilde{E}_1+\omega_1^{(m)}+\omega_2^{(m)})_{1\sim s_1}\omega_4^{(m)},$$

we obtain (6.6b) with $m+1$:

$$\mu_1^{(m+1)} = \omega_4^{(m+1)}\overline{G}_1^{(1)},$$

$$\|\omega_4^{(m+1)}\| \leqslant \left(1-\frac{\varepsilon}{2}\right)\|\omega_4^{(m)}\| \leqslant \left(1-\frac{\varepsilon}{2}\right)^m(1-\varepsilon).$$

From (5.6), (6.7) etc., we know that there is the following expression for $\boldsymbol{\mu}^{(m+1)}$:

$$\begin{aligned}\boldsymbol{\mu}^{(m+1)} &= \boldsymbol{F}_1^{(m+1)} - \Omega_1^{(m+1)}(\widetilde{E}_1+\omega_1^{(m)}+\omega_2^{(m)})\boldsymbol{F}^{(m)} \\ &= \boldsymbol{F}_1^{(m+1)} - \Omega_1^{(m+1)}(\widetilde{E}_1+\omega_1^{(m)}+\omega_2^{(m)})_{1\sim s_1}\boldsymbol{\mu}^{(m)} \\ &\quad - \Omega_1^{(m+1)}(\omega_1^{(m)}+\omega_2^{(m)})_{s_1+1\sim N}\begin{pmatrix}\boldsymbol{F}_2^{(m)} \\ \boldsymbol{F}_3^{(m)}\end{pmatrix}.\end{aligned}$$

Furthermore, by using

$$\|\boldsymbol{F}_1^{(m+1)}\| + \|\Omega_1^{(m+1)}(\omega_1^{(m)}+\omega_2^{(m)})\|\left\|\begin{pmatrix}\boldsymbol{F}_2^{(m)} \\ \boldsymbol{F}_3^{(m)}\end{pmatrix}\right\| \leqslant \|\boldsymbol{F}_1^{(m+1)}\| + \frac{\varepsilon_1}{2c_1}\left\|\begin{pmatrix}\boldsymbol{F}_2^{(m)} \\ \boldsymbol{F}_3^{(m)}\end{pmatrix}\right\| \leqslant c_2,$$

the estimate (6.6c)

$$\|\boldsymbol{\mu}^{(m+1)}\| \leqslant c_2 + \left(1-\frac{\varepsilon}{2}\right)\|\boldsymbol{\mu}^{(m)}\| \leqslant \frac{1-\left(1-\dfrac{\varepsilon}{2}\right)^{m+1}}{\dfrac{\varepsilon}{2}}c_2 \leqslant \frac{2}{\varepsilon}c_2$$

can be easily obtained.

From (6.6d) and (6.9), the relation

$$\nu_0^{(m)}\overline{\overline{G}}^{(m)-1} = \omega_5^{(m)}\overline{G}_3^{(m-1)}\overline{\overline{G}}^{(m)-1} = \omega_5^{(m)}(\widetilde{E}_3+\omega_3^{(m-1)}) \tag{6.12}$$

follows. Hence, according to (5.6), (5.5) etc., we have

$$\begin{aligned}\nu_0^{(m+1)} &= -\nu_0^{(m)}\overline{\overline{G}}^{(m)-1}H^{(m)} = -\omega_5^{(m)}(\widetilde{E}_3+\omega_3^{(m-1)})H^{(m)} \\ &= -\omega_5^{(m)}(\widetilde{E}_3+\omega_3^{(m-1)})_{s_2\sim N}\Omega_3^{(m)}\overline{G}_3^{(m)},\end{aligned}$$

and after letting

$$\omega_5^{(m+1)} = -\omega_5^{(m)}(\widetilde{E}_3+\omega_3^{(m-1)})_{s_2\sim N}\Omega_3^{(m)},$$

we obtain (6.6d) with $m+1$:

$$\nu_0^{(m+1)} = \omega_5^{(m+1)}\overline{G}_3^{(m)},$$

$$\|\omega_5^{(m+1)}\| \leqslant \|\omega_5^{(m)}\|\frac{1-\dfrac{\varepsilon}{2}}{1-\varepsilon}(1-\varepsilon) \leqslant \left(1-\frac{\varepsilon}{2}\right)^m(1-\varepsilon).$$

In the same way as we obtain the expression for $\nu_0^{(m+1)}$, we can derive the following expression for $\nu_1^{(m+1)}$:

$$\begin{aligned}\nu_1^{(m+1)} &= \nu_1^{(m)} - \nu_0^{(m)}\overline{\overline{G}}^{(m)^{-1}}\widetilde{\mu}_1^{(m)} = \nu_1^{(m)} - \omega_5^{(m)}(\widehat{E}_3 + \omega_3^{(m-1)})\widetilde{\mu}_1^{(m)} \\ &= \nu_1^{(m)} - \omega_5^{(m)}(\omega_3^{(m-1)})_{1\sim s_1}\mu_1^{(m)} \\ &= \overline{G}_3^{(0)} + (\omega_6^{(m)} - \omega_5^{(m)}(\omega_3^{(m-1)})_{1\sim s_1}\omega_4^{(m)})\overline{G}_1^{(1)},\end{aligned}$$

from which, after letting

$$\omega_6^{(m+1)} = \omega_6^{(m)} - \omega_5^{(m)}(\omega_3^{(m-1)})_{1\sim s_1}\omega_4^{(m)},$$

we obtain (6.6e) immediately:

$$\nu_1^{(m+1)} = \overline{G}_3^{(0)} + \omega_6^{(m+1)}\overline{G}_1^{(1)},$$

$$\begin{aligned}\|\omega_6^{(m+1)}\| &\leqslant \|\omega_6^{(m)}\| + \left(1-\frac{\varepsilon}{2}\right)^{2(m-1)}(1-\varepsilon)^2 c_0\delta_1 \\ &\leqslant (1-\varepsilon)^2 c_0 \delta_1 \sum_{l=0}^{m-1}\left(1-\frac{\varepsilon}{2}\right)^{2l} \leqslant \frac{\varepsilon_1(1-\varepsilon)}{\left(1-\frac{\varepsilon}{4}\right)2c_1}, \quad (m \geqslant 1).\end{aligned}$$

Also, we can derive the following expression for $\boldsymbol{\nu}^{(m+1)}$ in a similar way:

$$\boldsymbol{\nu}^{(m+1)} = \boldsymbol{\nu}^{(m)} - \nu_0^{(m)}\overline{\overline{G}}^{(m)^{-1}}\boldsymbol{F}^{(m)} = \boldsymbol{\nu}^{(m)} - \omega_5^{(m)}(\widehat{E}_3 + \omega_3^{(m-1)})\boldsymbol{F}^{(m)}.$$

Thus, we have

$$\|\boldsymbol{\nu}^{(m+1)}\| \leqslant \|\boldsymbol{\nu}^{(m)}\| + \left(1-\frac{\varepsilon}{2}\right)^m \|\boldsymbol{F}^{(m)}\|.$$

Moreover, from $\|\boldsymbol{\mu}^{(m)}\| \leqslant \frac{2}{\varepsilon} c_2$, we obtain

$$\|\boldsymbol{F}^{(m)}\| \leqslant \|\boldsymbol{\mu}^{(m)}\| + \|\boldsymbol{F}_2^{(m)}\| + \|\boldsymbol{F}_3^{(m)}\| \leqslant c_3.$$

Therefore, the estimate

$$\|\boldsymbol{\nu}^{(m+1)}\| \leqslant \frac{1-\left(1-\frac{\varepsilon}{2}\right)^{m+1}}{\frac{\varepsilon}{2}} c_3 \leqslant \frac{2}{\varepsilon} c_3$$

follows immediately. All the relations (6.6a—f) are proved.

It is easy to prove from these relations that the procedure of elimination and the procedure of calculation of the unknowns are stable, and that the iteration (5.7) is convergent.

To begin with, we prove that the process of calculation of $\mu_0^{(m)}$, $\mu_1^{(m)}, \cdots, \boldsymbol{\nu}^{(m)}$ is stable. In order to prove that, we investigate the relations between the set of variations $\delta\widetilde{\omega}_1^{(m)} \equiv (\delta\mu_0^{(m)})\overline{\overline{G}}^{(m)^{-1}}$, $\delta\mu_1^{(m)}$, $\delta\boldsymbol{\mu}^{(m)}$, $\delta\widetilde{\omega}_5^{(m)} \equiv (\delta\nu_0^{(m)})\overline{\overline{G}}^{(m)^{-1}}$, $\delta\nu_1^{(m)}, \delta\boldsymbol{\nu}^{(m)}$ and the set of variations $\delta\widetilde{\omega}_1^{(m+1)}, \cdots, \delta\boldsymbol{\nu}^{(m+1)}$. Taking a variation for every relation of (5.6), noticing the relations (6.7), (6.12) and

$$\delta\overline{\overline{G}}^{(m)^{-1}} = -\overline{\overline{G}}^{(m)^{-1}}\delta\overline{\overline{G}}^{(m)}\overline{\overline{G}}^{(m)^{-1}} = -\overline{\overline{G}}^{(m)^{-1}}\begin{pmatrix}\delta\mu_0^{(m)} \\ 0 \\ 0\end{pmatrix}\overline{\overline{G}}^{(m)^{-1}},$$

and using the following relations derived from them

$$-\overline{G}_1^{(m+1)}\delta\overline{\overline{G}}^{(m)^{-1}}=(\overline{G}_1^{(m+1)}\overline{\overline{G}}^{(m)^{-1}})_{1\sim s_1}\delta\mu_0^{(m)}\overline{\overline{G}}^{(m)^{-1}}=(\widetilde{E}_1+\omega_1^{(m)}+\omega_2^{(m)})_{1\sim s_1}\delta\widetilde{\omega}_1^{(m)};$$

$$-\nu_0^{(m)}\delta\overline{\overline{G}}^{(m)^{-1}}=\omega_5^{(m)}(\widetilde{E}_3+\omega_3^{(m-1)})_{1\sim s_1}\delta\mu_0^{(m)}\overline{\overline{G}}^{(m)^{-1}}=\omega_5^{(m)}(\widetilde{E}_3+\omega_3^{(m-1)})_{1\sim s_1}\delta\widetilde{\omega}_1^{(m)},$$

we obtain

$$\begin{cases}\delta\mu_0^{(m+1)}=\Omega_1^{(m+1)}(\widetilde{E}_1+\omega_1^{(m)}+\omega_2^{(m)})_{1\sim s_1}\delta\widetilde{\omega}_1^{(m)}H^{(m)},\\ \delta\mu_1^{(m+1)}=\Omega_1^{(m+1)}(\widetilde{E}_1+\omega_1^{(m)}+\omega_2^{(m)})_{1\sim s_1}\delta\widetilde{\omega}_1^{(m)}\widetilde{\mu}_1^{(m)}\\ \qquad -\Omega_1^{(m+1)}(\widetilde{E}_1+\omega_1^{(m)}+\omega_2^{(m)})_{1\sim s_1}\delta\mu_1^{(m)},\\ \delta\boldsymbol{\mu}^{(m+1)}=\Omega_1^{(m+1)}(\widetilde{E}_1+\omega_1^{(m)}+\omega_2^{(m)})_{1\sim s_1}\delta\widetilde{\omega}_1^{(m)}\boldsymbol{F}^{(m)}\\ \qquad -\Omega_1^{(m+1)}(\widetilde{E}_1+\omega_1^{(m)}+\omega_2^{(m)})_{1\sim s_1}\delta\boldsymbol{\mu}^{(m)},\\ \delta\nu_0^{(m+1)}=-\delta\widetilde{\omega}_5^{(m)}H^{(m)}+\omega_5^{(m)}(\widetilde{E}_3+\omega_3^{(m-1)})_{1\sim s_1}\delta\widetilde{\omega}_1^{(m)}H^{(m)},\\ \delta\nu_1^{(m+1)}=\delta\nu_1^{(m)}-\delta\widetilde{\omega}_5^{(m)}\widetilde{\mu}_1^{(m)}+\omega_5^{(m)}(\widetilde{E}_3+\omega_3^{(m-1)})_{1\sim s_1}\delta\widetilde{\omega}_1^{(m)}\widetilde{\mu}_1^{(m)}\\ \qquad -\omega_5^{(m)}(\widetilde{E}_3+\omega_3^{(m-1)})_{1\sim s_1}\delta\mu_1^{(m)},\\ \delta\boldsymbol{\nu}^{(m+1)}=\delta\boldsymbol{\nu}^{(m)}-\delta\widetilde{\omega}_5^{(m)}\boldsymbol{F}^{(m)}+\omega_5^{(m)}(\widetilde{E}_3+\omega_3^{(m-1)})_{1\sim s_1}\delta\widetilde{\omega}_1^{(m)}\boldsymbol{F}^{(m)}\\ \qquad -\omega_5^{(m)}(\widetilde{E}_3+\omega_3^{(m-1)})_{1\sim s_1}\delta\boldsymbol{\mu}^{(m)}.\end{cases}\tag{6.13}$$

Furthermore, because we can derive from (6.9)

$$H^{(m)}=\begin{pmatrix}0\\0\\\Omega_3^{(m)}\overline{G}_3^{(m)}\end{pmatrix}=\begin{pmatrix}0\\0\\\Omega_3^{(m)}(\widetilde{E}_3+\omega_3^{(m)})\end{pmatrix}\overline{\overline{G}}^{(m+1)},$$

the first and fourth relations of (6.13) can be rewritten as

$$\begin{cases}\delta\widetilde{\omega}_1^{(m+1)}=\Omega_1^{(m+1)}(\widetilde{E}_1+\omega_1^{(m)}+\omega_2^{(m)})_{1\sim s_1}(\delta\widetilde{\omega}_1^{(m)})_{s_2\sim N}\Omega_3^{(m)}(\widetilde{E}_3+\omega_3^{(m)}),\\ \delta\widetilde{\omega}_5^{(m+1)}=-(\delta\widetilde{\omega}_5^{(m)})_{s_2\sim N}\Omega_3^{(m)}(\widetilde{E}_3+\omega_3^{(m)})\\ \qquad +\omega_5^{(m)}(\widetilde{E}_3+\omega_3^{(m-1)})_{1\sim s_1}(\delta\widetilde{\omega}_1^{(m)})_{s_2\sim N}\Omega_3^{(m)}(\widetilde{E}_3+\omega_3^{(m)}).\end{cases}\tag{6.14}$$

Therefore, by using (6.6), (6.11) and the relation derived from (6.10)

$$\|\Omega_3^{(m)}(\widetilde{E}_3+\omega_3^{(m)})\|\leqslant 1-\frac{\varepsilon}{2},$$

we can get the following estimates

$$\begin{cases}\|\delta\widetilde{\omega}_1^{(m+1)}\|\leqslant\left(1-\frac{\varepsilon}{2}\right)^2\|\delta\widetilde{\omega}_1^{(m)}\|,\\ \|\delta\mu_1^{(m+1)}\|\leqslant c_4\|\delta\widetilde{\omega}_1^{(m)}\|+\left(1-\frac{\varepsilon}{2}\right)\|\delta\mu_1^{(m)}\|,\\ \|\delta\boldsymbol{\mu}^{(m+1)}\|\leqslant c_4\|\delta\widetilde{\omega}_1^{(m)}\|+\left(1-\frac{\varepsilon}{2}\right)\|\delta\boldsymbol{\mu}^{(m)}\|,\\ \|\delta\widetilde{\omega}_5^{(m+1)}\|\leqslant c_4\|\delta\widetilde{\omega}_1^{(m)}\|+\left(1-\frac{\varepsilon}{2}\right)\|\delta\widetilde{\omega}_5^{(m)}\|,\\ \|\delta\nu_1^{(m+1)}\|\leqslant c_4[\|\delta\widetilde{\omega}_1^{(m)}\|+\|\delta\mu_1^{(m)}\|+\|\delta\widetilde{\omega}_5^{(m)}\|]+\|\delta\nu_1^{(m)}\|,\\ \|\delta\boldsymbol{\nu}^{(m+1)}\|\leqslant c_4[\|\delta\widetilde{\omega}_1^{(m)}\|+\|\delta\boldsymbol{\mu}^{(m)}\|+\|\delta\widetilde{\omega}_5^{(m)}\|]+\|\delta\boldsymbol{\nu}^{(m)}\|,\end{cases}$$

which can be rewritten as

$$Z^{(m+1)} \leqslant S Z^{(m)}, \quad m=1, 2, \cdots, M-1,$$

where

$$Z^{(m)} = \begin{pmatrix} \|\delta\tilde{\omega}_1^{(m)}\| \\ \|\delta\mu_1^{(m)}\| \\ \|\delta\boldsymbol{\mu}^{(m)}\| \\ \|\delta\tilde{\omega}_5^{(m)}\| \\ \|\delta\nu_1^{(m)}\| \\ \|\delta\boldsymbol{\nu}^{(m)}\| \end{pmatrix},$$

$$S = \begin{pmatrix} \left(1-\frac{\varepsilon}{2}\right)^2 & 0 & 0 & 0 & 0 & 0 \\ c_4 & 1-\frac{\varepsilon}{2} & 0 & 0 & 0 & 0 \\ c_4 & 0 & 1-\frac{\varepsilon}{2} & 0 & 0 & 0 \\ c_4 & 0 & 0 & 1-\frac{\varepsilon}{2} & 0 & 0 \\ c_4 & c_4 & 0 & c_4 & 1 & 0 \\ c_4 & 0 & c_4 & c_4 & 0 & 1 \end{pmatrix},$$

c_4 being a bounded positive constant independent of m. It is easy to see that the eigenvalues of the matrix S are $\left(1-\frac{\varepsilon}{2}\right)^2$, $\left(1-\frac{\varepsilon}{2}\right)$ and 1, and that the elementary divisors corresponding to the two eigenvalues equal to 1 are linear because there are two linearly independent eigenvectors corresponding to them. Moreover, every element of this matrix is nonnegative. Therefore, using Lemma 6.1, and noticing $\overline{\overline{G}}^{(m)}$ and $\overline{\overline{G}}^{(m)-1}$ to be uniformly bounded with respect to m, we can see that the process of calculation of $\mu_0^{(m)}, \mu_1^{(m)}, \cdots, \boldsymbol{\nu}^{(m)}$ is stable.

Now let us prove that the procedure of the inverse sweep is also stable. We consider the effect of $\boldsymbol{U}_0$ and $\boldsymbol{U}_M$ on $\boldsymbol{U}_m$, $m=M-1, \cdots, 1$.

Consider the following process of calculation

$$\begin{cases} \boldsymbol{U}_m = \overline{\overline{G}}^{(m)-1}\boldsymbol{F}^{(m)} - \overline{\overline{G}}^{(m)-1}H^{(m)}\boldsymbol{U}_{m+1} - \overline{\overline{G}}^{(m)-1}\tilde{\mu}_1^{(m)}\boldsymbol{U}_0, \\ \boldsymbol{U}_0 = \boldsymbol{U}_0, \quad m=M-1, M-2, \cdots, 1. \end{cases} \tag{6.15}$$

We can rewrite the first formula above as

$$\overline{\overline{G}}^{(m)}\boldsymbol{U}_m = \boldsymbol{F}^{(m)} - H^{(m)}\boldsymbol{U}_{m+1} - \tilde{\mu}_1^{(m)}\boldsymbol{U}_0,$$

from which the following formula among the variations of $\boldsymbol{U}_{m+1}$, $\boldsymbol{U}_m$, $\boldsymbol{U}_0$ can be obtained:

$$\begin{aligned} \overline{\overline{G}}^{(m)}\delta\boldsymbol{U}_m &= -\begin{pmatrix} O \\ \Omega_3^{(m)}\overline{G}_3^{(m)} \end{pmatrix}\delta\boldsymbol{U}_{m+1} - \begin{pmatrix} \mu_1^{(m)} \\ O \end{pmatrix}\delta\boldsymbol{U}_0 \\ &= -\begin{pmatrix} O \\ \Omega_3^{(m)}(\tilde{E}_3 + \omega_3^{(m)}) \end{pmatrix}\overline{\overline{G}}^{(m+1)}\delta\boldsymbol{U}_{m+1} - \begin{pmatrix} \mu_1^{(m)} \\ O \end{pmatrix}\delta\boldsymbol{U}_0. \end{aligned}$$

Hence, we have

$$\|\bar{\bar{G}}^{(m)}\delta U_m\| \leqslant (1-\varepsilon)\frac{1-\frac{\varepsilon}{2}}{1-\varepsilon}\|\bar{\bar{G}}^{(m+1)}\delta U_{m+1}\| + c_4\|\delta U_0\|$$

$$= \left(1-\frac{\varepsilon}{2}\right)\|\bar{\bar{G}}^{(m+1)}\delta U_{m+1}\| + c_4\|\delta U_0\|,$$

which can be rewritten as

$$Z^{(m)} \leqslant S Z^{(m+1)}, \quad m = M-1,\ M-2,\ \cdots,\ 1,$$

where

$$Z^{(m)} = \begin{pmatrix} \|\bar{\bar{G}}^{(m)}\delta U_m\| \\ \|\delta U_0\| \end{pmatrix},$$

$$S = \begin{pmatrix} 1-\frac{\varepsilon}{2} & c_4 \\ 0 & 1 \end{pmatrix}.$$

This nonnegative matrix S satisfies the conditions of Lemma 6.1, so we can see by using Lemma 6.1 that the process of calculation of $\bar{\bar{G}}^{(m)}U_m$, U_0 is stable. Moreover, $\bar{\bar{G}}^{(m)}$ and $\bar{\bar{G}}^{(m)-1}$ are bounded matrices, so the process (6.15) of computing $\{U_m, U_0\}$ from $\{U_{m+1}, U_0\}$ for successive $m = M-1$, $M-2, \cdots, 1$ is stable. That is, the process of computing U_m from $\{U_{m+1}, U_0\}$ for successive $m = M-1, M-2, \cdots, 1$ is stable.

We have considered the effect of δU_0 and δU_M on δU_m. If we also want to consider the effect of $\delta\mu_0^{(m)}, \cdots, \delta\nu^{(m)}$, or the effect of $\delta\mu_0^{(1)}, \cdots, \delta\nu^{(1)}$, then similar results can be proved. In fact, their effect on δU_m is similar to the effect of δU_0 on δU_m. Thus, by using the fact that the norms of $U_0, U_1, \cdots, U_M$ have a uniform upper bound with respect to M (it is easy to prove this fact), we can derive the following estimate

$$\|\delta U_m\| \leqslant c[\|\delta U_M\| + \|\delta U_0\| + \|\delta\mu_0^{(1)}\| + \cdots + \|\delta\nu^{(1)}\|],$$

that is to say, the procedure of calculation of the unknowns is stable.

Lastly, we shall prove that the iteration (5.7) is convergent. In the case with one interval, (5.7) can be written as

$$B(Y^{(k+1)}) = C Y^{(k)} + D,$$

where

$$Y = \begin{pmatrix} U_0 \\ x_0' \\ U_M \\ x_1' \end{pmatrix}, \quad B(Y) = \begin{pmatrix} B_0(U_0, x_0') \\ \bar{G}_2^{(0)} U_0 \\ \nu_1^{(M)} U_0 \\ \mu_0^{(M)} U_M \\ \bar{G}_2^{(M)} U_M \\ B_1(U_M, x_1') \end{pmatrix}, \quad CY = \begin{pmatrix} 0 \\ 0 \\ -\nu_0^{(M)} U_M \\ -\mu_1^{(M)} U_0 \\ 0 \\ 0 \end{pmatrix}, \quad D = \begin{pmatrix} 0 \\ F_2^{(0)} \\ \nu^{(M)} \\ \mu^{(M)} \\ F_2^{(M)} \\ 0 \end{pmatrix}.$$

We know from (6.6) that $\mu_0^{(M)}=\overline{G}_1^{(M)}+O(\varepsilon_1)$ and $\nu_1^{(M)}=\overline{G}_3^{(0)}+O(\varepsilon_1)$. Moreover, when $\Delta\xi$ is sufficiently small and $\frac{\Delta t}{\Delta\xi}$ is bounded, ε_1 can be sufficiently small. Thus,

$$\mu_0^{(M)}\approx\overline{G}_1^{(M)}\approx\overline{G}_{1,M};\quad \nu_1^{(M)}\approx\overline{G}_3^{(0)}\approx\overline{G}_{3,0}.$$

Also, there are the following results: $\overline{G}_2^{(0)}\approx\overline{G}_{2,0}$ and $\overline{G}_2^{(M)}\approx\overline{G}_{2,M}$. Therefore, if the foregoing condition (5) holds, the norm of the Jacobi determinant of $\boldsymbol{B}(\boldsymbol{Y})$ is greater than a positive constant. On the other hand, for any $\boldsymbol{Y}^{(k+1)}$, $\boldsymbol{Y}^{(k)}$, we can always find a matrix $\widetilde{B}_{(k)}$ such that

$$\widetilde{B}_{(k)}(\boldsymbol{Y}^{(k+1)}-\boldsymbol{Y}^{(k)})=\boldsymbol{B}(\boldsymbol{Y}^{(k+1)})-\boldsymbol{B}(\boldsymbol{Y}^{(k)}),$$

where $\widetilde{B}_{(k)}$ may not be a Jacobi matrix of $\boldsymbol{B}$, but it is sufficiently close to a Jacobi matrix if $\boldsymbol{Y}^{(k+1)}$ is sufficiently close to $\boldsymbol{Y}^{(k)}$. Thus, there exists $\varepsilon_2>0$ such that when $\|\boldsymbol{Y}^{(k+1)}-\boldsymbol{Y}^{(k)}\|\leqslant\varepsilon_2$, $\widetilde{B}_{(k)}$ must have a bounded inverse matrix. Therefore, from

$$\boldsymbol{B}(\boldsymbol{Y}^{(k+1)})-\boldsymbol{B}(\boldsymbol{Y}^{(k)})=\widetilde{B}_{(k)}(\boldsymbol{Y}^{(k+1)}-\boldsymbol{Y}^{(k)})=C(\boldsymbol{Y}^{(k)}-\boldsymbol{Y}^{(k-1)}),$$

we can obtain

$$\|\boldsymbol{Y}^{(k+1)}-\boldsymbol{Y}^{(k)}\|<\|\widetilde{B}_{(k)}^{-1}\|\,\|C\|\,\|\boldsymbol{Y}^{(k)}-\boldsymbol{Y}^{(k-1)}\|.$$

We know from (6.6b) and (6.6d) that $\mu_1^{(M)}$ and $\nu_0^{(M)}\to 0$ as $M\to\infty$. Hence, if $\Delta\xi=\frac{1}{M}$ is sufficiently small, $\|\widetilde{B}_{(k)}^{-1}\|\,\|C\|<1$ will hold. Therefore, the iteration is convergent if $\|\boldsymbol{Y}^{(2)}-\boldsymbol{Y}^{(1)}\|\leqslant\varepsilon_2$. The inequality is guaranteed if the solution of $\boldsymbol{B}(\boldsymbol{Y})=\boldsymbol{D}$ depends continuously on the terms on the right hand side. In fact,

$$\boldsymbol{B}(\boldsymbol{Y}^{(1)})=C\boldsymbol{Y}^{(0)}+\boldsymbol{D},\quad \boldsymbol{B}(\boldsymbol{Y}^{(2)})=C\boldsymbol{Y}^{(1)}+\boldsymbol{D},$$

and $\|C\|$ is very small, so the terms on the right–hand side of the two equations are always very close, whether $\boldsymbol{Y}^{(0)}$ is close to $\boldsymbol{Y}^{(1)}$ or not. Hence, if the solution depends continuously on the terms on the right hand side, then $\boldsymbol{Y}^{(2)}$ must be close to $\boldsymbol{Y}^{(1)}$.

Because the Jacobi determinant of $\boldsymbol{B}(\boldsymbol{Y})$ is not equal to zero, the continuous dependence of the solution of $\boldsymbol{B}(\boldsymbol{Y})=\boldsymbol{D}$ on $\boldsymbol{D}$ can be guaranteed. Of course, if the equation $\boldsymbol{B}(\boldsymbol{Y})=\boldsymbol{D}$ has several solutions, in our discussion we suppose that all $\boldsymbol{Y}^{(k)}$, $k=0, 1, \cdots$, belong to the same "branch" of solutions.

Therefore, the iteration (5.7) is convergent, if the conditions (1)—(5) hold. The proof of this theorem is completed.

We have discussed the stability of the block–double–sweep method for "incomplete" linear algebraic systems. The stability of the block–double–sweep method for "complete" linear algebraic systems can be discussed in the same way. For example, by using certain results of Theorem 6.1 and the method of proving Theorem 6.1, the following result can be immediately obtained.

Theorem 6.2 If the conditions (1)—(5) hold, then the process of computing $\mu_0^{(m)}$, $\boldsymbol{\mu}^{(m)}$ of (5.14)′ for $m=1, 2, \cdots, M$ successively, and the process of computing $\boldsymbol{U}_m$ for $m=M-1, M-2, \cdots, 0$ successively by using (5.13) are stable.

In order to prove this theorem, we need to notice only the following fact. The relation (5.14)′

$$\tilde{\mu}_0^{(m+1)}\boldsymbol{U}_{m+1}=\tilde{\boldsymbol{\mu}}^{(m+1)}$$

can be obtained from the relation (5.4)′, the equations (5.1b) with $m=0$ and (5.11), i.e., from

$$\begin{cases} \mu_0^{(m+1)}\boldsymbol{U}_{m+1}+\mu_1^{(m+1)}\boldsymbol{U}_0=\boldsymbol{\mu}^{(m+1)}, \\ \nu_0^{(m+1)}\boldsymbol{U}_{m+1}+\nu_1^{(m+1)}\boldsymbol{U}_0=\boldsymbol{\nu}^{(m+1)}, \end{cases} \tag{5.4$'$}$$

$$\overline{G}_2^{(0)}\boldsymbol{U}_0=\boldsymbol{F}_2^{(0)}, \tag{5.1b}$$

$$\tilde{\mu}_0^{(0)}\boldsymbol{U}_0=\tilde{\boldsymbol{\mu}}^{(0)}. \tag{5.11}$$

Hence, there is an invertible matrix α_{m+1} such that

$$\mu_0^{(m+1)}-\mu_1^{(m+1)}\begin{pmatrix} \tilde{\mu}_0^{(0)} \\ \overline{G}_2^{(0)} \\ \nu_1^{(m+1)} \end{pmatrix}^{-1}\begin{pmatrix} O \\ O \\ \nu_0^{(m+1)} \end{pmatrix}=\alpha_{m+1}\tilde{\mu}_0^{(m+1)},$$

$$\boldsymbol{\mu}^{(m+1)}-\mu_1^{(m+1)}\begin{pmatrix} \tilde{\mu}_0^{(0)} \\ \overline{G}_2^{(0)} \\ \nu_1^{(m+1)} \end{pmatrix}^{-1}\begin{pmatrix} \tilde{\boldsymbol{\mu}}^{(0)} \\ \boldsymbol{F}_2^{(0)} \\ \boldsymbol{\nu}^{(m+1)} \end{pmatrix}=\alpha_{m+1}\tilde{\boldsymbol{\mu}}^{(m+1)}.$$

In order to distinguish $\mu_0^{(m)}$, $\boldsymbol{\mu}^{(m)}$ in the two procedures, we shall change $\mu_0^{(m)}$, $\boldsymbol{\mu}^{(m)}$ in (5.11) and (5.14)′ into $\tilde{\mu}_0^{(m)}$, $\tilde{\boldsymbol{\mu}}^{(m)}$.

By using the above fact, this theorem can be proved, but to save space we shall not give the proof here.

Appendix 1
Stability of Difference Schemes for Pure–Initial–Value Problems with Variable Coefficients[1)]

Introduction

Lax et al.[1]–[4], and Kreiss[5] have discussed the stability of difference schemes for pure–initial–value problems with variable coefficients, and have developed some theorems. However, concerning this subject, there still remain problems to be solved.

This paper discusses the stability of difference schemes for hyperbolic

1) This paper is an English translation of the paper in "Mathematicae Numericae Sinica, 1978, No. 1, 33—43".

systems with two independent variables. For any explicit and implicit horizontal–three–point schemes and for several horizontal–multi–point explicit schemes, in which the Rusanov[6] third–order scheme, and the Burstein–Mirin[7] third–order scheme are included (i.e., for almost all schemes[6]–[15] being used in practical work), we present sufficient conditions of stability of schemes with variable coefficients. These are: (i) the von Neumann condition (4); (ii) condition (5), which guarantees the difference equations to be well–conditioned, and which is automatically fulfilled for explicit schemes; (iii) the condition that the coefficients of the schemes are smooth functions of x and t. Conditions (4) and (5) are also necessary for schemes with constant coefficients. We do not require that the schemes be dissipative, nor do we require that the operators be symmetric. These conditions can be applied conveniently to both explicit and implicit schemes.

1. *General Results*

We consider the following initial–value problem of hyperbolic systems:

$$\begin{cases} \dfrac{\partial \boldsymbol{U}}{\partial t} + A(x, t)\dfrac{\partial \boldsymbol{U}}{\partial x} = 0, \\ \boldsymbol{U}(x, 0) = \boldsymbol{f}(x), \end{cases} \tag{1}$$

where $\boldsymbol{U}(x, t)$ and $\boldsymbol{f}(x)$ are N–dimensional vectors, and $A(x, t)$ is an $N\times N$–matrix. $A(x, t)$ has N real eigenvalues and N linearly independent eigenvectors, i.e., there is a matrix G such that

$$A = G^{-1}\Lambda G, \tag{2}$$

where $\Lambda(x, t)$ is a real diagonal matrix. We discuss the following horizontal $(H+1)$–point difference scheme:

$$\sum_{h=0}^{H} R_{h,m}^{k}(\Delta)\,\boldsymbol{U}_{m+h}^{k+1} = \sum_{h=0}^{H} S_{h,m}^{k}(\Delta)\,\boldsymbol{U}_{m+h}^{k}, \tag{3}$$

where we adopt the following notation:

$$\boldsymbol{U}_m^k = \boldsymbol{U}(m\Delta x, k\Delta t);$$

$$R_{h,m}^k(\Delta) = R_h(m\Delta x, k\Delta t, \Delta x, \Delta t);$$

$$S_{h,m}^k(\Delta) = S_h(m\Delta x, k\Delta t, \Delta x, \Delta t);$$

$R_{h,m}^k(\Delta)$ and $S_{h,m}^k(\Delta)$ are $N\times N$–matrices.

We also introduce other notation as follows: $R_{h,m}^k$, $S_{h,m}^k$ denote $R_{h,m}^k(\Delta)\,|_{\Delta x=\Delta t=0}$, $S_{h,m}^k(\Delta)\,|_{\Delta x=\Delta t=0}$ respectively; $\overline{F}$ denotes $(F_0, F_1, \cdots, F_H)$, where F_h is an $N\times N$–matrix, $h=0, 1, \cdots, H$; $\overline{F}^*$ is the conjugate transposition of $\overline{F}$; and $F(\theta) \equiv \sum_{h=0}^{H} F_h e^{ih\theta}$. Moreover, the superscript k and subscript m will be omitted, if no confusion results. We will discuss problems in L_2 space and suppose $\Delta t/\Delta x$ is bounded.

First, we give a lemma.

Lemma 1 If $\sum_l \left(\sum_{h=0}^{H} D_{h,l} e^{ih\theta}\right)^* E_l \left(\sum_{h=0}^{H} D_{h,l} e^{ih\theta}\right) = 0$, then every sum[1)] of all "N-matrix-elements" located on a diagonal line of $Q = \sum_l \overline{D}_l^* E_l \overline{D}_l$ is a null-matrix, where E_l and $D_{h,l}$ are $N\times N$-matrices, and where the matrix Q is an $N(H+1)\times N(H+1)$-matrix.

Proof From

$$\sum_l \left(\sum_{h=0}^{H} D_{h,l} e^{ih\theta}\right)^* E_l \left(\sum_{h=0}^{H} D_{h,l} e^{ih\theta}\right)$$

$$= \sum_l \sum_{h=-H}^{H} \sum_j D_{j,l}^* E_l D_{j+h,l} e^{ih\theta} = \sum_{h=-H}^{H} \sum_l \sum_j D_{j,l}^* E_l D_{j+h,l} e^{ih\theta} = 0,$$

we obtain

$$\sum_l \sum_j D_{j,l}^* E_l D_{j+h,l} = 0, \quad h = 0, \pm 1, \pm 2, \cdots, \pm H,$$

where j satisfies $0 \leqslant j \leqslant H$ and $0 \leqslant j+h \leqslant H$. However, $\sum_l \sum_j D_{j,l}^* E_l D_{j+h,l}$ is the sum of all "N-matrix-elements" located on "the h-th diagonal line" of $\sum_l \overline{D}_l^* E_l \overline{D}_l$. Therefore, we obtain the desired result.

Then, we give two theorems.

Theorem 1 If for any m, k, and θ, the following conditions are fulfilled:

(i) there exist two invertible matrices N_m^k and G_m^k such that

$$N_m^k R_m^k(\theta) G_m^{k-1} = \Lambda_{1,m}^k(\theta), \quad N_m^k S_m^k(\theta) G_m^{k-1} = \Lambda_{2,m}^k(\theta),$$

and

$$\Lambda_{1,m}^{k*}(\theta)\Lambda_{1,m}^k(\theta) - \Lambda_{2,m}^{k*}(\theta)\Lambda_{2,m}^k(\theta) \geqslant 0, \tag{4}$$

where $\Lambda_{1,m}^k(\theta)$ and $\Lambda_{2,m}^k(\theta)$ are diagonal matrices, and $\geqslant 0$ denotes that the matrix on its left side is nonnegative definite;

(ii)
$$\Lambda_{1,m}^{k*}(\theta)\Lambda_{1,m}^k(\theta) - c_1 I \geqslant 0, \tag{5}$$
where c_1 is a positive constant and I is an $N\times N$-unit matrix;

(iii) R_h, S_h, N, N^{-1}, G and G^{-1} satisfy the Lipschitz condition with respect to x and t, i.e., for every element f of these matrices, the following relations

$$|f_{m+1}^k - f_m^k| \leqslant c_2 \Delta x, \quad |f_m^{k+1} - f_m^k| \leqslant c_2 \Delta t$$

---

1) Let $A_{i,j}$ be an $N\times N$-matrix. We call

$$\sum_{0\leqslant\{i,i+h\}\leqslant H} A_{i,i+h}$$

the sum of all "N-matrix-elements" located on the h-th diagonal line of

$$\begin{pmatrix} A_{00}, A_{01}, \cdots, A_{0H} \\ A_{10}, A_{11}, \cdots, A_{1H} \\ \cdots\cdots\cdots\cdots\cdots\cdots \\ A_{H0}, A_{H1}, \cdots, A_{HH} \end{pmatrix},$$

where $0\leqslant\{i, i+h\}\leqslant H$ denotes the fact that $0\leqslant i\leqslant H$ and $0\leqslant i+h\leqslant H$.

are fulfilled, and these elements are bounded;

(iv) among R_h, S_h and $R_h(\Delta)$, $S_h(\Delta)$, there are the relations

$$\|R_h(\Delta)-R_h\|\leqslant c_2\Delta x,$$
$$\|S_h(\Delta)-S_h\|\leqslant c_2\Delta x;$$

(v)
$$V(\theta)\equiv R^*(\theta)N^*NR(\theta)-S^*(\theta)N^*NS(\theta)$$
$$=G^*\Lambda_1^*(\theta)\Lambda_1(\theta)G-G^*\Lambda_2^*(\theta)\Lambda_2(\theta)G$$

can be rewritten as

$$V(\theta)=\sum_l D_l^*(\theta)M_lD_l(\theta), \tag{6}$$

where M_l and $D_{h,l}$ are $N\times N$-matrices, satisfying the Lipschitz condition, and M_l is a nonnegative definite matrix,

then the scheme (3) is stable in the space L_2, i. e., there exists a constant c_3 such that

$$\|U^k\|^2\leqslant c_3\|U^0\|^2,\quad 0\leqslant k\Delta t\leqslant T_1,$$

where T_1 is a bounded constant and

$$\|U^k\|^2=\sum_m U^*(m\Delta x,\ k\Delta t)\,U(m\Delta x,\ k\Delta t)\,\Delta x.$$

Proof We use the energy method. We take

$$T^k=\sum_m\left(N_m^k\sum_{h=0}^{H}R_{h,m}^{k-1}(\Delta)\,U_{m+h}^k\right)^*\left(N_m^k\sum_{h=0}^{H}R_{h,m}^{k-1}(\Delta)\,U_{m+h}^k\right)\Delta x$$
$$=\sum_m(\overline{U},\ \overline{R}^*N^*N\overline{R}\overline{U})_m^k\Delta x+O(\Delta x)\,\|U^k\|^2,$$

as the energy sum, where $\overline{U}_m^*=(U_m^*,\ U_{m+1}^*,\ \cdots,\ U_{m+H}^*)$.

First, we prove that from conditions (i), (iii)—(v) we can obtain the inequality

$$T^{k+1}-T^k\leqslant c_4\Delta x\|U^k\|^2, \tag{7}$$

where c_4 is a constant. In fact, by using (3) and conditions (iii), (iv), we have

$$T^{k+1}-T^k=\sum_m\Bigg[\left(N_m^k\sum_{h=0}^{H}S_{h,m}^kU_{m+h}^k\right)^*\left(N_m^k\sum_{h=0}^{H}S_{h,m}^kU_{m+h}^k\right)$$
$$-\left(N_m^k\sum_{h=0}^{H}R_{h,m}^kU_{m+h}^k\right)^*\left(N_m^k\sum_{h=0}^{H}R_{h,m}^kU_{m+h}^k\right)\Bigg]\Delta x$$
$$+O(\Delta x)\,\|U^k\|^2$$
$$=\sum_m(\overline{U},\ (\overline{S}^*N^*N\overline{S}-\overline{R}^*N^*N\overline{R})\overline{U})_m^k\Delta x+O(\Delta x)\,\|U^k\|^2.$$

According to conditions (iii) and (v) and by using Lemma 1, we know that every sum of all "N-matrix-elements" located on a "diagonal line" of the matrix

$$Q_1=\overline{S}^*N^*N\overline{S}-\overline{R}^*N^*N\overline{R}+\sum_l\overline{D}_l^*M_l\overline{D}_l$$

is a null-matrix of order N, and that every element of Q_1 satisfies the Lipschitz condition with respect to x. Thus, from the fact that $\sum_l \overline{D}_l^* M_l \overline{D}_l$ is nonnegative definite, we can obtain inequality (7) immediately.

Then, from conditions (ii) and (iii), we derive

$$c_5 \| U^k \|^2 \leqslant T^k \leqslant c_5^{-1} \| U^k \|^2, \tag{8}$$

where c_5 is a positive constant. It is easy to obtain the right half of the inequality. In the following we derive the left half. By using condition (ii) and the Fejér–Reisz theorem[17], we know that there exists a diagonal matrix-polynomial $J(\theta) = \sum_{h=0}^{H} J_h e^{ih\theta}$ such that

$$R^*(\theta) N^* N R(\theta) - G^* \frac{c_1}{2} G$$

$$= G^* \Lambda_1^*(\theta) \Lambda_1(\theta) G - G^* \frac{c_1}{2} G = G^* J^*(\theta) J(\theta) G.$$

Moreover, because $J^*(\theta) J(\theta) \geqslant (c_1/2) I > 0$, from the fact that the elements of $\Lambda_1(\theta)$ satisfy the Lipschitz condition, we can see that the elements of $J(\theta)$ also satisfy the Lipschitz condition [2]. Thus, according to Lemma 1, every sum of all "N-matrix-elements" located on a "diagonal line" of the matrix

$$Q_2 = \overline{R}^* N^* N \overline{R} - \overline{G}^* \frac{c_1}{2} \overline{G} - \overline{J}_G^* \overline{J}_G$$

is a null-matrix of order N, where

$$\overline{G} = (G, \underbrace{0, \cdots, 0}_{H}), \quad 0 \text{ being an } N \times N\text{-null-matrix},$$

$$\overline{J}_G = (J_0 G, J_1 G, \cdots, J_H G).$$

In addition,

$$\left(\overline{U}, \overline{G}^* \frac{c_1}{2} \overline{G} \overline{U} \right)_m = \left(U, G^* \frac{c_1}{2} G U \right)_m \geqslant \frac{c_1}{2 \| G_m^{-1} \|^2} (U, U)_m.$$

Therefore, from the fact that $\overline{J}_G^* \overline{J}_G$ is nonnegative definite, we can derive the left half of (8) immediately. Obviously, the inequality

$$T^{k+1} - T^k \leqslant c_4 \Delta x \| U^k \|^2 \leqslant \frac{c_4}{c_5} \Delta x T^k$$

follows from (7) and (8) immediately. Furthermore, we can obtain the following inequalities:

$$T^{k+1} \leqslant \left(1 + \frac{c_4}{c_5} \Delta x \right) T^k \leqslant \left(1 + \frac{c_4}{c_5} \Delta x \right)^{k+1} T^0,$$

and

$$\| U^{k+1} \|^2 \leqslant c_5^{-2} \left(1 + \frac{c_4}{c_5} \Delta x \right)^{k+1} \| U^0 \|^2.$$

Therefore, when $\Delta x/\Delta t$ is bounded, there is a positive constant c_3 such that

$$\|U^k\|^2 \leqslant c_3 \|U^0\|^2,$$

i.e., the conclusion of the theorem is proved.

Definition Let $\widetilde{U}^k = (\cdots, U_0^{k*}, U_1^{k*}, \cdots, U_m^{k*}, \cdots)^*$ be the solution of the system of equations

$$B\widetilde{U}^k = C.$$

If there exists a positive constant such that the inequality

$$\|U^k\|^2 \leqslant c_6 \|C\|^2$$

is fulfilled for any solution $\widetilde{U}^k$, then we say that the system is well-conditioned in L_2, where $\|C\|^2 = \sum_j |c_j|^2 \Delta x$ and c_j is an element of C. Obviously, if the left half of (8) is fulfilled and $\|N_m^k\|$ is bounded, then the difference equations in (3) are well-conditioned. Therefore, we can obtain the following theorem:

Theorem 2 If for any m and θ, there are two invertible matrices N and G such that $NR(\theta)G^{-1} = \Lambda_1(\theta)$ and

$$\Lambda_1^*(\theta)\Lambda_1(\theta) - c_1 I \geqslant 0, \tag{5}$$

and if conditions (iii) and (iv) of Theorem 2 are fulfilled, then the difference equations in (3) are well-conditioned. In the case with constant coefficients, if there exist two invertible matrices N and G such that all NR_hG^{-1} are diagonal matrices, then condition (5) is also necessary.

Proof From the proof of Theorem 1, we know that (8) also is fulfilled under the conditions of this theorem. Therefore the first part of the conclusion is proved.

If (5) is not fulfilled, then there exists a number θ^* such that the j-th element of $\Lambda_1(\theta) = NR(\theta)G^{-1}$ is equal to zero when $\theta = \theta^*$. Therefore, in the case of constant coefficients, the equations

$$N\sum_{h=0}^{H} R_h U_{m+h}^k = \sum_{h=0}^{H} NR_h U_{m+h}^k = 0, \quad m = \cdots, 0, 1, 2, \cdots,$$

have a nontrivial solution:

$$U_{m+h}^k = G^{-1}\begin{pmatrix} 0 \\ \vdots \\ 0 \\ e^{i(m+h)\theta^*} \\ 0 \\ \vdots \\ 0 \end{pmatrix} \text{—— the } j\text{-th element.}$$

Therefore, the equations are not well conditioned, i.e., the second part of the theorem is proved.

From Theorem 2, we know that the Fourier method can be used for studying the properties of difference equations.

2. *Applications*

Before we apply the results in Section 1 to some concrete schemes, we point out the following fact: for quite a few difference schemes which can be written in the form (3), there exists an invertible matrix $\widetilde{H}^k_m$ such that $R^k_{h,m}$ and $S^k_{h,m}$ can be rewitten as

$$\begin{cases} R^k_{h,m} = [\widetilde{H} \sum_j c_{j,h} A^j]^k_m = [\widetilde{H} G^{-1} (\sum_j c_{j,h} A^j) G]^k_m, \\ S^k_{h,m} = [\widetilde{H} \sum_j b_{j,h} A^j]^k_m = [\widetilde{H} G^{-1} (\sum_j b_{j,h} A^j) G]^k_m, \end{cases} \tag{9}$$

where $c_{j,h}$ and $b_{j,h}$ are scalars. The C-I-R scheme[8], the Lax scheme[9], the Richtmyer two step scheme[14], the Wendroff scheme[10], and the Thomée scheme[11], $\cdots$, have this feature, and $\widetilde{H}^k_m = I$ or G^k_m. For these schemes, there are the following results:

Lemma 2 If the coefficients of scheme (3) satisfy (9), then there exists N such that

$$R^*(\theta) N^* N R(\theta) = G^* \Big(\sum_{h=0}^{H} d_h \sin^{2h} \frac{\theta}{2} \Big) G, \tag{10}$$

$$R^*(\theta) N^* N R(\theta) - S^*(\theta) N^* N S(\theta) = G^* \Big(\sum_{h=0}^{H} a_h \sin^{2h} \frac{\theta}{2} \Big) G, \tag{11}$$

where a_h and d_h are real diagonal matrices (for their concrete expressions see (14) and (13)). If condition (iii) of Theorem 1 is also fulfilled, then every element of a_h satisfies the Lipschitz condition with respect to x and t.

Proof Let $N = G\widetilde{H}^{-1}$. From (9), we have

$$\begin{aligned} G^{-1*} R^*(\theta) N^* N R(\theta) G^{-1} &= \Big(\sum_{h=0}^{H} \sum_j c_{j,h} A^j e^{ih\theta} \Big)^* \Big(\sum_{h=0}^{H} \sum_j c_{j,h} A^j e^{ih\theta} \Big) \\ &= \sum_{h=-H}^{H} \widetilde{d}_h e^{ih\theta}, \end{aligned}$$

where $\widetilde{d}_h = \sum_{0 \le (i, i+h) \le H} (\sum_j c_{j,i} A^j)(\sum_j c_{j,i+h} A^j)$. Obviously, $\widetilde{d}_h$ is a real diagonal matrix, and $\widetilde{d}_h = \widetilde{d}_{-h}$. Moreover, for any integer k, there is the expression

$$\cos k\theta = \sum_{j=0}^{k} e_{k,j} \sin^{2j} \frac{\theta}{2},$$

where any $e_{k,j}$ is a real constant. (For $k=1, 2, 3, 4$, the concrete expressions are as follows:

$$\begin{cases} \cos \theta = 1 - 2 \sin^2 \frac{\theta}{2}, \\ \cos 2\theta = 1 - 8 \sin^2 \frac{\theta}{2} + 8 \sin^4 \frac{\theta}{2}, \\ \cos 3\theta = 1 - 18 \sin^2 \frac{\theta}{2} + 48 \sin^4 \frac{\theta}{2} - 32 \sin^6 \frac{\theta}{2}, \\ \cos 4\theta = 1 - 32 \sin^2 \frac{\theta}{2} + 160 \sin^4 \frac{\theta}{2} - 256 \sin^6 \frac{\theta}{2} + 128 \sin^8 \frac{\theta}{2}). \end{cases} \tag{12}$$

Therefore, we have (10):

$$R^*(\theta)N^*NR(\theta)=G^*\Big(\tilde{d}_0+2\sum_{h=1}^{H}\tilde{d}_h\cos h\theta\Big)G=G^*\Big(\sum_{h=0}^{H}d_h\sin^{2h}\frac{\theta}{2}\Big)G,$$

where

$$\begin{cases} d_0=\tilde{d}_0+2\sum\limits_{k=1}^{H}e_{k,0}\tilde{d}_k=\Big(\sum\limits_{h=0}^{H}\sum\limits_{j}c_{j,h}A^j\Big)^2, \quad h=1,2,\cdots H, \\ d_h=2\sum\limits_{k=h}^{H}e_{k,h}\tilde{d}_k=2\sum\limits_{k=h}^{H}e_{k,h}\sum\limits_{0<i<i+k<H}(\sum\limits_{j}c_{j,i}A^j)(\sum\limits_{j}c_{j,i+k}A^j). \end{cases} \tag{13}$$

We can also obtain a similar expression for $S^*(\theta)N^*NS(\theta)$ in which the first term is $\Big(\sum\limits_{h=0}^{H}\sum\limits_{j}b_{j,h}A^j\Big)^2$. Moreover,

$$\sum_{h=0}^{H}\sum_{j}c_{j,h}A^j=\sum_{h=0}^{H}\sum_{j}b_{j,h}A^j.$$

Therefore, we obtain the equality (11), and a_h has the following expression:

$$\begin{aligned} a_h=2\sum_{k=h}^{H}e_{k,h}\sum_{0<i<i+k<H}&[(\sum_{j}c_{j,i}A^j)(\sum_{j}c_{j,i+k}A^j) \\ &-(\sum_{j}b_{j,i}A^j)(\sum_{j}b_{j,i+k}A^j)]. \end{aligned} \tag{14}$$

Furthermore, it is obvious that every element of a_h satisfies the Lipschitz condition if condition (iii) is fulfilled. Thus, we have obtained all conclusions of this theorem.

Lemma 3 Let $f_0+f_1\sin^2\frac{\theta}{2}+f_2\sin^4\frac{\theta}{2}$ be a scalar quadratic polynomial in $\sin^2\frac{\theta}{2}$. Necessary and sufficient conditions for $f_0+f_1\sin^2\frac{\theta}{2}+f_2\sin^4\frac{\theta}{2}\geqslant 0$ are

$$f_0\geqslant 0, \quad 2f_0+f_1+2(f_0(f_0+f_1+f_2))^{1/2}\geqslant 0, \quad f_0+f_1+f_2\geqslant 0. \tag{15}$$

Proof According to the equality

$$\begin{aligned} &f_0+f_1\sin^2\frac{\theta}{2}+f_2\sin^4\frac{\theta}{2} \\ &=f_0\cos^4\frac{\theta}{2}+(2f_0+f_1)\sin^2\frac{\theta}{2}\cos^2\frac{\theta}{2}+(f_0+f_1+f_2)\sin^4\frac{\theta}{2} \\ &=\Big[f_0^{1/2}\cos^2\frac{\theta}{2}-(f_0+f_1+f_2)^{1/2}\sin^2\frac{\theta}{2}\Big]^2 \\ &\quad+[2f_0+f_1+2f_0^{1/2}(f_0+f_1+f_2)^{1/2}]\sin^2\frac{\theta}{2}\cos^2\frac{\theta}{2}, \end{aligned}$$

we know that if

$$f_0\geqslant 0, \quad 2f_0+f_1+2f_0^{1/2}(f_0+f_1+f_2)^{1/2}\geqslant 0, \quad \text{and} \quad f_0+f_1+f_2\geqslant 0,$$

then

$$f_0+f_1\sin^2\frac{\theta}{2}+f_2\sin^4\frac{\theta}{2}\geqslant 0,$$

i.e., (15) is a sufficient condition. Furthermore, if we let $\sin^2\frac{\theta}{2}\approx 0$, $\cos^2\frac{\theta}{2}\approx 0$ and $f_0^{1/2}\cos^2\frac{\theta}{2}-(f_0+f_1+f_2)^{1/2}\sin^2\frac{\theta}{2}=0$, and observe the right-hand side of the equality, then we can find that (15) is also a necessary condition.

From Lemma 2 we know that if the coefficients of a scheme satisfy equality (9), then conditions (i) and (ii) are reduced to

$$\sum_{h=1}^{H}a_h\sin^{2h}\frac{\theta}{2}\geqslant 0 \quad\text{and}\quad \sum_{h=0}^{H}d_h\sin^{2h}\frac{\theta}{2}-c_1I\geqslant 0$$

respectively. Furthermore, because a_h and d_h are real diagonal matrices, these inequalities can be reduced to some inequalities on scalar polynomials in $\sin^2\frac{\theta}{2}$. From Lemma 3 we know that if H is not too large, these conditions are further reduced to some inequalities on the coefficients of schemes $c_{j,i}$ and $b_{j,i}$, and the eigenvalues λ_n of A. In the following, we shall prove that the other conditions of Theorem 1 guarantee that condition (v) is fulfilled in a series of cases. Therefore, from Theorem 1, we shall obtain some stability criteria which are convenient in applications. Moreover, for constant coefficients, these conditions are necessary.

Theorem 3 For a horizontal three-point scheme with (9), if

(i) $$a_1\geqslant 0,\quad a_1+a_2\geqslant 0, \tag{16}$$

(ii) there exists $c_1>0$ such that

$$\begin{cases} d_0-c_1I\geqslant 0,\\ 2(d_0-c_1I)+d_1+2(d_0-c_1I)^{1/2}(d_0-c_1I+d_1+d_2)^{1/2}\geqslant 0,\\ d_0-c_1I+d_1+d_2\geqslant 0,\end{cases} \tag{17}$$

and if conditions (iii) and (iv) of Theorem 1 are fulfilled, then the scheme is stable.

Proof According to Lemmas 2 and 3, conditions (i) and (ii) here guarantee that conditions (i) and (ii) of Theorem 1 are fulfilled. Therefore, the proof of this theorem is reduced to proving that condition (v) of Theorem 1 is fulfilled. In fact, we have the equality

$$\begin{aligned}V(\theta)&=R^*(\theta)N^*NR(\theta)-S^*(\theta)N^*NS(\theta)=G^*\Big(\sum_{h=1}^{2}a_h\sin^{2h}\frac{\theta}{2}\Big)G\\ &=G^*\Big(a_1\sin^2\frac{\theta}{2}\cos^2\frac{\theta}{2}+(a_1+a_2)\sin^4\frac{\theta}{2}\Big)G\\ &=[(1-e^{i2\theta})G]^*\frac{a_1}{16}[(1-e^{i2\theta})G]\\ &\quad+[(1-2e^{i\theta}+e^{i2\theta})G]^*\frac{a_1+a_2}{16}[(1-2e^{i\theta}+e^{i2\theta})G].\end{aligned}$$

Moreover, condition (iii) guarantees that a_1 and a_2 satisfy the Lipschitz condition with respect to x and t, Therefore, condition (v) is fulfilled, i.e., the scheme is stable.

These schemes in [8]—[16] are horizontal three-point schemes, and satisfy equalities (9). Therefore, we can discuss their stability by using Theorem 3. We give the following corollaries.

Corollary 1 When the C-I-R scheme[8], the Lax scheme[9] and the Lax-Wendroff second-order scheme[13] (or the Richtmyer[14] and MacCormack[15] two-step L-W scheme) are applied to (1), if

(1) $$\|A\| \frac{\Delta t}{\Delta x} \leqslant 1, \tag{18}$$

i.e., for every element λ_n of A,

$$|\lambda_n| \frac{\Delta t}{\Delta x} \leqslant 1;$$

(2) $A(x, t)$ and $G(x, t)$ satisfy the Lipschitz condition, i.e., for every element f of A and G,

$$|f(x+\Delta x, t+\Delta t)-f(x, t)| \leqslant c_2(|\Delta x|+|\Delta t|), \tag{19}$$

and

$$|G|^2 \geqslant \varepsilon \geqslant 0, \tag{20}$$

where ε is a positive constant.

then the schemes are stable.

Corollary 2 When the Wendroff[10] and Thomée[11] scheme is applied to (1), if

$$\|A\| \geqslant \varepsilon > 0, \tag{21}$$

and (19) and (20) are fulfilled, then the scheme is stable.

Corollary 3 When the Keller-Thomée[12] scheme and the scheme

$$U_m^{k+1}-U_m^k+\frac{\Delta t}{\Delta x} A_m^{k+1/2} \cdot \frac{1}{4}(U_{m+1}^{k+1}-U_{m-1}^{k+1}+U_{m+1}^k-U_{m-1}^k)=0,$$

which is similar to the Crank-Nicolson[16] scheme, are applied to (1), if (19) and (20) are fulfilled, then the schemes are stable.

In order to prove these corollaries, we only need to prove that (16) and (17) are fulfilled. This is easy. In fact, from the coefficients $c_{j,h}$ and $b_{j,h}$ of these schemes, we may easily obtain a_1, a_2, d_0, d_1 and d_2 by using (12), (13), and (14), and prove immediately that (16) and (17) are fulfilled. We shall not give the proof for all schemes. In the following, taking the L-W scheme and the scheme similar to the Crank-Nicolson scheme as examples, we shall explain the procedure of the proof. For the L-W scheme, we have

$$\sum_j c_{j,0}A^j=0, \quad \sum_j c_{j,1}A^j=I, \quad \sum_j c_{j,2}A^j=0,$$

$$\sum_j b_{j,0}A^j=\frac{1}{2}\frac{\Delta t}{\Delta x}A+\frac{1}{2}\left(\frac{\Delta t}{\Delta x}A\right)^2, \quad \sum_j b_{j,1}A^j=I-\left(\frac{\Delta t}{\Delta x}A\right)^2,$$

$$\sum_j b_{j,2}A^j=-\frac{1}{2}\frac{\Delta t}{\Delta x}A+\frac{1}{2}\left(\frac{\Delta t}{\Delta x}A\right)^2.$$

According to (12),

$$e_{1,0}=1, \quad e_{1,1}=-2,$$

$$e_{2,0}=1, \quad e_{2,1}=-8, \quad e_{2,2}=8.$$

Therefore, using (13) and (14), we obtain

$$d_0=I, \quad d_1=0, \quad d_2=0,$$

$$a_1=0, \quad a_2=4\left[I-\left(\frac{\Delta t}{\Delta x}A\right)^2\right]\left(\frac{\Delta t}{\Delta x}A\right)^2,$$

i.e., (16) is reduced to (18), and (17) is always fulfilled. Thus, it is easy to prove that the conditions of Theorem 3 are fulfilled if (18)—(20) are fulfilled. This is the desired conclusion. For the scheme similar to the Crank–Nicolson scheme[16],

$$\sum_j c_{j,0}A^j=-\frac{1}{4}\frac{\Delta t}{\Delta x}A, \quad \sum_j c_{j,1}A^j=I, \quad \sum_j c_{j,2}A^j=\frac{1}{4}\frac{\Delta t}{\Delta x}A,$$

$$\sum_j b_{j,0}A^j=\frac{1}{4}\frac{\Delta t}{\Delta x}A, \quad \sum_j b_{j,1}A^j=I, \quad \sum_j b_{j,2}A^j=-\frac{1}{4}\frac{\Delta t}{\Delta x}A.$$

Moreover, by using (13) and (14), we obtain

$$d_0=I, \quad d_1=\left(\frac{\Delta t}{\Delta x}A\right)^2, \quad d_2=-\left(\frac{\Delta t}{\Delta x}A\right)^2,$$

$$a_1=0, \quad a_2=0.$$

Thus, (16) and (17) are always fulfilled, and it is easy to obtain the conclusion we want.

Lemma 4 For a horizontal five-point explicit scheme with (9), if the order of accuracy of the scheme is greater than 2, then $a_1=0$.

Proof The horizontal five-point explicit scheme can be rewritten in the following form:

$$U^{k+1}_{m+2+j}=S_0(\Delta)U^k_m+S_1(\Delta)U^k_{m+1}+S_2(\Delta)U^k_{m+2}+S_3(\Delta)U^k_{m+3}+S_4(\Delta)U^k_{m+4},$$

where j is equal to any one among -2, -1, 0, 1, and 2. Because the scheme is at least of the second order accuracy, we have the following relations:

$$\begin{cases} S_0+S_1+S_2+S_3+S_4=I, \\ -2S_0-S_1+S_3+2S_4=-\left(\frac{\Delta t}{\Delta x}A-jI\right), \\ 4S_0+S_1+S_3+4S_4=\left(\frac{\Delta t}{\Delta x}A-jI\right)^2. \end{cases}$$

These expressions can be rewritten as

$$\begin{cases} S_1=\frac{1}{2}\left[\left(\frac{\Delta t}{\Delta x}A-jI\right)^2+\left(\frac{\Delta t}{\Delta x}A-jI\right)\right]-3S_0-S_4, \\ S_2=I-\left(\frac{\Delta t}{\Delta x}A-jI\right)^2+3S_0+3S_4, \\ S_3=\frac{1}{2}\left[\left(\frac{\Delta t}{\Delta x}A-jI\right)^2-\left(\frac{\Delta t}{\Delta x}A-jI\right)\right]-S_0-3S_4. \end{cases} \tag{22}$$

In addition, according to (14) and (12), in the case here,

$$a_1=-2[-2(\tilde{S}_0\tilde{S}_1+\tilde{S}_1\tilde{S}_2+\tilde{S}_2\tilde{S}_3+\tilde{S}_3\tilde{S}_4)-8(\tilde{S}_0\tilde{S}_2+\tilde{S}_1\tilde{S}_3+\tilde{S}_2\tilde{S}_4)$$
$$-18(\tilde{S}_0\tilde{S}_3+\tilde{S}_1\tilde{S}_4)-32\tilde{S}_0\tilde{S}_4],\quad \tilde{S}_k=GS_kG^{-1}.$$

By using these expressions, it is easy to prove the conclusion of this lemma. In fact, putting (22) into the expression for a_1, we immediately know $a_1=0$.

From this lemma and the other results mentioned above, we obtain the following result:

Theorem 4 When the Rusanov[6] and the Burstein–Mirin[7] third-order schemes are applied to (1), if (18)—(20) and

$$4\left(\lambda_n\frac{\Delta t}{\Delta x}\right)^2-\left(\lambda_n\frac{\Delta t}{\Delta x}\right)^4\leqslant\omega\leqslant 3 \tag{23}$$

are fulfilled, then the schemes are stable. (ω is a parameter of the Rusanov[6] scheme.)

Proof The schemes are third order, horizontal five-point explicit schemes, and (9) is fulfilled, so $a_1=0$. Therefore, according to Lemma 2,

$$V(\theta)=G^*\left(a_2\sin^4\frac{\theta}{2}+a_3\sin^6\frac{\theta}{2}+a_4\sin^8\frac{\theta}{2}\right)G$$
$$=G^*\left\{\frac{1}{256}|1-2e^{i\theta}+e^{i2\theta}|^2[a_2|1+2e^{i\theta}+e^{i2\theta}|^2\right.$$
$$\left.+(2a_2+a_3)|1-e^{i2\theta}|^2+(a_2+a_3+a_4)|1-2e^{i\theta}+e^{i2\theta}|^2]\right\}G.$$

Moreover, it can be verified that for the schemes, only when

$$a_2\geqslant 0,\quad 2a_2+a_3\geqslant 0,\quad a_2+a_3+a_4\geqslant 0,$$

condition (4) can be fulfilled. Therefore, when conditions (i), (iii) and (iv) of Theorem 1 are fulfilled, condition (v) of Theorem 1 is also fulfilled, i.e., in order to apply Theorem 1, we only need to prove that (i)—(iv) are fulfilled. Obviously, if (19) and (20) are fulfilled, then (iii) and (iv) are fulfilled. For explicit schemes, (ii) is always fulfilled. Moreover (18) and (23) guarantee that (i) is fulfilled. Therefore, Theorem 1 can be applied, i.e., the schemes are stable.

Theorem 5 For a second-order or third-order explicit scheme with (9) and $H=3$, if

(i) $$a_2\geqslant 0,\quad a_2+a_3\geqslant 0, \tag{24}$$

and if conditions (iii) and (iv) of Theorem 1 are fulfilled, then the scheme is stable.

Proof By using Lemma 4 and the method for proving Theorem 3, the result can be obtained. The concrete proof is omitted.

3. *Conclusions*

From the above results, we see that in a series of cases, conditions (i)—(iv) of Theorem 1 guarantee that condition (v) of Theorem 1 is fulfilled, i.e., if (i)—(iv) of Theorem 1 are fulfilled, then the schemes for pure-initial-value problems with variable coefficients are stable. Moreover, we also see that in a series of cases, if the von Neumann condition (4), condition (5) guaranteeing that the difference equations are well conditioned, and condition (iv) are fulfilled, and if G and A are smooth functions, then the schemes are stable. Condition (5) is always fulfilled for explicit schemes. In general, if A and G are smooth, then condition (iv) is fulfilled. Thus for a series of explicit schemes, if the von Neumann condition is fulfilled, and if G and A are smooth, then the schemes with variable coefficients are stable; for a series of implicit schemes, if these two conditions and condition (5) are fulfilled, then the schemes are also stable. Therefore, for a series of explicit schemes, the conditions of the papers [2] and [5] can be weakened. In [4], those results of [2] and [5] have been improved. The paper [4] also has discussed the stability of implicit schemes. However, the stability criterion in this paper seems to be more convenient. Moreover, this paper gives certain conditions which do not contain functions of θ, for example, conditions (16), (17) and (24). Therefore the results of this paper are more useful in practical applications.

References

[1] P. D. Lax, The scope of the energy method, *Bull. of the Amer. Math. Soc.*, **66** (1960), 1, 32—35.

[2] P. D. Lax, On the stability of difference approximations to solutions of hyperbolic equations with variable coefficients, *Comm. Pure Appl. Math.*, **14** (1961), 3, 497—520.

[3] P. D. Lax and B. Wendroff, On the stability of difference schemes, *Comm. Pure Appl. Math.*, **15** (1962), 4, 363—371.

[4] P. D. Lax and L. Nirenberg, On stability for difference schemes: A sharp form of Gårding's inequality, *Comm. Pure Appl. Math.*, **19** (1966), 4, 473—492.

[5] H.-O. Kreiss, On difference approximations of the dissipative type for hyperbolic differential equations, *Comm. Pure Appl. Math.*, **17** (1964), 3, 335—353.

[6] V. V. Rusanov, Third-order difference schemes for "through"-calculation of discontinuous solutions, D. A. N. USSR, **180** (1968), 6, 1303—1305 (in Russian).

[7] S. Z. Burstein and A. A. Mirin, Third order difference methods for hyperbolic equations, *J. Comp. Phys.*, **5** (1970), 3, 547—571.

[8] R. Courant, E. Isaacson and M. Rees, On the solution of nonlinear hyperbolic differential equations by finite differences, *Comm. Pure Appl. Math.*, **5** (1952), 3, 243—255.

[9] P. D. Lax, Weak solution of nonlinear hyperbolic equations and their numerical computation, *Comm. Pure Appl. Math.*, **7** (1954), 1, 159—193.

[10] B. Wendroff, On centered difference equations for hyperbolic systems, *J. Soc. Indust. Appl. Math.*, **8** (1960), 3, 549—555.

[11] V. Thomée, A stable difference scheme for the mixed boundary problem for a hyperbolic first order system in two dimensions. *J. Soc. Indust. Appl. Math.*, **10** (1962), 2, 229—245.

[12] H. B. Keller and V. Thomée, Unconditionally stable difference methods for mixed problems for quasi-linear hyperbolic systems in two dimensions, *Comm. Pure Appl. Math.*, **15** (1962), 1, 63—73.

[13] P. D. Lax and B. Wendroff, Systems of conservation laws, *Comm. Pure Appl. Math.*, **13**

(1960), 2, 217—237.

[14] R. D. Richtmyer, A survey of difference methods for nonsteady fluid dynamics, NCAR TN63-2, 1962.

[15] R. W. MacCormack, The effects of viscosity in hypervelocity impact cratering, AIAA Paper No. 69-354, 1969.

[16] J. Crank and P. Nicolson, A practical method for numerical evaluation of solutions of partial differential equations of the heat-conduction type, Proc. Cambridge Philos. Soc., **43** (1947), Part I, 50—67.

[17] N. I. Ahiezer, Lectures on theory of approximation, Gostehizdat, Moscow, 1947 (in Russian).

Appendix 2

A Block-Double-Sweep Method for "Incomplete" Linear Algebraic Systems and Its Stability[1)]

Abstract

The block-double-sweep methods for linear algebraic systems have been discussed by many authors. However, sometimes we may meet a system which consists of many linear and a few nonlinear equations, and in which the number of unknowns in the linear equations is greater than the number of linear equations. In this situation, these linear equations constitute an "incomplete" system. This paper presents a block-double-sweep method which can be used to solve the "incomplete" systems of linear equations and which is stable under very weak conditions. In addition, this paper points out that for linear tridiagonal systems we can obtain the stability conditions which are weaker than the conditions obtained previously.

Introduction

Rusanov's paper [1] discusses a block-double-sweep method——a direct method for solving a system in the following form:

$$b_i\boldsymbol{x}_i + a_{i+1}\boldsymbol{x}_{i+1} = \boldsymbol{c}_{i+1}, \quad i=0, 1, \cdots, M-1, \tag{1}$$

$$\begin{cases} g_0\boldsymbol{x}_0 = \boldsymbol{d}_0, \\ h_M\boldsymbol{x}_M = \boldsymbol{d}_M, \end{cases} \tag{2}$$

where b_i and a_{i+1} are $n\times n$-matrices; $\boldsymbol{x}_i$ and $\boldsymbol{c}_i$ are n-dimensional vectors; g_0 and h_M are $s\times n$- and $(n-s)\times n$-matrices respectively; $\boldsymbol{d}_0$ and $\boldsymbol{d}_M$ are s- and $(n-s)$-dimensional vectors respectively; and $0\leqslant s\leqslant n$. We may meet this type of system when solving differential equations numerically. In this case, equations (1) are difference equations approximating differential equations; and every equation of (2) is either a boundary condition for differential equations or a difference equation. Of course, this type of system can also appear in other problems. [2] discusses

1) This paper is an English translation of the paper in "Mathematicae Numericae Sinica, 1978, No. 3, 1—27".

block-tridiagonal equations. The idea of [2] is similar to that of [1]. Moreover, the block-tridiagonal equations can be rewritten in the form (1)—(2). Therefore, the two methods belong to the same type. This kind of block-double-sweep method has been applied to solving differential equations numerically and to other calculations. However, if "the boundary-condition-equations" for x_0 and x_M are nonlinear, then these methods cannot be applied. Moreover, if we need to solve many systems which consist of the same equations (1) and different equations (2), then it is uneconomical to solve these systems by using the method in [1]. We have discussed how to solve these problems by block-double-sweep methods: in [3] we have presented a method for solving these problems with a special type of system (1); and in this paper we generalize that method, i.e., present a method which can solve these problems with a general type of system (1). In addition, when the method in this paper is used, the equations (1) can be divided into several parts which can then be solved at the same time. Therefore, this method is more suitable for parallel computers than the usual block-double-sweep methods.

1. *A Description of the Method*

It is well known that the "double sweep" method is an elimination method. Eliminating unknowns can still be done for incomplete equations (1). In fact, if the rank of the $nM\times n(M-1)$-matrix

$$C=\begin{pmatrix} a_1 & & & \\ b_1 & a_2 & & 0 \\ & \ddots & \ddots & \\ & & b_{M-2} & a_{M-1} \\ 0 & & & b_{M-1} \end{pmatrix}$$

is $n(M-1)$, which is the coefficient matrix of $x_1, x_2, \cdots, x_{M-1}$ in (1), then we can always eliminate $x_1, x_2, \cdots, x_{M-1}$ from (1), and obtain n equations which contain x_0 and x_M only. After that, we can simultaneously solve the n equations and the boundary-condition-equations, which also contain x_0 and x_M only, and obtain x_0 and x_M. Finally, by using x_0, x_M and certain expressions which are obtained in the elimination procedure, we can obtain $x_1, x_2, \cdots, x_{M-1}$. Therefore, the elimination method can be applied to solving (1). The difficulties in constructing this type of method are how to guarantee that the procedure of elimination (the direct sweep) and the procedure of calculation of the unknowns $x_1, x_2, \cdots, x_{M-1}$ (the inverse sweep) are stable, and how to reduce the time of computation.

In the following, we present a double sweep method which is economical. and whose two procedures of calculation, the direct and the inverse sweeps, are stable under very weak conditions. It contains the following four steps:

(i) Some components of $\boldsymbol{x}_0$ or $\boldsymbol{x}_1$ are eliminated from some equations of $b_0\boldsymbol{x}_0+a_1\boldsymbol{x}_1=c_1$, i.e., we obtain a system of the following form from $b_0\boldsymbol{x}_0+a_1\boldsymbol{x}_1=c$:

$$\begin{pmatrix} E_{1,1} & \nu_{1,1} \\ 0 & \nu_{2,1} \end{pmatrix} F_{0,1}\boldsymbol{x}_0+\begin{pmatrix} \mu_{1,1} & 0 \\ \mu_{2,1} & E_{2,1} \end{pmatrix} F_{1,1}\boldsymbol{x}_1=\begin{pmatrix} d_{1,1} \\ d_{2,1} \end{pmatrix}, \tag{1.1}$$

where the two matrices and the vector are "$(n-s_1)\sim s_1$-broken", i.e., in every matrix, the upper-left, upper-right, lower-left and lower-right submatrices are $(n-s_1)\times(n-s_1)$-, $(n-s_1)\times s_1$-, $s_1\times(n-s_1)$- and $s_1\times s_1$-matrices respectively, and in the vector, $d_{1,1}$ and $d_{2,1}$ are $n-s_1$- and s_1-dimensional vectors respectively; $F_{0,1}$ and $F_{1,1}$ are $n\times n$-permutation matrices where one and only one element in every row and every column is 1, and the others are zero; $E_{1,1}$ and $E_{2,1}$ are unit matrices; and 0 denotes a matrix where every element is zero. This elimination procedure can be realized by a pivoting-elimination method similar to that in [4]. The concrete steps are as follows: First, we find the "pivot-element" in (b_0, a_1), i.e., the element which has the maximal modulus; every element which is on the same row as the pivot-element is divided by the pivot-element; and the unknown, corresponding to the pivot-element, is eliminated from the other equations. Then, in the remaining equations, i.e., in those equations which do not contain the unknown corresponding to the pivot-element, a new pivot-element is selected; every element of the equation in which the new pivot-element is located is divided by it; the unknown corresponding to it is eliminated from all the other equations. The step of eliminating an unknown is done n times. Finally, after making exchanges of rows, we obtain (1.1).

(ii) For $i=1, 2, \cdots, M-1$, the following procedure is carried out: the $\boldsymbol{x}_i$ and some components of $\boldsymbol{x}_0$ and $\boldsymbol{x}_{i+1}$ are eliminated from some of the equations

$$\begin{cases} \begin{pmatrix} E_{1,i} & \nu_{1,i} \\ 0 & \nu_{2,i} \end{pmatrix} F_{0,i}\boldsymbol{x}_0+\begin{pmatrix} \mu_{1,i} & 0 \\ \mu_{2,i} & E_{2,i} \end{pmatrix} F_{i,i}\boldsymbol{x}_i=\begin{pmatrix} \boldsymbol{d}_{1,i} \\ \boldsymbol{d}_{2,i} \end{pmatrix}, & (1.2.1) \\ \begin{pmatrix} b_{1,i} & b_{2,i} \\ b_{3,i} & b_{4,i} \end{pmatrix} F_{i,i}\boldsymbol{x}_i+\begin{pmatrix} a_{1,i+1} & a_{2,i+1} \\ a_{3,i+1} & a_{4,i+1} \end{pmatrix} F_{i,i}\boldsymbol{x}_{i+1}=\begin{pmatrix} \boldsymbol{c}_{1,i+1} \\ \boldsymbol{c}_{2,i+1} \end{pmatrix}. & (1.2.2) \end{cases}$$

and we thus obtain the following equations:

$$\begin{cases} \begin{pmatrix} E_{1,i+1} & \nu_{1,i+1} \\ 0 & \nu_{2,i+1} \end{pmatrix} F_{0,i+1}\boldsymbol{x}_0+\begin{pmatrix} \mu_{1,i+1} & 0 \\ \mu_{2,i+1} & E_{2,i+1} \end{pmatrix} F_{i+1,i+1}\boldsymbol{x}_{i+1}=\begin{pmatrix} \boldsymbol{d}_{1,i+1} \\ \boldsymbol{d}_{2,i+1} \end{pmatrix}, & (1.3.1) \\ (0 \quad \alpha_{1,i+1})F_{0,i+1}\boldsymbol{x}_0+(E_{1,i} \quad 0)F_{i,i}\boldsymbol{x}_i+(\beta_{1,i+1} \quad 0)F_{i+1,i+1}\boldsymbol{x}_{i+1} & \\ \qquad =\boldsymbol{\gamma}_{1,i+1}, & (1.3.2) \\ (0 \quad \nu_{2,i})F_{0,i}\boldsymbol{x}_0+(\mu_{2,i} \quad E_{2,i})F_{i,i}\boldsymbol{x}_i=\boldsymbol{d}_{2,i}. & (1.3.3) \end{cases}$$

Here, the matrices and the vectors in (1.2.1) and (1.2.2) are $(n-s_i)\sim s_i$-broken; the matrices and vectors in (1.3.1) are $(n-s_{i+1})\sim s_{i+1}$-broken; $\alpha_{1,i+1}$ and $\beta_{1,i+1}$ are $(n-s_i)\times s_{i+1}$ and $(n-s_i)\times(n-s_{i+1})$-matrices respectively; $\gamma_{1,i+1}$ is an $(n-s_i)$-dimensional vector; the equations (1.2.2) are the equations in (1) which contain $\boldsymbol{x}_i$ and $\boldsymbol{x}_{i+1}$; (1.3.3) is the second group of equations in (1.2.1). The concrete steps of calculation for obtaining (1.3.1), (1.3.2) are as follows:

First, by using (1.3.3), the last s_i components of $F_{i,i}\boldsymbol{x}_i$ are eliminated from (1.2.2), and we thus obtain the equations

$$\begin{pmatrix} 0 & b^*_{2,i} \\ 0 & b^*_{4,i} \end{pmatrix} F_{0,i}\boldsymbol{x}_0 + \begin{pmatrix} b^*_{1,i} & 0 \\ b^*_{3,i} & 0 \end{pmatrix} F_{i,i}\boldsymbol{x}_i + \begin{pmatrix} a_{1,i+1} & a_{2,i+1} \\ a_{3,i+1} & a_{4,i+1} \end{pmatrix} F_{i,i}\boldsymbol{x}_{i+1} = \begin{pmatrix} \boldsymbol{c}^*_{1,i+1} \\ \boldsymbol{c}^*_{2,i+1} \end{pmatrix}. \tag{1.3.4}$$

Then, selecting the pivot-elements in $\begin{pmatrix} \mu_{1,i} \\ b^*_{1,i} \\ b^*_{3,i} \end{pmatrix}$, eliminating the corresponding unknowns from all the other equations of

$$\begin{pmatrix} E_{1,i} & \nu_{1,i} \\ 0 & b^*_{2,i} \\ 0 & b^*_{4,i} \end{pmatrix} F_{0,i}\boldsymbol{x}_0 + \begin{pmatrix} \mu_{1,i} & 0 \\ b^*_{1,i} & 0 \\ b^*_{3,i} & 0 \end{pmatrix} F_{i,i}\boldsymbol{x}_i + \begin{pmatrix} 0 & 0 \\ a_{1,i+1} & a_{2,i+1} \\ a_{3,i+1} & a_{4,i+1} \end{pmatrix} F_{i,i}\boldsymbol{x}_{i+1} = \begin{pmatrix} \boldsymbol{d}_{1,i} \\ \boldsymbol{c}^*_{1,i+1} \\ \boldsymbol{c}^*_{2,i+1} \end{pmatrix} \tag{1.3.5$'$}$$

and making exchanges of rows, we obtain the following equations:

$$\begin{pmatrix} E^*_{1,i+1} & \nu^*_{1,i+1} \\ e^*_{1,i+1} & \alpha^*_{1,i+1} \\ e^*_{2,i+1} & \nu^*_{2,i+1} \end{pmatrix} F_{0,i}\boldsymbol{x}_0 + \begin{pmatrix} 0 & 0 \\ E_{1,i} & 0 \\ 0 & 0 \end{pmatrix} F_{i,i}\boldsymbol{x}_i + \begin{pmatrix} \mu^*_{1,i+1} & \bar{a}_{2,i+1} \\ \beta^*_{1,i+1} & a^*_{2,i+1} \\ \mu^*_{2,i+1} & a^*_{4,i+1} \end{pmatrix} F_{i,i}\boldsymbol{x}_{i+1} = \begin{pmatrix} \boldsymbol{d}^*_{1,i+1} \\ \boldsymbol{\gamma}^*_{1,i+1} \\ \boldsymbol{d}^*_{2,i+1} \end{pmatrix}. \tag{1.3.5}$$

Finally, selecting the pivot-elements in the first and third groups of (1.3.5), and eliminating the corresponding unknowns from all the other equations of (1.3.5), we obtain (1.3.1) and (1.3.2).

If the equation $|\lambda b_i + a_{i+1}| = 0$ always has s roots whose moduli are greater than $1/(1-2\varepsilon)$, and $n-s$ roots whose moduli are less than $1-2\varepsilon$, where $\varepsilon \in \left(0, \dfrac{1}{2}\right)$, then, as will be pointed out in the next section, s_i is always equal to s, and $\mu_{1,i}$ and $\nu_{2,i}$ are close to null-matrices when i is sufficiently large. Obviously, at the same time, $E_{1,i}$ is always an $(n-s)\times(n-s)$–unit matrix, and $E_{2,i}$ is always an $s\times s$–unit matrix. Moreover, because $\mu_{1,i}$ is almost a null–matrix, the pivot–elements must be in the lower n rows of (1.3.5)′ when they are selected from $\begin{pmatrix}\mu_{1,i}\\ b^*_{1,i}\\ b^*_{3,i}\end{pmatrix}$, i.e., they must be in $\begin{pmatrix}b^*_{1,i}\\ b^*_{3,i}\end{pmatrix}$. Therefore, if we want to save time in computation then after this case appears, we can carry out the calculation in the following way: After (1.3.4) is obtained, selecting the pivot–elements in $\begin{pmatrix}b^*_{1,i}\\ b^*_{3,i}\end{pmatrix}$, and eliminating some components of $\boldsymbol{x}_i$ from some equations of (1.3.4), we obtain the second and third groups of (1.3.5) with $e^*_{1,i+1}=0$ and $e^*_{2,i+1}=0$. Then, selecting the pivot–elements in $(\mu^*_{2,i+1},\ a^*_{4,i+1})$, and eliminating the corresponding unknowns from some equations of the second and third groups of (1.3.5), we obtain (1.3.2) and the second group of (1.3.1). Finally, eliminating the upper $n-s$ unknowns of $F_{i,i}\boldsymbol{x}_i$ from the first group of (1.2.1) by using (1.3.2), we obtain the first group of equations in (1.3.1). In this way, we also obtain (1.3.1) and (1.3.2).

Steps (i) and (ii) are called the Procedure of Elimination (the Direct Sweep). After it is completed, we may carry out the following calculation:

(iii) $\boldsymbol{x}_0$ and $\boldsymbol{x}_M$ are solved. We suppose the boundary–condition–equations are nonlinear, i.e., they are

$$\begin{cases}\boldsymbol{B}_1(\boldsymbol{x}_0)=0,\\ \boldsymbol{B}_2(\boldsymbol{x}_M)=0,\end{cases}\tag{1.4}$$

where $\boldsymbol{B}_1$ and $\boldsymbol{B}_2$ are s and $(n-s)$–dimensional vectors respectively, and every element is a nonlinear function of $\boldsymbol{x}_0$ or $\boldsymbol{x}_M$. From the proof in the next section, we know that if M is sufficiently large, in general, those matrices in (1.3.1) with $i=M-1$ are $(n-s)\sim s$–broken. Thus, it is reasonable to adopt the following iteration in order to solve $\boldsymbol{x}_0$ and $\boldsymbol{x}_M$:

$$\begin{cases}\boldsymbol{B}_1(\boldsymbol{x}_0^{k+1})=0,\\ (E_{1,M}\ \ \nu_{1,M})F_{0,M}\boldsymbol{x}_0^{k+1}=-(\mu_{1,M}\ \ 0)F_{M,M}\boldsymbol{x}_M^k+\boldsymbol{d}_{1,M},\end{cases}\tag{1.5.1}$$

$$\begin{cases}(\mu_{2,M}\ \ E_{2,M})F_{M,M}\boldsymbol{x}_M^{k+2}=-(0\ \ \nu_{2,M})F_{0,M}\boldsymbol{x}_0^{k+1}+\boldsymbol{d}_{2,M},\\ \boldsymbol{B}_2(\boldsymbol{x}_M^{k+2})=0.\end{cases}\tag{1.5.2}$$

The iteration is, in general, convergent. Therefore, we can obtain $\boldsymbol{x}_0$ and $\boldsymbol{x}_M$ when we have a method to solve $\boldsymbol{x}_0^{k+1}$ and $\boldsymbol{x}_M^{k+2}$ from (1.5.1) and (1.5.2) respectively.

For a problem with several groups of boundary–condition–equations, we need to solve the following equations:

$$\begin{cases} g_0^h \boldsymbol{x}_0 = \boldsymbol{d}_0^h, \\ h_M^h \boldsymbol{x}_M = \boldsymbol{d}_M^h, \end{cases} \quad h=1, 2, \cdots,$$
$$\begin{pmatrix} E_{1,M} & \nu_{1,M} \\ 0 & \nu_{2,M} \end{pmatrix} F_{0,M}\boldsymbol{x}_0 + \begin{pmatrix} \mu_{1,M} & 0 \\ \mu_{2,M} & E_{2,M} \end{pmatrix} F_{M,M}\boldsymbol{x}_M = \begin{pmatrix} \boldsymbol{d}_{1,M} \\ \boldsymbol{d}_{2,M} \end{pmatrix}.$$

(iv) After obtaining $\boldsymbol{x}_0$ and $\boldsymbol{x}_M$, we calculate $\boldsymbol{x}_{M-1}, \boldsymbol{x}_{M-2}, \cdots, \boldsymbol{x}_1$ by using (1.3.2), (1.3.3) as follows:

$$\begin{cases} (E_{1,i}\ 0)F_{i,i}\boldsymbol{x}_i = -(\beta_{1,i+1}\ 0)F_{i+1,i+1}\boldsymbol{x}_{i+1} \\ \qquad\qquad -(0\ \alpha_{1,i+1})F_{0,i+1}\boldsymbol{x}_0 + \boldsymbol{\gamma}_{1,i+1}, \\ (0\ E_{2,i})F_{i,i}\boldsymbol{x}_i = -(\mu_{2,i}\ 0)F_{i,i}\boldsymbol{x}_i - (0\ \nu_{2,i})F_{0,i}\boldsymbol{x}_0 + \boldsymbol{d}_{2,i}, \\ \qquad\qquad i = M-1, M-2, \cdots, 1. \end{cases} \tag{1.6}$$

We call this step the Procedure of Calculation of the Unknowns (the Inverse Sweep).

Finally, we point out that the elimination procedure here may start by eliminating $\boldsymbol{x}_1$ or $\boldsymbol{x}_{M-1}$, even by eliminating a middle $\boldsymbol{x}_i$. In addition, if some rows of a_1 are null vectors, then the equations which contain $\boldsymbol{x}_0$ only have already been of the desired type. Thus it is unnecessary to do any calculation on them in the elimination procedure and some rows of $\mu_{1,i}$ are always null-vectors. In this case, if the elimination procedure starts by eliminating $\boldsymbol{x}_1$, then the computing-time and the number of memory units will be less. If some rows of b_{M-1} are null-vectors, then it is reasonable to start the procedure by eliminating $\boldsymbol{x}_{M-1}$.

2. *The Stability of the Method*

In order to explain that the method in Section 1 is reasonable, we shall discuss the stability of this method. We shall prove that the procedure of elimination and the procedure of calculation of the unknowns are stable if the following conditions are fulfilled:

(1) For all i, the pencil of matrices $a_{i+1}+\lambda b_i$ is regular. Moreover, for all i, the equation $|a_{i+1}+\lambda b_i|=0$ has $n-s$ roots whose moduli are less than $1-2\varepsilon$, and the equation $|\mu a_{i+1}+b_i|=0$ has s roots whose moduli are less than $1-2\varepsilon$, where ε is a constant, $\varepsilon\in\left(0, \frac{1}{2}\right)$. As is well known from the theory of matrices[5], in this case we can find the matrices P_i and S_i such that

$$a_{i+1}+\lambda b_i = P_i\left[\begin{pmatrix} \lambda_{1,i} & 0 \\ 0 & E_2 \end{pmatrix} + \lambda\begin{pmatrix} E_1 & 0 \\ 0 & \lambda_{2,i} \end{pmatrix}\right]S_i,$$

where $\lambda_{1,i}$ and $\lambda_{2,i}$ are respectively $(n-s)\times(n-s)$ and $s\times s$-matrices of the Jordan form, the moduli of whose eigenvalues are less than $1-2\varepsilon$. Because

$$\begin{pmatrix} a & 1 & & 0 \\ & a & \ddots & \\ & & \ddots & 1 \\ 0 & & & a \end{pmatrix}$$

is similar to

$$\begin{pmatrix} a & \varepsilon & & 0 \\ & a & \ddots & \\ & & \ddots & \varepsilon \\ 0 & & & a \end{pmatrix} = \begin{pmatrix} \varepsilon^{-1} & & & 0 \\ & \varepsilon^{-2} & & \\ & & \ddots & \\ 0 & & & \varepsilon^{-k} \end{pmatrix} \begin{pmatrix} a & 1 & & 0 \\ & a & \ddots & \\ & & \ddots & 1 \\ 0 & & & a \end{pmatrix} \begin{pmatrix} \varepsilon^{1} & & & 0 \\ & \varepsilon^{2} & & \\ & & \ddots & \\ 0 & & & \varepsilon^{k} \end{pmatrix},$$

the constant on the off-diagonal line of a Jordan canonical matrix may be any constant except zero. Therefore, taking ε as the constant, we have

$$\|\lambda_{1,i}\| \leqslant 1-\varepsilon, \quad \|\lambda_{2,i}\| \leqslant 1-\varepsilon.$$

(2) There are three constants c_1, c_2 and c_3 such that

$$\|S_{i+1}-S_i\| \leqslant \frac{c_1}{M}, \quad i=0, 1, \cdots, M-1,$$

$$0 < c_2 \leqslant \Gamma(S_i) \leqslant c_3, \quad i=0, 1, \cdots, M,$$

where $\Gamma(S_i)$ is the Gram determinant of S_i, and let $S_M \equiv S_{M-1}$.

(3) c_1/M is sufficiently small (M is sufficiently large), i.e., the increment of S_i is very small when i changes.

Before proving our conclusion, we point out that the conditions here are weaker. In fact, if we let b_i and a_{i+1} be constant matrices, then the above conditions are reduced to:

That the pencil of matrices $a+\lambda b$ is regular, and that the equation $|a+\lambda b|=0$ does not have any root whose modulus is equal to 1.

Therefore, in general, the conditions are fulfilled. In the case where a_{i+1} and b_i change with i, if these functions are "continuous" with respect to i, then the most important condition is that the equation $|a_{i+1}+\lambda b_i|=0$ does not have any root whose modulus is equal to 1.

In the case where the equation $|a_{i+1}+\lambda b_i|=0$ has roots whose moduli approach 1, it is possible that all determinants of order $n(M-1)$ in C go to zero as $M\to\infty$. In this paper we shall not discuss the stability of the direct method in this case.

In short, the conditions are quite weak, but further problems must be considered.

Proof The proof is divided into two parts. First, the stability of the elimination procedure is proved. Then the stability of the procedure of calculation of the unknowns is proved. The definition of stability is as follows: Suppose the error of the initial data is less than ε^*. If the error of all data obtained in a procedure is less than $c^*\varepsilon^*$, c^* being a positive constant independent of M, then we say that the procedure is stable.

The structure of proof on the stability of the elimination procedure is as follows: First, we prove a fictitious procedure to be stable. Then, we prove the procedure in Section 1 to be stable.

According to the above conditions, equations (1) are equivalent to the following equations:

$$\begin{pmatrix} E_1 & 0 \\ 0 & \lambda_{2,i} \end{pmatrix} S_i \boldsymbol{x}_i + \begin{pmatrix} \lambda_{1,i} & 0 \\ 0 & E_2 \end{pmatrix} S_i \boldsymbol{x}_{i+1} = \begin{pmatrix} \boldsymbol{f}_{1,i+1} \\ \boldsymbol{f}_{2,i+1} \end{pmatrix}, \quad i=0, 1, \cdots, M-1.$$

Because $\|S_i - S_{i+1}\| \leqslant c_1/M$, $S_i = S_{i+1} + (S_i - S_{i+1}) S_{i+1}^{-1} S_{i+1}$, and $\|S_{i+1}^{-1}\|$ is bounded, we have

$$\begin{pmatrix} \lambda_{1,i} & 0 \\ 0 & E_2 \end{pmatrix} S_i = \begin{pmatrix} \lambda_{1,i}(E_1 + O_{1,i}) & \lambda_{1,i} O_{2,i} \\ O_{3,i} & E_2 + O_{4,i} \end{pmatrix} S_{i+1},$$

where $O_{1,i}$, $O_{2,i}$, $O_{3,i}$ and $O_{4,i}$ are $(n-s)\times(n-s)$-, $(n-s)\times s$-, $s\times(n-s)$- and $s\times s$- matrices respectively, and $\|O_{j,i}\| \leqslant c_4/M$, $j=1, 2, 3, 4$, c_4 being a constant. Therefore, the above equations can be rewritten as

$$\begin{pmatrix} E_1 & 0 \\ 0 & \lambda_{2,i} \end{pmatrix} \boldsymbol{Y}_i + \begin{pmatrix} \lambda_{1,i}(E_1 + O_{1,i}) & \lambda_{1,i} O_{2,i} \\ O_{3,i} & E_2 + O_{4,i} \end{pmatrix} \boldsymbol{Y}_{i+1} = \begin{pmatrix} \boldsymbol{f}_{1,i+1} \\ \boldsymbol{f}_{2,i+1} \end{pmatrix}, \tag{2.1}$$
$$i=0, 1, \cdots, M-1,$$

where $\boldsymbol{Y}_i$ denotes $S_i \boldsymbol{x}_i$. In the following, for $m=1, 2, \cdots, M$, we prove that the equations

$$\begin{pmatrix} E_1 & 0 \\ 0 & \lambda_{2,i} \end{pmatrix} \boldsymbol{Y}_i + \begin{pmatrix} \lambda_{1,i}(E_1 + O_{1,i}) & \lambda_{1,i} O_{2,i} \\ O_{3,i} & E_2 + O_{4,i} \end{pmatrix} \boldsymbol{Y}_{i+1} = \begin{pmatrix} \boldsymbol{f}_{1,i+1} \\ \boldsymbol{f}_{2,i+1} \end{pmatrix} \tag{2.1}$$
$$i=0, 1, \cdots, m-1$$

are equivalent to the equations

$$\begin{cases} \begin{pmatrix} E_1 & 0 \\ O_{7,i} & E_2 + O_{8,i} \end{pmatrix}^{-1} \begin{pmatrix} 0 & 0 \\ 0 & \omega_i \end{pmatrix} \boldsymbol{Y}_0 + \boldsymbol{Y}_i + \begin{pmatrix} E_1 & 0 \\ O_{7,i} & E_2 + O_{8,i} \end{pmatrix}^{-1} \\ \quad \times \begin{pmatrix} \lambda_{1,i}(E_1 + O_{1,i}) & \lambda_{1,i} O_{2,i} \\ 0 & 0 \end{pmatrix} \boldsymbol{Y}_{i+1} = \begin{pmatrix} \boldsymbol{g}_{1,i} \\ \boldsymbol{g}_{2,i} \end{pmatrix}, & (2.2.1) \\ \qquad i=1, \cdots, m-1 \text{ (when } m=1\text{, this is empty)}, \\ \begin{pmatrix} E_1 & O_{5,m} \\ 0 & \omega_m \end{pmatrix} \boldsymbol{Y}_0 + \begin{pmatrix} \bar{\omega}_m & \bar{\omega}_m O_{6,m} \\ O_{7,m} & E_2 + O_{8,m} \end{pmatrix} \boldsymbol{Y}_m = \begin{pmatrix} \boldsymbol{h}_{1,m} \\ \boldsymbol{h}_{2,m} \end{pmatrix}; & (2.2.2) \end{cases}$$

that $O_{5,i}$, $O_{6,i}$, $O_{7,i}$, $O_{8,i}$, ω_i, $\bar{\omega}_i$, $\boldsymbol{h}_{1,i}$, $\boldsymbol{h}_{2,i}$, $\boldsymbol{g}_{1,i}$ and $\boldsymbol{g}_{2,i}$ satisfy the following inequalities respectively:

$$\begin{cases} \|O_{5,i}\| \leqslant \dfrac{4c_4}{M} \dfrac{1-\left(1-\dfrac{\varepsilon}{2}\right)^{2i}}{\varepsilon\left(1-\dfrac{\varepsilon}{4}\right)} = \dfrac{4c_4}{M} \displaystyle\sum_{j=0}^{i-1} \left(1-\dfrac{\varepsilon}{2}\right)^{2j}, \\ \|O_{6,i}\| \leqslant 2c_4/M, \end{cases}$$

$$\begin{cases} \|O_{7,i}\| \leqslant \dfrac{c_4}{M} \dfrac{1-\left(1-\dfrac{\varepsilon}{2}\right)^{2i}}{\varepsilon\left(1-\dfrac{\varepsilon}{4}\right)}, \quad i=1, 2, \cdots, m, \\ \|O_{8,i}\| \leqslant 2c_4/M, \end{cases} \tag{2.2.3}$$

$$\begin{cases} \|\omega_i\| \leqslant \left(1-\dfrac{\varepsilon}{2}\right)^i, \\ \|\bar{\omega}_i\| \leqslant \left(1-\dfrac{\varepsilon}{2}\right)^i, \end{cases} \quad i=1, 2, \cdots, m,$$

$$\begin{cases} \|\boldsymbol{h}_{1,i}\| \leqslant \dfrac{\varepsilon}{2} c_5 \sum\limits_{j=0}^{i-1} \left(1-\dfrac{\varepsilon}{2}\right)^j = c_5\left[1-\left(1-\dfrac{\varepsilon}{2}\right)^i\right], \\ \|\boldsymbol{h}_{2,i}\| \leqslant \dfrac{\varepsilon}{2} c_5 \sum\limits_{j=0}^{i-1} \left(1-\dfrac{\varepsilon}{2}\right)^j = c_5\left[1-\left(1-\dfrac{\varepsilon}{2}\right)^i\right], \end{cases} \quad i=1, 2, \cdots, m, \tag{2.2.4}$$

$$\begin{cases} \|\boldsymbol{g}_{1,i}\| \leqslant c_5, \quad i=1, 2, \cdots, m-1, \\ \|\boldsymbol{g}_{2,i}\| \leqslant 2c_5, \end{cases}$$

where c_5 is a constant; and that the procedure of calculating $O_{5,i}$, $\bar{\omega}_i O_{6,i}$, $O_{7,i}$, $O_{8,i}$, ω_i, $\bar{\omega}_i$, $\boldsymbol{h}_{1,i}$, $\boldsymbol{h}_{2,i}$, $\boldsymbol{g}_{1,i}$, and $\boldsymbol{g}_{2,i}$ is stable. We shall use the inductive method. First, we prove that (2.1)′ is equivalent to (2.2.1)—(2.2.2), and that (2.2.3) and (2.2.4) are fulfilled. When $m=1$, the conclusion is obviously true. (Here, we suppose that M is so large that $(1-\varepsilon)\left(1+\dfrac{c}{M}\right) \leqslant \left(1-\dfrac{\varepsilon}{2}\right)$, where c is a constant. In the following proof, we shall use the fact several times.) We now suppose the conclusion is true for m, and we need to prove that it is true for $m+1$. In fact, the only things we need to prove are that the equations

$$\begin{cases} \begin{pmatrix} E_1 & O_{5,m} \\ 0 & \omega_m \end{pmatrix} \boldsymbol{Y}_0 + \begin{pmatrix} \bar{\omega}_m & \bar{\omega}_m O_{6,m} \\ O_{7,m} & E_2+O_{8,m} \end{pmatrix} \boldsymbol{Y}_m = \begin{pmatrix} \boldsymbol{h}_{1,m} \\ \boldsymbol{h}_{2,m} \end{pmatrix}, \\ \begin{pmatrix} E_1 & 0 \\ 0 & \lambda_{2,m} \end{pmatrix} \boldsymbol{Y}_m + \begin{pmatrix} \lambda_{1,m}(E_1+O_{1,m}) & \lambda_{1,m} O_{2,m} \\ O_{3,m} & E_2+O_{4,m} \end{pmatrix} \boldsymbol{Y}_{m+1} = \begin{pmatrix} \boldsymbol{f}_{1,m+1} \\ \boldsymbol{f}_{2,m+1} \end{pmatrix} \end{cases} \tag{2.3}$$

are equivalent to the equations

$$\begin{cases} \begin{pmatrix} E_1 & 0 \\ O_{7,m} & E_2+O_{8,m} \end{pmatrix}^{-1} \begin{pmatrix} 0 & 0 \\ 0 & \omega_m \end{pmatrix} \boldsymbol{Y}_0 + \boldsymbol{Y}_m + \begin{pmatrix} E_1 & 0 \\ O_{7,m} & E_2+O_{8,m} \end{pmatrix}^{-1} \\ \quad \times \begin{pmatrix} \lambda_{1,m}(E_1+O_{1,m}) & \lambda_{1,m} O_{2,m} \\ 0 & 0 \end{pmatrix} \boldsymbol{Y}_{m+1} = \begin{pmatrix} \boldsymbol{g}_{1,m} \\ \boldsymbol{g}_{2,m} \end{pmatrix}, \\ \begin{pmatrix} E_1 & O_{5,m+1} \\ 0 & \omega_{m+1} \end{pmatrix} \boldsymbol{Y}_0 + \begin{pmatrix} \bar{\omega}_{m+1} & \bar{\omega}_{m+1} O_{6,m+1} \\ O_{7,m+1} & E_2+O_{8,m+1} \end{pmatrix} \boldsymbol{Y}_{m+1} = \begin{pmatrix} \boldsymbol{h}_{1,m+1} \\ \boldsymbol{h}_{2,m+1} \end{pmatrix}, \end{cases} \tag{2.4}$$

and that (2.2.3) and (2.2.4) are fulfilled for $i=m+1$. Formula (2.3) can be rewritten as

$$\begin{cases}\begin{pmatrix}0 & 0\\ 0 & \omega_m\end{pmatrix}\boldsymbol{Y}_0+\begin{pmatrix}E_1 & 0\\ O_{7,m} & E_2+O_{8,m}\end{pmatrix}\boldsymbol{Y}_m\\ \quad+\begin{pmatrix}\lambda_{1,m}(E_1+O_{1,m}) & \lambda_{1,m}O_{2,m}\\ 0 & 0\end{pmatrix}\boldsymbol{Y}_{m+1}=\begin{pmatrix}\boldsymbol{f}_{1,m+1}\\ \boldsymbol{h}_{2,m}\end{pmatrix},\\ \begin{pmatrix}E_1 & O_{5,m}\\ 0 & 0\end{pmatrix}\boldsymbol{Y}_0+\begin{pmatrix}\overline{\omega}_m & \overline{\omega}_mO_{6,m}\\ 0 & \lambda_{2,m}\end{pmatrix}\boldsymbol{Y}_m\\ \quad+\begin{pmatrix}0 & 0\\ O_{3,m} & E_2+O_{4,m}\end{pmatrix}\boldsymbol{Y}_{m+1}=\begin{pmatrix}\boldsymbol{h}_{1,m}\\ \boldsymbol{f}_{2,m+1}\end{pmatrix}.\end{cases}$$

Therefore, when M is so large that $\begin{pmatrix}E_1 & 0\\ O_{7,m} & E_2+O_{8,m}\end{pmatrix}$ is invertible, we can obtain (2.4) from (2.3), and have the following expressions:

$$\begin{cases}\begin{pmatrix}E_1 & O_{5,m+1}\\ 0 & \omega_{m+1}\end{pmatrix}\\ \quad=\begin{pmatrix}E_1 & O_{5,m}\\ 0 & 0\end{pmatrix}-\begin{pmatrix}\overline{\omega}_m & \overline{\omega}_mO_{6,m}\\ 0 & \lambda_{2,m}\end{pmatrix}\begin{pmatrix}E_1 & 0\\ O_{7,m} & E_2+O_{8,m}\end{pmatrix}^{-1}\begin{pmatrix}0 & 0\\ 0 & \omega_m\end{pmatrix}\\ \quad=\begin{pmatrix}E_1 & O_{5,m}-\overline{\omega}_mO_{6,m}(E_2+O_{8,m})^{-1}\omega_m\\ 0 & -\lambda_{2,m}(E_2+O_{8,m})^{-1}\omega_m\end{pmatrix},\\ \begin{pmatrix}\overline{\omega}_{m+1} & \overline{\omega}_{m+1}O_{6,m+1}\\ O_{7,m+1} & E_2+O_{8,m+1}\end{pmatrix}=\begin{pmatrix}0 & 0\\ O_{3,m} & E_2+O_{4,m}\end{pmatrix}\\ \quad-\begin{pmatrix}\overline{\omega}_m & \overline{\omega}_mO_{6,m}\\ 0 & \lambda_{2,m}\end{pmatrix}\begin{pmatrix}E_1 & 0\\ O_{7,m} & E_2+O_{8,m}\end{pmatrix}^{-1}\begin{pmatrix}\lambda_{1,m}(E_1+O_{1,m}) & \lambda_{1,m}O_{2,m}\\ 0 & 0\end{pmatrix}\\ \quad=\begin{pmatrix}-\overline{\omega}_m[E_1-O_{6,m}(E_2+O_{8,m})^{-1}O_{7,m}]\lambda_{1,m}(E_1+O_{1,m})\\ O_{3,m}+\lambda_{2,m}(E_2+O_{8,m})^{-1}O_{7,m}\lambda_{1,m}(E_1+O_{1,m})\end{pmatrix.}\\ \qquad\begin{matrix}-\overline{\omega}_m[E_1-O_{6,m}(E_2+O_{8,m})^{-1}O_{7,m}]\lambda_{1,m}O_{2,m}\\ E_2+O_{4,m}+\lambda_{2,m}(E_2+O_{8,m})^{-1}O_{7,m}\lambda_{1,m}O_{2,m}\end{matrix}\Bigg).\end{cases}\tag{2.5}$$

Moreover, we can also obtain (2.3) from (2.4). Therefore, (2.3) is equivalent to (2.4). In the following, we shall prove that (2.2.3) and (2.2.4) are fulfilled. According to

$$\|O_{j,m}\|\leqslant\frac{c_4}{M},\quad j=1,\,2,\,3,\,4,$$

by using (2.5), we obtain

$$\begin{cases}\|O_{5,m+1}\|\leqslant\|O_{5,m}\|+\frac{4c_4}{M}\left(1-\frac{\varepsilon}{2}\right)^{2m}\\ \qquad\leqslant\frac{4c_4}{M}\sum_{j=0}^{m}\left(1-\frac{\varepsilon}{2}\right)^{2j}=\frac{4c_4}{M}\frac{1-\left(1-\frac{\varepsilon}{2}\right)^{2(m+1)}}{\varepsilon\left(1-\frac{\varepsilon}{4}\right)},\end{cases}$$

$$\begin{cases}
\|O_{6,m+1}\| = \|(E_1+O_{1,m})^{-1}O_{2,m}\| \leqslant \dfrac{2c_4}{M}, \\
\|O_{7,m+1}\| \leqslant \|O_{3,m}\| + \left(1-\dfrac{\varepsilon}{2}\right)^2 \|O_{7,m}\| \leqslant \dfrac{c_4}{M} \sum\limits_{j=0}^{m} \left(1-\dfrac{\varepsilon}{2}\right)^{2j} \\
\qquad = \dfrac{c_4}{M} \dfrac{1-\left(1-\dfrac{\varepsilon}{2}\right)^{2(m+1)}}{\varepsilon\left(1-\dfrac{\varepsilon}{4}\right)}, \\
\|O_{8,m+1}\| \leqslant \|O_{4,m}\| + \left(1-\dfrac{\varepsilon}{2}\right)^2 \dfrac{c_4}{M} \|O_{7,m}\| \\
\qquad \leqslant \dfrac{c_4}{M}\left[1+\left(1-\dfrac{\varepsilon}{2}\right)^2 \dfrac{c_4}{M} \dfrac{1}{\varepsilon\left(1-\dfrac{\varepsilon}{4}\right)}\right] \leqslant \dfrac{2c_4}{M}, \\
\|\omega_{m+1}\| \leqslant \|\omega_m\| \, \|\lambda_{2,m}(E_2+O_{8,m})^{-1}\| \leqslant \left(1-\dfrac{\varepsilon}{2}\right)^{m+1}, \\
\|\bar{\omega}_{m+1}\| \leqslant \|\bar{\omega}_m\| \, \|(E_1-O_{6,m}(E_2+O_{8,m})^{-1}O_{7,m})\lambda_{1,m}(E_1+O_{1,m})\| \\
\qquad \leqslant \left(1-\dfrac{\varepsilon}{2}\right)^{m+1}.
\end{cases}$$

In deriving these inequalities, we suppose that M is so large that $\left(1+\frac{c}{M}\right)\left(1+\left(\frac{c}{M}\right)^2\right)(1-\varepsilon)$, etc., are less than $1-\frac{\varepsilon}{2}$, where c is a constant. Using the following expressions

$$\begin{cases}
\begin{pmatrix} \boldsymbol{h}_{1,m+1} \\ \boldsymbol{h}_{2,m+1} \end{pmatrix} = \begin{pmatrix} \boldsymbol{h}_{1,m} \\ \boldsymbol{f}_{2,m+1} \end{pmatrix} - \begin{pmatrix} \bar{\omega}_m & \bar{\omega}_m O_{6,m} \\ 0 & \lambda_{2,m} \end{pmatrix} \\
\qquad \times \begin{pmatrix} E_1 & 0 \\ -(E_2+O_{8,m})^{-1}O_{7,m} & (E_2+O_{8,m})^{-1} \end{pmatrix} \begin{pmatrix} \boldsymbol{f}_{1,m+1} \\ \boldsymbol{h}_{2,m} \end{pmatrix} \\
\qquad = \begin{pmatrix} \boldsymbol{h}_{1,m} - \bar{\omega}_m(\boldsymbol{f}_{1,m+1} - O_{6,m}(E_2+O_{8,m})^{-1}(O_{7,m}\boldsymbol{f}_{1,m+1} - \boldsymbol{h}_{2,m})) \\ \boldsymbol{f}_{2,m+1} + \lambda_{2,m}(E_2+O_{8,m})^{-1}(O_{7,m}\boldsymbol{f}_{1,m+1} - \boldsymbol{h}_{2,m}) \end{pmatrix}, \\
\begin{pmatrix} \boldsymbol{g}_{1,m} \\ \boldsymbol{g}_{2,m} \end{pmatrix} = \begin{pmatrix} E_1 & 0 \\ -(E_2+O_{8,m})^{-1}O_{7,m} & (E_2+O_{8,m})^{-1} \end{pmatrix} \begin{pmatrix} \boldsymbol{f}_{1,m+1} \\ \boldsymbol{h}_{2,m} \end{pmatrix} \\
\qquad = \begin{pmatrix} \boldsymbol{f}_{1,m+1} \\ -(E_2+O_{8,m})^{-1}(O_{7,m}\boldsymbol{f}_{1,m+1} - \boldsymbol{h}_{2,m}) \end{pmatrix},
\end{cases} \tag{2.6}$$

and supposing

$$\max\{\|\boldsymbol{f}_{1,m+1}\|, \|\boldsymbol{f}_{2,m+1}\|\} \leqslant c_6, \quad m=0, 1, \cdots, M-1,$$

$$c_5 = \frac{4}{\varepsilon} c_6,$$

where c_6 is a constant, we can obtain the following inequalities very easily:

$$\begin{cases} \|\boldsymbol{h}_{1,m+1}\| \leqslant \|\boldsymbol{h}_{1,m}\| + \left(1-\frac{\varepsilon}{2}\right)^m \left[\left(1+\frac{c_7}{M^2}\right)\|\boldsymbol{f}_{1,m+1}\| + \frac{c_7}{M}\|\boldsymbol{h}_{2,m}\|\right] \\ \qquad \leqslant \|\boldsymbol{h}_{1,m}\| + 2\left(1-\frac{\varepsilon}{2}\right)^m c_6 \\ \qquad \leqslant 2c_6 \dfrac{1-\left(1-\frac{\varepsilon}{2}\right)^{m+1}}{\varepsilon/2} = c_5\left[1-\left(1-\frac{\varepsilon}{2}\right)^{m+1}\right], \\ \|\boldsymbol{h}_{2,m+1}\| \leqslant \|\boldsymbol{f}_{2,m+1}\| + \left(1-\frac{\varepsilon}{2}\right)\left[\frac{c_7}{M}\|\boldsymbol{f}_{1,m+1}\| + \|\boldsymbol{h}_{2,m}\|\right] \\ \qquad \leqslant \left(1+\frac{c_7}{M}\right)c_6 + \left(1-\frac{\varepsilon}{2}\right)\|\boldsymbol{h}_{2,m}\| \\ \qquad \leqslant \dfrac{1-\left(1-\frac{\varepsilon}{2}\right)^{m+1}}{\varepsilon/2} 2c_6 = c_5\left[1-\left(1-\frac{\varepsilon}{2}\right)^{m+1}\right], \\ \|\boldsymbol{g}_{1,m}\| = \|\boldsymbol{f}_{1,m+1}\| \leqslant c_6 < c_5, \\ \|\boldsymbol{g}_{2,m}\| \leqslant \frac{c_7}{M}\|\boldsymbol{f}_{1,m+1}\| + \left(1+\frac{c_7}{M}\right)\|\boldsymbol{h}_{2,m}\| \leqslant 2c_5, \end{cases}$$

where c_7 is a constant. Therefore, we derive all inequalities of (2.2.3)—(2.2.4). The fact that the procedure of calculating the coefficients is stable can be easily proved by using these expressions above. Let

$$O_{9,m} = (E_2+O_{8,m})^{-1}O_{7,m}, \quad \widehat{E}_{2,m} = (E_2+O_{8,m})^{-1},$$
$$O_{11,m} = \overline{\omega}_m O_{6,m}, \quad \delta O_{10,m} = \delta\overline{\omega}_m - \delta O_{11,m}O_{9,m}.$$

We observe the relations between the set of variations $\delta O_{9,m}$, $\delta\widehat{E}_{2,m}$, $\delta O_{10,m}$, $\delta O_{11,m}$, $\delta\omega_m$, $\delta O_{5,m}$, $\delta\boldsymbol{h}_{2,m}$, $\delta\boldsymbol{h}_{1,m}$ and the set of variations $\delta O_{9,m+1}$, $\delta\widehat{E}_{2,m+1}$, $\cdots$, $\delta\boldsymbol{h}_{1,m+1}$. According to (2.5) and (2.6), among these variations there are the following relations:

$$\begin{cases} \delta O_{9,m+1} = \delta[(E_2+O_{8,m+1})^{-1}O_{7,m+1}] \\ \qquad = -(E_2+O_{8,m+1})^{-1}\delta(E_2+O_{8,m+1})(E_2+O_{8,m+1})^{-1}O_{7,m+1} \\ \qquad\quad + (E_2+O_{8,m+1})^{-1}\delta O_{7,m+1} \\ \qquad = -(E_2+O_{8,m+1})^{-1}\lambda_{2,m}\delta O_{9,m}\lambda_{1,m}O_{2,m}(E_2+O_{8,m+1})^{-1}O_{7,m+1} \\ \qquad\quad + (E_2+O_{8,m+1})^{-1}\lambda_{2,m}\delta O_{9,m}\lambda_{1,m}(E_1+O_{1,m}) \\ \qquad = (E_2+O_{8,m+1})^{-1}\lambda_{2,m}\delta O_{9,m}\lambda_{1,m}(E_1+O_{1,m} \\ \qquad\quad - O_{2,m}(E_2+O_{8,m+1})^{-1}O_{7,m+1}), \\ \delta\widehat{E}_{2,m+1} = \delta(E_2+O_{8,m+1})^{-1} \\ \qquad = -(E_2+O_{8,m+1})^{-1}\lambda_{2,m}\delta O_{9,m}\lambda_{1,m}O_{2,m}(E_2+O_{8,m+1})^{-1}, \\ \delta O_{10,m+1} = \delta\overline{\omega}_{m+1} - \delta O_{11,m+1}O_{9,m+1} \\ \qquad = (-\delta O_{10,m} + O_{11,m}\delta O_{9,m})\lambda_{1,m}(E_1+O_{1,m}-O_{2,m}O_{9,m+1}), \\ \delta\boldsymbol{h}_{1,m+1} = \delta\boldsymbol{h}_{1,m} + \cdots. \end{cases}$$

Therefore, we have the following inequalities:

$$\begin{cases}\|\delta O_{9,m+1}\|\leqslant\left(1-\dfrac{\varepsilon}{2}\right)^2\|\delta O_{9,m}\|,\\ \|\delta\widetilde{E}_{2,m+1}\|\leqslant\dfrac{c_4}{M}\left(1-\dfrac{\varepsilon}{2}\right)^2\|\delta O_{9,m}\|,\\ \|\delta O_{10,m+1}\|\leqslant\left(1-\dfrac{\varepsilon}{2}\right)\|\delta O_{10,m}\|+\dfrac{2c_4}{M}\left(1-\dfrac{\varepsilon}{2}\right)^{m+1}\|\delta O_{9,m}\|,\\ \cdots.\end{cases}$$

These inequalities can be rewritten in the vectorial form:

$$\boldsymbol{Z}_{m+1}\equiv\begin{pmatrix}\|\delta O_9\|\\ \|\delta\widetilde{E}_2\|\\ \|\delta O_{10}\|\\ \|\delta O_{11}\|\\ \|\delta\omega\|\\ \|\delta O_5\|\\ \|\delta\boldsymbol{h}_2\|\\ \|\delta\boldsymbol{h}_1\|\end{pmatrix}_{m+1}\leqslant\begin{pmatrix}\left(1-\frac{\varepsilon}{2}\right)^2 & & & & & & & \\ c & 0 & & & & & 0 & \\ c & 0 & 1-\frac{\varepsilon}{2} & & & & & \\ c & 0 & c & 0 & & & & \\ 0 & c & 0 & 0 & 1-\frac{\varepsilon}{2} & & & \\ 0 & c & 0 & c & c & 1 & & \\ c & c & 0 & 0 & 0 & 0 & 1-\frac{\varepsilon}{2} & \\ c & c & c & c & c & 0 & c & 1\end{pmatrix}\boldsymbol{Z}_m,$$

where c is a bounded constant. The modulus of every eigenvalue of the matrix on the right hand side of the above vectorial inequality is less than or equal to 1, and the two eigenvectors corresponding to the double eigenvalue which is equal to 1 are linearly independent, so we have the following inequalities:

$$\boldsymbol{Z}_m\leqslant c_8'\boldsymbol{Z}_{m^*},\quad m=m^*,\ m^*+1,\ \cdots,\ M,$$

where c_8' is a matrix whose elements are nonnegative, bounded and independent of M, and m^* is any number among 1, 2, $\cdots$, M. After obtaining these inequalities, and using the relations

$$\|\delta O_{7,m+1}\|\leqslant c\|\delta O_{9,m}\|,\quad \|\delta O_{8,m+1}\|\leqslant c\|\delta O_{9,m}\|,$$
$$\|\delta\bar{\omega}_m\|\leqslant c[\|\delta O_{10,m}\|+\|\delta O_{11,m}\|],\quad \|\delta O_{9,m^*}\|\leqslant c[\|\delta O_{7,m^*}\|+\|\delta O_{8,m^*}\|],$$
$$\|\delta\widetilde{E}_{2,m^*}\|\leqslant c\|\delta O_{8,m^*}\|,\quad \|\delta O_{10,m^*}\|\leqslant c[\|\delta\bar{\omega}_{m^*}\|+\|\delta O_{11,m^*}\|],$$

where c is a constant independent of M, and noticing the expressions (2.6), we can obtain

$$\begin{aligned}\|\delta f\|\leqslant c_8[&\|\delta O_{5,m^*}\|+\|\delta O_{11,m^*}\|+\|\delta O_{7,m^*}\|+\|\delta O_{8,m^*}\|\\ &+\|\delta\omega_{m^*}\|+\|\delta\bar{\omega}_{m^*}\|+\|\delta\boldsymbol{h}_{1,m^*}\|+\|\delta\boldsymbol{h}_{2,m^*}\|],\end{aligned}\tag{2.7}$$

where f is any one among $O_{5,m}$, $O_{11,m}$, $\cdots$, $\boldsymbol{h}_{2,m}$, $\boldsymbol{g}_{1,m}$, and $\boldsymbol{g}_{2,m}$, m being equal to m^*, m^*+1, $\cdots$, M; and c_8 is a constant. That is, we have proved that the fictitious procedure is stable.

In order to prove that the real procedure is stable, we derive several relations between a certain set of quantities in the real procedure of calculation and a certain set of quantities in the fictitious procedure. Let us introduce the following notation:

$$\left\{\begin{aligned}
&H_{1,m}=\begin{pmatrix} E_1 & O_{5,m} \\ 0 & \omega_m \end{pmatrix} S_0 F_{0,m}^{-1}, \quad H_{2,m}=\begin{pmatrix} \bar{\omega}_m & \bar{\omega}_m O_{6,m} \\ O_{7,m} & E_2+O_{8,m} \end{pmatrix} S_m F_{m,m}^{-1}, \\
&G_{1,m}=F_{m-1,m-1} S_{m-1}^{-1} \begin{pmatrix} E_1 & 0 \\ O_{7,m-1} & E_2+O_{8,m-1} \end{pmatrix}^{-1} \begin{pmatrix} 0 & 0 \\ 0 & \omega_{m-1} \end{pmatrix} S_0 F_{0,m}^{-1}, \\
&G_{2,m}=F_{m-1,m-1} S_{m-1}^{-1} \begin{pmatrix} E_1 & 0 \\ O_{7,m-1} & E_2+O_{8,m-1} \end{pmatrix}^{-1} \\
&\qquad \times \begin{pmatrix} \lambda_{1,m-1}(E_1+O_{1,m-1}) & \lambda_{1,m-1} O_{2,m-1} \\ 0 & 0 \end{pmatrix} S_m F_{m,m}^{-1}.
\end{aligned}\right. \tag{2.8}$$

The equations (2.2.1), (2.2.2) are equivalent to (2.1)′, so they are also equivalent to the system which consists of the equations in (1) with $i=0$, $1, \cdots, m-1$. Therefore, from (2.2.1), (2.2.2) we can derive the equations (1.3.1)—(1.3.3) with $i=m-1$ which are obtained from the equations in (1) with $i=0, 1, \cdots, m-1$. Moreover, from the structure of (2.2.1), we can say with certainty that (1.3.1) with $i=m-1$ can be obtained from (2.2.2), and (1.3.2) with $i=m-1$ can be obtained from (2.2.1) with $i=m-1$ and (2.2.2). That is, there exist two matrices T_m and V_m, such that

$$\left\{\begin{aligned}
&T_m H_{1,m}=\begin{pmatrix} E_{1,m} & \nu_{1,m} \\ 0 & \nu_{2,m} \end{pmatrix}, \quad \left(\text{therefore, } T_m H_{1,m}\begin{pmatrix} E_{1,m} \\ 0 \end{pmatrix}=\begin{pmatrix} E_{1,m} \\ 0 \end{pmatrix}\right), \\
&T_m H_{2,m}=\begin{pmatrix} \mu_{1,m} & 0 \\ \mu_{2,m} & E_{2,m} \end{pmatrix}, \quad \left(\text{therefore, } T_m H_{2,m}\begin{pmatrix} 0 \\ E_{2,m} \end{pmatrix}=\begin{pmatrix} 0 \\ E_{2,m} \end{pmatrix}\right), \\
&T_m\begin{pmatrix} \boldsymbol{h}_{1,m} \\ \boldsymbol{h}_{2,m} \end{pmatrix}=\begin{pmatrix} \boldsymbol{d}_{1,m} \\ \boldsymbol{d}_{2,m} \end{pmatrix}, \\
&(E_{1,m-1}\ 0) G_{1,m}+V_m H_{1,m}=(0\ \alpha_{1,m}), \\
&\qquad \left(\text{thus, } (E_{1,m-1}\ 0) G_{1,m}\begin{pmatrix} E_{1,m} \\ 0 \end{pmatrix}+V_m H_{1,m}\begin{pmatrix} E_{1,m} \\ 0 \end{pmatrix}=0\right), \\
&(E_{1,m-1}\ 0) G_{2,m}+V_m H_{2,m}=(\beta_{1,m}\ 0), \\
&\qquad \left(\text{thus, } (E_{1,m-1}\ 0) G_{2,m}\begin{pmatrix} 0 \\ E_{2,m} \end{pmatrix}+V_m H_{2,m}\begin{pmatrix} 0 \\ E_{2,m} \end{pmatrix}=0\right), \\
&(E_{1,m-1}\ 0) F_{m-1,m-1} S_{m-1}^{-1}\begin{pmatrix} \boldsymbol{g}_{1,m-1} \\ \boldsymbol{g}_{2,m-1} \end{pmatrix}+V_m\begin{pmatrix} \boldsymbol{h}_{1,m} \\ \boldsymbol{h}_{2,m} \end{pmatrix}=\boldsymbol{\gamma}_{1,m}.
\end{aligned}\right. \tag{2.9}$$

Furthermore, we can obtain the following expressions:

$$\begin{cases} T_m = \left[H_{1,m}\begin{pmatrix} E_{1,m} & 0 \\ 0 & 0 \end{pmatrix} + H_{2,m}\begin{pmatrix} 0 & 0 \\ 0 & E_{2,m} \end{pmatrix} \right]^{-1}, \\ V_m = -(E_{1,m-1}\ 0)\left[G_{1,m}\begin{pmatrix} E_{1,m} & 0 \\ 0 & 0 \end{pmatrix} + G_{2,m}\begin{pmatrix} 0 & 0 \\ 0 & E_{2,m} \end{pmatrix} \right] T_m, \\ \begin{pmatrix} \mu_{1,m} & \nu_{1,m} \\ \mu_{2,m} & \nu_{2,m} \end{pmatrix} = T_m \left[H_{1,m}\begin{pmatrix} 0 & 0 \\ 0 & E_{2,m} \end{pmatrix} + H_{2,m}\begin{pmatrix} E_{1,m} & 0 \\ 0 & 0 \end{pmatrix} \right], \\ \begin{pmatrix} \boldsymbol{d}_{1,m} \\ \boldsymbol{d}_{2,m} \end{pmatrix} = T_m \begin{pmatrix} \boldsymbol{h}_{1,m} \\ \boldsymbol{h}_{2,m} \end{pmatrix}, \\ (\beta_{1,m}\ \ \alpha_{1,m}) = (E_{1,m-1}\ 0)\left\{ G_{1,m}\begin{pmatrix} 0 & 0 \\ 0 & E_{2,m} \end{pmatrix} + G_{2,m}\begin{pmatrix} E_{1,m} & 0 \\ 0 & 0 \end{pmatrix} \right\} \\ \qquad + V_m \left[H_{1,m}\begin{pmatrix} 0 & 0 \\ 0 & E_{2,m} \end{pmatrix} + H_{2,m}\begin{pmatrix} E_{1,m} & 0 \\ 0 & 0 \end{pmatrix} \right], \\ \boldsymbol{\gamma}_{1,m} = (E_{1,m-1}\ 0) F_{m-1,m-1} S_{m-1}^{-1} \begin{pmatrix} \boldsymbol{g}_{1,m-1} \\ \boldsymbol{g}_{2,m-1} \end{pmatrix} + V_m \begin{pmatrix} \boldsymbol{h}_{1,m} \\ \boldsymbol{h}_{2,m} \end{pmatrix}. \end{cases} \tag{2.10}$$

By using these expressions, we can obtain the desired conclusion. In preparation for obtaining the conclusion, we first prove that the norms of T_m and T_m^{-1} are less than a constant independent of M. In order to prove that, we observe the matrices

$$\begin{cases} Q_m = \left(\begin{pmatrix} E_1 & O_{5,m} \\ 0 & \omega_m \end{pmatrix} S_0 \quad \begin{pmatrix} \bar{\omega}_m & \bar{\omega}_m O_{6,m} \\ O_{7,m} & E_2 + O_{8,m} \end{pmatrix} S_m \right), \\ Q_m^* = \left(\begin{pmatrix} E_{1,m} & \nu_{1,m} \\ 0 & \nu_{2,m} \end{pmatrix} F_{0,m} \quad \begin{pmatrix} \mu_{1,m} & 0 \\ \mu_{2,m} & E_{2,m} \end{pmatrix} F_{m,m} \right). \end{cases} \tag{2.11}$$

Because S_0 is invertible, the rank of Q_m is equal to the rank of the matrix

$$\bar{Q}_m = \left(\begin{pmatrix} E_1 & O_{5,m} \\ 0 & \omega_m \end{pmatrix} \quad \begin{pmatrix} \bar{\omega}_m & 0 \\ 0 & E_2 \end{pmatrix} \begin{pmatrix} E_1 & O_{6,m} \\ O_{7,m} & E_2 + O_{8,m} \end{pmatrix} S_m S_0^{-1} \right).$$

It is easy to prove that the rank of $\bar{Q}_m$ is n, so the rank of Q_m also is n. In fact, if the horizontal vector $(\bar{c}_1, \bar{c}_2, \cdots, \bar{c}_n)$ satisfies the condition

$$(\bar{c}_1, \bar{c}_2, \cdots, \bar{c}_n)\bar{Q}_m = (0, 0, \cdots, 0),$$

then according to the structure of the first $(n-s)$ columns of $\bar{Q}_m$, we can certainly say that $\bar{c}_i = 0$, $i = 1, \cdots, n-s$. Moreover, when M is sufficiently large,

$$\begin{pmatrix} E_1 & O_{6,m} \\ O_{7,m} & E_2 + O_{8,m} \end{pmatrix} S_m S_0^{-1}$$

is invertible. Therefore the lower s vectors in the right half of $\bar{Q}_m$ are linearly independent, i.e., according to $\bar{c}_1 = \cdots = \bar{c}_{n-s} = 0$, we can say with certainty that $\bar{c}_{n-s+1} = \cdots = \bar{c}_n = 0$. This means that the rank of $\bar{Q}_m$ is n, i.e., the rank of Q_m is n. In addition, those elements of Q_m are bounded, so the Gram determinant corresponding to the row vectors of Q_m, $\Gamma(Q_m)$, is greater than zero and bounded. Moreover, when m and M are sufficiently large, we have

$$Q_m \approx \left(\begin{pmatrix} E_1 & 0 \\ 0 & 0 \end{pmatrix} S_0 \quad \begin{pmatrix} 0 & 0 \\ 0 & E_2 \end{pmatrix} S_m\right),$$

i.e.,

$$\Gamma(Q_m) \approx \Gamma((E_1\ 0)S_0) \cdot \Gamma((0\ E_2)S_m).$$

Also, according to the inequality in condition (2)

$$0 < c_2 \leqslant \Gamma(S_i) \leqslant c_3, \quad i = 0, 1, \cdots, M,$$

we can derive a similar inequality for $\Gamma((0\ E_2)S_m)$. Therefore, we can find two constants a_Q and b_Q, independent of M, such that

$$0 < a_Q \leqslant \Gamma(Q_m) \leqslant b_Q, \quad m = 1, 2, \cdots, M.$$

In addition, we can prove that there are two constants a_Q^* and b_Q^*, independent of M, such that

$$0 < a_Q^* \leqslant \Gamma(Q_m^*) \leqslant b_Q^*, \quad m = 1, 2, \cdots, M.$$

In fact, $\Gamma(Q_m^*)$ is equal to the sum of all squares of determinants of order n in Q_m^*, and the sum of the orders of $E_{1,m}$ and $E_{2,m}$ is n, so we have

$$a_Q^* = 1 \leqslant \Gamma(Q_m^*).$$

Moreover, because the elimination method with pivot-elements is adopted, there is a constant b_Q^* such that $\Gamma(Q_m^*) \leqslant b_Q^*$.

According to these results, it is easy to prove the norms of T_m and T_m^{-1} are less than a constant independent of M. In fact,

$$T_m Q_m = Q_m^*, \quad |T_m|^2 \Gamma(Q_m) = \Gamma(Q_m^*),$$

so

$$|T_m|^2 = \frac{\Gamma(Q_m^*)}{\Gamma(Q_m)} \geqslant \frac{1}{b_Q} > 0.$$

Moreover, the lower bound of the modulus of the determinant of order n which has the maximal modulus in Q_m is greater than zero, and the moduli of elements of Q_m^* are uniformly bounded with respect to M. Therefore, according to the relation $T_m Q_m = Q_m^*$, the moduli of elements of T_m are uniformly bounded. From this result and the fact that $|T_m|^2 \geqslant 1/b_Q > 0$, it follows that the norms of T_m and T_m^{-1} are less than a constant independent of M.

Incidentally, we point out that because the rank of Q_m is always n, it is always possible to obtain the equations (1. 3. 1), and that the matrices in (1. 3. 1) are always $(n-s) \sim s$-broken when m is sufficiently large. Moreover, in this case, the rank of the left half of Q_m "approximates" $n-s$, i.e., there is at least one determinant of order $(n-s)$ in Q_m whose modulus is greater than a positive constant and all determinants of order $(n-s+1)$ are close to zero; and the rank of the right half "approximates" s. Therefore, the order of $E_{1,m}$ in Q_m^* is not greater than $n-s$, the order of $E_{2,m}$ is not greater than s, i.e., $n-s_m \leqslant n-s$ and $s_m \leqslant s$.

From these results, it follows that $s=s_m$, i.e., those matrices in (1.3.1) must be $(n-s)\sim s$–broken.

In preparation for a discussion of the stability of the elimination procedure, we must also derive the expressions for $H_{1,m}$, $\cdots$, $G_{2,m}$ and T_m, $\cdots$, $\gamma_{1,m}$ with rather large M and m. In this case, $O_{5,m}$, $\cdots$, $O_{8,m}$, ω_m, and $\bar{\omega}_m$ are very small. For simplicity, we introduce some notation to express all the small quantities. δ denotes any quantity of $O(1/M)$, ε_m denotes any quantity of $O((1-\varepsilon/2)^m)$, and $\delta_m\equiv\delta+\varepsilon_m$. In addition, (1.3.1)—(1.3.5) must be $(n-s)\sim s$–broken, so we introduce the following $(n-s)\sim s$–broken expressions for S_iF^{-1} and $F_{i,j}S_i^{-1}$:

$$S_iF_{i,j}^{-1}=\begin{pmatrix} s_{i,j} & t_{i,j} \\ u_{i,j} & v_{i,j}\end{pmatrix},\quad (S_iF_{i,j}^{-1})^{-1}=F_{i,j}S_j^{-1}=\begin{pmatrix} r_{i,j} & l_{i,j} \\ m_{i,j} & n_{i,j}\end{pmatrix}.$$

According to these symbols, the expressions for $H_{1,m}$, $H_{2,m}$, $G_{1,m}$ and $G_{2,m}$ in (2.8) and the expressions for T_m, V_m, $\mu_{1,m}$, $\cdots$, and $\gamma_{1,m}$ in (2.10) can be rewritten as

$$\left\{\begin{aligned}
&H_{1,m}=\begin{pmatrix} s_{0,m}+\delta & t_{0,m}+\delta \\ \varepsilon_m & \varepsilon_m\end{pmatrix},\\
&H_{2,m}=\begin{pmatrix} \varepsilon_m & \varepsilon_m \\ u_{m,m}+\delta & v_{m,m}+\delta\end{pmatrix},\\
&G_{1,m}=\begin{pmatrix} r_{m-1,m-1}+\delta & l_{m-1,m-1}+\delta \\ m_{m-1,m-1}+\delta & n_{m-1,m-1}+\delta\end{pmatrix}\begin{pmatrix} 0 & 0 \\ \varepsilon_m & \varepsilon_m\end{pmatrix}=\begin{pmatrix} \varepsilon_m & \varepsilon_m \\ \varepsilon_m & \varepsilon_m\end{pmatrix},\\
&G_{2,m}=\begin{pmatrix} r_{m-1,m-1}\lambda_{1,m-1}s_{m,m}+\delta & r_{m-1,m-1}\lambda_{1,m-1}t_{m,m}+\delta \\ m_{m-1,m-1}\lambda_{1,m-1}s_{m,m}+\delta & m_{m-1,m-1}\lambda_{1,m-1}t_{m,m}+\delta\end{pmatrix},\\
&T_m=\begin{pmatrix} s_{0,m}+\delta & \varepsilon_m \\ \varepsilon_m & v_{m,m}+\delta\end{pmatrix}^{-1}=\begin{pmatrix} s_{0,m}^{-1}+\delta_m & \varepsilon_m \\ \varepsilon_m & v_{m,m}^{-1}+\delta_m\end{pmatrix},\\
&V_m=-(\varepsilon_m \quad r_{m-1,m-1}\lambda_{1,m-1}t_{m,m}+\delta)\begin{pmatrix} s_{0,m}^{-1}+\delta_m & \varepsilon_m \\ \varepsilon_m & v_{m,m}^{-1}+\delta_m\end{pmatrix}\\
&\quad\;=-(\varepsilon_m \quad r_{m-1,m-1}\lambda_{1,m-1}t_{m,m}v_{m,m}^{-1}+\delta_m),\\
&\begin{pmatrix} \mu_{1,m} & \nu_{1,m} \\ \mu_{2,m} & \nu_{2,m}\end{pmatrix}=\begin{pmatrix} s_{0,m}^{-1}+\delta & \varepsilon_m \\ \varepsilon_m & v_{m,m}^{-1}+\delta\end{pmatrix}\begin{pmatrix} \varepsilon_m & t_{0,m}+\delta \\ u_{m,m}+\delta & \varepsilon_m\end{pmatrix}\\
&\qquad\qquad=\begin{pmatrix} \varepsilon_m & s_{0,m}^{-1}t_{0,m}+\delta_m \\ v_{m,m}^{-1}u_{m,m}+\delta_m & \varepsilon_m\end{pmatrix},\\
&\begin{pmatrix} \boldsymbol{d}_{1,m} \\ \boldsymbol{d}_{2,m}\end{pmatrix}=T_m\begin{pmatrix} \boldsymbol{h}_{1,m} \\ \boldsymbol{h}_{2,m}\end{pmatrix}=\begin{pmatrix} s_{0,m}^{-1}\boldsymbol{h}_{1,m}+\delta_m \\ v_{m,m}^{-1}\boldsymbol{h}_{2,m}+\delta_m\end{pmatrix},\\
&(\beta_{1,m} \quad \alpha_{1,m})=(r_{m-1,m-1}\lambda_{1,m-1}s_{m,m}+\delta \quad \varepsilon_m)\\
&\qquad\qquad-(\varepsilon_m \quad r_{m-1,m-1}\lambda_{1,m-1}t_{m,m}v_{m,m}^{-1}+\delta_m)\begin{pmatrix} \varepsilon_m & t_{0,m}+\delta \\ u_{m,m}+\delta & \varepsilon_m\end{pmatrix}
\end{aligned}\right. \tag{2.12}$$

$$\left\{\begin{aligned}&=(r_{m-1,m-1}\lambda_{1,m-1}(s_{m,m}-t_{m,m}v_{m,m}^{-1}u_{m,m})+\delta_m\ \varepsilon_m)\\&=(r_{m-1,m-1}\lambda_{1,m}r_{m,m}^{-1}+\delta_m\ \varepsilon_m),\\ \boldsymbol{\gamma}_{1,m}&=r_{m-1,m-1}\boldsymbol{g}_{1,m-1}+l_{m-1,m-1}\boldsymbol{g}_{2,m-1}-r_{m-1,m-1}\lambda_{1,m-1}t_{m,m}v_{m,m}^{-1}\boldsymbol{h}_{2,m}+\delta_m.\end{aligned}\right.$$

When we derive these expressions, we use the following formula:

$$\begin{pmatrix} a & b \\ c & d \end{pmatrix}^{-1}=\begin{pmatrix} (a-bd^{-1}c)^{-1} & -(a-bd^{-1}c)^{-1}bd^{-1} \\ -(d-ca^{-1}b)^{-1}ca^{-1} & (d-ca^{-1}b)^{-1} \end{pmatrix},$$

where a and d are square matrics, and b and c are matrices.

In the following, we shall discuss the stability of the procedure of calculating $\mu_{1,i}$, $\cdots$, $\boldsymbol{d}_{2,i}$, $\alpha_{1,i}$, $\beta_{1,i}$ and $\boldsymbol{\gamma}_{1,i}$. According to the definition of stability, if the number of operations (addition, subtraction, multiplication, and division) is finite, and if the lower bound of all moduli of divisors is greater than zero, then the procedure must be stable. When calculating $\mu_{1,i}$, $\cdots$, $\boldsymbol{\gamma}_{1,i}$, we adopt the pivoting-elimination method. Moreover, from the above discussion, we know that the pivot elements are not equal to zero. Therefore, if M is finite, then the procedure must be stable. That is, for the method described in Section 1, the proof of stability is reduced to proving that there exists a bounded constant c^* such that the error for any M is less than c^* times initial error as $M\to\infty$. Therefore, in the following, only for $m\geqslant m^*$ where m^* is a positive constant, shall we discuss problems. By using the fact that the fictitious procedure is stable, we can prove that the real procedure is stable for $m\geqslant m^*$, where m^* is so large that T_m, V_m, $\cdots$, $\boldsymbol{\gamma}_{1,m}$ can be expressed in the form (2.12). For that, we derive the relations between the set of variations $\delta\mu_{1,m}$, $\cdots$, $\delta\boldsymbol{\gamma}_{1,m}$ and the set of variations $\delta O_{5,m}$, $\cdots$, $\delta\boldsymbol{h}_{2,m}$, $\delta\boldsymbol{g}_{1,m-1}$, $\delta\boldsymbol{g}_{2,m-1}$. By taking variations of equations (2.10) and (2.8), we can obtain the following expressions for $\delta\mu_{1,m}$, $\cdots$, $\delta\boldsymbol{\gamma}_{1,m}$ in terms of $\delta O_{5,m}$, $\cdots$, $\delta\boldsymbol{g}_{2,m-1}$:

$$\left\{\begin{aligned}&\begin{pmatrix}\delta\mu_{1,m} & \delta\nu_{1,m}\\ \delta\mu_{2,m} & \delta\nu_{2,m}\end{pmatrix}=\delta T_m\left[H_{1,m}\begin{pmatrix}0 & 0\\ 0 & E_{2,m}\end{pmatrix}+H_{2,m}\begin{pmatrix}E_{1,m} & 0\\ 0 & 0\end{pmatrix}\right]\\&\qquad\qquad+T_m\left[\delta H_{1,m}\begin{pmatrix}0 & 0\\ 0 & E_{2,m}\end{pmatrix}+\delta H_{2,m}\begin{pmatrix}E_{1,m} & 0\\ 0 & 0\end{pmatrix}\right],\\&\begin{pmatrix}\delta\boldsymbol{d}_{1,m}\\ \delta\boldsymbol{d}_{2,m}\end{pmatrix}=\delta T_m\begin{pmatrix}\boldsymbol{h}_{1,m}\\ \boldsymbol{h}_{2,m}\end{pmatrix}+T_m\begin{pmatrix}\delta\boldsymbol{h}_{1,m}\\ \delta\boldsymbol{h}_{2,m}\end{pmatrix},\\&(\delta\beta_{1,m}\ \delta\alpha_{1,m})=(E_{1,m-1}\ 0)\left[\delta G_{1,m}\begin{pmatrix}0 & 0\\ 0 & E_{2,m}\end{pmatrix}+\delta G_{2,m}\begin{pmatrix}E_{1,m} & 0\\ 0 & 0\end{pmatrix}\right]\\&\qquad\qquad+\delta V_m\left[H_{1,m}\begin{pmatrix}0 & 0\\ 0 & E_{2,m}\end{pmatrix}+H_{2,m}\begin{pmatrix}E_{1,m} & 0\\ 0 & 0\end{pmatrix}\right]\\&\qquad\qquad+V_m\left[\delta H_{1,m}\begin{pmatrix}0 & 0\\ 0 & E_{2,m}\end{pmatrix}+\delta H_{2,m}\begin{pmatrix}E_{1,m} & 0\\ 0 & 0\end{pmatrix}\right],\end{aligned}\right.\tag{2.13}$$

$$\begin{cases}\delta\boldsymbol{\gamma}_{1,m}=(E_{1,m-1}\ \ 0)F_{m-1,m-1}S_{m-1}^{-1}\begin{pmatrix}\delta\boldsymbol{g}_{1,m-1}\\ \delta\boldsymbol{g}_{2,m-1}\end{pmatrix}\\ \qquad +\delta V_m\begin{pmatrix}\boldsymbol{h}_{1,m}\\ \boldsymbol{h}_{2,m}\end{pmatrix}+V_m\begin{pmatrix}\delta\boldsymbol{h}_{1,m}\\ \delta\boldsymbol{h}_{2,m}\end{pmatrix},\end{cases}$$

where

$$\begin{cases}\delta T_m=-T_m\delta T_m^{-1}T_m=-T_m\left[\delta H_{1,m}\begin{pmatrix}E_{1,m} & 0\\ 0 & 0\end{pmatrix}+\delta H_{2,m}\begin{pmatrix}0 & 0\\ 0 & E_{2,m}\end{pmatrix}\right]T_m,\\ \delta V_m=-(E_{1,m-1}\ \ 0)\left[\delta G_{1,m}\begin{pmatrix}E_{1,m} & 0\\ 0 & 0\end{pmatrix}+\delta G_{2,m}\begin{pmatrix}0 & 0\\ 0 & E_{2,m}\end{pmatrix}\right]T_m\\ \qquad -(E_{1,m-1}\ \ 0)\left[G_{1,m}\begin{pmatrix}E_{1,m} & 0\\ 0 & 0\end{pmatrix}+G_{2,m}\begin{pmatrix}0 & 0\\ 0 & E_{2,m}\end{pmatrix}\right]\delta T_m,\\ \delta H_{1,m}=\begin{pmatrix}0 & \delta O_{5,m}\\ 0 & \delta\omega_m\end{pmatrix}S_0F_{0,m}^{-1},\\ \delta H_{2,m}=\begin{pmatrix}\delta\bar{\omega}_m & \delta(\bar{\omega}O_{6,m})\\ \delta O_{7,m} & \delta O_{8,m}\end{pmatrix}S_mF_{m,m}^{-1},\\ \delta G_{1,m}=-F_{m-1,m-1}S_{m-1}^{-1}\begin{pmatrix}E_1 & 0\\ O_{7,m-1} & E_2+O_{8,m-1}\end{pmatrix}^{-1}\begin{pmatrix}0 & 0\\ \delta O_{7,m-1} & \delta O_{8,m-1}\end{pmatrix}\\ \qquad\times\begin{pmatrix}E_1 & 0\\ O_{7,m-1} & E_2+O_{8,m-1}\end{pmatrix}^{-1}\begin{pmatrix}0 & 0\\ 0 & \omega_{m-1}\end{pmatrix}\\ \qquad\times S_0F_{0,m}^{-1}+F_{m-1,m-1}S_{m-1}^{-1}\begin{pmatrix}E_1 & 0\\ O_{7,m-1} & E_2+O_{8,m-1}\end{pmatrix}^{-1}\\ \qquad\times\begin{pmatrix}0 & 0\\ 0 & \delta\omega_{m-1}\end{pmatrix}S_0F_{0,m}^{-1},\\ \delta G_{2,m}=-F_{m-1,m-1}S_{m-1}^{-1}\begin{pmatrix}E_1 & 0\\ O_{7,m-1} & E_2+O_{8,m-1}\end{pmatrix}^{-1}\\ \qquad\times\begin{pmatrix}0 & 0\\ \delta O_{7,m-1} & \delta O_{8,m-1}\end{pmatrix}\begin{pmatrix}E_1 & 0\\ O_{7,m-1} & E_2+O_{8,m-1}\end{pmatrix}^{-1}\\ \qquad\times\begin{pmatrix}\lambda_{1,m-1}(E_1+O_{1,m-1}) & \lambda_{1,m-1}O_{2,m-1}\\ 0 & 0\end{pmatrix}S_mF_{m,m}^{-1}.\end{cases}$$

In order to obtain the desired conclusion, we also need to derive expressions for $\delta O_{5,m}$, $\cdots$, $\delta\boldsymbol{h}_{2,m}$ in terms of $\delta\mu_{1,m}$, $\cdots$, $\delta\boldsymbol{\gamma}_{1,m}$. (Indeed, $\delta O_{5,m}$, $\cdots$, $\delta\boldsymbol{h}_{2,m}$ depend on $\delta\mu_{1,m}$, $\delta\mu_{2,m}$, $\delta\nu_{1,m}$, $\delta\nu_{2,m}$, $\delta\boldsymbol{d}_{1,m}$ and $\delta\boldsymbol{d}_{2,m}$ only.) According to (2.9), we have

$$\begin{cases}\delta H_{1,m}=\delta T_m^{-1}\begin{pmatrix}E_{1,m} & \nu_{1,m}\\ 0 & \nu_{2,m}\end{pmatrix}+T_m^{-1}\begin{pmatrix}0 & \delta\nu_{1,m}\\ 0 & \delta\nu_{2,m}\end{pmatrix}\\ \qquad =\delta H_{1,m}\begin{pmatrix}E_{1,m} & \nu_{1,m}\\ 0 & 0\end{pmatrix}+\delta H_{2,m}\begin{pmatrix}0 & 0\\ 0 & \nu_{2,m}\end{pmatrix}+T_m^{-1}\begin{pmatrix}0 & \delta\nu_{1,m}\\ 0 & \delta\nu_{2,m}\end{pmatrix},\end{cases}$$

$$\begin{cases}\delta H_{2,m}=\delta T_m^{-1}\begin{pmatrix}\mu_{1,m} & 0\\ \mu_{2,m} & E_{2,m}\end{pmatrix}+T_m^{-1}\begin{pmatrix}\delta\mu_{1,m} & 0\\ \delta\mu_{2,m} & 0\end{pmatrix}\\ \qquad =\delta H_{1,m}\begin{pmatrix}\mu_{1,m} & 0\\ 0 & 0\end{pmatrix}+\delta H_{2,m}\begin{pmatrix}0 & 0\\ \mu_{2,m} & E_{2,m}\end{pmatrix}+T_m^{-1}\begin{pmatrix}\delta\mu_{1,m} & 0\\ \delta\mu_{2,m} & 0\end{pmatrix}.\end{cases}$$

When m is sufficiently large, $\begin{pmatrix}E_{1,m} & \nu_{1,m}\\ 0 & \nu_{2,m}\end{pmatrix}$ and $\begin{pmatrix}\mu_{1,m} & 0\\ \mu_{2,m} & E_{2,m}\end{pmatrix}$ are $(n-s)\sim s$–broken. Therefore, the above two expressions can be rewritten as

$$\begin{cases}\begin{pmatrix}\delta O_{5,m}u_{0,m} & \delta O_{5,m}v_{0,m}\\ \delta\omega_m u_{0,m} & \delta\omega_m v_{0,m}\end{pmatrix}=\begin{pmatrix}\delta O_{5,m}u_{0,m} & \delta O_{5,m}u_{0,m}\nu_{1,m}\\ \delta\omega_m u_{0,m} & \delta\omega_m u_{0,m}\nu_{1,m}\end{pmatrix}\\ \qquad\qquad +\delta H_{2,m}\begin{pmatrix}0 & 0\\ 0 & \nu_{2,m}\end{pmatrix}+T_m^{-1}\begin{pmatrix}0 & \delta\nu_{1,m}\\ 0 & \delta\nu_{2,m}\end{pmatrix},\\ \begin{pmatrix}\delta\overline{\omega}_m S_{m,m}+\delta(\overline{\omega}_m O_{6,m})u_{m,m} & \delta\overline{\omega}_m t_{m,m}+\delta(\overline{\omega}_m O_{6,m})v_{m,m}\\ \delta O_{7,m}S_{m,m}+\delta O_{8,m}u_{m,m} & \delta O_{7,m}t_{m,m}+\delta O_{8,m}v_{m,m}\end{pmatrix}\\ \quad =\begin{pmatrix}\delta O_{5,m}u_{0,m}\mu_{1,m} & 0\\ \delta\omega_m u_{0,m}\mu_{1,m} & 0\end{pmatrix}\\ \quad +\begin{pmatrix}(\delta\overline{\omega}_m t_{m,m}+\delta(\overline{\omega}O_{6,m})v_{m,m})\mu_{2,m} & \delta\overline{\omega}_m t_{m,m}+\delta(\overline{\omega}_m O_{6,m})v_{m,m}\\ (\delta O_{7,m}t_{m,m}+\delta O_{8,m}v_{m,m})\mu_{2,m} & \delta O_{7,m}t_{m,m}+\delta O_{8,m}v_{m,m}\end{pmatrix}\\ \quad +T_m^{-1}\begin{pmatrix}\delta\mu_{1,m} & 0\\ \delta\mu_{2,m} & 0\end{pmatrix}.\end{cases}$$

The expressions are not unique when $\delta O_{5,m}$, $\cdots$, $\delta\overline{\omega}_m$ are expressed in terms of $\delta\mu_{1,m}$, $\cdots$, $\delta\nu_{2,m}$ according to the above two expressions. We choose the following expressions which make $\delta\overline{\omega}_m t_{m,m}+\delta(\overline{\omega}_m O_{6,m})v_{m,m}$ and $\delta O_{7,m}t_{m,m}+\delta O_{8,m}v_{m,m}$ equal to zero:

$$\begin{cases}\begin{pmatrix}\delta O_{5,m}\\ \delta\omega_m\end{pmatrix}=T_m^{-1}\begin{pmatrix}\delta\nu_{1,m}\\ \delta\nu_{2,m}\end{pmatrix}(v_{0,m}-u_{0,m}\nu_{1,m})^{-1},\\ \begin{pmatrix}\delta\overline{\omega}_m & \delta(\overline{\omega}_m O_{6,m})\\ \delta O_{7,m} & \delta O_{8,m}\end{pmatrix}=\begin{pmatrix}\delta\overline{\omega}_m & \delta O_{11,m}\\ \delta O_{7,m} & \delta O_{8,m}\end{pmatrix}\\ \quad =\left(T_m^{-1}\begin{pmatrix}\delta\nu_{1,m}\\ \delta\nu_{2,m}\end{pmatrix}(v_{0,m}-u_{0,m}\nu_{1,m})^{-1}u_{0,m}\mu_{1,m}+T_m^{-1}\begin{pmatrix}\delta\mu_{1,m}\\ \delta\mu_{2,m}\end{pmatrix}\ \ 0\right)\\ \qquad\times\begin{pmatrix}s_{m,m}-t_{m,m}\mu_{2,m} & t_{m,m}\\ u_{m,m}-v_{m,m}\mu_{2,m} & v_{m,m}\end{pmatrix}^{-1}.\end{cases}$$

Because of

$$(v_{0,m}-u_{0,m}\nu_{1,m})^{-1}=(v_{0,m}-u_{0,m}s_{0,m}^{-1}t_{0,m}+\delta_m)^{-1}=n_{0,m}+\delta_m,$$

$$(s_{m,m}-t_{m,m}\mu_{2,m})^{-1}=(s_{m,m}-t_{m,m}v_m^{-1},_m u_{m,m}+\delta_m)^{-1}=r_{m,m}+\delta_m$$

and $u_{m,m}-v_{m,m}\mu_{2,m}=\delta_m$, all inverse matrices in the above two expressions exist and their norms are uniformly bounded with respect to M.

By using these results, it is easy to prove that the procedure of elimination is stable. From the above expressions, we derive

$$\|\delta f_{m^*}\| \leqslant c_9[\|\delta\mu_{1,m^*}\| + \|\delta\mu_{2,m^*}\| + \|\delta\nu_{1,m^*}\| + \|\delta\nu_{2,m^*}\|],$$

where f_{m^*} is any one among O_{5,m^*}, $\cdots$, and $\bar{\omega}_{m^*}$, and c_9 is a constant independent of M. Also, from (2.9), we derive

$$\begin{aligned}\begin{pmatrix}\delta \boldsymbol{h}_{1,m}\\ \delta \boldsymbol{h}_{2,m}\end{pmatrix} &= \delta T_m^{-1}\begin{pmatrix}\boldsymbol{d}_{1,m}\\ \boldsymbol{d}_{2,m}\end{pmatrix} + T_m^{-1}\begin{pmatrix}\delta\boldsymbol{d}_{1,m}\\ \delta\boldsymbol{d}_{2,m}\end{pmatrix}\\ &= \begin{pmatrix}0 & \delta O_{5,m}\\ 0 & \delta\omega_m\end{pmatrix}\begin{pmatrix}s_{0,m}\boldsymbol{d}_{1,m}\\ u_{0,m}\boldsymbol{d}_{1,m}\end{pmatrix} + \begin{pmatrix}\delta\bar{\omega}_m & \delta(\bar{\omega}_m O_{6,m})\\ \delta O_{7,m} & \delta O_{8,m}\end{pmatrix}\begin{pmatrix}t_{m,m}\boldsymbol{d}_{2,m}\\ v_{m,m}\boldsymbol{d}_{2,m}\end{pmatrix}\\ &\quad + T_m^{-1}\begin{pmatrix}\delta\boldsymbol{d}_{1,m}\\ \delta\boldsymbol{d}_{2,m}\end{pmatrix}.\end{aligned}$$

Therefore, we can obtain

$$\begin{aligned}\|\delta \tilde{f}_{m^*}\| \leqslant c_9'[&\|\delta\mu_{1,m^*}\| + \|\delta\mu_{2,m^*}\| + \|\delta\nu_{1,m^*}\| + \|\delta\nu_{2,m^*}\|\\ &+ \|\delta\boldsymbol{d}_{1,m^*}\| + \|\delta\boldsymbol{d}_{2,m^*}\|],\end{aligned} \tag{2.14}$$

where $\tilde{f}_{m^*}$ is any one among O_{5,m^*}, $\cdots$, $\bar{\omega}_{m^*}$, $\boldsymbol{h}_{1,m^*}$ and $\boldsymbol{h}_{2,m^*}$, and c_9' is a constant. In addition, from (2.13) we can obtain

$$\begin{aligned}\|\delta\tilde{\tilde{f}}_m\| \leqslant c_{10}[&\|\delta O_{5,m}\| + \|\delta O_{11,m}\| + \cdots + \|\delta\boldsymbol{h}_{2,m}\| + \|\delta\boldsymbol{g}_{1,m-1}\|\\ &+ \|\delta\boldsymbol{g}_{2,m-1}\| + \|\delta O_{7,m-1}\| + \|\delta O_{8,m-1}\| + \|\delta\omega_{m-1}\|],\\ &m = m^*+1,\ m^*+2,\ \cdots,\ M,\end{aligned} \tag{2.15}$$

where $\tilde{\tilde{f}}_m$ is any one among $\mu_{1,m}$, $\mu_{2,m}$, $\cdots$, and $\boldsymbol{\gamma}_{1,m}$, and c_{10} is a constant. Therefore, by using (2.7), we obtain

$$\begin{aligned}\|\delta\tilde{\tilde{f}}_m\| \leqslant c_{11}[&\|\delta\mu_{1,m^*}\| + \|\delta\mu_{2,m^*}\| + \|\delta\nu_{1,m^*}\| + \|\delta\nu_{2,m^*}\| + \|\delta\boldsymbol{d}_{1,m^*}\|\\ &+ \|\delta\boldsymbol{d}_{2,m^*}\|],\quad m = m^*+1,\ m^*+2,\ \cdots,\ M\end{aligned}$$

immediately, where c_{11} is a constant independent of M. That is, the elimination procedure, in which (1.3.1)—(1.3.3) are obtained from (1.2.1), (1.2.2), is stable.

In the following, we prove that the procedure of calculation of the unknowns (1.6) is stable. We observe the effect of $\delta\boldsymbol{x}_0$ and $\delta\boldsymbol{x}_M$ on $\delta\boldsymbol{x}_i$. From (1.6), we derive

$$\begin{aligned}(E_{1,i}\ \ 0)\delta(F_{i,i}\boldsymbol{x}_i) &= -(\beta_{1,i+1}\ \ 0)\delta(F_{i+1,i+1}\boldsymbol{x}_{i+1})\\ &\quad - (0\ \ \alpha_{1,i+1})\delta(F_{0,i+1}\boldsymbol{x}_0),\\ (0\ \ E_{2,i})\delta(F_{i,i}\boldsymbol{x}_i) &= -(\mu_{2,i}\ \ 0)\delta(F_{i,i}\boldsymbol{x}_i) - (0\ \ \nu_{2,i})\delta(F_{0,i}\boldsymbol{x}_0)\\ &= \mu_{2,i}(\beta_{1,i+1}\ \ 0)\delta(F_{i+1,i+1}\boldsymbol{x}_{i+1})\\ &\quad + (0\ \ \mu_{2,i}\alpha_{1,i+1})\delta(F_{0,i+1}\boldsymbol{x}_0) - (0\ \ \nu_{2,i})\delta(F_{0,i}\boldsymbol{x}_0).\end{aligned}$$

Let

$$-\beta_{2,i+1} = \mu_{2,i}\beta_{1,i+1},\quad \alpha_{i+1} = \begin{pmatrix}0 & \alpha_{1,i+1}\\ 0 & -\mu_{2,i}\alpha_{1,i+1}\end{pmatrix} + \begin{pmatrix}0 & 0\\ 0 & \nu_{2,i}\end{pmatrix}F_{0,i}F_{0,i+1}^{-1},$$

the above expressions can be rewritten as

$$\delta(F_{i,i}x_i) = -\begin{pmatrix} \beta_{1,i+1} & 0 \\ \beta_{2,i+1} & 0 \end{pmatrix}\delta(F_{i+1,i+1}x_{i+1}) - \alpha_{i+1}\delta(F_{0,i+1}x_0).$$

Furthermore, we can obtain

$$\delta(F_{i,i}x_i) = B(i+1,\ i+1)\delta(F_{i+1,i+1}x_{i+1}) - B(i+1,\ i)\alpha_{i+1}\delta(F_{0,i+1}x_0)$$
$$= B(i+1,\ q)\delta(F_{qq}x_q) - \sum_{j=i+1}^{q} B(i+1,\ j-1)\alpha_j\delta(F_{0,j}x_0).$$

In this expression, we adopt the following notation: $B(i+1,\ i)$ is an $n\times n$-unit matrix, and for $q\geqslant i+1$,

$$B(i+1,\ q) = (-1)^{q-i}\begin{pmatrix} \beta_{1,i+1} & 0 \\ \beta_{2,i+1} & 0 \end{pmatrix}\begin{pmatrix} \beta_{1,i+2} & 0 \\ \beta_{2,i+2} & 0 \end{pmatrix}\cdots\begin{pmatrix} \beta_{1,q} & 0 \\ \beta_{2,q} & 0 \end{pmatrix}.$$

Moreover, from (2.10) and $ur+vm=0$, i.e., $m=-v^{-1}ur$, we can obtain

$$\begin{pmatrix} \beta_{1,i+1} & 0 \\ \beta_{2,i+1} & 0 \end{pmatrix} = \begin{pmatrix} r_{i,i}\lambda_{1,i}r_{i+1,i+1}^{-1} + \delta_{i+1} & 0 \\ -v_{i,i}^{-1}u_{i,i}r_{i,i}\lambda_{1,i}r_{i+1,i+1}^{-1} + \delta_{i+1} & 0 \end{pmatrix}$$
$$= \begin{pmatrix} r_{i,i} & 0 \\ m_{i,i} & E_2 \end{pmatrix}\begin{pmatrix} \lambda_{1,i} + \delta_{i+1} & 0 \\ \delta_{i+1} & 0 \end{pmatrix}\begin{pmatrix} r_{i+1,i+1} & 0 \\ m_{i+1,i+1} & E_2 \end{pmatrix}^{-1}.$$

Therefore,

$$B(i+1,\ q) = (-1)^{q-i}\begin{pmatrix} r_{i,i} & 0 \\ m_{i,i} & E_2 \end{pmatrix}\begin{pmatrix} \lambda_{1,i} + \delta_{i+1} & 0 \\ \delta_{i+1} & 0 \end{pmatrix}$$
$$\cdots\begin{pmatrix} \lambda_{1,q-1} + \delta_q & 0 \\ \delta_q & 0 \end{pmatrix}\begin{pmatrix} r_{q,q} & 0 \\ m_{q,q} & E_2 \end{pmatrix}^{-1}.$$

In addition, when M and j are sufficiently large, the norm of $\begin{pmatrix} \lambda_{1,j} + \delta_{j+1} & 0 \\ \delta_{j+1} & 0 \end{pmatrix}$ is less than $1-\frac{\varepsilon}{2}$; every element of α_j is $O((1-\varepsilon/2)^j)$; and $\|\delta F_{jj}x_j\| = \|\delta x_j\|$. Therefore, we can obtain the following inequalities:

$$\|\delta x_i\| \leqslant c_9\left(1-\frac{\varepsilon}{2}\right)^{M-i}\|\delta x_M\| + c_9\frac{\left(1-\frac{\varepsilon}{2}\right)^{i+1}\left[1-\left(1-\frac{\varepsilon}{2}\right)^{2(M-i)}\right]}{1-\left(1-\frac{\varepsilon}{2}\right)^2}\|\delta x_0\|,$$
$$i = i_1,\ i_1+1,\ \cdots,\ M-1,$$

where c_9 is a positive constant independent of M, and i_1 is a number which makes $\left(1-\frac{\varepsilon}{2}\right)^{i_1}$ less than a certain number. That is, for $i=i_1$, i_1+1, $\cdots$, $M-1$, we have proved that the procedure of calculation of the unknowns is stable. Because i_1 is always a bounded number, and all elements of those matrices in (1.6) for $i=1,\ 2,\ \cdots,\ i_1-1$ are bounded, we obtain

$$\|\delta x_i\| \leqslant c^*[\|\delta x_0\| + \|\delta x_M\|],\quad i=1,\ 2,\ \cdots,\ M-1,$$

where c^* is a constant independent of M. That is, the whole procedure of calculating $\boldsymbol{x}_{M-1}$, $\boldsymbol{x}_{M-2}$, $\cdots$, and $\boldsymbol{x}_1$ is stable.

We must point out that the choice of formula used in the procedure of calculation of the unknowns is important for stability. In general, if (1) is used to calculate $\boldsymbol{x}_{M-1}$, $\boldsymbol{x}_{M-2}$, $\cdots$, $\boldsymbol{x}_1$, the procedure will not be stable. Also, it seems to be unreasonable that the second group of (1.3.5) and (1.3.3), i.e.,

$$\begin{cases}(E_{1,i} \quad 0)\boldsymbol{F}_{i,i}\boldsymbol{x}_i = -(\beta^*_{1,i+1} \quad a^*_{2,i+1})\boldsymbol{F}_{i,i}\boldsymbol{x}_{i+1} - (e^*_{1,i+1} \quad \alpha^*_{1,i+1})\boldsymbol{F}_{0,i}\boldsymbol{x}_0 + \boldsymbol{\gamma}^*_{1,i+1}, \\ (0, E_{2,i})\boldsymbol{F}_{i,i}\boldsymbol{x}_i = -(\mu_{2,i} \quad 0)\boldsymbol{F}_{i,i}\boldsymbol{x}_i - (0 \quad \nu_{2,i})\boldsymbol{F}_{0,i}\boldsymbol{x}_0 + \boldsymbol{d}_{2,i}\end{cases}$$

are chosen to calculate $\boldsymbol{x}_{M-1}$, $\boldsymbol{x}_{M-2}$, $\cdots$, $\boldsymbol{x}_1$. In order to explain this fact, we observe the expression

$$\begin{pmatrix}\beta^*_{1,i+1} & a^*_{2,i+1} \\ -\mu_{2,i}\beta^*_{1,i+1} & -\mu_{2,i}a^*_{2,i+1}\end{pmatrix}\boldsymbol{F}_{i,i}\boldsymbol{F}^{-1}_{i+1,i+1},$$

where i is rather large.

According to condition (1), b_i can be expressed as

$$b_i = P_i\begin{pmatrix}E_1 & 0 \\ 0 & \lambda_{2,i}\end{pmatrix}S_i.$$

Moreover, (1.3.4) is obtained by eliminating the last s elements of $\boldsymbol{F}_{i,i}\boldsymbol{x}_i$ from (1.2.2) through using (1.3.3). In addition, $\mu_{2,i} = v^{-1}_{i,i}u_{i,i} + \delta_i$ when i is rather large. Therefore, the expression for the coefficient matrix of $\boldsymbol{F}_{i,i}\boldsymbol{x}_i$ in (1.3.4) is

$$\begin{aligned}\begin{pmatrix}b^*_{1,i} & 0 \\ b^*_{3,} & 0\end{pmatrix} &= P_i\begin{pmatrix}E_1 & 0 \\ 0 & \lambda_{2,i}\end{pmatrix}S_iF^{-1}_{i,i} - P_i\begin{pmatrix}E_1 & 0 \\ 0 & \lambda_{2,i}\end{pmatrix}S_iF^{-1}_{i,i}\begin{pmatrix}0 \\ E_2\end{pmatrix}(v^{-1}_{i,i}u_{i,i} + \delta_i \quad E_2) \\ &= P_i\left[\begin{pmatrix}s & t \\ \lambda_2 u & \lambda_2 v\end{pmatrix}_i - \begin{pmatrix}t \\ \lambda_2 v\end{pmatrix}_i (v^{-1}u + \delta \quad E_2)_i\right] \\ &= P_i\begin{pmatrix}s - tv^{-1}u + \delta & 0 \\ \delta & 0\end{pmatrix}_i \\ &= P_i\begin{pmatrix}s - tv^{-1}u + \delta & 0 \\ \delta & E_2\end{pmatrix}_i\begin{pmatrix}E_1 & 0 \\ 0 & 0\end{pmatrix}.\end{aligned}$$

In order to change $\begin{pmatrix}b^*_{1,i} \\ b^*_{3,i}\end{pmatrix}$ into $\begin{pmatrix}E_1 \\ 0\end{pmatrix}$, it should be multiplied from the left by $\begin{pmatrix}E_1 & H_1 \\ 0 & H_2\end{pmatrix}\begin{pmatrix}s - tv^{-1}u + \delta & 0 \\ \delta & E_2\end{pmatrix}^{-1}_i P^{-1}_i$, where H_1 and H_2 are any matrices. (Of course, they are determinate when the pivoting-elimination method is used.)

In addition, because

$$a_{i+1} = P_i\begin{pmatrix}\lambda_1 & 0 \\ 0 & E_2\end{pmatrix}_i S_i \quad \text{and} \quad (s - tv^{-1}u)^{-1} = \gamma,$$

and because the second group of equations in (1.3.5) is obtained from the second and third groups of equations in (1.3.5)′ when i is rather large, we have

$$\begin{cases} (\beta^*_{1,i+1} \quad a^*_{2,i+1})F_{i,i}F^{-1}_{i+1,i+1} \\ \quad =(E_1 \quad 0)\begin{pmatrix} E_1 & H_1 \\ 0 & H_2\end{pmatrix}_i \begin{pmatrix} r+\delta & 0 \\ \delta & E_2\end{pmatrix}_i P_i^{-1}P_i\begin{pmatrix}\lambda_1 & 0 \\ 0 & E_2\end{pmatrix}_i S_iF^{-1}_{i+1,i+1} \\ \quad =(E_1 \quad H_1)_i\begin{pmatrix} r+\delta & 0 \\ \delta & E_2\end{pmatrix}_i\begin{pmatrix}\lambda_1 & 0 \\ 0 & E_2\end{pmatrix}_i\begin{pmatrix} s_{i+1,i+1}+\delta & t_{i+1,i+1}+\delta \\ u_{i+1,i+1}+\delta & v_{i+1,i+1}+\delta\end{pmatrix} \\ \quad =(r_{i,i}\lambda_{1,i}s_{i+1,i+1}+H_{1,i}u_{i+1,i+1}+\delta_i \quad r_{i,i}\lambda_{1,i}t_{i+1,i+1}+H_{1,i}v_{i+1,i+1}+\delta_i), \\ (\beta^*_{2,i+1} \quad \bar{a}_{4,i+1})F_{i,i}F^{-1}_{i+1,i+1}\equiv-\mu_{2,i}(\beta^*_{1,i+1} \quad a^*_{2,i+1})F_{i,i}F^{-1}_{i+1,i+1} \\ \quad =-(v^{-1}_{i,i}u_{i,i}+\delta_i)(r_{i,i}\lambda_{1,i}s_{i+1,i+1} \\ \qquad +H_{1,i}u_{i+1,i+1}+\delta_i \quad r_{i,i}\lambda_{1,i}t_{i+1,i+1}+H_{1,i}v_{i+1,i+1}+\delta_i), \end{cases}$$

i.e.,

$$\begin{pmatrix}\beta^*_{1,i+1} & a^*_{2,i+1} \\ \beta^*_{2,i+1} & \bar{a}_{4,i+1}\end{pmatrix}F_{i,i}F^{-1}_{i+1,i+1}$$

$$=\begin{pmatrix} r_{i,i}\lambda_{1,i}(s-tv^{-1}u)_{i+1,i+1}+\delta_i & 0 \\ -(v^{-1}ur)_{i,i}\lambda_{1,i}(s-tv^{-1}u)_{i+1,i+1}+\delta_i & 0\end{pmatrix}$$

$$+\begin{pmatrix} r_{i,i}\lambda_{1,i}(tv^{-1}u)_{i+1,i+1} + H_{1,i}u_{i+1,i+1}+\delta_i & r_{i,i}\lambda_{1,i}t_{i+1,i+1} + H_{1,i}v_{i+1,i+1}+\delta_i \\ -v^{-1}_{i,i}u_{i,i}(r_{i,i}\lambda_{1,i}(tv^{-1}u)_{i+1,i+1} + H_{1,i}u_{i+1,i+1})+\delta_i & -v^{-1}_{i,i}u_{i,i}(r_{i,i}\lambda_{1,i}t_{i+1,i+1} + H_{1,i}v_{i+1,i+1})+\delta_i\end{pmatrix}$$

$$=\begin{pmatrix}\beta_{1,i+1} & 0 \\ \beta_{2,i+1} & 0\end{pmatrix}$$

$$+\begin{pmatrix} r_{i,i}\lambda_{1,i}(tv^{-1}u)_{i+1,i+1} + H_{1,i}u_{i+1,i+1}+\delta_i & r_{i,i}\lambda_{1,i}t_{i+1,i+1} + H_{1,i}v_{i+1,i+1}+\delta_i \\ m_{i,i}\lambda_{1,i}(tv^{-1}u)_{i+1,i+1} - v^{-1}_{i,i}u_{i,i}H_{1,i}u_{i+1,i+1}+\delta_i & m_{i,i}\lambda_{1,i}t_{i+1,i+1} - v^{-1}_{i,i}u_{i,i}H_{1,i}v_{i+1,i+1}+\delta_i\end{pmatrix}.$$

When we derive these expressions, the following relations are used:

$$\begin{cases} r^{-1}_{i,i}=(s-tv^{-1}u)_{i,i}, \\ m_{i,i}=-v^{-1}_{i,i}u_{i,i}r_{i,i}. \end{cases}$$

We have pointed out that

$$\begin{pmatrix}\beta_{1,i+1} & 0 \\ \beta_{2,i+1} & 0\end{pmatrix}=\begin{pmatrix} r_{i,i} & 0 \\ m_{i,i} & E_2\end{pmatrix}\begin{pmatrix}\lambda_{1,i}+\delta_{i+1} & 0 \\ \delta_{i+1} & 0\end{pmatrix}\begin{pmatrix} r_{i+1,i+1} & 0 \\ m_{i+1,i+1} & E_2\end{pmatrix}^{-1}.$$

However, from the expression for $\begin{pmatrix}\beta^*_{1,i+1} & a^*_{2,i+1} \\ \beta^*_{2,i+1} & \bar{a}_{4,i+1}\end{pmatrix}F_{i,i}F^{-1}_{i+1,i+1}$, we know that, because of the second term, in general, the matrix cannot be expressed as

$$A_i\ B_i\ A^{-1}_{i+1},$$

where the norm of B_i is less than $1-\varepsilon/2$. Therefore, stability cannot be guaranteed. From this result, we know that: for the elimination procedure, it is not necessary to eliminate some variables from the second group of (1.3.5) and to obtain (1.3.2); but for the procedure of calculation of the unknowns, it is necessary, because the second term of $\begin{pmatrix} \beta^*_{1,i+1} & a^*_{2,i+1} \\ \beta^*_{2,i+1} & \bar{a}_{4,i+1} \end{pmatrix} F_{i,i} F^{-1}_{i+1,i+1}$ is just eliminated and the stability of the procedure of calculation of the unknowns is guaranteed.

Incidentally, we discuss the convergence of the iteration (1.5) very briefly. From (2.12), we know that

$$\begin{cases} \|\mu_{1,M}\| \leqslant O\left(\left(1-\frac{\varepsilon}{2}\right)^M\right), \\ \|\nu_{2,M}\| \leqslant O\left(\left(1-\frac{\varepsilon}{2}\right)^M\right), \end{cases}$$

i.e., they are very small when M is very large. According to this fact, by using the method in Section 4 of [3], we can prove that the iteration procedure (1.5.1), (1.5.2) is convergent under quite weak conditions.

Finally, we point out the following three facts:

(i) When conditions (1)—(3) are fulfilled, the rank of the matrix which consists of the coefficients of $\boldsymbol{x}_1$, $\boldsymbol{x}_2$, $\cdots$, and $\boldsymbol{x}_{M-1}$ in (1) is $n(M-1)$. It is this fact that guarantees that no difficulty will appear in the elimination procedure here.

(ii) When conditions (1)—(3) are fulfilled, in order that (1) and (1.4) may determine a reasonable solution, (1.4) should satisfy the following conditions:

(a) $\boldsymbol{B}_1(\boldsymbol{x}_0)$ and $\boldsymbol{B}_2(\boldsymbol{x}_M)$ are s- and $(n-s)$-dimensional vectors respectively.

(b) $\begin{cases} \boldsymbol{B}_1(\boldsymbol{x}_0)=0, \\ (E_1\ \ 0)S_0\boldsymbol{x}_0=\boldsymbol{h}_{1,M}, \end{cases}$ and $\begin{cases} (0\ \ E_2)S_M\boldsymbol{x}_M=\boldsymbol{h}_{2,M}, \\ \boldsymbol{B}_2(\boldsymbol{x}_M)=0 \end{cases}$

have reasonable solutions. That is, only the "boundary condition equations" which satisfy conditions (a) and (b) are suitable to (1). As is pointed out in [1], "multi-diagonal" equations can be rewritten in the form (1). Therefore, by using this result here, we can discuss how to give "boundary condition equations" in that case. For practical work, it is very important to ascertain this problem.

(iii) Under the conditions in this section, the block-double-sweep method[4] with pivot-elements for complete linear algebraic equations is stable. This conclusion can be obtained by using certain results in this section. However, we shall not give a concrete proof here.

3. *Applications and Generalization for Multidiagonal Equations*

(1) We consider the following system of hyperbolic differential

equations:

$$\frac{\partial \boldsymbol{x}}{\partial t}+A\frac{\partial \boldsymbol{x}}{\partial \xi}=\boldsymbol{f},$$

where $\boldsymbol{x}$ and $\boldsymbol{f}$ are n-dimensional vectors and A is an $n\times n$-matrix. For this system, we adopt the following difference scheme:

$$\frac{\boldsymbol{x}_{i+1}^{k+1}+\boldsymbol{x}_i^{k+1}-\boldsymbol{x}_{i+1}^{k}-\boldsymbol{x}_i^{k}}{2\Delta t}+A_{i+\frac{1}{2}}^{k+\alpha}\left(\alpha\frac{\boldsymbol{x}_{i+1}^{k+1}-\boldsymbol{x}_i^{k+1}}{\Delta\xi}+(1-\alpha)\frac{\boldsymbol{x}_{i+1}^{k}-\boldsymbol{x}_i^{k}}{\Delta\xi}\right)$$
$$=\boldsymbol{f}_{i+\frac{1}{2}}^{k+\alpha},\quad i=0,1,\cdots,M-1.$$

It can be rewritten as

$$\left(E-A_{i+\frac{1}{2}}^{k+\alpha}\alpha\frac{\Delta t^*}{\Delta\xi}\right)\boldsymbol{x}_i^{k+1}+\left(E+A_{i+\frac{1}{2}}^{k+\alpha}\alpha\frac{\Delta t^*}{\Delta\xi}\right)\boldsymbol{x}_{i+1}^{k+1}$$
$$=\Delta t^*\boldsymbol{f}_{i+\frac{1}{2}}^{k+\alpha}+\left(E+A_{i+\frac{1}{2}}^{k+\alpha}(1-\alpha)\frac{\Delta t^*}{\Delta\xi}\right)\boldsymbol{x}_i^{k}+\left(E-A_{i+\frac{1}{2}}^{k+\alpha}(1-\alpha)\frac{\Delta t^*}{\Delta\xi}\right)\boldsymbol{x}_{i+1}^{k}.\quad(3.1)$$

In these expressions, F_i^k denotes $F(k\Delta t, i\Delta\xi)$ and Δt^* denotes $2\Delta t$. We suppose that except for $\boldsymbol{x}_i^{k+1}$, $i=0, 1, \cdots, M$, all other elements in (3.1) are given. Then, the system is in the form (1), and

$$b_i=\left(E-A_{i+\frac{1}{2}}^{k+\alpha}\alpha\frac{\Delta t^*}{\Delta\xi}\right),$$

$$a_{i+1}=\left(E+A_{i+\frac{1}{2}}^{k+\alpha}\alpha\frac{\Delta t^*}{\Delta\xi}\right).$$

If λ_{A_i} denotes an eigenvalue of $A_{i+\frac{1}{2}}^{k+\alpha}$, then

$$\lambda_i=-\frac{1+\lambda_{A_i}\alpha\dfrac{\Delta t^*}{\Delta\xi}}{1-\lambda_{A_i}\alpha\dfrac{\Delta t^*}{\Delta\xi}}$$

is a solution of $|\lambda_i b_i+a_{i+1}|=0$. If $\lambda_{A_i}\in[\delta, c]$ and $\alpha\frac{\Delta t^*}{\Delta\xi}\in[\delta, c]$, where δ and c are constants and $c>\delta>0$, then $|\lambda_i|>\frac{1}{1-2\varepsilon}$, where $\varepsilon>0$; if $\lambda_{A_i}\in[-c, -\delta]$ and $\alpha\frac{\Delta t^*}{\Delta\xi}\in[\delta, c]$, then $|\lambda_i|<1-2\varepsilon$. Therefore, if $A_{i+\frac{1}{2}}^{k+\alpha}$ has s eigenvalues which are greater than δ, and $n-s$ eigenvalues which are less than $-\delta$, if the moduli are bounded, and if other conditions in Section 2 are fulfilled, then the procedure is stable when the method in Section 1 is applied to this system, and equations (1.3) are $(n-s)\sim s$-broken. When λ_{A_i} may be equal to zero, the procedure of elimination and the procedure of calculation of the unknowns are still stable if the scheme in [3] is adopted.

(2) We consider the equations

$$a_{i-1}\boldsymbol{x}_{i-1}+b_i\boldsymbol{x}_i+c_{i+1}\boldsymbol{x}_{i+1}=\boldsymbol{d}_i,\quad i=1, 2, \cdots, M-1,\quad(3.2)$$

where $\boldsymbol{x}_i$ and $\boldsymbol{d}_i$ are n-dimensional vectors, and a_{i-1}, b_i and c_{i+1} are $n\times n$-matrices. We rewrite them as

$$\begin{pmatrix} a_{i-1} & \frac{1}{2}b_i \\ 0 & E_n \end{pmatrix}\begin{pmatrix} x_{i-1} \\ x_i \end{pmatrix}+\begin{pmatrix} \frac{1}{2}b_i & c_{i+1} \\ -E_n & 0 \end{pmatrix}\begin{pmatrix} x_i \\ x_{i+1} \end{pmatrix}=\begin{pmatrix} d_i \\ 0 \end{pmatrix}, \qquad (3.2)'$$

$$i=1, 2, \cdots, M-1,$$

where E_n is an $n\times n$–unit matrix. Let $X_i=\begin{pmatrix} x_i \\ x_{i+1} \end{pmatrix}$, the equations (3.2)′ have the same form as (1). Therefore, the method described in Section 1 can be applied to (3.2)′. If conditions (1)—(3) of Section 2 are fulfilled, then the procedure of elimination and the procedure of calculation of the unknowns are stable. In those conditions, the most important thing is the character of the eigenvalues. In this case, we should investigate the character of the roots of the equation

$$\left|\lambda\begin{pmatrix} a_{i-1} & \frac{1}{2}b_i \\ 0 & E_n \end{pmatrix}+\begin{pmatrix} \frac{1}{2}b_i & c_{i+1} \\ -E_n & 0 \end{pmatrix}\right|=\begin{vmatrix} \lambda a_{i-1}+\frac{1}{2}b_i & \frac{\lambda}{2}b_i+c_{i+1} \\ -E_n & \lambda E_n \end{vmatrix}$$

$$=\begin{vmatrix} \lambda a_{i-1}+b_i+\frac{1}{\lambda}c_{i+1} & 0 \\ -E_n & \lambda E_n \end{vmatrix}=|\lambda^2 a_{i-1}+\lambda b_i+c_{i+1}|=0. \qquad (3.3)$$

If it has s roots whose moduli are greater than $1/(1-2\varepsilon)$, and $2n-s$ roots whose moduli are less than $1-2\varepsilon$, then the stability is generally guaranteed, and the boundary-condition-equations of (3.2) should consist of s equations for x_0 and x_1, and $2n-s$ equations for x_{M-1} and x_M.

For $n=1$, we need to consider the character of the roots of the quadratic equation $\lambda^2 a_{i-1}+\lambda b_i+c_{i+1}=0$. The condition

$$|b_i|>|a_{i-1}+c_{i+1}|+\delta, \quad \delta>0 \qquad (3.4)$$

is necessary and sufficient for the fact that one of the moduli of the roots is greater than $1/(1-2\varepsilon)$ and the other is less than $1-2\varepsilon$[1]. (We shall prove this conclusion later.) Therefore, if (3.4) is fulfilled, then the procedure of elimination and the procedure of calculation of the unknowns are stable, and those matrices are $1\sim1$–broken when i is very large.

For $n\neq1$, if $|\lambda^2 a_{i-1}+\lambda b_i+c_{i+1}|=\prod_{k=1}^{n}|\lambda^2 a_{i-1,k}+\lambda b_{i,k}+c_{i+1,k}|$, where $a_{i-1,k}$, $b_{i,k}$ and $c_{i+1,k}$ are scalars, and if for every k,

$$|b_{i,k}|>|a_{i-1,k}+c_{i+1,k}|+\delta,$$

then the double sweep method in Section 1 is stable, and those matrices are $n\sim n$–broken when i is very large.

Of course, for tridiagonal equations, it is not necessary that they be rewritten as (3.2)′ before the equations are solved. For example, if the

1) When δ is a constant, and b_i, a_{i-1}, and c_{i+1} are uniformly bounded with respect to i, ε will not depend on i.

equation $|\lambda^2 a_{i-1}+\lambda b_i+c_{i+1}|=0$ has n roots whose moduli are greater than $1/(1-2\varepsilon)$, and n roots whose moduli are less than $1-2\varepsilon$, then (3.2) can be directly solved in the following way:

(i) Eliminating $\boldsymbol{x}_1$ and $\boldsymbol{x}_2$ from some of the equations (3.2) with $i=1, 2$, we obtain

$$\begin{pmatrix} \tilde{\boldsymbol{\nu}}_{1,2} & E_2 \\ \tilde{\boldsymbol{\nu}}_{2,2} & 0 \end{pmatrix}\begin{pmatrix} \boldsymbol{x}_0 \\ \boldsymbol{x}_1 \end{pmatrix}+\begin{pmatrix} 0 & \tilde{\mu}_{1,2} \\ E_n & \tilde{\mu}_{2,2} \end{pmatrix}\begin{pmatrix} \boldsymbol{x}_2 \\ \boldsymbol{x}_3 \end{pmatrix}=\begin{pmatrix} \tilde{\boldsymbol{d}}_{1,2} \\ \tilde{\boldsymbol{d}}_{2,2} \end{pmatrix}. \tag{3.5}$$

In (3.5) (and the expressions (3.6), (3.7) below), the matrices are $n\sim n$-broken.

(ii) For $i=3, \cdots, M-1$, eliminating $\boldsymbol{x}_{i-1}$ from

$$\left\{\begin{array}{ll} \begin{pmatrix} \tilde{\boldsymbol{\nu}}_{1,i-1} & E_n \\ \tilde{\boldsymbol{\nu}}_{2,i-1} & 0 \end{pmatrix}\begin{pmatrix} \boldsymbol{x}_0 \\ \boldsymbol{x}_1 \end{pmatrix}+\begin{pmatrix} 0 & \tilde{\mu}_{1,i-1} \\ E_n & \tilde{\mu}_{2,i-1} \end{pmatrix}\begin{pmatrix} \boldsymbol{x}_{i-1} \\ \boldsymbol{x}_i \end{pmatrix}=\begin{pmatrix} \tilde{\boldsymbol{d}}_{1,i-1} \\ \tilde{\boldsymbol{d}}_{2,i-1} \end{pmatrix}, & (3.6.1) \\ a_{i-1}\boldsymbol{x}_{i-1}+b_i\boldsymbol{x}_i+c_{i+1}\boldsymbol{x}_{i+1}=\boldsymbol{d}_i, & (3.6.2) \end{array}\right.$$

we obtain

$$\begin{pmatrix} \tilde{\boldsymbol{\nu}}_{1,i} & E_n \\ \tilde{\boldsymbol{\nu}}_{2,i} & 0 \end{pmatrix}\begin{pmatrix} \boldsymbol{x}_0 \\ \boldsymbol{x}_1 \end{pmatrix}+\begin{pmatrix} 0 & \tilde{\mu}_{1,i} \\ E_n & \tilde{\mu}_{2,i} \end{pmatrix}\begin{pmatrix} \boldsymbol{x}_i \\ \boldsymbol{x}_{i+1} \end{pmatrix}=\begin{pmatrix} \tilde{\boldsymbol{d}}_{1,i} \\ \tilde{\boldsymbol{d}}_{2,i} \end{pmatrix}. \tag{3.7}$$

(iii) Solving (3.7) with $i=M-1$ and "boundary-condition-equations" simultaneously, we obtain $\boldsymbol{x}_0$, $\boldsymbol{x}_1$, $\boldsymbol{x}_{M-1}$ and $\boldsymbol{x}_M$. Then, we calculate $\boldsymbol{x}_{M-2}, \cdots, \boldsymbol{x}_2$ successively by using the lower n equations in (3.7), i.e., by using

$$\boldsymbol{x}_i=-\tilde{\boldsymbol{\nu}}_{2,i}\boldsymbol{x}_0-\tilde{\mu}_{2,i}\boldsymbol{x}_{i+1}+\tilde{\boldsymbol{d}}_{2,i}. \tag{3.8}$$

In the following, we shall briefly discuss the stability of the procedure (3.5)—(3.8). We suppose that

$$|\lambda_i^2 a_{i-1}+\lambda_i b_i+c_{i+1}|=0$$

has the following $2n$ roots:

$$|\lambda_{i,1}^*|\leqslant|\lambda_{i,2}^*|\leqslant\cdots\leqslant|\lambda_{i,n}^*|\leqslant 1-2\varepsilon,$$

$$\frac{1}{1-2\varepsilon}\leqslant|\lambda_{i,n+1}^*|\leqslant\cdots\leqslant|\lambda_{i,2n}^*|,$$

and that n-dimensional vectors $\boldsymbol{y}_{i,j}$ are the solutions for

$$(\lambda_{i,j}^{*2}a_{i-1}+\lambda_{i,j}^* b_i+c_{i+1})\boldsymbol{y}_{i,j}=0, \quad j=1, 2, \cdots, 2n.$$

Let

$$\lambda_{1,i}=\begin{pmatrix} \lambda_1^* & & 0 \\ & \lambda_2^* & \\ & & \ddots \\ 0 & & \lambda_n^* \end{pmatrix}, \quad \lambda_{2,i}=\begin{pmatrix} \lambda_{n+1}^* & & 0 \\ & \lambda_{n+2}^* & \\ & & \ddots \\ 0 & & \lambda_{2n}^* \end{pmatrix}^{-1},$$

$$Y_{1,i}=(\boldsymbol{y}_1, \boldsymbol{y}_2, \cdots, \boldsymbol{y}_n)_i, \quad Y_{2,i}=(\boldsymbol{y}_{n+1}, \boldsymbol{y}_{n+2}, \cdots, \boldsymbol{y}_{2n})_i,$$

then the above expressions can be rewritten as

$$a_{i-1}Y_{1,i}\lambda_{1,i}+b_iY_{1,i}+c_{i+1}Y_{1,i}\lambda_{1,i}^{-1}=0,$$

$$a_{i-1}Y_{2,i}\lambda_{2,i}^{-1}+b_iY_{2,i}+c_{i+1}Y_{2,i}\lambda_{2,i}=0,$$

or

$$\begin{pmatrix} a_{i-1} & \frac{1}{2}b_i \\ 0 & E_n \end{pmatrix}\begin{pmatrix} Y_1\lambda_1 & Y_2 \\ Y_1 & Y_2\lambda_2 \end{pmatrix}_i\begin{pmatrix} E_n & 0 \\ 0 & -\lambda_2^{-1} \end{pmatrix}_i$$

$$-\begin{pmatrix} \frac{1}{2}b_i & c_{i+1} \\ -E_n & 0 \end{pmatrix}\begin{pmatrix} Y_1\lambda_1 & Y_2 \\ Y_1 & Y_2\lambda_2 \end{pmatrix}_i\begin{pmatrix} -\lambda_1^{-1} & 0 \\ 0 & E_n \end{pmatrix}_i=0.$$

Therefore, (3.2)′ can be rewritten as

$$\begin{pmatrix} E_n & 0 \\ 0 & -\lambda_{2,i} \end{pmatrix}S_i\begin{pmatrix} \boldsymbol{x}_{i-1} \\ \boldsymbol{x}_i \end{pmatrix}+\begin{pmatrix} -\lambda_{1,i} & 0 \\ 0 & E_n \end{pmatrix}S_i\begin{pmatrix} \boldsymbol{x}_i \\ \boldsymbol{x}_{i+1} \end{pmatrix}=\begin{pmatrix} \boldsymbol{f}_{1,i} \\ \boldsymbol{f}_{2,i} \end{pmatrix},$$

where

$$S_i=\begin{pmatrix} Y_1\lambda_1 & Y_2 \\ Y_1 & Y_2\lambda_2 \end{pmatrix}_i^{-1}$$

$$=\begin{pmatrix} Y_1^{-1}Y_2\lambda_2Y_2^{-1}(Y_1\lambda_1Y_1^{-1}Y_2\lambda_2Y_2^{-1}-E_n)^{-1} & -Y_1^{-1}(Y_2\lambda_2Y_2^{-1}Y_1\lambda_1Y_1^{-1}-E_n)^{-1} \\ -Y_2^{-1}(Y_1\lambda_1Y_1^{-1}Y_2\lambda_2Y_2^{-1}-E_n)^{-1} & Y_2^{-1}Y_1\lambda_1Y_1^{-1}(Y_2\lambda_2Y_2^{-1}Y_1\lambda_1Y_1^{-1}-E_n)^{-1} \end{pmatrix}_i$$

and we suppose all inverse matrices here exist. Rewriting (3.7) in the form of (1.2.1), we immediately know

$$\begin{cases} F_{0,i}=\begin{pmatrix} 0 & E_n \\ E_n & 0 \end{pmatrix}, \quad F_{i,i}=\begin{pmatrix} 0 & E_n \\ E_n & 0 \end{pmatrix}, \\ \begin{pmatrix} \tilde{\nu}_{1,i} & \tilde{\mu}_{1,i} \\ \tilde{\nu}_{2,i} & \tilde{\mu}_{2,i} \end{pmatrix}=\begin{pmatrix} \nu_{1,i} & \mu_{1,i} \\ \nu_{2,i} & \mu_{2,i} \end{pmatrix}. \end{cases}$$

In addition, according to the expressions for $\mu_{1,i}$, $\mu_{2,i}$, $\nu_{1,i}$ and $\nu_{2,i}$ in (2.12), we obtain

$$\begin{cases} \tilde{\mu}_{2,j}=\mu_{2,j}=v_{i,i}^{-1}u_{i,i}+\delta_i \\ \quad =-(Y_2^{-1}(Y_1\lambda_1Y_1^{-1}Y_2\lambda_2Y_2^{-1}-E_n)^{-1})_i^{-1}(Y_2^{-1}Y_1\lambda_1Y_1^{-1})_i \\ \quad\quad \times(Y_2\lambda_2Y_2^{-1}Y_1\lambda_1Y_1^{-1}-E_n)_i^{-1}+\delta_i=-(Y_1\lambda_1Y_1^{-1})_i+\delta_i, \\ \tilde{\nu}_{1,i}=\nu_{1,i}=s_{0,i}^{-1}t_{0,i}+\delta_i \\ \quad =-(Y_1^{-1}(Y_2\lambda_2Y_2^{-1}Y_1\lambda_1Y_1^{-1}-E_n)^{-1})_0^{-1}(Y_1^{-1}Y_2\lambda_2Y_2^{-1})_0 \\ \quad\quad \times(Y_1\lambda_1Y_1^{-1}Y_2\lambda_2Y_2^{-1}-E_n)_0^{-1}+\delta_i=-(Y_2\lambda_2Y_2^{-1})_0+\delta_i, \\ \tilde{\mu}_{1,i}=\mu_{1,i}=\varepsilon_i, \\ \tilde{\nu}_{2,i}=\nu_{2,i}=\varepsilon_i. \end{cases}$$

Therefore, the procedure of elimination and the procedure of calculation of the unknowns in Section 1 with $F_{0,i}=F_{i,i}=\begin{pmatrix} 0 & E_n \\ E_n & 0 \end{pmatrix}$ are reasonable, and the procedures (3.5)—(3.7) are also reasonable. Moreover, because $\tilde{\mu}_{2,i}=-(Y_1\lambda_1Y_1^{-1})_i+\delta_i$ and the moduli of its eigenvalues are less than 1, the reasonableness of the procedure (3.8) is also guaranteed.

Incidentally, we should point out the following fact. According to the above results, we can definitely say that the double sweep method for complete tridiagonal equations is reasonable if conditions (1)—(3) in Section 2 are fulfilled. However, condition (3.4) is weaker than condition

$$b_i > -(a_{i-1}+c_{i+1})+\delta, \quad a_{i-1}<0,\ c_{i+1}<0,\ \delta>0, \tag{3.4$'$}$$

which is given as a condition for the stability of the double sweep method for "complete" tridiagonal equations with $n=1$ in [6]. That is, if a_i-a_{i-1}, b_i-b_{i-1} and c_i-c_{i-1} are quantities of $O\left(\frac{1}{M}\right)$, and M is sufficiently large, then condition (3.4)$'$ can be weakened into condition (3.4):

$$|b_i|>|a_{i-1}+c_{i+1}|+\delta,$$

where $|a_{i-1}|$, $|b_i|$ and $|c_{i+1}|$ are uniformly bounded with respect to i and δ is any positive constant.

For general tridiagonal equations, the following procedure seems to be reasonable.

(i) Eliminating some variables from the equations with $i=1$ and 2 in (3.2), we obtain

$$\begin{pmatrix} E_{1,2} & \tilde{\boldsymbol{\nu}}_{1,2} \\ 0 & \tilde{\boldsymbol{\nu}}_{2,2} \end{pmatrix} \tilde{F}_{0,2} \begin{pmatrix} \boldsymbol{x}_0 \\ \boldsymbol{x}_1 \end{pmatrix} + \begin{pmatrix} \tilde{\mu}_{1,2} & 0 \\ \tilde{\mu}_{2,2} & E_{2,2} \end{pmatrix} \tilde{F}_{2,2} \begin{pmatrix} \boldsymbol{x}_2 \\ \boldsymbol{x}_3 \end{pmatrix} = \begin{pmatrix} \tilde{\boldsymbol{d}}_{1,2} \\ \tilde{\boldsymbol{d}}_{2,2} \end{pmatrix}. \tag{3.5$'$}$$

(ii) For $i=3, \cdots, M-1$, finding the expression for $\boldsymbol{x}_{i-1}$ in terms of $\boldsymbol{x}_0$, $\boldsymbol{x}_1$, $\boldsymbol{x}_i$ and $\boldsymbol{x}_{i+1}$ from

$$\begin{cases} \begin{pmatrix} E_{1,i-1} & \tilde{\boldsymbol{\nu}}_{1,i-1} \\ 0 & \tilde{\boldsymbol{\nu}}_{2,i-1} \end{pmatrix} \tilde{F}_{0,i-1} \begin{pmatrix} \boldsymbol{x}_0 \\ \boldsymbol{x}_1 \end{pmatrix} + \begin{pmatrix} \tilde{\mu}_{1,i-1} & 0 \\ \tilde{\mu}_{2,i-1} & E_{2,i-1} \end{pmatrix} \tilde{F}_{i-1,i-1} \begin{pmatrix} \boldsymbol{x}_{i-1} \\ \boldsymbol{x}_i \end{pmatrix} \\ \quad = \begin{pmatrix} \tilde{\boldsymbol{d}}_{1,i-1} \\ \tilde{\boldsymbol{d}}_{2,i-1} \end{pmatrix}, \\ a_{i-1}\boldsymbol{x}_{i-1} + b_i\boldsymbol{x}_i + c_{i+1}\boldsymbol{x}_{i+1} = \boldsymbol{d}_i, \end{cases} \tag{3.6$'$}$$

and eliminating $\boldsymbol{x}_{i-1}$ from some equations of (3.6)$'$, we obtain

$$\begin{cases} \begin{pmatrix} E_{1,i} & \tilde{\boldsymbol{\nu}}_{1,i} \\ 0 & \tilde{\boldsymbol{\nu}}_{2,i} \end{pmatrix} \tilde{F}_{0,i} \begin{pmatrix} \boldsymbol{x}_0 \\ \boldsymbol{x}_1 \end{pmatrix} + \begin{pmatrix} \tilde{\mu}_{1,i} & 0 \\ \tilde{\mu}_{2,i} & E_{2,i} \end{pmatrix} \tilde{F}_{i,i} \begin{pmatrix} \boldsymbol{x}_i \\ \boldsymbol{x}_{i+1} \end{pmatrix} = \begin{pmatrix} \tilde{\boldsymbol{d}}_{1,i} \\ \tilde{\boldsymbol{d}}_{2,i} \end{pmatrix}, & (3.7)'_1 \\ (0 \quad \tilde{\alpha}_{1,i}) \tilde{F}_{0,i} \begin{pmatrix} \boldsymbol{x}_0 \\ \boldsymbol{x}_1 \end{pmatrix} + \boldsymbol{x}_{i-1} + (\tilde{\beta}_{1,i} \quad 0) \tilde{F}_{i,i} \begin{pmatrix} \boldsymbol{x}_i \\ \boldsymbol{x}_{i+1} \end{pmatrix} = \tilde{\boldsymbol{\gamma}}_{1,i}. & (3.7)'_2 \end{cases}$$

(iii) After $\boldsymbol{x}_0$, $\boldsymbol{x}_1$, $\boldsymbol{x}_{M-1}$ and $\boldsymbol{x}_M$ are obtained, by using

$$\boldsymbol{x}_{i-1} = -(0 \quad \tilde{\alpha}_{1,i}) \tilde{F}_{0,i} \begin{pmatrix} \boldsymbol{x}_0 \\ \boldsymbol{x}_1 \end{pmatrix} - (\tilde{\beta}_{1,i} \quad 0) \tilde{F}_{i,i} \begin{pmatrix} \boldsymbol{x}_i \\ \boldsymbol{x}_{i+1} \end{pmatrix} + \tilde{\boldsymbol{\gamma}}_{1,i}, \tag{3.8$'$}$$

we obtain $\boldsymbol{x}_{M-2}, \cdots, \boldsymbol{x}_2$ successively.

In order to obtain an intuitive idea about the above theoretical analysis, we observe (3.2) with $n=1$:

$$ax_{i-1}+bx_i+cx_{i+1}=d_i, \quad i=1, \cdots, M-1,$$

and we suppose a, b and c are constants. The system has the following general solution:

$$x_i = c_1 q_1^{-1} + c_2 q_2^{-1} + x_i^*,$$

where q_1 and q_2 are the roots of $\lambda^2 a + \lambda b + c = 0$, x_i^* is a particular solution. If (3.4) is fulfilled, then $|q_1| > 1/(1-2\varepsilon)$, and $|q_2| < 1-2\varepsilon$. c_1 and c_2 are two parameters; for example, we take

$$c_1 = \frac{(x_0 - x_0^*) q_2^{-i-1} - (x_{i+1} - x_{i+1}^*)}{q_2^{-i-1} - q_1^{-i-1}},$$

$$c_2 = \frac{-(x_0 - x_0^*) q_1^{-i-1} + (x_{i+1} - x_{i+1}^*)}{q_2^{-i-1} - q_1^{-i-1}}.$$

Therefore,

$$x_1 = (x_0 - x_0^*) \frac{q_1^{-1} - q_1^{-i-1} q_2^i}{1 - \left(\frac{q_1}{q_2}\right)^{-i-1}} + (x_{i+1} - x_{i+1}^*) \frac{q_2^i - q_1^{-1} q_2^{i+1}}{1 - \left(\frac{q_1}{q_2}\right)^{-i-1}} + x_1^*,$$

$$x_{i-1} = (x_0 - x_0^*) \frac{q_1^{-i+1} - q_1^{-i-1} q_2^2}{1 - \left(\frac{q_1}{q_2}\right)^{-i-1}} + (x_{i+1} - x_{i+1}^*) \frac{q_2^2 - q_1^{-i+1} q_2^{i+1}}{1 - \left(\frac{q_1}{q_2}\right)^{-i-1}} + x_{i-1}^*,$$

$$x_i = (x_0 - x_0^*) \frac{q_1^{-i} - q_1^{-i-1} q_2}{1 - \left(\frac{q_1}{q_2}\right)^{-i-1}} + (x_{i+1} - x_{i+1}^*) \frac{q_2 - q_1^{-i} q_2^{i+1}}{1 - \left(\frac{q_1}{q_2}\right)^{-i-1}} + x_i^*.$$

Rewriting these expressions in the form of (3.7) or (3.7)′, (3.8)′, we obtain

$$\widetilde{F}_{0,i} = \widetilde{F}_{i,i} = \begin{pmatrix} 0 & 1 \\ 1 & 0 \end{pmatrix},$$

$$\begin{pmatrix} \tilde{\nu}_{1,i} & \tilde{\mu}_{1,i} \\ \tilde{\nu}_{2,i} & \tilde{\mu}_{2,i} \\ \tilde{\alpha}_{1,i} & \tilde{\beta}_{1,i} \end{pmatrix} = \frac{-1}{1 - \left(\frac{q_1}{q_2}\right)^{-i-1}} \begin{pmatrix} q_1^{-1}(1 - q_1^{-i} q_2^i) & q_2^i (1 - q_1^{-1} q_2) \\ q_1^{-i}(1 - q_1^{-1} q_2) & q_2 (1 - q_1^{-i} q_2^i) \\ q_1^{-i+1}(1 - q_1^{-2} q_2^2) & q_2^2 (1 - q_1^{-i+1} q_2^{i-1}) \end{pmatrix},$$

$$\begin{pmatrix} \tilde{d}_{1,i} \\ \tilde{d}_{2,i} \\ \tilde{\gamma}_{1,i} \end{pmatrix} = -\frac{1}{1 - \left(\frac{q_1}{q_2}\right)^{-i-1}} \begin{pmatrix} q_1^{-1}(1 - q_1^{-i} q_2^i) x_0^* + q_2^i (1 - q_1^{-1} q_2) x_{i+1}^* \\ q_1^{-i}(1 - q_1^{-1} q_2) x_0^* + q_2 (1 - q_1^{-i} q_2^i) x_{i+1}^* \\ q_1^{-i+1}(1 - q_1^{-2} q_2^2) x_0^* + q_2^2 (1 - q_1^{-i+1} q_2^{i-1}) x_{i+1}^* \end{pmatrix} + \begin{pmatrix} x_1^* \\ x_i^* \\ x_{i-1}^* \end{pmatrix}.$$

From the character of these coefficients, we can easily understand why the procedure of calculating these coefficients is stable. In addition, because $|\tilde{\mu}_{2,i}|$ and $|\tilde{\beta}_{1,i}|$ are less than 1, the procedures (3.8) and (3.8)′ are also stable. In the case where a_{i-1}, b_i and c_{i+1} change with i, it is impossible to give any simple expressions for $\tilde{\nu}_{1,i}, \cdots, \tilde{\beta}_{1,i}$ like the above, but we can prove that these coefficients have similar characters.

In the following, we shall prove (3.4) is a necessary and sufficient condition for the fact that the quadratic equation $\lambda^2 a_{i-1} + \lambda b_i + c_{i+1} = 0$ has one root whose modulus is greater than $1/(1-\varepsilon)$, and one root whose modulus is less than $1-\varepsilon$. Obviously, we can always suppose $b_i>0$. When $a_{i-1}=0$, the conclusion is obviously true. In the following, we shall consider the case $a_{i-1}\neq 0$. If one of the moduli of the roots is greater than $1/(1-\varepsilon)$, and the other is less than $1-\varepsilon$, where $\varepsilon\in(0,1)$, then the two roots must be real roots. Therefore, we can rewrite these conditions about roots as

$$\begin{cases} 1+\varepsilon<\dfrac{1}{1-\varepsilon}<\dfrac{b_i+\sqrt{b_i^2-4a_{i-1}c_{i+1}}}{2|a_{i-1}|}, \\ -1+\varepsilon<\dfrac{b_i-\sqrt{b_i^2-4a_{i-1}c_{i+1}}}{2|a_{i-1}|}<1-\varepsilon, \end{cases}$$

i.e.,

$$\begin{cases} 2(1+\varepsilon)|a_{i-1}|-b_i<\sqrt{b_i^2-4a_{i-1}c_{i+1}}, \\ -2|a_{i-1}|(1-\varepsilon)<b_i-\sqrt{b_i^2-4a_{i-1}c_{i+1}}<2|a_{i-1}|(1-\varepsilon). \end{cases}$$

These inequalities are equivalent to the following three groups of inequalities:

(1) $\big|b_i-2|a_{i-1}|\big|+2|a_{i-1}|\varepsilon<\sqrt{b_i^2-4a_{i-1}c_{i+1}}<b_i+2|a_{i-1}|-2|a_{i-1}|\varepsilon$;

(2) $b_i^2-4|a_{i-1}|b_i+4a_{i-1}^2+4|a_{i-1}|\varepsilon[\,|b_i-2|a_{i-1}||+|a_{i-1}|\varepsilon]$
$<b_i^2-4a_{i-1}c_{i+1}<b_i^2+4|a_{i-1}|b_i+4|a_{i-1}|^2$
$-4|a_{i-1}|\varepsilon[b_i+2|a_{i-1}|-|a_{i-1}|\varepsilon]$;

(3) $\begin{cases} (\operatorname{sign} a_{i-1})\cdot(a_{i-1}+c_{i+1})+\varepsilon[\,|b_i-2|a_{i-1}||+|a_{i-1}|\varepsilon]<b_i, \\ -(\operatorname{sign} a_{i-1})\cdot(a_{i-1}+c_{i+1})+\varepsilon[b_i+2|a_{i-1}|-|a_{i-1}|\varepsilon]<b_i. \end{cases}$

Let $\delta=\min\{\varepsilon[\,|b_i-2|a_{i-1}||+|a_{i-1}|\varepsilon],\ \varepsilon[b_i+2|a_{i-1}|-|a_{i-1}|\varepsilon]\}>0$, from the last group of inequalities, we can obtain

$$b_i>|a_{i-1}+c_{i+1}|+\delta,$$

that is, we have proved (3.4) is a necessary condition. We now turn to proving (3.4) is a sufficient condition. We shall prove that if $b_i>|a_{i-1}+c_{i+1}|+\delta^*$, $\delta^*>0$ and $\max\{b_i, |a_{i-1}|, |c_{i+1}|\}\leqslant d$, then one of the moduli of the roots is less than $1-\varepsilon^{**}$ and the other is greater than $1-\varepsilon^{**}$, where $\varepsilon^{**}=\varepsilon^*/(1+\varepsilon^*)$, ε^* being equal to $\min\left\{\dfrac{\delta^*}{3d},\ \dfrac{\delta^*}{2d\sqrt{1+\delta^*/d}}\right\}$.

We observe the characters of the functions $f_1(\varepsilon)=\varepsilon\cdot[\,|b_i-2|a_{i-1}||+|a_{i-1}|\varepsilon]$ and $f_2(\varepsilon)=\varepsilon[b_i+2|a_{i-1}|-|a_{i-1}|\varepsilon]$. (i) When $\varepsilon\geqslant 0$, $f_1(\varepsilon)\leqslant d(\varepsilon^2+2\varepsilon)$. Moreover, the positive root $\varepsilon^+=-1+\sqrt{1+\delta^*/d}$ of the equation $d(\varepsilon^2+2\varepsilon)=\delta^*$ is greater than or equal to $\delta^*/(2d\sqrt{1+\delta^*/d})\equiv\bar{\varepsilon}^*$. Therefore, when $\bar{\varepsilon}^*\geqslant\varepsilon\geqslant 0$, $f_1(\varepsilon)\leqslant\delta^*$. (ii) When $\varepsilon\geqslant 0$, $f_2(\varepsilon)\leqslant 3d\varepsilon$. Therefore, when $\delta^*/3d\geqslant\varepsilon\geqslant 0$, $f_2(\varepsilon)\leqslant\delta^*$. According to these characters, after letting

$$\varepsilon^*=\min\left\{\frac{\delta^*}{3d},\ \frac{\delta^*}{2d\sqrt{1+\delta^*/d}}\right\},$$

we obtain the following inequalities:

$$\begin{cases} b_i > |a_{i-1}+c_{i+1}|+\delta^* \geqslant |a_{i-1}+c_{i+1}|+f_1(\varepsilon^*), \\ b_i > |a_{i-1}+c_{i+1}|+\varepsilon^* \geqslant |a_{i-1}+c_{i+1}|+f_2(\varepsilon^*). \end{cases}$$

Because every one of these inequalities, which have been used to prove that (3.14) is a necessary condition, is equivalent to each other, we easily know from the above results that one of the moduli of the roots is greater than $1+\varepsilon^*$, and the other is less than $1-\varepsilon^*$ when (3.4) is fulfilled. Let $1+\varepsilon^*=1/(1-\varepsilon^{**})$. By using the fact that $\varepsilon^{**}<\varepsilon^*$, we can obtain the desired conclusion.

In the following, we shall briefly discuss the application of the method in this section to solving parabolic and hyperbolic equations. We consider the equation

$$\frac{\partial x}{\partial t}+\alpha\frac{\partial x}{\partial \xi}-\beta\frac{\partial^2 x}{\partial \xi^2}=f, \quad \beta\geqslant 0,$$

where x, α, β and f are scalars. When the equation is approximated by

$$\begin{aligned} \Delta t\Bigg\{&\frac{x_i^{k+1}-x_i^k}{\Delta t}+\alpha_i^{k+\frac{1}{2}}\left(\frac{x_{i+1}^{k+1}-x_{i-1}^{k+1}+x_{i+1}^k-x_{i-1}^k}{4\Delta\xi}\right) \\ &-\beta_i^{k+\frac{1}{2}}\frac{x_{i+1}^{k+1}-2x_i^{k+1}+x_{i-1}^{k+1}+x_{i+1}^k-2x_i^k+x_{i-1}^k}{2\Delta\xi^2}\Bigg\}=f_i^{k+\frac{1}{2}}\Delta t, \qquad (3.9) \\ &i=1,\ 2,\ \cdots,\ M-1, \end{aligned}$$

the difference equations have the same form as (3.2), and the coefficients are

$$\begin{cases} a_{i-1}=-\alpha_i^{k+\frac{1}{2}}\dfrac{\Delta t}{4\Delta\xi}-\beta_i^{k+\frac{1}{2}}\dfrac{\Delta t}{2\Delta\xi^2}, \\ b_i=1+\beta_i^{k+\frac{1}{2}}\dfrac{\Delta t}{\Delta\xi^2}, \\ c_{i+1}=\alpha_i^{k+\frac{1}{2}}\dfrac{\Delta t}{4\Delta\xi}-\beta_i^{k+\frac{1}{2}}\dfrac{\Delta t}{2\Delta\xi^2}. \end{cases}$$

In this case, $b_i=|a_{i-1}+c_{i+1}|+1$, so the condition (3.4) is fulfilled. Therefore, it is reasonable to apply (3.5)—(3.8) to solving (3.9).

As mentioned above, when (3.4) is fulfilled, the double sweep method for "complete" systems is reasonable. (Of course, "boundary-condition-equations" must be reasonable.) When the differential equation is hyperbolic, i.e., $\beta=0$, a_{i-1} and c_{i+1} have different signs. Therefore condition (3.4)′ is not fulfilled (when Δt is rather large, even the inequality $b_i>|a_{i-1}|+|c_{i+1}|+\delta$ is not fulfilled). That is, (3.4)′ cannot be used to discuss the stability of the double sweep method for the "complete" systems, in which the equations for interior points are (3.9).

From this example, we know that it is valuable to weaken (3.4)′ into (3.4).

If the above differential equation is approximated by

$$\Delta t\left\{\frac{x_i^{k+1}-x_i^k}{\Delta t}+\alpha_i^k\frac{x_i^{k+1}-x_{i-1}^{k+1}}{\Delta\xi}-\beta_i^k\frac{x_{i+1}^{k+1}-2x_i^{k+1}+x_{i-1}^{k+1}}{\Delta\xi^2}\right\}=f_i^k\Delta t, \quad (3.10)$$

then

$$\begin{cases} a_{i-1}=-\alpha_i^k\dfrac{\Delta t}{\Delta\xi}-\beta_i^k\dfrac{\Delta t}{\Delta\xi^2},\\ b_i=1+\alpha_i^k\dfrac{\Delta t}{\Delta\xi}+2\beta_i^k\dfrac{\Delta t}{\Delta\xi^2},\\ c_{i+1}=-\beta_i^k\dfrac{\Delta t}{\Delta\xi^2}. \end{cases}$$

Therefore, when $\alpha_i^k\geqslant 0$, $b_i=|a_{i-1}+c_{i+1}|+1$, i.e., it is reasonable to solve (3.10) by using (3.5)—(3.8). When $\alpha_i^k\leqslant 0$, (3.4) is also fulfilled if $\partial x/\partial\xi$ is approximated by $(x_{i+1}^{k+1}-x_i^{k+1})/\Delta\xi$.

We consider the equations

$$\frac{\partial \boldsymbol{x}}{\partial t}+\alpha\frac{\partial \boldsymbol{x}}{\partial\xi}-\beta\frac{\partial^2\boldsymbol{x}}{\partial\xi^2}=f, \quad (3.11)$$

where α is an $n\times n$-matrix, and β is a nonnegative number. When (3.11) is approximated by (3.9), the coefficients of difference equations are

$$\begin{cases} a_{i-1}=-\alpha_i^{k+\frac12}\dfrac{\Delta t}{4\Delta\xi}-\beta_i^{k+\frac12}\dfrac{\Delta t}{2\Delta\xi^2}E_n,\\ b_i=E_n+2\beta_i^{k+\frac12}\dfrac{\Delta t}{2\Delta\xi^2}E_n,\\ c_{i+1}=\alpha_i^{k+\frac12}\dfrac{\Delta t}{4\Delta\xi}-\beta_i^{k+\frac12}\dfrac{\Delta t}{2\Delta\xi^2}E_n. \end{cases}$$

If α_i is similar to the diagonal matrix

$$\begin{pmatrix} \alpha_{i,1} & & 0\\ & \alpha_{i,2} & \\ & & \ddots \\ 0 & & \alpha_{i,n}\end{pmatrix},$$

then the condition (3.3) can be rewritten as

$$|\lambda^2 a_{i-1}+\lambda b_i+c_{i+1}|=\prod_{j=1}^n\left\{\lambda^2\left(-\alpha_{i,j}^{k+\frac12}\frac{\Delta t}{4\Delta\xi}-\beta_i^{k+\frac12}\frac{\Delta t}{2\Delta\xi^2}\right)\right.$$
$$\left.+\lambda\left(1+2\beta_i^{k+\frac12}\frac{\Delta t}{2\Delta\xi^2}\right)+\alpha_{i,j}^{k+\frac12}\frac{\Delta t}{4\Delta\xi}-\beta_i^{k+\frac12}\frac{\Delta t}{2\Delta\xi^2}\right\}=0.$$

Therefore, in this case, there are n roots whose moduli are greater than $1/(1-2\varepsilon)$, and there are n roots whose moduli are less than $1-2\varepsilon$. Furthermore, it is reasonable to solve these equations by using (3.5)—(3.8). In addition, in this case, when i is sufficiently large, $\tilde{\mu}_{1,i}$ and $\tilde{\nu}_{2,i}$ in (3.7) are close to zero. Therefore, n conditions should be given in each "boundary".

Lastly, we consider the equations

$$\frac{\partial \boldsymbol{x}}{\partial t}-\beta\frac{\partial^2\boldsymbol{x}}{\partial\xi^2}=f,$$

where β is a positive definite matrix. If it is approximated by (3.9), and β_i is similar to

$$\begin{pmatrix} \beta_{i,1} & & & 0 \\ & \beta_{i,2} & & \\ & & \ddots & \\ 0 & & & \beta_{i,n} \end{pmatrix},$$

then we have

$$|\lambda^2 a_{i-1}+\lambda b_i+c_{i+1}| = \prod_{j=1}^{n}\Big[\lambda^2\Big(-\beta_{i,j}^{k+\frac{1}{2}}\frac{\Delta t}{2\Delta\xi^2}\Big) + \lambda\Big(1+2\beta_{i,j}^{k+\frac{1}{2}}\frac{\Delta t}{2\Delta\xi^2}\Big)-\beta_{i,j}^{k+\frac{1}{2}}\frac{\Delta t}{2\Delta\xi^2}\Big]=0.$$

Therefore, in this case, there are n roots whose moduli are greater than $1/(1-2\varepsilon)$, and there are n roots whose moduli are less than $1-2\varepsilon$. That is, it is reasonable to apply (3.5)—(3.8) to this case.

(3) For general multi-diagonal equations

$$a_0^i \boldsymbol{x}_i + a_1^i \boldsymbol{x}_{i+1} + \cdots + a_m^i \boldsymbol{x}_{i+m} = \boldsymbol{b}_i, \quad i=0, \cdots, M-m, \tag{3.12}$$

where a_j^i is an $n\times n$-matrix, $\boldsymbol{x}_i$ and $\boldsymbol{b}_i$ are n-dimensional vectors, the situation is similar to that for tridiagonal equations. They are equivalent to

$$\begin{pmatrix} a_0^i & \frac{1}{2}a_1^i & \frac{1}{2}a_2^i & \cdots & \frac{1}{2}a_{m-1}^i \\ 0 & E_n & 0 & \cdots & 0 \\ 0 & 0 & E_n & \cdots & 0 \\ \cdots & \cdots & \cdots & \cdots & \cdots \\ 0 & 0 & 0 & \cdots & E_n \end{pmatrix}\begin{pmatrix} \boldsymbol{x}_i \\ \boldsymbol{x}_{i+1} \\ \boldsymbol{x}_{i+2} \\ \vdots \\ \boldsymbol{x}_{i+m-1} \end{pmatrix} + \begin{pmatrix} \frac{1}{2}a_1^i & \frac{1}{2}a_2^i & \cdots & \frac{1}{2}a_{m-1}^i & a_m \\ -E_n & 0 & \cdots & 0 & 0 \\ 0 & -E_n & \cdots & 0 & 0 \\ \cdots & \cdots & \cdots & \cdots & \cdots \\ 0 & 0 & \cdots & -E_n & 0 \end{pmatrix}\begin{pmatrix} \boldsymbol{x}_{i+1} \\ \boldsymbol{x}_{i+2} \\ \boldsymbol{x}_{i+3} \\ \vdots \\ \boldsymbol{x}_{i+m} \end{pmatrix} = \begin{pmatrix} \boldsymbol{b}_i \\ 0 \\ 0 \\ \vdots \\ 0 \end{pmatrix}. \tag{3.13}$$

If the method in Section 1 is applied to (3.13), then the procedure is reasonable when conditions (1)—(3) in Section 2 are fulfilled. In this case, we need to consider the roots of the equation

$$\left|\lambda\begin{pmatrix} a_0^i & \frac{1}{2}a_1^i & \frac{1}{2}a_2^i & \cdots & \frac{1}{2}a_{m-1}^i \\ 0 & E_n & 0 & \cdots & 0 \\ 0 & 0 & E_n & \cdots & 0 \\ \cdots & \cdots & \cdots & \cdots & \cdots \\ 0 & 0 & 0 & \cdots & E_n \end{pmatrix} + \begin{pmatrix} \frac{1}{2}a_1^i & \frac{1}{2}a_2^i & \cdots & \frac{1}{2}a_{m-1}^i & a_m^i \\ -E_n & 0 & \cdots & 0 & 0 \\ 0 & -E_n & \cdots & 0 & 0 \\ \cdots & \cdots & \cdots & \cdots & \cdots \\ 0 & 0 & \cdots & -E_n & 0 \end{pmatrix}\right|$$

$$
=\begin{vmatrix}
\lambda a_0^i+\frac{1}{2}a_1^i & \frac{\lambda}{2}a_1^i+\frac{1}{2}a_2^i & \cdots & \frac{\lambda}{2}a_{m-2}^i+\frac{1}{2}a_{m-1}^i & \frac{\lambda}{2}a_{m-1}^i+a_m^i \\
-E_n & \lambda E_n & \cdots & 0 & 0 \\
0 & -E_n & \cdots & 0 & 0 \\
\cdots & \cdots & \cdots & \cdots & \cdots \\
0 & 0 & \cdots & -E_n & \lambda E_n
\end{vmatrix}
$$

$$
=\begin{vmatrix}
\lambda a_0^i+\frac{1}{2}a_1^i & \frac{\lambda}{2}a_1^i+\frac{1}{2}a_2^i & \cdots & \frac{\lambda}{2}a_{m-2}^i+a_{m-1}^i+\frac{1}{\lambda}a_m^i & \frac{\lambda}{2}a_{m-1}^i+a_m^i \\
-E_n & \lambda E_n & \cdots & 0 & 0 \\
0 & -E_n & \cdots & 0 & 0 \\
\cdots & \cdots & \cdots & \cdots & \cdots \\
0 & 0 & \cdots & 0 & \lambda E_n
\end{vmatrix}
$$

$$
=\cdots=|\lambda^m a_0^i+\lambda^{m-1}a_1^i+\cdots+a_m^i|=0. \tag{3.14}
$$

If (3.14) has s roots whose moduli are greater than $1/(1-2\varepsilon)$, and $(mn-s)$ roots whose moduli are less than $1-2\varepsilon$, for $i=0, 1, \cdots, M-m$, then the stability of the double sweep method can be roughly guaranteed. Moreover, in order to obtain reasonable solutions, (3.12) should have s conditions about $\boldsymbol{x}_0, \boldsymbol{x}_1, \cdots, \boldsymbol{x}_{m-1}$, and $mn-s$ conditions about $\boldsymbol{x}_{M-m+1}, \cdots, \boldsymbol{x}_M$.

Of course, changing (3.12) into (3.13) is not always necessary. Some methods similar to (3.5)—(3.8) or (3.5)′—(3.8)′ can be used to solve (3.12). For example, firstly, we obtain some relations similar to (3.7) or $(3.7)'_1$ among $\boldsymbol{x}_0, \boldsymbol{x}_1, \cdots, \boldsymbol{x}_{m-1}, \boldsymbol{x}_j, \boldsymbol{x}_{j+1}, \cdots, \boldsymbol{x}_{j+m-1}$, $j=m, m+1, \cdots, M-m+1$. If necessary, we can obtain expressions for $\boldsymbol{x}_{j-1}$ in terms of $\boldsymbol{x}_0, \boldsymbol{x}_1, \cdots, \boldsymbol{x}_{m-1}, \boldsymbol{x}_j, \boldsymbol{x}_{j+1}, \cdots, \boldsymbol{x}_{j+m-1}$ like $(3.7)'_2$ for the tridiagonal equations. Secondly, the boundary values $\boldsymbol{x}_0, \boldsymbol{x}_1, \cdots, \boldsymbol{x}_{m-1}, \boldsymbol{x}_{M-m+1}, \boldsymbol{x}_{M-m+2}, \cdots, \boldsymbol{x}_M$ are calculated. Finally, $\boldsymbol{x}_{M-m}, \boldsymbol{x}_{M-m-1}, \cdots, \boldsymbol{x}_m$ are calculated successively by using the expressions similar to (3.8) or (3.8)′.

Finally, we should point out the following fact: For "complete" multidiagonal equations, we can construct a double sweep method for which the stability conditions are roughly the conditions of Section 2.

References

[1] Rusanov, V. V., On the stability of block-double-sweep methods, Symposium Mathematics of Computation, No. 6, The Academy of Sciences of USSR, Moscow, 1960 (in Russian).

[2] Gelfand, I. M. and Lokutsieuski, O. V., The double sweep method for solution of difference equations, appendix II of book [6].

[3] Zhu, Youlan, A numerical method for a class of initial-boundary-value problems of first order quasilinear hyperbolic systems with three independent variables, Technical Report of The Institute of Computing Technology, The Chinese Academy of Sciences, Beijing, 1973 (in Chinese).

[4] Rusanov, V. V., On solution of difference equations, Report of Academy of Sciences of USSR, 136 (1961), No. 1, 33—35 (in Russian).

[5] Gantmacher, F. R., The theory of matrices, Chelsea Publishing Co., New York, N. Y., 1959.

[6] Godunov, S. K. and Ryabenki, V. S., Theory of difference schemes——an introduction, North-Holland, Amsterdam and Interscience-Wiley, New York, 1964.

Appendix 3

Stability and Convergence of Difference Schemes for Linear Initial–Boundary–Value Problems

Abstract

In this paper, we discuss the stability of difference schemes for initial–boundary–value problems for general linear hyperbolic systems (LIBVP) with variable coefficients satisfying the Lipschitz condition, and prove under very weak conditoins that several classes of difference schemes are stable. In this paper, we also prove that if a q-th order scheme is stable, then the approximate solution obtained by using the scheme converges to the genuine solution, and the rate of convergence in L_2 is of order q when the solution and the coefficients of partial differential equations are sufficiently smooth.

Introduction

In [1] and [2], we discussed the stability of difference schemes for initial–boundary–value problems with two boundaries for linear hyperbolic systems with diagonal matrices, and proved that under very weak conditions, several classes of schemes are stable. In this paper, those results will be generalized to the situations with several internal boundaries and with general types of coefficient matrices. For several classes of schemes, we shall prove that if (i) the von Neumann condition is satisfied everywhere; (ii) the condition (2.12), which guarantees that the systems of difference equations are well–conditioned, is satisfied everywhere; (iii) coefficients of schemes satisfy the Lipschitz condition; and (iv) coefficients of partial differential equations satisfy the Lipschitz condition and certain quantities concerned with coefficients of partial differential equations are uniformly bounded with respect to x and t, then the schemes are stable. Condition (iii) is easily fulfilled when the coefficients of partial differential equations satisfy the Lipschitz condition. Therefore, by using these results, we can prove that several schemes are stable when they are applied to calculating initial–boundary–value problems for hyperbolic systems with coefficients satisfying the Lipschitz condition.

In this paper, we shall also prove that if schemes are stable and if solutions and coefficients of partial differential equations are sufficiently smooth, then the solutions of difference equations will converge to the genuine solutions of partial differential equations, and the q-th order schemes have a rate of convergence of order q in L_2. Therefore, the second order schemes for initial-boundary-value problems given in [1] have a rate of convergence of order 2.

1. *Formulation of Problems*

In this paper, we consider the following initial-boundary-value problem with $L+1$ boundaries for linear hyperbolic systems with variable coefficients.

(i) In regions $0\leqslant t\leqslant t^*$, $l-1\leqslant x\leqslant l$, $l=1, 2, \cdots, L$, a hyperbolic system

$$\frac{\partial \boldsymbol{U}}{\partial t}+A(x, t)\frac{\partial \boldsymbol{U}}{\partial x}+C(x, t)\boldsymbol{U}=\boldsymbol{F}(x, t) \tag{1.1}$$

is given, where x, t are scalars. $\boldsymbol{U}$, $\boldsymbol{F}$ are N-dimensional vectors, and A, C are $N\times N$-matrices. The matrix A has N real eigenvalues and a omplete system of eigenvectors, i.e., there are an invertible matrix G cand a real diagonal matrix Λ such that

$$A=G^{-1}\Lambda G.$$

(ii) At $x=0, 1, \cdots, L$, boundary conditions and internal boundary conditions

$$\begin{cases} B_0(t)\begin{pmatrix}\boldsymbol{U}_0^+\\ y_0\end{pmatrix}=\boldsymbol{E}_0(t), \\ B_l(t)\begin{pmatrix}\boldsymbol{U}_l^-\\ \boldsymbol{U}_l^+\\ y_{lt}\end{pmatrix}=\boldsymbol{E}_l(t), \quad l=1, 2, \cdots, L-1, \\ B_L(t)\begin{pmatrix}\boldsymbol{U}_L^-\\ y_{Lt}\end{pmatrix}=\boldsymbol{E}_L(t), \end{cases} \tag{1.2}$$

and the following ordinary differential equations

$$\frac{dy_l}{dt}-y_{lt}=0, \quad l=0, 1, \cdots, L \tag{1.3}$$

are given, where y_l and y_{lt} are scalar functions dependent on t. Because y_{lt} appear in boundary conditions, we call them boundary functions. $\boldsymbol{U}_l^\pm=\lim\limits_{\varepsilon\to 0}\boldsymbol{U}(l\pm\varepsilon, t)$, $\varepsilon>0$, that is, they are the values of $\boldsymbol{U}$ on the right-hand and the left-hand sides of the boundary $x=l$. We permit solutions to be discontinuous at boundaries, so the values on different sides are denoted by different symbols. B_0, $B_l\ (l=1, 2, \cdots, L-1)$, B_L are $\gamma_0\times(N+1)$-, $\gamma_l\times(2N+1)$-, $\gamma_L\times(N+1)$-matrices respectively, and $\boldsymbol{E}_l$ are γ_l-dimensional vectors.

(iii) At $t=0$, initial values of $\boldsymbol{U}$, $\overline{\boldsymbol{y}}$ and $\overline{\boldsymbol{y}}_t$,

$$\begin{cases} \boldsymbol{U}(x, 0)=\boldsymbol{D}_l(x), & l-1\leqslant x\leqslant l,\ l=1, 2, \cdots, L, \\ \overline{\boldsymbol{y}}(0)=\overline{\boldsymbol{y}}_0, \\ \overline{\boldsymbol{y}}_t(0)=\overline{\boldsymbol{y}}_{0t} \end{cases} \tag{1.4}$$

are given, where $\overline{\boldsymbol{y}}$, $\overline{\boldsymbol{y}}_t$, $\overline{\boldsymbol{y}}_0$, $\overline{\boldsymbol{y}}_{0t}$ are $(L+1)$-dimensional vectors, the elements of $\overline{\boldsymbol{y}}$ and $\overline{\boldsymbol{y}}_t$ being y_l and y_{lt} respectively, and $\boldsymbol{D}_l$ is an N-dimensional vector. Obviously, in order to guarantee that there exists a smooth solution with such a structure, the initial values have to satisfy the boundary conditions (1.2).

We need to determine $\boldsymbol{U}$ for $0\leqslant x\leqslant L$, $0\leqslant t\leqslant t^*$, and $\overline{\boldsymbol{y}}$ and $\overline{\boldsymbol{y}}_t$ for $0\leqslant t\leqslant t^*$.

Obviously, in order that the above problem has a unique solution, the boundary conditions should satisfy certain conditions. In the following, λ_n denotes an element of Λ, and suppose $\lambda_1\geqslant\lambda_2\geqslant\cdots\geqslant\lambda_N$. In this situation, these conditions can be described as follows: if

$$\begin{cases} \lambda^+_{N-p_l,l}>0\geqslant\lambda^+_{N-p_l+1,l}, & l=0, 1, \cdots, L-1, \\ \lambda^-_{q_l,l}\geqslant 0>\lambda^-_{q_l+1,l}, & l=1, 2, \cdots, L, \end{cases} \tag{1.5}$$

where $\lambda^{\pm}_{n,l}=\lambda_n|_{x=l\pm 0}$, i.e., $\lambda^{\pm}_{n,l}$ are the values of λ_n on the left-hand and the right-hand sides of the boundary $x=l$, then

(1)
$$\begin{cases} \gamma_0+p_0=N+1, \\ \gamma_l+p_l+q_l=2N+1, & l=1, 2, \cdots, L-1, \\ \gamma_L+q_L=N+1; \end{cases} \tag{1.6}$$

and

(2) when $\boldsymbol{F}^-_l$, $\boldsymbol{F}^+_{l-1}$, $l=1, 2, \cdots, L$ are given, the syste

$$\begin{cases} \begin{cases} B_0(t)\begin{pmatrix} \boldsymbol{U}^+_0 \\ y_{0t} \end{pmatrix}=\boldsymbol{E}_0(t), \\ \overline{G}^+_0\boldsymbol{U}^+_0=\boldsymbol{F}^+_0, \end{cases} \\ \begin{cases} \overline{G}^-_l\boldsymbol{U}^-_l=\boldsymbol{F}^-_l, \\ B_l(t)\begin{pmatrix} \boldsymbol{U}^-_l \\ \boldsymbol{U}^+_l \\ y_{lt} \end{pmatrix}=\boldsymbol{E}_l(t), \quad l=1, 2, \cdots, L-1, \\ \overline{G}^+_l\boldsymbol{U}^+_l=\boldsymbol{F}^+_l, \end{cases} \\ \begin{cases} \overline{G}^-_L\boldsymbol{U}^-_L=\boldsymbol{F}^-_L, \\ B_L(t)\begin{pmatrix} \boldsymbol{U}^-_L \\ y_{Lt} \end{pmatrix}=\boldsymbol{E}_L(t) \end{cases} \end{cases} \tag{1.7}$$

has a unique solution, where $\boldsymbol{F}^-_l$, $\boldsymbol{F}^+_l$ are q_l- and p_l-dimensional vectors respectively, and

$$\overline{G}^+_l=\begin{pmatrix} \boldsymbol{G}^*_{N-p_l+1} \\ \vdots \\ \boldsymbol{G}^*_N \end{pmatrix}\Bigg|_{x=l+0}, \quad \overline{G}^-_l=\begin{pmatrix} \boldsymbol{G}^*_1 \\ \vdots \\ \boldsymbol{G}^*_{q_l} \end{pmatrix}\Bigg|_{x=l-0},$$

G_n^* being the n-th row of G.

It can be easily seen that if the system (1.7) has a unique solution, then there exists a constant c_1 such that the following inequalities

$$\begin{cases} \sum_{n=1}^{N-p_0} |u_{n,0}^+|^2 + |y_{0t}|^2 \leqslant c_1 \left[\sum_{n=N-p_0+1}^{N} |u_{n,0}^+|^2 + \|E_0\|^2 \right], \\ \sum_{n=q_l+1}^{N} |u_{n,l}^-|^2 + \sum_{n=1}^{N-p_l} |u_{n,l}^+|^2 + |y_{lt}|^2 \\ \quad \leqslant c_1 \left[\sum_{n=1}^{q_l} |u_{n,l}^-|^2 + \sum_{n=N-p_l+1}^{N} |u_{n,l}^+|^2 + \|E_l\|^2 \right], \quad l=1, 2, \cdots, L-1, \\ \sum_{n=q_L+1}^{N} |u_{n,L}^-|^2 + |y_{Lt}|^2 \leqslant c_1 \left[\sum_{n=1}^{q_L} |u_{n,L}^-|^2 + \|E_L\|^2 \right] \end{cases} \tag{1.8}$$

are satisfied, where $u_n = G_n^* U$, the meanings of the subscript l and the superscript $\pm$ are the same as those for $U_l^\pm$, and $\|E_l\|^2 = \sum_i |E_{li}|^2$ is the square of the Euclidean vector norm of E_l, E_{li} being elements of E_l.

In the discussion below, we suppose that the following inequalities are satisfied:

$$\begin{cases} \sum_{n=1}^{N-p_0} |u_{n,0}^+|^2 + |y_{0t}|^2 \leqslant c_1 \left[\sum_{n=N-p_0+1}^{N} |\lambda_{n,0}^+| \, |u_{n,0}^+|^2 + \|E_0\|^2 \right], \\ \sum_{n=q_l+1}^{N} |u_{n,l}^-|^2 + \sum_{n=1}^{N-p_l} |u_{n,l}^+|^2 + |y_{lt}|^2 \\ \quad \leqslant c_1 \left[\sum_{n=1}^{q_l} |\lambda_{n,l}^-| \, |u_{n,l}^-|^2 + \sum_{n=N-p_l+1}^{N} |\lambda_{n,l}^+| \, |u_{n,l}^+|^2 + \|E_l\|^2 \right], \quad l=1,2,\cdots,L-1, \\ \sum_{n=q_L+1}^{N} |u_{n,L}^-|^2 + |y_{Lt}|^2 \leqslant c_1 \left[\sum_{n=1}^{q_L} |\lambda_{n,L}^-| \, |u_{n,L}^-|^2 + \|E_L\|^2 \right], \end{cases} \tag{1.9}$$

which are a little stronger than the condition (1.8).

Finally, we want to explain why the boundary functions $\bar{y}$ and $\bar{y}_t$ in the boundary conditions are introduced. This is because we hope that the results in this paper can be applied to proof of convergence of difference schemes for nonlinear problems. As is well-known, in nonlinear problems, the locations of boundaries in many situations are unknown and are determined through boundary conditions, where the locations of boundaries appear as unknown functions, and whose first variations are similar to (1.2)—(1.3).

2. *Proof of Stability and Convergence*

In [1] and [2], we presented several schemes for the above type of linear problems. This section will discuss stability of these schemes with respect to initial values, boundary values and nonhomogeneous terms of partial differential equations, convergence and rates of convergence in L_2.

First of all, we discuss the key problem among these——stability of schemes with respect to the initial values. Therefore, we suppose that $F(x, t)$ and $E_l(t)$ are equal to zero.

In this section, we use the following symbols. If f is an unknown, then $f_{l,m}^k$ denotes the approximate value of f at $x=l+m\Delta x$, $t=k\Delta t$; if f is given, then $f_{l,m}^k$ denotes the value of f at $x=l+m\Delta x$, $t=k\Delta t$. $\Delta x=\frac{1}{M}$ and Δt are respective lengths of the mesh in the x and t directions, and suppose $\Delta t/\Delta x=\text{constant}$.

In papers [1] and [2], when constructing difference schemes for initial–boundary–value problems, in order for schemes to apply to all situations, we first rewrite

$$\frac{\partial \boldsymbol{U}}{\partial t}+A\frac{\partial \boldsymbol{U}}{\partial x}+C\boldsymbol{U}=0 \tag{2.1}$$

as

$$\boldsymbol{G}_n^*\frac{\partial \boldsymbol{U}}{\partial t}+\lambda_n\boldsymbol{G}_n^*\frac{\partial \boldsymbol{U}}{\partial x}+\boldsymbol{G}_n^*C\boldsymbol{U}=0, \quad n=1,\ 2,\ \cdots,\ N, \tag{2.2}$$

and then establish difference equations approximating (2.2). It is easy to know that if the difference equation of a scheme for a single wave equation $\frac{\partial u}{\partial t}+\lambda\frac{\partial u}{\partial x}=0$ is

$$\sum_{h=H_1}^{H_2} r_{n,m,h}^k u_{n,m+h}^{k+1}=\sum_{h=H_1}^{H_2} s_{n,m,h}^k u_{n,m+h}^k,$$

where $r_{n,m,h}^k$, $s_{n,m,h}^k$ are scalar functions of λ_n, then the difference equation of the scheme for the equation (2.2) can always be written in the following form

$$\sum_{h=H_1}^{H_2}(r_{n,m,h}^k u_{n,m+h}^{k+1}+O_{r,h}^*(\Delta t)\widetilde{\boldsymbol{U}}_{m+h}^{k+1})=\sum_{h=H_1}^{H_2}(s_{n,m,h}^k u_{n,m+h}^k+O_{s,h}^*(\Delta t)\widetilde{\boldsymbol{U}}_{m+h}^k),$$

where $u_n=\boldsymbol{G}_n^*\boldsymbol{U}$, $\widetilde{\boldsymbol{U}}=\boldsymbol{G}\boldsymbol{U}$, and $O_{r,h}^*(\Delta t)$ and $O_{s,h}^*(\Delta t)$ denote N–dimensional row–vectors. Moreover, if G satisfies the Lipschitz condition with respect to x and t, and if the elements of G, G^{-1} and C are uniformly bounded with respect to x and t, then the norms of elements of the row–vectors are less than $c\Delta t$, c being a constant independent of Δt. (In the above formula, we omit the subscript l for all quantities.)

Therefore, we discuss stability of the following difference scheme for the homogeneous initial–boundary–value problem (1.1)—(1.4):

$$\begin{cases}\begin{cases} y_0^{k+1}+O(\Delta t)y_{0t}^{k+1}=y_0^k+O(\Delta t)y_{0t}^k, \\ \widetilde{B}_0^{k+1}\begin{pmatrix}\widetilde{\boldsymbol{U}}_{l,0}^{k+1}\\ y_{0t}^{k+1}\end{pmatrix}=0, \\ \left[\sum_{h=H_1}^{H_2}(r_{n,m,h}^k u_{n,m+h}^{k+1}+O_{r,h}^*(\Delta t)\widetilde{\boldsymbol{U}}_{m+h}^{k+1})\right]_0 \\ \quad=\left[\sum_{h=H_1}^{H_2}(s_{n,m,h}^k u_{n,m+h}^k+O_{s,h}^*(\Delta t)\widetilde{\boldsymbol{U}}_{m+h}^k)\right]_0, \\ \quad m\in G([\lambda_n]_0), \quad n=1,\ 2,\cdots,N, \end{cases}\end{cases}$$

$$
\left\{\begin{array}{l}
\left\{\begin{array}{l}
y_l^{k+1}+O(\Delta t)y_{lt}^{k+1}=y_l^k+O(\Delta t)y_{lt}^k, \\
\widetilde{B}_l^{k+1}\begin{pmatrix}\widetilde{U}_{l-1,M}^{k+1}\\ \widetilde{U}_{l,0}^{k+1}\\ y_{lt}^{k+1}\end{pmatrix}=0, \\
\left[\sum_{h=H_1}^{H_2}(r_{n,m,h}^k u_{n,m+h}^{k+1}+O_{r,h}^*(\Delta t)\widetilde{U}_{m+h}^{k+1})\right]_l \\
\quad =\left[\sum_{h=H_1}^{H_2}(s_{n,m,h}^k u_{n,m+h}^k+O_{s,h}^*(\Delta t)\widetilde{U}_{m+h}^k)\right]_l, \\
\quad m\in G([\lambda_n]_l),\ n=1,2,\cdots,N,\ l=1,2,\cdots,L-1,
\end{array}\right. \\
\left\{\begin{array}{l}
y_L^{k+1}+O(\Delta t)y_{Lt}^{k+1}=y_L^k+O(\Delta t)y_{Lt}^k, \\
\widetilde{B}_L^{k+1}\begin{pmatrix}\widetilde{U}_{L-1,M}^{k+1}\\ y_{Lt}^{k+1}\end{pmatrix}=0,
\end{array}\right.
\end{array}\right. \tag{2.3}
$$

where $[\cdots]_l$ denotes that every quantity in brackets has the subscript l. $G([\lambda_n]_l)$ is a set of integers, which indicates the range of m, which depends on the values of λ_n at $x=l$ and $l+1$, and whose definition is given in [1], [2]. $H_1\leqslant 0$, $H_2\geqslant 0$, and they depend on n, l, m, and satisfy the relation $0\leqslant H_1(m)+m\leqslant H_2(m)+m\leqslant M$. $\widetilde{B}_0$, $\widetilde{B}_l$ $(l=1,2,\cdots,L-1)$, and $\widetilde{B}_L$ denote the following matrices

$$
\widetilde{B}_0=B_0\begin{pmatrix}G_{0,0}^{-1} & 0\\ 0 & 1\end{pmatrix},\quad \widetilde{B}_l=B_l\begin{pmatrix}G_{l-1,M}^{-1} & & 0\\ & G_{l,0}^{-1} & \\ 0 & & 1\end{pmatrix},\quad \widetilde{B}_L=B_L\begin{pmatrix}G_{L-1,M}^{-1} & 0\\ 0 & 1\end{pmatrix}.
$$

The system (2.3) can be rewritten in the following form

$$
(\bar{R}^k+\bar{O}_1^k(\Delta t))\bar{U}^{k+1}=(\bar{S}^k+\bar{O}_2^k(\Delta t))\bar{U}^k. \tag{2.4}
$$

Here $\bar{U}$ denotes a vector whose elements are y_l, y_{lt} $(l=0,1,\cdots,L)$, and $u_{n,l,m}$ $(n=1,2,\cdots,N,\ m=0,1,\cdots,M,\ l=0,1,\cdots,L-1)$. $\bar{O}_1^k(\Delta t)$ and $\bar{O}_2^k(\Delta t)$ are two matrices which correspond to the coefficients $O_{r,h}^*(\Delta t)$, $O_{s,h}^*(\Delta t)$ and $O(\Delta t)$ in (2.3). Obviously, the norms of their elements are less than $c\Delta t$ and there are only several non-zero elements in every row and in every column. Therefore, we have the following estimates for them

$$
\|\bar{O}_1^k(\Delta t)\|\leqslant c\Delta t,\quad \|\bar{O}_2^k(\Delta t)\|\leqslant c\Delta t. \tag{2.5}
$$

where c is a constant independent of Δt.

Kreiss and Strang[3] pointed out that if operators $\bar{D}^k$ satisfy the relation

$$
\left\|\prod_{k=k_1}^{k_2}\bar{D}^k\right\|\leqslant c_2,\quad 0\leqslant k_1\Delta t\leqslant k_2\Delta t\leqslant t^*,\ c_2>1,
$$

and if operators $\bar{O}^k(\Delta t)$ satisfy the relation

$$
\|\bar{O}^k(\Delta t)\|\leqslant c_3\Delta t,\quad 0\leqslant k\Delta t\leqslant t^*,
$$

where c_2, c_3 are independent of Δt, then

$$\left\| \prod_{k=k_1}^{k_2} (\bar{D}^k + \bar{O}^k(\Delta t)) \right\| \leqslant c_2 e^{c_2 c_3 t^*} = c_4.$$

According to this result, we know that if

$$\left\| \prod_{k=k_1}^{k_2} (\bar{R}^k)^{-1} \bar{S}^k \right\| \leqslant c_2, \quad 0 \leqslant k_1 \Delta t \leqslant k_2 \Delta t \leqslant t^*, \tag{2.6}$$

and

$$\| (\bar{R}^k + \bar{O}_1^k(\Delta t))^{-1} (\bar{S}^k + \bar{O}_2^k(\Delta t)) - (\bar{R}^k)^{-1} \bar{S}^k \| \leqslant c_3 \Delta t, \tag{2.7}$$

then the inequality

$$\left\| \prod_{k=k_1}^{k_2} (\bar{R}^k + \bar{O}_1^k(\Delta t))^{-1} (\bar{S}^k + \bar{O}_2^k(\Delta t)) \right\| \leqslant c_4, \quad 0 \leqslant k_1 \Delta t \leqslant k_2 \Delta t \leqslant t^* \tag{2.8}$$

is fulfilled.

It is easy to know that if (2.5) is satisfied and if

$$\| (\bar{R}^k)^{-1} \| \leqslant c_2, \quad \| \bar{S}^k \| \leqslant c_2, \tag{2.9}$$

where c_2 is a constant independent of k, then (2.7) is fulfilled.

In fact,

$$(\bar{R}^k + \bar{O}_1^k(\Delta t))^{-1} = (\bar{R}^k)^{-1} - (\bar{R}^k)^{-1} \bar{O}_1^k(\Delta t) (\bar{R}^k + \bar{O}_1^k(\Delta t))^{-1},$$

and

$$\begin{aligned} &(\bar{R}^k + \bar{O}_1^k(\Delta t))^{-1} (\bar{S}^k + \bar{O}_2^k(\Delta t)) - (\bar{R}^k)^{-1} \bar{S}^k \\ &= (\bar{R}^k)^{-1} \bar{O}_2^k(\Delta t) - (\bar{R}^k)^{-1} \bar{O}_1^k(\Delta t) (\bar{R}^k + \bar{O}_1^k(\Delta t))^{-1} (\bar{S}^k + \bar{O}_2^k(\Delta t)), \end{aligned}$$

so, by using the inequality

$$\| (\bar{R}^k + \bar{O}_1^k(\Delta t))^{-1} \| \leqslant \frac{\| (\bar{R}^k)^{-1} \|}{1 - \| O_1^k(\Delta t) \| \, \| (\bar{R}^k)^{-1} \|},$$

and the fact that $\| (\bar{R}^k)^{-1} \|$ and $\| \bar{S}^k \|$ are uniformly bounded with respect to Δt and k, (2.7) can be derived. Therefore, in order to prove (2.8), the only thing we need to do is to prove that the inequalities (2.6) and (2.9) are fulfilled. In fact, every equation in the system

$$\bar{R}^k \bar{U}^{k+1} = \bar{S}^k \bar{U}^k \tag{2.10}$$

approximates a single wave equation $\dfrac{\partial u_n}{\partial t} + \lambda_n \dfrac{\partial u_n}{\partial x} = 0$. Moreover, the forms of the inequalities (1.9) are similar to those of (4.6) in [1]. Therefore, by using the method in [1], [2], we can prove that the results in [1], [2] are still correct for the system (2.10). Therefore, (2.6) and (2.9) are fulfilled for a series of situations. However, for the situation in this paper, we should take

$$\begin{aligned} \tilde{T} = \sum_{l=0}^{L-1} \left[\sum_{n=1}^{N} \tilde{T}_n \right]_l + \tilde{c}_2^2 \Bigg\{ & \left\| \tilde{B}_0 \begin{pmatrix} \tilde{U}_{0,0} \\ y_{0t} \end{pmatrix} \right\|^2 + |y_0|^2 \\ & + \sum_{l=1}^{L-1} \left(\left\| \tilde{B}_l \begin{pmatrix} \tilde{U}_{l-1,M} \\ \tilde{U}_{l,0} \\ Y_{lt} \end{pmatrix} \right\|^2 + |Y_l|^2 \right) + \left\| \tilde{B}_L \begin{pmatrix} \tilde{U}_{L-1,M} \\ y_{Lt} \end{pmatrix} \right\|^2 + |y_L|^2 \Bigg\} \Delta x \end{aligned}$$

as the energy sum, where the definition of $\widetilde{T}_n$ is the same as that in [1] and [2].

That is, when horizontal–two–point schemes, horizontal–three–point explicit schemes, second order horizontal–four–point explicit schemes (using horizontal–three–point schemes in the regions near the boundaries), and section–wise mixed schemes obtained by combining the above schemes (suppose that the number of subintervals is finite and that $|\lambda|\geqslant\varepsilon>0$ at the borders of two subintervals) are used for the initial–boundary–value problem (1.1)—(1.4) with the condition (1.9), if

(i) the von Neumann condition

$$r^{k*}_{n,l,m}(\theta)r^{k}_{n,l,m}(\theta)-s^{k*}_{n,l,m}(\theta)s^{k}_{n,l,m}(\theta)\geqslant 0, \tag{2.11}$$

$$m\in G([\lambda_n]_l),\ l=0,\ 1,\ \cdots,\ L-1,\ n=1,\ 2,\ \cdots,\ N$$

is satisfied everywhere, where $r^{k*}_{n,l,m}(\theta)$ and $s^{k*}_{n,l,m}(\theta)$ are conjugate complex numbers of $r^{k}_{n,l,m}(\theta)$ and $s^{k}_{n,l,m}(\theta)$ respectively, and

$$r^{k}_{n,l,m}(\theta)=\sum_{h=H_1}^{H_2}\lim_{\Delta t\to 0} r^{k}_{n,l,m,h}(\Delta t)e^{i\theta h},$$

$$s^{k}_{n,l,m}(\theta)=\sum_{h=H_1}^{H_2}\lim_{\Delta t\to 0} s^{k}_{n,l,m,h}(\Delta t)e^{i\theta h};$$

(ii) the following inequality

$$r^{k*}_{n,l,m}(\theta)r^{k}_{n,l,m}(\theta)\geqslant c_0>0 \tag{2.12}$$

is satisfied everywhere, where c_0 is a positive constant;

(iii) $r^{k}_{n,l,m,h}$ and $s^{k}_{n,l,m,h}$ satisfy the Lipschitz condition with respect to k, m,

$$|f^{k}_{m+1,h}-f^{k}_{m,h}|\leqslant c\Delta x,\quad |f^{k+1}_{m,h}-f^{k}_{m,h}|\leqslant c\Delta t$$

(or they satisfy these conditions section–wise), where f denotes $r_{n,l}$ or $s_{n,l}$, then, there exists a constant c_2 independent of Δt such that (2.6) and (2.9) are fulfilled.

In this paper, we shall not give any concrete proof, but only point out the structure of the proof. When conditions (i) and (iii) are satisfied, a diagonal matrix $\overline{H}^k$ uniformly bounded with respect to k can be found such that the inequality

$$\|\overline{H}^k\overline{R}^k\overline{U}^{k+1}\|^2\leqslant(1+\alpha\Delta t)\|\overline{H}^{k-1}\overline{R}^{k-1}\overline{U}^k\|^2 \tag{2.13}$$

is fulfilled, where α is a constant independent of Δt; when conditions (ii) and (iii) are satisfied, there exists a constant c such that

$$\|(\overline{H}^k\overline{R}^k)^{-1}\|\leqslant c. \tag{2.14}$$

Moreover, by using (2.13) and (2.14), we can obtain (2.6) and $\|(\overline{R}^k)^{-1}\|\leqslant c$ immediately, and $\|\overline{S}^k\|\leqslant c_2$ is obvious. Therefore, the conclusion above can be proved.

When (2.8) is fulfilled, we can obtain the following inequality

$$\|\overline{U}^k\|\leqslant c_3\|\overline{U}^0\|.$$

Furthermore, according to the uniform boundedness of G^{-1}, we can also obtain

$$\|\overline{\overline{U}}^k\| \leqslant c_5 \|\overline{\overline{U}}^0\|,$$

where $\overline{\overline{U}}$ is a vector whose elements are y_l, y_{lt} $(l=0, 1, \cdots, L)$ and $U_{l,m}$ $(l=0, 1, \cdots, L-1, m=0, 1, \cdots, M)$. That is, the scheme (2.3) is stable with respect to the initial values.

Therefore, we have the following conclusion: when the schemes mentioned above are applied to the problem (1.1)—(1.4) with (1.9), if conditions (i)—(iii) are fulfilled, and if

(iv) the elements of G satsify the Lipschitz condition with respect to x and t, and the elements of G, G^{-1} and c are uniformly bounded with respect to x, t, the inequality (2.8) is then fulfilled, and the procedure (2.3) is stable with respect to initial values.

Moreover, from (2.13) and (2.14), it is easily proved that the scheme is stable with respect to nonhomogeneous terms of boundary conditions. In fact, when the boundary conditions in (2.3) are changed to the form

$$\begin{cases} \widetilde{B}_0^{k+1}\begin{pmatrix} \widetilde{U}_{0,0}^{k+1} \\ y_{0t}^{k+1} \end{pmatrix} = E_0((k+1)\Delta t), \\ \widetilde{B}_l^{k+1}\begin{pmatrix} \widetilde{U}_{l-1,M}^{k+1} \\ \widetilde{U}_{l,0}^{k+1} \\ y_{lt}^{k+1} \end{pmatrix} = E_l((k+1)\Delta t), \quad l=1, 2, \cdots, L-1, \\ \widetilde{B}_L^{k+1}\begin{pmatrix} \widetilde{U}_{L-1,M}^{k+1} \\ y_{Lt}^{k+1} \end{pmatrix} = E_L((k+1)\Delta t), \end{cases}$$

the difference equations have the following form

$$\overline{R}_1^k \overline{U}^{k+1} = \overline{S}_1^k \overline{U}^k + \overline{E}^k,$$

where $\overline{R}_1^k$, $\overline{S}_1^k$ denote $\overline{R}^k + \overline{O}_1^k(\Delta t)$, $\overline{S}^k + \overline{O}_2^k(\Delta t)$ respectively. Moreover, $(\overline{S}_1^k \overline{U}^k, \overline{E}^k) = 0$. This is because at least one of the i-th elements of the two vectors $\overline{S}_1^k \overline{U}^k$ and $\overline{E}^k$ is equal to zero. Because $(\overline{S}_1^k \overline{U}^k, \overline{E}^k) = 0$ and $\overline{H}^k$ is a diagonal matrix, we have the relation

$$\|\overline{H}^k \overline{R}_1^k \overline{U}^{k+1}\|^2 = \|\overline{H}^k \overline{S}_1^k \overline{U}^k\|^2 + \|\overline{H}^k \overline{E}^k\|^2.$$

Moreover, the norms of $\overline{O}_1^k(\Delta t)$ and $\overline{O}_2^k(\Delta t)$ are $O(\Delta t)$, and using (2.13), we can thus obtain

$$\|\overline{H}^k \overline{S}_1^k (\overline{H}^{k-1} \overline{R}_1^{k-1})^{-1}\|^2 \leqslant 1 + \alpha_1 \Delta t.$$

Therefore, we have the relationship

$$\|\overline{H}^k \overline{R}_1^k \overline{U}^{k+1}\|^2 \leqslant (1+\alpha_1 \Delta t) \|\overline{H}^{k-1} \overline{R}_1^{k-1} \overline{U}^k\|^2 + \|\overline{H}^k \overline{E}^k\|^2.$$

Furthermore, we can obtain the relationship

$$\|\overline{U}^{k+1}\|^2 \leqslant c_6\left(\|\overline{U}^0\|^2 + \sum_{i=0}^{L} \sum_{k'=1}^{k+1} \|E_i(k'\Delta t)\|^2 \Delta t\right), \quad 0 \leqslant k\Delta t \leqslant t^*,$$

or

$$\|\overline{U}^{k+1}\| \leqslant c_7 \|\overline{U}^0\| + c_7 \sum_{i=0}^{L} \left(\sum_{k'=1}^{k+1} \|E_i(k'\Delta t)\|^2 \Delta t \right)^{1/2}, \tag{2.15}$$

that is, the procedure is stable with respect to both initial and boundary values.

We now prove that the scheme is stable with respect to nonhomogeneous terms of partial differential equations. In fact, the general form of difference equations approximating (1.1)—(1.4) is

$$\overline{R}_1^k \overline{U}^{k+1} = \overline{S}_1^k \overline{U}^k + \overline{E}^k + \overline{F}^k \Delta t, \quad k = 0, 1, \cdots, \tag{2.16}$$

where $\overline{E}^k$ comes from the nonhomogeneous terms of boundary conditions, $\overline{F}^k$ comes from the nonhomogeneous terms of partial differential equations, and every element of $\overline{F}^k$ is either a value approximating $F(x, t)$ or zero. Let $\overline{D}^k = (\overline{R}_1^k)^{-1} \overline{S}_1^k$, we can rewrite (2.16) as $\overline{U}^{k+1} = \overline{D}^k \overline{U}^k + (\overline{R}_1^k)^{-1} (\overline{E}^k + \overline{F}^k \Delta t)$. Hence, we have the relationship

$$\begin{aligned} \overline{U}^{k+1} &= \overline{D}^k \overline{U}^k + (\overline{R}_1^k)^{-1} (\overline{E}^k + \overline{F}^k \Delta t) \\ &= \overline{D}^k \overline{D}^{k-1} \cdots \overline{D}^0 \overline{U}^0 \\ &\quad + \overline{D}^k \overline{D}^{k-1} \cdots \overline{D}^1 (\overline{R}_1^0)^{-1} (\overline{E}^0 + \overline{F}^0 \Delta t) \\ &\quad + \cdots \\ &\quad + \overline{D}^k (\overline{R}_1^{k-1})^{-1} (\overline{E}^{k-1} + \overline{F}^{k-1} \Delta t) \\ &\quad + (\overline{R}_1^k)^{-1} (\overline{E}^k + \overline{F}^k \Delta t). \end{aligned} \tag{2.17}$$

On the other hand, (2.15) means that

$$\left\| \prod_{k=k_1}^{k_2} \overline{D}^k \right\| \leqslant c_7, \quad 0 \leqslant k_1 \Delta t \leqslant k_2 \Delta t \leqslant t^*,$$

and that

$$\begin{aligned} &\|\overline{D}^k \overline{D}^{k-1} \cdots \overline{D}^1 (\overline{R}_1^0)^{-1} \overline{E}^0 + \cdots + \overline{D}^k (\overline{R}_1^{k-1})^{-1} \overline{E}^{k-1} + (\overline{R}_1^k)^{-1} \overline{E}^k\| \\ &\leqslant c_7 \sum_{i=0}^{L} \left(\sum_{k'=1}^{k+1} \|E_i(k'\Delta t)\|^2 \Delta t \right)^{1/2}. \end{aligned}$$

Therefore, from (2.17), we have the following inequality

$$\|\overline{U}^{k+1}\| \leqslant c_7 \|U^0\| + c_7 \sum_{i=0}^{L} \left(\sum_{k'=1}^{k+1} \|E_i(k'\Delta t)\|^2 \Delta t \right)^{1/2} + c_8 \sum_{k'=0}^{k} \|\overline{F}^{k'}\| \Delta t, \tag{2.18}$$

that is, the procedure is also stable with respect to nonhomogeneous terms of partial differential equations.

By using (2.18), we can also prove that the schemes are convergent. If U, $\overline{y}$ and $\overline{y}_t$ satisfy relations (1.1)—(1.4), if they and the coefficients of partial differential equations are sufficiently smooth, and if a scheme is a q-th order one, then the genuine solution $\overline{U}(k\Delta t)$ corresponding to $\overline{U}^k$ satisfies the equation

$$\overline{R}_1^k \overline{U}((k+1)\Delta t) = \overline{S}_1^k \overline{U}(k\Delta t) + \overline{E}^k + \overline{F}^k \Delta t + \overline{E}_1^k(\Delta t^q) + \overline{E}_2^k(\Delta t^q) \Delta t,$$

where $\overline{E}_1^k(\Delta t^q)$ and $\overline{E}_2^k(\Delta t^q)$ are vectors whose elements are quantities of $O(\Delta t^q)$. $\overline{E}_1^k(\Delta t^q)$ has the same property as $\overline{E}^k$, that is, $(\overline{S}_1^k \overline{U}(k\Delta t), \overline{E}_1^k(\Delta t^q))$

$=0$. Therefore, the error $\overline{U}^k-\overline{U}(k\Delta t)$ satisfies the following equation

$$\overline{R}_1^k(\overline{U}^{k+1}-\overline{U}((k+1)\Delta t))=\overline{S}_1^k(\overline{U}^k-\overline{U}(k\Delta t))-\overline{E}_1^k(\Delta t^q)-\overline{E}_2^k(\Delta t^q)\Delta t,$$

and for the error, an inequality similar to (2.18) is fulfilled. Because of $\overline{U}^0-\overline{U}(0)=0$, according to that inequality similar to (2.18), we have the estimate $\|\overline{U}^{k+1}-\overline{U}((k+1)\Delta t)\|\leqslant c_9\Delta t^q$. Furthermore, G^{-1} is uniformly bounded with respect to x, t, thus, for the difference between $\overline{\overline{U}}^k$ and the corresponding genuine solution $\overline{\overline{U}}(k\Delta t)$, there is a similar estimate $\|\overline{\overline{U}}^{k+1}-\overline{\overline{U}}((k+1)\Delta t)\|\leqslant c_{10}\Delta t^q$, i.e., the approximate solution converges to the genuine solution and the rate of convergence is Δt^q in L_2. Particularly, when the unconditionally stable second order mixed scheme presented in [1] is applied to the computation of the problem (1.1)—(1.4) with (1.9), if G and Λ satisfy the Lipschitz condition, and if G, G^{-1}, Λ and C are uniformly bounded, then the approximate solution converges to the genuine solution, and the rate of convergence is Δt^2.

References

[1] Zhu Youlan, et al., Difference schemes for initial-boundary-value problems of hyperbolic systems and examples of application, Scientia Sinica, 1979, Special Issue (II), 261—280 (English Edition).

[2] Zhu Youlan, Difference schemes for initial-boundary-value problems for first order hyperbolic systems and their stability, Mathematicae Numericae Sinica, 1979, No. 1, 1—30.

[3] Richtmyer, R. D. and Morton, K. W., Difference methods for initial-value problems, Second Edition, Wiley, New York, 1967.

Chapter 2

Numerical Methods for a Certain Class of Initial-Boundary-Value Problems for the First Order Quasilinear Hyperbolic Systems in Three Independent Variables

Introduction

In this chapter, we consider the following initial-boundary-value problem for hyperbolic systems with three independent variables x, φ, t. The initial conditions are given in the t-direction, the boundary conditions and the internal boundary conditions in the x-direction, and the periodicity condition in the φ-direction. Because a periodicity condition is given in the φ-direction, the way of discretization for the pure-initial-value problems can be used in the φ-direction. Therefore, it is not difficult to extend the method of Chapter 1 to this class of initial-boundary-value problems with three independent variables.

Obviously, all the methods in Chapter 1 can be extended to this case. However, this chapter stresses practical application, instead of discussing the theoretical problems of difference methods. Therefore, we only give the scheme which is applied in Chapter 7, and almost no theoretical discussion will be carried out in this chapter.

§ 1 Formulation of Problems

We consider the following initial-boundary-value problem for hyperbolic systems with three independent variables.

I. In L regions

$$\begin{cases} x_{l-1}(\varphi, t) \leqslant x \leqslant x_l(\varphi, t), \\ 0 \leqslant \varphi \leqslant 2\pi, \\ t \geqslant 0, \end{cases}$$

$$l = 1, 2, \cdots, L,$$

a system of first order quasilinear hyperbolic equations

$$\frac{\partial \widetilde{\boldsymbol{U}}}{\partial t}+\widetilde{A}\frac{\partial \widetilde{\boldsymbol{U}}}{\partial x}+\widetilde{C}\frac{\partial \widetilde{\boldsymbol{U}}}{\partial \varphi}=\widetilde{\boldsymbol{F}}(\widetilde{\boldsymbol{U}}, x, \varphi, t) \tag{1.1}$$

is given.

II. At $t=0$, the initial conditions

$$\begin{cases}\widetilde{\boldsymbol{U}}(x, \varphi, 0)=\widetilde{\boldsymbol{D}}_l(x, \varphi), \\ \quad x_{l-1}(\varphi, 0)\leqslant x\leqslant x_l(\varphi, 0), \\ \qquad 0\leqslant\varphi\leqslant 2\pi, \quad l=1, 2,\cdots, L, \\ x_l(\varphi, 0)=c_{0,l}(\varphi), \quad x_{l,t}(\varphi, 0)=c_{1,l}(\varphi), \\ \qquad 0\leqslant\varphi\leqslant 2\pi, l=0, 1, \cdots, L\end{cases} \tag{1.2}$$

are given.

III. On the boundaries, the boundary and internal boundary conditions, compatible with the partial differential equations (1.1),

$$\begin{cases}\boldsymbol{B}_0(\widetilde{\boldsymbol{U}}_{x_0^+}, x_0, x_{0,\varphi}, x_{0,t}, \varphi, t)=0, \\ \boldsymbol{B}_l(\widetilde{\boldsymbol{U}}_{x_l^-}, \widetilde{\boldsymbol{U}}_{x_l^+}, x_l, x_{l,\varphi}, x_{l,t}, \varphi, t)=0, \\ \qquad l=1, 2, \cdots, L-1, \\ \boldsymbol{B}_L(\widetilde{\boldsymbol{U}}_{x_L^-}, x_L, x_{L,\varphi}, x_{L,t}, \varphi, t)=0\end{cases} \tag{1.3}$$

are given.

IV. In the φ-direction, the periodicity condition

$$f(\varphi)=f(\varphi+2\pi) \tag{1.4}$$

is given, where f stands for any unknown above.

We need to determine the shapes of boundary surfaces $x_l(\varphi, t)$ in the region $0\leqslant\varphi\leqslant 2\pi$, $t\geqslant 0$, $l=0, 1, \cdots, L$ and the unknown functions $\widetilde{\boldsymbol{U}}(x, \varphi, t)$ in the region $x_0(\varphi, t)\leqslant x\leqslant x_L(\varphi, t)$, $0\leqslant\varphi\leqslant 2\pi$, $t\geqslant 0$. In the above expressions, the following symbols are adopted: x, φ, t are independent variables; $\widetilde{\boldsymbol{U}}$, $\widetilde{\boldsymbol{F}}$, $\widetilde{\boldsymbol{D}}_l$ N-dimensional vectors; $\widetilde{A}$, $\widetilde{C}$ $N\times N$-matrices; $c_{0,l}$, $c_{1,l}$ scalars; $\boldsymbol{B}_l$ γ_l-dimensional vectors; $\widetilde{\boldsymbol{U}}_{x_l^+}$ and $\widetilde{\boldsymbol{U}}_{x_l^-}$ the respective values of $\boldsymbol{U}$ on the upper and the lower sides of the surface $x=x_l(\varphi, t)$; and $x_{l,\varphi}=\dfrac{\partial x_l}{\partial \varphi}$, $x_{l,t}=\dfrac{\partial x_l}{\partial t}$. Because the solution might be discontinuous, $\widetilde{\boldsymbol{U}}_{x_l^-}\neq\widetilde{\boldsymbol{U}}_{x_l^+}$ is possible. In the above problem, we require that the planes $t=$constant be space-like for the solution $\widetilde{\boldsymbol{U}}$. That is to say, for any real number a, the matrix $\widetilde{A}+a\widetilde{C}$ has N real bounded eigenvalues and N linearly independent and real eigenvectors. Therefore, for any real number a, there exist an invertible matrix

$$\widetilde{\widetilde{G}}=\begin{pmatrix}\widetilde{\boldsymbol{G}}_1^* \\ \widetilde{\boldsymbol{G}}_2^* \\ \vdots \\ \widetilde{\boldsymbol{G}}_N^*\end{pmatrix}$$

and a real diagonal matrix

$$\widetilde{A}=\begin{pmatrix}\widetilde{\lambda}_1 & & & 0\\ & \widetilde{\lambda}_2 & & \\ & & \ddots & \\ 0 & & & \widetilde{\lambda}_N\end{pmatrix},$$

such that
$$\widetilde{A}+a\widetilde{C}=\widetilde{\widetilde{G}}^{-1}\widetilde{\Lambda}\widetilde{\widetilde{G}}, \tag{1.5}$$
where $\widetilde{G}_n^*$ is the transpose of a column vector $\widetilde{G}_n$, and we suppose without any loss of generality that

$$\widetilde{\lambda}_1\geqslant\widetilde{\lambda}_2\geqslant\cdots\geqslant\widetilde{\lambda}_N.$$

We also assume that the boundary conditions and the internal boundary conditions are compatible with the partial differential equations (1.1). That is, if the eigenvalues of the matrix $\widetilde{A}-x_{l,\varphi}\widetilde{C}$ on the upper side of the surface x_l are $\widetilde{\lambda}_{n,x_l^+}$, and the eigenvalues on the lower side $\widetilde{\lambda}_{n,x_l^-}$ and if they satisfy the inequalities

$$\begin{cases}\widetilde{\lambda}_{1,x_l^+}\geqslant\cdots\geqslant\widetilde{\lambda}_{N-p_l,x_l^+}>x_{l,t}\geqslant\widetilde{\lambda}_{N-p_l+1,x_l^+}\geqslant\cdots\geqslant\widetilde{\lambda}_{N,x_l^+},\\ \qquad l=0,1,\cdots,L-1,\\ \widetilde{\lambda}_{1,x_l^-}\geqslant\cdots\geqslant\widetilde{\lambda}_{q_l,x_l^-}\geqslant x_{l,t}>\widetilde{\lambda}_{q_l+1,x_l^-}\geqslant\cdots\geqslant\widetilde{\lambda}_{N,x_l^-},\\ \qquad l=1,2,\cdots,L,\end{cases} \tag{1.6}$$

then the following two conditions are satisfied.

(1)[1]

$$\begin{cases}\gamma_0+p_0=N+1,\\ \gamma_l+p_l+q_l=2N+1, \quad l=1,2,\cdots,L-1,\\ \gamma_L+q_L=N+1.\end{cases} \tag{1.7}$$

(2) When x_l, $x_{l,\varphi}$ $(l=0,1,\cdots,L)$, φ, t, F_l^- and F_{l-1}^+ $(l=1,2,\cdots,L)$ are correctly given, the systems

$$\begin{aligned}&\begin{cases}B_0(\widetilde{U}_{x_0^+},x_{0,t})=0,\\ \widetilde{\widetilde{G}}_0^+\widetilde{U}_{x_0^+}=F_0^+,\end{cases}\\ &\begin{cases}\widetilde{\widetilde{G}}_l^-\widetilde{U}_{x_l^-}=F_l^-,\\ B_l(\widetilde{U}_{x_l^-},\widetilde{U}_{x_l^+},x_{l,t})=0,\\ \widetilde{\widetilde{G}}_l^+\widetilde{U}_{x_l^+}=F_l^+, \quad l=1,2,\cdots,L-1,\end{cases}\\ &\begin{cases}\widetilde{\widetilde{G}}_L^-\widetilde{U}_{x_L^-}=F_L^-,\\ B_L(\widetilde{U}_{x_L^-},x_{L,t})=0\end{cases}\end{aligned} \tag{1.8}$$

always have reasonable solutions $\widetilde{U}_{x_0^+}$, $x_{0,t}$; $\widetilde{U}_{x_l^-}$, $\widetilde{U}_{x_l^+}$, $x_{l,t}$ $(l=1,2,\cdots,L-1)$; and $\widetilde{U}_{x_L^-}$, $x_{L,t}$, where

$$\widetilde{\widetilde{G}}_l^+=\begin{pmatrix}\widetilde{G}^*_{N-p_l+1}\\ \vdots\\ \widetilde{G}^*_N\end{pmatrix}\Bigg|_{a=-x_{l,\varphi},x=x_l^+(\varphi,t)}, \quad \widetilde{\widetilde{G}}_l^-=\begin{pmatrix}\widetilde{G}^*_1\\ \vdots\\ \widetilde{G}^*_{q_l}\end{pmatrix}\Bigg|_{a=-x_{l,\varphi},x=x_l^-(\varphi,t)}$$

1) Indeed the condition (2) contains the condition (1) implicitly. However, we list the condition (1) here in order to make this clearer.

$\boldsymbol{F}_l^+$, $\boldsymbol{F}_l^-$ are respectively p_l and q_l–dimensional vectors, and for simplicity, we have not written down the dependence of $\boldsymbol{B}_l$ on x_l, $x_{l,\varphi}$, φ, t.

In the following, we shall explain the meaning of the compatibility conditions (1.7) and (1.8) and why these restrictions should be put on the boundary conditions (1.3).

Firstly, we point out the following fact. $\widetilde{\lambda}_{n,x\xi}$ is the slope $\left(\frac{df_{n,x\xi}}{dt}\right)\Big|_{t=t_k}$ of the line $x=f_{n,x\xi}(t)$ at $t=t_k$ on which a plane $\varphi=$constant intersects the characteristic surface $\Pi_{n,x\xi}$ through the curve $x=x_l(\varphi, t_k)$.

If a surface Π is characteristic, and its normal vector in the coordinate system $\{x, \varphi, t\}$ is $(n_1, n_2, n_3)^*$, then we can always find an N–dimensional non-zero vector $\boldsymbol{\Omega}$, such that there are actually only some inner derivatives in the characteristic surface in the equation

$$\boldsymbol{\Omega}^* \frac{\partial \widetilde{\boldsymbol{U}}}{\partial t}+\boldsymbol{\Omega}^*\widetilde{A}\frac{\partial \widetilde{\boldsymbol{U}}}{\partial x}+\boldsymbol{\Omega}^*\widetilde{C}\frac{\partial \widetilde{\boldsymbol{U}}}{\partial \varphi}=\boldsymbol{\Omega}^*\widetilde{\boldsymbol{F}},$$

i.e., such that the relation

$$(n_1\widetilde{A}^*+n_2\widetilde{C}^*+n_3E)\boldsymbol{\Omega}=0$$

holds, E being a unit matrix of order N. On the other hand, the normal line should be perpendicular to the tangent vector

$$\left(\frac{df_{n,x\xi}}{dt}, 0, 1\right)$$

for the curve $x=f_{n,x\xi}(t)$, and to the tangent vector

$$(x_{l,\varphi}, 1, 0)$$

for the curve $x=x_l(\varphi, t_k)$. That is, there are the following relationships

$$\begin{cases} n_1\dfrac{df_{n,x\xi}}{dt}+n_3=0, \\ n_1x_{l,\varphi}+n_2=0. \end{cases}$$

Therefore, we have

$$\left(\widetilde{A}^*-x_{l,\varphi}\widetilde{C}^*-\frac{df_{n,x\xi}}{dt}E\right)\boldsymbol{\Omega}=0,$$

which means that $\dfrac{df_{n,x\xi}}{dt}$ is an eigenvalue of $\widetilde{A}^*-x_{l,\varphi}\widetilde{C}^*$ and $\widetilde{A}-x_{l,\varphi}\widetilde{C}$.

$\widetilde{\lambda}_{n,x\xi}$ has the above physical meaning, and $x_{l,t}$ is the slope of the curve on which the surface x_l intersects the plane $\varphi=$constant, so (1.6) has the following meaning: through the curve $x=x_l(\varphi, t_k)$ one can draw the backward characteristic surfaces corresponding to $\widetilde{\lambda}_{1,x\xi}, \cdots, \widetilde{\lambda}_{q_l,x\xi}$ in the l–th region and the characteristic surfaces corresponding to $\widetilde{\lambda}_{N,x\xi}, \cdots, \widetilde{\lambda}_{N-p_l+1,x\xi}$ in the $(l+1)$–th region. This means that when the values on the boundary need to be determined, the partial differential equations can provide q_l and p_l compatibility relations respectively in the l–th and $(l+1)$–th regions. Obviously, there are the relationships:

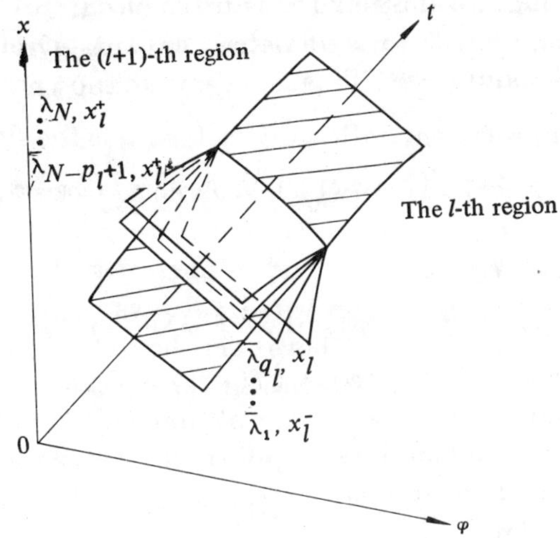

Figure 1.1

$$\begin{cases} x_l = \int x_{l,t}\, dt, \\ x_{l,\varphi} = \int \dfrac{\partial^2 x_l}{\partial\varphi\,\partial t}\, dt = \int \dfrac{\partial x_{l,t}}{\partial\varphi}\, dt \end{cases} \tag{1.9}$$

among x_l, $x_{l,\varphi}$ and $x_{l,t}$. Moreover, $\boldsymbol{B}_l$ $(l=1, \cdots, L-1)$ has $2N+3$ unknowns, which are $\widehat{U}_{x l}$, $\widehat{U}_{x l}$, x_l, $x_{l,\varphi}$, $x_{l,t}$. Therefore the dimensionality γ_l of $\boldsymbol{B}_l$ should be

$$2N+3-p_l-q_l-2=2N+1-p_l-q_l.$$

For the boundaries x_0 and x_L, the situation is similar. Hence, the condition (1.7) should be satisfied. Condition (2) says that not only the dimensionality of $\boldsymbol{B}_l$ should satisfy (1.7), but also by using (1.8), we can determine unique physical solutions for the shapes of the boundaries and the other quantities on the boundaries if the p_l+q_l compatibility relations are given.

Is the above initial-boundary-value problem for quasilinear hyperbolic systems well-posed? Does a solution exist? Is the solution unique?···. All these problems have not been solved, as far as we know. However there exists a number of practical problems whose formulations can be reduced to this type. For example, the formulation of problems in steady supersonic flow is in this form. Hence we can suppose that this formulation is probably correct. Of course, in the case of three independent variables, besides this type of initial-boundary-value problems, there are some other types.

We also point out that in certain physical problems, there exists a special case of this problem. Instead of the condition IV in the above formulation, a stronger condition IV′ is introduced.

IV′. Every function is a symmetric or anti-symmetric function with respect to $\varphi=0$ and $\varphi=\pi$. This is, every unknown f satisfies

$$\begin{cases} f(\varphi)=f(-\varphi), \\ f(\pi-\varphi)=f(\pi+\varphi), \quad \text{i.e., } f(\varphi)=f(2\pi-\varphi) \end{cases} \tag{1.10a}$$

or

$$\begin{cases} f(\varphi)=-f(-\varphi), \\ f(\pi-\varphi)=-f(\pi+\varphi), \quad \text{i.e., } f(\varphi)=-f(2\pi-\varphi). \end{cases} \tag{1.10b}$$

Evidently the relationship $f(\varphi)=\pm f(-\varphi)=f(2\pi+\varphi)$ holds in this case, i.e., this condition is a special case of the condition (1.4). Therefore, any method for the problem I—IV applies to the problem I—III, IV′. However, because of its particularity, we need to solve the problem only in the following region:

$$\begin{cases} x_0(\varphi, t)\leqslant x\leqslant x_L(\varphi, t), \\ 0\leqslant\varphi\leqslant\pi, \\ t\geqslant 0. \end{cases}$$

§ 2 Numerical Methods

In order to give an accurate and convenient treatment for the boundaries, we introduce the following curvilinear coordinate system $\{\xi, \varphi, t\}$:

$$\begin{cases} \xi=\dfrac{x-x_{l-1}(\varphi, t)}{x_l(\varphi, t)-x_{l-1}(\varphi, t)}+l-1, & \text{if } x_{l-1}(\varphi, t)\leqslant x\leqslant x_l(\varphi, t), \\ \varphi=\varphi, & 0\leqslant\varphi\leqslant 2\pi, \\ t=t, & l=1, 2,\cdots,L, \end{cases} \tag{2.1}$$

which makes all boundaries $x_l(\varphi, t)$ become coordinate surfaces. In this case, the surfaces $x=x_l(\varphi, t)$ become $\xi=l$, the region where we should determine the solution becomes $0\leqslant\xi\leqslant L$, $0\leqslant\varphi\leqslant 2\pi$, $t\geqslant 0$, and the boundary and internal boundary conditions are given on $\xi=$constant. Moreover, for $\tilde{\boldsymbol{U}}(x, \varphi, t)=\boldsymbol{U}(\xi, \varphi, t)$, there are the relations

$$\begin{cases} \dfrac{\partial\tilde{\boldsymbol{U}}}{\partial t}=\dfrac{\partial\boldsymbol{U}}{\partial t}+\dfrac{\partial\xi}{\partial t}\dfrac{\partial\boldsymbol{U}}{\partial\xi}, \\ \dfrac{\partial\tilde{\boldsymbol{U}}}{\partial x}=\dfrac{\partial\xi}{\partial x}\dfrac{\partial\boldsymbol{U}}{\partial\xi}, \\ \dfrac{\partial\tilde{\boldsymbol{U}}}{\partial\varphi}=\dfrac{\partial\xi}{\partial\varphi}\dfrac{\partial\boldsymbol{U}}{\partial\xi}+\dfrac{\partial\boldsymbol{U}}{\partial\varphi}, \end{cases}$$

where

$$\begin{cases} \dfrac{\partial \xi}{\partial t} = \dfrac{-x_{l-1,t} - (\xi - l + 1)(x_{l,t} - x_{l-1,t})}{x_l - x_{l-1}}, \\ \dfrac{\partial \xi}{\partial x} = \dfrac{1}{x_l - x_{l-1}}, \\ \dfrac{\partial \xi}{\partial \varphi} = \dfrac{-x_{l-1,\varphi} - (\xi - l + 1)(x_{l,\varphi} - x_{l-1,\varphi})}{x_l - x_{l-1}}, \end{cases} \tag{2.2}$$

if $l-1 \leqslant \xi \leqslant l \ (l=1, 2, \cdots, L), \ 0 \leqslant \varphi \leqslant 2\pi, \ t \geqslant 0$.

Consequently, the equation (1.1) can be rewritten as

$$\begin{cases} \dfrac{\partial \boldsymbol{U}}{\partial t} + A \dfrac{\partial \boldsymbol{U}}{\partial \xi} + C \dfrac{\partial \boldsymbol{U}}{\partial \varphi} = \boldsymbol{F}, \\ l-1 \leqslant \xi \leqslant l, \quad l=1, 2, \cdots, L, \\ 0 \leqslant \varphi \leqslant 2\pi, \ t \geqslant 0, \end{cases}$$

where

$$A(\boldsymbol{U}, \boldsymbol{X}, \boldsymbol{X}_\varphi, \boldsymbol{X}_t, \xi, \varphi, t) = \frac{\partial \xi}{\partial t} E + \frac{\partial \xi}{\partial x} \widetilde{A} + \frac{\partial \xi}{\partial \varphi} \widetilde{C},$$

$$C(\boldsymbol{U}, \boldsymbol{X}, \xi, \varphi, t) = \widetilde{C}(\widetilde{\boldsymbol{U}}(x, \varphi, t), x, \varphi, t),$$

$$\boldsymbol{F}(\boldsymbol{U}, \boldsymbol{X}, \xi, \varphi, t) = \widetilde{\boldsymbol{F}}(\widetilde{\boldsymbol{U}}(x, \varphi, t), x, \varphi, t),$$

E is an $N \times N$ unit matrix,

$$\boldsymbol{X}^* = (x_0, x_1, \cdots, x_L),$$

$$\boldsymbol{X}^*_\varphi = (x_{0,\varphi}, x_{1,\varphi}, \cdots, x_{L,\varphi}),$$

$$\boldsymbol{X}^*_t = (x_{0,t}, x_{1,t}, \cdots, x_{L,t}).$$

Therefore, the problem (1.1)—(1.4) of § 1 can be rewitten as follows:

I. In L regions, $t \geqslant 0, \ l-1 \leqslant \xi \leqslant l, \ 0 \leqslant \varphi \leqslant 2\pi, \ l=1, 2, \cdots, L$, the system of equations

$$\frac{\partial \boldsymbol{U}}{\partial t} + A \frac{\partial \boldsymbol{U}}{\partial \xi} + C \frac{\partial \boldsymbol{U}}{\partial \varphi} = \boldsymbol{F} \tag{2.3}$$

is given.

II. At $t=0$, the initial conditions

$$\begin{cases} \boldsymbol{U}(\xi, \varphi, 0) = \boldsymbol{D}_l(\xi, \varphi), \quad l-1 \leqslant \xi \leqslant l, \ 0 \leqslant \varphi \leqslant 2\pi, \\ \qquad\qquad l=1, 2, \cdots, L, \\ \boldsymbol{X}(\varphi, 0) = \boldsymbol{C}_0(\varphi), \quad \boldsymbol{X}_t(\varphi, 0) = \boldsymbol{C}_1(\varphi), \quad 0 \leqslant \varphi \leqslant 2\pi \end{cases} \tag{2.4}$$

are given.

III. On the boundaries, the boundary and internal boundary conditions, compatible with the partial differential equations (2.3),

$$\begin{cases} \boldsymbol{B}_0(\boldsymbol{U}_{0^+}, x_0, x_{0,\varphi}, x_{0,t}, \varphi, t) = 0, \\ \boldsymbol{B}_l(\boldsymbol{U}_{l^-}, \boldsymbol{U}_{l^+}, x_l, x_{l,\varphi}, x_{l,t}, \varphi, t) = 0, \\ \qquad l=1, 2, \cdots, L-1, \\ \boldsymbol{B}_L(\boldsymbol{U}_{L^-}, x_L, x_{L,\varphi}, x_{L,t}, \varphi, t) = 0 \end{cases} \tag{2.5}$$

are given.

IV. In the φ-direction, the periodicity condition

$$f(\varphi)=f(\varphi+2\pi) \tag{2.6}$$

is given, where f is any dependent variable. We need to determine $\boldsymbol{X}(\varphi, t)$ in the region $t\geqslant 0$, $0\leqslant\varphi\leqslant 2\pi$ and $\boldsymbol{U}(\xi, \varphi, t)$ in the region $t\geqslant 0$, $0\leqslant\xi\leqslant L$, $0\leqslant\varphi\leqslant 2\pi$. In the above expressions, $\boldsymbol{C}_0(\varphi)$ and $\boldsymbol{C}_1(\varphi)$ stand for $(L+1)$-dimensional vectors, whose elements are respectively $c_{0,l}(\varphi)$, $c_{1,l}(\varphi)$; $\boldsymbol{D}_l(\xi, \varphi)=\widetilde{\boldsymbol{D}}_l(x(\xi, \varphi, 0), \varphi)$; and $\boldsymbol{U}_{l^+}$ and $\boldsymbol{U}_{l^-}$ denote respective values of $\boldsymbol{U}$ on the upper and the lower sides of the plane $\xi=l$.

By using (1.5) with $a=\frac{\partial\xi}{\partial\varphi}\Big/\frac{\partial\xi}{\partial x}$, A can be expressed as

$$A=\frac{\partial\xi}{\partial x}\left[\frac{\frac{\partial\xi}{\partial t}}{\frac{\partial\xi}{\partial x}}E+\widetilde{A}+\frac{\frac{\partial\xi}{\partial\varphi}}{\frac{\partial\xi}{\partial x}}\widetilde{C}\right]$$

$$=\overline{\overline{G}}^{-1}\frac{\partial\xi}{\partial x}\left[\frac{\frac{\partial\xi}{\partial t}}{\frac{\partial\xi}{\partial x}}E+\widetilde{\Lambda}\right]\overline{\overline{G}}=\overline{G}^{-1}\Lambda\overline{G}, \tag{2.7}$$

where

$$\overline{G}(\boldsymbol{U}, \boldsymbol{X}, \boldsymbol{X}_\varphi, \xi, \varphi, t)=\overline{\overline{G}}(\widehat{\boldsymbol{U}}, x, \varphi, t, a)\big|_{a=\frac{\partial\xi}{\partial\varphi}/\frac{\partial\xi}{\partial x}},$$

$$\Lambda=\frac{\partial\xi}{\partial x}\left[\frac{\frac{\partial\xi}{\partial t}}{\frac{\partial\xi}{\partial x}}E+\widetilde{\Lambda}\right]=\begin{pmatrix}\lambda_1 & & & 0\\ & \lambda_2 & & \\ & & \ddots & \\ 0 & & & \lambda_N\end{pmatrix}\equiv\Lambda(\boldsymbol{U}, \boldsymbol{X}, \boldsymbol{X}_\varphi, \boldsymbol{X}_t, \xi, \varphi, t),$$

i.e.,

$$\lambda_n=\frac{\partial\xi}{\partial x}\left(\frac{\frac{\partial\xi}{\partial t}}{\frac{\partial\xi}{\partial x}}+\widetilde{\lambda}_n\right).$$

Because $\frac{\partial\xi}{\partial x}=\frac{1}{x_l-x_{l-1}}>0$ when $l-1\leqslant\xi\leqslant l$, and because the relation $\widetilde{\lambda}_1\geqslant\widetilde{\lambda}_2\geqslant\cdots\geqslant\widetilde{\lambda}_N$ is assumed, we can obtain

$$\lambda_1\geqslant\lambda_2\geqslant\cdots\geqslant\lambda_N. \tag{2.8}$$

Moreover, we have from (2.2)

$$\left.\frac{\frac{\partial\xi}{\partial t}}{\frac{\partial\xi}{\partial x}}\right|_{\xi=l^\pm}=-x_{l,t}, \quad \left.\frac{\frac{\partial\xi}{\partial\varphi}}{\frac{\partial\xi}{\partial x}}\right|_{\xi=l^\pm}=-x_{l,\varphi},$$

$$l=0, 1, \cdots, L.$$

Consequently, we obtain

$$\Lambda|_{\xi=l^\pm}=\left(\frac{\partial\xi}{\partial x}\right)_{l^\pm}(\widetilde{\Lambda}_{x\xi}-x_{l,t}E),$$

where $\Lambda_{x\xi}$ denotes the Jordan matrix of $\widetilde{A}-x_{l,\varphi}\widetilde{C}$ at $x=x_l^\pm$. Therefore, the inequality (1.6) can be rewritten as

$$\begin{cases}\lambda_{1,l^+}\geqslant\cdots\geqslant\lambda_{N-p_l,l}>0\geqslant\lambda_{N-p_l+1,l^+}\geqslant\cdots\geqslant\lambda_{N,l^+},\\ \qquad l=0,1,\cdots,L-1,\\ \lambda_{1,l^-}\geqslant\cdots\geqslant\lambda_{q_l,l^-}\geqslant0>\lambda_{q_l+1,l^-}\geqslant\cdots\geqslant\lambda_{N,l^-},\\ \qquad l=1,2,\cdots,L.\end{cases}\tag{2.9}$$

In this case, the compatibility conditions of (2.5) with the equations (2.3) are still (1.7) and (1.8). However in the symbols here, (1.8) can be rewritten as follows: when $\boldsymbol{X}$, $\boldsymbol{X}_\varphi$, φ, t, $\boldsymbol{F}_l^-$ and $\boldsymbol{F}_{l-1}^+$ $(l=1,2,\cdots,L)$ are properly given, the systems

$$\begin{cases}\boldsymbol{B}_0(\boldsymbol{U}_{0^+},x_{0,t})=0,\\ \overline{G}_0^+\boldsymbol{U}_{0^+}=\boldsymbol{F}_0^+,\end{cases}$$

$$\begin{cases}\overline{G}_l^-\boldsymbol{U}_{l^-}=\boldsymbol{F}_l^-,\\ \boldsymbol{B}_l(\boldsymbol{U}_{l^-},\boldsymbol{U}_{l^+},x_{l,t})=0,\\ \overline{G}_l^+\boldsymbol{U}_{l^+}=\boldsymbol{F}_l^+,\quad l=1,2,\cdots,L-1.\end{cases}\tag{2.10}$$

$$\begin{cases}\overline{G}_L^-\boldsymbol{U}_{L^-}=\boldsymbol{F}_L^-,\\ \boldsymbol{B}_L(\boldsymbol{U}_{L^-},x_{L,t})=0\end{cases}$$

always have reasonable solutions $\boldsymbol{U}_{0^+}$, $x_{0,t}$; $\boldsymbol{U}_{l^-}$, $\boldsymbol{U}_{l^+}$, $x_{l,t}$ $(l=1,2,\cdots,L-1)$; and $\boldsymbol{U}_{L^-}$, $x_{L,t}$ respectively, where

$$\overline{G}_l^+=\left.\begin{pmatrix}\boldsymbol{G}_{N-p_l+1}^*\\ \vdots\\ \boldsymbol{G}_N^*\end{pmatrix}\right|_{\xi=l^+},\quad \overline{G}_l^-=\left.\begin{pmatrix}\boldsymbol{G}_1^*\\ \vdots\\ \boldsymbol{G}_{q_l}^*\end{pmatrix}\right|_{\xi=l^-}.$$

As mentioned above, we require that the initial condtions be given on the plane $t=$constant and that the plane $t=$constant be space-like. If t is the time, these features hold naturally; if t is not the time, these features do not always hold. However, if the initial surface is space-like, and if there is a family of space-like surfaces each of which does not intersect any other one in the computational region, then the problem can be reduced into the above normal problem by using the coordinate transformation. A certain concrete skill is given in Chapter 7.

We solve the problem (2.3)—(2.6) in the following way.

In the region $l-1\leqslant\xi\leqslant l$, $0\leqslant\varphi\leqslant2\pi$, $t\geqslant0$, we construct a rectangular network with steps $\Delta\xi_l=\dfrac{1}{M_l}$, $\Delta\varphi=\dfrac{\pi}{J}$, Δt $(l=1,2,\cdots,L)$. $f_{l,m,j}^k$ stands for the value of f at the point $\xi=m\Delta\xi_l+(l-1)$, $\varphi=j\Delta\varphi$, $t=k\Delta t$. (If no confusion results, some superscripts and subscripts will be omitted.) The solution is obtained step by step, i.e., from the values at $t=t_k$:

$$\begin{cases} U^k_{l,m,j}, & m=0, 1, \cdots, M_l,\ l=1, 2, \cdots, L, \\ x^k_{l,j},\ x^k_{l,\varphi,j},\ x^k_{l,t,j}, & l=0, 1, \cdots, L, \end{cases} \qquad j=0, 1, \cdots, 2J-1,$$

the values at $t=(k+1)\Delta t$ are computed, $k=0, 1, \cdots$. In the process of computation, a scheme with an interim level is adopted. The first step is to compute the values at the interim level $t=\left(k+\frac{1}{2}\right)\Delta t$,

$$\begin{cases} U^{k+\frac{1}{2}}_{l,m,j+\frac{1}{2}}, & m=0, 1, \cdots, M_l,\ l=1, 2, \cdots, L, \\ x^{k+\frac{1}{2}}_{l,j+\frac{1}{2}},\ x^{k+\frac{1}{2}}_{l,\varphi,j+\frac{1}{2}},\ x^{k+\frac{1}{2}}_{l,t,j+\frac{1}{2}}, & l=0, 1, \cdots, L, \end{cases} \qquad j=0, 1, \cdots, 2J-1$$

from the values at the k-th level. The second step is to compute the values at the level $t=(k+1)\Delta t$ from the values at the k-th level and at the interim level.

Generally speaking, the process of calculation at each step is as follows: The equations (2.3) are rewritten as

$$G^*_n\left(\frac{\partial U}{\partial t}+\lambda_n\frac{\partial U}{\partial \xi}\right)+C^*_n\frac{\partial U}{\partial \varphi}=f_n, \quad n=1, 2, \cdots, N. \tag{2.11}$$

Then some difference equations approximating (2.11) are obtained by using the schemes (2.13)/(2.14) and (2.15)/(2.18) (or (2.19)/(2.21)) listed below. At the same time, the equations (1.9) are approximated by the scheme (2.22)/(2.23) listed below. (These schemes are respectively certain generalizations of the schemes (3.24)/(3.25), (3.7)/(3.12), (3.28)/(3.29) of Chapter 1.) By using these difference equations and the boundary conditions (2.5), we can determine all the quantities we need.

We shall use the following symbols:

$$\begin{cases} \sigma=\lambda\dfrac{\Delta t}{\Delta \xi}, \quad \tau=\dfrac{\Delta t}{\Delta \varphi}, \\ \Delta_m U_{m,j}=U_{m+\frac{1}{2},j}-U_{m-\frac{1}{2},j}, \\ \Delta_{m+} U_{m,j}=U_{m+1,j}-U_{m,j}, \\ \Delta_{m-} U_{m,j}=U_{m,j}-U_{m-1,j}, \\ \Delta_j U_{m,j}=U_{m,j+\frac{1}{2}}-U_{m,j-\frac{1}{2}}, \\ \mu_m f_{m,j}=\dfrac{1}{2}(f_{m+\frac{1}{2},j}+f_{m-\frac{1}{2},j}), \\ \mu_{m\pm} f_{m,j}=\dfrac{1}{2}(f_{m,j}+f_{m\pm1,j}), \\ \mu_j f_{m,j}=\dfrac{1}{2}(f_{m,j+\frac{1}{2}}+f_{m,j-\frac{1}{2}}), \\ \mu f_{m,j}=\mu_m\mu_j f_{m,j}. \end{cases} \tag{2.12}$$

As a generalization of (3.24)/(3.25) of Chapter 1 with $\alpha=1/2$, we construct the following second order scheme.

The first step of discretization, called the interim step of Scheme I, is that the equation (2.11) (omitting the subscript n) is approximated by the following difference equation

$$\mu G^{*k}_{m\mp\frac{1}{2},j+\frac{1}{2}}\left(\mu_m U^{k+\frac{1}{2}}_{m\mp\frac{1}{2},j+\frac{1}{2}}+\frac{1}{2}\mu\sigma^k_{m\mp\frac{1}{2},j+\frac{1}{2}}\Delta_m U^{k+\frac{1}{2}}_{m\mp\frac{1}{2},j+\frac{1}{2}}\right)$$
$$=\mu G^{*k}_{m\mp\frac{1}{2},j+\frac{1}{2}}\,\mu U^{k}_{m\mp\frac{1}{2},j+\frac{1}{2}}-\frac{1}{2}\tau\mu C^{*k}_{m\mp\frac{1}{2},j+\frac{1}{2}}\mu_m\Delta_j U^k_{m\mp\frac{1}{2},j+\frac{1}{2}}$$
$$+\frac{1}{2}\Delta t\mu f^k_{m\mp\frac{1}{2},j+\frac{1}{2}}. \tag{2.13}$$

(From now on, when "$\pm$" or "$\mp$" appears in this equation, we take the upper sign if $\sigma>0$, and the lower sign if $\sigma<0$.)

The second step, called the normal step of Scheme I, is that (2.11) is approximated by the difference equation

$$\mu G^{*k+\frac{1}{2}}_{m\mp\frac{1}{2},j}\left(\mu_m U^{k+1}_{m\mp\frac{1}{2},j}+\frac{1}{2}\mu\sigma^{k+\frac{1}{2}}_{m\mp\frac{1}{2},j}\Delta_m U^{k+1}_{m\mp\frac{1}{2},j}\right)$$
$$=\mu G^{*k+\frac{1}{2}}_{m\mp\frac{1}{2},j}\left(\mu_m U^k_{m\mp\frac{1}{2},j}-\frac{1}{2}\mu\sigma^{k+\frac{1}{2}}_{m\mp\frac{1}{2},j}\Delta_m U^k_{m\mp\frac{1}{2},j}\right)$$
$$-\tau\mu C^{*k+\frac{1}{2}}_{m\mp\frac{1}{2},j}\mu_m\Delta_j U^{k+\frac{1}{2}}_{m\mp\frac{1}{2},j}+\Delta t\,\mu f^{k+\frac{1}{2}}_{m\mp\frac{1}{2},j} \tag{2.14}$$

In the following, by (2.13)/(2.14) we mean that (2.13) is adopted in the first step and (2.14) is adopted in the second step, and we call it Scheme I.

As a generalization of (3.7)/(3.12) in Chapter 1, with $\alpha=1/2$, we construct the following scheme:

The first step is that the equation (2.11) is approximated by

$$\mu_j G^{*k}_{m,j+\frac{1}{2}}U^{k+\frac{1}{2}}_{m,j+\frac{1}{2}}=\mu_j G^{*k}_{m,j+\frac{1}{2}}\left(\mu_j U^k_{m,j+\frac{1}{2}}-\frac{1}{2}\mu_i\sigma^k_{m,j+\frac{1}{2}}\Delta_{m\mp}\mu_j U^k_{m,j+\frac{1}{2}}\right)$$
$$-\frac{1}{2}\tau\mu_j C^{*k}_{m,j+\frac{1}{2}}\mu_{m\mp}\Delta_j U^k_{m,j+\frac{1}{2}}+\frac{1}{2}\Delta t\mu_j f^k_{m,j+\frac{1}{2}}, \tag{2.15}$$

which is called the interim step of Scheme II′.

The second step, called the normal step of Scheme II′, is based on the following two schemes.

(i) The equation (2.11) is rewritten as

$$G^*\left[\frac{\partial U}{\partial t}\mp\frac{\Delta\xi}{\Delta t}\frac{\partial U}{\partial\xi}+\left(\lambda\pm\frac{\Delta\xi}{\Delta t}\right)\frac{\partial U}{\partial\xi}\right]+C^*\frac{\partial U}{\partial\varphi}=f.$$

Then, it is approximated by

$$\mu G^{*k+\frac{1}{2}}_{m\pm\frac{1}{2},j}U^{k+1}_{m,j}=\mu G^{*k+\frac{1}{2}}_{m\pm\frac{1}{2},j}\Big[U^k_{m\pm1,j}\pm\frac{1}{4}\Delta_j^2\Delta_m U^k_{m\pm\frac{1}{2},j}$$
$$-\left(\mu\sigma^{k+\frac{1}{2}}_{m\pm\frac{1}{2},j}\pm1\right)\mu_j\Delta_m U^{k+\frac{1}{2}}_{m\pm\frac{1}{2},j}\Big]$$
$$-\tau\mu C^{*k+\frac{1}{2}}_{m\pm\frac{1}{2},j}\mu_m\Delta_j U^{k+\frac{1}{2}}_{m\pm\frac{1}{2},j}+\Delta t\,\mu f^{k+\frac{1}{2}}_{m\pm\frac{1}{2},j}\equiv S_1U^k_{m,j}. \tag{2.16}$$

In order for the scheme to be almost the same as Richtmyer's, we add a third order term $\pm\frac{1}{4}\Delta_j^2\Delta_m \boldsymbol{U}^k_{m\pm\frac{1}{2},j}$ in the above expression.

(ii) The equation (2.11) is rewritten as

$$\boldsymbol{G}^*\left[\frac{\partial \boldsymbol{U}}{\partial t}\pm\frac{\Delta\xi}{\Delta t}\frac{\partial \boldsymbol{U}}{\partial \xi}+\left(\lambda\mp\frac{\Delta\xi}{\Delta t}\right)\frac{\Delta \boldsymbol{U}}{\Delta\xi}\right]+\boldsymbol{C}^*\frac{\partial \boldsymbol{U}}{\partial\varphi}=\boldsymbol{f}.$$

Then, it is approximated by

$$\begin{aligned}\mu \boldsymbol{G}^{*k+\frac{1}{2}}_{m\mp\frac{1}{2},j}\boldsymbol{U}^{k+1}_{m,j}=\mu \boldsymbol{G}^{*k+\frac{1}{2}}_{m\mp\frac{1}{2},j}\Big[\boldsymbol{U}^k_{m\mp1,j}-\Big(\mu\sigma^{k+\frac{1}{2}}_{m\mp\frac{1}{2},j}\mp1\Big)\mu_j\Delta_m \boldsymbol{U}^{k+\frac{1}{2}}_{m\mp\frac{1}{2},j}\Big]\\ -\tau\mu \boldsymbol{C}^{*k+\frac{1}{2}}_{m\mp\frac{1}{2},j}\mu_m\Delta_j \boldsymbol{U}^{k+\frac{1}{2}}_{m\mp\frac{1}{2},j}+\Delta t\,\mu\, \boldsymbol{f}^{k+\frac{1}{2}}_{m\mp\frac{1}{2},j}\equiv S_2\boldsymbol{U}_{m,j}.\end{aligned}\tag{2.17}$$

Using these two schemes as the basis, we can construct the following difference scheme for initial-boundary-value problems:

$$\begin{cases}\Big[\beta_{1,m}\mu \boldsymbol{G}^{*k+\frac{1}{2}}_{m\pm\frac{1}{2},j}+(1-\beta_{1,m})\mu \boldsymbol{G}^{*k+\frac{1}{2}}_{m\mp\frac{1}{2},j}\Big]\boldsymbol{U}^{k+1}_{m,j}\\ \quad=\beta_{1,m}S_1\boldsymbol{U}^k_{m,j}+(1-\beta_{1,m})S_2\boldsymbol{U}^k_{m,j},\quad m\in[m_0,\ m_1],\\ \mu \boldsymbol{G}^{*k+\frac{1}{2}}_{m\pm\frac{1}{2},j}\boldsymbol{U}^{k+1}_{m,j}=S_1\boldsymbol{U}^k_{m,j},\quad m\in[m_1+1,\ m_2-1],\\ \Big[(1-\beta_{2,m})\mu \boldsymbol{G}^{*k+\frac{1}{2}}_{m\pm\frac{1}{2},j}+\beta_{2,m}\mu \boldsymbol{G}^{*k+\frac{1}{2}}_{m\mp\frac{1}{2},j}\Big]\boldsymbol{U}^{k+1}_{m,j}\\ \quad=(1-\beta_{2,m})S_1\boldsymbol{U}^k_{m,j}+\beta_{2,m}S_2\boldsymbol{U}^k_{m,j},\quad m\in[m_2,\ m_3],\end{cases}\tag{2.18}$$

where m_0, m_1, m_2, m_3, $\beta_{1,m}, \beta_{2,m}$ are chosen in the same way as § 3 of Chapter 1.

However, we want to point out that the generalization of the scheme (3.12) of Chapter 1 with $\alpha=1/2$ in three independent variables is not unique. For example, we can construct another scheme which is also a generalization of (3.12) of Chapter 1.

The first step called the interim step of Scheme II is

$$\begin{aligned}\mu_j \boldsymbol{G}^{*k}_{m,j+\frac{1}{2}}\boldsymbol{U}^{k+\frac{1}{2}}_{m,j+\frac{1}{2}}=\mu_j \boldsymbol{G}^{*k}_{m,j+\frac{1}{2}}\Big(\mu_j \boldsymbol{U}^k_{m,j+\frac{1}{2}}-\frac{1}{2}\mu_j\sigma^k_{m,j+\frac{1}{2}}\Delta_{m\mp}\mu_j \boldsymbol{U}^k_{m,j+\frac{1}{2}}\Big)\\ -\frac{1}{2}\tau\mu_j \boldsymbol{C}^{*k}_{m,j+\frac{1}{2}}\Delta_j \boldsymbol{U}^k_{m,j+\frac{1}{2}}+\frac{1}{2}\Delta t\mu_j \boldsymbol{f}^k_{m,j+\frac{1}{2}}.\end{aligned}\tag{2.19}$$

The second step called the normal step of Scheme II is

$$\begin{aligned}\mu_j\tilde{\mu}_{m'}\boldsymbol{G}^{*k+\frac{1}{2}}_{m',j}\boldsymbol{U}^{k+1}_{m,j}=\mu_j\tilde{\mu}_{m'}\boldsymbol{G}^{*k+\frac{1}{2}}_{m',j}\mu^{(\prime)}_{m''}\boldsymbol{U}^k_{m'',j}\\ -\tau\mu_j\tilde{\mu}_{m'}\boldsymbol{C}^{*k+\frac{1}{2}}_{m',j}\tilde{\mu}_{m'}\Delta_j \boldsymbol{U}^{k+\frac{1}{2}}_{m',j}+\Delta t\mu_j\tilde{\mu}_{m'}\boldsymbol{f}^{k+\frac{1}{2}}_{m',j}\\ \equiv\tilde{S}_i\boldsymbol{U}^k_{m,j},\quad i=a,\ b.\end{aligned}\tag{2.20}$$

In the expression, m' satisfies the relationship

$$\frac{(m-m')\Delta\xi}{\frac{\Delta t}{2}}=(1\mp(m-m'))\,\mu_j\lambda^{k+\frac{1}{2}}_{m,j}\pm(m-m')\,\mu_j\lambda^{k+\frac{1}{2}}_{m\mp1,j}.$$

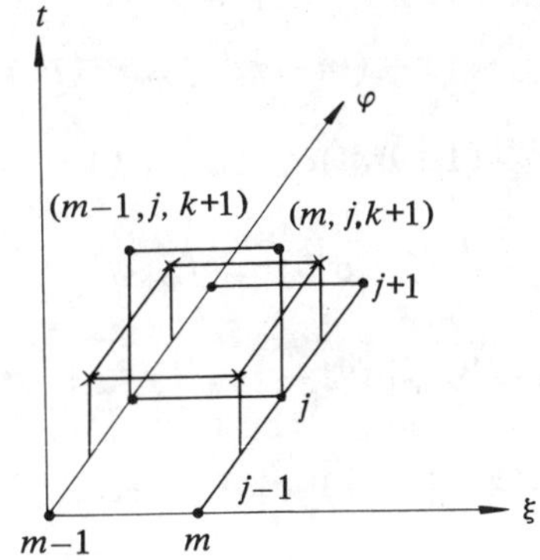

Scheme (2.13)/(2.14) $\sigma>0$

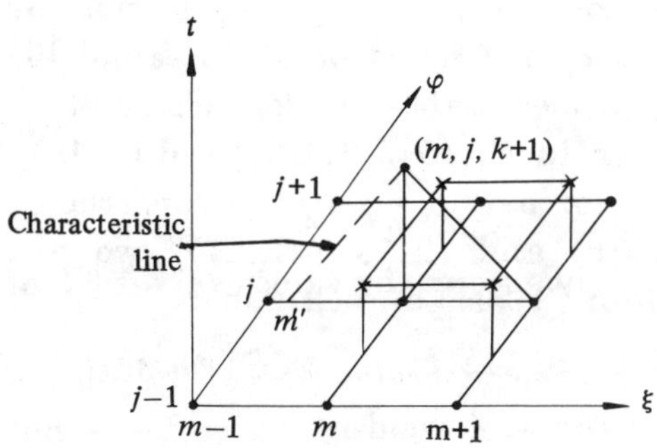

Scheme (2.15)/(2.16)
and scheme(2.19)/(2.20a)
$\sigma>0$

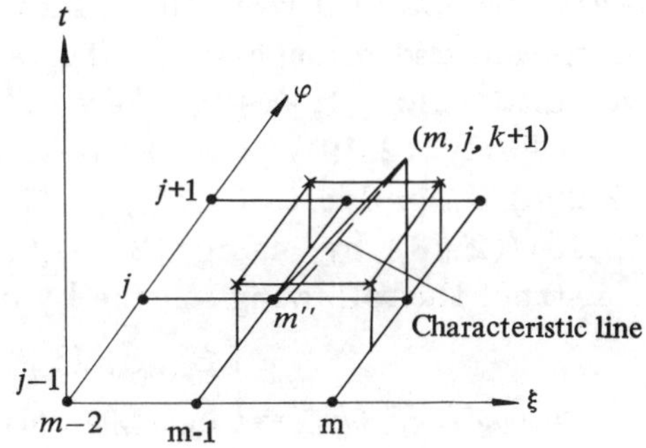

Scheme (2.15)/(2.17)
and scheme(2.19)/(2.20b)
$\sigma>0$

Figure 2.1 Mnemonic diagrams of several schemes

×——The points at the interim level, ●——The points at the normal level.

That is, we choose m' such that the characteristic line at the point $t=\left(k+\frac{1}{2}\right)\Delta t$, $\xi=m'\Delta\xi$ passes through the point $t=(k+1)\Delta t$, $\xi=m\Delta\xi$ on the t-ξ plane, the slope of the characteristic line at $m=m'$ being determined by the linear interpolation. Obviously, the above relationship can be rewritten as

$$m'=m-\frac{\mu_j\lambda_{m,j}^{k+\frac{1}{2}}\dfrac{\Delta t}{2\Delta\xi}}{1\mp\left(\mu_j\lambda_{m\mp1,j}^{k+\frac{1}{2}}-\mu_j\lambda_{m,j}^{k+\frac{1}{2}}\right)\dfrac{\Delta t}{2\Delta\xi}}.$$

In (2.20), m'' and m' satisfy the relation

$$m''=m-2(m-m').$$

That is, the characteristic line intersects the line $t=k\Delta t$, $\varphi=j\Delta\varphi$ at $\xi=m''\Delta\xi$. In (2.20), $\tilde{\mu}_{m'}f_{m'}$ and $\mu^{(i)}_{m''}f_{m''}$ $(i=a, b)$ denote respectively the linear and the quadric interpolations of f:

$$\tilde{\mu}_{m'}f_{m'}=(1\mp(m-m'))f_m\pm(m-m')f_{m\mp1},$$

$$\mu^{(a)}_{m''}f_{m''}=\frac{1}{2}(1+\sigma_{m'})\sigma_{m'}f_{m-1}+(1-\sigma_{m'})(1+\sigma_{m'})f_m$$

$$-\frac{1}{2}(1-\sigma_{m'})\sigma_{m'}f_{m+1},$$

$$\mu^{(b)}_{m''}f_{m''}=\frac{1}{2}(2-|\sigma_{m'}|)(1-|\sigma_{m'}|)f_m+|\sigma_{m'}|(2-|\sigma_{m'}|)f_{m\mp1}$$

$$-\frac{1}{2}(1-|\sigma_{m'}|)|\sigma_{m'}|f_{m\mp2},$$

$$\sigma_{m'}=m-m''=2(m-m').$$

Moreover, $\mu^{(a)}_{m''}f_{m''}$ stands for a centered interpolation: the value at $m''(m)$ is obtained from the values at $m-1$, m, $m+1$; $\mu^{(b)}_{m''}f_{m''}$ stands for an uncentered interpolation: the value at $m''(m)$ is obtained from the values at m, $m\pm1$, $m\pm2$, where $m''(m)$ means that m'' depends on m.

Clearly (2.19)/(2.20a)[1] is similar to (2.15)/(2.16), and (2.19)/(2.20b) is similar to (2.15)/(2.17). Moreover, as we have constructed (2.15)/(2.18) by using (2.15)/(2.16) and (2.15)/(2.17), we can construct the following scheme by using (2.20a) and (2.20b):

$$\mu_j\tilde{\mu}_{m'}\overset{*}{G}{}^{k+\frac{1}{2}}_{m',j}U^{k+1}_{m,j}=\begin{cases}\beta_{1,m}\tilde{S}_aU^k_{m,j}+(1-\beta_{1,m})\tilde{S}_bU^k_{m,j}, & m\in[m_0, m_1],\\ \tilde{S}_aU^k_{m,j}, & m\in[m_1+1, m_2-1],\\ (1-\beta_{2,m})\tilde{S}_aU^k_{m,j}+\beta_{2,m}\tilde{S}_bU^k_{m,j}, & m\in[m_2, m_3].\end{cases} \tag{2.21}$$

The scheme (2.19)/(2.21) has been applied to practical computation in Chapter 7, but (2.15)/(2.18) has not. However, we think that (2.15)/(2.18) has its own advantages. Therefore, we give two generalizations of the scheme (3.7)/(3.12) of Chapter 1 with $\alpha=1/2$.

We also want to point out that the coefficients of equations become infinite when $x_l-x_{l-1}=0$ because there is a factor $1/(x_l-x_{l-1})$ in $\frac{\partial\xi}{\partial t}$, $\frac{\partial\xi}{\partial x}$, $\frac{\partial\xi}{\partial\varphi}$. In order to avoid the appearance of ∞, we may carry out the calculation in the following way. We choose a net such that the relationship $x_l-x_{l-1}=0$ appears only at the normal level. Moreover, $x^k_l-x^k_{l-1}$ in the coefficients is changed into $x^{k+\frac{1}{2}}_l-x^{k+\frac{1}{2}}_{l-1}$ while the values at the interim level are computed. Because $x^{k+\frac{1}{2}}_l$ can be computed by using the

1) (2.20a) means that $i=a$ in (2.20) and (2.20b) is the formula for $i=b$.

scheme (2.22) below before the other quantities are obtained, no new iteration is needed when the above modification is adopted.

As a generalization of (3.28)/(3.29) approximating (1.9) of Chapter 1, we construct the following scheme approximating (1.9) of this chapter

$$\begin{cases} x^{k+\frac{1}{2}}_{l,j+\frac{1}{2}} = \mu_j x^k_{l,j+\frac{1}{2}} + \frac{1}{2}\mu_j x^k_{l,t,j+\frac{1}{2}}\Delta t, \\ x^{k+\frac{1}{2}}_{l,\varphi,j+\frac{1}{2}} = \mu_j x^k_{l,\varphi,j+\frac{1}{2}} + \frac{1}{2}\tau\Delta_j x^k_{l,t,j+\frac{1}{2}}, \end{cases} \tag{2.22}$$

$$\begin{cases} x^{k+1}_{l,j} = x^k_{l,j} + \mu_j x^{k+\frac{1}{2}}_{l,t,j}\Delta t, \\ x^{k+1}_{l,\varphi,j} = x^k_{l,\varphi,j} + \tau\Delta_j x^{k+\frac{1}{2}}_{l,t,j}. \end{cases} \tag{2.23}$$

We call (2.22) the interim step of the scheme for x and x_φ, and (2.23) the normal step of the scheme.

In § 2 of Chapter 1, we have pointed out that if $p_{l-1}+q_l<N$ in a region, we can introduce several new boundaries in the region such that $p_{l-1}+q_l\geqslant N$ in every new region. Therefore, it is enough for practical applications to construct a method for the initial-boundary-value problem (2.3)—(2.6) with the inequality $p_{l-1}+q_l\geqslant N$ in every region.

We shall now describe how a difference system is established for an initial-boundary-value problem with the inequality $p_{l-1}+q_l\geqslant N$ in every region.

In every region, the equations (2.11) are divided into three groups in the following way dependent on p_{l-1} and q_l. The equations with $n=1$ through $N-p_{l-1}$ are called the first group, the equations with $n=N-p_{l-1}+1$ through q_l the second group, and the equations with $n=q_l+1$ through N the third group. Then the three groups of equations are approximated in different ways described below. (The interim steps of Schemes I and II (II′) are adopted when the values at an interim level are computed, and the normal steps are adopted when a normal level is computed.)

Every equation of the first group in the l-th region is approximated at the points $j=0, 1, \cdots, 2J-1$, $m=1, 2, \cdots, M_l$. If $|\lambda|>\varepsilon$ at a point, ε being a given constant and satisfying the inequality $0<\varepsilon\leqslant\frac{\Delta\xi}{\Delta t}$, then we use Scheme I at the point; otherwise, we use Scheme II (or II′). In this case, $M_l(N-p_{l-1})$ difference equations are obtained for every pair of l and j, either when the values at an interim level need to be determined or when the values at a normal level need to be computed.

Every equation of the second group in the l-th region is approximated at the points $j=0, 1,\cdots, 2J-1$, $m=0, 1,\cdots, M_l$ by using scheme II (or II$'$). In this case, $(M_l+1)(p_{l-1}+q_l-N)$ difference equations are obtained for every pair of l and j.

Every equation of the third group in the l–th region is approximated at the points $j=0, 1, \cdots, 2J-1$, and $m=0, 1, \cdots, M_l-1$. If $|\lambda|>\varepsilon$ at a point, then we use Scheme I at the point; otherwise, we use Scheme II (or II$'$). In this case, $M_l(N-q_l)$ equations are obtained for every pair of l and j.

In addition to the difference equations approximating (2.11) in every region, we should also obtain two difference equations approximating (1.9) for every l. (1.9) are approximated by using (2.22) when the quantities at an interim level are evaluated, and by using (2.23) when the quantities at a normal level are evaluated, $j=0, 1, \cdots, 2J-1$, $l=0, 1, \cdots, L$. Hence, we obtain $2(L+1)$ difference equations. Of course, if $x_l(\varphi, t)$ is given for a boundary, then we can obtain x_l, $x_{l,\varphi}$ at the boundary directly.

Therefore, either for an interim level or for a normal level, the total number of difference equations in every φ–plane is

$$\sum_{l=1}^{L}[M_l(N-p_{l-1})+(M_l+1)(p_{l-1}+q_l-N)+M_l(N-q_l)]+2(L+1)$$
$$=N\sum_{l=1}^{L}(M_l+1)+\sum_{l=1}^{L}(p_{l-1}+q_l-2N)+2(L+1).$$

On the other hand, we know from (1.7) that the total number of boundary conditions in (2.5) is

$$\sum_{l=0}^{L}\gamma_l=N+1-p_0+\sum_{l=1}^{L-1}(2N+1-p_l-q_l)+N+1-q_L$$
$$=\sum_{l=1}^{L}(2N-p_{l-1}-q_l)+L+1.$$

Consequently, we have $N\sum_{l=1}^{L}(M_l+1)+3(L+1)$ equations in every φ–plane.

Clearly, the treatment in the φ–direction in (2.13)—(2.23) is explicit, i.e., all the unknowns in every equation belong to the same φ–plane. Therefore, the unknowns on every φ–plane can be determined independently, i.e., in order to determine the $N\sum_{l=1}^{L}(M_l+1)+3(L+1)$ unknowns on a φ–plane, namely, $\boldsymbol{U}_{l,m}$, $m=0, 1, \cdots, M_l$, $l=1, 2, \cdots, L$ and x_l, $x_{l,\varphi}$, $x_{l,t}$, $l=0, 1, \cdots, L$, the established difference equations and the boundary conditions only on this φ–plane are needed. If only Scheme II (or II$'$) is adopted, it is easy to compute $\boldsymbol{U}_{l,m}$, x_l, $x_{l,\varphi}$, $x_{l,t}$ by using these equations. If Schemes I and II (or II$'$) are used mixedly, we need to solve a system which is composed of a large number of linear algebraic equations and a few nonlinear equations (boundary conditions). The system can be solved by using the block–double–sweep method for "incomplete" systems of linear algebraic equations in § 5 of Chapter 1.

As mentioned above, the computation can be done φ–plane by φ–plane because an explicit treatment is adopted in the φ–direction. Moreover, only the values on three lines ($t=t_k\equiv k\Delta t$, $\varphi=\varphi_j\equiv j\Delta\varphi$, $\varphi=\varphi_{j+1}$ and $t=t_{k+\frac{1}{2}}$, $\varphi=\varphi_{j+\frac{1}{2}}$) appear in the schemes for interim levels (2.13), (2.15), (2.19), (2.22); and only the values on six lines ($t=t_k$, $\varphi=\varphi_{j-1}$, φ_j, φ_{j+1}; $t=t_{k+\frac{1}{2}}$, $\varphi=\varphi_{j-\frac{1}{2}}$, $\varphi_{j+\frac{1}{2}}$ and $t=t_{k+1}$, $\varphi=\varphi_j$) appear in the schemes (2.14), (2.18), (2.21), (2.23). Therefore, in order to save the memory capacity of keeping the quantities at an interim level and other quantities, we may carry out the calculation in the following way.

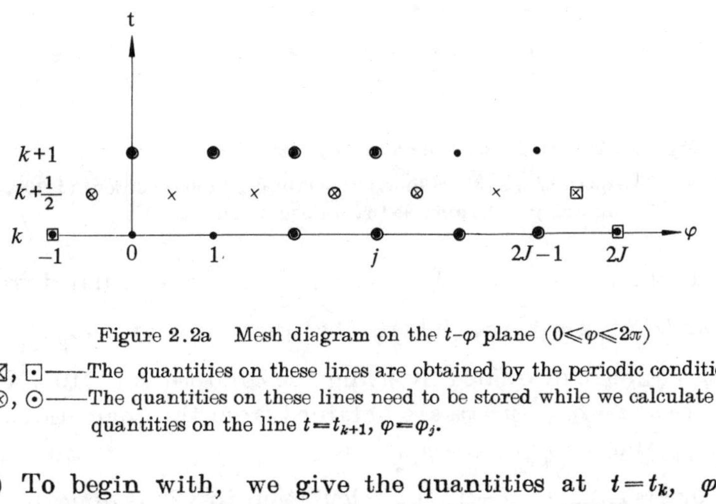

Figure 2.2a Mesh diagram on the t–φ plane ($0\leqslant\varphi\leqslant 2\pi$)

⊠, ⊡——The quantities on these lines are obtained by the periodic condition,
⊗, ⊙——The quantities on these lines need to be stored while we calculate the quantities on the line $t=t_{k+1}$, $\varphi=\varphi_j$.

(1) To begin with, we give the quantities at $t=t_k$, $\varphi=\varphi_{-1}$ by using the periodicity condition (2.6), next we obtain the quantities at $t=t_{k+\frac{1}{2}}$, $\varphi=\varphi_{-\frac{1}{2}}$ from the values at $t=t_k$, $\varphi=\varphi_{-1}$, φ_0, using the schemes for an interim level, and finally we give the quantities at $t=t_{k+\frac{1}{2}}$, $\varphi=\varphi_{2J-\frac{1}{2}}$ by using the periodicity eondition (2.6).

(2) For $j=0, 1, \cdots, 2J-2$, the following computation is done. (i) The quantities at $t=t_{k+\frac{1}{2}}$, $\varphi=\varphi_{j+\frac{1}{2}}$ are calculated by using the quantities at $t=t_k$, $\varphi=\varphi_j$, φ_{j+1}; (ii) The quantities at $t=t_{k+1}$, $\varphi=\varphi_j$ are calculated by using the quantities at $t=t_k$, $\varphi=\varphi_{j-1}$, φ_j, φ_{j+1} and at $t=t_{k+\frac{1}{2}}$, $\varphi=\varphi_{j-\frac{1}{2}}$, $\varphi_{j+\frac{1}{2}}$.

(3 After the quantities at $t=t_k$, $\varphi=\varphi_{2J}$ are obtained by using the

periodicity condition (2.6), we calculate the quantities at $t=t_{k+1}$, $\varphi=\varphi_{2J-1}$ by using the quantities at $t=t_k$, $\varphi=\varphi_{2J-2}$, φ_{2J-1}, φ_{2J} and $t=t_{k+\frac{1}{2}}$, $\varphi=\varphi_{2J-\frac{3}{2}}$, $\varphi_{2J-\frac{1}{2}}$.

Of course, if the conditions (1.10a, b) hold, then the region where we need to solve the problem is $\varphi=0$ through π. In this case, the calculation can be done in the following way.

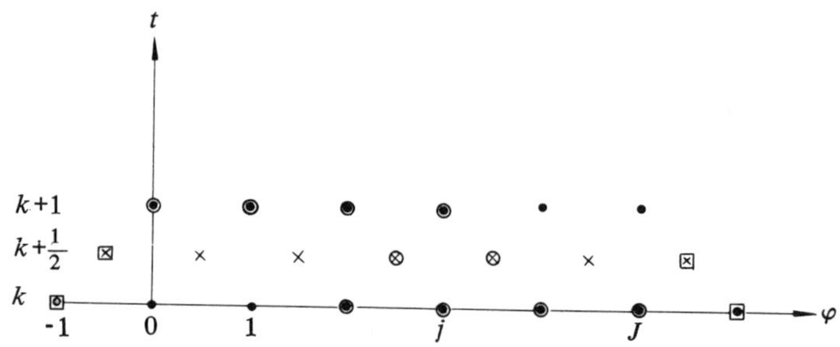

Figure 2.2b Mesh diagram on the t-φ plane $(0\leqslant\varphi\leqslant\pi)$

⊠, ⊡——The quantities on these lines are obtained by the conditions (1.10),

⊗, ⊙——The meaning is the same as that in Fig. 2.2a.

(1)′ The quantities at $t=t_{k+\frac{1}{2}}$, $\varphi=\varphi_{\frac{1}{2}}$ can be calculated from the quantities at $t=t_k$, $\varphi=\varphi_0$, φ_1. Then, the quantities at $t=t_{k+\frac{1}{2}}$, $\varphi=\varphi_{-\frac{1}{2}}$ and $t=t_k$, $\varphi=\varphi_{-1}$ are evaluated by using the conditions (1.10). Finally, the quantities at $t=t_{k+1}$, $\varphi=\varphi_0$ are obtained from the quantities at $t=t_k$, $\varphi=\varphi_{-1}$, φ_0, φ_1 and $t=t_{k+\frac{1}{2}}$, $\varphi=\varphi_{\pm\frac{1}{2}}$.

(2)′ For $j=1, 2, \cdots, J-1$, the calculation in (2) is done.

(3)′ After we obtain the quantities at $t=t_{k+\frac{1}{2}}$, $\varphi=\varphi_{J+\frac{1}{2}}$ and $t=t_k$, $\varphi=\varphi_{J+1}$ by using the conditions (1.10), the values at $t=t_{k+1}$, $\varphi=\varphi_J$ are calculated.

When the calculation is done in the above manner, we do not need to keep all the quantities at the interim level and at the old normal level. Hence, the memory capacity needed in the process of calculation is less (see Figs. 2.2a, b).

In the following, we shall carry out a general discussion on the stability of the above schemes. We shall show under what conditions the von Neumann condition holds when the above schemes are applied to the equation

$$\begin{cases} \dfrac{\partial u}{\partial t}+\lambda\dfrac{\partial u}{\partial \xi}+c\dfrac{\partial u}{\partial \varphi}=0, \\ u(\xi,\varphi,t)=u(\xi+2\pi,\varphi,t),\ u(\xi,\varphi,t)=u(\xi,\varphi+2\pi,t), \end{cases} \tag{2.24}$$

where λ and c are constants.

To begin with, we shall discuss the scheme (2.13)/(2.14). The difference equations corresponding to the equation (2.24) are

$$\mu_m u^{k+\frac{1}{2}}_{m\mp\frac{1}{2},j+\frac{1}{2}}+\frac{1}{2}\sigma\Delta_m u^{k+\frac{1}{2}}_{m\mp\frac{1}{2},j+\frac{1}{2}}=\mu u^k_{m\mp\frac{1}{2},j+\frac{1}{2}}-\frac{1}{2}\sigma_1\mu_m\Delta_j u^k_{m\mp\frac{1}{2},j+\frac{1}{2}},$$

$$\mu_m u^{k+1}_{m\mp\frac{1}{2},j}+\frac{1}{2}\sigma\Delta_m u^{k+1}_{m\mp\frac{1}{2},j}=\mu_m u^k_{m\mp\frac{1}{2},j}-\frac{1}{2}\sigma\Delta_m u^k_{m\mp\frac{1}{2},j}-\sigma_1\mu_m\Delta_j u^{k+\frac{1}{2}}_{m\mp\frac{1}{2},j},$$

$$m=0,1,\cdots,M-1,$$
$$j=0,1,\cdots,2J-1,$$

where $\sigma=\lambda\dfrac{\Delta t}{\Delta\xi}$, $\sigma_1=c\dfrac{\Delta t}{\Delta\varphi}$, $\Delta\xi=\dfrac{2\pi}{M}$, $\Delta\varphi=\dfrac{\pi}{J}$. Hence, for the scheme (2.13)/(2.14), the amplification factor is

$$\begin{aligned} g(\theta,\theta_1)=&\Big[\cos\frac{\theta}{2}-i\sigma\sin\frac{\theta}{2}-i2\sigma_1\cos\frac{\theta}{2}\sin\frac{\theta_1}{2} \\ &\times\left(\cos\frac{\theta}{2}\cos\frac{\theta_1}{2}-i\sigma_1\cos\frac{\theta}{2}\sin\frac{\theta_1}{2}\right) \\ &\times\left(\cos\frac{\theta}{2}+i\sigma\sin\frac{\theta}{2}\right)^{-1}\Big]\left(\cos\frac{\theta}{2}+i\sigma\sin\frac{\theta}{2}\right)^{-1} \\ =&\left(1+\sigma^2\operatorname{tg}^2\frac{\theta}{2}-2\sigma_1^2\sin^2\frac{\theta_1}{2}-i2\sigma_1\sin\frac{\theta_1}{2}\cos\frac{\theta_1}{2}\right) \\ &\times\left(1+i\sigma\operatorname{tg}\frac{\theta}{2}\right)^{-2}, \end{aligned}$$

where

$$\theta=m\Delta\xi,\quad \theta_1=j\Delta\varphi.$$

Therefore, by introducing the symbol

$$\tilde{\sigma}=\sigma\operatorname{tg}\frac{\theta}{2},$$

the von Neumann condition can be written in the following form

$$\begin{aligned} 0\geqslant&\left(1+\tilde{\sigma}^2-2\sigma_1^2\sin^2\frac{\theta_1}{2}\right)^2+4\sigma_1^2\sin^2\frac{\theta_1}{2}\left(1-\sin^2\frac{\theta_1}{2}\right)-(1+\sigma^2)^2 \\ =&-4(1+\tilde{\sigma}^2)\sigma_1^2\sin^2\frac{\theta_1}{2}+4\sigma_1^4\sin^4\frac{\theta_1}{2}+4\sigma_1^2\sin^2\frac{\theta_1}{2}-4\sigma_1^2\sin^4\frac{\theta_1}{2} \\ =&-4\tilde{\sigma}^2\sigma_1^2\sin^2\frac{\theta_1}{2}+4(\sigma_1^2-1)\sigma_1^2\sin^4\frac{\theta_1}{2}. \end{aligned}$$

This means that the von Neumann condition is equivalent to

$$\sigma_1^2\leqslant 1. \tag{2.25}$$

We shall now discuss the stability of the schemes (2.15)/(2.16) and (2.15)/(2.17).

When the equation (2.24) is approximated by (2.15)/(2.16), the difference equation is

$$
\begin{aligned}
u_{m,j}^{k+1} &= u_{m\pm1,j}^{k} \pm \frac{1}{4}\, \Delta_j^2\, \Delta_m u_{m\pm\frac{1}{2},j}^{k} - [(\sigma\pm1)\mu_j\Delta_m + \sigma_1\mu_m\Delta_j]\, u_{m\pm\frac{1}{2},j}^{k+\frac{1}{2}} \\
&= u_{m\pm1,j}^{k} \pm \frac{1}{4}\, \Delta_j^2 \Delta_{m\pm} u_{m,j}^{k} - [(\sigma\pm1)\mu_j\Delta_{m\pm} + \sigma_1\mu_{m\pm}\Delta_j]\, u_{m,j}^{k+\frac{1}{2}} \\
&= u_{m\pm1,j}^{k} \pm \frac{1}{4}\, \Delta_j^2 \Delta_{m\pm} u_{m,j}^{k} - [(\sigma\pm1)\mu_j\Delta_{m\pm} + \sigma_1\mu_{m\pm}\Delta_j] \\
&\quad \times \left(\mu_j - \frac{1}{2}\,\sigma\Delta_{m\mp}\mu_j - \frac{1}{2}\,\sigma_1\mu_{m\mp}\Delta_j\right) u_{m,j}^{k}.
\end{aligned}
$$

Hence, the amplification factor is

$$
\begin{aligned}
g(\theta, \theta_1) &= e^{\pm i\theta} + \left(i \sin\frac{\theta_1}{2}\right)^2 (e^{\pm i\theta}-1) \\
&\quad - \left[\mp(\sigma\pm1)\cos\frac{\theta_1}{2}(1-e^{\pm i\theta}) + \sigma_1(1+e^{\pm i\theta})\, i \sin\frac{\theta_1}{2}\right] \\
&\quad \times \left[\cos\frac{\theta_1}{2} \mp \frac{1}{2}\,\sigma\cos\frac{\theta_1}{2}(1-e^{\mp i\theta}) \right.\\
&\quad \left. - \frac{1}{2}\,\sigma_1(1+e^{\mp i\theta})\, i \sin\frac{\theta_1}{2}\right] \\
&= e^{\pm i\theta} - \sin^2\frac{\theta_1}{2}(e^{\pm i\theta}-1) + \cos^2\frac{\theta_1}{2}(1-e^{\pm i\theta}) \\
&\quad \pm \cos^2\frac{\theta_1}{2}\left[1-e^{\pm i\theta} - \frac{1}{2}(1-e^{\pm i\theta})\,(1-e^{\mp i\theta})\right]\sigma \\
&\quad - i\sin\frac{\theta_1}{2}\,\cos\frac{\theta_1}{2}\left[1+e^{\pm i\theta} + \frac{1}{2}(1-e^{\pm i\theta})(1+e^{\mp i\theta})\right]\sigma_1 \\
&\quad + 2i^2\left(\sigma\cos\frac{\theta_1}{2}\sin\frac{\theta}{2} + \sigma_1\sin\frac{\theta_1}{2}\cos\frac{\theta}{2}\right)^2 \\
&= 1 - 2i\cos\frac{\theta}{2}\,\cos\frac{\theta_1}{2}\left(\sigma\cos\frac{\theta_1}{2}\sin\frac{\theta}{2} + \sigma_1\sin\frac{\theta_1}{2}\cos\frac{\theta}{2}\right) \\
&\quad - 2\left(\sigma\cos\frac{\theta_1}{2}\sin\frac{\theta}{2} + \sigma_1\sin\frac{\theta_1}{2}\cos\frac{\theta}{2}\right)^2,
\end{aligned}
$$

and the von Neumann condition is

$$
(1-2\mu^2)^2 + 4\cos^2\frac{\theta}{2}\cos^2\frac{\theta_1}{2}\,\mu^2 \leqslant 1,
$$

where

$$
\mu = \sigma\cos\frac{\theta_1}{2}\sin\frac{\theta}{2} + \sigma_1\sin\frac{\theta_1}{2}\cos\frac{\theta}{2}.
$$

The above inequality is equivalent to

$$
1 \geqslant \mu^2 + \cos^2\frac{\theta}{2}\cos^2\frac{\theta_1}{2} = \cos^2\frac{\theta}{2}\cos^2\frac{\theta_1}{2}\left[1 + \left(\sigma \operatorname{tg}\frac{\theta}{2} + \sigma_1 \operatorname{tg}\frac{\theta_1}{2}\right)^2\right],
$$

$$\left(\sigma \operatorname{tg} \frac{\theta}{2}+\sigma_1 \operatorname{tg} \frac{\theta_1}{2}\right)^2 \leqslant \frac{1}{\cos^2 \frac{\theta}{2} \cos^2 \frac{\theta_1}{2}}-1$$
$$=\operatorname{tg}^2 \frac{\theta}{2}+\operatorname{tg}^2 \frac{\theta_1}{2}+\operatorname{tg}^2 \frac{\theta}{2} \operatorname{tg}^2 \frac{\theta_1}{2},$$

which can be rewritten as

$$(\sigma \sin \varphi+\sigma_1 \cos \varphi)^2 \leqslant \sin^2 \varphi \cos^2 \varphi+1, \tag{2.26}$$

where

$$\sin \varphi=\operatorname{tg} \frac{\theta}{2} \Big/ \sqrt{\operatorname{tg}^2 \frac{\theta}{2}+\operatorname{tg}^2 \frac{\theta_1}{2}},$$
$$\cos \varphi=\operatorname{tg} \frac{\theta_1}{2} \Big/ \sqrt{\operatorname{tg}^2 \frac{\theta}{2}+\operatorname{tg}^2 \frac{\theta_1}{2}}.$$

Therefore, noticing

$$(\sigma \sin \varphi+\sigma_1 \cos \varphi)^2 \leqslant (\sigma \sin \varphi+\sigma_1 \cos \varphi)^2+(\sigma \cos \varphi-\sigma_1 \sin \varphi)^2$$
$$\leqslant \sigma^2+\sigma_1^2,$$

we can see that the von Neumann condition holds if

$$\sigma^2+\sigma_1^2 \leqslant 1. \tag{2.27}$$

When (2.24) is approximated by (2.15)/(2.17), the difference equation is

$$u_{m,j}^{k+1}=u_{m\mp1,j}^k-\left[(\sigma\mp1)\mu_j \Delta_m+\sigma_1 \mu_m \Delta_j\right] u_{m\mp\frac{1}{2},j}^{k+\frac{1}{2}}$$
$$=u_{m\mp1,j}^k-\left[(\sigma\mp1)\mu_j \Delta_{m\mp}+\sigma_1 \mu_{m\mp} \Delta_j\right] u_{m,j}^{k+\frac{1}{2}}$$
$$=u_{m\mp1,j}^k-\left[(\sigma\mp1)\mu_j \Delta_{m\mp}+\sigma_1 \mu_{m\mp} \Delta_j\right]$$
$$\times\left(\mu_j-\frac{1}{2} \sigma \Delta_{m\mp} \mu_j-\frac{1}{2} \sigma_1 \mu_{m\mp} \Delta_j\right) u_{m,j}^k,$$

and the amplification factor is

$$g(\theta, \theta_1)=e^{\mp i\theta}-\left[\pm(\sigma\mp1) \cos \frac{\theta_1}{2}(1-e^{\mp i\theta})+\sigma_1(1+e^{\mp i\theta}) i \sin \frac{\theta_1}{2}\right]$$
$$\times\left[\cos \frac{\theta_1}{2} \mp \frac{1}{2} \sigma(1-e^{\mp i\theta}) \cos \frac{\theta_1}{2}-\frac{1}{2} \sigma_1(1+e^{\mp i\theta}) i \sin \frac{\theta_1}{2}\right]$$
$$=e^{\mp i\theta}\left\{1-\left[\pm(|\sigma|-1) 2i \sin \frac{\theta}{2} \cos \frac{\theta_1}{2}+\sigma_1 2i \cos \frac{\theta}{2} \sin \frac{\theta_1}{2}\right]\right.$$
$$\left.\times\left[\cos \frac{\theta}{2} \cos \frac{\theta_1}{2}-i\left(\pm(|\sigma|-1) \sin \frac{\theta}{2} \cos \frac{\theta_1}{2}+\sigma_1 \cos \frac{\theta}{2} \sin \frac{\theta_1}{2}\right)\right]\right\}$$
$$=e^{\mp i\theta}\left\{1-2i\mu_1 \cos \frac{\theta}{2} \cos \frac{\theta_1}{2}-2\mu_1^2\right\},$$

where

$$\mu_1= \pm(|\sigma|-1) \sin \frac{\theta}{2} \cos \frac{\theta_1}{2}+\sigma_1 \cos \frac{\theta}{2} \sin \frac{\theta_1}{2}.$$

Therefore, the von Neumann condition is equivalent to

$$[(|\sigma|-1)\sin\varphi+\sigma_1\cos\varphi]^2\leqslant\sin^2\varphi\cos^2\varphi+1, \qquad (2.28)$$

from which we can further see that the von Neumann condition holds if

$$(|\sigma|-1)^2+\sigma_1^2\leqslant 1. \qquad (2.29)$$

Evidently, if both (2.15)/(2.16) and (2.15)/(2.17) satisfy the von Neumann condition, and if $0\leqslant\beta_{1,m}\leqslant 1$, then the scheme (2.15)/(2.18) satisfies the von Neumann condition. That is to say, if both (2.26) and (2.28) (or both (2.27) and (2.29)) hold, then the scheme (2.15)/(2.18) satisfies the von Neumann condition.

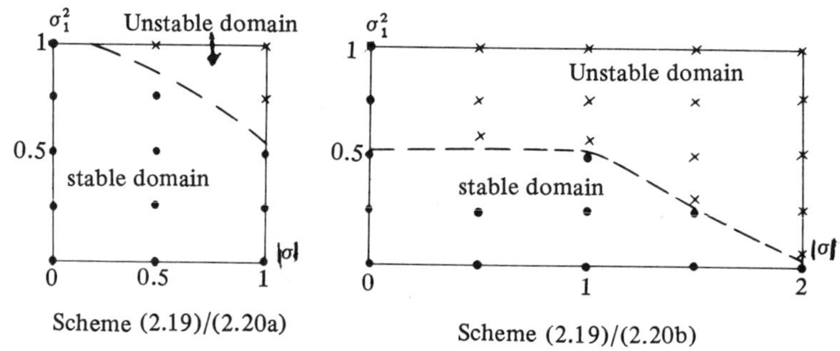

Figure 2.3 The stable domain of Scheme II

● —— $\max|g(\theta,\theta_1)|=1$ at that point, × —— $\max|g(\theta,\theta_1)|>1$ at that point. When calculating, we take $\Delta\frac{\theta}{2}=\Delta\frac{\theta_1}{2}=\frac{\pi}{64}$ for scheme (2.19)/(2.20a), and $\Delta\frac{\theta}{2}=\Delta\frac{\theta_1}{2}=\frac{\pi}{16}$ for scheme (2.19)/(2.20b).

We have not found the ranges on the $|\sigma|-\sigma_1^2$ plane where the schemes (2.19)/(2.20a) and (2.19)/(2.20b) satisfy the von Neumann condition, by using the analytical method. However, we have determined the rough ranges on the $|\sigma|-\sigma_1^2$ plane where the norm of the amplification factor is less than or equal to 1, by using the numerical method of computing the amplification factor.

We know from Fig. 2.3 that if

$$|\sigma|<1, \quad \sigma_1^2\leqslant 0.5,$$

then both (2.19)/(2.20a) and (2.19)/(2.20b) satisfy the von Neumann condition. Clearly, (2.19)/(2.21) also satisfies the von Neumann condition in this case if $0\leqslant\beta_{1,m}\leqslant 1$.

The above analysis on stability is still valid for a system with constant coefficients

$$\frac{\partial U}{\partial t}+A\frac{\partial U}{\partial \xi}+C\frac{\partial U}{\partial \varphi}=0$$

if A and C can be diagonalized by the same matrices, i.e., if A and C are commutative matrices. Therefore, if one of the above conditions is satisfied for every couple of eigenvalues of A and C, the corresponding scheme is stable. However, if A and C are not commutative, this analysis on stability is not valid. Moreover, here we discuss only the von Neumann condition. Therefore, there is a wide gap between the results here and the analysis of the stability of difference schemes for initial–boundary–value problems with variable coefficients. This is only a preliminary and general discussion on stability.

Chapter 3

Numerical Schemes for Certain Boundary–Value Problems of Mixed–Type and Elliptical Equations

§ 1 Formulation of Problems

In this chapter, we shall consider the following boundary–value problems. Suppose that $\boldsymbol{X}$ satisfies the first–order quasilinear system of equations

$$A\frac{\partial \boldsymbol{X}}{\partial r}+B\frac{\partial \boldsymbol{X}}{\partial s}+C\frac{\partial \boldsymbol{X}}{\partial \varphi}=\boldsymbol{H} \tag{1.1}$$

in the domain L:

$$\begin{cases} r_0(s,\ \varphi)\leqslant r\leqslant r_1(s,\ \varphi), \\ s_0\leqslant s\leqslant s_1, \\ 0\leqslant\varphi\leqslant 2\pi, \end{cases}$$

where A, B, and C are $n\times n$ matrixes, $\boldsymbol{X}$ and $\boldsymbol{H}$ are n–dimensional vectors, A, B, C, and $\boldsymbol{H}$ are the functions of $\boldsymbol{X}$, r, s and φ. $\boldsymbol{X}$, r_0 and r_1 also satisfy the following conditions:

(1) On the boundaries $r=r_0(s,\ \varphi)$ and $r=r_1(s,\ \varphi)$, there are the following relationships

$$\boldsymbol{B}_0(\boldsymbol{X}_0,\ r_0,\ r_{0,s},\ r_{0,\varphi},\ s,\ \varphi)=0, \tag{1.2}$$

$$\boldsymbol{B}_1(\boldsymbol{X}_1,\ r_1,\ r_{1,s},\ r_{1,\varphi},\ s,\ \varphi)=0, \tag{1.3}$$

where $\boldsymbol{B}_1$ is a t_1–dimensional vector, $\boldsymbol{B}_0$ is a t_0–dimensional vector, $\boldsymbol{X}_0$ and $\boldsymbol{X}_1$ are respective values of $\boldsymbol{X}$ on r_0 and r_1, and r_s, r_φ are $\frac{\partial r}{\partial s}, \frac{\partial r}{\partial \varphi}$. t_0 and t_1 satisfy

$$t_0+t_1=n+\Delta_0+\Delta_1, \tag{1.4}$$

where

$$\Delta_i=\begin{cases} 0, \text{ if } r_i(s,\ \varphi) \text{ is known,} \\ 1, \text{ if } r_i(s,\ \varphi) \text{ is unknown.} \end{cases} \quad (i=0,\ 1)$$

(2) All the functions satisfy the periodicity condition for φ, i.e.,

$$F(\varphi)=F(\varphi+2\pi). \tag{1.5}$$

Therefore, the problem to be solved is to find $\boldsymbol{X}$, r_0, r_1 in L, which satisfy (1.1), (1.2), (1.3), (1.4) and (1.5).

For convenience, we introduce the following coordinate transformation:

$$\begin{cases} \xi = \dfrac{r(s, \varphi) - r_0(s, \varphi)}{r_1(s, \varphi) - r_0(s, \varphi)}, \\ s = s, \\ \varphi = \varphi. \end{cases} \tag{1.6}$$

In the new coordinates, the boundaries $r_1(s, \varphi)$ and $r_0(s, \varphi)$ become the planes $\xi=1$ and $\xi=0$ respectively. Equations (1.1) become

$$\bar{A}\frac{\partial \boldsymbol{X}}{\partial \xi} + B\frac{\partial \boldsymbol{X}}{\partial s} + C\frac{\partial \boldsymbol{X}}{\partial \varphi} = \boldsymbol{H}, \tag{1.7}$$

where

$$\bar{A} = A\frac{\partial \xi}{\partial r} + B\frac{\partial \xi}{\partial s} + C\frac{\partial \xi}{\partial \varphi}.$$

Therefore the formulation of the above problem changes into the following. In the region

$$\begin{cases} 0 \leqslant \xi \leqslant 1, \\ s_0 \leqslant s \leqslant s_1, \\ 0 \leqslant \varphi \leqslant 2\pi, \end{cases} \tag{1.8}$$

$\boldsymbol{X}$ satisfies (1.7); on $\xi=0$ and $\xi=1$, $\boldsymbol{X}$, r_0 and r_1 satisfy (1.2) and (1.3); and they all satisfy (1.5) for φ. What we need is to find $\boldsymbol{X}$, r_0 and r_1.

Of course, it is required that the problem be properly posed. Here we do not intend to discuss this problem. We only point out that when equations (1.7) are of a mixed-type they may be properly posed under certain circumstances. For example, the problem of supersonic flow around blunt bodies discussed in Chapter 5 belongs to this case. There $r=r_1(s, \varphi)$ is the shock, $r=r_0(s, \varphi)$ is the body, $s=s_1$ is a space-like surface, and $s=s_0$ is the rotation axis. Therefore, the domain where we shall solve the problem is a cylinder bounded by a shock ($\xi=1$), a body ($\xi=0$), and a space-like surface ($s=s_1$) as shown in Fig. 1.1. The condition (1.3) is the Rankine-Hugoniot relations ($t_1=n$, $\Delta_1=1$), the condition (1.2) is the zero flux condition on the body ($t_0=1$, $\Delta_0=0$), and there is no condition on the space-like surface. Equations (1.7) are elliptic in the subsonic region—upsteam of the sonic surface $g(s, r, \varphi)$, and hyperbolic in the supersonic region—downstream of $g(s, r, \varphi)$.

In addition, there are several special cases in problems (1.2)—(1.8). For example, $\dfrac{\partial \boldsymbol{X}}{\partial s} \equiv 0$, $r_s \equiv 0$, or $\dfrac{\partial \boldsymbol{X}}{\partial \varphi} \equiv 0$, $r_\varphi \equiv 0$. As $\dfrac{\partial \boldsymbol{X}}{\partial s} \equiv 0$, $r_s \equiv 0$, the problem is as follows: Suppose that $\boldsymbol{X}$ satisfies

$$\bar{A}\frac{\partial \boldsymbol{X}}{\partial \xi} + C\frac{\partial \boldsymbol{X}}{\partial \varphi} = \boldsymbol{H} \tag{1.9}$$

in the domain L:

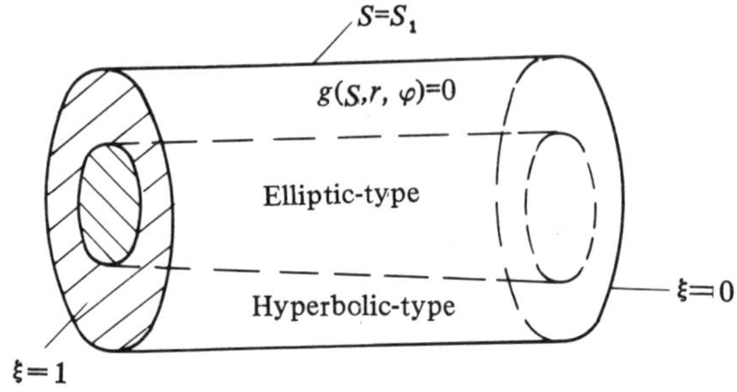

Figure 1.1 Domain of definition for the boundary-value problem of mixed-type equations

$$\begin{cases} 0\leqslant\xi\leqslant 1, \\ 0\leqslant\varphi\leqslant 2\pi \end{cases} \tag{1.10}$$

and satisfies

$$\begin{cases} \boldsymbol{B}_0(\boldsymbol{X}_0,\ r_0,\ r_{0,\varphi},\ \varphi)=0, \\ \boldsymbol{B}_1(\boldsymbol{X}_1,\ r_1,\ r_{1,\varphi},\ \varphi)=0, \end{cases} \tag{1.11}$$

on $\xi=0$ and $\xi=1$, where the dimensions of $\boldsymbol{B}_0$ and $\boldsymbol{B}_1$ satisfy (1.4) and all the functions satisfy (1.5). We need to solve $\boldsymbol{X}$ in the domain (1.10) and r_0, r_1. The problem of supersonic conical flow discussed in Chapter 6 belongs to this case. There $r=r_1(s)$ and $r=r_0(s)$ stand for the shock and the body respectively, and the equations pertain to the elliptical or mixed-type. As $\dfrac{\partial\boldsymbol{X}}{\partial\varphi}\equiv 0$, $r_\varphi\equiv 0$ the formulation of the problem changes into the following: suppose that $\boldsymbol{X}$ satisfies

$$\overline{A}\,\frac{\partial\boldsymbol{X}}{\partial\xi}+B\,\frac{\partial\boldsymbol{X}}{\partial s}=\boldsymbol{H} \tag{1.12}$$

in the domain L:

$$\begin{cases} 0\leqslant\xi\leqslant 1, \\ s_0\leqslant s\leqslant s_1 \end{cases} \tag{1.13}$$

and satisfies

$$\begin{cases} \boldsymbol{B}_0(\boldsymbol{X}_0,\ r_0,\ r_{0,s},\ s)=0, & \text{on } \xi=0, \\ \boldsymbol{B}_1(\boldsymbol{X}_1,\ r_1,\ r_{1,s},\ s)=0, & \text{on } \xi=1, \end{cases} \tag{1.14}$$

where the dimensions of $\boldsymbol{B}_0$ and $\boldsymbol{B}_1$ satisfy (1.4). We need to find $\boldsymbol{X}$ in the domain (1.13) and r_0, r_1. The problem of axisymmetric supersonic flow around blunt bodies in Chapter 5 belongs to this case. For the problem of supersonic conical flow in Chapter 6, if there exists a subdomain: $0\leqslant\xi\leqslant 1$, $\varphi_0\leqslant\varphi\leqslant\varphi_1$ where the problem can be solved alone, then the problem in this subdomain also belongs to this case.

§ 2 Numerical Schemes

There are many methods of solving the above boundary-value problem. For example, we can construct an initial-boundary-value problem of hyperbolic equations and let its asymptotic solution be the solution of the original problem. This is called the time-dependent method or the unsteady state method. Another kind of method is called the steady state, in which the boundary-value problem is directly solved. In the latter kind of method, the solution of the boundary-value problem is usually reduced to the solution of a series of initial-value problems and obtained by an iteration method. In the following we shall give two schemes using this method——one is called the explicit scheme, the other is called the implicit scheme. However, this method will give rise to an improperly posed problem because of encountering the Cauchy problem of elliptic equations. Some remarks about this will be made in the last section of this chapter.

Let us introduce some lines in the domain L: $0\leqslant\xi\leqslant 1$, $s_0\leqslant s\leqslant s_1$, $0\leqslant\varphi\leqslant 2\pi$. For example, we make k_1+1 coordinate surfaces with $s=s_i=$ constant and k_2 coordinate surfaces with $\varphi=\varphi_j=$constant. Then the intersection of these coordinate surfaces forms $(k_1+1)\,k_2$ lines. Set $\boldsymbol{X}_{i,j}$ to be the value of $\boldsymbol{X}$ at the line $s=s_i$, $\varphi=\varphi_j$. For a fixed ξ, we use some kinds of differences in place of differentials with respect to s, φ, e.g., we construct interpolation polynomials of $\boldsymbol{X}$ by the use of $\boldsymbol{X}_{i,j}$ and determine $\dfrac{\partial \boldsymbol{X}}{\partial s}$, $\dfrac{\partial \boldsymbol{X}}{\partial \varphi}$ by the polynomials (see § 4). Let equations (1.7) hold for all the lines, then we obtain a system of ordinary differential equations

$$\bar{A}\left(\frac{d\boldsymbol{X}}{d\xi}\right)_{i,j}+B\left(\frac{\partial \boldsymbol{X}}{\partial s}\right)_{i,j}+C\left(\frac{\partial \boldsymbol{X}}{\partial \varphi}\right)_{i,j}=\boldsymbol{H}_{i,j}. \tag{2.1}$$

The original boundary-value problem becomes a boundary-value one of ordinary differential equations. Owing to the different treatment of the terms $\left(\dfrac{d\boldsymbol{X}}{d\xi}\right)_{i,j}$, $\left(\dfrac{\partial \boldsymbol{X}}{\partial s}\right)_{i,j}$, $\left(\dfrac{\partial \boldsymbol{X}}{\partial \varphi}\right)_{i,j}$, we get different schemes.

1. *An Explicit Scheme*

Suppose that $\boldsymbol{X}^k_{i,j}$ are known at $\xi=\xi_k$. In order to get $\boldsymbol{X}^{k+1}_{i,j}$ at $\xi=\xi_{k+1}=\xi_k+\Delta\xi$ the equations (2.1) can be written as

$$\left(\frac{d\boldsymbol{X}}{d\xi}\right)_{i,j}=-\left(\bar{A}^{-1}B\left(\frac{\partial \boldsymbol{X}}{\partial s}\right)_{i,j}+\bar{A}^{-1}C\left(\frac{\partial \boldsymbol{X}}{\partial \varphi}\right)_{i,j}\right)+\bar{A}^{-1}\boldsymbol{H}_{i,j}. \tag{2.2}$$

This is a system of ordinary differential equations for $\boldsymbol{X}_{i,j}$. The standard numerical method, e.g., the Runge-Kutta method, can be used to integrate these equations.

However, it often happens in practice that $\overline{A}$ is singular at some points of the domain where we shall solve tho problem. For example, $\overline{A}$ is singular on the body for the supersonic flow around blunt bodies. Obviously, in this case (2.2) cannot be used to obtain the solution on the body. One way to get the solution is to use extrapolation. In general, when functions vary slowly, it is suitable to do so, but when functions vary rapidly, this will bring about large errors. In order to avoid the singularity, we put forward an implicit scheme.

2. An Implicit Scheme

For equations (2.1), the following two-step scheme is used:
For the first step,

$$\overline{A}^k \frac{\widetilde{\boldsymbol{X}}^{k+1}-\boldsymbol{X}^k}{\Delta\xi}+B^k\frac{\left(\frac{\partial\widetilde{\boldsymbol{X}}}{\partial s}\right)^{k+1}+\left(\frac{\partial\boldsymbol{X}}{\partial s}\right)^k}{2}+C^k\frac{\left(\frac{\partial\widetilde{\boldsymbol{X}}}{\partial\varphi}\right)^{k+1}+\left(\frac{\partial\boldsymbol{X}}{\partial\varphi}\right)^k}{2}=\boldsymbol{H}^k, \tag{2.3}$$

and for the second step,

$$\frac{\widetilde{\overline{A}}^{k+1}+\overline{A}^k}{2}\frac{\boldsymbol{X}^{k+1}-\boldsymbol{X}^k}{\Delta\xi}+\frac{1}{2}(\widetilde{B}^{k+1}+B^k)\frac{\left(\frac{\partial\boldsymbol{X}}{\partial s}\right)^{k+1}+\left(\frac{\partial\boldsymbol{X}}{\partial s}\right)^k}{2}+\frac{1}{2}(\widetilde{C}^{k+1}+C^k)\frac{\left(\frac{\partial\boldsymbol{X}}{\partial\varphi}\right)^{k+1}+\left(\frac{\partial\boldsymbol{X}}{\partial\varphi}\right)^k}{2}=\frac{1}{2}(\widetilde{\boldsymbol{H}}^{k+1}+\boldsymbol{H}^k), \tag{2.4}$$

where subscripts i, j are omitted, the quantities with superscript k donote the values at ξ_k, and the values with a tilde are those obtained at the first step.

Clearly, the terms $\left(\frac{\partial\boldsymbol{X}}{\partial s}\right)^{k+1}$ and $\left(\frac{\partial\boldsymbol{X}}{\partial\varphi}\right)^{k+1}$ involve the values at different points, therefore equations (2.3) and (2.4) must be solved simultaneously. The usual numerical methods, e.g., the Gauss elimination, can be used.

As mentioned above, equations (2.3) and (2.4) must be solved simultaneously, therefore the amount of calculation will be much greater than that for the explicit scheme. In order to reduce the amount of calculation the combined scheme can be used, that is, the explicit scheme is used when $\overline{A}$ is not singular, and the implicit scheme is used when $\overline{A}$ is singular.

In addition, when this scheme is applied to practical problems it is necessary to do some special treatment to some points. The details are omitted here (see Chapter 5).

Finally, we would like to point out that the above scheme is only suitable for some special kinds of boundary-value problems of mixed-type equations. In order to get a scheme suitable for general cases, the following way is effective. First we transform the equations

$$A\frac{\partial \boldsymbol{X}}{\partial r}+B\frac{\partial \boldsymbol{X}}{\partial s}=\boldsymbol{H}$$

into the "characteristic form"

$$G\frac{\partial \boldsymbol{X}}{\partial r}+\Lambda G\frac{\partial \boldsymbol{X}}{\partial s}=\overline{\boldsymbol{H}}, \qquad \Lambda=\begin{pmatrix}\lambda_1 & & 0\\ & \ddots & \\ 0 & & \lambda_n\end{pmatrix},$$

where λ_i and the elements of G may be complex. Then we construct difference equations according to the feature of λ_i. The details are omitted here.

§ 3 Iteration Methods

The procedure of solution is as follows. According to (1.4), the dimension of the vector B_1 in (1.3) should be $t_1=n+\Delta_0+\Delta_1-t_0$. If r_1 on each line is known, $r_{1,s}$, $r_{1,\varphi}$ can be determined by the interpolation polynomial. Thus $\boldsymbol{X}_1$ can be determined by (1.3) if on each line r_1 and $t_0-\Delta_0-\Delta_1$ components of $\boldsymbol{X}_1$ are given. That is, if on each line t_0 parameters——$t_0-\Delta_0-\Delta_1$ components of $\boldsymbol{X}_1$, r_1 and r_0 (if unknown)——are given, then $\boldsymbol{X}_1$, r_1, r_0 are determined. Once $\boldsymbol{X}_1$, r_1, r_0 are obtained, the successive $\boldsymbol{X}$ at $\xi_k=1-k\Delta\xi$, $k=1, 2, \cdots, \frac{1}{\Delta\xi}$ can be obtained by using the previous schemes. After getting $\boldsymbol{X}$ at $\xi=0$ we should check whether they satisfy (1.2) or not. If they do not satisfy (1.2), we adjust the parameters until (1.2) is satisfied within a tolerant error. Suppose that the total number of lines is $(l+1)/t_0$. Then the total number of parameters is $l+1$. We denote these parameters by f_i ($i=0, 1, \cdots, l$).

Obviously, the above procedure can be considered as a procedure for solving a system of transcendental equations, i.e. for solving equations

$$q_j(f_0, f_1, \cdots, f_l)=0, \quad j=0, 1, \cdots, l, \tag{3.1}$$

where $q_j=0$ is the boundary condition (1.2). There are many methods for solving (3.1), among which the Newton-iteration method is more convenient. That is, the increments δf_i of f_i satisfy

$$\sum_{i=0}^{l}\frac{\partial q_j}{\partial f_i}\,\delta f_i=-q_j, \quad j=0, 1, \cdots, l. \tag{3.2}$$

It is equivalent to the iteration formula

$$\boldsymbol{F}_{m+1}=\boldsymbol{F}_m-Q_m'^{-1}\boldsymbol{Q}_m, \quad m=0, 1, \cdots, \tag{3.3}$$

where

$$\boldsymbol{F}=\begin{pmatrix} f_0 \\ f_1 \\ \vdots \\ f_l \end{pmatrix}, \quad \boldsymbol{Q}=\begin{pmatrix} q_0 \\ q_1 \\ \vdots \\ q_l \end{pmatrix},$$

$$Q'=\begin{pmatrix} \frac{\partial q_0}{\partial f_0} & \frac{\partial q_0}{\partial f_1} & \cdots & \frac{\partial q_0}{\partial f_l} \\ \vdots & & & \vdots \\ \frac{\partial q_l}{\partial f_0} & \cdots\cdots & \cdots\cdots & \frac{\partial q_l}{\partial f_l} \end{pmatrix}.$$

For determining Q', we usually use numerical methods, i.e., we take

$$\frac{\partial q_j}{\partial f_i}=\frac{q_j(f_0, f_1, \cdots, f_i+\Delta f_i, \cdots, f_l)-q_j(f_0, f_1, \cdots, f_i, \cdots, f_l)}{\Delta f_i},$$

$$j=0, 1, \cdots, l, \quad i=0, 1, \cdots, l. \tag{3.4}$$

Set

$$\widetilde{F}_m=\Delta F\cdot E,$$

$$\widetilde{Q}_m=(\boldsymbol{Q}_m^{(1)}-\boldsymbol{Q}_m, \cdots, \boldsymbol{Q}_m^{(l+1)}-\boldsymbol{Q}_m), \tag{3.5}$$

where E is an identity matrix of order $l+1$, ΔF is a scalar and $\boldsymbol{Q}_m^{(i)}=\boldsymbol{Q}(\boldsymbol{F}_m+\Delta F\boldsymbol{e}_i)$, $\boldsymbol{e}_i$ being an $(l+1)$–dimensional vector whose i–th element is 1 and the remaining elements are 0. Then Q'_m can be replaced by $\widetilde{Q}_m\widetilde{F}_m^{-1}$, and (3.3) can be written into the form

$$\boldsymbol{F}_{m+1}=\boldsymbol{F}_m-\widetilde{F}_m\widetilde{Q}_m^{-1}\boldsymbol{Q}_m. \tag{3.6}$$

When (3.6) is used, (2.1) is integrated $l+2$ times from $\xi=1$ to $\xi=0$ in each iteration, therefore it takes a lot of time. In order to save time, we can use the simplified Newton–iteration method, i.e.,

$$\boldsymbol{F}_{m+1}=\boldsymbol{F}_m-\widetilde{F}_0\widetilde{Q}_0^{-1}\boldsymbol{Q}_m. \tag{3.7}$$

But the rate of convergence in this case may be slower. We have put forward a method which can improve the rate of convergence but does not increase the running time too much. If $\boldsymbol{Q}_{m-l-1}, \boldsymbol{Q}_{m-l}, \cdots, \boldsymbol{Q}_m(\boldsymbol{Q}_i=\boldsymbol{Q}(\boldsymbol{F}_i))$ are known for $\boldsymbol{F}=\boldsymbol{F}_{m-l-1}, \boldsymbol{F}_{m-l}, \cdots, \boldsymbol{F}_m$, Q'_m at $\boldsymbol{F}_m$ can be approximately determined under certain circumstances. In fact, there is the expansion expression

$$\boldsymbol{Q}_i\approx\boldsymbol{Q}_m+Q'_m(\boldsymbol{F}_i-\boldsymbol{F}_m), \quad i=m-l-1, m-l, \cdots, m-1, \tag{3.8}$$

i.e.,

$$(\boldsymbol{Q}_{m-l-1}-\boldsymbol{Q}_m, \boldsymbol{Q}_{m-l}-\boldsymbol{Q}_m, \cdots, \boldsymbol{Q}_{m-1}-\boldsymbol{Q}_m)$$

$$\approx Q'_m(\boldsymbol{F}_{m-l-1}-\boldsymbol{F}_m, \cdots, \boldsymbol{F}_{m-1}-\boldsymbol{F}_m). \tag{3.9}$$

Thus if $(\boldsymbol{Q}_{m-l-1}-\boldsymbol{Q}_m, \cdots, \boldsymbol{Q}_{m-1}-\boldsymbol{Q}_m)$ is invertible, then we can get

$$Q_m'^{-1}=(\boldsymbol{F}_{m-l-1}-\boldsymbol{F}_m, \cdots, \boldsymbol{F}_{m-1}-\boldsymbol{F}_m)$$

$$\times(\boldsymbol{Q}_{m-l-1}-\boldsymbol{Q}_m, \cdots, \boldsymbol{Q}_{m-1}-\boldsymbol{Q}_m)^{-1}. \tag{3.10}$$

Thus (3.3) can be rewritten into the form

$$\boldsymbol{F}_{m+1}=\boldsymbol{F}_m-(\boldsymbol{F}_{m-l-1}-\boldsymbol{F}_m,\ \cdots,\ \boldsymbol{F}_{m-1}-\boldsymbol{F}_m)$$
$$\times(\boldsymbol{Q}_{m-l-1}-\boldsymbol{Q}_m,\ \cdots,\ \boldsymbol{Q}_{m-1}-\boldsymbol{Q}_m)^{-1}\boldsymbol{Q}_m,$$
$$m=l+1,\ l+2,\ \cdots. \tag{3.11}$$

Obviously, when the above formula is used, it is necessary at the beginning to integrate $l+2$ times, but later on to integrate only one time.

Now, we proceed to estimate the rate of convergence. Since

$$\widetilde{Q}_m=Q'_m\widetilde{F}_m+O(\Delta F^2),$$

we have

$$Q'^{-1}_m=\widetilde{F}_m\widetilde{Q}^{-1}_m+O(\Delta F^2)\widetilde{Q}^{-1}_m=\widetilde{F}_m\widetilde{Q}^{-1}_m+O(\Delta F). \tag{3.12}$$

Then, for the difference form of the Newton formula there is the following expression of estimation:

$$\boldsymbol{F}^*-\boldsymbol{F}_{m+1}=\boldsymbol{F}^*-\boldsymbol{F}_m+\widetilde{F}_m\widetilde{Q}^{-1}_m\boldsymbol{Q}_m$$
$$=\boldsymbol{F}^*-\boldsymbol{F}_m+Q'^{-1}_m\boldsymbol{Q}_m+(\widetilde{F}_m\widetilde{Q}^{-1}_m-Q'^{-1}_m)\boldsymbol{Q}_m$$
$$=O(\|\boldsymbol{F}^*-\boldsymbol{F}_m\|^2)+O(\Delta F\|\boldsymbol{Q}_m\|), \tag{3.13}$$

where $\boldsymbol{F}^*$ is the exact solution, and $\|\boldsymbol{F}\|$ denotes the norm of $\boldsymbol{F}$. The above expression can be rewritten as

$$\|\boldsymbol{F}^*-\boldsymbol{F}_{m+1}\|\leqslant A_1\|\boldsymbol{F}^*-\boldsymbol{F}_m\|^2+A_2\Delta F\|\boldsymbol{Q}_m\|,$$
$$m=0,\ 1,\cdots, \tag{3.14}$$

where A_1 and A_2 are proper constants. Similarly, for the difference form of the simplified Newton formula (3.7), there is the expression of estimation

$$\boldsymbol{F}^*-\boldsymbol{F}_{m+1}=\boldsymbol{F}^*-\boldsymbol{F}_m+\widetilde{F}_0\widetilde{Q}^{-1}_0\boldsymbol{Q}_m$$
$$=\boldsymbol{F}^*-\boldsymbol{F}_m+Q'^{-1}_m\boldsymbol{Q}_m+(\widetilde{F}_m\widetilde{Q}^{-1}_m-Q'^{-1}_m)\boldsymbol{Q}_m$$
$$+(\widetilde{F}_0\widetilde{Q}^{-1}_0-\widetilde{F}_m\widetilde{Q}^{-1}_m)\boldsymbol{Q}_m,$$
$$\widetilde{F}_0\widetilde{Q}^{-1}_0-\widetilde{F}_m\widetilde{Q}^{-1}_m=Q'^{-1}_0-Q'^{-1}_m+O(\Delta F)=O(\|\boldsymbol{F}_0-\boldsymbol{F}_m\|)+O(\Delta F).$$

Therefore there exist constants A_1, A_2, A_3 such that the following inequality holds:

$$\|\boldsymbol{F}^*-\boldsymbol{F}_{m+1}\|\leqslant A_1\|\boldsymbol{F}^*-\boldsymbol{F}_m\|^2+A_2\Delta F\|\boldsymbol{Q}_m\|+A_3\|\boldsymbol{F}_0-\boldsymbol{F}_m\|\,\|\boldsymbol{Q}_m\|,$$
$$m=0,\ 1,\ 2,\ \cdots. \tag{3.15}$$

Since

$$(\boldsymbol{Q}_{m-l-1}-\boldsymbol{Q}_m,\ \cdots,\ \boldsymbol{Q}_{m-1}-\boldsymbol{Q}_m)=Q'_m(\boldsymbol{F}_{m-l-1}-\boldsymbol{F}_m,\ \cdots,\ \boldsymbol{F}_{m-1}-\boldsymbol{F}_m)$$
$$+O(\sup_{1\leqslant i\leqslant l+1}\|\boldsymbol{F}_{m-i}-\boldsymbol{F}_m\|^2),$$

for the iteration formula (3.11), we have

$$(\boldsymbol{F}_{m-l-1}-\boldsymbol{F}_m,\ \cdots,\ \boldsymbol{F}_{m-1}-\boldsymbol{F}_m)(\boldsymbol{Q}_{m-l-1}-\boldsymbol{Q}_m,\ \cdots,\ \boldsymbol{Q}_{m-1}-\boldsymbol{Q}_m)^{-1}$$
$$=Q'^{-1}_m+O(\sup_{1\leqslant i\leqslant l+1}\|\boldsymbol{F}_{m-i}-\boldsymbol{F}_m\|).$$

Then

$$F^* - F_{m+1} = F^* - F_m + Q_m'^{-1} Q_m + O(\sup_{1\leqslant i\leqslant l+1} \|F_{m-i} - F_m\|) Q_m.$$

That is, there exist constants A_1 and A_4 such that the following inequality holds:

$$\|F^* - F_{m+1}\|^2 \leqslant A_1 \|F^* - F_m\|^2 + A_4 \sup_{1\leqslant i\leqslant l+1} \|F_{m-i} - F_m\| \|Q_m\|,$$

$$m = l+1,\ l+2,\ \cdots. \tag{3.16}$$

It is easy to see that when ΔF is in the same order of magnitude as $\|F^* - F_m\|$ or in less order than $\|F^* - F_m\|$, the rate of convergence of the difference form of the Newton formula is close to that of the Newton formula. However because $\|F_0 - F_m\|$ usually increases with m, the simplified Newton method is of a lower rate of convergence. Since $\sup_{1\leqslant i\leqslant l+1} \|F_{m-i} - F_m\|$ usually decreases as m increases, the rate of convergence of (3.11) may be faster than that of the simplified Newton method. This has been shown by our experience.

§ 4 Interpolation Polynomials

When using the method of lines, we make use of the interpolation polynomial to determine the derivatives needed. For example, we construct an interpolation polynomial $g = \sum_{i=0}^{n} a_i s^i$ by use of the values g_k at points $s = s_k$ on the lines which are formed by the intersection of coordinate planes with ξ = constant and coordinate planes with φ = constant. Then the derivative $\frac{\partial g}{\partial s}$ is determined by $\frac{\partial g}{\partial s} = \sum_{i=0}^{n} a_i i s^{i-1}$. Similarly, we can determine $\frac{\partial g}{\partial \varphi}$. However, since the function is periodic for φ, it is better to use the trigonometric interpolation polynomial, e.g., for even functions we take $g = \sum_{j=0}^{n} b_j \cos^j \varphi$.

In order to save computer time, we do not calculate the coefficients a_i and b_j, but use the following procedure. As we know, the values of functions and their derivatives at interpolated points can be expressed by the linear combination of the values at the interpolation nodes. Moreover, the coefficients only depend on the positions of interpolation nodes and interpolated points, that is, once the nodes and the interpolated points are given, the coefficients may be determined. Thus when the values of the functions and their derivatives at interpolated points are required, the only thing we need to do is to take the inner product of these coefficients and the corresponding values of functions at the nodes. This fact can be explained as follows. Suppose that the interpolation polynomial is

$$g(s)=\sum_{i=0}^{n} a_i s^i, \tag{4.1}$$

and the nodes are s_k ($k=0, 1, \cdots, n$). The value of a function at a point s is required. At the nodes there are conditions

$$\sum_{i=0}^{n} a_i s_k^i = g_k, \quad k=0, 1, \cdots, n,$$

where $g_k=g(s_k)$. By matrix notation, they can be written as

$$\boldsymbol{M}\boldsymbol{a}=\boldsymbol{g}, \tag{4.2}$$

where

$$\boldsymbol{M}=\begin{pmatrix} 1 & s_0 & s_0^2 & \cdots & s_0^n \\ 1 & s_1 & s_1^2 & \cdots & s_1^n \\ \vdots & \vdots & \vdots & & \vdots \\ 1 & s_n & s_n^2 & \cdots & s_n^n \end{pmatrix},$$

$$\boldsymbol{a}=\begin{pmatrix} a_0 \\ a_1 \\ \vdots \\ a_n \end{pmatrix}, \quad \boldsymbol{g}=\begin{pmatrix} g_0 \\ g_1 \\ \vdots \\ g_n \end{pmatrix}.$$

Then we have $\boldsymbol{a}=\boldsymbol{M}^{-1}\boldsymbol{g}$. Thus the formula

$$g(s)=\sum_{i=0}^{n} a_i s^i = \boldsymbol{d}_0^* \boldsymbol{a} \tag{4.3}$$

can be rewritten into

$$g(s)=\boldsymbol{d}_0^*(\boldsymbol{M}^{-1}\boldsymbol{g})=(\boldsymbol{M}^{-1*}\boldsymbol{d}_0)^*\boldsymbol{g}=(\boldsymbol{M}^{*-1}\boldsymbol{d}_0)^*\boldsymbol{g}=\boldsymbol{b}^*\boldsymbol{g}, \tag{4.4}$$

where $\boldsymbol{d}_0^*=(1, s, \cdots, s^n)=\{s^i\}_{i=0,1,\cdots,n}$, $\boldsymbol{b}=\boldsymbol{M}^{*-1}\boldsymbol{d}_0$, which can be obtained by solving the linear system of equations

$$\boldsymbol{M}^*\boldsymbol{b}=\boldsymbol{d}_0.$$

Clearly, once the nodes and the interpolated point are given, $\boldsymbol{M}^*$ and $\boldsymbol{d}_0$ are determined, and $\boldsymbol{b}$ can then be obtained. Therefore, when $g(s)$ is required, what we need to do is to make an inner product by (4.4).

Since

$$g'(s)=\sum_{i=0}^{n} a_i i s^{i-1}=\boldsymbol{d}_1^*\boldsymbol{a}=(\boldsymbol{M}^{*-1}\boldsymbol{d}_1)^*\boldsymbol{g}, \tag{4.5}$$

$$g''(s)=\sum_{i=0}^{n} a_i (i(i-1)s^{i-2})=\boldsymbol{d}_2^*\boldsymbol{a}=(\boldsymbol{M}^{*-1}\boldsymbol{d}_2)^*\boldsymbol{g}, \tag{4.6}$$

where

$$\boldsymbol{d}_1^*=\{is^{i-1}\}_{i=0,1,\cdots,n},$$
$$\boldsymbol{d}_2^*=\{i(i-1)s^{i-2}\}_{i=0,1,\cdots,n},$$

the derivatives can be obtained in the same way.

For the trigonometric interpolation polynomial, the procedure is similar and is omitted here (for a concrete procedure see Chapters 5 and 6).

§ 5 Remarks on Improperly Posed Problems

As mentioned previously, when the above method is used, we shall meet the Cauchy problem of elliptic equations. In general, this problem is improperly posed in the Hadamard sense, i.e., there exists such a solution that its initial value is arbitrarily small but it may be arbitrarily large somewhere in a domain containing the initial-value line. In other words, the solution does not depend continuously on the initial value. The improperly posed example is presented first by Hadamard[30] for the Laplace equation. Since then, many people have done a great deal of work and have pointed out that if the set of solutions is contracted then the improperly posed problem in the Hadamard sense may become properly posed in the Tihonov sense (for definition of a properly posed problem in the Tihonov sense please refer to [29], p. 4). The main results are as follows.

Lavrent'ev[26] proved the following theorem for the Laplace equation:

Suppose that s is a single connected bounded domain in x, y plane, and $\varphi(x, y)$ is a continously differentiable function of $\bar{s}$ ($\bar{s}=s\cup\partial s$), and is positive in s and equal to zero on the boundary ∂s of s. Σ is a surface $z=-\varphi(x, y)$, $(x, y)\in\bar{s}$ in the x, y, z space, and D is an open domain bounded by Σ and $z=0$. Suppose that $u(x, y, z)$ is harmonic in D and equal to zero on Σ and satisfies inequalities

$$\iint_{\Sigma} \mathrm{grad}^2 u(x, y, z)\,d\sigma\leqslant\varepsilon,$$

$$\iint_{s} u^2(x, y, 0)\,dx\,dy\leqslant M,$$

where ε and M are arbitrary constants. Let s_t be the intersection of D and plane $z=t$, and φ_0 be the maximum of φ in s, then for an arbitrary t in the interval $(-\varphi_0, 0)$ the following inequality holds

$$\iint_{s_t} u^2(x, y, t)\,dx\,dy\leqslant 2C^{\frac{\varphi_0^3}{8}} M^{1+\frac{t}{\varphi_0}}\varepsilon^{-\frac{t}{\varphi_0}},$$

C being a constant. From the above theorem we know that in the set of bounded functions the Cauchy problem of the Laplace equation whose initial value is given on Σ is properly posed in the sense of L_2-norm. [31—33] gave similar theorems for the initial-value problem of the Cauchy-Riemann equations and the inverse heat conduction equation.

Zhang[27] discussed the following initial-value problem in domain G:

$$\begin{cases} \dfrac{\partial\bar{u}}{\partial y}=AD^{p}\bar{u}, \\ \bar{u}(0, x)=\bar{u}_0(x), \end{cases} \tag{5.1}$$

where $\bar{u}$ is a vector function, A a constant matrix, p an integer, G the domain: $y \in [0, Y]$, $x \in (-\infty, \infty)$, and $D^p = \dfrac{\partial^p}{\partial x^p}$. He proved the following theorem:

If $\bar{u}(y, x)$ is a square integrable solution of (5.1), i.e., for an arbitrary $y \in [0, Y]$,

$$\|\bar{u}(y)\| = \left[\int_{-\infty}^{\infty} |\bar{u}(y, x)|^2 dx\right]^{\frac{1}{2}} < \infty,$$

then the following inequality holds:

$$\|\bar{u}(y)\| \leqslant \tilde{c} \|\bar{u}(Y)\|^{\frac{y}{Y}} \|\bar{u}(0)\|^{1-\frac{y}{Y}},$$

where $\tilde{c}$ is a constant suitable for all solutions $\bar{u}(y, x)$. From this we can directly get the following theorem:

The initial problem (5.1) is properly posed in the Tihonov sense in the set of uniformly bounded solutions.

At the same time some other authors have developed many stable and convergent algorithms for this initial-value problem. For example, for the Cauchy problem of the Laplace equation:

$$u_{xx} + u_{yy} = 0,$$
$$u(x, 0) = u_0(x), \quad u_y(x, 0) = u_1(x), \quad 0 \leqslant x \leqslant 1,$$

Pucci[28] proved the following theorem:

Suppose that $u(x, y)$ is a harmonic function in the domain D, the interval $(0, 1)$ on x lies in D, and h is the distance of this interval from the boundary ∂D. Suppose that $\alpha_i^{(n)}$, $\beta_i^{(n)}$ ($i = 0, 1, \cdots, 4n$) are the approximate values of $u(x, y)$, $u_y(x, y)$ at $4n+1$ points: $\left(\dfrac{i}{4n}, 0\right)$ $(i = 0, 1, \cdots, 4n)$, with an error E_n, i.e., there are the following relationships

$$\left|u\left(\frac{i}{4n}, 0\right) - \alpha_i^{(n)}\right| \leqslant E_n,$$
$$\left|u_y\left(\frac{i}{4n}, 0\right) - \beta_i^{(n)}\right| \leqslant E_n.$$

Furthermore, we can construct the following approximate function

$$u_n(x, y) = \sum_{s=0}^{n} \frac{y^{2s}}{(2s)!}(-1)^s \Delta^{2s} \alpha_{i(x)}^{(n)} + \sum_{s=0}^{n} \frac{y^{2s+1}}{(2s+1)!}(-1)^s \Delta^{2s} \beta_{i(x)}^{(n)},$$

where

$$\Delta = \begin{cases} \Delta_+, & \text{for } i \leqslant 2n, \\ \Delta_-, & \text{for } i > 2n, \end{cases}$$
$$\Delta^s = \Delta(\Delta^{s-1}),$$
$$\Delta_+ \alpha_i^{(n)} = 4n(\alpha_{i+1}^{(n)} - \alpha_i^{(n)}),$$
$$\Delta_- \alpha_i^{(n)} = 4n(\alpha_i^{(n)} - \alpha_{i-1}^{(n)}),$$

and $i(x)$ is the integer which makes $|4nx - i|$ take minimum. The conclusion is that if

$$\lim_{n\to\infty} E_n e^{8nh}=0, \tag{5.2}$$

then the sequence $\{u_n(x, y)\}$ converges uniformly to $u(x, y)$ in the rectangle $R\equiv[0\leqslant x\leqslant 1,\ |y|<h']$, where h' is any positive constant less than h.

Some convergent difference schemes for the initial–value problems of the Laplace equation and the Cauchy–Riemann equations are developed in [27].

From the above discussion, it is seen that even though the problem is improperly posed, the reasonable solution may be obtained as long as the growth of error is strictly controlled by certain conditions, such as (5.2).

If we consider the initial–value problem of the Laplace equation in the set of uniformly bounded analytic functions, as pointed out above, the problem is properly posed. When the method of lines is used to solve this problem, we get an initial–value problem of ordinary differential equations. It can be shown that when $n\to\infty$, the solution of ordinary differential equations converges to that of the original partial differential equations. However, generally, there is no explicit analytic solution for these ordinary equations, thus we use the numerical methods. Therefore there exist truncation and rounding errors. The truncation error can be controlled, but the rounding error is random. Thus we have to consider the stability of the method of lines. For this purpose, we consider the Dirichlet problem of the Laplace equation in the rectangular domain:

$$\begin{cases} u_{xx}+u_{yy}=0, \\ u(0, y)=u(1, y)=0, \\ u(x, b)=u(x, -b)=\sin \pi x. \end{cases} \tag{5.3}$$

Obviously, the exact solution is

$$u(x, y)=\frac{\operatorname{ch} \pi y \sin \pi x}{\operatorname{ch} \pi b}. \tag{5.4}$$

Owing to symmetry, only the domain $0\leqslant x\leqslant\frac{1}{2}$, $0\leqslant y\leqslant b$ needs to be considered. We make $N-1$ lines parallel to the axis x:

$$y_n=nh=nb/N.$$

The symmetry conditions are

$$u_x\left(\frac{1}{2}, y\right)=0,$$

$$u(x, y)=u(x, -y).$$

For simplicity, we use the three point formula to approximate u_{yy}, i.e.,

$$u_{yy}=\frac{u(x, y+\delta y)-2u(x, y)+u(x, y-\delta y)}{\delta y^2}+O(\delta y^2).$$

Substituting the above equation into (5.3), we get

$$\begin{aligned}
&u_n''+(u_{n+1}-2u_n+u_{n-1})/h^2=0,\\
&u_n(0)=0, \quad u_n'\left(\frac{1}{2}\right)=0,\\
&u_N=\sin \pi x, \quad u_{-n}(x)=u_n(x),\\
&(n=0, 1, \cdots, N-1),
\end{aligned} \tag{5.5}$$

where $u_n(x)$ is the approximation of $u(x, y_n)$, the quantities with a prime denote the derivatives with respect to x. It can be proved that the general solution of (5.5) is

$$u_n(x)=\sum_{m=1,3}^{2N-1} T_n(\theta_m)(A_m e^{\mu_m x}+B_m e^{-\mu_m x})+\frac{\operatorname{ch} nz}{\operatorname{ch} Nz} \sin \pi x. \tag{5.6}$$

where

$$\begin{aligned}
&\mu_m=(2N/b)\sin(m\pi/4N),\\
&\theta_m=\cos(m\pi/2N),\\
&z=\operatorname{ch}^{-1}\left(1+\frac{1}{2}(\pi b)^2/N^2\right),
\end{aligned}$$

and $T_n(\theta_m)$ is a Chebyshev polynomial of order n. Now, $u_n(0)=u_n(1)=0$, so $A_m=B_m=0$. The second part of (5.6) is a particular solution. The inverse hyperbolic cosine function z can be expanded to

$$z=\frac{\pi b}{N}\left[\left(1-\frac{\pi^2 b^2}{24N^2}\right)+O(N^{-4})\right].$$

Then we get

$$v_n(x)=\frac{\operatorname{ch} \pi y_n}{\operatorname{ch} \pi b} \sin \pi x\left[1+\left(\tan h\pi b-\frac{y_n}{b} \tan h\pi y_n\right)\left(\frac{\pi^3 b^3}{24N^2}+O(N^{-4})\right)\right]. \tag{5.7}$$

It is seen that the solution of the ordinary differential equations obtained by the method of lines consists of two parts. The first part is zero for the present boundary conditions and the second part is a particular solution which converges to the exact solution (5.4) of the original differential equations with error $O(N^{-2})$.

However, it must be noticed that if A_m and B_m are not equal to zero (even very small), the first part of (5.7) may be predominant when x and N are large enough. Therefore (5.7) will not converge to (5.4). This is very important. In fact, even though the initial values $u_n(0)$ and $u_n'(0)$ are exact, they are no longer exact after integrating one step δx because of the truncation and rounding errors. Therefore the first part of (5.7) occurs and grows exponentially with x and N. We can conclude that, on the one hand, in order to reduce the truncation error it is necessary to use as many lines as possible, on the other hand, in order to reduce the growth of rounding error, it is necessary to use as few lines as possible. This means that when the method of lines is used for either mixed–type equations or elliptic equations, the number of lines must be properly selected.

PART II

INVISCID SUPERSONIC FLOW AROUND BODIES

Introduction to Part II

§ 1 Outline of Part II

In Part II we shall put emphasis on the application of the methods described in Part I to the inviscid steady supersonic flow around bodies. In addition, we shall give some supplementary information required for computation and also present some computed results. Occasionally, we shall also discuss some mechanical phenomena.

The problem referred to as the steady inviscid supersonic flow around bodies is as follows. Suppose that there exists a blunt–nosed body or a sharp–nosed body in a supersonic flow. We consider an inviscid and non–heat conducting flow. Thus the governing equations are a system of the first order quasilinear partial differential equations. Because the flow is supersonic, there is a shock in front of the bodies, which divides the flow field into two regions, one undisturbed by bodies and the other disturbed by bodies. Then the determination of the disturbed flow will be reduced to solving a boundary value problem with the boundary conditions given at the shock and the body. In the vicinity of the blunt nose the flow is usually subsonic and transonic, which means the governing equations are of the mixed–type. The flow over the afterbody is supersonic, which means the governing equations are hyperbolic. Therefore, the solution of the equations can be obtained in two steps. (i) The first step is to solve a boundary value problem of mixed–type equations in the subsonic–transonic region, and (ii) the second step is to solve an initial–boundary value problem of hyperbolic equations in the supersonic region with the initial conditions given on a space–like surface and the boundary conditions given at the shock and the body. The initial conditions are obtained from solutions at the first step. For the sharp–nosed body the situation is rather different. When the vertex angle of cones is small, the shock is attached. In this case the vertex is a singular point, and the solution of this problem is reduced to a boundary value problem of the elliptical or mixed–type equations, which represents a conical flow. Once the conical flow is obtained, the procedure to get the solution for the flow over the afterbody is similar to that for the supersonic flow around the blunt body. When the vertex angle of the cones is large, the shock is detached. This case is similar to the flow around the blunt body. In principle the flow around the bodies should

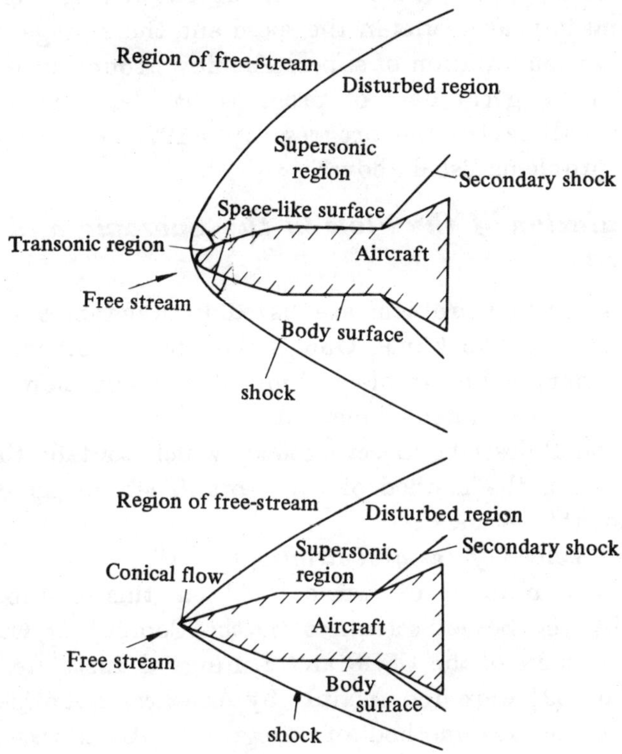

Figure II-1 Patterns of steady inviscid supersonic flow arounodised

include the base flow, but for the base flow we have to consider viscosity, which is beyond this book. Therefore we think that problems of the steady inviscid supersonic flow are composed of the following:

(1) to solve the subsonic and transonic flow in the vicinity of the blunt nose;

(2) to solve the conical flow;

(3) to solve the supersonic flow over the afterbody.

In Chapters 5, 6, 7 we shall describe the numerical methods for the three problems, respectively, and give some results.

Some preliminary knowledge, which is necessary for the application of the methods described in Part I to the flow around bodies and is not usually discussed in general books on mechanics, is given in Chapter 4. It includes the type of the governing equations, compatibility relations, weak discontinuities, strong discontinuities, singular lines, and the state equation of air etc.

§ 2 Literature Review

During the last two decades with the urgent demand for astronautic techniques and improvements in the speed and the storage of computer, many works on computation of supersonic flow around bodies have been carried out and a great deal of progress has been achieved. In the following we shall review the progress upto 1977 in foreign countries[1)] on the three problems listed above[2)].

1. *Calculation of the Flow in the Subsonic and Transonic Region*

The works on this problem are listed in References A. The works can be divided into two kinds. One is the direct methods[A1–121] which solve the boundary value problem of mixed-type equations directly. The other is the "time-dependent" methods [A122–183].

First we shall describe direct methods which contain the method of integral relations, the method of lines, etc. Then we shall review the "time-dependent" methods.

(i) *The Method of Integral Relations*

There are a number of works [A1–59] on this method. [A1—25] were done by the Soviet scientists in the Computing Center of the Academy of Sciences of the USSR and Leningrad Institute of Technical Physics. [A26—59] were done mainly by American scientists and others. The basic idea of the method of integral relations was proposed by Dorodnitsyn[A1]. Belotserkovskii[A2] applied successfully this method to the supersonic flow with a detached shock. In [A3, 4] he presented the detailed method and results for two-dimensional flow around bodies. In [A5—7] he considered the axisymmetrical flow around bodies and gave systematical results for the flow around blunt bodies in table form. He also gave some results for the flow around the flat cylindric body. Later he proposed the two other schemes[A8], Schemes II and III, and therefore he named the original one Seheme I. In [A9—19] he and his colleagues developed this method further, and applied it to the equilibrium air flow[A9, 11], the equilibrium CO_2 flow[A11], the nonequilibrium O_2 flow[A10, 12], the nonequilibrium air flow[A13], the flow with vibration and dissociation relaxation of O_2[A19], the flow with radiation[A15], and the flow around a body with a sonic corner[A16]. They also applied Scheme III to the low supersonic flow[A18]. Scheme II was extensively used[A10, 13, 15, 16]. The detailed formulas for these three schemes were given in [A17], and the

1) It means that the progress made in China is not included. However, we list the authors' papers at the end of the General References, which partially reflect the progress in China.

2) When we finished this book, we did not get all the literatures in 1977.

improved treatment on the sonic singularities was given in [A14]. On the basis of these works and others, in 1966 the Computing Center of Academy of Sciences of the USSR published a monograph: Supersonic Flow past Blunt Bodies. It was revised and republished in 1967. Later they calculated[A21, 22] the radiation flow with different physical models from that in [A15], and considered the nonequilibrium flow with the concentrations close to the equilibrium state[A23]. In several international conferences[A24, A25] Belotserkovskii summarized and discussed some of the results obtained by the method of integral relations and other methods.

Besides Belotserkovskii and his colleagues, various authors have done a great deal of work. [A26—29] are written by some scientist from USSR. [A26, 28] calculate the flow around bodies with corners, and [A27, 29] consider the flow around blunt bodies under a magnetic field. From the U. S. A. and other countries, we cite [A30—54]. Several of these[A30, 33, 34, 42, 49] use the method of integral relations to obtain some of their results. Some modify the method and calculate the flow around asymmetric bodies[A31, 37, 38, 50], bodies with a corner[A32, 51], and cones at large angles of attack[A53]. Some calculate the equilibrium flow[A43, 48, 52], nonequilibrium flow[A36, 39, 45], and radiation flow[A40, 46]. Papers [A31, 35, 41, 44, 50] discuss and modify the method. [A47] determines the body and its flow field, corresponding to a given pressure distribution along the body.

From the above it is obvious that the method of integral relations has been successfully applied to the two-dimensional and axisymmetric flow around blunt bodies. However, there are only few works on three-dimensional flow. [A55] attempts to apply the method of integral relations to three-dimensional flow, but has not carried it through. [A56] extends the method to three-dimensional flow and gives some results. [A58] has considered the equilibrium flow, but only used two meridional planes. The authors encountered some difficulties in extending the method to a general case. Similarly, [A56, 57] also consider only a first approximation in the φ direction. [A20] has extended the Scheme II to the three-dimensional case. [A59] considers the equilibrium and non-equilibrium flow, and gives some results and detailed formulas, but the scheme is not natural for the case with several φ planes. This is because the author did not realize how to establish the difference equations for the mixed-type equations with three variables.

(ii) *The Method of Lines*

The method of lines was proposed by Telenin and his colleagues at Moscow University ([A60—86]). The basic idea was proposed in [A63] in 1963. [A61] gives results for equilibrium air and CO_2 flow around a sphere. [A62] describes the method in detail and gives a new scheme for the perfect air flow, in which the singularities at the body are removed.

A series of results and mechanical discussions are given in [62—66]. In [A62, 63] the bodies under consideration include the concave body, the body with discontinuous curvature in the subsonic region, the disk-like body, the ellipsoid, and the body with a sonic corner at which the series expansion is used. Some results for low supersonic flow are given in [A62, 65] and for nonequilibrium flow in [A62, 64]. The detailed formulas for nonequilibrium flow are presented in [A64]. [A66] calculates the flow around a cylinder with an elliptic cross-section. In a later paper [A67], certain improvements are made to remove the singularities at the body for equilibrium flow. Furthermore, the method of lines has been applied to the various flows such as flow with vibration relaxation, nonequilibrium flow, nonequilibrium flow with rapid chemical reactions, radiation flow, flow with strong injection from the body, and nonequilibrium flow around the segmental body. In all these calculations the method has been modified, if necessary.

The method of lines is successful not only for two-dimensional but also for three-dimensional flow around blunt bodies. In 1964 Telenin et al. presented the results for three-dimensional flow around blunt bodies[A83], and in 1965 they gave a detailed method and analysis[A84]. [A85] calculates the flow around the segmental body at angle of attack. Also, Golomazov and Zyuzin[A86] use the method of lines with lower-order approximation and obtain good results.

(iii) *The Other Direct Methods*

In addition to the method of integral relations and the method of lines, there are various other methods. However they are not as systematic as the two previous methods. Therefore we shall put them in to one subsection.

The first one we should mention is the Van Dyke inverse method[A87–96]. The method was proposed in 1958[A87] by Van Dyke, who was at Ames Aeronautic Laboratory at that time. Some systematic results obtained by this method are found in [A88]. In 1960 Lick applied the same idea to nonequilibrium flow[A89]. Later the method was used for nonequilibrium flow with a more complex model[A93–95]. Inouye et al., of the Ames Research Center, put it into practical calculation and obtained a number of results[A90, 94–96], including some nonequilibrium flow.

[97, 98] list the inverse method for three-dimensional flow. Joss[A97] calculates the inverse problem, while Wobb and others[A98] calculate the direct problem for blunt body flow at angle of attack.

Fuller[A99] also uses the inverse method to get a number of results for two-dimensional flow, but uses a little different procedure from Van Dyke's.

Another inverse method was proposed by Vaglio-Laurin and Ferri in 1958 [A100]. A code is listed in [A101]. Some modification in terms of the PLK method is presented in [A102]. [A103] discusses application of the method given in [A100] to equilibrium air flow.

Mangler[A104] also proposed a scheme which has the same feature as the scheme by Vaglio-Laurin and Ferri, that is, the last marching line is coincident with the body. This is more convenient than the Van Dyke method. In the Van Dyke method the integration cannot be marched simultaneously to the body, which causes some inconvenience.

Swigart[A105-108] presents a method in which he uses series expansion to get a system of ordinary differential equations and integrates this system from the shock to the body. [A105—107] consider the inverse problem. [A108] considers the direct problem, and the equations contain second-order terms in angle of attack so that the method has a certain degree of accuracy for three-dimensional flow.

The Garabedian method is discussed in [A109—111]. In this method he uses the analytic continuation to extend the problem to a complex domain, and integrates the equations from the shock to the body. [A110] points out the instability of this continuation.

In the early stage there were the method of flux analysis by Uchida and Yasuhara[A112] and the stream-tube method discussed in [A113—115]. Van Tuyl[A116] used the rational approximation in the calculation of flows past blunt bodies. Jurak[A117] used the least-squares method for the hypersonic flow past blunt bodies. Severinov[A118] used the artificial viscosity in the marching process from the shock to the body. Bratos et al.[A119] used the Lax finite difference scheme in the marching process.

[A120] compares the current methods used in the U. S. A. for the blunt body flow. [A121] analyses the reasons which cause instability for the solution of the blunt body flow.

Except in a few methods, such as the Garabedian methed, Scheme III of the method of integral relations, and the least-squares method, in these direct methods the problem was reduced to an ill-posed initial value problem. Therefore it is necessary to control the growth of errors. In addition, the initial value problem can be divided into two kinds. In one kind the initial values are given at the shock and "marching" is from the shock to the body. In the other kind the initial values are given at the symmetric axis and "marching" is along the body. The former includes the Scheme II of the method of integral relations, the method of lines, the marching methods (the Van Dyke method and other inverse methods), the series expansion method, the method of flux analysis, and the stream-tube method. The latter mainly contains the Scheme I of the method of integral relations. There exists the "stream singularity" in the former, which is removed in some works, and there exists the "sonic singularity" in the latter, which has not been removed. From this point of view, the former is easier to apply than the latter. In addition, in some methods the body is coincident with a coordinate plane and in the others it is not so. The former methods are simpler because it is unnecessary to use extrapolation to obtain the values on the body. Therefore, the Vaglio–Laurin and Ferri method, the Mangler method and the method of lines without singularity etc. are more effective because they have two advantages:

(1) The shock and body are coincident with a coordinate plane so that it is convenient to march from the shock to the body.

(2) There is no singularity in the marching procedure. However, there still exists the instability since the problem is ill–posed. Thus it is necessary to control the growth of errors in the calculation.

(iv) *The Time–Dependent Method*

A different kind of method from that described above is the time–dependent method. This method is based on the following physical idea. Suppose that the body is fixed and the free stream flow does not vary with time. Thus, when time is long enough, the flow will tend to the steady state no matter what the initial value is. Instead of solving steady equations, we consider a time–dependent problem and want to find its asymptotic solution which is the solution of the steady equations. The time–dependent problem is an initial–boundary–value problem of the hyperbolic system. Mathematically, we have changed a boundary value problem of the mixed–type equations to an initial–boundary–value problem of the hyperbolic system. Compared with the direct methods, the dimension of the problem is raised by one, that is, the problem with two (or three) independent variables becomes one with three (or four) independent variables. Therefore the amount of calculation may increase, but the advantage is that the present initial–boundary–value problem is well–posed and unlike most of the direct methods there is no singularity. Therefore it is easier to implement. In addition, a number of direct methods require that the flow parameters vary smoothly, which restricts their applications. However the time–dependent method has no such restrictions.

[A122—183] are references which discuss the time-dependent method. [A122—158] relate to two–dimensional flow and [A159—183] relate to three–dimensional flow. The status of two–dimensional flow is as follows:

In 1961 Godunov et al. [A122] calculated the supersonic flow around blunt bodies by using the time–dependent method. The difference seheme was constructed by the physical idea of decomposition of discontinuities. [A129, 132] apply this method to calculation, and modify some procedures, thereby improving accuracy.

Babenko and Rusanov present a scheme and some results in [A123]. The scheme is implicit. The resulting difference equations are solved by an iteration in one direction and the double-sweep method in another direction. In fact, the scheme is the application of the method in [C62], which is used to calculate the steady flow, to the calculation of the time-dependent flow. The results in [A123] were presented earlier in [C71]. Later on this method was used by a number of people: Kosorukov[A137] used it to calculate the nonequilibrium flow, Rusanov analysed some results in [A152], Ivanova et al.[A154] used it to calculate the flow around the sharp-nosed body with a detached shock, Radvogin[A155] obtained some results by use of it, and Gilinskii et al.[A158] used it to calculate the flow around a blunt body with strong injection.

Bohachevsky and Rubin[124] calculated the nonequilibrium flow around blunt bodies by use of the Lax scheme. This method is different from those described above. It is a kind of shock-capturing method. Later Bohachevsky et al. calculated the supersonic flow over convex and concave bodies with radiation and ablation effects.

In 1966 Moretti and Abbett[A125] presented a time-dependent method for blunt body flows. In interior points the Lax-Wendroff scheme is used for the nonconservative equations[18,22], and at the shock and the body the characteristic method is used. Bastianon[A128] also used this method and found that the flow over concave bodies is unsteady and varies periodically. Kyriss[A136] modified this scheme and calculated the nonequilibrium flow. The modification is as follows: The equations are divided into two groups. The above scheme is used for the equations determining the pressure and velocity, while the implicit scheme is used for the stream-characteristic compatibility relations. D'Souza et al.[A142] used this method to calculate hypersonic internal flow over blunt leading edges and two blunt bodies. In their work some procedures, such as the arrangement of grid points, were modified. This method was also used by Lin et al.[A156] to calculate flow with nonuniform free stream. In [A130] Moretti discussed some procedures relating to boundary conditions and their effects on calculation and concluded that his procedures are correct.

Burstein used the Lax-Wendroff scheme to calculate the flow around blunt bodies in [A126]. Lapidus[A127] did similar work by using Richtmyer's two-step L-W scheme and a curvilinear coordinate. Unlike [A125], these two works adopt the shock-capturing technique. [A60] mentions that Roslyakov and Chudov calculated the blunt body flow by the alternative use of the Lax scheme and the "Cross" scheme in 1960, but did not give any concrete results.

Barnwell calculated the blunt body flow by using a first-order scheme and considered the radiation in [A131]. The scheme is similar to the Lax scheme, but does not use the "half-step" points. Later he adopted another method[A134,135] in which the second-order two-step scheme was used for interior points and the characteristic method for the shock points. In these two works some blunt body flow with sharp corners are calculated, and the sharp corners may be sonic points or in the subsonic region. In addition, this method was also used to calculate the flow around sphere–cones with a large vertex angle by Rao[A148] et al.

In 1970 Vliegenthart[A133] adopted the Shuman filtering technique in calculation of the shock waves. Harten and Zwas[A145,146] proposed the switched Shuman filtering technique in order to reduce the big error inherent in the Vliegenthart's technique, and achieved some improvement. They all took the flat cylinder as an example.

In 1970 Belotserkovskii et al. used the "large particle" method to calculate the blunt body flow [A138]. The method is presented in detail in [A140]. Later, some results obtained by using this method were presented in [A141]. [A150] gave the complete description of this method and its results.

Coakley and Porter calculated the MHD blunt body flow in [A139]. They used the Lax scheme[17] in the interior points and the characteristic method at the boundary points.

In [A143] Nichols used the Lax[17] scheme and the Rusanov scheme[20] for blunt body flows and made a comparison between the two. He also considered a modified Lax scheme.

In [A147] MacCormack and Paullay calculated the supersonic flow by using time–splitting of difference operators. In [A149] they also consider the effect of arrangement of the grid points on accuracy of results, when using shock–capturing technique. They focus on the study of methods. Actually most problems presented in these works can be solved by the steady state method for two–dimensional flow.

In 1975 Hsieh[A151] calculated the low supersonic flow by the MacCormack scheme.

In 1975, Krasil'nikov et al. [A153] used the Magomedov–Kholodov[C99] scheme to calculate some flows around sphere–cones with a large vertex angle.

In 1977, Srinivas and Gururaja[A157] calculated the supersonic flow around flat–nose bodies by their local stability scheme.

There are a number of works on the three–dimensional flow.

In 1966 Bohackevsky and Mates[A159] extended the work of [A124] to the three–dimensional case, and calculated the supersonic flow over sphere segment with sharp corners at angle of attack. The accuracy of the results was not satisfactory because of the unsatisfactory treatments at sharp corners. Later Xerikos and Anderson calculated some flows over other blunt bodies by a similar method[A163].

In 1966 Rusanov[A160] calculated the three–dimensional flow. The scheme is an extension of that in [A123], details of the method are presented in [A164] and the analysis of results is given in [A165]. In

[A167] they made a detailed theoretical study of this method and analyzed the results. This work is the most complete in theoretical analysis and the most thorough in its presentation of results. Later, Förster et al. also used this method and proposed a technique to save running time. The work by Lipnitskii et al.[A175] is again related to this method. They modified the scheme as follows: the original scheme is used for the wave-characteristic compatibility relation and another scheme is used for the stream-characteristic compatibility relation. They obtained some results by this modified scheme. Later Savinov et al.[A182] used this scheme to calculate the nonequilibrium flow.

In 1967 Moretti and Bleich presented some results on three-dimensional blunt body flows. The method is an extension of that in [A125], i.e., the scheme similar to the Lax-Wendroff[18, 22] scheme is used for interior points and the four-dimensional characteristic method for boundary points. The details of the method are given in GASL TR-637. In [A162] the authors changed the cylindrical coordinate used in [A161] to the spherical coordinate and made some calculations. In [A171, 172, 177] Li made some modifications to this method and calculated the nonequilibrium flow. As for the calculation of boundary points by the characteristic method, Porter and Coakley have also given a detailed discussion in [A174].

In [A166] Cohen et al. apply the Godunov method[A122] to three-dimensional flow problems. They took only two planes in the φ direction. Ivanov [A181] develops a further series of results by using a version of the Godunov method.

In 1971 Bornwell extended his works[A134, 135] to the three-dimensional case. For the calculation of body points he considered two methods: the characteristic method and the difference method. This work was further elaborated in [A169, 170] and a FORTRAN IV code was given.

In 1972 Moretti et al.[A173] revised the scheme given in [A162]. They used the MacCormack scheme at interior points and the predictor-corrector method at boundary points. This paper mainly describes the calculation of flow in the supersonic region, and mentions the revision for computation of the blunt body flow as well.

In 1973 Rizzi and Inouye[A176] presented a time-split finite-volume method for blunt body flow. The method is a kind of shock-capturing method. The scheme is an extension of that used by MacCormack in [A147]. Later this method was used to calculate the nonequilibrium flow[A179].

Belotserkovskii et al.[A178] applied the scheme presented by Magomedov[C99] to the calculation of radiation flow. Kostrykin et al.[A183] did some further work in this respect.

As mentioned early, to calculate supersonic blunt body flows by the time-dependent method is actually to solve an unsteady flow problem. Therefore each scheme used for multi-dimensional unsteady flows can also be used for the blunt body flow. In fact, almost all the schemes in common use have been used for the blunt body flow, for example, the Godunov scheme, the Lax scheme, the Lax-Wendroff scheme, the Richtmyer two-step L-W scheme, the Rusanov scheme, the MacCormack predictor-corrector two-step scheme, etc. In addition, there are other schemes, e.g., the BVLR scheme (the scheme presented by Babenko, Rusanov et al.), the modified scheme of Lipnitskii et al., the Barnwell two-step scheme, the Magomedov-Kholodov scheme, the "large particle" scheme of Belotserkovskii, the time-splitting difference method of MacCormack et al., and so on.

Although there are various schemes, they can be catalogued into two kinds. One is the shock-capturing technique which is similar to the artificial viscosity method. The other is the shock-separating method in which the shock is treated as a discontinuity. The works by Bohachevsky et al., Burstein, Lapidus, Vleigenthart et al., Belotserkovskii et al., MacCormack et al., Nichols, Srinivas et al., belong to the first kind. The works by Babenko and Rusanov et al., Moretti et al., Barnwell, and Coakley and Porter, belong to the second kind. Methods of the first kind are rather simple since they do not need any special treatment for the shock. But for them a careful consideration of schemes is necessary in order to reduce oscillation in the vicinity of the shock. The common problem encountered in this kind is that the calculation of the body points needs to be separately considered. There are several methods, but it seems that the correctness of some procedures is doubtful. In addition, compared with the second kind, the first kind usually needs more grid points, therefore consumes much running time. Methods of the second kind are not so simple as the first, since the shock has to be considered separately. However, not much consideration is needed for constructing schemes for the interior points since the functions at these points are rather smooth. The characteristic methods (or an implicit one such as the BVLR scheme) are usually used for the shock and body points. Hence these treatments are rather reliable. Of course, whether they are correct or not needs to be further analysed. If there are many discontinuities, the second kind will be more complicated in dealing with these discontiunities.

2. *Calculation of Conical Flow*

The works on this problem are listed in References B. Like the calculation in the subsonic and transonic region the methods can be divided into two kinds: the direct method and the method of stabilization.

[B1—45] are the works on the direct method and [B46—63] are the works on the method of stabilization.

(i) *The Direct Method*

Among the numerical methods for solving the quasilinear equations of conical flows, the earlier work was done by Maslen[B1]. There he considered the nonlinear equations, rather than the equations derived from the approximate linearized theory. But he supposed that the change of entropy was small, therefore some terms were neglected. The method is a combination of the characteristic method and the relaxation method. He calculated the conical flow over delta wings at the angle of attack, but did not consider the embedded shock in the leeward region.

The successive works on conical flows are as follows:

In 1959 Briggs[B2] applied the inverse method which was used by Van Dyke to calculate the blunt body flow to calculation of conical flows. He calculated the conical flow with shock waves shaped like elliptical cones. Later he calculated the conical flow with shock waves shaped like elliptical cones at angle of attack. He only considered the inverse problem. In 1962 Stocker and Mauger[B6] also used the inverse method in which they gave the shock and then marched from the shock to the body and determined the whole flow. However there is a big difference between this method and the method used by Briggs. Besides the inverse problem, the direct one——the conical flow over cones at small angle of attack——has also been calculated. Eastman and Omar[B14] use the inverse method to calculate the windward part of conical flow over cones at large angle of attack and consider the equilibrium air flow. In [B11] Syagaev proposes an inverse method and calculates the conical flow over cones and elliptical cones at small angle of attack. In [B15] he and Makhin modify the original scheme, eliminating the unnecessary assumption and introducing the transformation of independent variables. Therefore they greatly improve the method.

In 1960 Chushkin and Schennikov calculated the conical flow around cones at small angle of attack by the method of integral relations[B4]. They used one strip and integrated equations along the body. Brook[B5] discussed the application of the method of integral relations with one strip to the conical flow over the elliptic cones and delta wings, but did not do any calculation. Later he[B12] reconsidered it, but he still did not give any results. In 1963 Kennet calculated the windward flow over the delta wing [B10] by using the method of integral relations. Akinrelere [B22] also calculated the same kind of flow by the method of integral relations, but unlike Kennet he thought that the shock was attached to the leading edge, rather than detached. Therefore the two schemes were different. In [B13] Melnik calculated the conical flow over elliptical cones. His code was originally designed to use any number of strips, but there were some difficulties in the calculation. Therefore only the result obtained by one strip was given. In 1962 Belotserkovskii and Chushkin[B9] proposed another scheme of the method of integral relations in which the integration was carried out from the shock to the body, but they did not put it into effect until 1970. In 1970 Chushkin[B24] gave the details of the scheme and some results about the conical flow over cones at small angle of attack.

In 1962 Babaev proposed a method of calculating the flow over delta wings in the leeward[B7] and the windward regions[B8]. It is a difference method. First the differential equations are differenced to a system of nonlinear equations, then this system is solved by a specific method. In [B7, 8] some results are given.

In 1967 Bazzhin and Chelysheva [B16] applied the method of lines to the conical flow over cones at large angle of attack. They calculated the windward flow over cones and elliptical cones at large angle of attack. In 1968 Jones[B18] proposed another scheme for the method of lines and calculated the conical flow over cones, elliptic cones, and other bodies at small angle of attack. In 1969 he[B19] presented a quite complete table about the conical flow over cones at small angle of attack, and within an extensive range of Mach number and vertex angle of cones, investigated the values of α/θ_c at which the governing equations become the mixed-type. Here α is the angle of attack and θ_c is the vertex angle of cones. At the present this table is the most complete and accurate. Several other works on conical flow, using the method of lines, are the following. Ndefo[B20] calculated the conical flow over cones at small angle of attack. Later he and Holt presented a number of their results in [B25]. South and Klunker[B21] calculated the windward flow over elliptic cones, conical wings with a cross section consisting of a parabolic segment or a circle, and delta wings. In [B26] they presented other results at small angle of attack. The authors of [B27], [B18] and [B21] discussed the application of the method of lines to elliptic differential equations, and presented results. Nakao[B30] calculated the conical flow at large angle of attack and roughly considered the effect of separation by modifying the body.

In 1968 Bazzhin et al. [B17] calculated the hyperbolic region in conical flow by the characteristic method, and gave some results for cones and elliptical cones at large angle of attack. Later they gathered these results and those of [B16] together in [B23], with presentation of analysis and discussion.

In 1975 Fletcher [B28, 29] calculated the conical flow over cones at large angle of attack by the method of lines and the characteristic method. He precisely determined the embedded shock and the position of the vortical singularity (the vortical singularity is detached from the body in the given results). In his calculation the method of lines is used in the windward region, the characteristic method is used in the hyperbolic region, and the modified method of lines is used in the leeward region.

From the above it can be seen that the methods, which are used for the conical flow over bodies whose shapes are close to circular cones at small angle of attack, the windward flow over bodies whose shapes are close to circular cones at large angle of attack, and the windward flow over delta wings, are of four kinds: the inverse method, the method of integral relations, the method of lines, and the difference method. In fact the first three methods are used for the conical flow after they have been successfully used for the blunt body flow. It is evident that the works on the conical flow are greatly influenced by the works on the blunt body flow. Almost all of the methods used for the latter have been used for the former. This is normal, since the two mathematical problems are similar. It is worthwhile noting that the method of lines is developed later, but it is the most successful and provides a number of results. Several people have studied the method of integral relations along the body (corresponding to the scheme I of the method of integral relations in the blunt body flow), but they only use one strip and have obtained few results with unsatisfactory accuracy.

The works about the conical flow over circular cones at zero angle of attack and about the conical flow over circular cones and bodies whose shapes are close to circular cones at small angle of attack, and concerned with the series expansion method, are listed in [B31—45].

[B31—33] by Kopal present the earliest tables of the conical flow over cones. In [B31] the results at zero angle of attack are listed. [B33] gives results obtained by Stone's first and second order expansions. Some of the results are drawn in curves in [B35]. In [B40, 41, 45] Sims presents similar results.

In [B34] Ferri points out the existence of vortical singularity in the conical flow. He uses his first order expansion method to calculate the conical flow over cones at angle of attack. In [B38, 42] Woods also uses the series expansion method to calculate the asymmetric conical flow.

Vasil'ev proposes his expansion method in [B43] and he expands the series up to second order terms. In [B44] Kurosaki outlines some good results obtained by the series expansion method.

[B36, 37, 39] present some results concerning equilibrium conical flow over cones at zero angle of attack. The results in [B37] and [B39] are quite systematic.

(ii) *The Method of Stabilization*

Besides the direct method there is the so-called method of stabilization, which can be described as follows: for a cone, no matter what its nose is, when it is long enough the flow over afterbody will be almost the same as that over a pointed cone. Therefore the calculation of the flow over a pointed cone can be reduced to that over a long cone with an arbitrary nose. That is, we can obtain a conical flow in the following way: we calculate a steady flow in the supersonic region until the flow is stabilized. Although the basic idea is as above, the variations can be allowed in its implementation.

Besides the above method, some people add the time-dependent terms to the governing equations of conical flows and stabilize the solution with time. This method has also succeeded[B58, B60].

The earliest work is that done by Babenko and Voskresenskii, in 1961[B46]. They calculated the conical flow over cones at small angle of attack by the method of stabilization. An implicit scheme is used. The equations are solved by the double-sweep method in one direction and by the iteration method in the other direction. A few results are given in [B46], but more systematic results are presented in [B47], where a table of conical flow over cones at small angle of attack is presented. Some results are discussed in [B48]. In addition, some results of conical flow over elliptical cones are also given in [B48]. In [B50] Gonidou used this method and got some results. He found that the vortical singularity is detached from the body. Later Voskresenskii used this method to calculate the windward flow over delta wings with an arbitrary cross section[B52]. He supposed that the shock is attached to the leading edge and modified the method in order to meet this physical model.

In 1965 Moretti[B49] calculated the conical flow by the method of stabilization. He used the Lax-Wendroff scheme at interior points and the characteristic method at boundary points, and obtained some results of conical flow over cones at small angle of attack. These results are given briefly in [B51].

In 1969 Beeman and Powers[B53] calculated the conical flow by the method of stabilization. They used the method of three-dimensional characteristics and calculated the windward and leeward flows over delta wings and wings with other cross sections. Later they presented more detailed results in [B56].

In 1970 Rakich[B54] calculated the conical flow over elliptic cones by another kind of method of three-dimensional characteristics.

In 1971 Kutler and Lomax[B55] calculated the flows over cones and wing-fuselage combined bodies at angle of attack. They used the MacCormack[23] scheme. They adopted Von Neumann's idea to smear out the shock and called this technique the "shock-capturing" method. Bazzhin[B60] used a similar method (the method of stabilization with time) to calculate supersonic flow over delta wings at small angle of attack.

In 1971 Lapygin[B57] calculated the flow over the V-type wings by the Rusanov scheme[20]. The shock is also smeared out.

In 1973 Bachmanova, Lapygin and Lipnitskii proposed a method of stabilization with time[B58]. They did extensive calculations of conical flow over cones with wide ranges of Mach number, vertex angle and angle of attack. They gave the position of the embedded shock in the leeward region for certain cases, therefore showed the complete picture of conical flows over cones at large angle of attack. The schemes consist of the implicit scheme used for blunt body flows, which is proposed by one of the authors[A175], and the Rusanov scheme[20]. The scheme in [A175] is a modified BVLR scheme. In their calculation a shock-separating implicit scheme is used for the flow in the windward region and the Rusanov scheme, which is not a shock-separating scheme, is used for the flow in the leeward region where the embedded shock may occur. The results are analysed in [B62].

In 1973 Ivanov and Kraiko[B59] calculated the conical flow over cones, elliptic cones and delta wings. Their method was proposed previously by the authors in [24, 25], and is a version of the Godunov method.

In 1975 Kosykh and Minailos[B61] calculated the conical flow by the method of stabilization. In their calculation the shock-capturing technique is used. The adopted scheme is of almost second order accuracy in space.

In 1977 Daywitt and Anderson[B63] calculated the supersonic flow over cones at large angle of attack. They adopted the method of stabilization with t in both the windward and leeward regions. The floating-shock-fitting technique[C115, 129] was used, the MacCormack scheme[23] being used at the interior points, and the method of Abbett[C51, 54], Kentzer[61] et al. being used at the boundary. They gave the positions of the embedded shock and the vortical singularity. When the "spherical velocity" was greater than the local sound speed, the calculation was carried out along the φ direction.

Thus there are a number of works and a number of results on the method of stabilization. As we see, the method of stabilization with time is similar to the method for blunt body flow, and the method of stabilization with length is similar to the method for the supersonic region. Brief comments on these methods can be found in subsections 1 and 3.

3. *Calculation of Flow in Supersonic Regions*

In Special References C we list papers on calculation of flow in supersonic regions. Papers [C1—59] discuss two-dimensional or axisymmetric flow, and papers [C60—159] discuss three-dimensional flow. The status of two-dimensional problems is first reviewed, and the status of three-dimensional problems is reviewed thereafter.

(i) *Two-dimensional Flow*

In 1946 Ferri[C1] computed supersonic rotational flow by the method of characteristics, which was later described in detail in [35].However, little computation was done in his paper. In 1959 Casaccio[C2] obtained several results of flow around a sphere-cylinder at $M_\infty=20$ by using high-speed computers. After that, a series of papers on the method of characteristics appeared.

In 1959 Ehlers[C3] investigated the problem of the application of the method of characteristics adapted to high-speed computers and obtained some results. Later the problem of equilibrium flow of an ideal dissociating gas and the problem of nonequilibrium flow were considered[C25, 30, 35], and the corresponding code written in FORTRAN was offered [C25, 35]. However, he only presented a few results:

In 1960 Chushkin[C4] obtained several results on blunt wedges and cones for several Mach numbers and vertex angles using the method of characteristics. Later a series of tables on computed results of blunt cones were presented[C9], and analysis of computed results was done[C10, 15]. In [C8] Katskova, Haumova et al. described systematically the method of characteristics which was developed in the Computing Center, Academy of Sciences, U. S. S. R. in the leter Fifties. In 1963 Katskova and Kraiko[C22, 27] extended this method to the case of nonequilibrium flow, and investigated the calculation of equations of physical-chemical reactions for the gas close to equilibrium state. Later Dushin[C38] calculated nonequilibrium flow by a method based on the scheme in [C8], and presented a scheme for the flow close to equilibrium. Moreover, Haumova[C23, 28] computed equilibrium flow by using the scheme in [C8], and Fomin and Shulishnina[C40] extended it to the flow with radiation.

In 1961 Capiaux[C5] et al. computed the flow fields around sphere-cylinders and sphere-cone-cylinders by the method of characteristics, and the flow field near an expansion corner was accurately calculated. Later[C16] nonequilibrium flow was considered.

In 1961 Kennedy[C6] et al. computed the flow fields of the perfect gas and an equilibrium gas around a blunted 9° cone by the method of characteristics. Vaglio-Laurin[C7] et al. calculated the flow of equilibrium gases around sphere-cone-cylinders.

In 1962 Inouye and Lomax[C11] dealt with flows around various bodies, including the bodies with expansion corners, by the method of characteristics. Later[C32] they further calculated the flow field around bodies with compression corners.

In 1962 Sedney[C12] et al. applied the method of characteristics to calculation of nonequilibrium flow. Later[C19, 36] Sedney and his colleagues did some research on the calculation of nonequilibrium flow by using the method of characteristics.

In 1962 Arkhinov and Khoroshko[C14] also calculated nonequilibrium flow using the method of characteristics.

In 1963 Powers[C17] et al. computed flow fields of equilibrium gas around cone–cylinders. Eastman[C18] et al. calculated flow around a cone–cylinder–flare, making certain treatments to the secondary shocks. Roslyakov[C24, 43] et al. also calculated flow fields around bodies with expansion corners and compression corners, and considered the intersection between two shocks belonging to the same family. However, they have not yet computed the down–stream flow field behind the point of intersection. In the paper [C42] they summarize works on this subject done at Moscow University. In addition, Thompson[C48] et al. also computed the flow fields around bodies with both expansion corners and compression corners.

Lunev[C37] calculates flow fields around sphere–cones by the method of characteristics. Later Znamenskii [C58] computed more complicated flow field past bodies by the method of characteristics.

Besides the above work on "the natural method of characteristics", in 1964 Katskova and Chushkin[C26] presented the net–characteristic method and computed several results. Later Chushkin[C47] applied the mothod to calculation of a combustible gas flow. However, their main purpose was to calculate three–dimensional problems by using that type of method. Therefore, they calculated a few problems in two–dimensions. Zapryanov[C39] also proposed a net–characteristic method. The only difference from the paper [C26], in which three–point–interpolation is used, is the use of a high–order–interpolation at the initial level. The scheme in [C39] is a special case of the scheme for three–dimensional problems given by the same author in an earlier paper [C75].

In addition to the method of characteristics, there are the method of integral relations and the difference method.

In 1962 Belotserkovskii and Chushkin discussed the application of the method of integral relations to calculation of flow in supersonic region in the comprehensive paper [C64] on the method of integral relations. However, no computation was done in [C64]. In 1962 Traugatt[C13] computed the flow in supersonic region by using the method of integral relations. He used only one strip in his calculation both in the subsonic and the supersonic regions. In addition, South and Newman[C20, 29, 31, 33], Chushkin and Li Likang[C34] also worked in this direction. In [C20, 31, 33] the authors have focused their attention on the flow of nonequilibrium gas, while in [C34] they focus on the flow around slender bodies. Hence each scheme has its own features.

In 1963 D'yakonov and Zaitseva[C21] solved a number of problems in two dimensions by using the difference method which was developed by Babenko[C62] et al. for calculation of three–dimensional problems. Later, D'yakonov[C41] et al. calculated several problems of equilibrium flow by using the same method. Moreover, this method was extended to calculation of nonequilibrium flow by Severinov[C44]. Rusanov[C45] computed flow fields of several slender bodies, including a slender sphere–cone with a length of 10,000 radii of the sphere,and he investigated asymptotic properties of those flow fields. In his calculations, the values on the body were modified. In addition, Rusanov and Lyubimov[C50] investigated the formation of internal shocks by using this method. Lebedev[C55], Vishnevetskii[C56] et al. also obtained some results for two-dimensional or axisymmetric supersonic flow by this method.

Kutler and Lomax[C49], Walkden and Caine[C52], Moretti[C53, 57] et al. have also worked on difference methods. In [C49] and [C52] discontinuities are smeared, i.e., the shock–capturing methods are used. In [C53] Moretti discusses how to avoid oscillations of results in an entropy layer. In [C57] the floating–shock–fitting method for two–dimensional unsteady flow is proposed, and certain embedded shocks in two–dimensional steady supersonic flow are presented. Abbett[C51, 54] studies the calculation of boundary points, and tests a number of methods for the boundary points. His conclusion is that the method of characteristics and the method of simple waves are the best. (In the method of simple waves the values are first obtained by using a difference scheme, and are then modified by the theory of simple waves.) In [C59], Rusanov presents a third order scheme for calculation of boundary points, and also does some tests.

We can see from the papers cited above that for two–dimensional problems, the method of characteristics has been widely used. It is used not only for bodies with simple shape, but also for bodies with expansion corners[C5, 7, 11, 17, 18, 24, 32, 43, 48] or with compression corners[C18, 32, 43, 48]. However, there are not many works on the method of integral relations and difference methods. In addition, almost every difference scheme is a special case of a scheme for three–dimensional problems.

(ii) *Three–dimensional Flow*

In 1960 Moretti[C60] presented a method for three-dimensional supersonic flow around bodies. In his paper certain relationships on intersection lines of characteristic surfaces and meridian planes are used to determine the values at mesh–points on a series of meridian planes. However, no results are given in that paper. In a later paper [C63] several results of flow around sphere–cones and blunt flat–plate delta wings at small angles of attack are given. In [C65] Moretti briefly summarizes these results. In addition, this method has been applied in the calculation of lower supersonic flow around bodies by Hsieh[C140].

In 1961 Fowell[C61] did some computation on rotational flow with a shock by the method of bicharacteristic–tetrahedrons. But Saverwein[C69] et al. pointed out that the scheme is not stable.

In 1961 Babenko and Voskresenskii[C62] suggested an implicit scheme for calculation of steady supersonic flow, and computed flows around cones and pointed bodies of revolution. In their method the difference equations are solved by the double-sweep method in one direction and by the iteration method in the other direction. A summary of that method is given in [C66], and several other results are given and analysed in [C71]. A detailed discussion about that method is given in [C72]: the "condition" of the system of difference equations, the stability of the double-sweep method, and the stability of the difference scheme for initial-boundary-value problems are discussed; calculation of equilibrium gas flow and conical flow is considered; and a large number of results, mainly, a large number of results of flow around circular cones at small angles of attack, are presented and analysed. In addition, several results of pointed bodies are given in [C86]. In [C91] and [C98] Rusanov and Lyubimov also present and analyse several results of blunt bodies. Later a book on flow around blunt bodies was published[C108], in which results of various flows, for example, equilibrium air flow around bodies at zero angle of attack, flow around blunt cones at angle of attack, and flow around bodies with analytic generators at angle of attack are given and analysed. For equilibrium air flow, in addition to pressure, density and velocity, concentrations of species of equilibrium air are given. In that book, the length of bodies is usually about 50, and sometimes greater than 50 radii of the nose.

Besides Babenko, Voskresenskii, Lyubimov and Rusanov of the Institute of Applied Mathematics, Academy of Sciences of the U.S.S.R., D'yakonov and his colleagues at Moscow University have computed supersonic flow around bodies by using the above method. In [C73, 74] they calculate equilibrium air and perfect gas flows around blunt cones. In [C106] the scheme is slightly modified by using a special double-sweep method in which the direct sweep is carried out at different directions in different regions. In [C107], that method is applied to the calculation of flow around bodies with expansion edges and compression edges. In [C120] they compute flow around bodies at large angle of attack after slightly modifying that method. Also, Antonets[C104] calculates flow around complicated bodies and flow with contact discontinuities by using a modified scheme. Later he applies that method to nonequilibrium flow[C136] and flow around bodies with elliptic cross sections[C149]. In addition, Burdel'nyi and Minostsev[C119] compute nonequilibrium flow around bow-shaped bodies at angle of attack using a modified scheme. The scheme is modified in the same way as Mikhailov et al. do in [A175]: partial differential equations are first rewritten in the characteristic form; then for the wave-characteristic compatibility relations the "four-point" scheme is still adopted on the z-ξ plane, and for the stream-characteristic compatibility relations a "six–point" scheme is adopted on the z–ξ plane.

Among methods applicable for supersonic flow around bodies, this method has the strongest theoretical basis. Moreover, a large number of results are obtained by using it.

In paper [C64] the authors suggest that the method of integral relations be applied to three–dimensional supersonic flow. That suggestion was realized by Chushkin[C94] in 1968. The basic idea of that method is as follows: In the φ–direction, trigonometric polynomials are used, and in the ξ–direction, polynomials in ξ are used. However, the integration is used only in the ξ–direction. Thus, strictly speaking, that scheme is a combination of the integral relation method and the line method.

In 1963 Mikhailov[C67] investigated the structure of the flow field near a spatial supersonic singular line by using the series expansion. Later the results on singular lines were applied in numerical calculation.

Burnat et al. present a method called the characteristic–tetrahedron method in [C68]. However, no results are given in that paper.

In 1964 Zapryanov and Minostsev[C75] proposed a type of characteristic method. The main feature of that scheme is to use the backward characteristic–tetrahedrons and to adopt a "through approximation" in every initial plane. The idea of "through approximation" comes from the method of lines, and is suitable to the case where functions are quite smooth. In [C87] Minostsev gave the details of the method, and corrected the. method for shock–points, using one compatibility relation instead of two. He also presented several results for smooth bodies, including a bow–shaped body. Those results are analysed in [C93].

In [C76] and [C26] Chushkin considers the possibility of extending the scheme of the paper [C26] to three–dimensional problems, which was realized by Katskova and Chushkin[C78] in 1965. They computed equilibrium air flow. Later [C88, 95, 101, 102, 105] the scheme was further applied to various cases, including nonequilibrium gas[C95, 105] and combustible gas [C101, 102]. In the scheme all functions are approximated by polynomials in the φ–direction. Thus the problem is reduced to a system in two independent variables which can be solved by using the net–characteristic method. A similar method is presented by Camarero[C147]. In that method the Chebyshev approximation is used in the ξ–direction, and the spline interpolation is used in the φ–direction.

Pridmore Brown and Franks[C77] outline a method called the characteristic–pentahedron method, and also present some tests.

In 1965 Podladchikov [C79, 80] presented a characteristic–tetrahedron method. The way of marching is similar to the way of taking the second family of characteristics as marching–curves in two–dimensional problems. He obtained several results, including results of flow fields around bodies with expansion edges.

In 1966 Magomedov[C81] also proposed a type of characteristic method, in which all quantities at interior points are determined by certain relations along backward stream lines and bicharacteristics. He pointed out that there are some problems if two wave-compatibility relations are used for a shock point, and doubted the correctness of the algorithms for shock-points in the papers [C61, 75, 80]. He presented an algorithm for shock-points, using only one wave-compatibility relation, and carried out some theoretical analysis on the methods for shock-points [C82]. Later [C89] he computed equilibrium gas flow by a slightly modified scheme, gave several results of blunt cones at angle of attack, and showed the possibility of the appearance of a shock on the leeward side of the flow field around a cylinder at angle of attack. He also computed flow around blunt delta wings[C90]. In 1968 Magomedov, Lunev et al. wrote a book [C92], in which several tables of results of equilibrium gas flow fields around blunt cones at angle of attack are given. The angles of attack are usually five degrees, the lengths of bodies are usually 20—30 radii of the sphere, and the maximal length is about 44.

In 1967 Chu[C83] presented a type of characteristic method. The flow parameters at new mesh-points were determined by "the least square method of compatibility relations" instead of certain compatibility relations. When his scheme was derived, the method in paper [C70] for obtaining compatibility relations of a system of partial differential equations was used. In paper [C83] results of blunt cone flow fields are shown. In papers [C111, 117, 126] the scheme is slightly modified, and results of flow fields around swept-back delta wings, space shuttles and slab delta wings are presented.

In 1967 Rakich [C84] calculated sphere-cone flow fields by a method of characteristics, which is similar to the scheme in the paper [C78]. In paper [C96] detailed formulae are presented, and the entropy layers are computed after introducing artificial viscosity. In later papers, flow fields around noncircular bodies at angle of attack [C97], flow fields around space shuttles[C114, 118] and nonequilibrium flow[C128, 139] are computed. In addition, Lewis, Black et al.[C112, 125] have computed flow fields around bodies in nonuniform freestreams by using this method.

In 1967 Strom [C85] also presented a method of characteristics, in which the intersection points of stream lines with marching-surfaces are taken as mesh-points. When an interior point is calculated, four bicharacteristics are used in the following way. They are divided into two groups in a certain way, either of which contains three bicharacteristics. Two groups of values of unknowns at each interior point can be obtained by using the two groups of bicharacteristics. Then the averages are taken as final solutions. The algorithms for body points and shock points are similar. Some results are given in his paper.

In 1969 Magomedov and Kholodov[C99] presented an explicit scheme. In the process of constructing the scheme, discretized compatibility relations are used in order to eliminate "space derivatives" of unknowns. Both smooth body flow fields and sphere–cylinder–flare flow fields are computed. Kosarev[C109] applies the scheme to computation of flow fields containing embedded shocks. The result is not good if the embedded shock is strong. In paper [C130] a corresponding conservative scheme is constructed, and some results on embedded shocks are presented.

In 1969 Grigor'ev and Magomedov[C100] proposed a method called the method of characteristic surfaces. It is a direct extension of the method of characteristics in two-dimensional problems to three-dimensional problems. In this method, the characteristic surfaces take the place of the characteristic curves in two-dimensional problems. The flow fields around bodies with expansion edges at angle of attack have been computed by using this method.

In 1971 Kutler and Lomax[C103] computed steady supersonic flow fields by a "shock-capturing" finite difference method, the adopted scheme being the MacCormack scheme[23]. The method was further developed in [C114, 116, 118, 121—123]. Moreover, a third-order non-centered difference scheme was presented by them and by Warming[C116,121]. In that method the values of unknowns at body points are first predicted by difference schemes, and are then corrected by the compression wave and expansion wave theory. In addition, the reflection theory is also occasionally used. Later Shankar, Kutler, Sanders, Chaussee et al.[C137,138,141,145,148] applied the scheme to the computation of inviscid supersonic corner flows and several other flows. Also, Schiff[C113] computed supersonic flow fields past bodies in coning motion by using the scheme.

The MacCormack scheme has been applied to the shock-fitting method as well as to the shock-capturing method. Concerning this subject, we may cite the following works:

(1) Thomas [C110] et al. applied the MacCormack scheme to computation of interior points. However, they used a special algorithm for boundary points: in order to determine the values on boundary points, they chose a linearly independent subset of equations from the boundary conditions and the partial differential equations. For body points, they used the boundary conditions on bodies and several combinations of partial differential equations. For shock points, they used the jump conditions on shocks and one partial differential equation. Kutler, Davy and others[C141,142] computed the main shock by the same algorithm. (It

was a nonequilibrium flow that Davy and others[C142] computed, and for the equations of chemical reactions, the implicit schemes were adopted.) Solomon and others[C150] also used that algorithm for practical calculation. However, they found that some nonphysical oscillations appear if flow fields around long bodies were computed. In [C150], besides the Thomas algorithm, the method of characteristics was used. For the method of characteristics, nonphysical oscillations did not appear.

(2) Moretti[C115] et al. also applied the MacCormack scheme to the computation of interior points. In order to determine the values on shock points they used the jump conditions on shocks and one wave-compatibility condition, as they had done in the computation of conical flows by the method of stabilization. However, a small modification was made——they adopted a predictor-corrector scheme. For body points a similar algorithm was used. In [C115] there are two points worth noticing: ① the authors have accurately dealt with embedded shocks, and ② the conformal transformation is used when they arrange mesh-points, which improve results in the region near the leading edge of a wing. The method of [C115] is further developed in [C124], where the flow field around an aeroplane is computed. On the basis of [C115] and [C124], Moretti[C129] developed the floating-shock-fitting technique, which was later used to compute complicated flow fields by Moretti, Marconi and others[C133,143]. In [C143] the intersection between two shocks was computed. Unfortunately, this problem has not yet been solved completely——the contact discontinuity and the reflected expansion wave have not been considered.

The splitting method was used by MacCormack, Warming, Rizzi and others[C127,144,146]. In [C146], where nonequilibrium flow is considered, not only the split of space but also the split of fluid dynamics equations and chemical reaction equations are adopted, and an implicit scheme is used for chemical reaction equations. In addition, the surface-marching technique and the mesh-fitting technique for shock points are presented in the above papers.

Walkden[C132,135]and others presented a scheme,which was an extension of the Courant-Isaacson-Rees scheme[16],and they computed the flow fields around delta-wings without using the shock-fitting technique.

D'attorro[C134] and others proposed their own two-step L-W scheme, and used it to calculate flow fields around aeroplanes.

The papers [C151—155] are earlier work on the calculation of steady supersonic flow fields. In these papers either only irrotational flow is considered or only methods are given. Thus we do not introduce their contents, only listing them at the end of the Reference C.

From what we have mentioned above, we can see that there are a number of methods for three-dimensional supersonic flow field computation, which can be grouped in the method of "characteristics", the method of integral relations, and the difference method. The method of "characteristics" contains the method of characteristics on reference planes, the characteristic - tetrahedron method, the characteristic - pentahedron method, the characteristic–multihedron method and the characteristic–surface method. The difference method can further be divided into the shock–capturing method and the shock–fitting method. In the former, shocks are smeared in computed results, and in the latter, shocks are accurately determined in computation.

Both the system of steady supersonic flow equations and the system used for subsonic–transonic flow computation by the time–dependent method are hyperbolic. Therefore research work on numerical methods for the two systems has several common points. Of course, they have their own features. The first feature of research work on numerical methods for steady supersonic flow is that the percentage of papers on the method of "characteristics" is quite large. The reason for this is probably that the method of characteristics is widely used in computation of two–dimensional supersonic flow, so a number of scientists try to extend the method of characteristics to three–dimensional problems. The second feature is that the shock–fitting difference method is greatly developed. The work of Babenko and his colleagues and the work of Moretti and others are the most important works on this subject. The prominent characteristic of the work of Babenko and others is that it has quite a strong theoretical basis and has been widely applied. The prominent characteristic of the work of Moretti and others is that they have computed complicated flow fields with several shocks successfully, and have developed the floating–shock–fitting technique. The third feature is that many algorithms for boundary points have been tested. The tests show that the algorithms based on the characteristic theory or implicitly based on it give good results, and that other algorithms often, but not always, give somewhat unreasonable results.

Chapter 4

Inviscid Steady Flow

§ 1 The System of Fundamental Differential Equations and Its Characteristics

1. *Differential Equations*

Let us consider steady flow, and assume that the medium is a gas whose viscosity and heat conductivity are negligible. The external force acting on the gas will also be neglected. Under these conditions we can derive the basic flow equations in the integral form as follows[36]:

$$\begin{cases} \displaystyle\iint_\sigma \rho V_n \boldsymbol{V} d\sigma + \iint_\sigma p\boldsymbol{n}\, d\sigma = 0, \\ \displaystyle\iint_\sigma \rho V_n d\sigma = 0, \\ \displaystyle\iint_\sigma \rho V_n \left(e + \frac{|\boldsymbol{V}|^2}{2}\right) d\sigma + \iint_\sigma pV_n d\sigma = \iint_\sigma \rho V_n \left(h + \frac{|\boldsymbol{V}|^2}{2}\right) d\sigma = 0. \end{cases} \tag{1.1}$$

Here σ is any closed surface in space; $\boldsymbol{V}$ is the velocity vector of the moving gas, p the pressure, ρ the density, e the specific internal energy (i.e., the internal energy per unit mass of gas), and $h = e + \frac{p}{\rho}$ the specific enthalpy; $\boldsymbol{n}$ is the unit outward normal to the surface σ; $V_n = \boldsymbol{V}\cdot\boldsymbol{n}$ is the normal component of the velocity.

It is easy to see that these equations represent the laws of conservation of momentum, mass and energy in steady flow. In fact, the quantity $\rho V_n d\sigma$ is just the transport of mass of the gas across the surface element $d\sigma$ per unit time ($V_n < 0$: inflow; $V_n > 0$: outflow). Therefore the second equation represents the law of conservation of mass, that is, net mass flux through any fixed closed surface in space is equal to zero. Similarly, the first equation states that the rate of change of momentum in the volume enclosed by the surface σ equals the resultant of pressure forces acting on the surface, and the third one states that the increase of internal energy and kinetic energy in the volume equals the work done by pressure forces on it over time. In other words, the equations represent the conservation of momentum and of energy for steady flow.

Using the Gauss formula, from (1.1) we can derive the following equations in the differential form:

$$\begin{cases} \boldsymbol{V}\cdot\nabla\boldsymbol{V}+\dfrac{1}{\rho}\nabla p=0, \\ \boldsymbol{V}\cdot\nabla\rho+\rho\nabla\cdot\boldsymbol{V}=0, \\ \boldsymbol{V}\cdot\nabla\left(h+\dfrac{|\boldsymbol{V}|^2}{2}\right)=\boldsymbol{V}\cdot\nabla h-\dfrac{1}{\rho}\boldsymbol{V}\cdot\nabla p=0. \end{cases} \tag{1.2}$$

Furthermore, if the enthalpy can be expressed as a function of pressure and density, i.e.,

$$h=h(p, \rho), \tag{1.3}$$

then by using the identity

$$\boldsymbol{V}\cdot\nabla h-\frac{1}{\rho}\boldsymbol{V}\cdot\nabla p=\frac{\partial h}{\partial\rho}\boldsymbol{V}\cdot\nabla\rho+\left(\frac{\partial h}{\partial p}-\frac{1}{\rho}\right)\boldsymbol{V}\cdot\nabla p$$

we can rewrite the third equation of (1.2) in the form

$$\boldsymbol{V}\cdot\nabla p-a^2\boldsymbol{V}\cdot\nabla\rho=0,$$

where

$$a^2=-\frac{\partial h}{\partial\rho}\Big/\left(\frac{\partial h}{\partial p}-\frac{1}{\rho}\right). \tag{1.4}$$

The positive quantity a is usually referred to as the speed of sound. Thus, under the assumption of (1.3), the fundamental differential equations take the following form:

$$\begin{cases} \boldsymbol{V}\cdot\nabla\boldsymbol{V}+\dfrac{1}{\rho}\nabla p=0, \\ \boldsymbol{V}\cdot\nabla\rho+\rho\nabla\cdot\boldsymbol{V}=0, \\ \boldsymbol{V}\cdot\nabla p-a^2\boldsymbol{V}\cdot\nabla\rho=0, \end{cases} \tag{1.5}$$

which is a closed system of equations for the basic flow properties $\boldsymbol{V}$, p, and ρ. The relation (1.3) is applicable to a perfect gas as well as to an equilibrium gas (see § 4).

However, the assumption (1.3) is no longer true for a non-equilibrium gas. In fact, for a gas mixture with multiple components, when non-equilibrium chemical reactions and vibrational relaxation are involved, we should use, instead of (1.3), a more general relation:

$$h=h(p, \rho, c_1, \cdots, c_{m-5}), \tag{1.6}$$

where c_i is the concentration of the i-th component in the mixture, $i=1, 2, \cdots, m-5$. Here we assume that the mixture consists of $m-5$ components. Unlike the gas mixture in a state of equilibrium, h now cannot be determined only by p and ρ. To complete the system we shall add the following equations:

$$\boldsymbol{V}\cdot\nabla c_i=-d_{i+5}, \quad i=1, 2, \cdots, m-5 \tag{1.7}$$

(for details, see § 5). Then, since

$$\boldsymbol{V}\cdot\nabla h=\frac{\partial h}{\partial p}\,\boldsymbol{V}\cdot\nabla p+\frac{\partial h}{\partial \rho}\,\boldsymbol{V}\cdot\nabla \rho+\sum_{i=1}^{m-5}\frac{\partial h}{\partial c_i}\,\boldsymbol{V}\cdot\nabla c_i$$
$$=\frac{\partial h}{\partial p}\,\boldsymbol{V}\cdot\nabla p+\frac{\partial h}{\partial \rho}\,\boldsymbol{V}\cdot\nabla \rho-\sum_{i=1}^{m-5}\frac{\partial h}{\partial c_i}\,d_{i+5},$$

the third equation of (1.2) can be reduced to

$$\boldsymbol{V}\cdot\nabla p-a^2\boldsymbol{V}\cdot\nabla\rho=-d_5,$$
$$d_5\equiv\frac{-1}{\dfrac{\partial h}{\partial p}-\dfrac{1}{\rho}}\sum_{i=1}^{m-5}\frac{\partial h}{\partial c_i}d_{i+5},$$

where a^2 is still defined by (1.4), and h is defined by (1.6). Thus, differential equations for the flow properties $\boldsymbol{V}$, p, ρ, c_1, $\cdots$, c_{m-5} now become

$$\begin{cases}\boldsymbol{V}\cdot\nabla\boldsymbol{V}+\dfrac{1}{\rho}\,\nabla p=0,\\ \boldsymbol{V}\cdot\nabla\rho+\rho\nabla\cdot\boldsymbol{V}=0,\\ \boldsymbol{V}\cdot\nabla p-a^2\boldsymbol{V}\cdot\nabla\rho=-d_5,\\ \boldsymbol{V}\cdot\nabla c_i=-d_{i+5},\quad i=1,\,2,\,\cdots,\,m-5.\end{cases}\tag{1.8}$$

Note that the equations (1.5) are a special case of the equations (1.8), in which $d_{i+5}=0$, $i=1,\,2,\,\cdots,\,m-5$.

In a system of orthogonal curvilinear coordinates x_1, x_2, x_3, the equations (1.8) take the following form:

$$\begin{cases}\dfrac{u_1}{H_1}\dfrac{\partial u_1}{\partial x_1}+\dfrac{u_2}{H_2}\dfrac{\partial u_1}{\partial x_2}+\dfrac{u_3}{H_3}\dfrac{\partial u_1}{\partial x_3}+\dfrac{1}{\rho H_1}\dfrac{\partial p}{\partial x_1}+d_1=0,\\ \dfrac{u_1}{H_1}\dfrac{\partial u_2}{\partial x_1}+\dfrac{u_2}{H_2}\dfrac{\partial u_2}{\partial x_2}+\dfrac{u_3}{H_3}\dfrac{\partial u_2}{\partial x_3}+\dfrac{1}{\rho H_2}\dfrac{\partial p}{\partial x_2}+d_2=0,\\ \dfrac{u_1}{H_1}\dfrac{\partial u_3}{\partial x_1}+\dfrac{u_2}{H_2}\dfrac{\partial u_3}{\partial x_2}+\dfrac{u_3}{H_3}\dfrac{\partial u_3}{\partial x_3}+\dfrac{1}{\rho H_3}\dfrac{\partial p}{\partial x_3}+d_3=0,\\ \dfrac{\rho}{H_1}\dfrac{\partial u_1}{\partial x_1}+\dfrac{\rho}{H_2}\dfrac{\partial u_2}{\partial x_2}+\dfrac{\rho}{H_3}\dfrac{\partial u_3}{\partial x_3}+\dfrac{u_1}{H_1}\dfrac{\partial \rho}{\partial x_1}+\dfrac{u_2}{H_2}\dfrac{\partial \rho}{\partial x_2}\\ \quad+\dfrac{u_3}{H_3}\dfrac{\partial \rho}{\partial x_3}+d_4=0,\\ \dfrac{u_1}{H_1}\dfrac{\partial p}{\partial x_1}+\dfrac{u_2}{H_2}\dfrac{\partial p}{\partial x_2}+\dfrac{u_3}{H_3}\dfrac{\partial p}{\partial x_3}\\ \quad-a^2\left(\dfrac{u_1}{H_1}\dfrac{\partial \rho}{\partial x_1}+\dfrac{u_2}{H_2}\dfrac{\partial \rho}{\partial x_2}+\dfrac{u_3}{H_3}\dfrac{\partial \rho}{\partial x_3}\right)+d_5=0,\\ \dfrac{u_1}{H_1}\dfrac{\partial c_i}{\partial x_1}+\dfrac{u_2}{H_2}\dfrac{\partial c_i}{\partial x_2}+\dfrac{u_3}{H_3}\dfrac{\partial c_i}{\partial x_3}+d_{i+5}=0,\\ \qquad i=1,\,2,\,\cdots,\,m-5,\end{cases}\tag{1.9}$$

where u_1, u_2 and u_3 are the projections of the velocity vector $\boldsymbol{V}$ in the x_1, x_2 and x_3–coordinate line directions respectively; and H_1, H_2 and H_3 are the Lamé coefficients corresponding to the x_1, x_2 and x_3–coordinates; and

$$\begin{cases} d_1 = \dfrac{u_2}{H_1H_2}\left(u_1\dfrac{\partial H_1}{\partial x_2} - u_2\dfrac{\partial H_2}{\partial x_1}\right) + \dfrac{u_3}{H_1H_3}\left(u_1\dfrac{\partial H_1}{\partial x_3} - u_3\dfrac{\partial H_3}{\partial x_1}\right), \\ d_2 = \dfrac{u_3}{H_2H_3}\left(u_2\dfrac{\partial H_2}{\partial x_3} - u_3\dfrac{\partial H_3}{\partial x_2}\right) + \dfrac{u_1}{H_2H_1}\left(u_2\dfrac{\partial H_2}{\partial x_1} - u_1\dfrac{\partial H_1}{\partial x_2}\right), \\ d_3 = \dfrac{u_1}{H_3H_1}\left(u_3\dfrac{\partial H_3}{\partial x_1} - u_1\dfrac{\partial H_1}{\partial x_3}\right) + \dfrac{u_2}{H_3H_2}\left(u_3\dfrac{\partial H_3}{\partial x_2} - u_2\dfrac{\partial H_2}{\partial x_3}\right), \\ d_4 = \dfrac{\rho}{H_1H_2H_3}\left(u_1\dfrac{\partial H_2H_3}{\partial x_1} + u_2\dfrac{\partial H_3H_1}{\partial x_2} + u_3\dfrac{\partial H_1H_2}{\partial x_3}\right). \end{cases} \tag{1.10}$$

These equations can be obtained directly from (1.8) through the use of the expressions of the gradient of a scalar function and of the divergence of a vectorial function in the x_1, x_2, x_3–coordinate system.

The equations (1.9) can be written in the vectorial form

$$A_1\frac{\partial \boldsymbol{U}}{\partial x_1} + A_2\frac{\partial \boldsymbol{U}}{\partial x_2} + A_3\frac{\partial \boldsymbol{U}}{\partial x_3} + \boldsymbol{D} = 0, \tag{1.11}$$

where

$$A_j = \frac{1}{H_j}\begin{pmatrix} u_j & 0 & 0 & \delta_{1,j}/\rho & 0 & & & & \\ 0 & u_j & 0 & \delta_{2,j}/\rho & 0 & & \mathbf{0} & & \\ 0 & 0 & u_j & \delta_{3,j}/\rho & 0 & & & & \\ \rho\delta_{1,j} & \rho\delta_{2,j} & \rho\delta_{3,j} & 0 & u_j & & & & \\ 0 & 0 & 0 & u_j & -a^2u_j & & & & \\ & & & & & u_j & & & \\ & \mathbf{0} & & & & & u_j & & \\ & & & & & & & \ddots & \\ & & & & & & & & u_j \end{pmatrix},$$

$$\boldsymbol{U} = \begin{pmatrix} u_1 \\ u_2 \\ u_3 \\ p \\ \rho \\ c_1 \\ c_2 \\ \vdots \\ c_{m-5} \end{pmatrix} \equiv \begin{pmatrix} u_1 \\ u_2 \\ u_3 \\ u_4 \\ \vdots \\ u_m \end{pmatrix}, \quad \boldsymbol{D} = \begin{pmatrix} d_1 \\ d_2 \\ d_3 \\ d_4 \\ \vdots \\ d_m \end{pmatrix}.$$

(1.11) are usually called the inviscid steady flow equations. As indicated above, by putting the values $m=5$ and $d_5=0$ in (1.11), we immediately obtain the vectorial form of (1.5) in an orthogonal curvilinear coordinate system.

Now we will consider several commonly used systems of orthogonal coordinates and write the expressions for d_1, d_2, d_3 and d_4.

(i) The cylindrical coordinates z, r, φ

Setting $x_1=z$, $x_2=r$ and $x_3=\varphi$, we have

$$H_1=1, \quad H_2=1, \quad H_3=r.$$

If we use u, v and w in place of u_1, u_2 and u_3 respectively, then, according to (1.10), we have

$$\begin{cases} d_1=0, \\ d_2=-\dfrac{w^2}{r}, \\ d_3=\dfrac{vw}{r}, \\ d_4=\dfrac{\rho v}{r}. \end{cases} \tag{1.12}$$

(ii) The spherical coordinates R, θ, φ

Let $x_1=R$, $x_2=\theta$ and $x_3=\varphi$. The Lamé coefficients are

$$H_1=1, \quad H_2=R, \quad H_3=R\sin\theta.$$

Again, using the notation u, v and w instead of u_1, u_2 and u_3, we have

$$\begin{cases} d_1=-\dfrac{v^2+w^2}{R}, \\ d_2=\dfrac{1}{R}(uv-w^2\operatorname{ctg}\theta), \\ d_3=\dfrac{1}{R}(uw+vw\operatorname{ctg}\theta), \\ d_4=\dfrac{\rho}{R}(2u+v\operatorname{ctg}\theta). \end{cases} \tag{1.13}$$

(iii) The "boundary-layer" coordinates S, n, φ

Suppose that there exists a surface of revolution with non-concave and smooth generators and an axis of symmetry called the z-axis. We now define the boundary-layer coordinate system as follows. Let P be any point in space. Denote by Φ the plane through the axis z and the point P. On this plane we draw through P a straight line which is normal to the generator and which intersects it at the point p, as shown in Fig. 1.1. Denote the arc $\overset{\frown}{OP}$ by S and the directed segment $\overline{pP}$ by n. Let φ denote the directed angle between the plane Φ and a fixed plane through the axis z. Then we define the numbers S, n and φ as the coordinates of the point P. It is easy to see that a system of coordinates so defined is an orthogonal curvilinear coordinate system, and the corresponding Lamé coefficients are

$$H_1=1+\frac{n}{R}, \quad H_2=1, \quad H_3=r,$$

where $R=R(S)$ is the radius of curvature of the generator at the point $(S, 0, \varphi)$ and $r=r(S, n)$ is the distance from the point (S, n, φ) to the

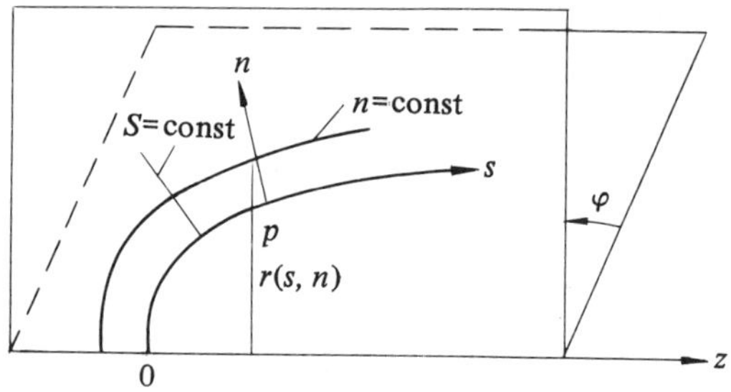

Figure 1.1 The "boundary-layer" coordinates

axis z. Thus, according to (1.10), we have

$$\begin{cases} d_1 = \dfrac{uv}{\left(1+\dfrac{n}{R(S)}\right)R(S)} - \dfrac{w^2}{\left(1+\dfrac{n}{R(S)}\right)r}\dfrac{\partial r}{\partial S}, \\ d_2 = -\dfrac{w^2}{r}\dfrac{\partial r}{\partial n} - \dfrac{u^2}{1+\dfrac{n}{R(S)}}\dfrac{1}{R(S)}, \\ d_3 = \dfrac{uw}{\left(1+\dfrac{n}{R(S)}\right)r}\dfrac{\partial r}{\partial S} + \dfrac{vw}{r}\dfrac{\partial r}{\partial n}, \\ d_4 = \dfrac{\rho}{\left(1+\dfrac{n}{R(S)}\right)r}\left[u\dfrac{\partial r}{\partial S} + v\left(\dfrac{r}{R(S)} + \left(1+\dfrac{n}{R(S)}\right)\dfrac{\partial r}{\partial n}\right)\right], \end{cases} \tag{1.14}$$

where u, v and w are the S, n and φ-components of the flow velocity vector respectively.

Incidentally, we notice that both the cylindrical and the spherical coordinate systems are special cases of the "boundary-layer" coordinate system.

For practical applications, it is sometimes more convenient to use some other set of variables as the independent variables in the differential equations. Now we represent the fundamental differential equations in an arbitrary set of independent variables, say ξ_1, ξ_2 and ξ_3. Suppose that there are the following relations between $\{\xi_1, \xi_2, \xi_3\}$ and a set of orthogonal coordinate variables $\{x_1, x_2, x_3\}$:

$$x_j = x_j(\xi_1, \xi_2, \xi_3), \quad j=1, 2, 3. \tag{1.15}$$

Then for any U we have

$$\frac{\partial U}{\partial x_j} = \frac{\partial \xi_1}{\partial x_j}\frac{\partial U}{\partial \xi_1} + \frac{\partial \xi_2}{\partial x_j}\frac{\partial U}{\partial \xi_2} + \frac{\partial \xi_3}{\partial x_j}\frac{\partial U}{\partial \xi_3}.$$

Using this formula, we can transform (1.11) into the form

$$A_1 \frac{\partial \boldsymbol{U}}{\partial x_1}+A_2 \frac{\partial \boldsymbol{U}}{\partial x_2}+A_3 \frac{\partial \boldsymbol{U}}{\partial x_3}+\boldsymbol{D}$$
$$=A_{\xi_1}\frac{\partial \boldsymbol{U}}{\partial \xi_1}+A_{\xi_2}\frac{\partial \boldsymbol{U}}{\partial \xi_2}+A_{\xi_3}\frac{\partial \boldsymbol{U}}{\partial \xi_3}+\boldsymbol{D}=0, \tag{1.16}$$

where

$$A_{\xi_j}=\frac{\partial \xi_j}{\partial x_1} A_1+\frac{\partial \xi_j}{\partial x_2} A_2+\frac{\partial \xi_j}{\partial x_3} A_3,$$

or, in detail,

$$A_{\xi_j}=\begin{pmatrix} V_{\xi_j} & 0 & 0 & \dfrac{\frac{\partial \xi_j}{\partial x_1}}{\rho H_1} & 0 & & & & \\ 0 & V_{\xi_j} & 0 & \dfrac{\frac{\partial \xi_j}{\partial x_2}}{\rho H_2} & 0 & & & \mathbf{0} & \\ 0 & 0 & V_{\xi_j} & \dfrac{\frac{\partial \xi_j}{\partial x_3}}{\rho H_3} & 0 & & & & \\ \dfrac{\rho\frac{\partial \xi_j}{\partial x_1}}{H_1} & \dfrac{\rho\frac{\partial \xi_j}{\partial x_2}}{H_2} & \dfrac{\rho\frac{\partial \xi_j}{\partial x_3}}{H_3} & 0 & V_{\xi_j} & & & & \\ 0 & 0 & 0 & V_{\xi_j} & -a^2 V_{\xi_j} & & & & \\ & & & & & V_{\xi_j} & & & \\ & & & & & & V_{\xi_j} & & \\ & \mathbf{0} & & & & & & \ddots & \\ & & & & & & & & V_{\xi_j} \end{pmatrix}$$

with

$$V_{\xi_j}=\boldsymbol{V}\cdot\boldsymbol{\xi}_j, \qquad \boldsymbol{\xi}_j=\begin{pmatrix} \dfrac{1}{H_1}\dfrac{\partial \xi_j}{\partial x_1} \\ \dfrac{1}{H_2}\dfrac{\partial \xi_j}{\partial x_2} \\ \dfrac{1}{H_3}\dfrac{\partial \xi_j}{\partial x_3} \end{pmatrix}. \tag{1.17}$$

2. *Characteristic Surfaces and Compatibility Relations*

In this subsection we discuss the characteristics of the fundamental equations (1.11). We first give the definition of characteristics[37, 38].

Suppose that there is a smooth surface in space, which is given in an η_1, η_2, η_3-coordinate system by

$$\Phi(\eta_1, \eta_2, \eta_3)=0, \tag{1.18}$$

and that the pseudo-linear differential equations

$$B_1 \frac{\partial \boldsymbol{U}}{\partial \eta_1} + B_2 \frac{\partial \boldsymbol{U}}{\partial \eta_2} + B_3 \frac{\partial \boldsymbol{U}}{\partial \eta_3} + \boldsymbol{D} = 0 \tag{1.19}$$

have a solution $\boldsymbol{U}$. If for this solution the equality

$$\left| \frac{\partial \Phi}{\partial \eta_1} B_1 + \frac{\partial \Phi}{\partial \eta_2} B_2 + \frac{\partial \Phi}{\partial \eta_3} B_3 \right| = 0 \tag{1.20}$$

holds at every point on the surface (1.18), then it is called a characteristic surface of (1.19) associated with the given solution. Here $|A|$ denotes the determinant of the matrix A.

Let

$$Q = \frac{\partial \Phi}{\partial \eta_1} B_1 + \frac{\partial \Phi}{\partial \eta_2} B_2 + \frac{\partial \Phi}{\partial \eta_3} B_3,$$

which is called the characteristic matrix of (1.19). Clearly, when the condition (1.20) holds, the algebraic system

$$\boldsymbol{\omega}^* Q = 0 \tag{1.21}$$

must have a non-trivial solution vector $\boldsymbol{\omega}$. It is called a characteristic-combination vector of (1.19). Here $\boldsymbol{\omega}^*$ is the transpose of the column vector $\boldsymbol{\omega}$. Then, the equation

$$\boldsymbol{\omega}^* \left(B_1 \frac{\partial \boldsymbol{U}}{\partial \eta_1} + B_2 \frac{\partial \boldsymbol{U}}{\partial \eta_2} + B_3 \frac{\partial \boldsymbol{U}}{\partial \eta_3} + \boldsymbol{D} \right) = 0 \tag{1.22}$$

is called a compatibility relation, or a characteristic-compatibility relation, of the equations (1.19) on the characteristic surface (1.18), where $\boldsymbol{\omega}$ is a solution of the equations (1.21).

It is important to notice the invariance of characteristics of differential equations with respect to any transformation of the independent variables, by which we mean that under a given transformation of the independent variables, the characteristic surfaces and the associated compatibility relations of the original differential equations will be transformed into the characteristic surfaces and the associated compatibility relations of the new differential equations.

In fact, in the new coordinate system $\{\xi_1, \xi_2, \xi_3\}$, connected with the old coordinate system $\{\eta_1, \eta_2, \eta_3\}$ by the relations

$$\eta_j = \eta_j(\xi_1, \xi_2, \xi_3), \quad j = 1, 2, 3, \tag{1.23}$$

the equation of the surface (1.18) assumes the form

$$\tilde{\Phi}(\xi_1, \xi_2, \xi_3) = \Phi(\eta_1(\xi_1, \xi_2, \xi_3), \eta_2(\xi_1, \xi_2, \xi_3), \eta_3(\xi_1, \xi_2, \xi_3)) = 0, \tag{1.24}$$

and because

$$B_1 \frac{\partial \boldsymbol{U}}{\partial \eta_1} + B_2 \frac{\partial \boldsymbol{U}}{\partial \eta_2} + B_3 \frac{\partial \boldsymbol{U}}{\partial \eta_3} = B_{\xi_1} \frac{\partial \boldsymbol{U}}{\partial \xi_1} + B_{\xi_2} \frac{\partial \boldsymbol{U}}{\partial \xi_2} + B_{\xi_3} \frac{\partial \boldsymbol{U}}{\partial \xi_3},$$

the original differential equations (1.19) are transformed into the form

$$B_{\xi_1}\frac{\partial U}{\partial \xi_1}+B_{\xi_2}\frac{\partial U}{\partial \xi_2}+B_{\xi_3}\frac{\partial U}{\partial \xi_3}+D=0, \tag{1.25}$$

where

$$B_{\xi_j}=\frac{\partial \xi_j}{\partial \eta_1}B_1+\frac{\partial \xi_j}{\partial \eta_2}B_2+\frac{\partial \xi_j}{\partial \eta_3}B_3, \quad j=1, 2, 3.$$

Then, using

$$\frac{\partial \tilde{\Phi}}{\partial \xi_l}=\sum_{j=1}^{3}\frac{\partial \Phi}{\partial \eta_j}\frac{\partial \eta_j}{\partial \xi_l}, \quad l=1, 2, 3,$$

we obtain

$$\begin{aligned}
&\frac{\partial \tilde{\Phi}}{\partial \xi_1}B_{\xi_1}+\frac{\partial \tilde{\Phi}}{\partial \xi_2}B_{\xi_2}+\frac{\partial \tilde{\Phi}}{\partial \xi_3}B_{\xi_3}\\
&=\sum_l\frac{\partial \tilde{\Phi}}{\partial \xi_l}B_{\xi_l}=\sum_l\sum_j\sum_k\frac{\partial \Phi}{\partial \eta_j}\frac{\partial \eta_j}{\partial \xi_l}\frac{\partial \xi_l}{\partial \eta_k}B_k\\
&=\sum_j\sum_k\frac{\partial \Phi}{\partial \eta_j}\delta_{j,k}B_k=\sum_j\frac{\partial \Phi}{\partial \eta_j}B_j\\
&=\frac{\partial \Phi}{\partial \eta_1}B_1+\frac{\partial \Phi}{\partial \eta_2}B_2+\frac{\partial \Phi}{\partial \eta_3}B_3,
\end{aligned}$$

which shows that when (1.20) holds, the equality

$$\left|\frac{\partial \tilde{\Phi}}{\partial \xi_1}B_{\xi_1}+\frac{\partial \tilde{\Phi}}{\partial \xi_2}B_{\xi_2}+\frac{\partial \tilde{\Phi}}{\partial \xi_3}B_{\xi_3}\right|=0$$

holds. In other words, if the surface (1.18) is a characteristic surface of the original equations (1.19), then it (now given by the equation (1.24)) is also a characteristic surface of the new differential equations (1.25). Furthermore, the compatibility relation (1.22) is transformed into

$$\omega^*\left(B_{\xi_1}\frac{\partial U}{\partial \xi_1}+B_{\xi_2}\frac{\partial U}{\partial \xi_2}+B_{\xi_3}\frac{\partial U}{\partial \xi_3}+D\right)=0,$$

which is the compatibility relation of the new differential equations (1.25), because the solution ω of the algebraic equations (1.21) also satisfies the algebraic equations

$$\omega^*\left(\frac{\partial \tilde{\Phi}}{\partial \xi_1}B_{\xi_1}+\frac{\partial \tilde{\Phi}}{\partial \xi_2}B_{\xi_2}+\frac{\partial \tilde{\Phi}}{\partial \xi_3}B_{\xi_3}\right)=0.$$

The proof is completed. The above statements allow us to investigate the characteristics of differential equations in terms of any independent variables chosen.

We further point out that a compatibility relation actually represents an "interior" relation in the characteristic surface associated with it, that is, it contains the directional derivatives of the dependent variables only along interior directions in the characteristic surface. This can be proven as follows. Blocking by columns the matrix B_j, which is assumed to be $m\times m$-dimensional, and denoting the i-th column by $B_{j,i}$ $(i=1, 2, \cdots, m, j=1, 2, 3)$, we can write the equations (1.21) and (1.22) in the forms

$$\boldsymbol{\omega}^*\left(\frac{\partial\Phi}{\partial\eta_1}B_{1,i}+\frac{\partial\Phi}{\partial\eta_2}B_{2,i}+\frac{\partial\Phi}{\partial\eta_3}B_{3,i}\right)$$
$$=\tilde{\omega}_{i,1}\frac{\partial\Phi}{\partial\eta_1}+\tilde{\omega}_{i,2}\frac{\partial\Phi}{\partial\eta_2}+\tilde{\omega}_{i,3}\frac{\partial\Phi}{\partial\eta_3}=0,$$
$$i=1, 2, \cdots, m \tag{1.26}$$

and

$$\sum_{i=1}^{m}\left(\boldsymbol{\omega}^*B_{1,i}\frac{\partial u_i}{\partial\eta_1}+\boldsymbol{\omega}^*B_{2,i}\frac{\partial u_i}{\partial\eta_2}+\boldsymbol{\omega}^*B_{3,i}\frac{\partial u_i}{\partial\eta_3}\right)+\boldsymbol{\omega}^*\boldsymbol{D}$$
$$=\sum_{i=1}^{m}\left(\tilde{\omega}_{i,1}\frac{\partial u_i}{\partial\eta_1}+\tilde{\omega}_{i,2}\frac{\partial u_i}{\partial\eta_2}+\tilde{\omega}_{i,3}\frac{\partial u_i}{\partial\eta_3}\right)+\boldsymbol{\omega}^*\boldsymbol{D}=0$$

respectively, where u_i is the i–th component of the vector $\boldsymbol{U}$, and

$$\tilde{\omega}_{i,j}=\boldsymbol{\omega}^*B_{j,i}.$$

Since $\tilde{\omega}_{i,j}$ satisfies (1.26), the vector $(\tilde{\omega}_{i,1}, \tilde{\omega}_{i,2}, \tilde{\omega}_{i,3})$ must lie on the tangent plane to the characteristic surface (1.18). This shows the derivative

$$\tilde{\omega}_{i,1}\frac{\partial u_i}{\partial\eta_1}+\tilde{\omega}_{i,2}\frac{\partial u_i}{\partial\eta_2}+\tilde{\omega}_{i,3}\frac{\partial u_i}{\partial\eta_3}$$

is an interior derivative of u_i in the surface, which is the desired result.

In what follows we shall discuss the differential equations (1.11) in an orthogonal curvilinear coordinate system. Let

$$\Phi(x_1, x_2, x_3)=0 \tag{1.27}$$

be a characteristic surface of (1.11). By definition, it must satisfy the equation

$$|Q|=\left|\frac{\partial\Phi}{\partial x_1}A_1+\frac{\partial\Phi}{\partial x_2}A_2+\frac{\partial\Phi}{\partial x_3}A_3\right|=0, \tag{1.28}$$

where

$$Q=\begin{pmatrix}
\psi & 0 & 0 & N_1/\rho & 0 & & & & \\
0 & \psi & 0 & N_2/\rho & 0 & & & \mathbf{0} & \\
0 & 0 & \psi & N_3/\rho & 0 & & & & \\
N_1\rho & N_2\rho & N_3\rho & 0 & \psi & & & & \\
0 & 0 & 0 & \psi & -a^2\psi & & & & \\
 & & & & & \psi & & & \\
 & \mathbf{0} & & & & & \psi & & \\
 & & & & & & & \ddots & \\
 & & & & & & & & \psi
\end{pmatrix}$$

with

$$N_j=\frac{1}{H_j}\frac{\partial\Phi}{\partial x_j}, \quad j=1, 2, 3,$$

$$\psi = \boldsymbol{V}\cdot\boldsymbol{N}, \quad \boldsymbol{N}=\begin{pmatrix} N_1 \\ N_2 \\ N_3 \end{pmatrix}.$$

Clearly, (1.28) can be rewritten as

$$\psi^{m-2}(\psi^2 - a^2|\boldsymbol{N}|^2) = 0, \tag{1.29}$$

where

$$|\boldsymbol{N}|^2 = N_1^2 + N_2^2 + N_3^2.$$

Now we introduce the concepts of characteristic normal cones, characteristic cones, space–like surfaces and others. Suppose there exists a solution $\boldsymbol{U}$ for (1.11) in the neighborhood of a given point P. The definitions to be given below are all related to the given solution $\boldsymbol{U}$.

Every solution vector $\boldsymbol{N}$ of the equation (1.29) is called a characteristic normal of the differential equations (1.11). Obviously, the direction normal to a characteristic surface is a characteristic normal. Taking all possible characteristic normal vectors passing through the given point P, we obtain a conical surface with the vertex at P, because when $\boldsymbol{N}$ is a solution of (1.29), $t\boldsymbol{N}$ is also its solution, t being any real number. It is called the characteristic normal cone. The plane passing through the point P and perpendicular to a characteristic normal at P is called a characteristic plane. Through the point P there is a family of such planes. Their envelope is also a conical surface. It is called the characteristic cone at P. The direction of each generator of this cone is called a characteristic direction. A spatial curve is called a bicharacteristic curve if the tangent direction to it at every point is a characteristic direction. All possible bicharacteristic curves through a given point also form a conical surface, called the characteristic conoid at this point.

Equation (1.29) can be split in two: $\psi^{m-2}=0$ and $\psi^2 - a^2|\boldsymbol{N}|^2=0$, which are mutually independent unless $a=0$ (we are not interested in this case). Accordingly, the characteristic normal cone as well as the characteristic cone and the characteristic conoid at a point can be broken into two branches, either of which is a conical surface. For the purpose of clarity, where necessary, we shall add the prefix "stream" or "wave" to a term when the term is associated with the equation

$$\psi^{m-2}=0 \quad \text{or} \quad \psi^2 - a^2|\boldsymbol{N}|^2=0.$$

For example, we may say stream-characteristic normals, wave-characteristic cones, and so on.

In what follows we shall discuss the geometrical and physical nature of the stream–characteristics and the wave–characteristics.

By definition, a stream–characteristic normal, denoted by $\boldsymbol{N}_s$, satisfies the equation

$$\psi_s = \boldsymbol{V}\cdot\boldsymbol{N}_s = u_1 N_{s,1} + u_2 N_{s,2} + u_3 N_{s,3} = 0. \tag{1.30}$$

It is obvious that every solution vector $\boldsymbol{N}_s$ of the equation lies on the plane orthogonal to the corresponding velocity vector $\boldsymbol{V}$. Thus we immediately conclude that, at any given point in space, the stream–characteristic normal cone degenerates into the plane that is perpendicular to the velocity vector $\boldsymbol{V}$; any plane through $\boldsymbol{V}$ is a stream–characteristic plane; the stream–characteristic cone is the straight line parallel to $\boldsymbol{V}$; the direction of the velocity is stream–characteristic; the streamline (a curve is called a streamline if the velocity vector is tangent to it at every point) through the given point is the stream–bicharacteristic curve and the stream–characteristic conoid as well.

We now observe the wave–characteristics. The equation determining a wave–characteristic normal, denoted by $\boldsymbol{N}_W$, is

$$\psi_W^2 - a^2 |\boldsymbol{N}_W|^2 = 0$$

with

$$\psi_W = \boldsymbol{V} \cdot \boldsymbol{N}_W.$$

Obviously, without loss of generality, we can require $|\boldsymbol{N}_W| = 1$. Thus we have

$$\begin{cases} |\psi_W| = |\boldsymbol{V} \cdot \boldsymbol{N}_W| = a, \\ |\boldsymbol{N}_W| = 1. \end{cases} \tag{1.31}$$

Note that the quantity ψ_W is the projection of the velocity vector in the direction $\boldsymbol{N}_W$. It is apparent that the equation (1.31) has a real solution if and only if

$$|\boldsymbol{V}| \geqslant a. \tag{1.32}$$

For the time being we suppose that this condition holds at points to be considered. It is easily seen from (1.31) that the wave–characteristic normal cone is a circular cone with the semi–open angle equal to $\arccos(a/|\boldsymbol{V}|)$ and with the tip at the considered point P, and the wave–characteristic cone is a circular cone with the semi–open angle equal to $\alpha = \arcsin(a/|\boldsymbol{V}|)$ and with the tip at the same point. Given a generator of the normal cone, that is, a characteristic normal $\boldsymbol{N}_W$, we can determine a corresponding characteristic plane, which touches the wave–characteristic cone along a generator of it. If we denote the direction of this generator by $\boldsymbol{\tau}$, it is easy to see that there is a relation $\boldsymbol{\tau} = \boldsymbol{V} \mp a\boldsymbol{N}_W$ between the characteristic direction $\boldsymbol{\tau}$ and the corresponding characteristic normal $\boldsymbol{N}_W$. The wave–characteristic conoid is often called the Mach cone, its semi–open angle the Mach angle, and the quantity $M = |\boldsymbol{V}|/a$ the Mach number. Between the Mach angle α and the Mach number M there exists the relation

$$\alpha = \arcsin\frac{1}{M}. \tag{1.33}$$

Geometrically these are illustrated in Fig. 1.2. Both the wave–characteristic cone and the Mach cone at a point have two "leaves": the leave

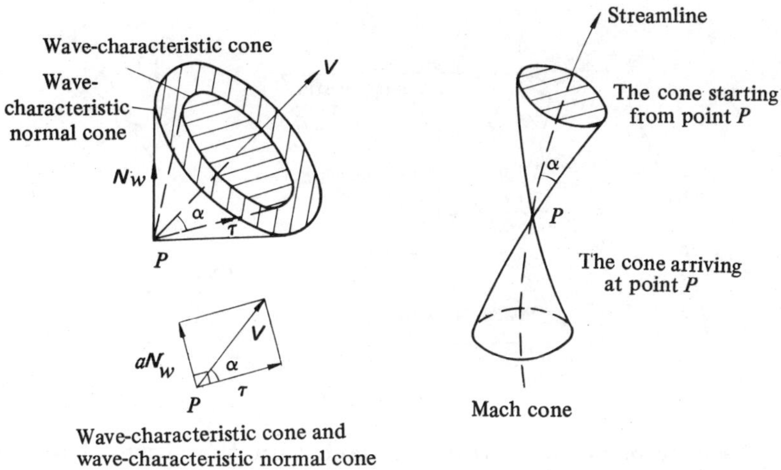

Figure 1.2

which is "expanding" along the direction of the velocity $\boldsymbol{V}$ is referred to as the "issuing" cone, and the other is referred to as the "arriving" cone, with reference to the point considered. From the theory of partial differential equations[37], it is known that the disturbance of the solution at a given point influences the solution only at those points in the Mach cone issuing from the given point and hence, in a supersonic flow the downstream has no influence on the upstream.

A surface is space–like if the issuing wave–characteristic cone from every point on it lies on the same side of it. As indicated above, the flow on the downstream side of a space–like surface has no influence on the flow on the upstream side, and therefore the flow field on the upstream may be calculated independently of the downstream. It is on this basis that, in solving the problem of a supersonic flow past a body, one can compute its solution first in the subsonic–transonic region in front of a certain space–like surface and then from one space–like surface to another in the supersonic region. It is easy to prove that the necessary and sufficient condition for a surface to be space–like is

$$|\boldsymbol{V}\cdot\boldsymbol{N}|>a, \tag{1.34}$$

where $\boldsymbol{N}$ is the unit normal to the surface. The condition (1.34) says that the projection of the velocity in the normal direction is greater than the speed of sound at any point of a space–like surface. This can be concluded from the fact that the normal vectors to a space–like surface always fall into the wave–characteristic normal cones and $|\boldsymbol{V}\cdot\boldsymbol{N}_W|$ is equal to a (see Fig. 1.3).

Given a closed curve on a given space–like surface, we can draw a

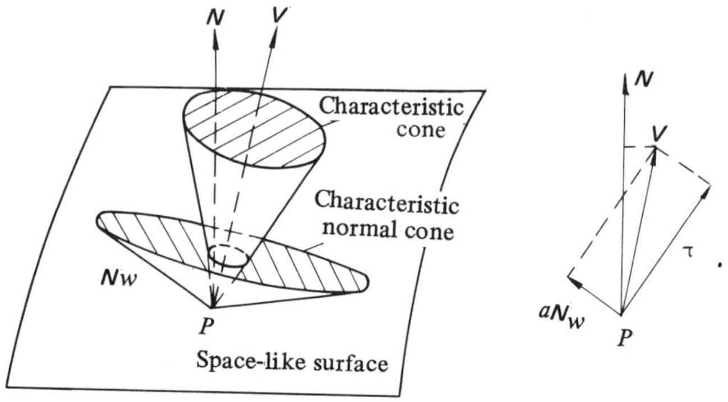

Figure 1.3

family of issuing Mach cones from the points on the curve. This family has two envelopes, which are the wave–characteristic surfaces (see Fig. 1.4). As indicated above, the disturbance from the curve should influence only the flow field between these two characteristic surfaces. In other words, the wave–characteristic surfaces are the boundary surfaces of propagation of the disturbance from the curve. In addition, we can draw a streamline from each point on the given curve. All the streamlines form a stream–surface which is the stream–characteristic surface through the curve. Thus we have seen that through any curve on a given space–like surface we can draw, on each side of the surface, e.g., on the downstream side, two wave–characteristic surfaces and one stream–characteristic surface between the two wave–characteristic surfaces.

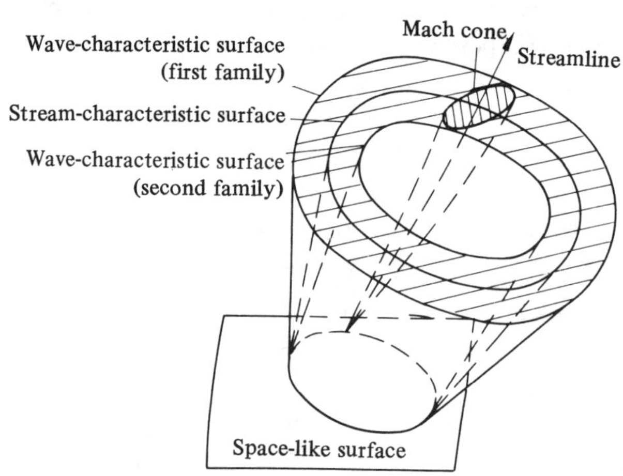

Figure 1.4

In what follows we shall give some possible choices of vector $\boldsymbol{\omega}$, which is used to construct a compatibility relation associated with a wave- or stream-characteristic surface (see (1.21) for the definition of $\boldsymbol{\omega}$). For a stream-characteristic normal $\boldsymbol{N}_s$ satisfying the equation (1.30), the corresponding characteristic matrix is

$$Q_s=\begin{pmatrix} 0 & 0 & 0 & N_{s,1}/\rho & \\ 0 & 0 & 0 & N_{s,2}/\rho & \\ 0 & 0 & 0 & N_{s,3}/\rho & 0 \\ \rho N_{s,1} & \rho N_{s,2} & \rho N_{s,3} & 0 & \\ & 0 & & & 0 \end{pmatrix}. \tag{1.35}$$

Then the system of equations

$$\boldsymbol{\omega}^* Q_s=0 \tag{1.36}$$

has $m-2$ linearly independent solution vectors:

$$\begin{cases} \boldsymbol{\omega}_2^*=(\omega_{2,1},\ \omega_{2,2},\ \omega_{2,3},\ 0,\ 0,\ \cdots,\ 0,\ 0), \\ \boldsymbol{\omega}_3^*=(\omega_{3,1},\ \omega_{3,2},\ \omega_{3,3},\ 0,\ 0,\ \cdots,\ 0,\ 0), \\ \boldsymbol{\omega}_4^*=(0,\ 0,\ 0,\ 0,\ 1,\ 0,\ \cdots,\ 0,\ 0), \\ \vdots \\ \boldsymbol{\omega}_{m-1}^*=(0,\ 0,\ 0,\ 0,\ 0,\ 0,\ \cdots,\ 0,\ 1), \end{cases} \tag{1.37}$$

where $(\omega_{2,1},\ \omega_{2,2},\ \omega_{2,3})$ and $(\omega_{3,1},\ \omega_{3,2},\ \omega_{3,3})$ satisfy the linear equations

$$\begin{cases} \omega_{2,1}N_{s,1}+\omega_{2,2}N_{s,2}+\omega_{2,3}N_{s,3}=0, \\ \omega_{3,1}N_{s,1}+\omega_{3,2}N_{s,2}+\omega_{3,3}N_{s,3}=0, \end{cases} \tag{1.38}$$

and are required to be mutually independent. We can, for example, choose

$$\begin{pmatrix} \omega_{2,1} \\ \omega_{2,2} \\ \omega_{2,3} \end{pmatrix}=\boldsymbol{V},\quad \begin{pmatrix} \omega_{3,1} \\ \omega_{3,2} \\ \omega_{3,3} \end{pmatrix}=\boldsymbol{V}\times\boldsymbol{N}_s\equiv\boldsymbol{T}\equiv\begin{pmatrix} t_1 \\ t_2 \\ t_3 \end{pmatrix}. \tag{1.39}$$

From (1.37) it is seen that the fifth and all succeeding equations in (1.11) have been stream-compatibility relations (in any stream-characteristic surface of (1.11)). We are not surprised if noting that each of them contains only the derivatives along the streamlines. Notice, also, that any linear combination of the stream-compatibility relations in a stream-characteristic surface is still a stream-compatibility relation in the same surface. Thus the choice of the compatibility relations is not unique, in particular, the choice of $\boldsymbol{\omega}_2$ and $\boldsymbol{\omega}_3$ is not unique.

Now let $\boldsymbol{N}_W$ be a wave-characteristic normal which satisfies the equation (1.31). The corresponding characteristic matrix is

$$Q_W=\begin{pmatrix} \psi_W & 0 & 0 & N_{W,1}/\rho & 0 & & & \\ 0 & \psi_W & 0 & N_{W,2}/\rho & 0 & & \mathbf{0} & \\ 0 & 0 & \psi_W & N_{W,3}/\rho & 0 & & & \\ N_{W,1}\rho & N_{W,2}\rho & N_{W,3}\rho & 0 & \psi_W & & & \\ 0 & 0 & 0 & \psi_W & -a^2\psi_W & & & \\ & & & & & \psi_W & & \\ & \mathbf{0} & & & & & \ddots & \\ & & & & & & & \psi_W \end{pmatrix}.$$

Then the equations

$$\boldsymbol{\omega}^* Q_W = 0$$

have a solution, which can be written as

$$\boldsymbol{\omega}^* = \left(\frac{N_{W,1}\rho a^2}{\psi_W}, \frac{N_{W,2}\rho a^2}{\psi_W}, \frac{N_{W,3}\rho a^2}{\psi_W}, -a^2, -1, 0, \cdots, 0\right). \quad (1.40)$$

The corresponding wave-compatibility relation is

$$\boldsymbol{\omega}^* \left(A_1 \frac{\partial \boldsymbol{U}}{\partial x_1} + A_2 \frac{\partial \boldsymbol{U}}{\partial x_2} + A_3 \frac{\partial \boldsymbol{U}}{\partial x_3} + \boldsymbol{D}\right) = 0. \quad (1.41)$$

For practical applications it may be useful to find all the characteristic surfaces through a given spacial curve and all the compatibility relations on them. Suppose we have a curve given by the equations

$$\begin{cases} \xi_1(x_1, x_2, x_3) = \text{constant}, \\ \xi_2(x_1, x_2, x_3) = \text{constant} \end{cases} \quad (1.42)$$

in an orthogonal curvilinear coordinate system x_1, x_2, x_3. The normal, denoted by $\boldsymbol{\xi}_j$, to each surface $\xi_j(x_1, x_2, x_3) = \text{constant}$ $(j=1, 2)$ is

$$\boldsymbol{\xi}_j = \begin{pmatrix} \dfrac{1}{H_1} \dfrac{\partial \xi_j}{\partial x_1} \\ \dfrac{1}{H_2} \dfrac{\partial \xi_j}{\partial x_2} \\ \dfrac{1}{H_3} \dfrac{\partial \xi_j}{\partial x_3} \end{pmatrix}.$$

Therefore a tangent to the curve (1.42) has the direction given by

$$\boldsymbol{\xi}_1 \times \boldsymbol{\xi}_2.$$

Since a normal vector $\boldsymbol{N}$ to a characteristic surface through the given curve is orthogonal to $\boldsymbol{\xi}_1 \times \boldsymbol{\xi}_2$, it can be expressed in the form

$$\boldsymbol{N} = a_1 \boldsymbol{\xi}_1 + a_2 \boldsymbol{\xi}_2. \quad (1.43)$$

Let $\boldsymbol{N}_s$ be a stream-characteristic normal, which, according to (1.43), has the form

$$\boldsymbol{N}_s = a_{s,1} \boldsymbol{\xi}_1 + a_{s,2} \boldsymbol{\xi}_2$$

and should satisfy the equation (1.30), i.e.,

$$\boldsymbol{N}_s \cdot \boldsymbol{V} = a_{s,1} V_{\xi_1} + a_{s,2} V_{\xi_2} = 0. \tag{1.44}$$

From it we have

$$-\frac{a_{s,1}}{a_{s,2}} = \frac{V_{\xi_2}}{V_{\xi_1}},$$

and $\boldsymbol{N}_s$ can be represented as

$$\boldsymbol{N}_s = V_{\xi_2} \boldsymbol{\xi}_1 - V_{\xi_1} \boldsymbol{\xi}_2. \tag{1.45}$$

Thus, using (1.37) and (1.39) with the substitution of (1.45) for $\boldsymbol{N}_s$ and noting (1.16), we obtain the following stream-compatibility relations:

$$\begin{cases}
\boldsymbol{\omega}_2^* \left(A_1 \dfrac{\partial \boldsymbol{U}}{\partial x_1} + A_2 \dfrac{\partial \boldsymbol{U}}{\partial x_2} + A_3 \dfrac{\partial \boldsymbol{U}}{\partial x_3} + \boldsymbol{D} \right) \\
\quad = \boldsymbol{\omega}_2^* \left(A_{\xi_1} \dfrac{\partial \boldsymbol{U}}{\partial \xi_1} + A_{\xi_2} \dfrac{\partial \boldsymbol{U}}{\partial \xi_2} + A_{\xi_3} \dfrac{\partial \boldsymbol{U}}{\partial \xi_3} + \boldsymbol{D} \right) \\
\quad = V_{\xi_1} \boldsymbol{V} \cdot \dfrac{\partial \boldsymbol{V}}{\partial \xi_1} + V_{\xi_2} \boldsymbol{V} \cdot \dfrac{\partial \boldsymbol{V}}{\partial \xi_2} + V_{\xi_3} \boldsymbol{V} \cdot \dfrac{\partial \boldsymbol{V}}{\partial \xi_3} \\
\qquad + \dfrac{1}{\rho} \left(V_{\xi_1} \dfrac{\partial p}{\partial \xi_1} + V_{\xi_2} \dfrac{\partial p}{\partial \xi_2} + V_{\xi_3} \dfrac{\partial p}{\partial \xi_3} \right) \\
\qquad + \boldsymbol{V} \cdot \boldsymbol{D}_{1-3} = 0, \\
\boldsymbol{\omega}_3^* \left(A_1 \dfrac{\partial \boldsymbol{U}}{\partial x_1} + A_2 \dfrac{\partial \boldsymbol{U}}{\partial x_2} + A_3 \dfrac{\partial \boldsymbol{U}}{\partial x_3} + \boldsymbol{D} \right) \\
\quad = \boldsymbol{\omega}_3^* \left(A_{\xi_1} \dfrac{\partial \boldsymbol{U}}{\partial \xi_1} + A_{\xi_2} \dfrac{\partial \boldsymbol{U}}{\partial \xi_2} + A_{\xi_3} \dfrac{\partial \boldsymbol{U}}{\partial \xi_3} + \boldsymbol{D} \right) \\
\quad = V_{\xi_1} \boldsymbol{T} \cdot \dfrac{\partial \boldsymbol{V}}{\partial \xi_1} + V_{\xi_2} \boldsymbol{T} \cdot \dfrac{\partial \boldsymbol{V}}{\partial \xi_2} + V_{\xi_3} \boldsymbol{T} \cdot \dfrac{\partial \boldsymbol{V}}{\partial \xi_3} \\
\qquad + \dfrac{1}{\rho} \left(\boldsymbol{T} \cdot \boldsymbol{\xi}_1 \dfrac{\partial p}{\partial \xi_1} + \boldsymbol{T} \cdot \boldsymbol{\xi}_2 \dfrac{\partial p}{\partial \xi_2} + \boldsymbol{T} \cdot \boldsymbol{\xi}_3 \dfrac{\partial p}{\partial \xi_3} \right) \\
\qquad + \boldsymbol{T} \cdot \boldsymbol{D}_{1-3} = 0, \\
\boldsymbol{\omega}_4^* \left(A_1 \dfrac{\partial \boldsymbol{U}}{\partial x_1} + A_2 \dfrac{\partial \boldsymbol{U}}{\partial x_2} + A_3 \dfrac{\partial \boldsymbol{U}}{\partial x_3} + \boldsymbol{D} \right) \\
\quad = V_{\xi_1} \dfrac{\partial p}{\partial \xi_1} + V_{\xi_2} \dfrac{\partial p}{\partial \xi_2} + V_{\xi_3} \dfrac{\partial p}{\partial \xi_3} \\
\qquad - a^2 \left(V_{\xi_1} \dfrac{\partial \rho}{\partial \xi_1} + V_{\xi_2} \dfrac{\partial \rho}{\partial \xi_2} + V_{\xi_3} \dfrac{\partial \rho}{\partial \xi_3} \right) + d_5 = 0, \\
\boldsymbol{\omega}_{i+4}^* \left(A_1 \dfrac{\partial \boldsymbol{U}}{\partial x_1} + A_2 \dfrac{\partial \boldsymbol{U}}{\partial x_2} + A_3 \dfrac{\partial \boldsymbol{U}}{\partial x_3} + \boldsymbol{D} \right) \\
\quad = V_{\xi_1} \dfrac{\partial c_i}{\partial \xi_1} + V_{\xi_2} \dfrac{\partial c_i}{\partial \xi_2} + V_{\xi_3} \dfrac{\partial c_i}{\partial \xi_3} + d_{i+5} = 0, \\
\qquad i = 1, 2, \cdots, m-5,
\end{cases} \tag{1.46}$$

where

$$\boldsymbol{D}_{1-3} = \begin{pmatrix} d_1 \\ d_2 \\ d_3 \end{pmatrix}. \tag{1.47}$$

From (1.10) we can see that $\boldsymbol{V}\cdot\boldsymbol{D}_{1-3}\equiv 0$. Thus, introducing the notation

$$\begin{cases} \sigma=-\dfrac{a_{s,1}}{a_{s,2}}=\dfrac{V_{\xi_2}}{V_{\xi_1}}, \quad m_2=0,\ m_3=\boldsymbol{T}\cdot\boldsymbol{D}_{1-3}, \\ m_{i+4}=d_{i+5}, \quad i=0,\ 1,\ 2,\ \cdots,\ m-5, \end{cases} \tag{1.48}$$

and noting that

$$\begin{cases} \boldsymbol{T}\cdot\boldsymbol{\xi}_1=\boldsymbol{V}\times\boldsymbol{N}_s\cdot\boldsymbol{\xi}_1=\boldsymbol{V}\times(V_{\xi_2}\boldsymbol{\xi}_1-V_{\xi_1}\boldsymbol{\xi}_2)\cdot\boldsymbol{\xi}_1 \\ \qquad =V_{\xi_1}\boldsymbol{V}\times\boldsymbol{\xi}_1\cdot\boldsymbol{\xi}_2, \\ \boldsymbol{T}\cdot\boldsymbol{\xi}_2=\boldsymbol{V}\times\boldsymbol{N}_s\cdot\boldsymbol{\xi}_2=V_{\xi_2}\boldsymbol{V}\times\boldsymbol{\xi}_1\cdot\boldsymbol{\xi}_2, \end{cases} \tag{1.49}$$

we can rewrite (1.46) in the form

$$\begin{cases} V_{\xi_1}\boldsymbol{V}\cdot\left(\dfrac{\partial\boldsymbol{V}}{\partial\xi_1}+\sigma\dfrac{\partial\boldsymbol{V}}{\partial\xi_2}\right)+\dfrac{1}{\rho}V_{\xi_1}\left(\dfrac{\partial p}{\partial\xi_1}+\sigma\dfrac{\partial p}{\partial\xi_2}\right) \\ \quad +V_{\xi_3}\left(\boldsymbol{V}\cdot\dfrac{\partial\boldsymbol{V}}{\partial\xi_3}+\dfrac{1}{\rho}\dfrac{\partial p}{\partial\xi_3}\right)=-m_2, \\ V_{\xi_1}\boldsymbol{T}\cdot\left(\dfrac{\partial\boldsymbol{V}}{\partial\xi_1}+\sigma\dfrac{\partial\boldsymbol{V}}{\partial\xi_2}\right)+\dfrac{1}{\rho}V_{\xi_1}\boldsymbol{V}\times\boldsymbol{\xi}_1\cdot\boldsymbol{\xi}_2\left(\dfrac{\partial p}{\partial\xi_1}+\sigma\dfrac{\partial p}{\partial\xi_2}\right) \\ \quad +V_{\xi_3}\boldsymbol{T}\cdot\dfrac{\partial\boldsymbol{V}}{\partial\xi_3}+\dfrac{1}{\rho}\boldsymbol{T}\cdot\boldsymbol{\xi}_3\dfrac{\partial p}{\partial\xi_3}=-m_3, \\ V_{\xi_1}\left(\dfrac{\partial p}{\partial\xi_1}+\sigma\dfrac{\partial p}{\partial\xi_2}\right)-a^2V_{\xi_1}\left(\dfrac{\partial\rho}{\partial\xi_1}+\sigma\dfrac{\partial\rho}{\partial\xi_2}\right) \\ \quad +V_{\xi_3}\left(\dfrac{\partial p}{\partial\xi_3}-a^2\dfrac{\partial\rho}{\partial\xi_3}\right)=-m_4, \\ V_{\xi_1}\left(\dfrac{\partial c_i}{\partial\xi_1}+\sigma\dfrac{\partial c_i}{\partial\xi_2}\right)+V_{\xi_3}\dfrac{\partial c_i}{\partial\xi_3}=-m_{i+4}, \\ \qquad i=1,\ 2,\ \cdots,\ m-5. \end{cases} \tag{1.50}$$

Now let $\boldsymbol{N}_W$ be a wave-characteristic normal to a wave-characteristic surface through the curve (1.42). Obviously, according to (1.43), it can be expressed in the form

$$\boldsymbol{N}_W=a_{W,1}\boldsymbol{\xi}_1+a_{W,2}\boldsymbol{\xi}_2$$

and should satisfy (1.31), that is,

$$\begin{cases} |\boldsymbol{V}\cdot\boldsymbol{N}_W|=|a_{W,1}V_{\xi_1}+a_{W,2}V_{\xi_2}|=a, \\ |\boldsymbol{N}_W|=|a_{W,1}\boldsymbol{\xi}_1+a_{W,2}\boldsymbol{\xi}_2|=1, \end{cases}$$

or

$$\begin{cases} a_{W,1}^2V_{\xi_1}^2+2a_{W,1}a_{W,2}V_{\xi_1}V_{\xi_2}+a_{W,2}^2V_{\xi_2}^2=a^2, \\ a_{W,1}^2\boldsymbol{\xi}_1\cdot\boldsymbol{\xi}_1+2a_{W,1}a_{W,2}\boldsymbol{\xi}_1\cdot\boldsymbol{\xi}_2+a_{W,2}^2\boldsymbol{\xi}_2\cdot\boldsymbol{\xi}_2=1. \end{cases}$$

Subtracting from the first equation the second one multiplied by a^2, we obtain

$$a_{W,1}^2(V_{\xi_1}^2-a^2\boldsymbol{\xi}_1\cdot\boldsymbol{\xi}_1)+2a_{W,1}a_{W,2}(V_{\xi_1}V_{\xi_2}-a^2\boldsymbol{\xi}_1\cdot\boldsymbol{\xi}_2)$$
$$+a_{W,2}^2(V_{\xi_2}^2-a^2\boldsymbol{\xi}_2\cdot\boldsymbol{\xi}_2)=0,$$

or, we can rewrite it as the following equation for $-a_{W,1}/a_{W,2}$ when $a_{W,2}\neq 0$,

$$\beta_1\left(-\frac{a_{W,1}}{a_{W,2}}\right)^2-2\beta_2\left(-\frac{a_{W,1}}{a_{W,2}}\right)+\beta_3=0 \tag{1.51}$$

with

$$\begin{cases}\beta_1=V_{\xi_1}^2-a^2\boldsymbol{\xi}_1\cdot\boldsymbol{\xi}_1,\\ \beta_2=V_{\xi_1}V_{\xi_2}-a^2\boldsymbol{\xi}_1\cdot\boldsymbol{\xi}_2,\\ \beta_3=V_{\xi_2}^2-a^2\boldsymbol{\xi}_2\cdot\boldsymbol{\xi}_2.\end{cases} \tag{1.52}$$

Obviously, when $\beta_2^2-\beta_1\beta_3\geqslant 0$, the equation has real solutions:

$$\sigma_\pm\equiv\left(-\frac{a_{W,1}}{a_{W,2}}\right)_\pm=\frac{\beta_2\pm\sqrt{\beta_2^2-\beta_1\beta_3}}{\beta_1}. \tag{1.53}$$

In the following discussion, for convenience, we consider only the wave-characteristic normals $\boldsymbol{N}_W$ which satisfy $\psi_W=\boldsymbol{V}\cdot\boldsymbol{N}_W=a$. Then we can write

$$\boldsymbol{N}_W=\frac{a}{a_{W,1}V_{\xi_1}+a_{W,2}V_{\xi_2}}(a_{W,1}\boldsymbol{\xi}_1+a_{W,2}\boldsymbol{\xi}_2) \tag{1.54}$$

in place of $\boldsymbol{N}_W=a_{W,1}\boldsymbol{\xi}_1+a_{W,2}\boldsymbol{\xi}_2$, and using

$$\frac{a_{W,1}\boldsymbol{\xi}_1+a_{W,2}\boldsymbol{\xi}_2}{a_{W,1}V_{\xi_1}+a_{W,2}V_{\xi_2}}=\frac{(-a_{W,1}/a_{W,2})\boldsymbol{\xi}_1-\boldsymbol{\xi}_2}{(-a_{W,1}/a_{W,2})V_{\xi_1}-V_{\xi_2}},$$

we finally obtain two solutions for $\boldsymbol{N}_W$,

$$\boldsymbol{N}_W^\pm=\frac{a}{\sigma_\pm V_{\xi_1}-V_{\xi_2}}(\sigma_\pm\boldsymbol{\xi}_1-\boldsymbol{\xi}_2), \tag{1.55}$$

which means that in the case of $\beta_2^2-\beta_1\beta_3>0$ there are two wave-characteristic surfaces through the curve (1.42), their normals being given, respectively, by $\boldsymbol{N}_W^+$ and $\boldsymbol{N}_W^-$. According to (1.40) and (1.41) we obtain

$$\begin{cases}\boldsymbol{\omega}_1^*=(N_{W,1}^+\rho a,\ N_{W,2}^+\rho a,\ N_{W,3}^+\rho a,\ -a^2,\ -1,\ 0,\ \cdots,\ 0),\\ \boldsymbol{\omega}_m^*=(N_{W,1}^-\rho a,\ N_{W,2}^-\rho a,\ N_{W,3}^-\rho a,\ -a^2,\ -1,\ 0,\ \cdots,\ 0),\end{cases} \tag{1.56}$$

and the corresponding wave-compatibility relations are

$$\begin{aligned}
&\boldsymbol{\omega}_i^*\left(A_1\frac{\partial\boldsymbol{U}}{\partial x_1}+A_2\frac{\partial\boldsymbol{U}}{\partial x_2}+A_3\frac{\partial\boldsymbol{U}}{\partial x_3}+\boldsymbol{D}\right)\\
&=\boldsymbol{\omega}_i^*\left(A_{\xi_1}\frac{\partial\boldsymbol{U}}{\partial\xi_1}+A_{\xi_2}\frac{\partial\boldsymbol{U}}{\partial\xi_2}+A_{\xi_3}\frac{\partial\boldsymbol{U}}{\partial\xi_3}+\boldsymbol{D}\right)\\
&=\rho a\boldsymbol{N}_W^\pm\cdot\left(V_{\xi_1}\frac{\partial\boldsymbol{V}}{\partial\xi_1}+V_{\xi_2}\frac{\partial\boldsymbol{V}}{\partial\xi_2}+V_{\xi_3}\frac{\partial\boldsymbol{V}}{\partial\xi_3}\right)\\
&\quad+a\boldsymbol{N}_W^\pm\cdot\left(\boldsymbol{\xi}_1\frac{\partial p}{\partial\xi_1}+\boldsymbol{\xi}_2\frac{\partial p}{\partial\xi_2}+\boldsymbol{\xi}_3\frac{\partial p}{\partial\xi_3}\right)\\
&\quad+\rho a\boldsymbol{N}_W^\pm\cdot\boldsymbol{D}_{1-3}-a^2\rho\left(\boldsymbol{\xi}_1\cdot\frac{\partial\boldsymbol{V}}{\partial\xi_1}+\boldsymbol{\xi}_2\cdot\frac{\partial\boldsymbol{V}}{\partial\xi_2}+\boldsymbol{\xi}_3\cdot\frac{\partial\boldsymbol{V}}{\partial\xi_3}\right)\\
&\quad-a^2\left(V_{\xi_1}\frac{\partial\rho}{\partial\xi_1}+V_{\xi_2}\frac{\partial\rho}{\partial\xi_2}+V_{\xi_3}\frac{\partial\rho}{\partial\xi_3}\right)
\end{aligned}$$

$$-a^2d_4-\left(V_{\xi_1}\frac{\partial p}{\partial \xi_1}+V_{\xi_2}\frac{\partial p}{\partial \xi_2}+V_{\xi_3}\frac{\partial p}{\partial \xi_3}\right)$$

$$+a^2\left(V_{\xi_1}\frac{\partial \rho}{\partial \xi_1}+V_{\xi_2}\frac{\partial \rho}{\partial \xi_2}+V_{\xi_3}\frac{\partial \rho}{\partial \xi_3}\right)-d_5$$

$$=\rho a(V_{\xi_1}\boldsymbol{N}_W^{\pm}-a\boldsymbol{\xi}_1)\cdot\frac{\partial \boldsymbol{V}}{\partial \xi_1}+\rho a(V_{\xi_2}\boldsymbol{N}_W^{\pm}-a\boldsymbol{\xi}_2)\cdot\frac{\partial \boldsymbol{V}}{\partial \xi_2}$$

$$+\rho a(V_{\xi_3}\boldsymbol{N}^{\pm}-a\xi_3)\cdot\frac{\partial \boldsymbol{V}}{\partial \xi_3}+(a\boldsymbol{N}_W^{\pm}\cdot\boldsymbol{\xi}_1-V_{\xi_1})\frac{\partial p}{\partial \xi_1}$$

$$+(a\boldsymbol{N}_W^{\pm}\cdot\boldsymbol{\xi}_2-V_{\xi_2})\frac{\partial p}{\partial \xi_2}+(a\boldsymbol{N}_W^{\pm}\cdot\boldsymbol{\xi}_3-V_{\xi_3})\frac{\partial p}{\partial \xi_3}$$

$$+\rho a\boldsymbol{N}_W^{\pm}\cdot\boldsymbol{D}_{1-3}-a^2d_4-d_5=0,$$

where $i=1$, m, and $i=1$ corresponds to $\boldsymbol{N}_W^{+}$ and $i=m$ to $\boldsymbol{N}_W^{-}$. Futhermore, by virtue of

$$V_{\xi_2}\boldsymbol{N}_W^{\pm}-a\boldsymbol{\xi}_2=\frac{a}{\sigma_\pm V_{\xi_1}-V_{\xi_2}}\,[V_{\xi_2}(\sigma_\pm\boldsymbol{\xi}_1-\boldsymbol{\xi}_2)-(\sigma_\pm V_{\xi_1}-V_{\xi_2})\boldsymbol{\xi}_2]$$

$$=\frac{a}{\sigma_\pm V_{\xi_1}-V_{\xi_2}}\,(V_{\xi_2}\sigma_\pm\boldsymbol{\xi}_1-V_{\xi_1}\sigma_\pm\boldsymbol{\xi}_2)=\frac{\sigma_\pm a\boldsymbol{N}_s}{\sigma_\pm V_{\xi_1}-V_{\xi_2}}$$

$$=\sigma_\pm\frac{a}{\sigma_\pm V_{\xi_1}-V_{\xi_2}}\,[V_{\xi_1}(\sigma_\pm\boldsymbol{\xi}_1-\boldsymbol{\xi}_2)-(\sigma_\pm V_{\xi_1}-V_{\xi_2})\boldsymbol{\xi}_1]$$

$$=\sigma_\pm(V_{\xi_1}\boldsymbol{N}_W^{\pm}-a\boldsymbol{\xi}_1),$$

$$a\boldsymbol{\xi}_2\cdot\boldsymbol{N}_W^{\pm}-V_{\xi_2}=\frac{a^2}{\sigma_\pm V_{\xi_1}-V_{\xi_2}}\,(\sigma_\pm\boldsymbol{\xi}_2\cdot\boldsymbol{\xi}_1-\boldsymbol{\xi}_2\cdot\boldsymbol{\xi}_2)-V_{\xi_2}$$

$$=\frac{1}{\sigma_\pm V_{\xi_1}-V_{\xi_2}}\,(\sigma_\pm a^2\boldsymbol{\xi}_1\cdot\boldsymbol{\xi}_2-a^2\boldsymbol{\xi}_2\cdot\boldsymbol{\xi}_2-\sigma_\pm V_{\xi_1}V_{\xi_2}+V_{\xi_2}^2)$$

$$=\frac{1}{\sigma_\pm V_{\xi_1}-V_{\xi_2}}\,(-\sigma_\pm\beta_2+\beta_3)=\frac{1}{\sigma_\pm V_{\xi_1}-V_{\xi_2}}(-\beta_1\sigma_\pm^2+\beta_2\sigma_\pm)$$

$$=\frac{\pm\sigma_\pm}{V_{\xi_2}-\sigma_\pm V_{\xi_1}}\sqrt{\beta_2^2-\beta_1\beta_3}$$

$$=\frac{\sigma_\pm}{\sigma_\pm V_{\xi_1}-V_{\xi_2}}\,[-\sigma_\pm(V_{\xi_1}^2-a^2\boldsymbol{\xi}_1\cdot\boldsymbol{\xi}_1)+V_{\xi_1}V_{\xi_2}-a^2\boldsymbol{\xi}_1\cdot\boldsymbol{\xi}_2]$$

$$=\sigma_\pm\frac{1}{\sigma_\pm V_{\xi_1}-V_{\xi_2}}\,[a^2\boldsymbol{\xi}_1\cdot(\sigma_\pm\boldsymbol{\xi}_1-\boldsymbol{\xi}_2)-V_{\xi_1}(\sigma_\pm V_{\xi_1}-V_{\xi_2})]$$

$$=\sigma_\pm(a\boldsymbol{\xi}_1\cdot\boldsymbol{N}_W^{\pm}-V_{\xi_1}),$$

the compatibility relations can be rewritten in the form

$$\boldsymbol{G}^\pm\cdot\left(\frac{\partial \boldsymbol{V}}{\partial \xi_1}+\sigma_\pm\frac{\partial \boldsymbol{V}}{\partial \xi_2}\right)+g_4^\pm\left(\frac{\partial p}{\partial \xi_1}+\sigma_\pm\frac{\partial p}{\partial \xi_2}\right)$$

$$+\rho a(V_{\xi_3}\boldsymbol{N}_W^{\pm}-a\boldsymbol{\xi}_3)\cdot\frac{\partial \boldsymbol{V}}{\partial \xi_3}+(a\boldsymbol{N}_W^{\pm}\cdot\boldsymbol{\xi}_3-V_{\xi_3})\frac{\partial p}{\partial \xi_3}=-m_i,$$

where

$$\begin{cases} \boldsymbol{G}^{\pm}=\rho a(V_{\xi_1}\boldsymbol{N}_W^{\pm}-a\boldsymbol{\xi}_1)=\rho a^2\boldsymbol{N}_s/(\sigma_{\pm}V_{\xi_1}-V_{\xi_2}), \\ g_4^{\pm}=a\boldsymbol{N}_W^{\pm}\cdot\boldsymbol{\xi}_1-V_{\xi_1}=\pm\sqrt{\beta_2^2-\beta_1\beta_3}/(V_{\xi_2}-\sigma_{\pm}V_{\xi_1}), \\ m_i=\rho a\boldsymbol{N}_W^{\pm}\cdot\boldsymbol{D}_{1-3}-a^2d_4-d_5. \end{cases} \tag{1.57}$$

Thus, we have found all the characteristic surfaces (namely, one stream- and two wave-characteristic surfaces) through the given curve (1.42) and the independent compatibility relations on them. These relations are

$$\begin{cases} \boldsymbol{G}^{+}\cdot\left(\dfrac{\partial \boldsymbol{V}}{\partial\xi_1}+\sigma_{+}\dfrac{\partial \boldsymbol{V}}{\partial\xi_2}\right)+g_4^{+}\left(\dfrac{\partial p}{\partial\xi_1}+\sigma_{+}\dfrac{\partial p}{\partial\xi_2}\right) \\ \quad +\rho a(V_{\xi_1}\boldsymbol{N}_W^{+}-a\boldsymbol{\xi}_3)\cdot\dfrac{\partial \boldsymbol{V}}{\partial\xi_3}+(a\boldsymbol{N}_W^{+}\cdot\boldsymbol{\xi}_3-V_{\xi_3})\dfrac{\partial p}{\partial\xi_3}=-m_1, \\ V_{\xi_1}\boldsymbol{V}\cdot\left(\dfrac{\partial \boldsymbol{V}}{\partial\xi_1}+\sigma\dfrac{\partial \boldsymbol{V}}{\partial\xi_2}\right)+\dfrac{1}{\rho}V_{\xi_1}\left(\dfrac{\partial p}{\partial\xi_1}+\sigma\dfrac{\partial p}{\partial\xi_2}\right) \\ \quad +V_{\xi_3}\left(\boldsymbol{V}\cdot\dfrac{\partial \boldsymbol{V}}{\partial\xi_3}+\dfrac{1}{\rho}\dfrac{\partial p}{\partial\xi_3}\right)=-m_2, \\ V_{\xi_1}\boldsymbol{T}\cdot\left(\dfrac{\partial \boldsymbol{V}}{\partial\xi_1}+\sigma\dfrac{\partial \boldsymbol{V}}{\partial\xi_2}\right)+\dfrac{1}{\rho}V_{\xi_1}\boldsymbol{V}\times\boldsymbol{\xi}_1\cdot\boldsymbol{\xi}_2\left(\dfrac{\partial p}{\partial\xi_1}+\sigma\dfrac{\partial p}{\partial\xi_2}\right) \\ \quad +V_{\xi_3}\boldsymbol{T}\cdot\dfrac{\partial \boldsymbol{V}}{\partial\xi_3}+\dfrac{1}{\rho}\boldsymbol{T}\cdot\boldsymbol{\xi}_3\dfrac{\partial p}{\partial\xi_3}=-m_3, \\ V_{\xi_1}\left(\dfrac{\partial p}{\partial\xi_1}+\sigma\dfrac{\partial p}{\partial\xi_2}\right)-a^2V_{\xi_1}\left(\dfrac{\partial \rho}{\partial\xi_1}+\sigma\dfrac{\partial \rho}{\partial\xi_2}\right) \\ \quad +V_{\xi_3}\left(\dfrac{\partial p}{\partial\xi_3}-a^2\dfrac{\partial \rho}{\partial\xi_3}\right)=-m_4, \\ V_{\xi_1}\left(\dfrac{\partial c_i}{\partial\xi_1}+\sigma\dfrac{\partial c_i}{\partial\xi_2}\right)+V_{\xi_3}\dfrac{\partial c_i}{\partial\xi_3}=-m_{i+4}, \quad i=1,2,\cdots,m-5, \\ \boldsymbol{G}^{-}\cdot\left(\dfrac{\partial \boldsymbol{V}}{\partial\xi_1}+\sigma_{-}\dfrac{\partial \boldsymbol{V}}{\partial\xi_2}\right)+g_4^{-}\left(\dfrac{\partial p}{\partial\xi_1}+\sigma_{-}\dfrac{\partial p}{\partial\xi_2}\right) \\ \quad +\rho a(V_{\xi_1}\boldsymbol{N}_W^{-}-a\boldsymbol{\xi}_3)\cdot\dfrac{\partial \boldsymbol{V}}{\partial\xi_3}+(a\boldsymbol{N}_W^{-}\cdot\boldsymbol{\xi}_3-V_{\xi_3})\dfrac{\partial p}{\partial\xi_3}=-m_m, \end{cases} \tag{1.58}$$

which are called the $\xi_1-\xi_2$ characteristic form of (1.11), or shortly, the characteristic form of (1.11).

We notice that under certain conditions this set of equations is equivalent to the original system (1.11) or (1.16). In fact, since the system (1.58) was obtained by multiplying the system (1.11) with the matrix

$$\boldsymbol{\Omega}=\begin{pmatrix} \omega_1^* \\ \omega_2^* \\ \vdots \\ \omega_m^* \end{pmatrix}, \tag{1.59}$$

we need only to look for the conditions under which the matrix is invertible. To this end we now calculate its determinant:

$$\begin{vmatrix} \rho a N^+_{W,1} & \rho a N^+_{W,2} & \rho a N^+_{W,3} & -a^2 & -1 & 0 & \cdots & 0 & 0 \\ u_1 & u_2 & u_3 & 0 & 0 & 0 & \cdots & 0 & 0 \\ t_1 & t_2 & t_3 & 0 & 0 & 0 & \cdots & 0 & 0 \\ 0 & 0 & 0 & 0 & 1 & 0 & \cdots & 0 & 0 \\ \cdots & \cdots & \cdots & \cdots & \cdots & \cdots & \cdots & \cdots & \cdots \\ 0 & 0 & 0 & 0 & 0 & 0 & \cdots & 0 & 1 \\ \rho a N^-_{W,1} & \rho a N^-_{W,2} & \rho a N^-_{W,3} & -a^2 & -1 & 0 & \cdots & 0 & 0 \end{vmatrix}$$
$$=(-1)^{m+3}\rho a^3(\boldsymbol{N}^+_W-\boldsymbol{N}^-_W)\cdot\boldsymbol{V}\times\boldsymbol{T}.$$

It is not equal to zero if and only if $(\boldsymbol{N}^+_W-\boldsymbol{N}^-_W)\cdot\boldsymbol{V}\times\boldsymbol{T}\neq 0$. Furthermore, using (1.55) and (1.53) we have

$$\begin{aligned}
\boldsymbol{N}^\pm_W &= \frac{a(\sigma_\pm\boldsymbol{\xi}_1-\boldsymbol{\xi}_2)}{\sigma_\pm V_{\xi_1}-V_{\xi_2}} = \frac{a[(\beta_2\pm\sqrt{\beta_2^2-\beta_1\beta_3})\boldsymbol{\xi}_1-\beta_1\boldsymbol{\xi}_2]}{(\beta_2\pm\sqrt{\beta_2^2-\beta_1\beta_3})V_{\xi_1}-\beta_1 V_{\xi_2}} \\
&= a[\beta_2\boldsymbol{\xi}_1-\beta_1\boldsymbol{\xi}_2\pm\sqrt{\beta_2^2-\beta_1\beta_3}\,\boldsymbol{\xi}_1] \\
&\quad\times\frac{[\beta_2 V_{\xi_1}-\beta_1 V_{\xi_2}\mp\sqrt{\beta_2^2-\beta_1\beta_3}\,V_{\xi_1}]}{(\beta_2 V_{\xi_1}-\beta_1 V_{\xi_2})^2-(\beta_2^2-\beta_1\beta_3)V_{\xi_1}^2} \\
&= a\,\frac{(\beta_3 V_{\xi_1}-\beta_2 V_{\xi_2})\boldsymbol{\xi}_1-(\beta_2 V_{\xi_1}-\beta_1 V_{\xi_2})\boldsymbol{\xi}_2}{(\beta_3 V_{\xi_1}-\beta_2 V_{\xi_2})V_{\xi_1}-(\beta_2 V_{\xi_1}-\beta_1 V_{\xi_2})V_{\xi_2}} \\
&\quad\pm a\sqrt{\beta_2^2-\beta_1\beta_3}\,\frac{-V_{\xi_2}\boldsymbol{\xi}_1+V_{\xi_1}\boldsymbol{\xi}_2}{(\beta_3 V_{\xi_1}-\beta_2 V_{\xi_2})V_{\xi_1}-(\beta_2 V_{\xi_1}-\beta_1 V_{\xi_2})V_{\xi_2}} \\
&= a\,\frac{-(V_{\xi_2}\boldsymbol{\xi}_1-V_{\xi_1}\boldsymbol{\xi}_2)\cdot\boldsymbol{\xi}_2\boldsymbol{\xi}_1+(V_{\xi_2}\boldsymbol{\xi}_1-V_{\xi_1}\boldsymbol{\xi}_2)\cdot\boldsymbol{\xi}_1\boldsymbol{\xi}_2}{|V_{\xi_2}\boldsymbol{\xi}_1-V_{\xi_1}\boldsymbol{\xi}_2|^2} \\
&\quad\pm\frac{\sqrt{\beta_2^2-\beta_1\beta_3}}{a}\,\frac{V_{\xi_2}\boldsymbol{\xi}_1-V_{\xi_1}\boldsymbol{\xi}_2}{|V_{\xi_2}\boldsymbol{\xi}_1-V_{\xi_1}\boldsymbol{\xi}_2|^2} \\
&= a\,\frac{-\boldsymbol{N}_s\cdot\boldsymbol{\xi}_2\boldsymbol{\xi}_1+\boldsymbol{N}_s\cdot\boldsymbol{\xi}_1\boldsymbol{\xi}_2}{|\boldsymbol{N}_s|^2}\pm\frac{\sqrt{\beta_2^2-\beta_1\beta_3}}{a}\,\frac{\boldsymbol{N}_s}{|\boldsymbol{N}_s|^2},
\end{aligned}\tag{1.60}$$

and therefore

$$\boldsymbol{N}^+_W-\boldsymbol{N}^-_W=\frac{2\sqrt{\beta_2^2-\beta_1\beta_3}}{a|\boldsymbol{N}_s|^2}\,\boldsymbol{N}_s,\tag{1.61}$$

which implies that $\boldsymbol{N}^+_W-\boldsymbol{N}^-_W$ has the same direction as $\boldsymbol{N}_s$. Since $\boldsymbol{N}_s$ is orthogonal to $\boldsymbol{V}$, the vectors $\boldsymbol{N}^+_W-\boldsymbol{N}^-_W$, $\boldsymbol{V}$ and $\boldsymbol{T}=\boldsymbol{V}\times\boldsymbol{N}_s$ are mutually orthogonal. Hence the matrix Ω is invertible when

$$\beta_2^2-\beta_1\beta_3\neq 0,\quad |\boldsymbol{N}_s|\neq 0,\quad |\boldsymbol{V}|\neq 0.\tag{1.62}$$

In addition, according to (1.52) we have

$$\begin{aligned}
\beta_2^2-\beta_1\beta_3 &= V_{\xi_1}^2V_{\xi_2}^2-2V_{\xi_1}V_{\xi_2}a^2\boldsymbol{\xi}_1\cdot\boldsymbol{\xi}_2+a^4|\boldsymbol{\xi}_1\cdot\boldsymbol{\xi}_2|^2 \\
&\quad -V_{\xi_1}^2V_{\xi_2}^2-a^4|\boldsymbol{\xi}_1|^2|\boldsymbol{\xi}_2|^2+V_{\xi_1}^2a^2|\boldsymbol{\xi}_2|^2+V_{\xi_2}^2a^2|\boldsymbol{\xi}_1|^2 \\
&= a^2[|V_{\xi_2}\boldsymbol{\xi}_1-V_{\xi_1}\boldsymbol{\xi}_2|^2-a^2(|\boldsymbol{\xi}_1|^2|\boldsymbol{\xi}_2|^2-|\boldsymbol{\xi}_1\cdot\boldsymbol{\xi}_2|^2)],
\end{aligned}\tag{1.63}$$

and therefore the inequality $\beta_2^2-\beta_1\beta_3>0$ implies $|N_s|=|V_{\xi_2}\xi_1-V_{\xi_1}\xi_2|\neq 0$ and $|V|\neq 0$. Thus we conclude that the equivalency of (1.58) and (1.11) holds if

$$\beta_2^2-\beta_1\beta_3>0, \tag{1.64}$$

or equivalently,

$$\frac{|V_{\xi_2}\xi_1-V_{\xi_1}\xi_2|^2}{|\xi_1|^2|\xi_2|^2-|\xi_1\cdot\xi_2|^2}>a^2, \tag{1.65}$$

where we suppose that ξ_1 is not parallel to ξ_2.

The condition (1.65) physically signifies that the projection of the velocity V in the normal X to the plane on which the vectors N_s and $\xi_1\times\xi_2$ lie, is greater than the speed of sound a. In fact, since X is perpendicular to $\xi_1\times\xi_2$, it can be represented in the form

$$X=x_1\xi_1+x_2\xi_2;$$

and since X is also perpendicular to N_s, we have

$$X\cdot N_s=x_1\xi_1\cdot N_s+x_2\xi_2\cdot N_s=0.$$

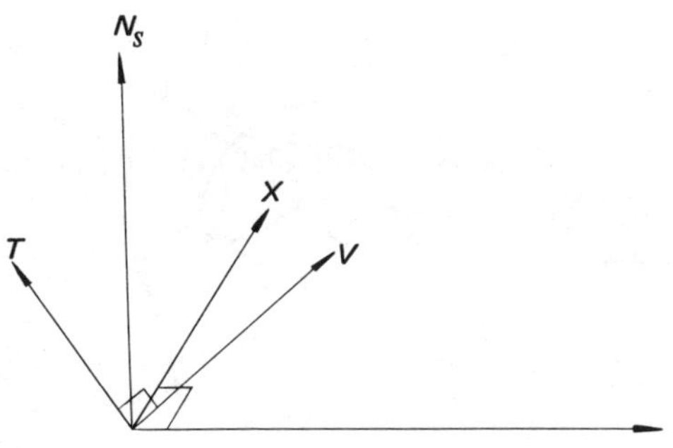

Figure 1.5

From these we can represent the unit vector X in the form

$$X=\frac{(\xi_2\cdot N_s)\xi_1-(\xi_1\cdot N_s)\xi_2}{|(\xi_2\cdot N_s)\xi_1-(\xi_1\cdot N_s)\xi_2|}$$

$$=\frac{\xi_2\cdot(V_{\xi_2}\xi_1-V_{\xi_1}\xi_2)\xi_1-\xi_1\cdot(V_{\xi_2}\xi_1-V_{\xi_1}\xi_2)\xi_2}{|\xi_2\cdot(V_{\xi_2}\xi_1-V_{\xi_1}\xi_2)\xi_1-\xi_1\cdot(V_{\xi_2}\xi_1-V_{\xi_1}\xi_2)\xi_2|}.$$

Therefore

$$V_X^2\equiv(V\cdot X)^2=\frac{|V_{\xi_2}\xi_1-V_{\xi_1}\xi_2|^4}{|\xi_2\cdot(V_{\xi_2}\xi_1-V_{\xi_1}\xi_2)\xi_1-\xi_1\cdot(V_{\xi_2}\xi_1-V_{\xi_1}\xi_2)\xi_2|^2}.$$

Noticing

$$|\xi_2\cdot(V_{\xi_2}\xi_1-V_{\xi_1}\xi_2)\xi_1-\xi_1\cdot(V_{\xi_2}\xi_1-V_{\xi_1}\xi_2)\xi_2|^2$$

$$=(V_{\xi_2}\xi_1\cdot\xi_2-V_{\xi_1}|\xi_2|^2)^2|\xi_1|^2-2(V_{\xi_2}\xi_1\cdot\xi_2-V_{\xi_1}|\xi_2|^2)$$

$$\times(V_{\xi_2}|\boldsymbol{\xi}_1|^2-V_{\xi_1}\boldsymbol{\xi}_1\cdot\boldsymbol{\xi}_2)\boldsymbol{\xi}_1\cdot\boldsymbol{\xi}_2+(V_{\xi_2}|\boldsymbol{\xi}_1|^2-V_{\xi_1}\boldsymbol{\xi}_1\cdot\boldsymbol{\xi}_2)^2|\boldsymbol{\xi}_2|^2$$
$$=(V_{\xi_1}^2|\boldsymbol{\xi}_2|^2-2V_{\xi_1}V_{\xi_2}\boldsymbol{\xi}_1\cdot\boldsymbol{\xi}_2+V_{\xi_2}^2|\boldsymbol{\xi}_1|^2)[|\boldsymbol{\xi}_1|^2|\boldsymbol{\xi}_2|^2-(\boldsymbol{\xi}_1\cdot\boldsymbol{\xi}_2)^2]$$
$$=|V_{\xi_2}\boldsymbol{\xi}_1-V_{\xi_1}\boldsymbol{\xi}_2|^2[|\boldsymbol{\xi}_1|^2|\boldsymbol{\xi}_2|^2-|\boldsymbol{\xi}_1\cdot\boldsymbol{\xi}_2|^2],$$

we can further obtain

$$V_X^2=\frac{|V_{\xi_2}\boldsymbol{\xi}_1-V_{\xi_1}\boldsymbol{\xi}_2|^2}{|\boldsymbol{\xi}_1|^2|\boldsymbol{\xi}_2|^2-(\boldsymbol{\xi}_1\cdot\boldsymbol{\xi}_2)^2},$$

which validates the above statement.

It is clear that the vectors $\boldsymbol{X}$, $\boldsymbol{V}$ and $\boldsymbol{\xi}_1\times\boldsymbol{\xi}_2$ all lie on the same plane. Consequently, $\boldsymbol{X}$ is the normal to $\boldsymbol{\xi}_1\times\boldsymbol{\xi}_2$ in the plane on which $\boldsymbol{V}$ and $\boldsymbol{\xi}_1\times\boldsymbol{\xi}_2$ lie. Thus if the curve (1.42) lies on a space-like surface, then the projection of $\boldsymbol{V}$ in the direction $\boldsymbol{X}$ is greater than the sound speed, i.e., the condition (1.65) holds (see Fig. 1.6), which means that the system (1.58) is equivalent to (1.11) if the curve (1.42) lies on a space-like surface.

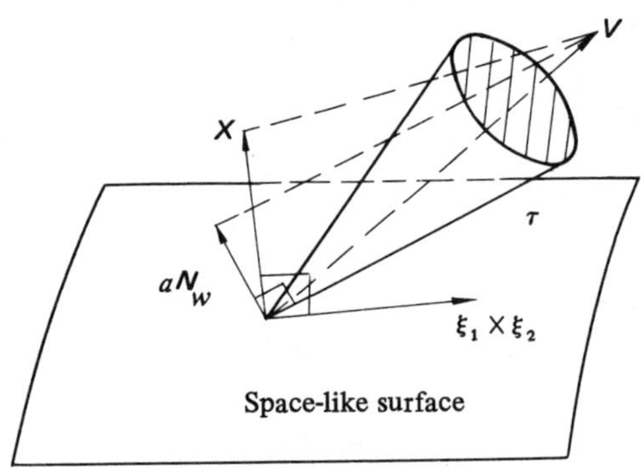

Figure 1.6

As indicated above, the characteristics for a given physical problem are invariant with respect to any transformation of coordinates. This means that the equations (1.58) can be derived directly from the equations (1.16). We now verify this result. In fact, in the ξ_1, ξ_2, ξ_3-space, the equation defining the curve (1.42) takes the form

$$\begin{cases}\xi_1=\text{constant},\\ \xi_2=\text{constant},\end{cases} \tag{1.66}$$

and the tangent vector to it can be represented in the form

$$\tilde{\boldsymbol{\tau}}=\begin{pmatrix}0\\0\\1\end{pmatrix}.$$

Then if we start with the equation

$$A_{\xi_1}\frac{\partial \boldsymbol{U}}{\partial \xi_1}+A_{\xi_2}\frac{\partial \boldsymbol{U}}{\partial \xi_2}+A_{\xi_3}\frac{\partial \boldsymbol{U}}{\partial \xi_3}+\boldsymbol{D}=0$$

and find all its characteristic normals, which may be expressed in the form

$$\tilde{\boldsymbol{N}}=\begin{pmatrix}a_1\\a_2\\0\end{pmatrix}, \tag{1.67}$$

and the compatibility relations on the corresponding characteristic surfaces through the given curve, we shall obtain the equations (1.58). In fact, $\tilde{\boldsymbol{N}}$ satisfies

$$|a_1A_{\xi_1}+a_2A_{\xi_2}|=0 \tag{1.68}$$

and the corresponding compatibility relations can be written as

$$\boldsymbol{\omega}^*\left(A_{\xi_1}\frac{\partial \boldsymbol{U}}{\partial \xi_1}+A_{\xi_2}\frac{\partial \boldsymbol{U}}{\partial \xi_2}+A_{\xi_3}\frac{\partial \boldsymbol{U}}{\partial \xi_3}+\boldsymbol{D}\right)=0, \tag{1.69}$$

where $\boldsymbol{\omega}$ is a solution of the equation

$$\boldsymbol{\omega}^*(a_1A_{\xi_1}+a_2A_{\xi_2})=0. \tag{1.70}$$

Noting (see the deduction of (1.16))

$$\begin{aligned}&A_1\frac{\partial \boldsymbol{U}}{\partial x_1}+A_2\frac{\partial \boldsymbol{U}}{\partial x_2}+A_3\frac{\partial \boldsymbol{U}}{\partial x_3}+\boldsymbol{D}\\&=A_{\xi_1}\frac{\partial \boldsymbol{U}}{\partial \xi_1}+A_{\xi_2}\frac{\partial \boldsymbol{U}}{\partial \xi_2}+A_{\xi_3}\frac{\partial \boldsymbol{U}}{\partial \xi_3}+\boldsymbol{D},\end{aligned}$$

we see that to validate the above statement it suffices to verify that a_1 and a_2 defined in (1.43) and the corresponding $\boldsymbol{\omega}$ satisfy (1.68) and (1.70). In fact, using (1.43), we have

$$\begin{aligned}&\frac{\partial \Phi}{\partial x_1}A_1+\frac{\partial \Phi}{\partial x_2}A_2+\frac{\partial \Phi}{\partial x_3}A_3\\&=N_1H_1A_1+N_2H_2A_2+N_3H_3A_3\\&=a_1\sum_i\frac{\partial \xi_1}{\partial x_i}A_i+a_2\sum_i\frac{\partial \xi_2}{\partial x_i}A_i=a_1A_{\xi_1}+a_2A_{\xi_2},\end{aligned}$$

and therefore, since $\boldsymbol{N}$ satisfies the equation (1.28), we are led to

$$|a_1A_{\xi_1}+a_2A_{\xi_2}|=0$$

and correspondingly

$$\boldsymbol{\omega}^*(a_1A_{\xi_1}+a_2A_{\xi_2})=0,$$

as what we want to prove.

We have proved that the system (1.58), a set of compatibility relations on the characteristic surfaces through the given curve (1.42) in the original coordinate space, is a set of compatibility relations on the characteristics through the straightline (1.66) in the new coordinate space. This is the reason why we are interested in (1.58). In fact, as will be seen in the next three chapters, by virtue of this property of (1.58), the use of the new coordinates and of the corresponding differential equations (1.58) is beneficial to constructing difference schemes. The equations (1.58) could be derived directly from (1.16). But we started rather from (1.11) in physical space so that the physical meaning might be observed more easily and intuitively.

Incidentally, it can be seen from (1.68) and (1.70) that $\sigma_\pm = -(a_{W,1}/a_{W,2})_\pm$ and $\sigma = -a_{s,1}/a_{s,2}$ are actually the eigenvalues of the matrix $A_{\xi_1}^{-1}A_{\xi_2}$ and the vector $\boldsymbol{\omega}^* A_{\xi_1}$ is a left-eigenvector of it, that is, for each $\boldsymbol{\omega}$ above there is a certain σ_i such that

$$\boldsymbol{\omega}^* A_{\xi_1}(\sigma_i I - A_{\xi_1}^{-1}A_{\xi_2}) = 0.$$

We point out that if $\beta_1 = V_{\xi_1}^2 - a^2|\boldsymbol{\xi}_1|^2 > 0$ (i.e. the surface $\xi_1(x_1, x_2, x_3) =$ constant is space-like) and $V_{\xi_1} > 0$, we have

$$\sigma_+ > \sigma > \sigma_-. \tag{1.71}$$

Hence, as we usually distinguish between two different families of characteristic curves in the case of two-dimensional flow, we sometimes classify the wave-characteristic surfaces through given curves into two families: those corresponding to σ_+ belong to the first family and those corresponding to σ_- belong to the second one. The proof of inequality (1.71) is simple. In fact, according to the definitions of σ_+, σ, and σ_-, we may write (1.71) in another form:

$$\frac{\beta_2 + (\beta_2^2 - \beta_1\beta_3)^{\frac{1}{2}}}{\beta_1} > \frac{V_{\xi_2}}{V_{\xi_1}} > \frac{\beta_2 - (\beta_2^2 - \beta_1\beta_3)^{\frac{1}{2}}}{\beta_1}.$$

Since $\beta_1 > 0$ and $V_{\xi_1} > 0$, the above inequality is equivalent to

$$\begin{cases} V_{\xi_1}(\beta_2^2 - \beta_1\beta_3)^{\frac{1}{2}} > \beta_1 V_{\xi_2} - \beta_2 V_{\xi_1}, \\ V_{\xi_1}(\beta_2^2 - \beta_1\beta_3)^{\frac{1}{2}} > \beta_2 V_{\xi_1} - \beta_1 V_{\xi_2}, \end{cases}$$

or

$$V_{\xi_1}(\beta_2^2 - \beta_1\beta_3)^{\frac{1}{2}} > |\beta_2 V_{\xi_1} - \beta_1 V_{\xi_2}| = a^2 |V_{\xi_1}\boldsymbol{\xi}_1\cdot\boldsymbol{\xi}_2 - V_{\xi_2}\boldsymbol{\xi}_1\cdot\boldsymbol{\xi}_1|.$$

Squaring each side of the inequality and in view of (1.63), we obtain

$$\begin{aligned} &a^2 V_{\xi_1}^2 [|V_{\xi_2}\boldsymbol{\xi}_1 - V_{\xi_1}\boldsymbol{\xi}_2|^2 - a^2(|\boldsymbol{\xi}_1|^2|\boldsymbol{\xi}_2|^2 - |\boldsymbol{\xi}_1\cdot\boldsymbol{\xi}_2|^2)] \\ &\quad > a^4(V_{\xi_1}^2(\boldsymbol{\xi}_1\cdot\boldsymbol{\xi}_2)^2 - 2V_{\xi_1}V_{\xi_2}\boldsymbol{\xi}_1\cdot\boldsymbol{\xi}_2|\boldsymbol{\xi}_1|^2 + V_{\xi_2}^2|\boldsymbol{\xi}_1|^4), \end{aligned}$$

or, in another form,

$$a^2(V_{\xi_1}^2-a^2|\xi_1|^2)|V_{\xi_2}\xi_1-V_{\xi_1}\xi_2|^2>0.$$

This inequality holds under the condition given above and therefore the proof is completed.

As stated above, in the case of $\beta_2^2-\beta_1\beta_3>0$, we can derive from (1.16) the equations (1.58) by finding all the roots σ_i of the algebraic equation

$$|\sigma_i A_{\xi_1}-A_{\xi_2}|=0$$

and the corresponding solution vectors of the linear system

$$\omega_i^*(\sigma_i A_{\xi_1}-A_{\xi_2})=0$$

and then by multiplying from the left the equations (1.16) with the obtained matrix

$$\begin{pmatrix}\omega_1^*\\ \vdots\\ \omega_m^*\end{pmatrix}.$$

These operations could also occur even for the case of $\beta_2^2-\beta_1\beta_3<0$. However in that case $\sigma_\pm$ as well as ω_1 and ω_m would not be real numbers anymore, and therefore the corresponding equations (the first and the m-th) could not be the compatibility relations. Even so, the use of (1.58) instead of (1.16) is of benefit for practical applications. This is because the equations from the second to the $(m-1)$-th in (1.58) possess the property of hyperbolic equations and the other two possess the property of elliptical equations. Such properties should be considered when we use difference methods for solving the differential equations. Of course, the operations involving the complex are not convenient. To avoid such complication we may take the real part and the imaginary part of the first equation instead of taking the first and the m-th equations. We now derive these two equations. From (1.60), the real part and the imaginary part of N_W^+ have the forms

$$\begin{cases}\operatorname{Re} N_W^+=-a\,\dfrac{(N_s\cdot\xi_2)\xi_1-(N_s\cdot\xi_1)\xi_2}{|N_s|^2},\\[2ex] \operatorname{Im} N_W^+=\sqrt{\beta_1\beta_3-\beta_2^2}\,\dfrac{N_s}{a|N_s|^2},\end{cases}\tag{1.72}$$

where

$$N_s=V_{\xi_2}\xi_1-V_{\xi_1}\xi_2.$$

Consequently, the corresponding real-part equation can be written as

$$(\rho a^2/|N_s|^2)\Big(-V_{\xi_1}N_s\cdot\xi_2\xi_1\cdot\frac{\partial V}{\partial\xi_1}+V_{\xi_1}N_s\cdot\xi_1\xi_2\cdot\frac{\partial V}{\partial\xi_1}$$

$$-V_{\xi_2}N_s\cdot\xi_2\xi_1\cdot\frac{\partial V}{\partial\xi_2}+V_{\xi_2}N_s\cdot\xi_1\xi_2\cdot\frac{\partial V}{\partial\xi_2}-V_{\xi_3}N_s\cdot\xi_2\xi_1\cdot\frac{\partial V}{\partial\xi_3}$$

$$+V_{\xi_3}\boldsymbol{N}_s\cdot\boldsymbol{\xi}_1\boldsymbol{\xi}_2\cdot\frac{\partial \boldsymbol{V}}{\partial\xi_3}-|\boldsymbol{N}_s|^2\left[\boldsymbol{\xi}_1\cdot\frac{\partial \boldsymbol{V}}{\partial\xi_1}+\boldsymbol{\xi}_2\cdot\frac{\partial \boldsymbol{V}}{\partial\xi_2}+\boldsymbol{\xi}_3\cdot\frac{\partial \boldsymbol{V}}{\partial\xi_3}\right]\Big)$$

$$+(a^2/|\boldsymbol{N}_s|^2)\Big[(-\boldsymbol{N}_s\cdot\boldsymbol{\xi}_2\boldsymbol{\xi}_1\cdot\boldsymbol{\xi}_1+\boldsymbol{N}_s\cdot\boldsymbol{\xi}_1\boldsymbol{\xi}_2\cdot\boldsymbol{\xi}_1)\frac{\partial p}{\partial\xi_1}$$

$$+(-\boldsymbol{N}_s\cdot\boldsymbol{\xi}_2\boldsymbol{\xi}_1\cdot\boldsymbol{\xi}_2+\boldsymbol{N}_s\cdot\boldsymbol{\xi}_1\boldsymbol{\xi}_2\cdot\boldsymbol{\xi}_2)\frac{\partial p}{\partial\xi_2}$$

$$+(-\boldsymbol{N}_s\cdot\boldsymbol{\xi}_2\boldsymbol{\xi}_1\cdot\boldsymbol{\xi}_3+\boldsymbol{N}_s\cdot\boldsymbol{\xi}_1\boldsymbol{\xi}_2\cdot\boldsymbol{\xi}_3)\frac{\partial p}{\partial\xi_3}\Big]-V_{\xi_1}\frac{\partial p}{\partial\xi_1}-V_{\xi_1}\frac{\partial p}{\partial\xi_2}$$

$$-V_{\xi_1}\frac{\partial p}{\partial\xi_3}+(\rho a^2/|\boldsymbol{N}_s|^2)(-\boldsymbol{N}_s\cdot\boldsymbol{\xi}_2\boldsymbol{\xi}_1+\boldsymbol{N}_s\cdot\boldsymbol{\xi}_1\boldsymbol{\xi}_2)\cdot\boldsymbol{D}_{1-3}$$

$$-a^2d_4-d_5=0,$$

or, after reduction,

$$-(\rho a^2\boldsymbol{N}_s\cdot\boldsymbol{\xi}_1/|\boldsymbol{N}_s|^2)\boldsymbol{N}_s\cdot\left(\frac{\partial \boldsymbol{V}}{\partial\xi_1}+\sigma_f\frac{\partial \boldsymbol{V}}{\partial\xi_2}\right)$$

$$+g_5V_{\xi_1}\left(\frac{\partial p}{\partial\xi_1}+\sigma\frac{\partial p}{\partial\xi_2}\right)+\frac{\rho a^2}{|\boldsymbol{N}_s|^2}(V_{\xi_1}\boldsymbol{S}-|\boldsymbol{N}_s|^2\boldsymbol{\xi}_3)\cdot\frac{\partial \boldsymbol{V}}{\partial\xi_3}$$

$$+\left(\frac{a^2}{|\boldsymbol{N}_s|^2}\boldsymbol{S}\cdot\boldsymbol{\xi}_3-V_{\xi_3}\right)\frac{\partial p}{\partial\xi_3}=-\widetilde{m}_1,$$

where

$$\begin{cases}\sigma_f=\dfrac{\boldsymbol{N}_s\cdot\boldsymbol{\xi}_2}{\boldsymbol{N}_s\cdot\boldsymbol{\xi}_1}, \quad g_5=\dfrac{a^2}{|\boldsymbol{N}_s|^2}[|\boldsymbol{\xi}_1|^2|\boldsymbol{\xi}_2|^2-(\boldsymbol{\xi}_1\cdot\boldsymbol{\xi}_2)^2]-1,\\ \boldsymbol{S}=-\boldsymbol{N}_s\cdot\boldsymbol{\xi}_2\boldsymbol{\xi}_1+\boldsymbol{N}_s\cdot\boldsymbol{\xi}_1\boldsymbol{\xi}_2,\\ \widetilde{m}_1=\dfrac{\rho a^2}{|\boldsymbol{N}_s|^2}\boldsymbol{S}\cdot\boldsymbol{D}_{1-3}-a^2d_4-d_5.\end{cases} \tag{1.73}$$

The imaginary-part equation divided by $\rho a\sqrt{\beta_1\beta_3-\beta_2^2}/(a|\boldsymbol{N}_s|^2)$ is

$$V_{\xi_1}\boldsymbol{N}_s\cdot\frac{\partial \boldsymbol{V}}{\partial\xi_1}+V_{\xi_2}\boldsymbol{N}_s\cdot\frac{\partial \boldsymbol{V}}{\partial\xi_2}+V_{\xi_3}\boldsymbol{N}_s\cdot\frac{\partial \boldsymbol{V}}{\partial\xi_3}+\frac{1}{\rho}\boldsymbol{N}_s\cdot\boldsymbol{\xi}_1\frac{\partial p}{\partial\xi_1}$$

$$+\frac{1}{\rho}\boldsymbol{N}_s\cdot\boldsymbol{\xi}_2\frac{\partial p}{\partial\xi_2}+\frac{1}{\rho}\boldsymbol{N}_s\cdot\boldsymbol{\xi}_3\frac{\partial p}{\partial\xi_3}+\boldsymbol{N}_s\cdot\boldsymbol{D}_{1-3}=0,$$

or

$$V_{\xi_1}\boldsymbol{N}_s\cdot\left(\frac{\partial \boldsymbol{V}}{\partial\xi_1}+\sigma\frac{\partial \boldsymbol{V}}{\partial\xi_2}\right)+\frac{\boldsymbol{N}_s\cdot\boldsymbol{\xi}_1}{\rho}\left(\frac{\partial p}{\partial\xi_1}+\sigma_f\frac{\partial p}{\partial\xi_2}\right)$$

$$+V_{\xi_3}\boldsymbol{N}_s\cdot\frac{\partial \boldsymbol{V}}{\partial\xi_3}+\frac{1}{\rho}\boldsymbol{N}_s\cdot\boldsymbol{\xi}_3\frac{\partial p}{\partial\xi_3}=-\widetilde{m}_m,$$

where

$$\widetilde{m}_m=\boldsymbol{N}_s\cdot\boldsymbol{D}_{1-3}. \tag{1.74}$$

Finally, in the case of $\beta_2^2-\beta_1\beta_3<0$, the equations which are derived from (1.16) in the way stated above have the form

$$
\left\{
\begin{array}{l}
-\dfrac{\rho a^2 \boldsymbol{N}_s\cdot\boldsymbol{\xi}_1}{|\boldsymbol{N}_s|^2}\boldsymbol{N}_s\cdot\left(\dfrac{\partial \boldsymbol{V}}{\partial\xi_1}+\sigma_f\dfrac{\partial \boldsymbol{V}}{\partial\xi_2}\right)+g_5 V_{\xi_1}\left(\dfrac{\partial p}{\partial\xi_1}+\sigma\dfrac{\partial p}{\partial\xi_2}\right)\\
\quad+\dfrac{\rho a^2}{|\boldsymbol{N}_s|^2}(V_{\xi_1}\boldsymbol{S}-|\boldsymbol{N}_s|^2\boldsymbol{\xi}_3)\cdot\dfrac{\partial \boldsymbol{V}}{\partial\xi_3}\\
\quad+\left(\dfrac{a^2}{|\boldsymbol{N}_s|^2}\boldsymbol{S}\cdot\boldsymbol{\xi}_3-V_{\xi_1}\right)\dfrac{\partial p}{\partial\xi_3}=-\widetilde{m}_1,\\
V_{\xi_1}\boldsymbol{V}\cdot\left(\dfrac{\partial \boldsymbol{V}}{\partial\xi_1}+\sigma\dfrac{\partial \boldsymbol{V}}{\partial\xi_2}\right)+\dfrac{1}{\rho}V_{\xi_1}\left(\dfrac{\partial p}{\partial\xi_1}+\sigma\dfrac{\partial p}{\partial\xi_2}\right)\\
\quad+V_{\xi_1}\left(\boldsymbol{V}\cdot\dfrac{\partial \boldsymbol{V}}{\partial\xi_3}+\dfrac{1}{\rho}\dfrac{\partial p}{\partial\xi_3}\right)=-m_2,\\
V_{\xi_1}\boldsymbol{T}\cdot\left(\dfrac{\partial \boldsymbol{V}}{\partial\xi_1}+\sigma\dfrac{\partial \boldsymbol{V}}{\partial\xi_2}\right)+\dfrac{1}{\rho}V_{\xi_1}\boldsymbol{V}\times\boldsymbol{\xi}_1\cdot\boldsymbol{\xi}_2\left(\dfrac{\partial p}{\partial\xi_1}+\sigma\dfrac{\partial p}{\partial\xi_2}\right)\\
\quad+V_{\xi_1}\boldsymbol{T}\cdot\dfrac{\partial \boldsymbol{V}}{\partial\xi_3}+\dfrac{1}{\rho}\boldsymbol{T}\cdot\boldsymbol{\xi}_3\dfrac{\partial p}{\partial\xi_3}=-m_3,\\
V_{\xi_1}\left(\dfrac{\partial p}{\partial\xi_1}+\sigma\dfrac{\partial p}{\partial\xi_2}\right)-a^2V_{\xi_1}\left(\dfrac{\partial \rho}{\partial\xi_1}+\sigma\dfrac{\partial \rho}{\partial\xi_2}\right)\\
\quad+V_{\xi_1}\left(\dfrac{\partial p}{\partial\xi_3}-a^2\dfrac{\partial \rho}{\partial\xi_3}\right)=-m_4,\\
V_{\xi_1}\left(\dfrac{\partial c_i}{\partial\xi_1}+\sigma\dfrac{\partial c_i}{\partial\xi_2}\right)+V_{\xi_1}\dfrac{\partial c_i}{\partial\xi_3}=-m_{i+4},\quad i=1,2,\cdots,m-5,\\
V_{\xi_1}\boldsymbol{N}_s\cdot\left(\dfrac{\partial \boldsymbol{V}}{\partial\xi_1}+\sigma\dfrac{\partial \boldsymbol{V}}{\partial\xi_2}\right)+\dfrac{\boldsymbol{N}_s\cdot\boldsymbol{\xi}_1}{\rho}\left(\dfrac{\partial p}{\partial\xi_1}+\sigma_f\dfrac{\partial p}{\partial\xi_2}\right)\\
\quad+V_{\xi_1}\boldsymbol{N}_s\cdot\dfrac{\partial \boldsymbol{V}}{\partial\xi_3}+\dfrac{1}{\rho}\boldsymbol{N}_s\cdot\boldsymbol{\xi}_3\dfrac{\partial p}{\partial\xi_3}=-\widetilde{m}_m.
\end{array}
\right.
\tag{1.75}
$$

This system of equations can be obtained by left-multiplying (1.16) with the matrix

$$
\widetilde{\boldsymbol{\Omega}}=\begin{pmatrix}\widetilde{\boldsymbol{\omega}}_1^*\\ \boldsymbol{\omega}_2^*\\ \vdots\\ \boldsymbol{\omega}_{m-1}^*\\ \widetilde{\boldsymbol{\omega}}_m^*\end{pmatrix},
$$

where

$$
\widetilde{\boldsymbol{\omega}}_1^*=\left(\rho a^2\frac{\boldsymbol{S}^*}{|\boldsymbol{N}_s|^2},\ -a^2,\ -1,\ 0,\ \cdots,\ 0\right),
$$

$$
\widetilde{\boldsymbol{\omega}}_m^*=(\boldsymbol{N}_s^*,\ 0,\ \cdots,\ 0).
$$

Its determinant is equal to $(-1)^m a^2\boldsymbol{V}\times\boldsymbol{T}\cdot\boldsymbol{N}_s$. But, as proved previously, the vectors $\boldsymbol{V}$, $\boldsymbol{T}$ and $\boldsymbol{N}_s$ are mutually orthogonal, and therefore the equations (1.75) are equivalent to (1.16) where $|V|\neq 0$ and $|\boldsymbol{N}_s|\neq 0$, no matter whether $\beta_2^2-\beta_1\beta_3$ is positive or not.

At the end of this section we discuss the "type" of differential equations. Consider the system (1.19)

$$B_1\frac{\partial U}{\partial\eta_1}+B_2\frac{\partial U}{\partial\eta_2}+B_3\frac{\partial U}{\partial\eta_3}+D=0,$$

where B_1, B_2 and B_3 are real square matrices of order m, and U and D are vectors with m components.

The system (1.19) is said to be hyperbolic at a given point in space if there exists a transformation of independent variables given by

$$\eta_j=\eta_j(\xi_1,\ \xi_2,\ \xi_3),\quad j=1,\ 2,\ 3$$

such that for properly chosen real numbers α_2 and α_3 the algebraic equation of the m-th degree for α_1

$$|\alpha_1\beta_{\xi_1}+\alpha_2B_{\xi_2}+\alpha_3B_{\xi_3}|=0 \tag{1.76}$$

has m real roots at the considered point, and that for every root α_1 the linear system of equations for $\boldsymbol{\omega}$

$$\boldsymbol{\omega}^*(\alpha_1B_{\xi_1}+\alpha_2B_{\xi_2}+\alpha_3B_{\xi_3})=0 \tag{1.77}$$

has real solutions and among all such solution vectors there are just m linearly independent vectors. Here

$$B_{\xi_j}=\frac{\partial\xi_j}{\partial\eta_1}B_1+\frac{\partial\xi_j}{\partial\eta_2}B_2+\frac{\partial\xi_j}{\partial\eta_3}B_3\quad(j=1,\ 2,\ 3).$$

Note that α_1, α_2 and α_3 satisfying (1.76) are, in fact, three components $\partial\tilde{\Phi}/\partial\xi_1$, $\partial\tilde{\Phi}/\partial\xi_2$ and $\partial\tilde{\Phi}/\partial\xi_3$, of the normal to the characteristic surface $\tilde{\Phi}(\xi_1,\ \xi_2,\ \xi_3)=0$ in the ξ_1, ξ_2, ξ_3-space,

By this definition, the system (1.11) is hyperbolic at points where $|\boldsymbol{V}|>a$. In fact, in this case we can choose a system of coordinates ξ_1, ξ_2, ξ_3 such that $\boldsymbol{\xi}_1\times\boldsymbol{\xi}_2$ is perpendicular to $\boldsymbol{V}$. This leads to $V_X=|\boldsymbol{V}|>a$ and therefore the equations (1.30) and (1.31) always have real solutions. As proved previously, the matrix Ω (1.59) is invertible, that is, the vectors $\boldsymbol{\omega}_1$, $\boldsymbol{\omega}_2$, $\cdots$, $\boldsymbol{\omega}_m$ are linearly independent.

However, if $|\boldsymbol{V}|<a$, then the equation (1.31) has no real solutions and therefore the equation (1.29) must have some complex solutions, which means the system (1.11) is not hyperbolic. In that case, the system is known as elliptical.

Note that where $|\boldsymbol{V}|<a$ the equation (1.29) for $\partial\Phi/\partial x_1$, given $\partial\Phi/\partial x_2$ and $\partial\Phi/\partial x_3$, has a pair of conjugate complex roots and one $(m-2)$-multiple real root, and that where $|\boldsymbol{V}|=0$, it has infinitely many real roots so that the system (1.11) has singularities there.

Thus, for steady flow the domain of definition for the solution of differential equations can be divided into two parts: the hyperbolic region where $|\boldsymbol{V}|>a$ and the elliptical region where $|\boldsymbol{V}|<a$. The "type" of the system changes across the surface where $|\boldsymbol{V}|=a$. Noting that the type is independent of the choice of independent variables, the above

statement applies in the case of equations (1.16) as well as in the case of equations (1.11).

If for a chosen set of independent variables, ξ_1, ξ_2 and ξ_3, we have $\frac{\partial U}{\partial \xi_3}=0$ everywhere, then the problem is reduced to a problem with two independent variables. Thus, two-dimensional steady flows (e.g., the plane flows) lead to such problems. Moreover, the spacial steady supersonic flow past a cone also leads to such a problem. This is based on the fact that the variables $\boldsymbol{V}$, p, ρ, etc. remain constant along each of the rays from the tip of the cone. In that case, if we choose a system of coordinates ξ_1, ξ_2, ξ_3 such that the rays are the ξ_3-coordinate lines, then $\frac{\partial U}{\partial \xi_3}=0$ holds identically. In the coordinates ξ_1, ξ_2, ξ_3 the basic differential equations for the conical flow have the form (1.16) with $\frac{\partial U}{\partial \xi_3}=0$, and the system (1.58) or (1.75), when setting $\frac{\partial U}{\partial \xi_3}=0$, is the "characteristic form" for the conical flow. As indicated before, the system (1.58) can be derived from (1.16) only if it is hyperbolic, i.e., if the condition (1.65) holds. In the case of the conical flow, this condition signifies that the projection V_X of the velocity on the tangent plane to the spherical surface with the center at the tip of the cone is greater than the speed of sound. The quantity V_X may be referred to as the "spherical velocity". Thus the conical flow equations are hyperbolic when the spherical velocity is greater than the speed of sound, and elliptical when less than the speed of sound.

§ 2 Discontinuities, Singularities, and the Intersection and Reflection of Strong Discontinuities

1. *Strong Discontinuities*

As is well-known, in an inviscid flow the variables $\boldsymbol{V}$, p, ρ, etc. (the "flow properties") or their derivatives possibly suffer discontinuities. The surfaces across which the discontinuities occur are called discontinuity surfaces. Moreover, a surface is called a strong discontinuity surface if the flow properties themselves are discontinuous, and is called a weak discontinuity surface if the derivatives of the flow properties are discontinuous while the flow properties are continuous.

In this subsection we shall observe the strong discontinuities and derive the "jump conditions", i.e., the conditions that the flow properties must satisfy across the discontinuity surfaces.

As stated in § 1, the flow properties should satisfy the following laws of conservation of momentum, mass, and energy:

$$\begin{cases} \iint\limits_\sigma \rho V_n \boldsymbol{V} \, d\sigma + \iint\limits_\sigma p\boldsymbol{n} \, d\sigma = 0, \\ \iint\limits_\sigma \rho V_n \, d\sigma = 0, \\ \iint\limits_\sigma \rho V_n \left(h + \frac{|\boldsymbol{V}|^2}{2}\right) d\sigma = 0, \end{cases}$$

in which there is a relation among h, p and ρ

$$h = h(p, \rho, c_1, \cdots, c_{m-5})$$

and c_i satisfies (1.7), i.e.,

$$\boldsymbol{V} \cdot \nabla c_i = -d_{i+5}, \quad i = 1, 2, \cdots, m-5.$$

Using the second equation of (1.5), the so-called "continuity equation"

$$\operatorname{div} \rho \boldsymbol{V} = 0,$$

we can replace the above equation by

$$\operatorname{div} c_i \rho \boldsymbol{V} = -\rho d_{i+5}, \quad i = 1, 2, \cdots, m-5.$$

Again we can rewrite it in the integral form

$$\iiint\limits_\tau \operatorname{div} c_i \rho \boldsymbol{V} \, d\tau = \iint\limits_\sigma c_i \rho \boldsymbol{V} \cdot \boldsymbol{n} \, d\sigma = -\iiint\limits_\tau \rho d_{i+5} \, d\tau, \tag{2.1}$$

where τ is the volume enclosed by the surface σ.

Now let Π be a strong discontinuity surface and P be a point on it. Take as σ any closed surface around the point P. Then the discontinuity surface Π cuts the closed surface σ and the volume τ enclosed by it into two parts. They are denoted, respectively, by σ_1 and τ_1, σ_2 and τ_2. We denote the intersection of Π and τ by Π_τ and the unit normal to Π_τ by $\boldsymbol{n}$ (see Fig. 2.1).

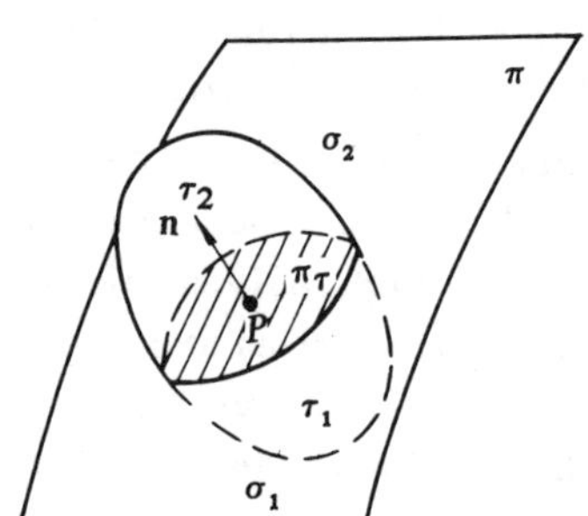

Figure 2.1

Then for any given law of conservation of the form

$$\iint_{\sigma} \boldsymbol{A}\cdot\boldsymbol{n}\,d\sigma=\iiint_{\tau} b\,d\tau,$$

we have

$$\begin{cases}\displaystyle\iint_{\sigma_1+\sigma_2} \boldsymbol{A}\cdot\boldsymbol{n}\,d\sigma=\iiint_{\tau_1+\tau_2} b\,d\tau,\\ \displaystyle\iint_{\sigma_1} \boldsymbol{A}\cdot\boldsymbol{n}\,d\sigma\pm\iint_{\Pi_\tau}(\boldsymbol{A}\cdot\boldsymbol{n})_1 d\sigma=\iiint_{\tau_1} b\,d\tau,\\ \displaystyle\iint_{\sigma_2} \boldsymbol{A}\cdot\boldsymbol{n}\,d\sigma\mp\iint_{\Pi_\tau}(\boldsymbol{A}\cdot\boldsymbol{n})_2 d\sigma=\iiint_{\tau_2} b\,d\tau.\end{cases}$$

Here the choice between "+" and "−" depends on the direction of $\boldsymbol{n}$ chosen, but in any case we can rewrite it as

$$\iint_{\Pi_\tau}(\boldsymbol{A}_1\cdot\boldsymbol{n}-\boldsymbol{A}_2\cdot\boldsymbol{n})\,d\sigma=0.$$

From this, using the arbitrariness of τ, we conclude that at the point P the equality

$$\boldsymbol{A}_1\cdot\boldsymbol{n}=\boldsymbol{A}_2\cdot\boldsymbol{n}$$

holds, where $\boldsymbol{n}$ is the unit normal to the discontinuity surface at the point in consideration, and $\boldsymbol{A}_1$ and $\boldsymbol{A}_2$ are the values of $\boldsymbol{A}$ on two different sides of the discontinuity surface at the point. Applying this process to the above integral equations we obtain the following set of relations

$$\begin{cases}(\rho V_n\boldsymbol{V})_1+p_1\boldsymbol{n}=(\rho V_n\boldsymbol{V})_2+p_2\boldsymbol{n},\\ (\rho V_n)_1=(\rho V_n)_2,\\ (\rho V_n)_1\left(h+\dfrac{|\boldsymbol{V}|^2}{2}\right)_1=(\rho V_n)_2\left(h+\dfrac{|\boldsymbol{V}|^2}{2}\right)_2,\\ (\rho V_n c_i)_1=(\rho V_n c_i)_2,\quad i=1,2,\cdots,m-5,\end{cases}\tag{2.2}$$

which the flow must satisfy on a strong discontinuity surface.

There are two cases in which the set of relations will hold: either $V_{n,1}=0$ or $V_{n,1}\neq0$. In the first case, when $V_{n,1}=0$, we obtain $\rho_2V_{n,2}=0$ and $p_2=p_1$ from (2.2). Thus, on the assumption that $\rho_2\neq0$, we have

$$\begin{cases}p_2=p_1,\\ V_{n,2}=0,\\ V_{n,1}=0.\end{cases}\tag{2.3}$$

Conversely, when (2.3) holds, the jump conditions given by (2.2) are satisfied no matter what ρ, c_i and the tangent-component of the velocity are on both sides of the discontinuity surfaces. Such discontinuities are called contact discontinuities. Since $V_{n,2}=V_{n,1}=0$ means that the moving gas does not pass through the discontinuity surfaces, we know that contact discontinuity surfaces must be streamplanes and, hence, must be characteristic surfaces.

Now we turn to the case where $V_{n,1}\neq 0$. Choosing the unit vectors $\boldsymbol{t}_1$ and $\boldsymbol{t}_2$ such that $\boldsymbol{n}$, $\boldsymbol{t}_1$ and $\boldsymbol{t}_2$ are orthogonal to each other, we can rewrite the first equation of (2.2) in its component form

$$p_1+(\rho V_n^2)_1=p_2+(\rho V_n^2)_2,$$

$$(\rho V_n \boldsymbol{V}\cdot\boldsymbol{t}_1)_1=(\rho V_n \boldsymbol{V}\cdot\boldsymbol{t}_1)_2,$$

$$(\rho V_n \boldsymbol{V}\cdot\boldsymbol{t}_2)_1=(\rho V_n \boldsymbol{V}\cdot\boldsymbol{t}_2)_2.$$

By using these equalities, it is easy to derive from (2.2) the following relations:

$$\begin{cases} \boldsymbol{V}_2=\boldsymbol{V}_1+\dfrac{p_1-p_2}{\rho_1 V_{n,1}}\boldsymbol{n}=\boldsymbol{V}_1-V_{n,1}\left(1-\dfrac{\rho_1}{\rho_2}\right)\boldsymbol{n}, \\ p_2=p_1+\rho_1 V_{n,1}^2-\rho_2 V_{n,2}^2=p_1+\rho_1 V_{n,1}^2\left(1-\dfrac{\rho_1}{\rho_2}\right), \\ h_2+\dfrac{V_{n,2}^2}{2}=h_1+\dfrac{V_{n,1}^2}{2}, \\ c_{i,2}=c_{i,1}, \quad i=1,\,2,\,\cdots,\,m-5. \end{cases} \tag{2.4}$$

These are equivalent to (2.2) based on the assumption that $V_{n,1}\neq 0$. The discontinuity surface on which the relations (2.4) are satisfied is called a shock surface or, in short, a shock. This is a kind of strong discontinuity different from contact discontinuity. The gas will pass through a shock surface, and the normal velocity, the pressure, the density and other thermodynamic quantities will all be discontinuous on it. In what follows we shall assume that the gas crosses the shock from τ_1 towards τ_2. The side next to τ_1 is referred to as the front side of the shock, and the other, the back side. The conditions (2.4), which relate the flow properties on the different sides with the normal direction to the shock, are called the shock relations, or the Rankine–Hugoniot conditions.

It should be noted that not all the strong discontinuity surfaces satisfying the shock relations can exist physically. For their existence, certain "stability conditions" should be added. In the case of non–steady one–dimensional flow, it is known from [52, 62, 63] that the contact discontinuities are always stable, and the stability conditions for the shocks are given by

$$\omega_+(U_1,\,U^{(1)})\leqslant\omega_+(U_1,\,U_2)\leqslant\omega_+(U^{(2)},\,U_2), \tag{2.5}$$

where

$$\omega_+(U_1,\,U^{(1)})\equiv u_1+\frac{1}{\rho_1}\left(\frac{p^{(1)}-p_1}{\dfrac{1}{\rho_1}-\dfrac{1}{\rho^{(1)}}}\right)^{\frac{1}{2}},$$

$$\omega_+(U^{(2)}, U_2) \equiv u_2 + \frac{1}{\rho_2}\left(\frac{p^{(2)}-p_2}{\dfrac{1}{\rho_2}-\dfrac{1}{\rho^{(2)}}}\right)^{\frac{1}{2}},$$

$$U=\begin{pmatrix} u \\ p \\ \rho \end{pmatrix}.$$

Here u is the speed of the moving gas in the Euler coordinate frame, p the pressure and ρ the density. The indices 1 and 2 refer to the front side and the back side of the shock respectively. $p^{(i)}$ and $\rho^{(i)}$ $(i=1, 2)$ stand for the values of p and ρ at a point between the points (p_1, ρ_1) and (p_2, ρ_2) on the curve in the p, ρ-plane given by

$$h(p, \rho)-h(p_i, \rho_i)-\frac{1}{2}(p-p_i)\left(\frac{1}{\rho}+\frac{1}{\rho_i}\right)=0, \tag{2.6}$$

which is usually called the shock adiabatic curve or the Hugoniot curve through (p_i, ρ_i). The inequality (2.5) can also be rewritten in the form

$$\frac{p^{(1)}-p_1}{\dfrac{1}{\rho_1}-\dfrac{1}{\rho^{(1)}}} \leqslant \frac{p_2-p_1}{\dfrac{1}{\rho_1}-\dfrac{1}{\rho_2}} \leqslant \frac{p^{(2)}-p_2}{\dfrac{1}{\rho_2}-\dfrac{1}{\rho^{(2)}}}. \tag{2.7}$$

It may be supposed that these results are applicable to the spatial steady flow. Thus the condition (2.7) is necessary for a shock to be stable, but a contact discontinuity is always stable.

Now we shall study some properties of shock and the Hugoniot curve, and we shall show that the stability condition (2.7) can be reduced to a simpler form in some cases. Introducing the notation $\widetilde{V}=\dfrac{1}{\rho}$ and denoting the specific entropy by S and the specific internal energy by e, we assume

$$e_p \equiv \left(\frac{\partial e}{\partial p}\right)_{\widetilde{V}}>0, \quad e_{\widetilde{V}} \equiv \left(\frac{\partial e}{\partial \widetilde{V}}\right)_p>0, \tag{2.8}$$

and therefore

$$a^2=\frac{-\dfrac{\partial h}{\partial \rho}}{\dfrac{\partial h}{\partial p}-\dfrac{1}{\rho}}=\frac{\widetilde{V}^2 h_{\widetilde{V}}}{h_p-\widetilde{V}}=\frac{\widetilde{V}^2(e_{\widetilde{V}}+p)}{e_p}=-\widetilde{V}^2\left(\frac{\partial p}{\partial \widetilde{V}}\right)_s>0.$$

We first derive a few relations which hold along the Hugoniot curve. Using the thermodynamic relation

$$T\,dS=dh-\widetilde{V}\,dp=de+p\,d\widetilde{V}=e_p\,dp+(e_{\widetilde{V}}+p)\,d\widetilde{V},$$

it follows from (2.6) that

$$\left\{\begin{aligned}
& dh-\frac{1}{2}(\tilde{V}_i+\tilde{V})dp-\frac{1}{2}(p-p_i)d\tilde{V} \\
& \quad =T\,dS-\frac{1}{2}(\tilde{V}_i-\tilde{V})dp-\frac{1}{2}(p-p_i)d\tilde{V}=0, \\
& \frac{dp}{d\tilde{V}}=\left(\frac{\partial p}{\partial \tilde{V}}\right)_s+\left(\frac{\partial p}{\partial S}\right)_{\tilde{V}}\frac{dS}{d\tilde{V}}=\left(\frac{\partial p}{\partial \tilde{V}}\right)_s+\frac{T}{e_p}\frac{dS}{d\tilde{V}} \\
& \quad =\left(\frac{\partial p}{\partial \tilde{V}}\right)_s+\frac{1}{2e_p}(\tilde{V}_i-\tilde{V})\left(\frac{dp}{d\tilde{V}}+\frac{p-p_i}{\tilde{V}_i-\tilde{V}}\right) \\
& \quad =\left(\frac{\partial p}{\partial \tilde{V}}\right)_s+\frac{1}{2e_p}(\tilde{V}_i-\tilde{V})^2\frac{d}{d\tilde{V}}\left(\frac{p-p_i}{\tilde{V}_i-\tilde{V}}\right), \\
& \frac{d^2p}{d\tilde{V}^2}=\left(\frac{\partial^2 p}{\partial \tilde{V}^2}\right)_s+\frac{\partial^2 p}{\partial \tilde{V}\,\partial S}\frac{dS}{d\tilde{V}}+\frac{1}{2}(\tilde{V}_i-\tilde{V})\left(\frac{dp}{d\tilde{V}}+\frac{p-p_i}{\tilde{V}_i-\tilde{V}}\right) \\
& \quad \times\frac{d\left(\dfrac{1}{e_p}\right)}{d\tilde{V}}+\frac{\tilde{V}_i-\tilde{V}}{2e_p}\frac{d^2p}{d\tilde{V}^2}.
\end{aligned}\right. \tag{2.9}$$

Hence we have at the point $(p_i, \tilde{V}_i)$

$$\left\{\begin{aligned}
& \frac{dp}{d\tilde{V}}=\left(\frac{\partial p}{\partial \tilde{V}}\right)_s, \\
& \frac{d^2p}{d\tilde{V}^2}=\left(\frac{d^2p}{\partial \tilde{V}^2}\right)_s,
\end{aligned}\right.$$

which implies that the Hugoniot curve is tangent to the constant entropy curve at this point and that their second derivatives are equal. Rewriting the Hugoniot equation (2.6) in the form

$$e(p, \tilde{V})-e(p_i, \tilde{V}_i)+\frac{1}{2}(\tilde{V}-\tilde{V}_i)(p+\quad =0,$$

we know that $p=p_i$ when $\tilde{V}=\tilde{V}_i$ (otherwise, a contradiction with $e_p>0$ would result, see (2.8)). In addition, it is easily seen from (2.9) that at any point other than at the point $(p_i, \tilde{V}_i)$ on the Hugoniot curve, the following relations can be derived from each other:

$$\left\{\begin{aligned}
& \frac{dp}{d\tilde{V}}=\left(\frac{\partial p}{\partial \tilde{V}}\right)_s, \\
& \frac{dS}{d\tilde{V}}=0, \\
& \frac{dp}{d\tilde{V}}=-\frac{p-p_i}{\tilde{V}_i-\tilde{V}}, \\
& \frac{d}{d\tilde{V}}\left(\frac{p-p_i}{\tilde{V}_i-\tilde{V}}\right)=0.
\end{aligned}\right. \tag{2.10}$$

Based on these results, we shall show that if, in addition to the assumption of $\left(\frac{\partial p}{\partial \tilde{V}}\right)_s<0$, the condition

$$\left(\frac{\partial^2 p}{\partial \tilde{V}^2}\right)_s > 0 \quad \left(\text{or} \left(\frac{\partial^2 p}{\partial \tilde{V}^2}\right)_s < 0\right) \tag{2.11}$$

holds everywhere, then the condition (2.7) can be replaced by

$$p_2 > p_1 \quad (\text{or } p_2 < p_1). \tag{2.12}$$

To this end we must first prove that under the given conditions the equality

$$\frac{d}{d\tilde{V}}\left(\frac{p-p_i}{\tilde{V}_i-\tilde{V}}\right)=0 \tag{2.13}$$

never occurs on the Hugoniot curve. In fact, from the expression

$$\frac{d}{d\tilde{V}}\left[\frac{p-p_i}{\tilde{V}-\tilde{V}_i}\left(\frac{\partial p}{\partial \tilde{V}}\right)_s^{-1}\right]=\left(\frac{\partial p}{\partial \tilde{V}}\right)_s^{-1}\frac{d}{d\tilde{V}}\left(\frac{p-p_i}{\tilde{V}-\tilde{V}_i}\right)$$
$$-\frac{p-p_i}{\tilde{V}-\tilde{V}_i}\left(\frac{\partial p}{\partial \tilde{V}}\right)_s^{-2}\left[\left(\frac{\partial^2 p}{\partial \tilde{V}^2}\right)_s+\frac{\partial^2 p}{\partial \tilde{V}\,\partial S}\,\frac{dS}{d\tilde{V}}\right],$$

we know that if there were a point, say $(p^*, \tilde{V}^*)$, such that the equality (2.13) holds at that point, then, from (2.10), we would obtain

$$\frac{d}{d\tilde{V}}\left[\frac{p-p_i}{\tilde{V}-\tilde{V}_i}\left(\frac{\partial p}{\partial \tilde{V}}\right)_s^{-1}\right]=-\left(\frac{\partial p}{\partial \tilde{V}}\right)_s^{-1}\left(\frac{\partial^2 p}{\partial \tilde{V}^2}\right)_s$$

at $(p^*, \tilde{V}^*)$. On the other hand, from

$$\lim_{\tilde{V}\to\tilde{V}_i}\frac{d}{d\tilde{V}}\left(\frac{p-p_i}{\tilde{V}-\tilde{V}_i}\right)=\lim_{\tilde{V}\to\tilde{V}_i}\frac{\dfrac{dp}{d\tilde{V}}-\dfrac{p-p_i}{\tilde{V}-\tilde{V}_i}}{\tilde{V}-\tilde{V}_i}$$
$$=\frac{d^2p}{d\tilde{V}^2}-\lim_{\tilde{V}\to\tilde{V}_i}\frac{d}{d\tilde{V}}\left(\frac{p-p_i}{\tilde{V}-\tilde{V}_i}\right)$$

we obtain

$$\frac{d}{d\tilde{V}}\left(\frac{p-p_i}{\tilde{V}-\tilde{V}_i}\right)=\frac{1}{2}\,\frac{d^2p}{d\tilde{V}^2}=\frac{1}{2}\left(\frac{\partial^2 p}{\partial \tilde{V}^2}\right)_s$$

and therefore

$$\frac{d}{d\tilde{V}}\left[\frac{p-p_i}{\tilde{V}-\tilde{V}_i}\left(\frac{\partial p}{\partial \tilde{V}}\right)_s^{-1}\right]=-\frac{1}{2}\left(\frac{\partial p}{\partial \tilde{V}}\right)_s^{-1}\left(\frac{\partial^2 p}{\partial \tilde{V}^2}\right)_s$$

at $(p_i, \tilde{V}_i)$. Thus this derivative at $(p_i, \tilde{V}_i)$ has the same sign as that at $(p^*, \tilde{V}^*)$. However, at $(p_i, \tilde{V}_i)$ we have

$$\frac{p-p_i}{\tilde{V}-\tilde{V}_i}\left(\frac{\partial p}{\partial \tilde{V}}\right)_s^{-1}=1,$$

and at $(p^*, \tilde{V}^*)$, where (2.13) holds, we also have

$$\frac{p-p_i}{\tilde{V}-\tilde{V}_i}\left(\frac{\partial p}{\partial \tilde{V}}\right)_s^{-1}=1.$$

Furthermore, without loss of generality, we may assume that between $(p_i, \tilde{V}_i)$ and $(p^*, \tilde{V}^*)$ on the Hugoniot curve there are no other points where (2.13) holds, that is, there are no other points at which the function $\dfrac{p-p_i}{\widetilde{V}-\widetilde{V}_i}\left(\dfrac{\partial p}{\partial \widetilde{V}}\right)_s^{-1}$ has a value of 1. Hence, the values of the derivative of this function at the points $(p_i, \widetilde{V}_i)$ and $(p^*, \widetilde{V}^*)$ should not have the same sign. Thus, we have proved that under the assumption that (2.11) holds everywhere, the equality (2.13) never takes place on the Hugoniot curve. This means that $\dfrac{d}{d\widetilde{V}}\left(\dfrac{p-p_i}{\widetilde{V}_i-\widetilde{V}}\right)$ is either positive everywhere or negative everywhere. Further, since at $(p_i, \widetilde{V}_i)$ we have

$$\frac{d}{d\widetilde{V}}\left(\frac{p-p_i}{\widetilde{V}_i-\widetilde{V}}\right)=-\frac{1}{2}\left(\frac{\partial^2 p}{\partial \widetilde{V}^2}\right)_s,$$

we know that if $\left(\dfrac{\partial^2 p}{\partial \widetilde{V}^2}\right)_s>0$ (or <0) everywhere, then it follows that $\dfrac{d}{d\widetilde{V}}\left(\dfrac{p-p_i}{\widetilde{V}_i-\widetilde{V}}\right)<0$ (or >0) everywhere.

Based on this fact we can prove that the condition (2.7) is satisfied if

$$\widetilde{V}_2\leqslant\widetilde{V}_1, \quad \text{when } \left(\frac{\partial^2 p}{\partial \widetilde{V}^2}\right)_s>0,$$

or if

$$\widetilde{V}_2\geqslant\widetilde{V}_1, \quad \text{when } \left(\frac{\partial^2 p}{\partial \widetilde{V}^2}\right)_s<0.$$

This is illustrated in Fig. 2. 2. In fact, in either case we have

$$\operatorname{tg}\theta_1\leqslant\operatorname{tg}\theta\leqslant\operatorname{tg}\theta_2,$$

where

$$\operatorname{tg}\theta_1=\frac{p^{(1)}-p_1}{\widetilde{V}_1-\widetilde{V}^{(1)}}, \quad \operatorname{tg}\theta=\frac{p_2-p_1}{\widetilde{V}_1-\widetilde{V}_2}, \quad \operatorname{tg}\theta_2=\frac{p^{(2)}-p_2}{\widetilde{V}_2-\widetilde{V}^{(2)}}.$$

What we now need to do is to write the above conditions in the desired form. As described above, on the Hugoniot curve there are no points, other than $(p_i, \widetilde{V}_i)$, where $\widetilde{V}=\widetilde{V}_i$. Similarly, it can be proven by the use of (2.6) and (2.8) that on the curve there is only one point where $p=p_i$. Since $\dfrac{dp}{d\widetilde{V}}=\left(\dfrac{\partial p}{\partial \widetilde{V}}\right)_s<0$ at $(p_i, \widetilde{V}_i)$, the inequality $\dfrac{p-p_i}{\widetilde{V}_i-\widetilde{V}}>0$ holds everywhere on the curve. It follows that the condition $\widetilde{V}_2\leqslant\widetilde{V}_1$ (or $\widetilde{V}_2\geqslant\widetilde{V}_1$) is equivalent to

$$p_2\geqslant p_1 \quad (\text{or } p_2\leqslant p_1).$$

This ends the proof.

For a perfect gas,

$$e=h-p\widetilde{V}=\frac{\gamma}{\gamma-1}p\widetilde{V}-p\widetilde{V}=\frac{1}{\gamma-1}p\widetilde{V},$$

where γ is a constant greater than 1. Hence,

$$\left(\frac{\partial p}{\partial \widetilde{V}}\right)_s=-\frac{e_{\widetilde{V}}+p}{e_p}=-\frac{\dfrac{1}{\gamma-1}p+p}{\dfrac{1}{\gamma-1}\widetilde{V}}=-\gamma\frac{p}{\widetilde{V}}<0,$$

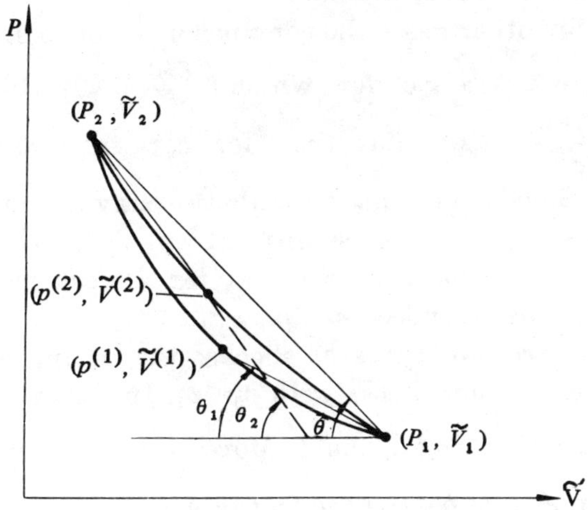

(a) If $\frac{d}{d\tilde{V}}\left(\frac{p-p_i}{\tilde{V}_i-\tilde{V}}\right)<0$

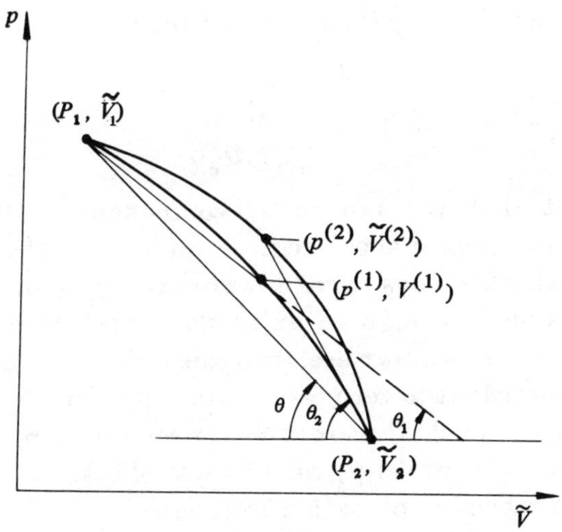

(b) If $\frac{d}{d\tilde{V}}\left(\frac{p-p_i}{\tilde{V}_i-\tilde{V}}\right)>0$

Figure 2.2 Shock-adiabatic curves

$$\left(\frac{\partial^2 p}{\partial \tilde{V}^2}\right)_s=-\gamma\left[\left(\frac{\partial p}{\partial \tilde{V}}\right)_s\Big/\tilde{V}-p/\tilde{V}^2\right]=\gamma(\gamma+1)\frac{p}{\tilde{V}^2}>0.$$

Consequently, the stability condition for a shock in the case of a perfect gas is

$$p_2 \geqslant p_1,$$

that is, the pressure on the back side of the shock should be greater than the pressure on the front side.

However, for other gases the conclusion is not so simple. We do not know if there exists a gas for which $\left(\frac{\partial^2 p}{\partial \tilde{V}^2}\right)_s < 0$ always holds. It is possible that $\left(\frac{\partial^2 p}{\partial \tilde{V}^2}\right)_s < 0$ holds true for certain gases under certain conditions[54]. Since the shock adiabatic curves touch the constant entropy curves up to the second order derivative, as indicated previously, it is not hard to see that for a weak shock the stability condition in that case is indeed $p_2 \leqslant p_1$.

Thus, there are two types of shocks: the compression shocks with $p_2 \geqslant p_1$ and the expansion shocks with $p_2 \leqslant p_1$. In either case the relations $\frac{p_2 - p_1}{\tilde{V}_1 - \tilde{V}_2} > 0$ and $\rho_1 V_{n,1} = \rho_2 V_{n,2}$ hold. Moreover, it follows from them that the compression shocks possess the features

$$\begin{cases} p_2 \geqslant p_1, \\ \rho_2 \geqslant \rho_1, \\ V_{n,2} \leqslant V_{n,1}, \end{cases} \tag{2.14}$$

and the expansion shocks possess the features

$$\begin{cases} p_2 \leqslant p_1, \\ \rho_2 \leqslant \rho_1, \\ V_{n,2} \geqslant V_{n,1}. \end{cases} \tag{2.15}$$

Using the fact that the tangential component of the flow velocity remains constant across a shock surface, we can see that the flow through a compression shock always turns towards the tangent plane to the shock surface and the flow through an expansion shock always turns towards the normal to the shock surface, provided that the tangential velocity component is not equal to zero, as is illustrated in Fig. 2.3.

From now on we shall mainly be concerned with compression shocks and, for the sake of simplicity, we will use shocks to mean compression shocks unless a special explanation is given.

In addition to (2.14) we can prove the following inequalities

$$a_1^2 \leqslant V_{n,1}^2, \quad V_{n,2}^2 \leqslant a_2^2,$$

or, when we choose the normal $\boldsymbol{n}$ directed to the back side (which means that $V_{n,1} > 0$ and $V_{n,2} > 0$), then

$$a_1 \leqslant V_{n,1}, \quad V_{n,2} \leqslant a_2. \tag{2.16}$$

In fact, they can be derived from (2.7) and the following formulae:

$$a^2 = -\tilde{V}^2 \left(\frac{\partial p}{\partial \tilde{V}}\right)_s,$$

$$V_{n,i}^2 = \frac{1}{\rho_i^2}\left(\frac{p_2-p_1}{\dfrac{1}{\rho_1}-\dfrac{1}{\rho_2}}\right) = \tilde{V}_i^2\,\frac{p_2-p_1}{\tilde{V}_1-\tilde{V}_2},$$

$$\lim_{\substack{p^{(i)}\to p_i\\ \tilde{V}^{(i)}\to\tilde{V}_i}} \frac{p^{(i)}-p_i}{\tilde{V}_i-\tilde{V}^{(i)}} = -\left(\frac{\partial p}{\partial \tilde{V}}\right)_s\Bigg|_{p=p_i,\,\tilde{V}=\tilde{V}_i} = \frac{a_i^2}{\tilde{V}_i^2}.$$

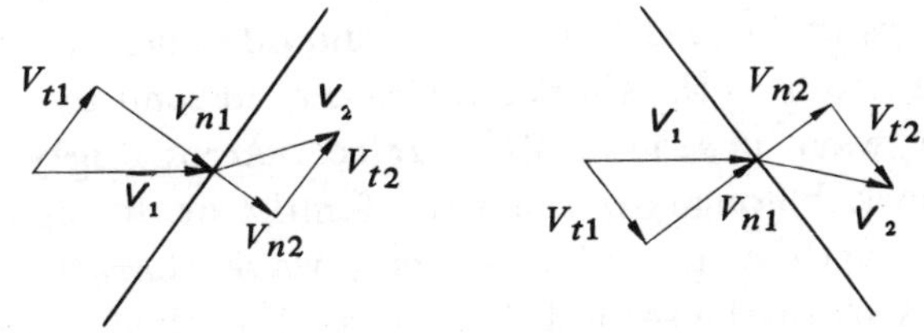

First family Second family

(a) Compression shock ($p_2 \geqslant p_1$)

(If $\left(\frac{\partial^2 p}{\partial \tilde{V}^2}\right)_s$ is always greater than zero, then only the compression shock appears.)

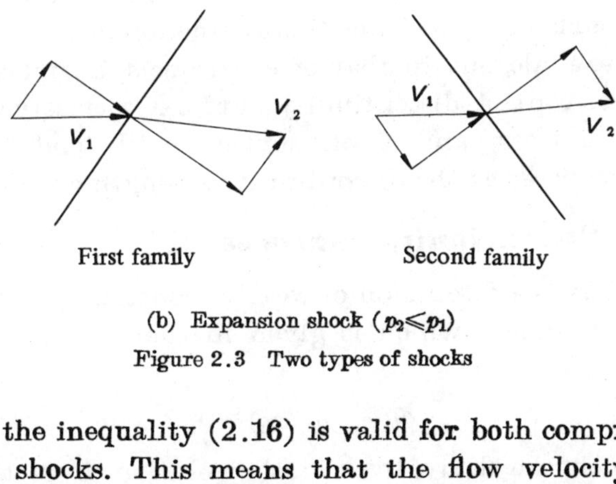

First family Second family

(b) Expansion shock ($p_2 \leqslant p_1$)

Figure 2.3 Two types of shocks

Note that the inequality (2.16) is valid for both compression shocks and expansion shocks. This means that the flow velocity in front of shocks must be greater than the speed of sound. Therefore, a stationary shock may appear only in a supersonic flow. In addition, it is easily seen that the Mach cones arriving at the front side of a shock must entirely be in front of the shock. Therefore, the flow properties on the front side of a shock surface are determined only by the flow in front of it as if the shock did not exist. On the other hand, the flow properties on the back side can be uniquely determined by the flow properties on the front side and the normal direction to the shock. When the coming flow in front of the shock is fixed, the flow properties on the back side depend only on the normal direction to the shock. We can measure the strength of a shock by the "jump" of any flow property, such as $|V_{n,2}-V_{n,1}|$, $|p_2-p_1|$. If the direction of the coming flow is perpendicular to the shock surface, then there is the maximum of shock strength. At that time, as shown by (2.16), the flow speed on the back side must be subsonic. As the shock surface gets oblique, the shock strength will weaken and the flow speed on the back may be supersonic. In the limiting case, when the strength approaches zero, that is, $V_{n,2}\to V_{n,1}$ and $a_2\to a_1$, it follows from (2.16) that $V_{n,1}$ should equal a_1. This means that the normal direction to the shock surface should tend towards the normal direction to a wave-characteristic surface. Accordingly, we sometimes say that a shock belongs to the first family or to the second family, depending on whether it tend towards a wave-characteristic surface of the first family or of the second family as the shock strength becomes infinitely small.

We conclude this subsection with the following remarks. In supersonic flow there can be two types of strong discontinuities, contact discontinuities and shocks, but in subsonic flow there can be only contact discontinuities. As pointed out in § 1, in supersonic flow there exist two types of characteristic surfaces, stream-characteristic surfaces and wave-characteristic surfaces, whereas in subsonic flow only stream-characteristic surfaces exist. Thus this distinction of strong discontinuity surfaces is quite analogous to that of characteristic surfaces. Moreover, as stated above, contact discontinuity surfaces themselves are stream-characteristic surfaces, and shock surfaces will tend towards wave-characteristic surfaces as the discontinuity strength goes to zero.

2. *Weak Discontinuity Surfaces*

What follows is a discussion of weak discontinuity surfaces. Suppose we have such a surface which is given in the η_1, η_2, η_3-space by the equation

$$\Phi(\eta_1, \eta_2, \eta_3)=0 \tag{2.17}$$

and across which the derivatives of the solution $\boldsymbol{U}$ of the differential equations (1.19) are discontinuous, but the solution itself is continuous. Let $\left(\dfrac{\partial \boldsymbol{U}}{\partial \eta_j}\right)_1$ and $\left(\dfrac{\partial \boldsymbol{U}}{\partial \eta_j}\right)_2$ denote the derivatives on the different sides of the surface (2.17). Then from (1.19) we should have

$$\begin{cases} B_1\left(\dfrac{\partial \boldsymbol{U}}{\partial \eta_1}\right)_1+B_2\left(\dfrac{\partial \boldsymbol{U}}{\partial \eta_2}\right)_1+B_3\left(\dfrac{\partial \boldsymbol{U}}{\partial \eta_3}\right)_1+\boldsymbol{D}=0,\\ B_1\left(\dfrac{\partial \boldsymbol{U}}{\partial \eta_1}\right)_2+B_2\left(\dfrac{\partial \boldsymbol{U}}{\partial \eta_2}\right)_2+B_3\left(\dfrac{\partial \boldsymbol{U}}{\partial \eta_3}\right)_2+\boldsymbol{D}=0,\end{cases}$$

where the coefficient matrices in these two systems assume the same values on the surface because they depend only on the solution, not on its derivatives. Subtracting the second system from the first one, we obtain

$$B_1\left[\frac{\partial U}{\partial \eta_1}\right]+B_2\left[\frac{\partial U}{\partial \eta_2}\right]+B_3\left[\frac{\partial U}{\partial \eta_3}\right]=0,$$

where

$$\left[\frac{\partial U}{\partial \eta_j}\right]=\left(\frac{\partial U}{\partial \eta_j}\right)_1-\left(\frac{\partial U}{\partial \eta_j}\right)_2, \quad j=1,\ 2,\ 3.$$

Defining the new coordinates ζ_1, ζ_2, ζ_3 through the transformation

$$\begin{cases}\zeta_1=\zeta_1(\eta_1,\ \eta_2,\ \eta_3)\equiv\Phi(\eta_1,\ \eta_2,\ \eta_3),\\ \zeta_2=\zeta_2(\eta_1,\ \eta_2,\ \eta_3),\\ \zeta_3=\zeta_3(\eta_1,\ \eta_2,\ \eta_3),\end{cases}$$

we can transform the system into

$$B_{\zeta_1}\left[\frac{\partial U}{\partial \zeta_1}\right]+B_{\zeta_2}\left[\frac{\partial U}{\partial \zeta_2}\right]+B_{\zeta_3}\left[\frac{\partial U}{\partial \zeta_3}\right]=0,$$

where

$$B_{\zeta_j}=\frac{\partial \zeta_j}{\partial \eta_1}B_1+\frac{\partial \zeta_j}{\partial \eta_2}B_2+\frac{\partial \zeta_j}{\partial \eta_3}B_3, \quad j=1,\ 2,\ 3,$$

and the equation of the surface is transformed to $\zeta_1=0$. Clearly, we obtain

$$\left[\frac{\partial U}{\partial \zeta_2}\right]=\left[\frac{\partial U}{\partial \zeta_3}\right]=0,$$

because $\frac{\partial U}{\partial \zeta_2}$ and $\frac{\partial U}{\partial \zeta_3}$ represent the interior derivatives of U in the surface $\zeta_1=0$, and U is continuous across this surface. Thus, we obtain

$$B_{\zeta_1}\left[\frac{\partial U}{\partial \zeta_1}\right]=0. \tag{2.18}$$

However, from the assumption made for weak discontinuity, $\left[\frac{\partial U}{\partial \zeta_1}\right]\neq 0$, it is necessary to require that

$$|B_{\zeta_1}|=\left|\frac{\partial \Phi}{\partial \eta_1}B_1+\frac{\partial \Phi}{\partial \eta_2}B_2+\frac{\partial \Phi}{\partial \eta_3}B_3\right|=0. \tag{2.19}$$

Remembering the definition of the characteristic surfaces (see (1.20)), we conclude that a weak discontinuity surface must be a characteristic surface.

Applying this result to the problems of steady flows, we can say that in supersonic flow there are two types of weak discontinuities, "wave" weak discontinuities and "stream" weak discontinuities, and in subsonic flow there is only one type, "stream" weak discontinuities.

It should be noted that the jumps of the first derivatives across a weak discontinuity surface cannot be arbitrary but have to satisfy the equations (2.18). In the case of steady flow, taking orthogonal curvilinear coordinates x_1, x_2, x_3 as η_1, η_2, η_3 and considering the equations (1.11) in place of (1.19), we find that the matrix B_{ζ_1} in (2.18) is indeed the characteristic matrix of the equations (1.11), i.e., Q in (1.28). From (1.35) it can be seen that for a stream weak discontinuity surface the equations (2.18) take the form

$$\begin{cases} \left[\dfrac{\partial p}{\partial \zeta_1}\right]=0, \\ \boldsymbol{N}_s \cdot \left[\dfrac{\partial \boldsymbol{V}}{\partial \zeta_1}\right]=\left[\boldsymbol{N}_s \cdot \dfrac{\partial \boldsymbol{V}}{\partial \zeta_1}\right]=0, \end{cases} \tag{2.20}$$

where $\boldsymbol{N}_s$ is normal to the surface. It follows that the quantities $\dfrac{\partial p}{\partial \zeta_1}$ and $\boldsymbol{N}_s \cdot \dfrac{\partial \boldsymbol{V}}{\partial \zeta_1}$ are continuous across a stream weak discontinuity surface and among $\left[\dfrac{\partial u_1}{\partial \zeta_1}\right]$, $\left[\dfrac{\partial u_2}{\partial \zeta_1}\right]$, $\cdots$, $\left[\dfrac{\partial c_{m-5}}{\partial \zeta_1}\right]$, only $m-2$ quantities are independent. For example, given $\left[\dfrac{\partial \rho}{\partial \zeta_1}\right]$, $\left[\dfrac{\partial c_i}{\partial \zeta_1}\right]$, $i=1, 2, \cdots, m-5$, and given two of $\left[\dfrac{\partial u_1}{\partial \zeta_1}\right]$, $\left[\dfrac{\partial u_2}{\partial \zeta_1}\right]$ and $\left[\dfrac{\partial u_3}{\partial \zeta_1}\right]$, then $\left[\dfrac{\partial U}{\partial \zeta_1}\right]$ is uniquely determined. In the other case, let B_{ζ_1} be Q_W, given previously. Then equations (2.18) are equivalent to

$$\begin{cases} a\left[\dfrac{\partial u_1}{\partial \zeta_1}\right]+\dfrac{N^{\pm}_{W,1}}{\rho}\left[\dfrac{\partial p}{\partial \zeta_1}\right]=\left[a\dfrac{\partial u_1}{\partial \zeta_1}+\dfrac{N^{\pm}_{W,1}}{\rho}\dfrac{\partial p}{\partial \zeta_1}\right]=0, \\ a\left[\dfrac{\partial u_2}{\partial \zeta_1}\right]+\dfrac{N^{\pm}_{W,2}}{\rho}\left[\dfrac{\partial p}{\partial \zeta_1}\right]=\left[a\dfrac{\partial u_2}{\partial \zeta_1}+\dfrac{N^{\pm}_{W,2}}{\rho}\dfrac{\partial p}{\partial \zeta_1}\right]=0, \\ a\left[\dfrac{\partial u_3}{\partial \zeta_1}\right]+\dfrac{N^{\pm}_{W,3}}{\rho}\left[\dfrac{\partial p}{\partial \zeta_1}\right]=\left[a\dfrac{\partial u_3}{\partial \zeta_1}+\dfrac{N^{\pm}_{W,3}}{\rho}\dfrac{\partial p}{\partial \zeta_1}\right]=0, \\ \left[\dfrac{\partial p}{\partial \zeta_1}\right]-a^2\left[\dfrac{\partial \rho}{\partial \zeta_1}\right]=\left[\dfrac{\partial p}{\partial \zeta_1}-a^2\dfrac{\partial \rho}{\partial \zeta_1}\right]=0, \\ \left[\dfrac{\partial c_i}{\partial \zeta_1}\right]=0, \quad i=1, 2, \cdots, m-5, \end{cases} \tag{2.21}$$

where $(N^{\pm}_{W,1}, N^{\pm}_{W,2}, N^{\pm}_{W,3})$ is normal to the wave weak discontinuity surface considered and satisfies the relation

$$N^{\pm}_{W,1}u_1+N^{\pm}_{W,2}u_2+N^{\pm}_{W,3}u_3=a.$$

It follows that there is no discontinuity about the quantities $a\dfrac{\partial u_1}{\partial \zeta_1}+\dfrac{N^{\pm}_{W,1}}{\rho}\dfrac{\partial p}{\partial \zeta_1}$, $a\dfrac{\partial u_2}{\partial \zeta_1}+\dfrac{N^{\pm}_{W,2}}{\rho}\dfrac{\partial p}{\partial \zeta_1}$, $a\dfrac{\partial u_3}{\partial \zeta_1}+\dfrac{N^{\pm}_{W,3}}{\rho}\dfrac{\partial p}{\partial \zeta_1}$, $\dfrac{\partial p}{\partial \zeta_1}-a^2\dfrac{\partial \rho}{\partial \zeta_1}$, and $\dfrac{\partial c_i}{\partial \zeta_1}$ $(i=1, 2, \cdots, m-5)$, and among $\left[\dfrac{\partial u_1}{\partial \zeta_1}\right], \cdots, \left[\dfrac{\partial c_{m-5}}{\partial \zeta_1}\right]$ only one is independent. For example, $\left[\dfrac{\partial U}{\partial \zeta_1}\right]$ can be uniquely determined by $\left[\dfrac{\partial p}{\partial \zeta_1}\right]$.

We can show that if the weak discontinuity surfaces pass through the curve given by (1.42), then relations (2.20) and (2.21) can be written respectively as

$$\begin{cases} \left[\boldsymbol{G}^{+}\cdot\dfrac{\partial \boldsymbol{V}}{\partial \zeta_1}+g_4^{+}\dfrac{\partial p}{\partial \zeta_1}\right]=0, \\ \left[\boldsymbol{G}^{-}\cdot\dfrac{\partial \boldsymbol{V}}{\partial \zeta_1}+g_4^{-}\dfrac{\partial p}{\partial \zeta_1}\right]=0, \end{cases} \tag{2.22}$$

and

$$\begin{cases} \left[\boldsymbol{V}\cdot\dfrac{\partial \boldsymbol{V}}{\partial \zeta_1}+\dfrac{1}{\rho}\dfrac{\partial p}{\partial \zeta_1}\right]=0, \\ \left[\boldsymbol{T}\cdot\dfrac{\partial \boldsymbol{V}}{\partial \zeta_1}+\dfrac{1}{\rho}\,\boldsymbol{V}\times\boldsymbol{\xi}_1\cdot\boldsymbol{\xi}_2\,\dfrac{\partial p}{\partial \zeta_1}\right]=0, \\ \left[\dfrac{\partial p}{\partial \zeta_1}-a^2\dfrac{\partial \rho}{\partial \zeta_1}\right]=0, \\ \left[\dfrac{\partial c_i}{\partial \zeta_1}\right]=0, \quad i=1, 2, \cdots, m-5, \\ \left[\boldsymbol{G}^{\mp}\cdot\dfrac{\partial \boldsymbol{V}}{\partial \zeta_1}+g_4^{\mp}\dfrac{\partial p}{\partial \zeta_1}\right]=0, \end{cases} \tag{2.23}$$

where $\boldsymbol{G}^{\mp}$, $g_4^{\mp}$, $\boldsymbol{T}$, $\boldsymbol{\xi}_1$ and $\boldsymbol{\xi}_2$ have the same expressions as those in (1.58), but $\boldsymbol{G}^{\mp}$ and $g_4^{\mp}$ here may be complex numbers. Such relations are, in fact, generalization about the following results of homogeneous hyperbolic differential equations in two independent variables——the Riemann invariants along one characteristic curve issuing from a given point will not have discontinuities across the other characteristic curves.

It is easy to derive (2.22) from (2.20) because $\boldsymbol{G}^{\pm}$ is different from $\boldsymbol{N}_s$ only by a scalar factor. In order to derive (2.23) from (2.21) we make the inner products of the system consisting of the first three equations in (2.21) and each of the vectors $\boldsymbol{V}$, $\boldsymbol{T}$ and $\boldsymbol{G}^{\mp}$ and then divide the results by a. Finally, using (1.55), (1.57) and (1.60), we obtain

$$\begin{cases} \boldsymbol{V}\cdot\boldsymbol{N}_W^{\pm}=a, \\ \boldsymbol{T}\cdot\boldsymbol{N}_W^{\pm}=\boldsymbol{V}\times(V_{\xi_2}\boldsymbol{\xi}_1-V_{\xi_1}\boldsymbol{\xi}_2)\cdot\boldsymbol{N}_W^{\pm} \\ \qquad =\boldsymbol{V}\times(V_{\xi_2}\boldsymbol{\xi}_1-V_{\xi_1}\boldsymbol{\xi}_2)\cdot a(\sigma_{\pm}\boldsymbol{\xi}_1-\boldsymbol{\xi}_2)/(\sigma_{\pm}V_{\xi_1}-V_{\xi_2}) \\ \qquad =(-V_{\xi_1}a+V_{\xi_1}a\sigma_{\pm})\boldsymbol{V}\times\boldsymbol{\xi}_1\cdot\boldsymbol{\xi}_2/(\sigma_{\pm}V_{\xi_1}-V_{\xi_2})=a\boldsymbol{V}\times\boldsymbol{\xi}_1\cdot\boldsymbol{\xi}_2, \\ \boldsymbol{G}^{\mp}\boldsymbol{N}_W^{\pm}/\rho a=\dfrac{a\boldsymbol{N}_s}{\sigma_{\mp}V_{\xi_1}-V_{\xi_2}}\cdot\left(a\,\dfrac{-\boldsymbol{N}_s\cdot\boldsymbol{\xi}_2\boldsymbol{\xi}_1+\boldsymbol{N}_s\cdot\boldsymbol{\xi}_1\boldsymbol{\xi}_2}{\boldsymbol{N}_s\cdot\boldsymbol{N}_s}\pm\dfrac{\sqrt{\beta_2^2-\beta_1\beta_3}\,\boldsymbol{N}_s}{a\boldsymbol{N}_s\cdot\boldsymbol{N}_s}\right) \\ \qquad =\mp\sqrt{\beta_2^2-\beta_1\beta_3}/(V_{\xi_2}-\sigma_{\mp}V_{\xi_1})=g_4^{\mp}. \end{cases}$$

This can immediately lead to the desired result.

We can also derive (2.22) and (2.23) more directly in the following way. Let us start with the equations (1.16) or, equivalently, (1.58). Suppose that the surface $\widetilde{\Phi}(\xi_1, \xi_2, \xi_3)=0$ is a weak discontinuity surface through the straightline in the $\{\xi_1, \xi_2, \xi_3\}$–space,

$$\begin{cases} \xi_1=\text{constant}, \\ \xi_2=\text{constant}. \end{cases} \tag{2.24}$$

Along this straightline we have

$$\left(\frac{\partial\widetilde{\Phi}}{\partial\xi_1}, \frac{\partial\widetilde{\Phi}}{\partial\xi_2}, \frac{\partial\widetilde{\Phi}}{\partial\xi_3}\right)=(a_1, a_2, 0),$$

and B_{ζ_1} is equal to

$$\begin{pmatrix}
(\mathbf{G}^+)^*(a_1+a_2\sigma_+) & g_4^+(a_1+a_2\sigma_+) & 0 & 0 & \cdots & 0 \\
V_{\xi_1}\boldsymbol{V}^*(a_1+a_2\sigma) & V_{\xi_1}(a_1+a_2\sigma)/\rho & 0 & 0 & \cdots & 0 \\
V_{\xi_1}\boldsymbol{T}^*(a_1+a_2\sigma) & V_{\xi_1}\boldsymbol{V}\times\boldsymbol{\xi}_1\cdot\boldsymbol{\xi}_2(a_1+a_2\sigma)/\rho & 0 & 0 & \cdots & 0 \\
0 \quad 0 \quad 0 & V_{\xi_1}(a_1+a_2\sigma) & -a^2V_{\xi_1}(a_1+a_2\sigma) & 0 & \cdots & 0 \\
0 \quad 0 \quad 0 & 0 & 0 & V_{\xi_1}(a_1+a_2\sigma) & \cdots & 0 \\
\cdots & \cdots & \cdots & \cdots & \cdots & \cdots \\
0 \quad 0 \quad 0 & 0 & 0 & 0 & \cdots & V_{\zeta_1}(a_1+a_2\sigma) \\
(\mathbf{G}^-)^*(a_1+a_2\sigma_-) & g_4^-(a_1+a_2\sigma_-) & 0 & 0 & \cdots & 0
\end{pmatrix}.$$

Hence, for the stream weak discontinuity we obtain

$$a_1+a_2\sigma=0,$$

which leads to the relations (2.22). For the wave weak discontinuities we obtain

$$a_1+a_2\sigma_+=0 \quad \text{or} \quad a_1+a_2\sigma_-=0,$$

which leads to the relations (2.23), and for the wave weak discontinuity of the first family, the minus sign should be taken in (2.23) and for that of the second family the plus sign should be taken.

Obviously, our ability to derive the same result in different ways shows that such properties of the weak discontinuities are invariant with respect to the choice of coordinates.

3. *Singular Curves*

In the preceding subsections we studied the cases where the solutions of the differential equations or their derivatives may have two different values at each point on a surface. Sometimes, the solutions can have multiple values, even infinitely many values, at a point. Such a point is called a singular point. If every point on a curve is a singular point, the curve is called a singular curve.

In a steady flow there can be various singular curves and singular points. For example, the intersections of strong discontinuity surfaces are singular curves; the edges of the bodies in a flow are singular curves; the sharp points on the surface of a body in a flow are singular points. In the supersonic flow against a body with a sharp tip, the tip point is a singular point and the solutions at this point satisfy the conical flow equations under certain conditions. Thus, it can also be stated that the solution of the conical flow, which will be given in Chapter 6, is to deal with the solutions at a kind of singular point.

We shall now discuss the structure of the solution in the neighbourhood of a singular curve[1)].

For purposes of simplicity we will restrict our discussion to a consideration of the equations (1.11) with $m=5$, i.e., the equations (1.5). For a more general case the result is similar.

Let L be a curve in space which is given in the cylindrical coordinates z, r, φ by the equations

$$\begin{cases} z=z_0(\varphi), \\ r=r_0(\varphi). \end{cases} \tag{2.25}$$

We now introduce a new coordinate system $\bar{R}$, $\bar{\theta}$, Φ associated with the given curve as follows: in the plane $\varphi=\Phi$ we define the coordinate system $\bar{R}$, $\bar{\theta}$ as the polar coordinates with the pole Q on the curve which has the coordinates (z_0, r_0, Φ) in the original cylindrical coordinate system, where $\bar{R}$ is the radial coordinate of a point and $\bar{\theta}$ is the angle of the inclination of the radius vector to the axis z (see Fig. 2.4). It is clear that between the new and old coordinates there exist the relations

$$\begin{cases} z=\bar{R}\cos\bar{\theta}+z_0(\Phi), \\ r=\bar{R}\sin\bar{\theta}+r_0(\Phi), \\ \varphi=\Phi. \end{cases} \tag{2.26}$$

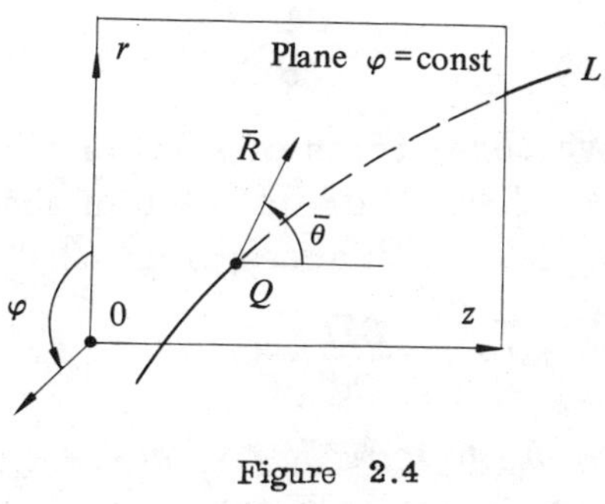

Figure 2.4

1) Part of those given below are taken from the paper [C67].

From (2.26) we have

$$\begin{pmatrix} \dfrac{\partial \bar{R}}{\partial z} & \dfrac{\partial \bar{\theta}}{\partial z} & \dfrac{\partial \Phi}{\partial z} \\ \dfrac{\partial \bar{R}}{\partial r} & \dfrac{\partial \bar{\theta}}{\partial r} & \dfrac{\partial \Phi}{\partial r} \\ \dfrac{\partial \bar{R}}{\partial \varphi} & \dfrac{\partial \bar{\theta}}{\partial \varphi} & \dfrac{\partial \Phi}{\partial \varphi} \end{pmatrix}$$

$$=\begin{pmatrix} \cos\bar{\theta} & -\sin\bar{\theta}/\bar{R} & 0 \\ \sin\bar{\theta} & \cos\bar{\theta}/\bar{R} & 0 \\ -\dfrac{dz_0}{d\Phi}\cos\bar{\theta}-\dfrac{dr_0}{d\Phi}\sin\bar{\theta} & \left(\dfrac{dz_0}{d\Phi}\sin\bar{\theta}-\dfrac{dr_0}{d\Phi}\cos\bar{\theta}\right)\Big/\bar{R} & 1 \end{pmatrix}. \tag{2.27}$$

From the first five equations of (1.9), using (1.12) and (2.26), we obtain the following steady flow equation,

$$A_{\bar{R}}\frac{\partial U}{\partial \bar{R}}+A_{\bar{\theta}}\frac{\partial U}{\partial \bar{\theta}}+A_{\Phi}\frac{\partial U}{\partial \Phi}+D=0,$$

where

$$\begin{cases} A_{\bar{R}}=\dfrac{\partial \bar{R}}{\partial z}A_1+\dfrac{\partial \bar{R}}{\partial r}A_2+\dfrac{\partial \bar{R}}{\partial \varphi}A_3, \\ A_{\bar{\theta}}=\dfrac{\partial \bar{\theta}}{\partial z}A_1+\dfrac{\partial \bar{\theta}}{\partial r}A_2+\dfrac{\partial \bar{\theta}}{\partial \varphi}A_3, \\ A_{\Phi}=\dfrac{\partial \Phi}{\partial z}A_1+\dfrac{\partial \Phi}{\partial r}A_2+\dfrac{\partial \Phi}{\partial \varphi}A_3. \end{cases} \tag{2.28}$$

Here $U^*=(u, v, w, p, \rho)$, and u, v and w are, respectively, the z–, r–, and φ–components of the velocity vector V. Note that we are concerned only with the flow in the vicinity of the points on the given curve. Suppose $\dfrac{\partial U}{\partial \bar{R}}$, $\dfrac{\partial U}{\partial \bar{\theta}}$, $\dfrac{\partial U}{\partial \Phi}$, A_1, A_2 and A_3 all are bounded when $\bar{R}\to 0$. Then, $A_{\bar{R}}$, A_{Φ} and D are also bounded, but $A_{\bar{\theta}}=O\left(\dfrac{1}{\bar{R}}\right)\to\infty$. Thereby we finally obtain

$$\bar{R}A_{\bar{\theta}}\frac{\partial U}{\partial \bar{\theta}}=0. \tag{2.29}$$

This equation can have three types of solutions, which describe three different flow patterns. We shall discuss each of them separately.

(i) Suppose $|\bar{R}A_{\bar{\theta}}|\neq 0$. In this case, from (2.29) we have

$$\frac{\partial U}{\partial \bar{\theta}}=0.$$

Thus if in the region $\bar{\theta}_0(\Phi)\leqslant\bar{\theta}\leqslant\bar{\theta}_1(\Phi)$, $\Phi_0\leqslant\Phi\leqslant\Phi_1$ that condition is satisfied, the solution of equations (2.29) in the region has the following form:

$$\begin{cases} u=u(\Phi), \\ v=v(\Phi), \\ w=w(\Phi), \\ p=p(\Phi), \\ \rho=\rho(\Phi). \end{cases} \tag{2.30}$$

Such a flow is called the "single-valued flow", by which we mean that the flow properties at a point on the given curve are independent of $\overline{\theta}$ for $\overline{\theta}_0 \leqslant \overline{\theta} \leqslant \overline{\theta}_1$.

Now we turn to a discussion of the case where $|\overline{R}A_{\bar{\theta}}|=0$. Recalling the definition of characteristic surfaces and noting the expression (2.28) for $A_{\bar{\theta}}$, we immediately find that where $|\overline{R}A_{\bar{\theta}}|=0$ the surface $\overline{\theta}=$constant must be a characteristic surface. On the curve L the normal to a surface $\overline{\theta}(z, r, \varphi)=$constant is

$$N=\begin{pmatrix} \overline{R}\dfrac{\partial\theta}{\partial z} \\ \overline{R}\dfrac{\partial\overline{\theta}}{\partial r} \\ \overline{R}\dfrac{1}{r_0}\dfrac{\partial\overline{\theta}}{\partial\varphi} \end{pmatrix}=\begin{pmatrix} -\sin\overline{\theta} \\ \cos\overline{\theta} \\ b \end{pmatrix},$$

where

$$b=\frac{1}{r_0}\frac{dz_0}{d\Phi}\sin\overline{\theta}-\frac{1}{r_0}\frac{dr_0}{d\Phi}\cos\overline{\theta}.$$

Thus there are two possibilities.

(ii) Suppose

$$\psi=\boldsymbol{V}\cdot\boldsymbol{N}=-u\sin\overline{\theta}+v\cos\overline{\theta}+wb=0 \tag{2.31}$$

in the region $\overline{\theta}_0(\Phi)\leqslant\overline{\theta}\leqslant\overline{\theta}_1(\Phi)\neq\overline{\theta}_0(\Phi)$, $\Phi_0\leqslant\Phi\leqslant\Phi_1$. Then every surface $\overline{\theta}=$constant in that region is a stream-characteristic surface and the equations (2.29) can be reduced to

$$\begin{cases} \dfrac{\partial p}{\partial\overline{\theta}}=0, \\ \boldsymbol{N}_s\cdot\dfrac{\partial\boldsymbol{V}}{\partial\overline{\theta}}=-\sin\overline{\theta}\dfrac{\partial u}{\partial\overline{\theta}}+\cos\overline{\theta}\dfrac{\partial v}{\partial\overline{\theta}}+b\dfrac{\partial w}{\partial\overline{\theta}}=0. \end{cases} \tag{2.32}$$

The flow described by (2.32) will be referred to as the "stream-singular flow".

In order to observe such a flow, we rewrite the second equation of (2.32) in the form

$$\boldsymbol{N}_s\cdot\frac{\partial\boldsymbol{V}}{\partial\overline{\theta}}=\frac{\partial\boldsymbol{V}\cdot\boldsymbol{N}_s}{\partial\overline{\theta}}-\boldsymbol{V}\cdot\frac{\partial\boldsymbol{N}_s}{\partial\overline{\theta}}=-\boldsymbol{V}\cdot\frac{\partial\boldsymbol{N}_s}{\partial\overline{\theta}}$$
$$=-\left(-u\cos\overline{\theta}-v\sin\overline{\theta}+w\left(\frac{1}{r_0}\frac{dz_0}{d\Phi}\cos\overline{\theta}+\frac{1}{r_0}\frac{dr_0}{d\Phi}\sin\overline{\theta}\right)\right)=0. \tag{2.33}$$

The tangent to the given curve L is

$$\boldsymbol{t}_3=\frac{1}{\sqrt{\left(\frac{1}{r_0}\frac{dz_0}{d\Phi}\right)^2+\left(\frac{1}{r_0}\frac{dr_0}{d\Phi}\right)^2+1}}\begin{pmatrix}\frac{1}{r_0}\frac{dz_0}{d\Phi}\\ \frac{1}{r_0}\frac{dr_0}{d\Phi}\\ 1\end{pmatrix}. \tag{2.34}$$

Obviously,

$$\boldsymbol{t}_3\cdot\frac{\partial \boldsymbol{N}_s}{\partial\overline{\theta}}=0,\quad \boldsymbol{t}_3\cdot\boldsymbol{N}_s=0. \tag{2.35}$$

Therefore $\boldsymbol{V}$ is parallel to $\boldsymbol{t}_3$. If we denote by $q(\Phi)$ the projection of the velocity vector on the plane normal to the given curve L, we have

$$\begin{cases}p=p(\Phi),\\ q(\Phi)=0.\end{cases} \tag{2.36}$$

Thus we can conclude that, in a stream-singular flow around the curve L, the pressure is independent of $\overline{\theta}$ and the flow velocity is directed along the curve. However the modulus of the velocity vector and the density change with $\overline{\theta}$. Hence the stream-singular curve is a curve along which a number of streamplanes "come together". Note that the stream-singular curves are distinct from the streamlines in an ordinary sense although through a streamline there are also many streamplanes. The "entropy singularity curve" is such a singular curve.

(iii) Suppose

$$|\boldsymbol{V}\cdot\boldsymbol{N}|^2/|\boldsymbol{N}|^2=|-u\sin\overline{\theta}+v\cos\overline{\theta}+wb|^2/(1+b^2)=a^2 \tag{2.37}$$

in the region $\overline{\theta}_0(\Phi)\leqslant\overline{\theta}\leqslant\overline{\theta}_1(\Phi)\neq\overline{\theta}_0(\Phi)$, $\Phi_0\leqslant\Phi\leqslant\Phi_1$. In this case every surface $\overline{\theta}=\text{constant}$ in the region is a wave-characteristic surface and therefore the equations (2.29) can be written in the form

$$\begin{cases}a\dfrac{\partial u}{\partial\overline{\theta}}+\dfrac{N_{W,1}}{\rho}\dfrac{\partial p}{\partial\overline{\theta}}=0,\\ a\dfrac{\partial u}{\partial\overline{\theta}}+\dfrac{N_{W,2}}{\rho}\dfrac{\partial p}{\partial\overline{\theta}}=0,\\ a\dfrac{\partial w}{\partial\overline{\theta}}+\dfrac{N_{W,3}}{\rho}\dfrac{\partial p}{\partial\overline{\theta}}=0,\\ \dfrac{\partial p}{\partial\overline{\theta}}-a^2\dfrac{\partial\rho}{\partial\overline{\theta}}=0,\end{cases} \tag{2.38}$$

where

$$N_W=\frac{a}{\tilde{\psi}_W}\begin{pmatrix}-\sin\overline{\theta}\\ \cos\overline{\theta}\\ b\end{pmatrix},\quad \tilde{\psi}_W=-u\sin\overline{\theta}+v\cos\overline{\theta}+wb.$$

Using the same process as has been used to derive (2.23) from (2.21), we can rewrite the equations (2.38) in the following form——the "stream-compatibility relations" and the "wave-compatibility relation" on the wave-characteristic surface of the other family:

$$\left\{\begin{array}{l}\boldsymbol{V}\cdot\dfrac{\partial \boldsymbol{V}}{\partial\overline{\theta}}+\dfrac{1}{\rho}\dfrac{\partial p}{\partial\overline{\theta}}=0,\\ \overline{R}\boldsymbol{T}\cdot\dfrac{\partial \boldsymbol{V}}{\partial\overline{\theta}}+\dfrac{1}{\rho}\boldsymbol{V}\times\overline{\boldsymbol{R}}\cdot\overline{R}\overline{\boldsymbol{\theta}}\dfrac{\partial p}{\partial\overline{\theta}}=0,\\ \dfrac{\partial p}{\partial\overline{\theta}}-a^2\dfrac{\partial \rho}{\partial\overline{\theta}}=0,\\ \overline{R}\boldsymbol{G}^{\mp}\cdot\dfrac{\partial \boldsymbol{V}}{\partial\overline{\theta}}+\overline{R}g_4^{\mp}\dfrac{\partial p}{\partial\overline{\theta}}=0,\end{array}\right. \tag{2.39}$$

in which the minus sign should appear when N_W in (2.38) corresponds to the first family, and the plus sign when N_W corresponds to the second family; $\boldsymbol{T}$, $\boldsymbol{G}^{\mp}$ and $g_4^{\mp}$ have the same forms as in (1.39) and (1.57), except that ξ_1 and ξ_2 are replaced respectively by

$$\overline{\boldsymbol{R}}=\begin{pmatrix}\cos\overline{\theta}\\ \sin\overline{\theta}\\ -\dfrac{1}{r_0}\left(\dfrac{dz_0}{d\Phi}\cos\overline{\theta}+\dfrac{dr_0}{d\Phi}\sin\overline{\theta}\right)\end{pmatrix}$$

and

$$\overline{\boldsymbol{\theta}}=\frac{1}{\overline{R}}\begin{pmatrix}-\sin\overline{\theta}\\ \cos\overline{\theta}\\ \dfrac{1}{r_0}\left(\dfrac{dz_0}{d\Phi}\sin\overline{\theta}-\dfrac{dr_0}{d\Phi}\cos\overline{\theta}\right)\end{pmatrix}.$$

We have multiplied the second and fourth equations of (2.39) by the factor $\overline{R}$ so that the coefficients are bounded.

The flow described by (2.38) or (2.39) is referred to as the "wave-singular flow".

In order to clarify the features of this flow we introduce a new variable θ in place of $\overline{\theta}$ through the relations

$$\left\{\begin{array}{l}\cos\theta=\boldsymbol{t}_1(\overline{\theta})\cdot\boldsymbol{t}_1(0),\\ \sin\theta=\boldsymbol{t}_1(\overline{\theta})\cdot\boldsymbol{t}_2(0),\end{array}\right. \tag{2.40}$$

where

$$\left\{\begin{array}{l}\boldsymbol{t}_2(\overline{\theta})=\dfrac{1}{\sqrt{1+b^2}}\begin{pmatrix}-\sin\overline{\theta}\\ \cos\overline{\theta}\\ b\end{pmatrix},\quad b=\dfrac{1}{r_0}\left(\dfrac{dz_0}{d\Phi}\sin\overline{\theta}-\dfrac{dr_0}{d\Phi}\cos\overline{\theta}\right),\\ \boldsymbol{t}_1(\overline{\theta})=\boldsymbol{t}_2(\overline{\theta})\times\boldsymbol{t}_3.\end{array}\right.$$

We first observe the geometrical meaning of θ. It is easily seen that $\boldsymbol{t}_2(\overline{\theta})$ is perpendicular to $\boldsymbol{t}_3$, by their definitions. Therefore $\boldsymbol{t}_1(\overline{\theta})$, $\boldsymbol{t}_2(\overline{\theta})$ and $\boldsymbol{t}_3$ are mutually orthogonal, which implies that $\boldsymbol{t}_1(\overline{\theta})$ and $\boldsymbol{t}_2(\overline{\theta})$ lie on the plane normal to $\boldsymbol{t}_3$. In addition, $\boldsymbol{t}_2(\overline{\theta})$ is also perpendicular to the $\overline{R}$-axis, i.e., perpendicular to the vector

$$\boldsymbol{t}_4(\bar{\theta}) = \begin{pmatrix} \cos\bar{\theta} \\ \sin\bar{\theta} \\ 0 \end{pmatrix},$$

so that $\boldsymbol{t}_1(\bar{\theta})$ lies on the plane determined by $\boldsymbol{t}_4(\bar{\theta})$ and $\boldsymbol{t}_3$ (see Fig. 2.5). Furthermore, from the first relation of (2.40), it is evident that θ is the angle between $\boldsymbol{t}_1(0)$ and $\boldsymbol{t}_1(\bar{\theta})$ and therefore is the projection of the angle $\bar{\theta}$ on the plane normal to $\boldsymbol{t}_3$, $\bar{\theta}$ being measured on the plane $\Phi=$ constant. Notice that $\boldsymbol{t}_1(\bar{\theta})$, $\boldsymbol{t}_1(0)$, $\boldsymbol{t}_2(0)$ all lie on the same plane and $\boldsymbol{t}_1(0)$ is perpendicular to $\boldsymbol{t}_2(0)$. Thus, if the first relation of (2.40) holds, then the left-hand side of the second relation of (2.40) must be equal to the right-hand side in the modulus. Thus the purpose of introducing the second relation of (2.40) is only to define which direction is positive when the angle θ is measured.

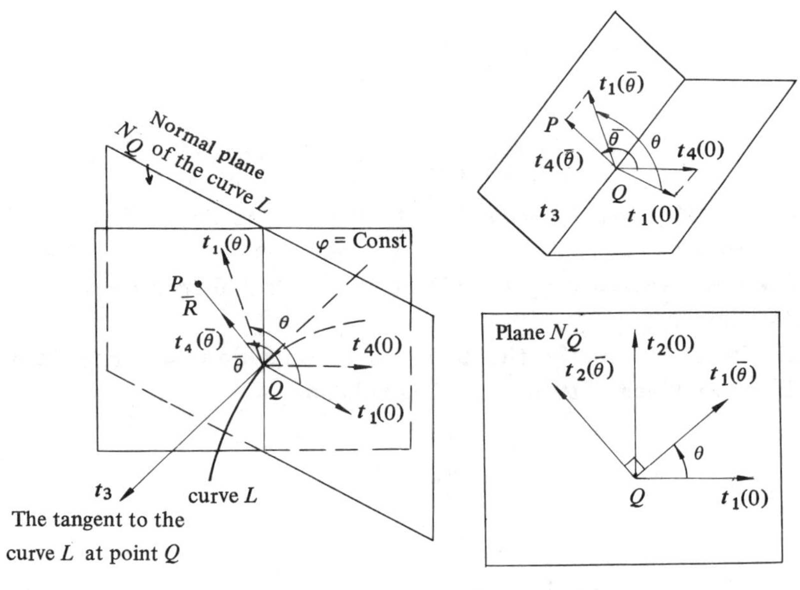

Figure 2.5

Since equations (2.38) are homogeneous, we can replace $\dfrac{\partial}{\partial\bar{\theta}}$ by $\dfrac{\partial}{\partial\theta}$.

Then, after multiplying the first three equations by $\dfrac{1}{a}\boldsymbol{t}_1$, $\dfrac{\tilde{\psi}_W}{a\sqrt{1+b^2}}\boldsymbol{t}_2$ and $\dfrac{1}{a}\boldsymbol{t}_3$ respectively, we obtain

$$\begin{cases} \boldsymbol{t}_1 \cdot \dfrac{\partial \boldsymbol{V}}{\partial \theta} = 0, \\ \boldsymbol{t}_2 \cdot \boldsymbol{V} \boldsymbol{t}_2 \cdot \dfrac{\partial \boldsymbol{V}}{\partial \theta} + \dfrac{1}{\rho} \dfrac{\partial p}{\partial \theta} = 0, \\ \boldsymbol{t}_3 \cdot \dfrac{\partial \boldsymbol{V}}{\partial \theta} = 0. \end{cases} \tag{2.41}$$

Let

$$V_1 = \boldsymbol{V} \cdot \boldsymbol{t}_1, \quad V_2 = \boldsymbol{V} \cdot \boldsymbol{t}_2, \quad V_3 = \boldsymbol{V} \cdot \boldsymbol{t}_3.$$

From

$$\begin{cases} \boldsymbol{t}_1(\bar{\theta}) = \cos\theta \boldsymbol{t}_1(0) + \sin\theta \boldsymbol{t}_2(0), \\ \boldsymbol{t}_2(\bar{\theta}) = -\sin\theta \boldsymbol{t}_1(0) + \cos\theta \boldsymbol{t}_2(0) \end{cases} \tag{2.42}$$

and the fact that $\boldsymbol{t}_3$ is independent of $\bar{\theta}$, we find

$$\frac{\partial \boldsymbol{t}_1(\bar{\theta})}{\partial \theta} = \boldsymbol{t}_2(\bar{\theta}), \quad \frac{\partial \boldsymbol{t}_2(\bar{\theta})}{\partial \theta} = -\boldsymbol{t}_1(\bar{\theta}), \quad \frac{\partial \boldsymbol{t}_3}{\partial \theta} = 0. \tag{2.43}$$

Therefore, from (2.38) and (2.37) we obtain

$$\begin{cases} \dfrac{\partial V_1}{\partial \theta} - V_2 = 0, \\ V_2\left(\dfrac{\partial V_2}{\partial \theta} + V_1\right) + \dfrac{1}{\rho} \dfrac{\partial p}{\partial \theta} = 0, \\ \dfrac{\partial V_3}{\partial \theta} = 0, \\ \dfrac{\partial p}{\partial \theta} - a^2 \dfrac{\partial \rho}{\partial \theta} = 0, \\ V_2 = \mp a. \end{cases} \tag{2.44}$$

Then, multiplying the system of the first two equations of (2.44) from the left by $(V_1, 1)$ and $(-V_2/(V_1^2+V_2^2),\ V_1/(V_2(V_1^2+V_2^2)))$ leads to

$$\begin{cases} \dfrac{1}{2} \dfrac{\partial}{\partial \theta}(V_1^2+V_2^2) + \dfrac{1}{\rho} \dfrac{\partial p}{\partial \theta} = 0, \\ \dfrac{\partial}{\partial \theta}\left(\operatorname{arc\,tg} \dfrac{V_2}{V_1} + \theta\right) + \dfrac{V_1}{\rho V_2 (V_1^2+V_2^2)} \dfrac{\partial p}{\partial \theta} \\ \quad = \dfrac{a}{V_2}\left[\dfrac{\partial}{\partial \theta}\left(\operatorname{arc\,tg} \dfrac{a}{V_1} + \dfrac{V_2}{a}(\theta - \pi)\right) + \dfrac{V_1}{\rho a (V_1^2+V_2^2)} \dfrac{\partial p}{\partial \theta}\right] = 0. \end{cases} \tag{2.45}$$

Now, introducing

$$\begin{cases} \beta = \operatorname{arc\,tg} \dfrac{a}{V_1} + \dfrac{V_2}{a}(\theta - \pi), \\ q^2 = V_1^2 + V_2^2, \end{cases} \tag{2.46}$$

we obtain

$$\begin{cases} \dfrac{1}{2}\dfrac{\partial q^2}{\partial\theta}+\dfrac{1}{\rho}\dfrac{\partial p}{\partial\theta}=0, \\ \dfrac{\partial V_3}{\partial\theta}=0, \\ \dfrac{\partial p}{\partial\theta}-a^2\dfrac{\partial\rho}{\partial\theta}=0, \\ \dfrac{\partial\beta}{\partial\theta}+\dfrac{V_1}{\rho a q^2}\dfrac{\partial p}{\partial\theta}=0, \\ V_2=\mp a, \end{cases} \tag{2.47}$$

or, in terms of the variable β,

$$\begin{cases} \dfrac{\partial q^2}{\partial\beta}=2aq^2/V_1, \\ \dfrac{\partial V_3}{\partial\beta}=0, \\ \dfrac{\partial\rho}{\partial\beta}=-\rho q^2/(aV_1), \\ \dfrac{\partial p}{\partial\beta}=-\rho a q^2/V_1, \\ V_2=\mp a, \end{cases} \tag{2.48}$$

with $|V_1|=\sqrt{q^2-a^2}$.

From Fig. 2.6 it is seen that β is the angle on the $(\boldsymbol{t}_1, \boldsymbol{t}_2)$-plane between $\boldsymbol{t}_1(\pi)$ and the projection $\boldsymbol{q}$ of the velocity vector $\boldsymbol{V}(\theta)$ when $V_1>0$. The angle β is measured clockwise when $V_2/a=-1$ and counterclockwise when $V_2/a=1$.

The equations (2.48) are quite similar to that for the Prandtl–Meyer flow in two-dimensional case. In the present case we have one additional equation, $\dfrac{\partial V_3}{\partial\beta}=0$, the second of (2.48), and q is the modulus of the projection of the velocity on the normal plane to the curve L, but not the modulus of the velocity. Therefore, we say that the equations

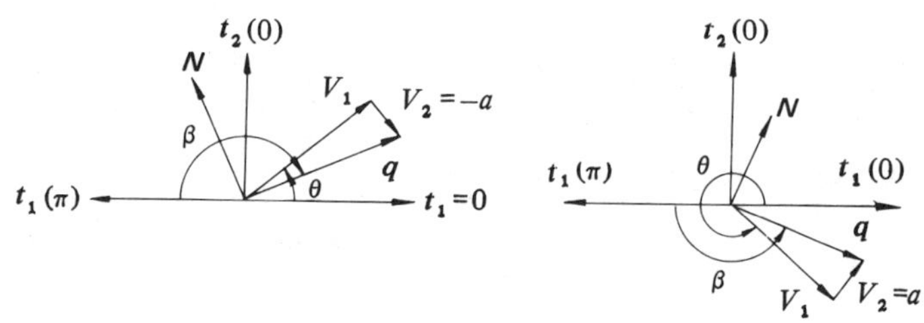

Figure 2.6

(2.48) describe a three-dimensional centered simple wave.

Consider the case where $V_1>0$ and $V_2=-a$ at first. From (2.48) we know that p decreases as β increases. Therefore, if θ decreases as β increases, the flow described by (2.48) is a centered expansion wave of the first family issuing from the curve L (see Fig. 2.7). Similarly, if β increases with the increase of θ when $V_1>0$ and $V_2=a$, the flow is a centered expansion wave of the second family. Now let us see what $\frac{d\theta}{d\beta}>0$ means when $V_1>0$ and $V_2=a$. From (2.46) and (2.48) we have

$$\begin{aligned}
d\beta &= \frac{1}{q^2}(V_1\,da - a\,dV_1) + d\theta \\
&= \frac{1}{q^2}\Big(V_1 da - \frac{a}{2V_1}(dq^2 - 2a\,da)\Big) + d\theta \\
&= \frac{1}{V_1}\,da - \frac{a^2}{V_1^2}\,d\beta + d\theta = \Big(\frac{1}{V_1}\,\frac{da}{d\rho}\,\frac{d\rho}{d\beta} - \frac{a^2}{V_1^2}\Big)d\beta + d\theta \\
&= \Big(-\frac{\rho q^2}{aV_1^2}\,\frac{da}{d\rho} - \frac{a^2}{V_1^2}\Big)d\beta + d\theta,
\end{aligned}$$

so

$$\frac{d\theta}{d\beta} = \Big(1 + \frac{a^2}{V_1^2} + \frac{\rho q^2}{aV_1^2}\,\frac{da}{d\rho}\Big) = \frac{\rho q^2}{aV_1^2}\Big(\frac{a}{\rho} + \frac{da}{d\rho}\Big).$$

$\Big($If $\Big(\frac{\partial^2 p}{\partial V^2}\Big)_s$ is always greater than zero, then only the centered expansion wave appears.$\Big)$

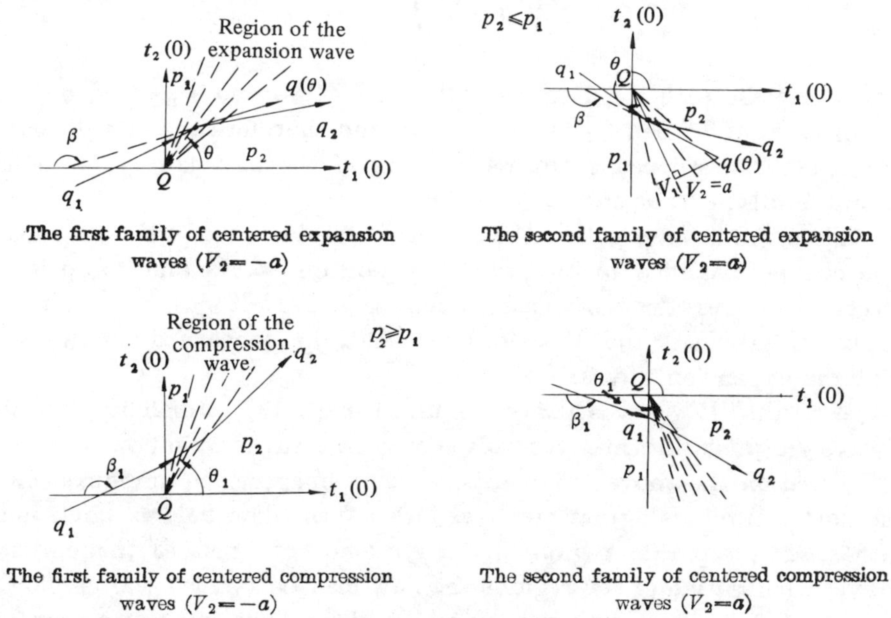

The first family of centered expansion waves ($V_2=-a$)

The second family of centered expansion waves ($V_2=a$)

The first family of centered compression waves ($V_2=-a$)

The second family of centered compression waves ($V_2=a$)

Figure 2.7 Two types of centered simple waves

On the other hand, we have

$$\frac{d}{d\left(\frac{1}{\rho}\right)}\left(\frac{dp}{d\left(\frac{1}{\rho}\right)}\right)=\frac{d}{d\left(\frac{1}{\rho}\right)}\left[\frac{dp}{d\rho}(-\rho^2)\right]=\frac{d^2p}{d\rho^2}\rho^4+\frac{dp}{d\rho}2\rho^3$$

$$=2a\frac{da}{d\rho}\rho^4+a^2 2\rho^3=2a\rho^4\left(\frac{da}{d\rho}+\frac{a}{\rho}\right).$$

Since the process under consideration is isentropic, we conclude that when $V_1>0$ and $V_2=a$ the inequality $\frac{d\theta}{d\beta}>0$ means that

$$\left(\frac{\partial^2 p}{\partial\left(\frac{1}{\rho}\right)^2}\right)_s>0, \tag{2.49}$$

in which the pressure p is considered as a function of the specific entropy S and the specific volume $1/\rho$. Also, we can show that when $V_1>0$ and $V_2=-a$, the inequality $\frac{d\theta}{d\beta}<0$ is equivalent to (2.49). Thus, we assert that the equations (2.48) under the condition (2.49) describe the centered expansion waves, which belong to the first family when $V_1>0$ and $V_2=-a$ or the second family when $V_1>0$ and $V_2=a$.

Similarly, we can find that under the condition

$$\left(\frac{\partial^2 p}{\partial\left(\frac{1}{\rho}\right)^2}\right)_s<0,$$

p will increase with the decrease of θ when $V_1>0$ and $V_2=-a$, or with the increase of θ when $V_1>0$ and $V_2=a$, and therefore the flow described by (2.48) is a centered compression wave of the first family, or of the second family, correspondingly.

Thus, we have seen that the centered simple waves, like the shocks, also can be classified in two types: expansion waves and compression waves. However for most gases, such as a perfect gas, the condition (2.49) is satisfied, and therefore we shall be concerned mainly with centered expansion waves.

Note that $|V_2|=a$ is always required in (2.48), which implies that a wave-singular flow may take place only in a supersonic flow.

Based on the above statements, we may imagine the configuration of the flow around a singular curve as follows: the flow field is partitioned into several separate regions by a sequence of surfaces through the curve; within each of the regions the flow may be a single-valued flow, a stream-singular flow, or a wave-singular flow. Those delimiting surfaces are either strong discontinuity surfaces(shocks or contact discontinuities) or characteristic surfaces (streamplanes or wave-characteristic surfaces), which are weak discontinuities, possibly with great discontinuity strength. Such configurations are, of course, physically possible only when the jump conditions and the stability conditions for the strong discontinuity surfaces are satisfied and the flow properties remain continuous across the weak discontinuity surfaces.

In conclusion, we would like to make a few remarks. First, the equations (2.44) can be obtained almost immediately from (1.11), if we use the local cylindrical coordinates with $\boldsymbol{t}_3$ as the axis of symmetry. However, it is often interesting to know the dependence of the flow properties on the angle $\bar{\theta}$. Now it is convenient to do so. Moreover, after knowing $U(\bar{\theta})$, using

$$\begin{pmatrix} u \\ v \\ w \end{pmatrix} = (\boldsymbol{t}_1 \quad \boldsymbol{t}_2 \quad \boldsymbol{t}_3) \begin{pmatrix} V_1 \\ V_2 \\ V_3 \end{pmatrix}, \tag{2.50}$$

we can also obtain the components of the velocity in the z, r, φ-coordinate directions. This may be important for certain practical calculations. Next, the fourth equation of (2.47) is, indeed, the fourth of (2.39), which is a wave-compatibility relation of the second family when $V_2 = -a$ or of the first family when $V_2 = a$. The first and the second equations of (2.47) can be derived from the first and the second equations of (2.39). Thus, the first three equations of (2.47) are stream-compatibility relations and the fourth is a wave-compatibility relation of "the other family". However, deriving (2.47) from (2.38) rather than from (2.39) is easier to understand. Finally, when we take the non-equilibrium process into account, the above conclusion on the flow configuration around a singular curve is still valid, but then we need to consider the equations for c_i, $i=1, 2, \cdots m-5$. It is easily seen that (i) for a single-valued flow we have $c_i = c_i(\Phi)$; (ii) for a stream-singular flow the result is still (2.36), that is, no constraint is imposed on c_i; (iii) for a centered simple wave we ought to add $\dfrac{\partial c_i}{\partial \beta} = 0$ to the system (2.48).

4. *Space-like Singular Curves and the Interaction between Strong Discontinuities*

In the present subsection we shall deal with several types of "space-like" singular curves, including those formed by the collisions and reflections of strong discontinuities. Here by "space-like" we mean that at any point on the singular curve the projection of every possible velocity value $\boldsymbol{V}(\theta)$ on the normal plane to the curve is greater than the corresponding speed of sound $a(\theta)$ in the modulus.

First of all we should point out that there is one degree of freedom for determining a shock through a given curve; that is, when the flow properties in front of a shock are known, the normal direction to the shock and flow properties can be determined uniquely by just one condition on them. This can be seen from (2.4). In the system (2.4), when U_1 is known, the number of unknowns is two more than the number of equations in system (2.4). (Note that a unit vector is uniquely defined by two components of it.) However, we have forced the shock surface to pass through the given curve, so we now lack only one condition. Similarly, there is also one degree of freedom for determining a centered expansion wave. This results from the fact that the region of the expansion is bounded by a characteristic surface through the given curve and the flow properties are always continuous. This can also be seen from the equations (2.48). When the flow in front of the expansion wave is known, determining it is actually to integrate the ordinary differential equations (2.48) from the given initial values. The integration will be terminated when the solutions satisfy a given condition, such as a given direction of flow. Finally, it is seen from (2.3) that we have $m-2$ degrees of freedom for determining a contact discontinuity through a given curve when the flow on one of the two sides of the discontinuity surface is known ($m+2$ unknowns need to be determined at that time).

As an example we may consider a supersonic flow past a sharp corner, which is assumed to be a "space-like" curve (see Fig. 2.8). To be clear about what will happen around the corner, for the moment, we might as well assume that the coming stream on the left is uniform and parallel to the solid surface. In general, we can further assume that, after the flow changes its direction, the flow direction should be parallel to the solid surface on the right. This is the only condition that we may impose on the flow on the right. According to the statements given above, it is evident that there should be a "singular flow" with one degree of freedom on the edge which coordinates the flows on two sides of the corner. Intuitively, this should be a shock wave when the corner is concave or an expansion wave when convex (assuming that the gas obeys the inequality (2.49)).

In what follows we shall deal with a variety of singular curves in detail.

(i) *Concave Corners*

Let us suppose that the flow field around the corner has this structure: two single-valued flow regions separated by an oblique shock surface issuing from the edge. The flow properties U_1 in front of the shock are assumed to be known and we want to determine the normal direction to the shock and the flow properties U_2 behind it.

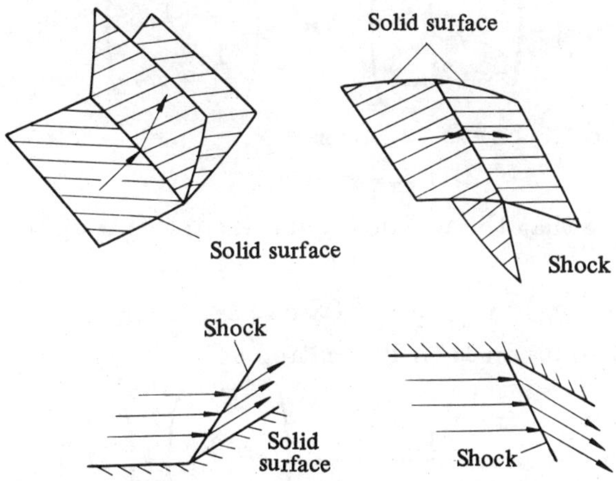

The first family of shock The second family of shock

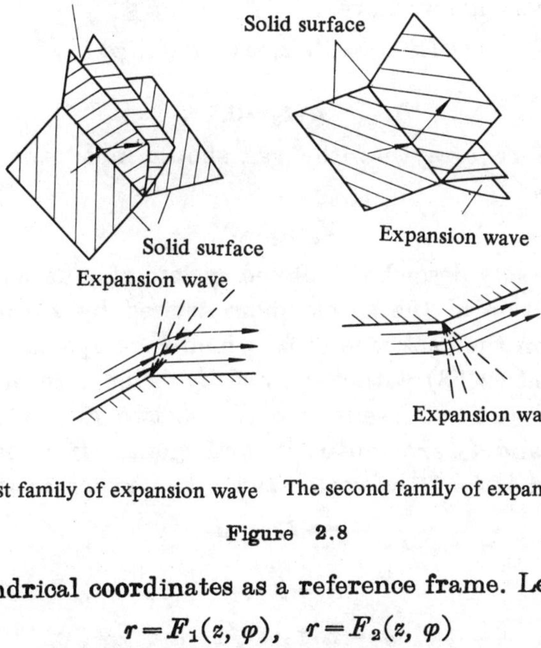

The first family of expansion wave The second family of expansion wave

Figure 2.8

Take cylindrical coordinates as a reference frame. Let

$$r=F_1(z, \varphi), \quad r=F_2(z, \varphi) \tag{2.51}$$

be the equations describing the solid surface before and after the edge, respectively. Their intersection, i.e., the edge of the body, is given by the equations

$$\begin{cases} r=F_1(z, \varphi), \\ r=F_2(z, \varphi). \end{cases} \tag{2.52}$$

Then the normals to the two surfaces in (2.51) are

$$n_1=\begin{pmatrix} F_{1,z} \\ -1 \\ \dfrac{1}{F_1}F_{1,\varphi} \end{pmatrix}, \quad n_2=\begin{pmatrix} F_{2,z} \\ -1 \\ \dfrac{1}{F_2}F_{2,\varphi} \end{pmatrix}, \tag{2.53}$$

and the tangent to the curve (2.52) is

$$t_3=n_1\times n_2. \tag{2.54}$$

We further assume that the unknown shock surface sought after is given by the equation

$$r=F(z,\ \varphi). \tag{2.55}$$

The unit normal to it can then be written as

$$n=\frac{1}{\sqrt{1+F_z^2+\left(\dfrac{1}{F}F_\varphi\right)^2}}\begin{pmatrix} F_z \\ -1 \\ \dfrac{1}{F}F_\varphi \end{pmatrix}=\begin{pmatrix} n_1 \\ n_2 \\ n_3 \end{pmatrix}. \tag{2.56}$$

Note that here we are concerned with the shape of the shock and the flow properties on both sides of it only at the points on the edge. Obviously, on the edge we have

$$F(z,\ \varphi)=F_1(z,\ \varphi)=F_2(z,\ \varphi)$$

and

$$n\cdot t_3=0. \tag{2.57}$$

In addition, we require that the gas should not penetrate the solid surface, that is,

$$V_2\cdot n_2=0. \tag{2.58}$$

Now we have already described a closed system of equations to determine the normal direction to the shock characterized by F_z and F_φ and the flow properties on the back side of it, which is composed of the equations (2.4), (2.57) and (2.58) with n defined by (2.56). In order to find the solutions of the equations easily, we choose two unit vectors t_1 and t_2 such that t_1, t_2 and t_3 are mutually orthogonal. It is obvious that by using (2.57) and $|n|=1$ the vector n can be written in the form

$$n=\alpha_1 t_1+\alpha_2 t_2, \tag{2.59}$$

where

$$\alpha_1=n\cdot t_1,$$
$$\alpha_2=n\cdot t_2=\sqrt{1-\alpha_1^2}.$$

From the first equation of (2.4) we obtain

$$\begin{cases} V_{2,1}=V_{1,1}-V_{n,1}\left(1-\dfrac{\rho_1}{\rho_2}\right)\alpha_1, \\ V_{2,2}=V_{1,2}-V_{n,1}\left(1-\dfrac{\rho_1}{\rho_2}\right)\alpha_2, \end{cases}$$

where

$$V_{i,j} = \boldsymbol{V}_i \cdot \boldsymbol{t}_j, \quad i=1, 2, \quad j=1, 2.$$

Thereby, the equation (2.58) becomes

$$\begin{aligned}\boldsymbol{V}_2 \cdot \boldsymbol{n}_2 &= V_{2,1}\boldsymbol{t}_1 \cdot \boldsymbol{n}_2 + V_{2,2}\boldsymbol{t}_2 \cdot \boldsymbol{n}_2 \\ &= V_{1,1}\boldsymbol{t}_1 \cdot \boldsymbol{n}_2 + V_{1,2}\boldsymbol{t}_2 \cdot \boldsymbol{n}_2 - V_{n,1}\left(1 - \frac{\rho_1}{\rho_2}\right)(\alpha_1 \boldsymbol{t}_1 \cdot \boldsymbol{n}_2 + \alpha_2 \boldsymbol{t}_2 \cdot \boldsymbol{n}_2) = 0.\end{aligned}$$

From this equation, using

$$\begin{aligned}V_{n,1} &= \boldsymbol{V}_1 \cdot \boldsymbol{n} = \alpha_1 V_{1,1} + \alpha_2 V_{1,2} \\ &= \alpha_1 V_{1,1} + \sqrt{1-\alpha_1^2}\, V_{1,2} \equiv V_{n,1}(\alpha_1), \end{aligned} \tag{2.60}$$

we can obtain the expression for ρ_2 as a function of α_1 as follows:

$$\rho_2(\alpha_1) = \rho_1 \left[1 - \frac{V_{1,1}\boldsymbol{t}_1 \cdot \boldsymbol{n}_2 + V_{1,2}\boldsymbol{t}_2 \cdot \boldsymbol{n}_2}{(\alpha_1 V_{1,1} + \sqrt{1-\alpha_1^2}\, V_{1,2})(\alpha_1 \boldsymbol{t}_1 \cdot \boldsymbol{n}_2 + \sqrt{1-\alpha_1^2}\, \boldsymbol{t}_2 \cdot \boldsymbol{n}_2)}\right]^{-1}. \tag{2.61}$$

Substituting (2.60) and (2.61) in the second equation of (2.4) and the relation $\rho_2 V_{n,2} = \rho_1 V_{n,1}$, we obtain

$$\begin{cases} p_2(\alpha_1) = p_1 + \rho_1 V_{n,1}^2(\alpha_1)\left(1 - \dfrac{\rho_1}{\rho_2(\alpha_1)}\right), \\ V_{n,2}(\alpha_1) = \rho_1 V_{n,1}(\alpha_1)/\rho_2(\alpha_1). \end{cases} \tag{2.62}$$

Furthermore, substituting (2.60)—(2.62) into the third equation of (2.4), we obtain the non-linear algebraic equation for α_1,

$$h(p_2(\alpha_1), \rho_2(\alpha_1)) + \frac{1}{2} V_{n,2}^2(\alpha_1) = h(p_1, \rho_1) + \frac{1}{2} V_{n,1}^2(\alpha_1). \tag{2.63}$$

Thus, our problem can be solved in the following way: by choosing the unit vectors $\boldsymbol{t}_1$ and $\boldsymbol{t}_2$, for example, $\boldsymbol{t}_1 = \boldsymbol{n}_1/|\boldsymbol{n}_1|$, $\boldsymbol{t}_2 = \boldsymbol{t}_3 \times \boldsymbol{t}_1/|\boldsymbol{t}_3 \times \boldsymbol{t}_1|$, by solving (2.63) for α_1 and by using (2.59)—(2.62) and the first and the fourth equations in (2.4), we can obtain the desired solutions for $\boldsymbol{n}$, $\boldsymbol{V}_2$, p_2, ρ_2 and $c_{i,2}$ ($i=1, 2, \cdots, m-5$). Then $\boldsymbol{F}_z$ and $\boldsymbol{F}_\varphi$ for the shock can be determined from (2.56), that is,

$$\begin{cases} \boldsymbol{F}_z = -n_1/n_2, \\ \boldsymbol{F}_\varphi = -F n_3/n_2, \end{cases} \tag{2.64}$$

where F is the r-coordinate of the point under consideration on the edge.

The non-linear algebraic equation (2.63) can be solved by the use of any method (e.g., by the method of iteration). But it should be noted that this equation can have more than one root (for example, three roots in the case of a perfect gas) and we must choose a root so that the shock stability condition (2.7) is satisfied and the flow behind the shock is supersonic.

(ii) *Convex Corners*

We now turn to deal with a supersonic flow around a convex corner. Suppose that a centered expansion wave will occur on the edge and the flow in front of it is known. Our problem is to determine the boundaries of the expansion wave and the flow properties within the wave and on its boundaries at the edge.

As indicated previously, the front boundary of the expansion wave is a characteristic surface through the edge of the body and therefore is determined in advance by the flow in front of it. The flow in the wave is governed by the differential equations (2.37) and (2.38), or equivalently by the equations (2.48). Thus, in order to determine the flow properties at each singular point on the edge we need only to integrate the ordinary differential equations (2.48) together with $\frac{\partial c_i}{\partial \beta}=0$ ($i=1, 2, \cdots, m-5$) from the initial values on the front boundary until the solution satisfies the condition (2.58). Here we again assume that the solid surfaces are given by the equations in (2.51), and $\boldsymbol{n}_2$ in (2.58) is the normal to the surface $r=F_2(z, \varphi)$. Once the condition (2.58) is satisfied, the back boundary of the expansion wave is determined. In fact, the condition (2.58) can be written (refer to Fig. 2.6) in the form

$$\begin{aligned}\boldsymbol{V}_2\cdot\boldsymbol{n}_2=&(\boldsymbol{V}_2\cdot\boldsymbol{t}_1(\pi))(\boldsymbol{t}_1(\pi)\cdot\boldsymbol{n}_2)+(\boldsymbol{V}_2\cdot\boldsymbol{t}_2(0))(\boldsymbol{t}_2(0)\cdot\boldsymbol{n}_2)\\&+(\boldsymbol{V}_2\cdot\boldsymbol{t}_3)(\boldsymbol{t}_3\cdot\boldsymbol{n}_2)=0.\end{aligned}$$

Noting that $\boldsymbol{t}_3\cdot\boldsymbol{n}_2=0$, from this we can obtain the formula for the angle β_2 corresponding to the back boundary as follows:

$$\operatorname{tg}\beta_2=-\frac{V_2}{a}\,\frac{\boldsymbol{V}_2\cdot\boldsymbol{t}_2(0)}{\boldsymbol{V}_2\cdot\boldsymbol{t}_1(\pi)}=\frac{V_2}{a}\,\frac{\boldsymbol{t}_1(\pi)\cdot\boldsymbol{n}_2}{\boldsymbol{t}_2(0)\cdot\boldsymbol{n}_2}. \tag{2.65}$$

Here we have introduced the factor $-V_2/a$ so as to be in accordance with the given rules of choosing the sign of β (we always have $|V_2|=a$!).

Now, we are able to determine the shape of the boundaries of the expansion wave on the edge. From (2.46) we can calculate the angles θ_1 and θ_2, respectively, for the front and back boundaries. It is seen from Fig. 2.6 that on the edge the normals to the boundary surfaces, denoted by $\boldsymbol{n}_3$ and $\boldsymbol{n}_4$, can be represented in the form

$$\begin{cases}\boldsymbol{n}_3=-\sin\theta_1\boldsymbol{t}_1(0)+\cos\theta_1\boldsymbol{t}_2(0),\\ \boldsymbol{n}_4=-\sin\theta_2\boldsymbol{t}_1(0)+\cos\theta_2\boldsymbol{t}_2(0).\end{cases}$$

Thus, letting the equations

$$r=F_3(z, \varphi), \qquad r=F_4(z, \varphi) \tag{2.66}$$

describe the front and back boundary surfaces respectively, we can find the values of the partial derivatives $F_{3,z}$, $F_{3,\varphi}$, $F_{4,z}$ and $F_{4,\varphi}$ on the edge by the formulae

$$\begin{cases}F_{i,z}=-n_{i,1}/n_{i,2},\\ F_{i,\varphi}=-F_i n_{i,3}/n_{i,2},\end{cases} \qquad i=3, 4, \tag{2.67}$$

where $n_{i,1}$, $n_{i,2}$, and $n_{i,3}$ are, respectively, the z, r, and φ–components of the vector $\boldsymbol{n}_i$ $(i=3, 4)$, and $F_3=F_4$ is the r–coordinate of the edge.

We notice that in the case of a perfect gas we can find the analytic expressions for the solution of the equations (2.48), but in general we should solve this system by a numerical method. In any case, if given a surface through the edge in the expansion wave to which the normal is denoted by $\boldsymbol{n}$, we can find the solutions on the edge corresponding to the normal direction to this surface by integrating the equations (2.48) and by taking the given relations on the edge

$$\boldsymbol{n}(\theta)=|\boldsymbol{n}(\theta)|(-\sin\theta\boldsymbol{t}_1(0)+\cos\theta\boldsymbol{t}_2(0)) \tag{2.68}$$

into account.

(iii) *Reflection of a Shock on a Wall*

Suppose that there is a shock wave hitting a rigid wall and the direction of the flow behind it is not parallel to the wall surface. Obviously, the flow should be along the wall, because the wall is assumed to be impenetrable. Such a conflict can be resolved if there occurs another shock, a "reflected" shock, which makes the flow go along the wall, as illustrated in Fig. 2.9.

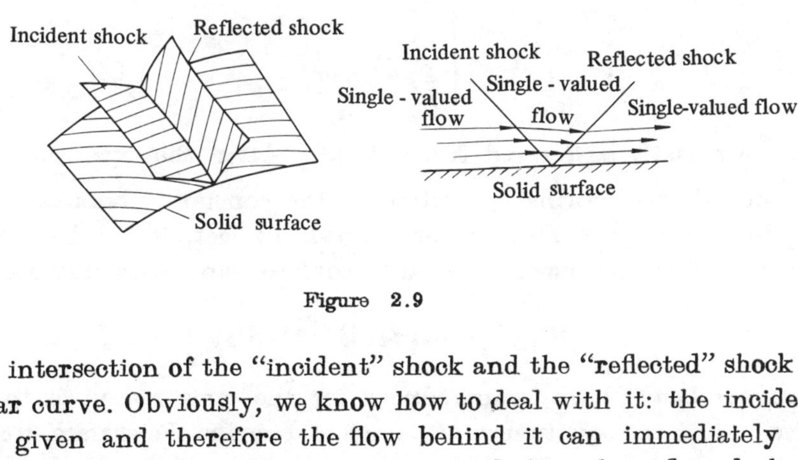

Figure 2.9

The intersection of the "incident" shock and the "reflected" shock is a singular curve. Obviously, we know how to deal with it: the incident shock is given and therefore the flow behind it can immediately be determined by the explicit shock relations, and then the reflected shock can be handled as in the case (i).

(iv) *Reflection of a Shock on a Surface of Constant Pressure*

Now we consider the case in which an "incident" shock hits a surface on which the pressure remains constant, as is illustrated in Fig. 2.10. Obviously, the pressure behind the "incident" shock is higher than that on the constant pressure surface. Therefore, transition of an expansion wave has to take place. This expansion wave can be dealt with in a way

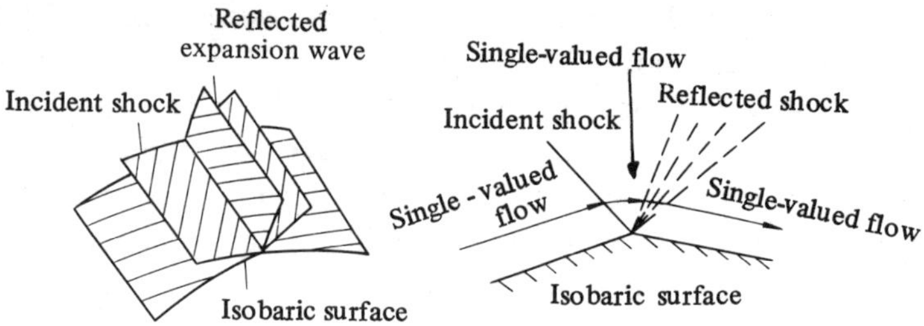

Figure 2.10

similar to that in case (ii), that is, the equations (2.48) are integrated until the pressure is equal to the given pressure on the constant pressure surface. However, it will be more convenient to integrate the differential equations in the independent variable p

$$\left\{\begin{array}{l}\dfrac{\partial q^2}{\partial p}=-\dfrac{2}{\rho},\\[2mm] \dfrac{\partial V_3}{\partial p}=0,\\[2mm] \dfrac{\partial \rho}{\partial p}=a^{-2},\\[2mm] \dfrac{\partial \beta}{\partial p}=\dfrac{-V_1}{\rho a q^2},\\[2mm] V_2=\mp a,\end{array}\right. \tag{2.69}$$

which can easily be derived from (2.48). After the expansion wave is determined, the normal direction to the constant pressure surface at the intersection can also be determined. In fact, it can be seen from Fig. 2.6 that the normal to this surface can be represented by

$$\boldsymbol{n}_5=\frac{V_2}{a}\sin\beta\boldsymbol{t}_1(0)-\cos\beta\boldsymbol{t}_2(0),$$

where the factor V_2/a is introduced for the same reason as in (2.65). Thus, using the equation $r=F_5(z, \varphi)$ to describe the surface, we have

$$\left\{\begin{array}{l}F_{5,z}=-n_{5,1}/n_{5,2},\\ F_{5,\varphi}=-F_5 n_{5,3}/n_{5,2},\end{array}\right.$$

where $n_{5,1}$, $n_{5,2}$ and $n_{5,3}$ are the components of the normal vector.

(v) *More General Cases of Intersection of Discontinuities*

Suppose that there are two strong discontinuity surfaces which are given by the equations $r=F_1(z, \varphi)$ and $r=F_2(z, \varphi)$ and which meet on the curve

$$\begin{cases} r = F_1(z, \varphi), \\ r = F_2(z, \varphi) \end{cases} \tag{2.70}$$

(see Fig. 2.11). Let us observe some of the possible flow patterns in an infinitely small neighborhood of the singular curve (2.70). In general, the flow properties, i.e., u, v, w, p, ρ, c_1, c_2, $\cdots$, c_{m-5}, on the back sides of these two discontinuity surfaces are not consistent. Thus, to coordinate them, there must occur a sequence of "reflected" waves (singularities), which possess m degrees of freedom in total. It is easily seen that the possible order in which they occur is such: a wave (shock or expansion wave) of the first (or, second) family, and then a contact discontinuity, and lastly a wave of the second (first) family. Moreover, there can be neither more than one wave which has the same family number nor more than one contact discontinuity. Furthermore, as indicated previously, a shock or an expansion wave possesses one degree of freedom and a contact discontinuity possesses $m-2$ degrees of freedom. Therefore, the flow around the singular curve, in general, has the configuration shown in Fig. 2.11.

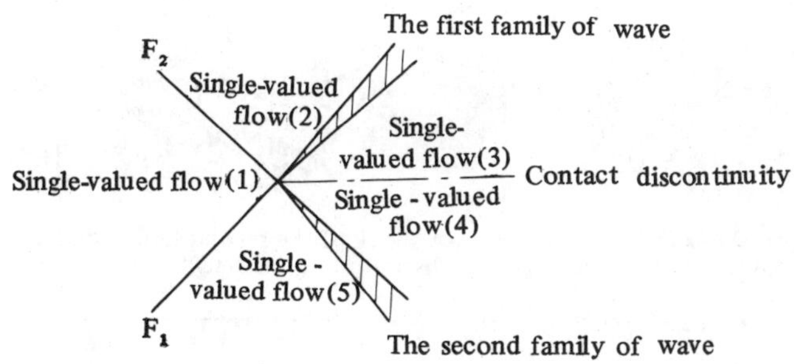

Figure 2.11 Possible flow patterns in the neighborhood of the singular curve at which two discontinuities intersect

Now we shall describe how to determine each of the reflected waves (see Fig. 2.11). Suppose the single–valued flows in zones 2 and 5 are given. Let $r = F_4(z, \varphi)$ be the equation describing the contact discontinuity surface, which passes through the curve (2.70). Clearly, the normal direction to it at a point on the curve can be determined by either of $F_{4,z}$ or $F_{4,\varphi}$, say $F_{4,z}$, if the curve (2.70) is not tangent to the surface φ=constant at the point under consideration. If a value of $F_{4,z}$ is given, the flows in zones 3 and 4 can then be determined by use of the methods given in case (i) or (ii). However, the results may not satisfy the condition $p_3 = p_4$, the first of the jump conditions (2.3) on contact discontinuities. The value of $F_{4,s}$ needs to be adjusted. This procedure is repeated until that condition is satisfied. In the end we can determine the whole configuration: the wave of the first family, the wave of the second one and the contact discontinuity, as well as the flow properties in zones 3 and 4 at the point under consideration on the singular curve (2.70). The process of adjusting $F_{4,s}$ is, indeed, a process of solving a non-linear algebraic equation of the form

$$p_4(F_{4,s}) = p_3(F_{4,s}).$$

Therefore, we may use any known method to correct the value of $F_{4,s}$.

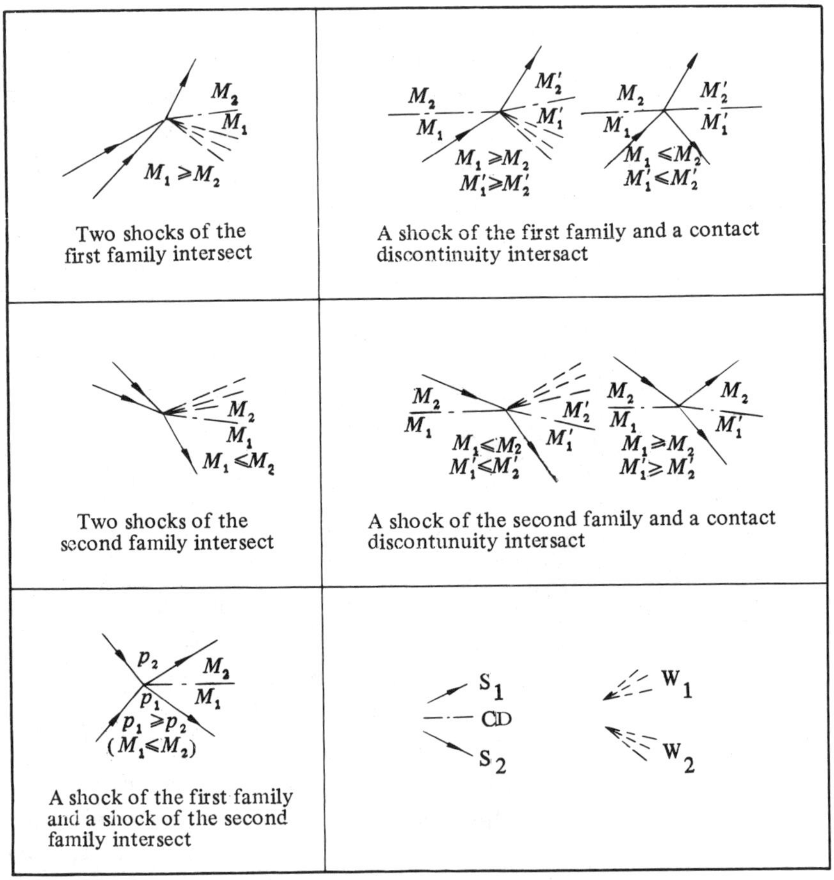

Figure 2.12 Possible flow patterns where two discontinuities intersect

(We suppose that the quantities $\frac{V^2}{2} + \gamma p/(\gamma-1)\rho$ on two sides of a contact discontinuity are equal, therefore the relation $M_1 \gtrless M_2$ is equivalent to $\rho_1 \gtrless \rho_2$. "S_1"——Shock of the first family, "CD"——Contact discontinuity, "S_2"——Shock of the second family, "W_1"——Expansion wave of the first family, "W_2"——Expansion wave of the second family.)

Of course, one can use the pressure $p_0=p_3=p_4$ as the object to be corrected instead of the slope $F_{4,z}$. At that time a procedure similar to that in case (iv) should be repeated for each of the first and the second family waves until the two "constant pressure" surfaces coincide and become the contact discontinuity surface we are looking for. Which type of wave will the desired wave of the first family (and the second family) be, a shock wave or an expansion wave? This can be determined by the procedure given above. It is, of course, much better if one can predict it in a specific environment. Such problems were discussed in [34, 39, 64]. In Fig. 2.12, some of the possible flow patterns are shown. Other possibilities also exist. For example, in the case where two shocks of the same family meet, there can occur two reflected shocks and one contact discontinuity[64].

The previous discussion has been based on the assumption that the state equation of the gas is convex, that is, on the assumption that the inequality $(\partial^2 p/\partial(1/\rho)^2)_s>0$ holds everywhere. The situation may be more complicated in cases where the state equation is not convex. Papers [52, 53] consider such cases for non-steady flow. They point out that a compressive process may be accomplished with a sequence of compression shocks and centered compression waves, and an expansive process with a sequence of expansion shocks and centered expansion waves.

For supersonic flows, in addition to "space-like" singular curves, there are other kinds of singular curves——the entropy-singularity curves, which are themselves streamlines. Such a singular curve will appear in a flow past a cone. In fact, it is a singularity of the conical flow equations. We shall not deal with it here.

§ 3 Boundary Conditions and Internal Boundary Conditions

As is well known, for the determination of specific solutions of partial differential equations it is necessary to add certain further conditions, such as initial conditions or boundary conditions. In this section we shall be concerned with the proper formulation of some boundary-value problems of the inviscid steady flow equations (1.11). We shall discuss what boundary conditions should be prescribed on different types of boundaries of the domain in which the solutions are sought. Moreover, in view of the fact that the solutions of (1.11) possibly suffer some kinds of discontinuities within the domain, or from other considerations, we shall also deal with the interface or internal boundary surfaces which divide the domain. The conditions that the solutions must satisfy on such surfaces will be called internal boundary conditions or interface conditions.

Now, we first discuss how many boundary conditions should be prescribed on a given boundary surface Π. Since the boundary surface itself may sometimes be unknown, we consider each boundary condition to be a relation among the flow parameters ($\boldsymbol{V}$, p, ρ, etc.) and the shape parameters of the boundary surface.

We shall see that the number of boundary conditions on the boundary surface depends on $V_n = \boldsymbol{V}\cdot\boldsymbol{n}$, the normal component of the velocity vector, where $\boldsymbol{n}$ is the unit outward normal to the boundary surface of the domain: this number will be $m+1$ when $V_n < -a$, m when $-a \leqslant V_n < 0$, 2 when $0 \leqslant V_n < a$, and 1 when $a \leqslant V_n$ (see the first row of Table 3.1). These conclusions are based on the observation of the characteristics of the differential equations in the vicinity of the boundary surface. (Gu and Chen studied the same problem in [65] from another point of view.)

Table 3.1

	$V_n < -a$	$-a \leqslant V_n < 0$	$0 \leqslant V_n < a$	$a \leqslant V_n$
Number of boundary conditions	$m+1$*	m	2	1
The number of equations which the differential equations can supply	0	1	$m-1$	m

* $m=5$ for three-dimensional flow of perfect gas and equilibrium air; $m=4$ for two-dimensional flow.

In what follows we briefly observe each case.

(i) When $V_n < 0$, the gas flows into the domain of solution, and the stream-characteristic surface through any curve on the boundary surface Π (only the characteristic surfaces in the domain of solution are our concern here) is "issuing" from Π (see Fig. 3.1). The corresponding stream-compatibility relations in (1.58) or (1.75) then represent how the flow parameters on the boundary influence the flow in the interior of the domain. Therefore, those relations cannot be used for determining the unknowns on the boundary. However, the situation is reversed when $V_n \geqslant 0$, i.e., when no gas goes into the domain. If $V_n > 0$, the stream-characteristic surface through a curve on the boundary is "arriving" and the corresponding compatibility relations describe the dependence of the boundary flow parameters on the internal flow parameters. If $V_n = 0$, then the boundary surface itself is a stream surface and the corresponding stream-compatibility relations connect the flow parameters at the points on the boundary surface. Thus, when $V_n \geqslant 0$, the $m-2$ equations of (1.58) or (1.75) can be used for the determination of the unknowns on the boundary.

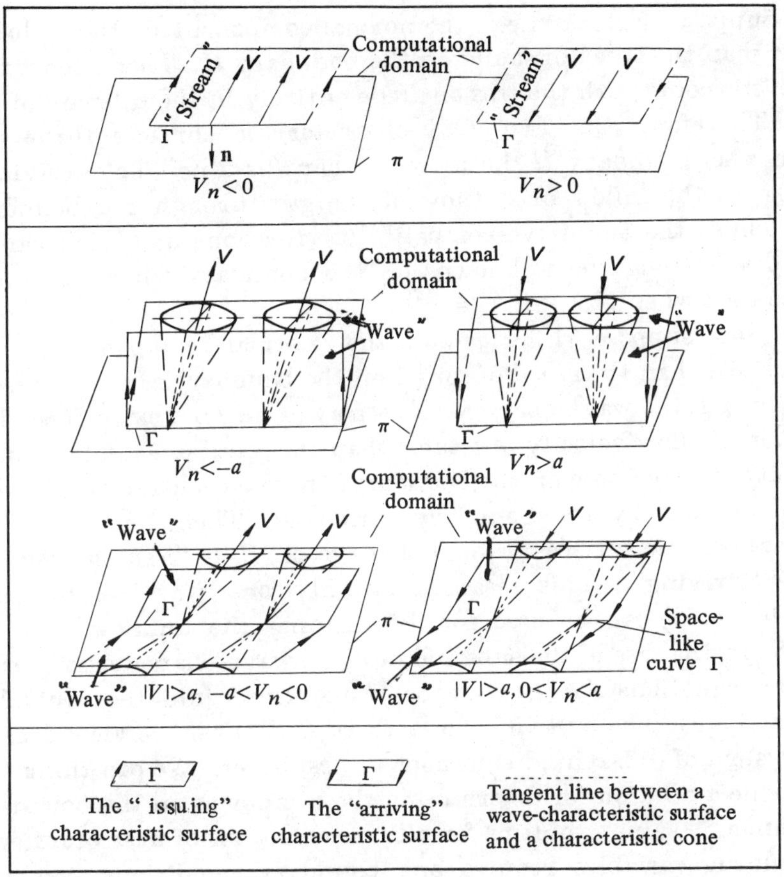

Figure 3.1(a)

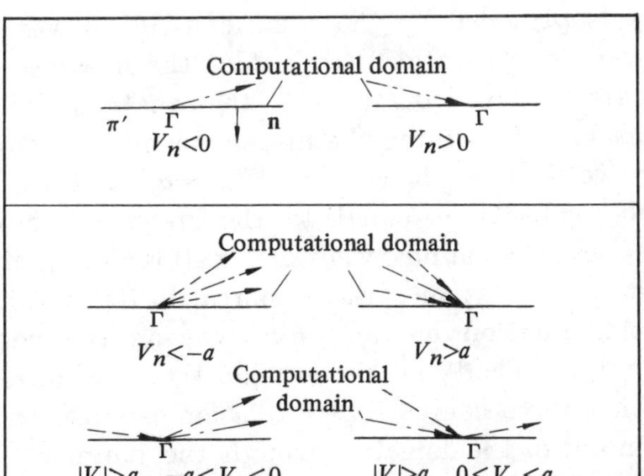

Figure 3.1(b) (A side view of. Fig 3.1(a))

(ii) Suppose $|V_n|>a$, i.e., the normal component of the velocity is greater than the speed of sound on the boundary Π. Then, the wave-characteristic cones with the tips on Π lie entirely in the interior of the domain. Therefore the two wave-characteristic surfaces through a curve on the boundary Π both are either "issuing" or "arriving", depending on the inflow or outflow of the gas through the boundary surface. Thus, the two wave-compatibility relations of (1.58) can be used for determining the unknowns on the boundary when $V_n>a$ and cannot be used when $V_n<-a$.

(iii) Now suppose $|V_n|<a$, i.e., the normal component of the velocity is less than the speed of sound on the boundary. In this case, as indicated in § 1, the wave-characteristics may or may not exist, depending on whether the flow velocity is greater than the speed of sound or not. If the velocity is supersonic on the boundary, the wave-characteristic cones exist and are cut by the boundary surface Π. Then, one of the two wave-characteristic surfaces through a curve on Π is "issuing" and the other is "arriving". This means that only one wave-compatibility relation in (1.58) can be used for determining the unknowns on the boundary. In the case of the subsonic flow, no wave-characteristics exist. Then, the equations (1.11) can be transformed into the equivalent equations (1.75). The first and the m-th of (1.75) can be viewed as an elliptic system of differential equations of first order. We can think that they provide a relation for determining the unknowns on the boundary. (The boundary-value problems for elliptic equations of first order with two unknown variables require one boundary condition; one may think the other relation needed for the determination of two unknowns on the boundary is provided by the equations themselves.) Thus, in the case of $|V_n|<a$, we can think that besides the stream-compatibility relations, which are derived from (1.11), the equations (1.11) provide one more relation for determining the unknowns on the boundary.

(iv) Finally we consider the case of $|V_n|=a$, that is, the normal component of the velocity is equal to the speed of sound on the boundary. In this case, the boundary surface itself is a wave-characteristic surface. Therefore, at least one wave-compatibility relation can be used for the determination of the unknowns on the boundary. If, furthermore, the flow velocity is greater than the speed of sound on the boundary, the other wave-compatibility relation can also be used when the gas flows from out of the domain through the boundary. Thus, one wave-compatibility relation can be used for the determination of the unknowns on the boundary in the case of the inflow ($V_n=-a$), and two wave-compatibility relations can be used in the case of the outflow ($V_n=a$).

Based on the above discussion, we may obtain the following conclusions:

(*a*) When $V_n < -a$ on the boundary, $m+1$ boundary conditions are necessary for the determination of the values of $m+1$ unknowns, namely, m flow parameters and one shape parameter of the boundary surface, because the differential equations (1.11) do not provide anything for the determination of them.

(*b*) When $-a \leqslant V_n < 0$, i.e., when the gas flows into the domain at a speed whose normal component is less than or equal to the speed of sound, the equations (1.11) "provide" one equation for determining the unknowns on the boundary and therefore m boundary conditions are required.

(*c*) When $V_n \geqslant a$, i.e., when the gas flows from out of the domain and the normal component of the velocity is greater than or equal to the speed of sound, the equations (1.11) "provide" m equations for the determination of the unknowns on the boundary and just one boundary condition is required.

(*d*) When $0 \leqslant V_n < a$, i.e., when the gas flows from out of the domain and the normal component of the velocity is less than the speed of sound, or when no gas passes through the boundary surface, the equations (1.11) "provide" $m-1$ relations for the determination of the

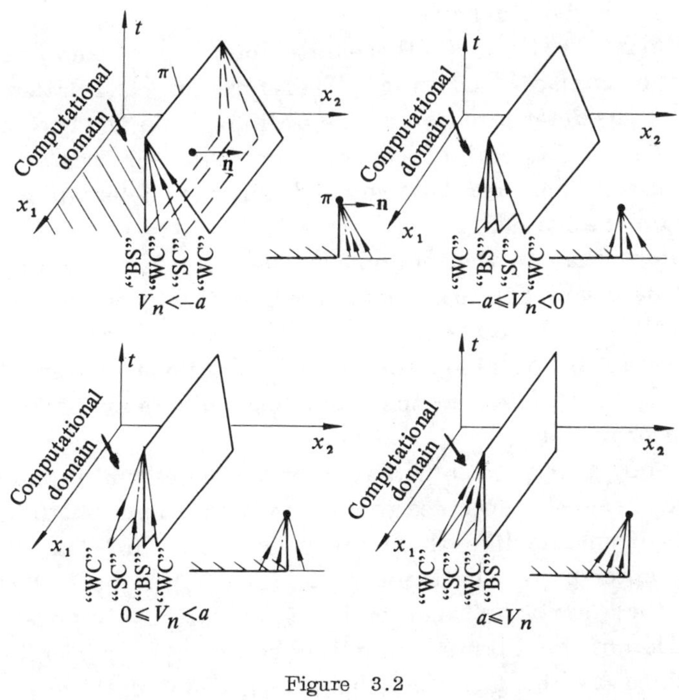

Figure 3.2

"BS"——Boundary surface, "WC"——Wave-characteristic surface,
"SC"——Stream-characteristic surface.

unknowns on the boundary and therefore two boundary conditions are necessary.

These results are listed in Table 3.1.

We can also obtain the above results by observing the formulation of the boundary conditions for the hyperbolic system of differential equations for non-steady flow. Following our intuition, we wish to consider only a two-dimensional non-steady flow. Suppose that the boundary curve in a physical plane does not move with time. The geometry of the boundary surface and the "arriving" characteristic surfaces is shown in Fig. 3.2. There are four different cases:

(*a*) When $V_n<-a$, where $V_n=\boldsymbol{V}\cdot\boldsymbol{n}$ is the normal component of the velocity on the boundary curve, there are no characteristic surfaces arriving at the boundary from the interior of the domain. Therefore, no relations are provided by the differential equations for the determination of the unknowns on the boundary.

(*b*) When $-a\leqslant V_n<0$, there is one arriving wave-characteristic surface. We notice that there is one compatibility relation on a wave-characteristic surface and there are $m-2$ compatibility relations on a stream-characteristic surface, even for non-steady flow, where m is assumed to be the number of unknown flow parameters. Thus one wave-compatibility relations can be used for the determination of the unknowns on the boundary.

(*c*) When $0\leqslant V_n<a$, there are one wave- and one stream-characteristic surfaces arriving. Therefore $m-1$ relations can be provided for the determination of the unknowns on the boundary.

(*d*) When $a\leqslant V_n$, all of the characteristic surfaces are arriving. Therefore m relations can be provided for the determination of the unknowns on the boundary.

It follows that if the boundary does not move with time, the number of necessary boundary conditions on it will be $m+1$, m, 2 and 1 respectively for the cases (*a*)—(*d*). It is easy to see that the same conclusions can be obtained for three-dimensional nonsteady flow as well as steady flow. Note that these conclusions here are compatible with those given previously.

Incidentally, we point out that in order to determine the unknowns on the boundary the specified boundary conditions should also meet further requirements like those given by (1.8) in Chapter 1 or by (1.8) in Chapter 2. We shall not write them down here. Besides, why are only the "arriving" compatibility relations applicable to the determination of the unknowns on the boundary? Physically, this can be interpreted by the fact that the propagation of the disturbance of the flow is directed, as indicated above. From the viewpoint of the numerical methods, when we use the difference equations to approximate the compatibility relations in the differential form, only the "arriving" relations can result in the stable computation, but the "issuing" compatibility relations cannot.

Now we deal with several kinds of boundary surfaces, on the basis of Table 3.1, which may be encountered when we solve the problems of steady flow.

(1) A shock surface with given flow parameters in front of it. Obviously, the boundary conditions on the shock boundary are the jump conditions on a shock surface, which are given by (2.4). In Table 3.1, we have assumed that the normal $\boldsymbol{n}$ to the boundary surface is chosen to be directed outward. In our case, this means that the direction of the normal $\boldsymbol{n}$ is towards the front side from the back side of the shock. Then, from the properties of a compression shock it follows that the inequality $-a \leqslant V_n < 0$ holds. Thus, according to Table 3.1, m boundary conditions are required. From (2.4) we know that the shock conditions contain exactly m equations.

(2) A wall boundary. In this case we have $V_n = 0$. However, according to Table 3.1 we should need two boundary conditions. Remember that the number of boundary conditions in Table 3.1 is obtained on the assumption that the shape of the boundary surface can be unknown, which make the number of unknowns on the boundary increase by one. In accordance with this convention, for a wall boundary we can write two boundary conditions as follows:

$$\begin{cases} f(x_1, x_2, x_3) = 0, \\ V_n = 0, \end{cases} \tag{3.1}$$

where the first one is the equation of the wall surface and is always given.

(3) A constant pressure surface or a stream surface on which the pressure at every point is known. Then, we have $V_n = 0$ on the boundary. The boundary conditions can be written in the form

$$\begin{cases} p = p^*, \\ V_n = 0, \end{cases} \tag{3.2}$$

where p^* is a given function. If this surface is an interface between the moving fluid under consideration and another fluid which is at rest and in which the pressure is known, then it is a special contact discontinuity and the conditions given by (3.2) are, in fact, those two equations in (2.3) which should be satisfied on the boundary by the flow under consideration.

(4) A stream surface on which the Mach number at every point is known. In this case the boundary conditions are

$$\begin{cases} \boldsymbol{M}=\boldsymbol{M}^*, \\ V_n=0, \end{cases} \tag{3.3}$$

where M is the Mach number $|\boldsymbol{V}|/a$, and M^* is a given function.

(5) A space-like surface where $a<V_n$. The only boundary condition is

$$f(x_1, x_2, x_3)=0, \tag{3.4}$$

which is the equation of the given surface. Of course, such a boundary should be specified so that the condition $a<V_n$ holds everywhere on it.

(6) A wave-characteristic surface where $a=V_n$. The boundary condition on such a boundary surface is obviously

$$V_n=a. \tag{3.5}$$

In addition, it may be necessary to know to which family the wave-characteristic surface belongs.

(7) A space-like surface where $V_n<-a$. On such a given surface the flow parameters are assumed to be known. The boundary conditions can be written in the form

$$\begin{cases} f(x_1, x_2, x_3)=0, \\ \boldsymbol{U}=\boldsymbol{U}^*, \end{cases} \tag{3.6}$$

for which $\boldsymbol{U}^*$ is a known function defined on the given boundary surface where $V_n<-a$.

(8) A wave-characteristic surface where $V_n=-a$. According to Table 3.1, m boundary conditions are required in this case. For the determination of the unknowns on the boundary what the differential equations can provide is only the wave-compatibility relation on the wave-characteristic surface under consideration; the m boundary conditions must be specified so that the $m+1$ unknowns on the boundary can be determined by them and by that compatibility relation. For example, as the boundary conditions we can specify

$$\begin{cases} V_n=-a, \\ u_i=u_i^*, \quad i=1, 2, 3, 5, \cdots, m, \end{cases} \tag{3.7}$$

where u_i^* is a known function. Here $i=4$ is excluded, that is, the pressure p is not given. The conditions (3.7) plus the compatibility relation on this wave-characteristic surface amount to giving this characteristic surface and all the flow parameters on it, that is, amount to giving

$$\begin{cases} f(x_1, x_2, x_3)=0, \\ \boldsymbol{U}=\boldsymbol{U}^*, \end{cases} \tag{3.8}$$

where $\boldsymbol{U}^*$ is a given function on the surface $f(x_1, x_2, x_3)=0$. The quantities defined by (3.8) should satisfy the condition $V_n=-a$ and the wave-compatibility relation, whereas the flow parameters in (3.6) are arbitrarily given to some extent.

(9) A given boundary surface where $-a<V_n<0$. According to Table 3.1, we need m boundary conditions. The possible boundary conditions are

$$\begin{cases} f(x_1, x_2, x_3)=0, \\ u_1t_{1,1}+u_2t_{1,2}+u_3t_{1,3}=u_2^*, \\ u_1t_{2,1}+u_2t_{2,2}+u_3t_{2,3}=u_3^*, \\ u_i=u_i^*, \quad i=4,\ 5,\ \cdots,\ m\,(u_4\equiv p,\ u_5\equiv\rho), \end{cases} \tag{3.9}$$

where u_i^* for $i=2,\ 3,\ \cdots,\ m$ is a known function on the boundary surface, and $(t_{1,1},\ t_{1,2},\ t_{1,3})$ and $(t_{2,1},\ t_{2,2},\ t_{2,3})$ are two unit vectors on the tangent plane to the surface. The conditions (3.9) explicitly specify the tangential component of the velocity and other flow parameters on the boundary.

(10) A given boundary surface where $0<V_n<a$. Two boundary conditions are required. They may be

$$\begin{cases} f(x_1, x_2, x_3)=0, \\ p=p^*, \end{cases} \tag{3.10}$$

where p^* is a known function on the given surface $f(x_1, x_2, x_3)=0$.

In what follows we shall be concerned with the interface conditions which should be satisfied by the solution on an interface within the domain under consideration. Such a surface can be a discontinuity surface, and therefore, in general, $2m+1$ unknown quantities need to be determined on it, including one shape parameter of the surface and $2m$ flow parameter values on the two sides of it. Obviously we should require that the sum of the number of interface conditions and the number of equations which can be "provided" by the differential equations for the determination of the unknowns on the surface be equal to $2m+1$.

We will now deal with two types of interface surfaces which will be encountered when we solve the problems of steady flows numerically in Chapter 7.

(1) Strong discontinuity surfaces. There are two kinds of strong discontinuities: contact discontinuities and shock waves. As shown in § 2, across a discontinuity surface the flow parameters should satisfy three conditions, given by (2.3), for a contact discontinuity or m conditions, given by (2.4), for a shock. It is obvious that they are enough to determine the discontinuity which appears in the interior of the domain under consideration. In fact, for a contact discontinuity we have $V_n=0$ on either side of the discontinuity surface. According to Table 3.1, there are altogether $2(m-1)$ relations provided by the

differential equations from the two sides of the discontinuity surface. Therefore for determining $2m+1$ unknowns we need just three more conditions. For a shock, we have $V_n \geqslant a$ on the front side and $-a \leqslant V_n < 0$ on the back side and therefore, from Table 3.1, we know that the differential equations provide $m+1$ relations,and m additional conditions should be required. The shock conditions (2.4) contain just m relations.

(2) Continuity surfaces. When a continuity surface is taken as an interface surface, the interface conditions are obviously the continuity conditions of the flow parameters, that is,

$$\boldsymbol{U}_1 = \boldsymbol{U}_2.$$

If the interface is not a characteristic surface, then, according to Table 3.1, we have m differential equations in total from the two sides of the surface. Thus we need $m+1$ more conditions. In this case the surface should be given. Therefore, the $m+1$ interface conditions required can be written in the form

$$\begin{cases} f(x_1, x_2, x_3) = 0, \\ \boldsymbol{U}_1 = \boldsymbol{U}_2, \end{cases} \tag{3.11}$$

where the first one is the equation of the surface. If the interface surface is stream–characteristic, then we have $V_n = 0$. In that case, according to Table 3.1, we have $m-1$ "provided" equations from each side of the surface and therefore we need three interface conditions. Note that among these $m-1$ equations there are $m-2$ stream–compatibility relations. Because of the continuity conditions these stream–compatibility relations on the different sides of the surface should be the same. Obviously, we can rewrite the $2(m-2)$ stream–compatibility relations plus the three interface conditions in the equivalent form:

$$\begin{cases} \boldsymbol{U}_1 = \boldsymbol{U}_2, \\ V_n = 0 \end{cases} \tag{3.12}$$

plus the $m-2$ stream–compatibility relations on either side of the surface. The other two equations needed for the determination of the $2m+1$ unknowns associated with the interface surface will be "provided" by the differential equations. Likewise, for an interface surface which is wave–characteristic, where $|V_n| = a$ holds, according to Table 3.1, we need m interface conditions. In place of them and the two wave–compatibility relations on the two sides of the surface, we can use

$$\begin{cases} \boldsymbol{U}_1 = \boldsymbol{U}_2, \\ |V_n| = a \end{cases} \tag{3.13}$$

plus one wave–compatibility relation on either side of the surface in the process of determining the unknowns on the boundary.

§ 4 Calculation of Thermodynamic Properties of Equilibrium Air

As is well-known, in the study of aerodynamics problems, air at a moderate range of temperature and pressure may be considered as a perfect gas. The equation of state describing a perfect gas is

$$\frac{p}{\rho}=\frac{RT}{\mu_0}, \tag{4.1}$$

where p is the pressure of the gas, ρ the density, T the absolute temperature, R the universal gas constant equal to 0.0820819 liter·atm/mol·deg or 0.831695×10^8 erg/mol·deg, and μ_0 the molecular weight of the gas. For "frozen" air, that is, air when no chemical reactions take place in it, the average molecular weight is approximately 28.97 gram/mol.

For a perfect gas, the internal energy and enthalpy are functions of temperature alone. In a moderate range of temperature the specific heats may be considered as constants. From the first law of thermodynamics and from the equation of state (4.1), we can obtain expressions for the specific internal energy e, the specific enthalpy h, and the speed of sound a, as follows:

$$e=C_{\bar{v}}T=\frac{p}{(\gamma-1)\rho}, \tag{4.2}$$

$$h=C_pT=\frac{\gamma p}{(\gamma-1)\rho}, \tag{4.3}$$

$$a^2\equiv-\frac{\left(\dfrac{\partial h}{\partial\rho}\right)_p}{\left(\dfrac{\partial h}{\partial p}\right)_\rho-\dfrac{1}{\rho}}=\frac{\gamma p}{\rho}, \tag{4.4}$$

where C_p and $C_{\bar{v}}$ are the specific heats at constant pressure and at constant volume, respectively, and γ is the ratio of the specific heats $C_p/C_{\bar{v}}$. For frozen air, we have $\gamma=1.4$.

As shown experimentally, when an aircraft flies at hypersonic speed, air behind the head shock wave is strongly compressed and can reach very high temperatures, for example, several thousands of degrees and even ten-thousands of degrees. In that case, chemical reactions among the components of air may take place and the internal degrees of freedom of motion of the molecules of gas may be excited. If T and p do not change with time, the system in question may reach an equilibrium state after these processes last for a period of time, usually referred to as the relaxation time. As we know, if the relaxation time is extremely short in comparison with the characteristic time of the flow problem considered, we may view the system as being in an equilibrium state. Otherwise, we have to deal with the non-equilibrium processes. In any event, for the case of high temperature we shall not be able to use the formulae (4.1)—(4.4) for the calculation of the thermodynamic quantities.

In this section we shall be concerned with the calculation of the thermodynamic properties of air in a state of equilibrium. The case of air in a non-equilibrium state will be discussed in the next section.

Consider a mixture of perfect gases. It is assumed to obey the perfect-gas equation of state:

$$\frac{p}{\rho}=\frac{R}{\mu}T. \tag{4.5}$$

Here

$$\mu=\sum_{i=1}^{l}x_i\mu_i,$$

where l is the number of components of the mixture, μ_i and x_i are the molecular weight and the concentration (the mole fraction) of the i-th component for $i=1, 2, \cdots, l$. The component concentrations $x_1, x_2, \cdots, x_l$ are not constants, but depend on two of the state variables, e.g., p and ρ, in the case of chemical equilibrium.

From the theory of chemical thermodynamics we know that a set of l non-linear algebraic equations for $x_1, x_2, \cdots, x_l$ can be constructed, depending on given models of chemical reactions. This set of equations, together with the equation (4.5), can then be used to determine the quantities $T(p, \rho)$, $x_1(p, \rho)$, $x_2(p, \rho)$, $\cdots$, $x_l(p, \rho)$ as functions of p and ρ. Moreover, by using the results obtained we may calculate all the thermodynamic properties of each component and of the mixture for the given p and ρ.

Thus, for the problems of equilibrium-gas flow, we generally need to solve a set of non-linear algebraic equations simultaneously while solving the basic system of partial differential equations. The work [C72] gives a method of solution of the algebraic equations. Another way to treat such problems is to solve simultaneously a set of ordinary differential equations derived from that set of algebraic equations instead of solving the latter directly. The common disadvantage of these two ways is that they require much computer time.

As is well known, the thermodynamic functions of equilibrium air have been calculated from a complicated set of algebraic equations and have been given in the form of tables or curves by many authors[40-47]. Apparently, these data can be stored in computers and used efficiently for solving the problems of aerodynamics. Unfortunately, this will require much computer storage capacity.

In view of these considerations, many authors propose a compromise, in which they construct simple analytic expressions to approximate those existing tables of thermodynamic functions and then use them, instead of directly using the tables or solving the algebraic or differential equations. Readers are referred to [48—51], [C63], [C73], also [A94], [A96] and [A98]. Among these works, some authors obtained quite accurate approximation expressions (e.g., [48]), but a greater amount of computation work is still required because of the complexity of the expressions; some authors simpler expressions, but the accuracy of these is not satisfactory. D'yakonov [C73] and Harry Bailey (refer to [A94]) used the simple piecewise analytic approximations, whose virtue is that those approximate expressions give quite accurate results and do not need too much computer time. Obviously, this is a practically acceptable method.

Following D'yakonov, the present authors have again obtained approximate expressions based on the tables given by [40] for several thermodynamic quantities, such as the sound speed $a(p, \rho)$, the specific enthalpy $h(p, \rho)$, the average molecular weight $\mu(p, \rho)$ and the temperature $T(p, \rho)$, where the pressure p and the density ρ are taken as the arguments. The actual calculation shows that the discrepancy does not exceed 1—2%. In these approximate formulae, besides one logarithmic term, only the calculation of the second-degree polynomials is involved. Therefore they require a lesser amount of computation in comparison with those given by [48]. When we apply them to the problems of equilibrium-air flows around a body, the total time of computation increases by about 5% compared with that in the case of frozen air.

We will now give the general description of the approximate expressions. Let us first introduce the transformation of the independent variables:

$$\begin{cases} \sigma = \log p, \\ q = (1 + 0.1667|\sigma|)^{-\operatorname{sign}\sigma}(Q - 0.45), \end{cases} \tag{4.6}$$

where

$$Q = \frac{p}{\rho} \times 10^{-4}. \tag{4.7}$$

Here the unit of the pressure p is atm, and the unit of the density ρ is g/cm^3. Figs. 4.1—4.6 show the dependences of the quantities f_a, f_h and μ, which will be defined later, on p and T, or on p and q. Evidently, when we take p and q as the arguments, the curves change comparatively smoothly and gently. For a fixed value of q the functions change slowly with p and are sometimes monotonic. Hence, these functions may be better approximated by the piecewise biquadratic polynomials in σ and q. At lower temperature, i.e., when $Q < 0.45$, we shall use the second-degree polynomials in Q or the biquadratic polynomials in σ and Q.

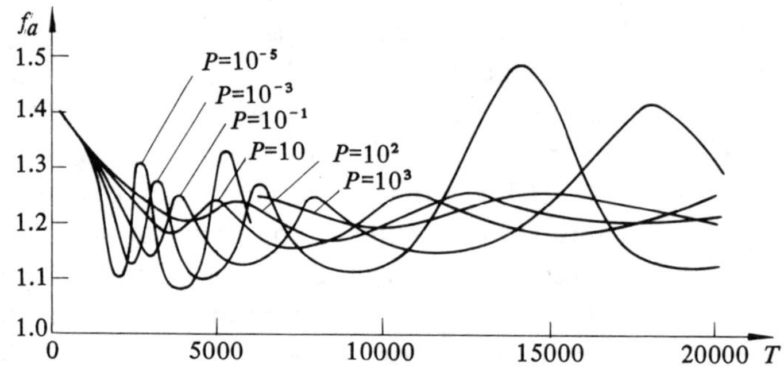

Figure 4.1 Variations of f_a with T

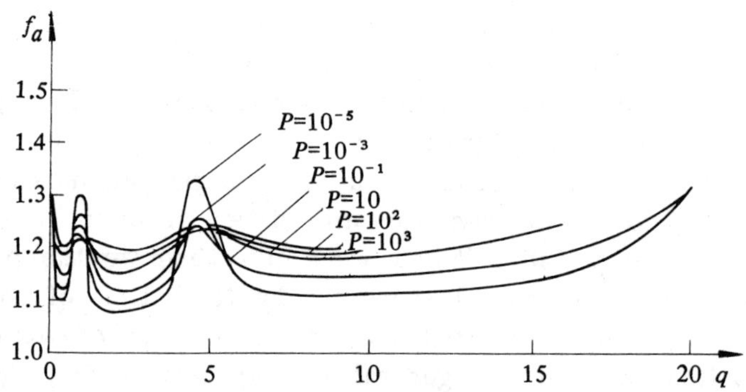

Figure 4.2 Variations of f_a with q

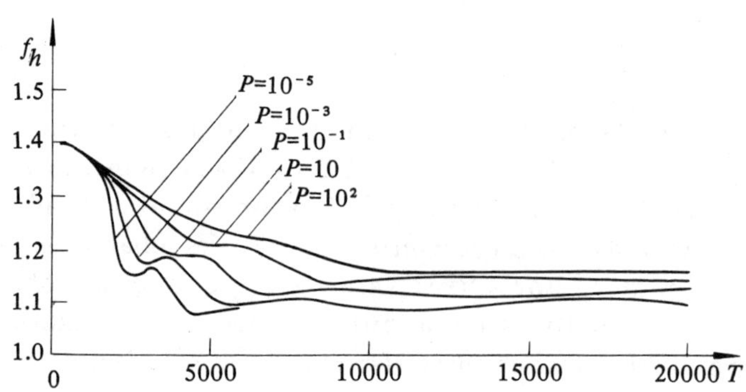

Figure 4.3 Variations of f_h with T

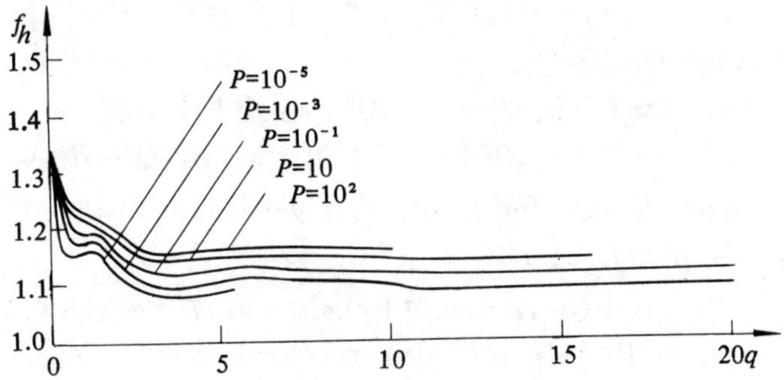

Figure 4.4 Variations of f_h with q

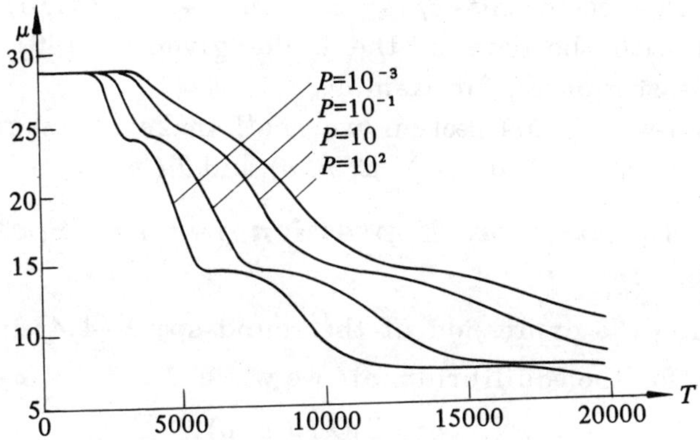

Figure 4.5 Variations of μ with T

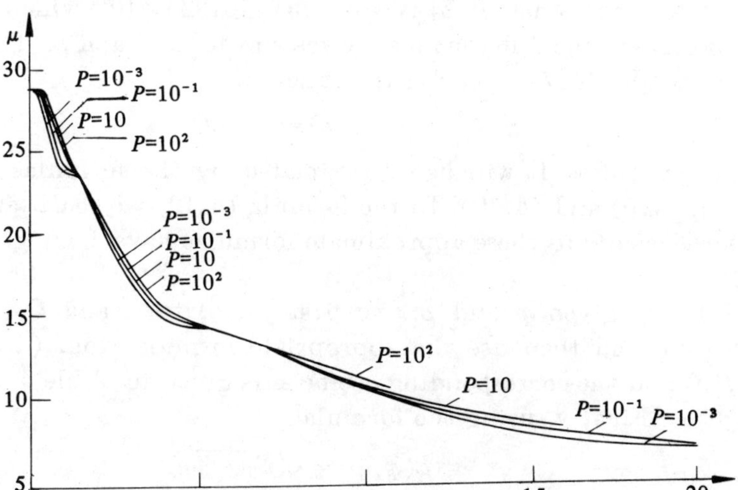

Figure 4.6 Variations of μ with q

Concretely, letting $f(p, \rho)$ be any of the functions to be approximated, then the approximate formula has the following forms:

(i) when $Q_0 \leqslant Q < Q_1$[1],

$$f(p, \rho) \approx f^*(Q) = \beta_{0,7}Q^2 + \beta_{0,8}Q + \beta_{0,9}; \tag{4.8}$$

(ii) when $Q_1 \leqslant Q < Q_2$,

$$f(p, \rho) \approx f^{**}(\sigma, Q) = (\beta_{1,1}Q^2 + \beta_{1,2}Q + \beta_{1,3})\sigma^2 + (\beta_{1,4}Q^2 + \beta_{1,5}Q + \beta_{1,6})\sigma + \beta_{1,7}Q^2 + \beta_{1,8}Q + \beta_{1,9}; \tag{4.9}$$

(iii) when $Q \geqslant Q_2$, for $q_j \leqslant q < q_{j+1}$, $j = 0, 1, \cdots, n (q_0 = 0)$,

$$f(p, \rho) \approx f_j(\sigma, q) = (\alpha_{j,1}q^2 + \alpha_{j,2}q + \alpha_{j,3})\sigma^2 + (\alpha_{j,4}q^2 + \alpha_{j,5}q + \alpha_{j,6})\sigma + \alpha_{j,7}q^2 + \alpha_{j,8}q + \alpha_{j,9}. \tag{4.10}$$

Here, we assume $10^{-5} \leqslant p \leqslant 10^3$ atm and $Q_0 = 0.056657224$, $Q_1 = 0.33990483$, and $Q_2 = 0.45$. The numbers n and $q_1, \cdots, q_n$ will take different values for the different approximated functions, depending on the accuracy required. The coefficients $\beta_{i,k}$, $\alpha_{j,k}$ in (4.8)—(4.10) can then be determined with the data in the tables given by [40], by using the method of least squares, for example.

In the rest of this section we shall make a few remarks on the calculation of $a(p, \rho)$, $h(p, \rho)$, $\mu(p, \rho)$ and $T(p, \rho)$.

1. *An Approximate Expression for the Speed of Sound* $a(p, \rho)$

Imitating the expression of the sound speed (4.4) in the case of a perfect gas, for the equilibrium air we write

$$a^2 = \Sigma_1 Q f_a(p, \rho).$$

Here m/sec, atm and g/cm³ are taken respectively as the units of a, p and ρ, and the constant Σ_1 is equal to 0.101325×10^7, which is introduced to coordinate the different metric systems for a, p and ρ. The quantity Q is defined by (4.7). Clearly, the function

$$f_a(p, \rho) = a^2 / \Sigma_1 Q \tag{4.11}$$

is dimensionless. It will be arproximated by the formulae of the forms (4.8), (4.9) and (4.10). In the formula (4.10) we shall take $n = 19$. All the coefficients in these approximate formulae have been given in Table 4.1.

Thus, given p and ρ, we first calculate σ and Q, and q when $Q > 0.45$, and then use the appropriate formula from (4.8), (4.9) or (4.10) and the corresponding coefficients given in Table 4.1 to calculate $f_a(p, \rho)$. Finally, using the formula

$$a(p, \rho) = \sqrt{\Sigma_1 Q f_a} \tag{4.12}$$

1) Even when $Q < Q_0$ (i.e., when $T < 200°\mathrm{K}$) the formula (4.8) is possibly valid.

we shall find the value of the sound speed.

It can be verified that the results given by this process differ by at most 1% from the tables[40] when pressure and temperature fall over the range given by [40], that is, when $10^{-5} \leqslant p \leqslant 10^3$ atm and $200 \leqslant T \leqslant 20{,}000$°K.

2. *An Approximate Expression for the Specific Enthalpy* $h(p, \rho)$

Let

$$h(p, \rho) = \Sigma_2 Q \frac{f_h(p, \rho)}{f_h(p, \rho) - 1}, \tag{4.13}$$

where the constant $\Sigma_2 = 0.2420106 \times 10^3$ is introduced so as to coordinate the different metric systems of the specific enthalpy h [cal/g], the pressure p [atm] and the density ρ [g/cm^3]. The dimensionless function

$$f_h(p, \rho) = \frac{1}{1 - \Sigma_2 Q/h} \tag{4.14}$$

will be approximated by formulae of the types given by (4.8)—(4.10). All the coefficients in the approximate formulae have been listed in Table 4.2. Thus, for given p and ρ, the specific enthalpy $h(p, \rho)$ can be calculated by the use of the proper approximate formula together with the expression (4.13). The deviation from [40] does not exceed 1.5%.

3. *Approximate Expressions for the Average Molecular Weight* $\mu(p, \rho)$ *and the Temperature* $T(p, \rho)$

As indicated above, the equation of state for equilibrium air has the form

$$\frac{p}{\rho} = \tilde{R} T(p, \rho)/\mu(p, \rho).$$

Here the units of the average molecular weight μ, the temperature T, the pressure p and the density ρ are g/mol, °K, atm and g/cm^3 respectively. Then the constant $\tilde{R}$ is equal to 0.820819×10^2. Obviously, one of $T(p, \rho)$ and $\mu(p, \rho)$ can be obtained immediately from the other by the use of the above equation.

Now let

$$f_T(p, \rho) = \frac{\tilde{R} \rho T(p, \rho)}{p} = \frac{\tilde{R} T(p, \rho)}{Q} \times 10^{-4}, \tag{4.15}$$

$$f_\mu(p, \rho) = \mu(p, \rho). \tag{4.16}$$

Then we may construct their approximate formulae of the types (4.8)—(4.10). The coefficients in the formulae can be determined, respectively, through the use of the values p, ρ, T and p, ρ, μ given by [40]. These coefficients have been listed in Tables 4.3 and 4.4.

Table 4.1 Coefficients of the approximate expression for $f_a(p, \rho)$

i	Q_i		$\beta_{i,1}$		$\beta_{i,2}$		$\beta_{i,3}$		$\beta_{i,4}$	
0	0.56657224	−1	0		0		0		0	
1	0.33990483	0	−0.28610491	0	0.21073710	0	−0.38575344	−1	−0.95497886	−1
2	0.450	0								
j	q_j		$\alpha_{j,1}$		$\alpha_{j,2}$		$\alpha_{j,3}$		$\alpha_{j,4}$	
0	0	0	0.15584849	0	−0.41147597	−1	−0.16798927	−2	0.29460989	0
1	0.2400	0	−0.56803366	−1	0.56195762	−1	−0.12793552	−1	−0.20263832	0
2	0.4500	0	0.28472436	−1	−0.23008340	−1	0.55799442	−2	−0.20535258	−1
3	0.6500	0	−0.11406494	0	0.18292781	0	−0.68056510	−1	−0.20418603	0
4	0.9000	0	0.71152276	−1	−0.16521297	0	0.95244244	−1	0.41920179	0
5	0.1100	1	0.16828039	−1	−0.41876785	−1	0.25306768	−1	−0.68953720	−1
6	0.1500	1	−0.10163628	−2	0.45859643	−2	−0.42374526	−2	−0.21767695	−2
7	0.2500	1	−0.44927450	−3	0.24534700	−2	−0.24505186	−2	−0.20691671	−2
8	0.3500	1	0.41173255	−2	−0.30600221	−1	0.57296549	−1	−0.16479645	−1
9	0.4500	1	−0.29829234	−2	0.27246393	−1	−0.59233174	−1	0.34988500	−1
10	0.5200	1	0.41237892	−2	−0.48354488	−1	0.14172590	0	−0.13817766	−1
11	0.6000	1	0.11580352	−3	−0.21598149	−2	0.88453460	−2	−0.20612466	−2
12	0.7000	1	0.76721297	−4	−0.12283571	−2	0.42401707	−2	0.30454001	−3
13	0.1200	2	0.10718431	−4	0.21796329	−4	−0.12572584	−2	−0.88750729	−4
14	0.1700	2	−0.29186444	−3	0.11846451	−1	−0.11482995	0	−0.25520100	−2
15	0.2200	2	0.10337060	−2	−0.50806272	−1	0.62195385	0	0.70777254	−2
16	0.2500	2	0.55375302	−3	−0.28198066	−1	0.35671935	0	0.19951484	−2
17	0.2700	2	−0.83369153	−4	0.52580239	−2	−0.82133017	−1	−0.20310540	−2
18	0.3050	2	−0.53143951	−3	0.33235825	−1	−0.51863851	0	−0.28866930	−2
19	0.3540	2								

Table 4.1 (continued)

i	$\beta_{i,5}$		$\beta_{i,6}$		$\beta_{i,7}$		$\beta_{i,8}$		$\beta_{i,9}$	
0	0		0		−0.34240645	0	−0.14405247	0	0.14102685	1
1	0.70341147	−1	−0.12875919	−1	0.19756667	1	−0.17077187	1	0.16739470	1
2										
j	$\alpha_{j,5}$		$\alpha_{j,6}$		$\alpha_{j,7}$		$\alpha_{j,8}$		$\alpha_{j,9}$	
0	−0.14914867	−1	−0.56072507	−3	−0.89587516	0	−0.94441167	−1	0.13055461	1
1	0.16829590	0	−0.15889813	−1	0.95605105	0	−0.96289779	0	0.14073047	1
2	0.11392743	−1	0.17840740	−1	0.35763465	0	−0.41500949	0	0.12819343	1
3	0.25079920	0	−0.60181008	−1	0.71883659	0	−0.92526251	0	0.14609910	1
4	−0.85169130	0	0.42711630	0	−0.12778888	1	0.27205046	1	−0.20285187	0
5	0.22097100	0	−0.16214405	0	−0.17119642	−1	−0.15503588	0	0.14347120	1
6	0.10093328	−1	0.39243181	−2	0.52255218	−1	−0.23269637	0	0.13951093	1
7	0.12390179	−1	−0.24903270	−2	0.16645041	−1	−0.76986621	−1	0.12283985	1
8	0.11385249	0	−0.18108004	0	0.12678688	−1	−0.39082864	−1	0.11443232	1
9	−0.32954082	0	0.77195988	0	−0.92942799	−1	0.89757542	0	−0.93180399	0
10	0.16788719	0	−0.49494432	0	0.20524333	−2	−0.62419027	−1	0.14914960	1
11	0.27397648	−1	−0.75241767	−1	0.93980779	−2	−0.13989485	0	0.16919078	1
12	−0.48642245	−2	0.34667798	−1	0.12501414	−2	−0.22775316	−1	0.12713199	1
13	0.40088899	−2	−0.15175708	−1	0.13203358	−2	−0.27467745	−1	0.13175211	1
14	0.90995726	−1	−0.78206998	0	0.91170584	−3	−0.86009882	−2	0.11148803	1
15	−0.34001800	0	0.40394401	1	−0.97907085	−2	0.44288794	0	−0.36379077	1
16	−0.86522374	−1	0.87865998	0	0.53414658	−2	−0.31135066	0	0.57604484	1
17	0.12301144	0	−0.18436514	1	0.26223076	−2	−0.16642565	0	0.38297395	1
18	0.18037855	0	−0.27973902	1	−0.32253297	−2	0.19498406	0	−0.17534920	1
19										

Note. i, j are subscripts; the coefficients are written in the normalized form, i.e., a number $A=x\cdot 10^{\pm n}$ is represented by $x \pm n$, where $|x|<1$; for xample, the coefficient $\alpha_{1,2}=0.56195762\times 10^{-1}$; the unit of p is atm; the unit of ρ is g/cm^3.

Table 4.2 Coefficients of the approximate expression for $f_h(p, \rho)$

i	Q_i		$\beta_{i,1}$		$\beta_{i,2}$		$\beta_{i,3}$		$\beta_{i,4}$	
0	0.56657224	−1	0		0		0		0	
1	0.33990483	0	−0.26882402	−1	0.19800847	−1	−0.36245373	−2	−0.13914086	−1
2	0.4500	0								
j	q_j		$\alpha_{j,1}$		$\alpha_{j,2}$		$\alpha_{j,3}$		$\alpha_{j,4}$	
0	0.0000	0	−0.36267775	−1	−0.41306761	−2	−0.15784262	−3	0.43045628	−1
1	0.1000	0	0.22291754	−1	−0.19788361	−1	0.82233057	−3	0.19589834	0
2	0.2000	0	0.70809125	−1	−0.41666533	−1	0.32572701	−2	0.13440650	0
3	0.3000	0	0.35082114	−1	−0.18675773	−1	−0.42452678	−3	−0.36946390	−1
4	0.4000	0	−0.49032691	−2	0.14502839	−1	−0.72983104	−2	−0.85613776	−1
5	0.5000	0	−0.85059207	−2	0.17286123	−1	−0.77892896	−2	−0.65232996	−1
6	0.6000	0	−0.67969194	−2	0.15398689	−1	−0.72720695	−2	−0.63050360	−1
7	0.7000	0	−0.11900312	−1	0.21884081	−1	−0.93111813	−2	−0.33985235	−1
8	0.9000	0	0.28930221	−2	−0.68179508	−2	0.45380463	−2	0.35456223	−1
9	0.1100	1	0.11663117	−2	−0.36068436	−2	0.30951480	−2	0.18096130	−1
10	0.1500	1	0.48800021	−3	−0.12166055	−2	0.10359918	−2	−0.15656251	−2
11	0.2000	1	−0.27006982	−3	0.16184308	−2	−0.16018007	−2	−0.24161106	−2
12	0.2500	1	−0.15485075	−3	0.10193826	−2	−0.82429937	−3	−0.95334904	−3
13	0.3000	1	−0.17011521	−3	0.11347514	−2	−0.10330258	−2	−0.68488368	−3
14	0.3500	1	−0.31579403	−4	0.15792136	−3	0.68881581	−3	−0.66968560	−3
15	0.4000	1	−0.41633355	−4	0.20635420	−3	0.65594771	−3	−0.43273568	−3
16	0.4500	1	0.81112793	−4	−0.95997505	−3	0.34188198	−2	0.71925708	−3
17	0.5200	1	0.94883446	−4	−0.10604033	−2	0.35686885	−2	0.93259029	−3
18	0.6000	1	−0.11085020	−4	0.18074367	−3	−0.63328787	−4	0.16440461	−3
19	0.7000	1	0.27564421	−5	0.36948800	−4	0.26500363	−3	−0.35778365	−4
20	0.1200	2	0.10258155	−5	0.20300884	−4	0.71398886	−3	−0.44130302	−5
21	0.1700	2	−0.17908744	−4	0.71195302	−3	−0.55720098	−2	−0.74224236	−4
22	0.2200	2	−0.10331104	−4	0.38222436	−3	−0.19855566	−2	−0.63162060	−4
23	0.2500	2	−0.16896145	−4	0.78587387	−3	−0.79736442	−2	−0.93382310	−4
24	0.2700	2	0.36756261	−5	−0.28459855	−3	0.59322900	−2	0.32971019	−4
25	0.3050	2	0.37304146	−4	−0.22812220	−2	0.35546376	−1	0.22317982	−3
26	0.3540	2								

Table 4.2 (continued)

i	$\beta_{i,5}$		$\beta_{i,6}$		$\beta_{i,7}$		$\beta_{i,8}$		$\beta_{i,9}$	
0	0		0		−0.38701791	0	0.37223599	−1	0.14005476	1
1	0.10248737	−1	−0.18760274	−2	0.17506582	0	−0.27942963	0	0.14432391	1
2										

j	$\alpha_{j,5}$		$\alpha_{j,6}$		$\alpha_{j,7}$		$\alpha_{j,8}$		$\alpha_{j,9}$	
0	−0.19661940	−1	−0.81697900	−4	0.11518052	0	−0.13282684	0	0.13529466	1
1	−0.55744489	−1	0.19980299	−2	−0.29090810	0	−0.32880712	−1	0.13470128	1
2	−0.33206930	−1	−0.49808355	−4	−0.51508158	0	0.61278692	−1	0.13371479	1
3	0.76479621	−1	−0.17534013	−1	−0.17244677	0	−0.15211018	0	0.13703274	1
4	0.11653269	0	−0.25768461	−1	0.25605463	0	−0.50209969	0	0.14417630	1
5	0.92454840	−1	−0.18824729	−1	0.35234688	0	−0.60019416	0	0.14667372	1
6	0.90367205	−1	−0.18357897	−1	0.28116526	0	−0.51649075	0	0.14421405	1
7	0.48613225	−1	−0.33720220	−2	0.24011624	0	−0.44850086	0	0.14146616	1
8	−0.83986757	−1	0.59720381	−1	0.62299591	−1	−0.12617525	0	0.12686000	1
9	−0.48885645	−1	0.42114870	−1	−0.76101106	−1	0.17664198	0	0.11029659	1
10	0.10548136	−1	−0.27968526	−2	0.11970922	−2	−0.62234635	−1	0.12873599	1
11	0.13156365	−1	−0.46113672	−2	0.21972622	−1	−0.13978911	0	0.13593667	1
12	0.59924080	−2	0.41562642	−2	0.13561071	−1	−0.97205613	−1	0.13054802	1
13	0.45828114	−2	0.59688657	−2	0.10737670	−1	−0.80504440	−1	0.12807873	1
14	0.44298683	−2	0.63179903	−2	0.83468442	−2	−0.64175148	−1	0.12529224	1
15	0.23719274	−2	0.10758555	−1	0.56815348	−2	−0.42524168	−1	0.12089634	1
16	−0.81742506	−2	0.34888503	−1	−0.58098679	−3	0.13907672	−1	0.10818362	1
17	−0.10608426	−1	0.41777685	−1	−0.32845298	−2	0.40281177	−1	0.10177978	1
18	−0.15149329	−2	0.14871411	−1	−0.15935155	−2	0.19572809	−1	0.10811714	1
19	0.12687510	−2	0.51945891	−2	0.22425863	−3	−0.67999070	−2	0.11767095	1
20	0.48646174	−3	0.10065452	−1	0.25478964	−3	−0.70469112	−2	0.11752771	1
21	0.29616621	−2	−0.11837516	−1	0.19179286	−3	−0.47825691	−2	0.11549894	1
22	0.25796372	−2	−0.87870614	−2	−0.14682626	−3	0.10032782	−1	0.99294332	0
23	0.43608711	−2	−0.34430252	−1	−0.43942766	−3	0.24820295	−1	0.80613134	0
24	−0.22801673	−2	0.52766208	−1	−0.33046281	−3	0.18957827	−1	0.88498262	0
25	−0.13621015	−1	0.22172033	0	0.42951341	−3	−0.27505305	−1	0.15951403	1
26										

Table 4.3 Coefficients of the approximate expression for $f_T(p, \rho)$

i	Q_i		$\beta_{i,1}$		$\beta_{i,2}$		$\beta_{i,3}$		$\beta_{i,4}$	
0	0.56657224	−1	0		0		0		0	
1	0.33990483	0	−0.19912751	0	0.14667191	0	−0.26848235	−1	−0.17353173	−1
2	0.4500	0								
j	q_j		$\alpha_{j,1}$		$\alpha_{j,2}$		$\alpha_{j,3}$		$\alpha_{j,4}$	
0	0.0000	0	−0.46055239	0	−0.18027963	−1	−0.11691964	−2	0.10856880	1
1	0.2400	0	0.84824209	0	−0.61917547	0	0.67719642	−1	0.17818967	1
2	0.4500	0	0.32058880	0	−0.13551977	0	−0.43075629	−1	−0.20693394	1
3	0.6500	0	−0.45765888	0	0.88114672	0	−0.37509920	0	−0.23034733	1
4	0.9000	0	−0.53631547	0	0.11047782	1	−0.51265573	0	−0.15058188	1
5	0.1100	1	0.16105761	−1	−0.88183342	−1	0.13117231	0	0.58076089	0
6	0.1500	1	0.14356809	−1	−0.56368462	−1	0.87385134	−1	0.50475570	−1
7	0.2500	1	−0.64293320	−2	0.39102923	−1	−0.21379944	−1	−0.41965840	−1
8	0.3500	1	−0.33165159	−2	0.18777003	−1	0.11628782	−1	−0.69783328	−1
9	0.4500	1	0.10676245	−2	−0.23015215	−1	0.11091491	0	0.45887327	−1
10	0.5200	1	−0.45915235	−1	0.53065998	0	−0.14977796	1	0.93771695	−1
11	0.6000	1	0.32805422	−2	−0.37934082	−1	0.14273682	0	0.22722054	−1
12	0.7000	1	−0.15897889	−2	0.28834428	−1	−0.85996522	−1	−0.68499253	−2
13	0.1200	2	0.40998557	−3	−0.14494374	−1	0.14598158	0	−0.95408967	−3
14	0.1700	2	0.12375792	−4	−0.14106993	−2	0.38468331	−1	−0.13336439	−2
15	0.2200	2	0.95095878	−2	−0.44989387	0	0.53084475	1	0.24625072	−1
16	0.2500	2	−0.32872338	−4	0.35163722	−4	0.24259215	−1	0.40832321	−3
17	0.2700	2	0.23241326	−3	−0.14109747	−1	0.21277859	0	0.26208780	−2
18	0.3050	2	−0.33876497	−3	0.19328992	−1	−0.27576440	0	−0.19515221	−2
19	0.3540	2								

Table 4.3 (continued)

i	$\beta_{i,5}$		$\beta_{i,6}$		$\beta_{i,7}$		$\beta_{i,8}$		$\beta_{i,9}$	
0	0		0		0.15344031	0	−0.49257688	−1	0.28977223	2
1	0.12781875	−1	−0.23397173	−2	0.23424115	1	−0.18699574	1	0.29343184	2
2										
j	$\alpha_{j,5}$		$\alpha_{j,6}$		$\alpha_{j,7}$		$\alpha_{j,8}$		$\alpha_{j,9}$	
0	−0.10457205	0	−0.10189083	−3	−0.17021984	1	0.38117527	0	0.28976042	2
1	−0.11582485	0	−0.37502844	−1	−0.14263670	2	0.60366961	1	0.28342257	2
2	0.32568222	1	−0.77531870	0	−0.34978085	1	−0.40135109	1	0.30684764	2
3	0.34455862	1	−0.79909371	0	0.81413195	1	−0.18977613	2	0.35493898	2
4	0.25839782	1	−0.66974667	0	0.48816926	1	−0.12827858	2	0.32599417	2
5	−0.18746521	1	0.17099852	1	−0.15711238	1	0.14439714	1	0.24708312	2
6	−0.15452672	0	0.32293914	0	−0.20925991	−1	−0.41414462	1	0.29598494	2
7	0.26735691	0	−0.15401111	0	0.86959541	0	−0.82284878	1	0.34250339	2
8	0.45531592	0	−0.47110342	0	0.75392493	0	−0.73912780	1	0.32737068	2
9	−0.57003844	0	0.18006605	1	0.15897995	0	−0.20748211	1	0.20860648	2
10	−0.10999221	1	0.32612623	1	0.12230733	0	−0.20189889	1	0.21561948	2
11	−0.26136554	0	0.78770986	0	−0.70222915	−1	0.18654699	0	0.15259821	2
12	0.15363867	0	−0.66829261	0	0.33123633	−1	−0.12895000	1	0.20528169	2
13	0.25932405	−1	0.15182176	−1	0.25181135	−1	−0.10735437	1	0.19080414	2
14	0.34521439	−1	−0.21140242	−1	0.16840815	−1	−0.78538455	0	0.16592060	2
15	−0.11788317	1	0.14108611	2	0.13177953	−1	−0.66716794	0	0.15764120	2
16	−0.35750944	−1	0.66705930	0	−0.52429928	−2	0.22180261	0	0.50529473	1
17	−0.15567892	0	0.22921623	1	−0.33923635	−2	0.11747617	0	0.65206527	1
18	0.11582492	0	−0.17352296	1	−0.37894661	−2	0.13181150	0	0.64528297	1
19										

Table 4.4 Coefficients of the approximate expression for $f_\mu(p, \rho)$

i	Q_i		$\beta_{i,1}$		$\beta_{i,2}$		$\beta_{i,3}$		$\beta_{i,4}$	
0	0.56657224	−1	0		0		0		0	
1	0.33990483	0	−0.14737770	0	0.10855441	0	−0.19870842	−1	0.10964714	0
2	0.4500	0								
j	q_j		$\alpha_{j,1}$		$\alpha_{j,2}$		$\alpha_{j,3}$		$\alpha_{j,4}$	
0	0.0000	0	−0.46932310	0	−0.16961777	−1	−0.86534241	−3	0.12157549	1
1	0.2400	0	0.85792100	0	−0.62529896	0	0.68686320	−1	0.17227704	1
2	0.4500	0	0.20873624	0	−0.19609962	−1	−0.72413813	−1	−0.21437404	1
3	0.6500	0	−0.49762647	0	0.94801414	0	−0.40293124	0	−0.22869024	1
4	0.9000	0	−0.59015173	0	0.12119468	1	−0.56552517	0	−0.15597252	1
5	0.1100	1	0.30255195	−1	−0.12303239	0	0.15225956	0	0.60391392	0
6	0.1500	1	0.15560954	−1	−0.61837227	−1	0.93528860	−1	0.52258998	−1
7	0.2500	1	−0.64834558	−2	0.39522839	−1	−0.22093743	−1	−0.48608036	−1
8	0.3500	1	−0.30436290	−2	0.15235626	−1	0.20773625	−1	−0.67922632	−1
9	0.4500	1	0.25249900	−2	−0.39199817	−1	0.15296858	0	0.49088223	−1
10	0.5200	1	−0.38903383	−1	0.44502745	0	−0.12447900	1	0.86596081	−1
11	0.6000	1	0.24555920	−2	−0.33649375	−1	0.13834786	0	0.19273290	−1
12	0.7000	1	−0.58848020	−3	0.93797361	−2	−0.13696384	−1	−0.34668117	−2
13	0.1200	2	0.41256937	−3	−0.13491715	−1	0.11660989	0	−0.72499257	−3
14	0.1700	2	−0.64654189	−4	0.26546307	−2	−0.19960377	−1	−0.14219984	−2
15	0.2200	2	−0.13569545	−3	0.43258032	−2	−0.22342200	−1	−0.15580950	−3
16	0.2500	2	−0.38008341	−1	0.19360844	1	−0.24645903	2	−0.16436495	0
17	0.2700	2	−0.14575358	−1	0.85922638	0	−0.12653382	2	−0.61839463	−1
18	0.3050	2	0.62871141	−3	−0.39125762	−1	0.60277300	0	0.44257743	−2
19	0.3540	2								

Table 4.4 (continued)

i	$\beta_{i,5}$		$\beta_{i,6}$		$\beta_{i,7}$		$\beta_{i,8}$		$\beta_{i,9}$	
0	0		0		−0.44703484	−7	0.26077032	−7	0.28970000	2
1	−0.80763101	−1	0.14783654	−1	0.19834964	1	−0.15490748	1	0.29267374	2
2										
j	$\alpha_{j,5}$		$\alpha_{j,6}$		$\alpha_{j,7}$		$\alpha_{j,8}$		$\alpha_{j,9}$	
0	−0.13348003	0	0.64380377	−3	−0.88812903	0	0.17384382	0	0.28971949	2
1	−0.78736333	−1	−0.41698773	−1	−0.14803835	2	0.64036251	1	0.28278346	2
2	0.33202408	1	−0.78827003	0	−0.28655140	1	−0.46703724	1	0.30844135	2
3	0.34324918	1	−0.80074728	0	0.85266253	1	−0.19610967	2	0.35742342	2
4	0.26884523	1	−0.72012524	0	0.48099216	1	−0.12679001	2	0.32514103	2
5	−0.19292184	1	0.17413092	1	−0.16961154	1	0.17623777	1	0.24500891	2
6	−0.16414238	0	0.33491874	0	−0.26981130	−1	−0.41134477	1	0.29559077	2
7	0.30745799	0	−0.21366322	0	0.86309984	0	−0.81880608	1	0.34182603	2
8	0.43632750	0	−0.42810271	0	0.76011115	0	−0.74384666	1	0.32820635	2
9	−0.61207121	0	0.19202217	1	0.14420887	0	−0.19361034	1	0.20532022	2
10	−0.10373148	1	0.31172761	1	0.89095565	−1	−0.16552447	1	0.20561820	2
11	−0.24027534	0	0.75865954	0	−0.69246463	−1	0.15799740	0	0.15382681	2
12	0.82933794	−1	−0.38953941	0	0.33449615	−1	−0.13087703	1	0.20617947	2
13	0.23162265	−1	−0.67103022	−1	0.26186605	−1	−0.11014661	1	0.19176170	2
14	0.43325886	−1	−0.20844991	0	0.17035770	−1	−0.78710017	0	0.16476541	2
15	−0.97662119	−2	0.34674084	0	0.13864538	−2	−0.10763539	0	0.91025847	1
16	0.83657401	1	−0.10641020	3	−0.16270034	0	0.82522191	1	−0.97339531	2
17	0.36448448	1	−0.53687110	2	−0.65546220	−1	0.37799682	1	−0.47414110	2
18	−0.27262453	0	0.41524663	1	0.63647859	−2	−0.49113216	0	0.15959237	2
19										

Thus, we may use either the approximate formulae of $f_T(p, \rho)$ or the approximate formulae of $f_\mu(p, \rho)$ to calculate the average molecular weight μ, and the temperature can then be obtained from the formula

$$T(p, \rho)=\frac{f_T Q}{\widetilde{R}}\times 10^4 \quad [°\mathrm{K}] \tag{4.17}$$

or

$$T(p, \rho)=\frac{f_\mu Q}{\widetilde{R}}\times 10^4 \quad [°\mathrm{K}]. \tag{4.18}$$

It can be imagined that for given p and ρ the values of μ or T calculated by the different approximate formulae mentioned above are not the same (even in the tables given by [40], the equality $\mu=\widetilde{R}\rho T/p$ holds only with a lower accuracy). Generally speaking, it is a little better to use Table 4.4 for calculating μ, while Table 4.3 may result in a slightly more accurate calculation of T. However, either way can be used to calculate μ and T with a deviation of not more than 2% from [40].

§ 5 A Non-equilibrium Model of Air

Much work has been devoted to the study of the non-equilibrium flow and a number of non-equilibrium models for air have been presented. In this section we shall present a model that we use for calculating the flow of non-equilibrium air around a body. A number of chemical reactions and ionization processes have been included in the model, though it could be simplified for some specific flow problems.

1. *Chemical Reactions and Ionization Reactions*

In our model of high-temperature air, thirteen components will be involved. They are O_2, N_2, NO, O, N, Ar, NO^+, e^-, O^+, N^+, O_2^+, N_2^+, O_2^-.

The chemical reactions considered include the dissociation-recombination reactions (taking the effect of the different collision species on dissociation into account), the atom exchange reactions, and the bimolecular reactions. The concrete reactions are listed as follows.

Dissociation-recombination reactions:

$$(1)\quad O_2+\begin{pmatrix} O_2^+, & N_2^+, & O_2^- & \\ NO^+, & e^-, & O^+, & N^+ \\ Ar, & N_2, & N, & NO \end{pmatrix}+5.12\,\mathrm{eV}$$

$$\underset{k_{r,1}}{\overset{k_{f,1}}{\rightleftarrows}} 2O+\begin{pmatrix} O_2^+, & N_2^+, & O_2^- & \\ NO^+, & e^-, & O^+, & N^+ \\ Ar, & N_2, & N, & NO \end{pmatrix},$$

$$(2)\quad O_2+(O_2)+5.12\,\mathrm{ev} \underset{k_{r,2}}{\overset{k_{f,2}}{\rightleftarrows}} 2O+(O_2),$$

$$(3)\ O_2+(O)+5.12\,eV \underset{k_{r,3}}{\overset{k_{f,3}}{\rightleftarrows}} 2O+(O),$$

$$(4)\ N_2+\begin{pmatrix} O_2^+ & N_2^+, & O_2^- & \\ NO^+, & e^-, & O^+, & N^+ \\ Ar, & O_2, & O, & NO \end{pmatrix}+9.76\,eV$$

$$\underset{k_{r,4}}{\overset{k_{f,4}}{\rightleftarrows}} 2N+\begin{pmatrix} O_2^+, & N_2^+, & O_2^- & \\ NO^+, & e^-, & O^+, & N^+ \\ Ar, & O_2, & O, & NO \end{pmatrix},$$

$$(5)\ N_2+(N_2)+9.76\,eV \underset{k_{r,5}}{\overset{k_{f,5}}{\rightleftarrows}} 2N+(N_2),$$

$$(6)\ N_2+(N)+9.76\,eV \underset{k_{r,6}}{\overset{k_{f,6}}{\rightleftarrows}} 2N+(N),$$

$$(7)\ NO+\begin{pmatrix} O_2^+, & N_2^+, & O_2^- & \\ Ar, & O_2, & N_2 & \\ NO^+, & e^-, & O^+, & N^+ \end{pmatrix}+6.49\,eV$$

$$\underset{k_{r,7}}{\overset{k_{f,7}}{\rightleftarrows}} N+O+\begin{pmatrix} O_2^+, & N_2^+, & O_2^- & \\ Ar, & O_2, & N_2 & \\ NO^+, & e^-, & O^+, & N^+ \end{pmatrix},$$

$$(8)\ NO+(O, N, NO)+6.49\,eV \underset{k_{r,8}}{\overset{k_{f,8}}{\rightleftarrows}} N+O+(O, N, NO).$$

Atom exchange reactions:

$$(9)\ N_2+O+3.27\,eV \underset{k_{r,9}}{\overset{k_{f,9}}{\rightleftarrows}} NO+N,$$

$$(10)\ NO+O+1.37\,eV \underset{k_{r,10}}{\overset{k_{f,10}}{\rightleftarrows}} O_2+N.$$

Bimolecular reactions:

$$(11)\ N_2+O_2+1.9\,eV \underset{k_{r,11}}{\overset{k_{f,11}}{\rightleftarrows}} 2NO.$$

The ionization processes have been analysed carefully in the paper [55]. When the flying Mach number is less than 30, the predominate electron production process would be atom–atom collision recombination ionization. For example, at a temperature lower than 8000°K, the major contribution to the electron production is due to the collisions between oxygen atoms and nitrogen atoms, and the contribution from other atom–atom collisions, photoionization, electron impact and atom–molecule and molecule–molecule collisions is insignificant. However, at higher temperature, the species N^+ and O^+ make even more contribution to the electron concentration than NO^+, so that it is necessary to consider those reactions producing N^+ and O^+. For high–temperature air, in addition to the atom (molecule)–atom (molecule) collision ionization and electron impact, the charge exchange reactions are sometimes important especially, when the rate of recombination of the positive ions with electrons is comparable with the rate of ionization. The reason for this is that since the different ions recombine with the electrons at different rates and since the charge exchange reactions continuously change the relative population between the atomic and the molecular ions, the charge exchange reactions may have a certain effect on the rate of electron production. Finally, electron attachment to oxygen molecules which have considerable electron affinity, is also involved. Thus the following ionization processes will be considered:

Atom–atom collision recombination ionization

$$(12)\quad N+O+2.76\,\mathrm{eV} \underset{k_{r,12}}{\overset{k_{f,12}}{\rightleftarrows}} NO^{+}+e^{-},$$

$$(13)\quad N+N+5.82\,\mathrm{eV} \underset{k_{r,13}}{\overset{k_{f,13}}{\rightleftarrows}} N_2^{+}+e^{-},$$

$$(14)\quad O+O+6.96\,\mathrm{eV} \underset{k_{r,14}}{\overset{k_{f,14}}{\rightleftarrows}} O_2^{+}+e^{-}.$$

Atom–atom (molecule) collision ionization

$$(15)\quad N+(X)+14.54\,\mathrm{eV} \underset{k_{r,15}}{\overset{k_{f,15}}{\rightleftarrows}} N^{+}+e^{-}+(X),$$

$$(16)\quad O+(X)+13.61\,\mathrm{eV} \underset{k_{r,16}}{\overset{k_{f,16}}{\rightleftarrows}} O^{+}+e^{-}+(X)$$

$$(X=O_2,\ N_2,\ NO,\ O,\ N,\ Ar,\ NO^+,\ O^+,\ N^+,\ O_2^+,\ N_2^+).$$

Electron–impact ionization

$$(17)\quad O+e^{-}+13.61\,\mathrm{eV} \underset{k_{r,17}}{\overset{k_{f,17}}{\rightleftarrows}} O^{+}+2e^{-},$$

$$(18)\quad N+e^{-}+14.54\,\mathrm{eV} \underset{k_{r,18}}{\overset{k_{f,18}}{\rightleftarrows}} N^{+}+2e^{-}.$$

Electron attachment reactions

$$(19)\quad O_2+e^{-}+(O_2)-0.44\,\mathrm{eV} \underset{k_{r,19}}{\overset{k_{f,19}}{\rightleftarrows}} O_2^{-}+(O_2),$$

$$(20)\quad O_2+e^{-}+(N_2)-0.44\,\mathrm{eV} \underset{k_{r,20}}{\overset{k_{f,20}}{\rightleftarrows}} O_2^{-}+(N_2).$$

Charge exchange reactions

$$(21)\ O_2^+ + N + 0.16\,\text{eV} \underset{k_{r,21}}{\overset{k_{f,21}}{\rightleftarrows}} O^+ + NO,$$

$$(22)\ N_2^+ + N - 1.04\,\text{eV} \underset{k_{r,22}}{\overset{k_{f,22}}{\rightleftarrows}} N^+ + N_2.$$

The general form of the reactions listed above may be written as

$$\sum_i \nu'_{i,j} M_i \underset{k_{r,j}}{\overset{k_{f,j}}{\rightleftarrows}} \sum_i \nu''_{i,j} M_i \quad (j=1,\ 2,\ \cdots,\ 22).$$

If the rate of reaction is measured as the mole fraction produced in unit volume per unit time, then the net rate of the j-th reaction is

$$\bar{\omega}_j = k_{f,j} \prod_{i=1}^{13} (\rho c_i)^{\nu'_{i,j}} - k_{r,j} \prod_{i=1}^{13} (\rho c_i)^{\nu''_{i,j}},$$

where the units of ρ, c_i and $\bar{\omega}_j$ are g/cm^3, mol/g and mol/cm^3 · sec, respectively.

The total rate of production of the i-th component can then be written as

$$\omega_i = \sum_{j=1}^{22} (\nu''_{i,j} - \nu'_{i,j}) \bar{\omega}_j = \sum_{j=1}^{22} (\nu''_{i,j} - \nu'_{i,j}) \left[k_{f,j} \prod_{i=1}^{13} (\rho c_i)^{\nu'_{i,j}} - k_{r,j} \prod_{i=1}^{13} (\rho c_i)^{\nu''_{i,j}} \right]. \quad (5.1)$$

The equation of conservation of mass for each component has the form

$$\frac{dc_i}{dt} = \frac{\omega_i}{\rho} \quad (i=1,\ 2,\ \cdots,\ 13), \quad (5.2)$$

where $c_i = \dfrac{\rho_i}{\rho W_i}$ is the number of moles of the i-th component contained in a gram of air mixture, W_i being the molecular weight of the i-th component.

The reaction rate constants have been listed in Table 5.1.

2. *The Vibrational Excitation Process*: *The CVDV Model*

The excitation processes of the internal degrees of freedom of molecules in high-temperature gases are quite complex. In general, the translational relaxation and the rotational relaxation of the molecules may be completed through just a few collisions between the molecules, while the relaxation of the vibrational degree of freedom may continue much longer, usually taking 10^4—10^5 collisions to reach the equilibrium state. When calculating the flow field around a body flying at a supersonic speed, we may assume that the translational/rotational degree of freedom is in equilibrium state immediately behind the strong head shock wave. However, we shall have to consider the relaxation process of the vibrational degree of freedom if the vibrational relaxation time is comparable with the flow characteristic time of the problem under consideration. In our model of non-equilibrium air we shall consider the vibrational relaxation of the molecules O_2, N_2 and NO.

Table 5.1

Reactions	(X)	$k_{f,j}$	$k_{r,j}$
1	O_2^+, N_2^+, O_2^- Ar, N_2, N, NO NO^+, e^-, O^+, N^+	$1.2\times10^{21}T^{-3/2}\exp(-59380/T)$	$1.0\times10^{18}T^{-1}$
2	O_2	$3.6\times10^{21}T^{-3/2}\exp(-59380/T)$	$3.0\times10^{18}T^{-1}$
3	O	$2.1\times10^{18}T^{-1/2}\exp(-59380/T)$	1.75×10^{15}
4	O_2^+, N_2^+, O_2^- Ar, O_2, O, NO NO^+, e^-, O^+, N^+	$1.9\times10^{17}T^{-1/2}\exp(-113260/T)$	$1.06\times10^{16}T^{-1/2}$
5	N_2	$4.8\times10^{17}T^{-1/2}\exp(-113260/T)$	$2.67\times10^{16}T^{-1/2}$
6	N	$4.3\times10^{22}T^{-3/2}\exp(-113260/T)$	$2.39\times10^{21}T^{-3/2}$
7	O_2^+, N_2^+, O_2^- Ar, O_2, N_2 NO^+, e^-, O^+, N^+	$4.0\times10^{20}T^{-3/2}\exp(-75490/T)$	$1.0\times10^{20}T^{-3/2}$
8	O, N_2, NO	$8.0\times10^{21}T^{-3/2}\exp(-75490/T)$	$2.0\times10^{21}T^{-3/2}$
9		$5.0\times10^{13}\exp(-37750/T)$	1.11×10^{13}
10		$2.4\times10^{11}T^{1/2}\exp(-19240/T)$	$1.0\times10^{12}T^{1/2}\exp(-3120/T)$

11		$9.1\times10^{24}T^{-5/2}\exp(-65000/T)$	$4.79\times10^{23}T^{-5/2}\exp(-43360/T)$
12		$3.0\times10^{12}T^{-1/2}\exp(-32500/T)$	$\dfrac{3\times10^{21}T^{-3/2}}{(1.4+1.2\times10^{-4}T+1.4\times10^{-8}T^2)}$
13		$8.5\times10^{(9\pm1)}T\exp(-67700/T)$	$5.0\times10^{(18\pm1)}T^{-1/2}$
14		$6.0\times10^{(8\pm1)}T^{1/2}\exp(-80800/T)$	$5.0\times10^{(19\pm1)}T^{-1}$
15	O_2, N_2, NO O, N, Ar, NO^+ O^+, N^+, O_2^+, N_2^+	$6.0\times10^{4}T^{3/2}\exp(-168800/T)$	1.2×10^{12}
16	Ditto	$4.0\times10^{4}T^{3/2}\exp(-157800/T)$	7.0×10^{11}
17		$1.63\times10^{(13\pm1)}T^{1/2}\exp(-157800/T)$	$2.8\times10^{(20\pm1)}T^{-1}$
18		$2.7\times10^{(13\pm1)}T^{1/2}\exp(-168800/T)$	$5.3\times10^{(20\pm1)}T^{-1}$
19	O_2	$5.0\times10^{(19\pm1)}T^{-1/2}$	$3.0\times10^{(11\pm1)}T\exp(-5110/T)$
20	N_2	$2.0\times10^{(18\pm1)}T^{-1/2}$	$1.2\times10^{(10\pm1)}T\exp(-5110/T)$
21		$9.0\times10^{(17\pm2)}T^{-3/2}\exp(-1800/T)$	$3.0\times10^{(13\pm2)}T^{-1/2}$
22		$5.0\times10^{(13\pm2)}T^{-1/2}$	$2.0\times10^{(19\pm2)}T^{-2}\exp(-13000/T)$

As indicated in many research reports, the vibrational relaxation of the molecules is related to the dissociation–recombination of them. The higher the molecules are vibrationally excited, the more easily they are dissociated. On the other hand, the dissociation process will reduce the number of energetic oscillators and hence will reduce the average vibrational energy remaining in the molecules. The recombination process will cause the production of new oscillators and hence will increase the average vibrational energy of the molecules. Therefore we should consider the effect of each on the other. For this we shall use the CVDV model introduced by the paper [56], where coupling between vibration and dissociation is considered.

In that model, the effect of vibrational relaxation on dissociation is usually included through a vibrational coupling factor

$$\nabla = k_f / k_{f,eq},$$

where k_f is the dissociation rate constant to be used and $k_{f,eq}$ stands for the value of the dissociation rate constant that would exist in the case of vibrational equilibrium. The experimental values of the dissociation rate constants listed in Table 5.1, which are measured in the vibrational "quasi–equilibrium" zone, are taken as $k_{f,eq}$ in our model. Thus, for the foregoing reactions, we have

$$(k_{f,1},\ k_{f,2},\ k_{f,3}) = \nabla_1 (k_{f,1,eq},\ k_{f,2,eq},\ k_{f,3,eq}),$$
$$(k_{f,4},\ k_{f,5},\ k_{f,6}) = \nabla_2 (k_{f,4,eq},\ k_{f,5,eq},\ k_{f,6,eq}),$$
$$(k_{f,7},\ k_{f,8}) = \nabla_3 (k_{f,7,eq},\ k_{f,8,eq}),$$
$$(k_{f,j} = k_{f,j,eq},\ j \geqslant 9),$$

where ∇_i is the coupling factor, $i=1$ for O_2, $i=2$ for N_2, and $i=3$ for NO, represented by[57]

$$\nabla_i = k_{f,j}/k_{f,j,eq} = \frac{Q(T)Q(T_{F_i})}{Q(T_{v_i})Q(-U_i)}$$
$$= \frac{\dfrac{1-\exp(-N_i\theta_i/T)}{1-\exp(-\theta_i/T)} \cdot \dfrac{1-\exp(-N_i\theta_i/T_{F_i})}{1-\exp(-\theta_i/T_{F_i})}}{\dfrac{1-\exp(-N_i\theta_i/T_{v_i})}{1-\exp(-\theta_i/T_{v_i})} \cdot \dfrac{1-\exp(N_i\theta_i/U_i)}{1-\exp(\theta_i/U_i)}}.$$

Here N_i, θ_i and T_{vi} are the total number of vibrational levels, the characteristic vibrational temperature, and the vibrational temperature of molecules, respectively, and furthermore

$$\frac{1}{T_{F_i}} = \frac{1}{T_{v_i}} - \frac{1}{T} - \frac{1}{U_i},$$
$$N_i = D_{0,i} k / \mu_i hc,$$
$$U_i = D_{0,i}/6,$$

where U_i is an adjustable parameter (in °K), $D_{0,i}$ is the dissociation temperature, k is the Boltzmann constant, h is the Planck constant, μ_i is the wave number, and c is the speed of light. Their values have been listed in Table 5.2.

Table 5.2

Species \ Parameters	N_i	U_i (°K)	θ_i (°K)	$D_{0,i}$ (°K)	μ_i (cm^{-1})
O_2	26	9894.67	2274	59368	1579.78
N_2	33	18878	3396	113268	2359.43
NO	27	12584.17	2742	75505	1905.6

When we consider the effect of the molecular dissociation-recombination on the vibrational relaxation, the vibrational relaxation equation can be written as [56]

$$\frac{de_{v_i}}{dt}=\frac{e_{v_i}(T)-e_{v_i}(T_{v_i})}{\tau_{v_i}}-\frac{\overline{E}_i(T,T_{v_i})-e_{v_i}(T_{v_i})}{c_i}\left(\frac{dc_i}{dt}\right)_f$$
$$+\frac{\overline{E}_i(T,T)-e_{v_i}(T_{v_i})}{c_i}\left(\frac{dc_i}{dt}\right)_r\equiv E_i\quad(i=1,2,3),\tag{5.3}$$

where $e_{v_i}(T)$——is the average vibrational energy at a temperature equal to the local translational temperature T,

$e_{v_i}(T_{v_i})$——is the average vibrational energy at the vibrational temperature T_{v_i},

$\overline{E}_i(T,T_{v_i})$——is the average energy lost from vibration in a dissociation,

$\overline{E}_i(T,T)$——is the average energy gained in a recombination,

$\left(\frac{dc_i}{dt}\right)_f$——is the rate at which the molecules are dissociating,

and

$\left(\frac{dc_i}{dt}\right)_r$——is the rate at which the molecules are being formed by recombination.

For a harmonic-oscillator model, we have the formulae

$$\begin{cases} e_{v_i}(T)=\dfrac{R\theta_i}{\exp(\theta_i/T)-1}, \\ e_{v_i}(T_{v_i})=\dfrac{R\theta_i}{\exp(\theta_i/T_{v_i})-1}, \\ \overline{E}_i(T,T_{v_i})=R\left[\dfrac{\theta_i}{\exp(\theta_i/T_{F_i})-1}-\dfrac{N_i\theta_i}{\exp(N_i\theta_i/T_{F_i})-1}\right], \\ \overline{E}_i(T,T)=R\left[\dfrac{\theta_i}{\exp(-\theta_i/U_i)-1}-\dfrac{N_i\theta_i}{\exp(-N_i\theta_i/U_i)-1}\right], \end{cases}\tag{5.4}$$

and

$$\begin{cases} \dfrac{1}{c_1}\left(\dfrac{dc_1}{dt}\right)_f = \rho\left(k_{f,1}\sum\limits_{i=2,3,5,6-13} c_i + k_{f,2}c_1 + k_{f,3}c_4\right), \\ \dfrac{1}{c_1}\left(\dfrac{dc_1}{dt}\right)_r = \dfrac{\rho^2 c_4^2}{c_1}\left(k_{r,1}\sum\limits_{i=2,3,5,6-13} c_i + k_{r,2}c_1 + k_{r,3}c_4\right), \\ \dfrac{1}{c_2}\left(\dfrac{dc_2}{dt}\right)_f = \rho\left(k_{f,4}\sum\limits_{i=1,3,4,6-13} c_i + k_{f,5}c_2 + k_{f,6}c_5\right), \\ \dfrac{1}{c_2}\left(\dfrac{dc_2}{dt}\right)_r = \dfrac{\rho^2 c_5^2}{c_2}\left(k_{r,4}\sum\limits_{i=1,3,4,6-13} c_i + k_{r,5}c_2 + k_{r,6}c_5\right), \\ \dfrac{1}{c_3}\left(\dfrac{dc_3}{dt}\right)_f = \rho\left(k_{f,7}\sum\limits_{i=1,2,6-13} c_i + k_{f,8}\sum\limits_{i=3,4,5} c_i\right), \\ \dfrac{1}{c_3}\left(\dfrac{dc_3}{dt}\right)_r = \dfrac{\rho^2 c_4 c_5}{c_3}\left(k_{r,7}\sum\limits_{i=1,2,6-13} c_i + k_{r,8}\sum\limits_{i=3,4,5} c_i\right), \\ \tau_{v_1}^{-1} = 0.723\times 10^{17}\rho T^{1/6}\exp(-218/T^{1/3})(1-\exp(-\theta_1/T)) \\ \qquad \times \left(3c_1 + 5c_4 + \sum\limits_{i=2,3,5,6-13} c_i\right), \\ \tau_{v_2}^{-1} = 0.13\times 10^{15}\rho T^{1/2}\exp(-193/T^{1/3})(1-\exp(-\theta_2/T)) \\ \qquad \times \left(2.5c_2 + 35c_5 + \sum\limits_{i=1,3,4,6-13} c_i\right), \\ \tau_{v_3}^{-1} = 0.127\times 10^{15}\rho T^{1/6}\exp(-84/T^{1/3})(1-\exp(-\theta_3/T)) \\ \qquad \times \left(0.05\left(\sum\limits_{i=1,2,6-13} c_i\right) + c_3 + c_4 + c_5\right). \end{cases} \tag{5.5}$$

where R is the universal gas constant.

3. *The Equation of State and Thermodynamic Functions*

As stated above, the non-equilibrium air has been assumed to be a gas mixture with n components. Each of the components will be assumed to be a thermodynamically perfect gas, that is, it satisfies the equation

$$p_i = \frac{\rho_i}{W_i} RT \quad (i=1, 2, \cdots, n),$$

where R is the universal gas constant equal to 0.83169×10^8 erg/mol·deg, and p_i, ρ_i and W_i are, respectively, the partial pressure, the density, and the molecular weight. The mixture, itself, will also be assumed to obey the perfect-gas equation of state

$$p = \frac{\rho}{W} RT, \tag{5.6}$$

where p and ρ are the total pressure and density of the mixture, respectively, and W is its average molecular weight, defined by the formula

$$W = \frac{1}{\sum\limits_{i=1}^{n} c_i},$$

with

$$c_i = \frac{\rho_i}{\rho W_i},$$

which is the concentration of the i-th component, that is, the mole number of the i-th component in a gram of mixture. For the total specific enthalpy h, we have the following expressions:

$$\begin{cases} h = \sum_{i=1}^{n} c_i h_i, \\ h_i = c_{p_i} T + e_{v_i} + h_i^0 \quad (i = 1, 2, \cdots, n), \end{cases} \tag{5.7}$$

where h_i, c_{p_i}, e_{v_i}, and h_i^0 are, respectively, the specific enthalpy, the specific heat at constant pressure, the molecular vibrational energy, and the generation enthalpy for the i-th component. The specific heat c_{p_i} will take the value $\frac{7}{2}R$ for a diatomic component or $\frac{5}{2}R$ for a uniatomic component. The values of h_i^0 for air are listed in Table 5.3.

Table 5.3 Chemical generating enthalpy h_i^0 (erg/mol)

Components	O_2	N_2	NO	O	N
h_i^0	0	0	9.16×10^{11}	2.47×10^{12}	4.7×10^{12}

Components	e^-	O^+	N^+	O_2^+
h_i^0	0	1.56×10^{13}	1.87×10^{13}	1.16×10^{18}

Components	N_2^+	Ar	NO^+	O_2^-
h_i^0	1.5×10^{13}	0	9.83×10^{12}	-4.24×10^{13}

It follows from (5.7) that

$$\begin{cases} \frac{\partial h}{\partial p} = \frac{c_p W}{\rho R}, \quad \frac{\partial h}{\partial \rho} = -\frac{c_p p W}{R\rho^2} = -\frac{c_p T}{\rho}, \\ \frac{\partial h}{\partial c_i} = h_i - c_p W T, \quad \frac{\partial h}{\partial e_{v_i}} = c_i, \\ a^2 = \dfrac{-\dfrac{\partial h}{\partial \rho}}{\dfrac{\partial h}{\partial p} - \dfrac{1}{\rho}} = \frac{c_p p W}{\rho(c_p W - R)}, \end{cases} \tag{5.8}$$

where

$$c_p = \sum_{i=1}^{n} c_i c_{p_i}.$$

If we make the thermodynamic quantities nondimensional in the following way:

$$h—V_\infty^2, \quad c_i—\frac{1}{W_\infty}, \quad h_i,\ h_i^0,\ e_{v_i}—W_\infty V_\infty^2,$$

$$p—\rho_\infty V_\infty^2, \quad \rho—\rho_\infty, \quad W—W_\infty, \quad T—W_\infty V_\infty^2/R,$$

where the quantities with the suffix ∞ denote the values of the flow parameters in the undisturbed region, and x–y means that y is the dimensionless factor for x, then in the nondimensional form the equation (5.6) will become

$$p=\frac{\rho}{W}T, \tag{5.9}$$

and the second expression of (5.7) will become

$$h_i=c_{p_i}T/R+e_{v_i}+h_i^0, \tag{5.10}$$

whereas the expressions (5.8) will remain unchanged.

Chapter 5

Calculation of Supersonic Flow around Blunt Bodies

§ 1 Introduction

In this chapter we will describe the numerical calculations of supersonic flow around blunt bodies. It is well-known that this problem is a boundary-value problem for the first-order quasilinear system of mixed-type equations.

Since the late fifties many numerical methods have been developed for solving this problem. They can be divided into two kinds, i.e., the time-dependent method and the steady state method. The latter covers the finite difference method, the method of integral relations, and the method of lines etc. For details, please refer to the introduction for the second part of this book. Here we mainly discuss the application of the method of lines, in Chapter 3, to the blunt body flow. The method of lines combines the advantages of simplicity, less storage and satisfactory accuracy etc. We apply the explicit scheme of this method in calculating three-dimensional and axisymmetric flow around blunt bodies, and also apply the implicit scheme in calculating axisymmetric flow around blunt bodies. The forms of bodies are ellipsoids with various axial ratios and somewhat like discs. The frozen, equilibrium and nonequilibrium air flows will be considered.

The accuracy of results is determined through two different methods. One method to determine the accuracy of results is to compare those results obtained by using different numbers of lines or different integration steps and to verify the validity of certain conditions(i.e.,to do some calculations relating to the conservation law of total energy, etc.). The other method is to check results by comparing the computed figures with those obtained by other methods or with experimental data. All these comparisons show that the method of lines can give us satisfactory results.

As to the improperly posed problem, our calculations show, as pointed out in § 5 of Chapter 3, that there is no need to worry about the serious growth of the rounding error as long as a proper number of lines is taken.

§ 2 Formulation of Problems

1. *Governing Equations*

We will consider inviscid, non-heat conducting, steady flow with chemical reactions. In orthogonal curvilinear coordinates, the fluid dynamic equations are of the same form (1.11) as those in Chapter 4. For convenience, we shall use slightly different notations. Suppose that in orthogonal curvilinear coordinates (s, n, φ) the shock is denoted by the equation $n=f_1(s, \varphi)$ and the body by the equation $n=f_2(s, \varphi)$. For the convenience of calculation, the following coordinate transformation is introduced:

$$\begin{cases} \xi=\dfrac{n-f_2(s, \varphi)}{f_1(s, \varphi)-f_2(s, \varphi)}=\dfrac{n-f_2(s, \varphi)}{\varepsilon(s, \varphi)}, \\ \eta=s, \\ \psi=\varphi, \end{cases}$$

where

$$\varepsilon(s, \varphi)=f_1(s, \varphi)-f_2(s, \varphi).$$

Obviously, with the new coordinates the shock and body lie in the respective planes $\xi=1$ and $\xi=0$.

From Chapter 4 it can be seen that for the coordinates (ξ, η, ψ) the fluid dynamic equations have the form (1.16). Since we have

$$\begin{cases} \dfrac{\partial\xi}{\partial s}=-\dfrac{\xi\left(\dfrac{\partial f_1}{\partial s}-\dfrac{\partial f_2}{\partial s}\right)+\dfrac{\partial f_2}{\partial s}}{\varepsilon}, \\ \dfrac{\partial\eta}{\partial s}=1, \\ \dfrac{\partial\psi}{\partial s}=0, \\ \dfrac{\partial\xi}{\partial n}=\dfrac{1}{\varepsilon}, \\ \dfrac{\partial\eta}{\partial n}=0, \\ \dfrac{\partial\psi}{\partial n}=0, \\ \dfrac{\partial\xi}{\partial\varphi}=-\dfrac{\xi\left(\dfrac{\partial f_1}{\partial\varphi}-\dfrac{\partial f_2}{\partial\varphi}\right)+\dfrac{\partial f_2}{\partial\varphi}}{\varepsilon}, \\ \dfrac{\partial\eta}{\partial\varphi}=0, \\ \dfrac{\partial\psi}{\partial\varphi}=1, \end{cases}$$

these equations can be written as:

$$\left\{\begin{array}{l}\dfrac{u}{h_s}\dfrac{\partial u}{\partial\eta}+\dfrac{c}{\varepsilon}\dfrac{\partial u}{\partial\xi}+\dfrac{w}{h_\varphi}\dfrac{\partial u}{\partial\psi}+\dfrac{1}{h_s\rho}\dfrac{\partial p}{\partial\eta}+\dfrac{\sigma}{\varepsilon\rho}\dfrac{\partial p}{\partial\xi}+a_s=0,\\ \dfrac{u}{h_s}\dfrac{\partial v}{\partial\eta}+\dfrac{c}{\varepsilon}\dfrac{\partial v}{\partial\xi}+\dfrac{w}{h_\varphi}\dfrac{\partial v}{\partial\psi}+\dfrac{\lambda}{\varepsilon\rho}\dfrac{\partial p}{\partial\xi}+a_n=0,\\ \dfrac{u}{h_s}\dfrac{\partial w}{\partial\eta}+\dfrac{c}{\varepsilon}\dfrac{\partial w}{\partial\xi}+\dfrac{w}{h_\varphi}\dfrac{\partial w}{\partial\psi}+\dfrac{1}{\rho h_\varphi}\dfrac{\partial p}{\partial\psi}+\dfrac{\beta}{\varepsilon\rho}\dfrac{\partial p}{\partial\xi}+a_\varphi=0,\\ \dfrac{u}{h_s}\dfrac{\partial\rho}{\partial\eta}+\dfrac{c}{\varepsilon}\dfrac{\partial\rho}{\partial\xi}+\dfrac{w}{h_\varphi}\dfrac{\partial\rho}{\partial\psi}+\rho\Big(\dfrac{1}{h_s}\dfrac{\partial u}{\partial\eta}+\dfrac{\sigma}{\varepsilon}\dfrac{\partial u}{\partial\xi}+\dfrac{\lambda}{\varepsilon}\dfrac{\partial v}{\partial\xi}\\ \qquad+\dfrac{\beta}{\varepsilon}\dfrac{\partial w}{\partial\xi}+\dfrac{1}{h_\varphi}\dfrac{\partial w}{\partial\psi}\Big)+b=0,\\ \dfrac{u}{h_s}\dfrac{\partial p}{\partial\eta}+\dfrac{c}{\varepsilon}\dfrac{\partial p}{\partial\xi}+\dfrac{w}{h_\varphi}\dfrac{\partial p}{\partial\psi}-a^2\Big(\dfrac{u}{h_s}\dfrac{\partial\rho}{\partial\eta}+\dfrac{c}{\varepsilon}\dfrac{\partial\rho}{\partial\xi}+\dfrac{w}{h_\varphi}\dfrac{\partial\rho}{\partial\psi}\Big)=g,\\ \dfrac{u}{h_s}\dfrac{\partial c_i}{\partial\eta}+\dfrac{c}{\varepsilon}\dfrac{\partial c_i}{\partial\xi}+\dfrac{w}{h_\varphi}\dfrac{\partial c_i}{\partial\psi}=\dfrac{\omega_i}{\rho}\quad(i=1,2,\cdots,l_1),\\ \dfrac{u}{h_s}\dfrac{\partial e_{v_i}}{\partial\eta}+\dfrac{c}{\varepsilon}\dfrac{\partial e_{v_i}}{\partial\xi}+\dfrac{w}{h_\varphi}\dfrac{\partial e_{v_i}}{\partial\psi}=E_i\quad(i=1,2,\cdots,l_2),\end{array}\right.\tag{2.1}$$

where

$$\left\{\begin{array}{l}c=\sigma u+\lambda v+\beta w,\\ \sigma=-\dfrac{1}{h_s}\Big(\xi\dfrac{\partial\varepsilon}{\partial s}+\dfrac{\partial f_2}{\partial s}\Big),\\ \lambda=\dfrac{1}{h_n},\\ \beta=-\dfrac{1}{h_\varphi}\Big(\xi\dfrac{\partial\varepsilon}{\partial\varphi}+\dfrac{\partial f_2}{\partial\varphi}\Big),\\ a_s=\dfrac{v}{h_sh_n}\Big(u\dfrac{\partial h_s}{\partial n}-v\dfrac{\partial h_n}{\partial s}\Big)+\dfrac{w}{h_sh_\varphi}\Big(u\dfrac{\partial h_s}{\partial\varphi}-w\dfrac{\partial h_\varphi}{\partial s}\Big),\\ a_n=\dfrac{w}{h_nh_\varphi}\Big(v\dfrac{\partial h_n}{\partial\varphi}-w\dfrac{\partial h_\varphi}{\partial n}\Big)+\dfrac{u}{h_nh_s}\Big(v\dfrac{\partial h_n}{\partial s}-u\dfrac{\partial h_s}{\partial n}\Big),\\ a_\varphi=\dfrac{u}{h_\varphi h_s}\Big(w\dfrac{\partial h_\varphi}{\partial s}-u\dfrac{\partial h_s}{\partial\varphi}\Big)+\dfrac{v}{h_\varphi h_n}\Big(w\dfrac{\partial h_\varphi}{\partial n}-v\dfrac{\partial h_n}{\partial\varphi}\Big),\\ b=\dfrac{\rho}{h_sh_nh_\varphi}\Big(u\dfrac{\partial(h_nh_\varphi)}{\partial s}+v\dfrac{\partial(h_sh_\varphi)}{\partial n}+w\dfrac{\partial(h_sh_n)}{\partial\varphi}\Big),\\ g=\dfrac{-1}{\dfrac{\partial h}{\partial p}-\dfrac{1}{\rho}}\Big[\displaystyle\sum_{i=1}^{l_1}\dfrac{\partial h}{\partial c_i}\dfrac{\omega_i}{\rho}+\sum_{i=1}^{l_2}\dfrac{\partial h}{\partial e_{v_i}}E_i\Big]\\ \quad=\dfrac{a^2}{\dfrac{\partial h}{\partial\rho}}\Big[\displaystyle\sum_{i=1}^{l_1}\dfrac{\partial h}{\partial c_i}\dfrac{\omega_i}{\rho}+\sum_{i=1}^{l_2}\dfrac{\partial h}{\partial e_{v_i}}E_i\Big].\end{array}\right.$$

Here u, v, w represent components of velocity vector $\boldsymbol{V}$ in the directions s, n, φ, respectively, h_s, h_n, h_φ represent the Lamé coefficients, p stands for pressure, ρ for density, c_i for the mass fraction of the i–th species, e_{v_i} for vibration energy of molecules, and h for specific enthalpy.

Also ω_i, E_i, $\frac{\partial h}{\partial \rho}$, $\frac{\partial h}{\partial c_i}$, $\frac{\partial h}{\partial e_{v_i}}$, a^2, etc. depend on the adopted model of fluid flow. If the nonequilibrium model described in § 5 of Chapter 4 is adopted, the expressions of ω_i, E_i, $\cdots$, a^2 will be (5.1), (5.3)—(5.5), and (5.8) in Chapter 4.

2. Boundary Conditions

(1) *Shock Conditions*

According to (2.4) in Chapter 4, on the shock $n=f_1(s, \varphi)$ the Rankine–Hugoniot relations must be satisfied. For frozen and equilibrium flows, the Rankine–Hugoniot relations are

$$\begin{cases} h+\dfrac{V_N^2}{2}=h_\infty+\dfrac{V_{\infty,N}^2}{2}, \\ p+\rho V_N^2=p_\infty+\rho_\infty V_{\infty,N}^2, \\ u=u_\infty-\left(1-\dfrac{\rho_\infty}{\rho}\right)n_1 V_{\infty,N}, \\ v=v_\infty-\left(1-\dfrac{\rho_\infty}{\rho}\right)n_2 V_{\infty,N}, \\ w=w_\infty-\left(1-\dfrac{\rho_\infty}{\rho}\right)n_3 V_{\infty,N}, \end{cases} \tag{2.2}$$

and for nonequilibrium flow, they are (2.2) and

$$\begin{cases} c_i=c_{i,\infty}, & i=1, 2, \cdots, l_1, \\ e_{v_i}=e_{v_i,\infty}, & i=1, 2, \cdots, l_2, \end{cases}$$

where the quantities with subscript ∞ denote those before the shock and the quantities without subscript denote those behind the shock. $\boldsymbol{N}$ is the normal unit vector to the shock and V_N the projection of the velocity vector on the normal to the shock, i.e., $V_N=\boldsymbol{V}\cdot\boldsymbol{N}$. n_1, n_2, n_3 are direction cosines of the shock normal, i.e,

$$\boldsymbol{N}=\begin{pmatrix} n_1 \\ n_2 \\ n_3 \end{pmatrix}=\frac{1}{\sqrt{1+\left(\dfrac{f'_{1s}h_n}{h_s}\right)^2+\left(\dfrac{f'_{1\varphi}h_n}{h_\varphi}\right)^2}}\begin{pmatrix} -\dfrac{f'_{1s}}{h_s}h_n \\ 1 \\ -\dfrac{f'_{1\varphi}h_n}{h_\varphi} \end{pmatrix},$$

where the quantities with a prime denote the derivatives with respect to the subindex.

(2) *Body Conditions*

According to (3.1) in Chapter 4, on the body $n=f_2(s, \varphi)$ the zero flux condition must be satisfied, i.e.,

$$q=-\frac{f'_{2s}}{h_s}u+\frac{v}{h_n}-\frac{f'_{2\varphi}}{h_\varphi}w=0, \tag{2.3}$$

where the quantities with a prime denote the derivatives with respect to the subindex.

Suppose that the dimensionless manner is

$$\begin{cases} u,\ v,\ w,\ V \sim V_\infty,\ \rho \sim \rho_\infty, \\ p \sim \rho_\infty V_\infty^2,\ l \sim l^*,\ h \sim V_\infty^2, \\ c_i \sim \dfrac{1}{W_\infty},\quad \omega_i \sim \dfrac{\rho_\infty V_\infty}{l^* W_\infty}, \\ e_{v_i} \sim W_\infty V_\infty^2, E_i \ \sim \dfrac{W_\infty V_\infty^3}{l^*}. \end{cases}$$

It is obvious that equations (2.1)—(2.3) are satisfied by the dimensionless quantities. In the above symbols the quantities with a subscript ∞ denote the free stream values, W is molecular weight, and l^* is a characteristic length.

3. *Formulation of Problems*

For supersonic flow around blunt bodies, the following pictorial description applies to many cases: a stand-off shock occurs in front of the body and the flow behind the shock and near the blunt nose is subsonic, whereas the downstream flow is supersonic. Thus it is easy to find a space-like surface Π (see Fig. 2.1). Because the flow upstream of Π is not influenced by the flow downstream of Π, the solution at the

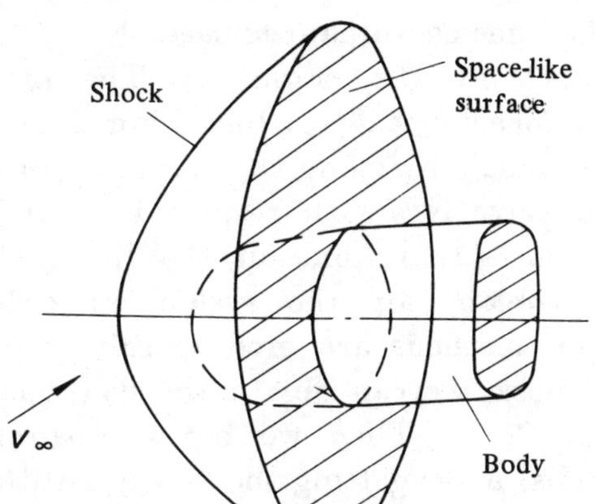

Figure 2.1

upstream region can be determined independently, Suppose that we take an orthogonal curvilinear coordinate system $(s,\ n,\ \varphi)$, where $s=0$ is a rotation axis, and the surface $s=\eta_1,\ f_2(s,\ \varphi) \leqslant n \leqslant f_1(s,\ \varphi),\ 0 \leqslant \varphi \leqslant 2\pi$ is taken as Π. Thus the problem we want to solve is reduced to the following: in the domain

$$\begin{cases} 0 \leqslant \xi \leqslant 1, \\ 0 \leqslant \eta \leqslant \eta_1, \\ 0 \leqslant \psi \leqslant 2\pi, \end{cases}$$

we determine the solution of equations (2.1) with the following boundary conditions: (1) on the shock and body, (2.2) and (2.3) are satisfied; (2) for ψ, there is the periodicity condition $f(\psi) = f(\psi + 2\pi)$, f denoting any function; (3) on the surface Π, the condition $V_N > a$ is satisfied, V_N being the projection of the velocity vector $\boldsymbol{V}$ on the outward unit normal to the surface Π. The condition (3) implies that Π is a space-like surface and that the fluid flows from out of the computational domain. Since equations (2.1) are elliptic in the subsonic region and hyperbolic in the supersonic region, the above problem is a boundary-value problem of mixed-type equations, and is coincident with that described in Chapter 3. Therefore the method discussed in Chapter 3 can be applied to this problem.

§ 3 Methods of Solution

We shall adopt the method of lines described in Chapter 3. The procedures can be outlined as follows: Suppose that the computational domain is L: $0 \leqslant \xi \leqslant 1$, $0 \leqslant \eta \leqslant \eta_1$, $0 \leqslant \psi \leqslant 2\pi$. We make some coordinate surfaces $\eta = \eta_i =$ constant and divide the interval $[0, \eta_1]$ into k_1 segments. Similarly, we make some coordinate surfaces $\psi = \psi_j =$ constant and divide the interval $[0, 2\pi]$ into $2k_2$ segments. The intersection of these coordinate surfaces forms $k_1 \times 2k_2 + 1$ lines. For a fixed ξ, we construct interpolation polynomials by using the values on these lines, and hence the corresponding derivatives with respect to η and ψ can be determined. Let equations (2.1) hold on the lines. Thus we obtain a boundary-value problem for the system of ordinary differential equations. Iteration methods are used to solve this boundary-value problem. Given a shock, we can obtain the flow quantities behind the shock by virtue of (2.2). Then we begin integrating the ordinary differential equations. After getting the flow quantities on the body, we check whether these quantities satisfy (2.3) or not. If they do not satisfy (2.3), then we adjust the shock and repeat the above procedure until (2.3) is satisfied within an acceptable tolerant error.

1. *Schemes*

Two integration schemes are used for the systems of ordinary differential equations described above.

I. *Explicit Scheme*

Rewriting equations (2.1) in the explicit form for $\frac{\partial u}{\partial \xi}, \frac{\partial v}{\partial \xi}, \cdots, \frac{\partial e_{v_i}}{\partial \xi}$, we obtain the following equations:

$$\left\{\begin{aligned}
&\frac{\partial p}{\partial \xi}=\frac{a^2[F_4c-\rho(\sigma F_1+\lambda F_2+\beta F_3)]+F_5c}{c^2-\tau^2a^2},\\
&\frac{\partial \rho}{\partial \xi}=\frac{1}{a^2}\left[-\frac{F_5}{c}+\frac{\partial p}{\partial \xi}\right],\\
&\frac{\partial u}{\partial \xi}=\frac{1}{c}\left[F_1-\frac{\sigma}{\rho}\frac{\partial p}{\partial \xi}\right],\\
&\frac{\partial v}{\partial \xi}=\frac{1}{c}\left[F_2-\frac{\lambda}{\rho}\frac{\partial p}{\partial \xi}\right],\\
&\frac{\partial w}{\partial \xi}=\frac{1}{c}\left[F_3-\frac{\beta}{\rho}\frac{\partial p}{\partial \xi}\right],\\
&\frac{\partial c_i}{\partial \xi}=\frac{1}{c}F_{6,i},\quad i=1,2,\cdots,l_1,\\
&\frac{\partial e_{v_i}}{\partial \xi}=\frac{1}{c}F_{7,i},\quad i=1,2,\cdots,l_2,
\end{aligned}\right. \tag{3.1}$$

where

$$\left\{\begin{aligned}
&\tau^2=\sigma^2+\lambda^2+\beta^2,\\
&F_1=-\varepsilon\left[\frac{u}{h_s}\frac{\partial u}{\partial \eta}+\frac{w}{h_\varphi}\frac{\partial u}{\partial \psi}+\frac{1}{h_s\rho}\frac{\partial p}{\partial \eta}+a_s\right],\\
&F_2=-\varepsilon\left[\frac{u}{h_s}\frac{\partial v}{\partial \eta}+\frac{w}{h_\varphi}\frac{\partial v}{\partial \psi}+a_n\right],\\
&F_3=-\varepsilon\left[\frac{u}{h_s}\frac{\partial w}{\partial \eta}+\frac{w}{h_\varphi}\frac{\partial w}{\partial \psi}+\frac{1}{\rho h_\varphi}\frac{\partial p}{\partial \psi}+a_\varphi\right],\\
&F_4=-\varepsilon\left[\rho\left(\frac{1}{h_s}\frac{\partial u}{\partial \eta}+\frac{1}{h_\varphi}\frac{\partial w}{\partial \psi}\right)+\frac{u}{h_s}\frac{\partial \rho}{\partial \eta}+\frac{w}{h_\varphi}\frac{\partial \rho}{\partial \psi}+b\right],\\
&F_5=-\varepsilon\left[\frac{u}{h_s}\left(\frac{\partial p}{\partial \eta}-a^2\frac{\partial \rho}{\partial \eta}\right)+\frac{w}{h_\varphi}\left(\frac{\partial p}{\partial \psi}-a^2\frac{\partial \rho}{\partial \psi}\right)-g\right],\\
&F_{6,i}=-\varepsilon\left[\frac{u}{h_s}\frac{\partial c_i}{\partial \eta}+\frac{w}{h_\varphi}\frac{\partial c_i}{\partial \psi}-\frac{\omega_i}{\rho}\right],\\
&F_{7,i}=-\varepsilon\left[\frac{u}{h_s}\frac{\partial e_{v_i}}{\partial \eta}+\frac{w}{h_\varphi}\frac{\partial e_{v_i}}{\partial \psi}-E_i\right].
\end{aligned}\right. \tag{3.2}$$

For integration of equations (3.1), the standard explicit methods for ordinary differential equations, e.g., the Runge–Kutta method, can be used.

Obviously, c in equations (3.1) vanishes on the body. Therefore, the integration cannot be marched to the body by using explicit schemes. In order to obtain the flow quantities on the body suitable procedure is extrapolation. That is, when integrating equations (3.1) to some ξ near the body (e.g., $\xi=0.1$) we use the values at several points, such as $\xi=0.3$, 0.2, 0.1, to extrapolate the values at $\xi=0$. If the flow quantities vary with ξ rather slowly, the extrapolation can give good results, but if the flow quantities vary with ξ rather rapidly, the extrapolation will produce some errors. In [A62], [A67] this singularity is removed by introducing the stream function, but the function is only applicable to the axisymmetric frozen and equilibrium flow and is not applicable to nonequilibrium flow. In order to be applicable to the general cases, we put forward an implicit scheme to remove the singularity.

II. *Implicit Scheme*

For simplicity, equations (2.1) are written in a matrix form:

$$A\frac{\partial \boldsymbol{X}}{\partial \xi}+B\frac{\partial \boldsymbol{X}}{\partial \eta}+D\frac{\partial \boldsymbol{X}}{\partial \psi}=\boldsymbol{H}, \tag{3.3}$$

where A, B, D are matrices, $\boldsymbol{X}$ and $\boldsymbol{H}$ are vectors. Let $\boldsymbol{X}_{i,j}$ denote the values of $\boldsymbol{X}$ on lines. Then on the lines equations (3.3) become

$$A_{i,j}\left(\frac{\partial \boldsymbol{X}}{\partial \xi}\right)_{i,j}+B_{i,j}\left(\frac{\partial \boldsymbol{X}}{\partial \eta}\right)_{i,j}+D_{i,j}\left(\frac{\partial \boldsymbol{X}}{\partial \psi}\right)_{i,j}=\boldsymbol{H}_{i,j}. \tag{3.4}$$

Suppose that $\boldsymbol{X}^k_{i,j}$ at $\xi=\xi_k$ are given. It is required to get $\boldsymbol{X}^{k+1}_{i,j}$ at $\xi=\xi_k+\Delta\xi$. The two-step implicit scheme (2.3)/(2.4) of Chapter 3 can be used. That is, in the first step we take

$$A^k\frac{\tilde{\boldsymbol{X}}^{k+1}-\boldsymbol{X}^k}{\Delta\xi}+B^k\left[\left(\frac{\partial \tilde{\boldsymbol{X}}}{\partial \eta}\right)^{k+1}+\left(\frac{\partial \boldsymbol{X}}{\partial \eta}\right)^{k}\right]\Big/2$$
$$+D^k\left[\left(\frac{\partial \tilde{\boldsymbol{X}}}{\partial \psi}\right)^{k+1}+\left(\frac{\partial \boldsymbol{X}}{\partial \psi}\right)^{k}\right]\Big/2=\boldsymbol{H}^k, \tag{3.5}$$

and in the second step we take

$$\frac{\tilde{A}^{k+1}+A^k}{2}\,\frac{\boldsymbol{X}^{k+1}-\boldsymbol{X}^k}{\Delta\xi}+\frac{\tilde{B}^{k+1}+B^k}{2}\left[\left(\frac{\partial \boldsymbol{X}}{\partial \eta}\right)^{k+1}+\left(\frac{\partial \boldsymbol{X}}{\partial \eta}\right)^{k}\right]\Big/2$$
$$+\frac{\tilde{D}^{k+1}+D^k}{2}\left[\left(\frac{\partial \boldsymbol{X}}{\partial \psi}\right)^{k+1}+\left(\frac{\partial \boldsymbol{X}}{\partial \psi}\right)^{k}\right]\Big/2=(\tilde{\boldsymbol{H}}^{k+1}+\boldsymbol{H}^k)/2, \tag{3.6}$$

where subscripts i, j are omitted, and the values with a tilde "~" are those obtained in the first step. If we examine the concrete form of A, B, and D, it is easy to see that, except for the stagnation point, the singularity does not occur when the scheme (3.5)/(3.6) is used. As for the stagnation point, it is necessary to do some special treatments, for example, we can let $\boldsymbol{H}$ include $\boldsymbol{X}^{k+1}$, and solve $\boldsymbol{X}^{k+1}$ from the equations. We shall present this kind of treatment for axisymmetric flow in the next section.

Clearly, when equation (3.5) or (3.6) is used, the difference equations must be solved simultaneously. Therefore the calculation will be much more complicated than that for explicit schemes. However, as pointed out above, the singularity occurs only near the body. In order to reduce the amount of calculation, a combined scheme can be used, that is, the implicit scheme is used near the body and the explicit scheme is used in the remainder of the domain.

Suppose that $s=0$ is a rotation axis in the orthogonal coordinates (s, n, φ). Then there are the terms 0/0 in equations (2.1) at this axis. The l'Hospitale rule can be used to obtain the equations needed. In addition, this axis is a singular line (φ is multivalued), therefore we need to perform some appropriate treatment, the details of which will be given later.

2. *Interpolation Polynomials*

In our calculations we construct interpolation polynomials of the functions in terms of values on the lines and determine the corresponding derivatives from the polynomials. In general, the procedure is as follows. For $\xi=$constant and $\psi=$constant, we construct the interpolation polynomials in η in terms of g_k which are values of g at η_k, and hence determine the derivative $\frac{\partial g}{\partial \eta}$. Similarly, the derivative $\frac{\partial g}{\partial \psi}$ can be determined. However, at present the functions are periodic with ψ, therefore the trigonometric interpolations in ψ are used.

In order to reduce the amount of computation, the technique discribed in § 4 of Chapter 3 is adopted. That is, we calculate directly the combination coefficients instead of a_i in equation (4.1), in § 4 of Chapter 3. The concrete formulae for different cases will be given in the next two sections.

3. *Iteration Method*

The iteration method described in § 3 of Chapter 3 is used to adjust the shock. The formula is

$$\begin{aligned} \boldsymbol{F}_{m+1}=\boldsymbol{F}_m-&(\boldsymbol{F}_{m-(l+1)}-\boldsymbol{F}_m, \cdots, \boldsymbol{F}_{m-1}-\boldsymbol{F}_m) \\ &\times(\boldsymbol{Q}_{m-(l+1)}-\boldsymbol{Q}_m, \cdots, \boldsymbol{Q}_{m-1}-\boldsymbol{Q}_m)^{-1}\boldsymbol{Q}_m, \\ &m=l+1,\ l+2, \cdots, \end{aligned} \tag{3.7}$$

where $l+1$ is the total number of lines. $\boldsymbol{F}$ and $\boldsymbol{Q}$ are vectors, i.e.,

$$\boldsymbol{F}=\begin{pmatrix} f_{1,0} \\ f_{1,1} \\ \vdots \\ f_{1,l} \end{pmatrix}, \quad \boldsymbol{Q}=\begin{pmatrix} q_0 \\ q_1 \\ \vdots \\ q_l \end{pmatrix},$$

where $f_{1,i}$ signifies the position of the shock on the lines, and q_i is the quantity characterizing the body condition on the lines, i.e.,

$$q=-f'_{2s}u/h_s+v/h_n+f'_{2\varphi}w/h_\varphi.$$

4. *Initial Values of the Shock*

When the above method is used, whether the initial values of the shock are good or not is important. To get good initial values of the shock, we adopt the method of successive transition for some parameters.

For example, when the result at some angle of attack α^* is required, the angle of attack α can be taken as the parameter. If the solution at α_0 is known, then it can be taken as the initial value at $\alpha_0+\Delta\alpha$ $\left(\Delta\alpha=\frac{\alpha^*-\alpha}{L}\right)$. When the solution at $\alpha_0+\Delta\alpha$ is obtained, the initial value at $\alpha_0+2\Delta\alpha$ can be extrapolated from the values at α_0 and $\alpha_0+\Delta\alpha$. Once the shocks are obtained at three successive angles of attack, say, at $\alpha_0+(\bar{l}-3)\Delta\alpha$, $\alpha_0+(\bar{l}-2)\Delta\alpha$, and $\alpha_0+(\bar{l}-1)\Delta\alpha$, then the initial value of the shock at $\alpha_0+\bar{l}\Delta\alpha$ can be extrapolated from the three successive values. This procedure is continued until $\bar{l}=L$. As long as L is large enough, the above method can give good initial values of the shock in order to guarantee the convergence of iteration. Moreover, in addition to α and M_∞, the geometric parameter of the body, etc., can be taken as parameters.

Obviously, when this method is used, it is necessary to know the initial value of the shock for a certain body and for a certain free stream condition. This is possible from the available results.

It is worthwhile pointing out that the increment of the parameter should be adequately selected so that computations can be carried out successfully in as little time as possible.

Our experience has shown that the above procedure is effective.

§ 4 Calculation of the Axisymmetric Flow

In this section we shall give detailed formulae for axisymmetric flow around some blunt bodies in the several coordinate systems and make some remarks about special treatments.

1. *Coordinates and Governing Equations*

We take the boundary layer coordinates for the axisymmetric flow.

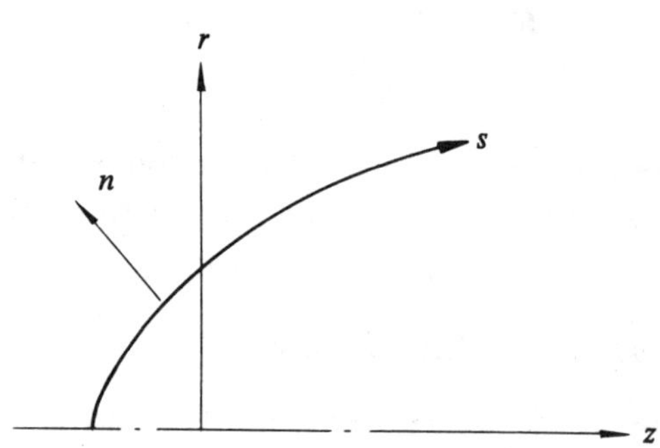

Figure 4.1

Thus the Lamé coefficients are

$$h_s = 1 + \frac{n}{R(s)}, \quad h_n = 1, \quad h_\varphi = r(s, n), \tag{4.1}$$

where $R(s)$ is the radius of curvature of the basic coordinate line and r is the distance of the point denoted by (s, n) to the symmetric axis. Suppose that the equation of the basic line of the boundary layer coordinate system is $r = r_0(z)$ in the cylindrical coordinates (r, z, φ). In the following, $\dot{r}_0$ denotes the derivative of r_0 with respect to z, $\ddot{r}_0$ the second derivative, and we assume that $\ddot{r} \leqslant 0$, and $\dot{r}_0^{-1}|_{r=0} = 0$. The coordinates are shown in Fig. 4.1. Consider the axisymmetric flow, which means the flow quantities are independent of φ and $w = 0$. Thus equations (3.1) can be written as

$$\left\{\begin{aligned} &\frac{\partial p}{\partial \xi} = \frac{a^2[F_4 c - \rho(\sigma F_1 + \lambda F_2)] + F_5 c}{c^2 - \tau^2 a^2}, \\ &\frac{\partial \rho}{\partial \xi} = \frac{1}{a^2}\left[-\frac{F_5}{c} + \frac{\partial p}{\partial \xi}\right], \\ &\frac{\partial u}{\partial \xi} = \frac{1}{c}\left[F_1 - \frac{\sigma}{\rho}\frac{\partial p}{\partial \xi}\right], \\ &\frac{\partial v}{\partial \xi} = \frac{1}{c}\left[F_2 - \frac{\lambda}{\rho}\frac{\partial p}{\partial \xi}\right], \\ &\frac{\partial c_i}{\partial \xi} = \frac{1}{c} F_{6,i}, \quad i = 1, 2, \cdots, l_1, \\ &\frac{\partial e_{v_i}}{\partial \xi} = \frac{1}{c} F_{7,i}, \quad i = 1, 2, \cdots, l_2, \end{aligned}\right. \tag{4.2}$$

where

$$\left\{\begin{aligned} &F_1 = -\varepsilon\left[\frac{1}{h_s}\left(u\frac{\partial u}{\partial \eta} + \frac{1}{\rho}\frac{\partial p}{\partial \eta}\right) + a_s\right], \\ &F_2 = -\varepsilon\left[\frac{u}{h_s}\frac{\partial v}{\partial \eta} + a_n\right], \\ &F_4 = -\varepsilon\left[\frac{1}{h_s}\left(\rho\frac{\partial u}{\partial \eta} + u\frac{\partial \rho}{\partial \eta}\right) + b\right], \\ &F_5 = -\varepsilon\left[\frac{u}{h_s}\left(\frac{\partial p}{\partial \eta} - a^2\frac{\partial \rho}{\partial \eta}\right) - g\right], \\ &F_{6,i} = -\varepsilon\left[\frac{u}{h_s}\frac{\partial c_i}{\partial \eta} - \frac{\omega_i}{\rho}\right], \\ &F_{7,i} = -\varepsilon\left[\frac{u}{h_s}\frac{\partial e_{v_i}}{\partial \eta} - E_i\right], \\ &\sigma = -\frac{1}{h_s}(\xi\varepsilon' + f_2'), \\ &\lambda = \frac{1}{h_n} = 1, \\ &\tau^2 = \sigma^2 + \lambda^2 = 1 + \sigma^2, \end{aligned}\right. \tag{4.3}$$

$$
\left\{
\begin{aligned}
& c=\sigma u+v, \\
& a_s=\frac{uv}{h_s R}, \\
& a_n=-\frac{u^2}{h_s R}, \\
& b=\frac{\rho}{h_s r}\left[u\frac{\partial r}{\partial s}+v\left(\frac{r}{R}+h_s\frac{\partial r}{\partial n}\right)\right], \\
& n=\xi\varepsilon+f_2=\xi(f_1-f_2)+f_2, \\
& r=r_0+\frac{n|\dot{r}_0^{-1}|}{\sqrt{1+(\dot{r}_0^{-1})^2}}=r_0+\frac{n}{\sqrt{1+\dot{r}_0^2}}, \\
& \frac{\partial r}{\partial s}=\frac{\operatorname{sign}\dot{r}_0^{-1}}{\sqrt{1+(\dot{r}_0^{-1})^2}}h_s=\frac{\dot{r}_0}{\sqrt{1+\dot{r}_0^2}}h_s, \\
& \frac{\partial r}{\partial n}=\frac{|\dot{r}_0^{-1}|}{\sqrt{1+(\dot{r}_0^{-1})^2}}=\frac{1}{\sqrt{1+\dot{r}_0^2}}, \\
& R^{-1}=\frac{-\ddot{r}_0}{[1+\dot{r}_0^2]^{3/2}}, \quad (\text{if } \ddot{r}_0\leqslant 0).
\end{aligned}
\right.
$$

Equations (3.3) become

$$A\frac{\partial \boldsymbol{X}}{\partial \xi}+B\frac{\partial \boldsymbol{X}}{\partial \eta}=\boldsymbol{H}, \tag{4.4}$$

where A, B, $\boldsymbol{X}$, $\boldsymbol{H}$ are

$$
\left\{
\begin{aligned}
& A=\begin{bmatrix}
V_\xi & 0 & \dfrac{\sigma}{\varepsilon\rho} & 0 & & \\
0 & V_\xi & \dfrac{1}{\varepsilon\rho} & 0 & \multicolumn{2}{c}{\mathbf{0}} \\
\dfrac{\rho\sigma}{\varepsilon} & \dfrac{\rho}{\varepsilon} & 0 & V_\xi & & \\
0 & 0 & V_\xi & -a^2V_\xi & & \\
 & \mathbf{0} & & & V_\xi & \\
 & & & & & \ddots \quad V_\xi
\end{bmatrix}, \\
& B=\begin{bmatrix}
V_\eta & 0 & \dfrac{1}{h_s\rho} & 0 & & \\
0 & V_\eta & 0 & 0 & \multicolumn{2}{c}{\mathbf{0}} \\
\dfrac{\rho}{h_s} & 0 & 0 & V_\eta & & \\
0 & 0 & V_\eta & -a^2V_\eta & & \\
 & \mathbf{0} & & & V_\eta & \\
 & & & & & \ddots \quad V_\eta
\end{bmatrix},
\end{aligned}
\right. \tag{4.5}
$$

$$
\boldsymbol{H}=\begin{bmatrix} -a_s \\ -a_n \\ -b \\ g \\ \omega_1/\rho \\ \vdots \\ \omega_{l_1}/\rho \\ E_1 \\ \vdots \\ E_{l_2} \end{bmatrix}, \quad \boldsymbol{X}=\begin{bmatrix} u \\ v \\ p \\ \rho \\ c_1 \\ \vdots \\ c_{l_1} \\ e_{v_1} \\ \vdots \\ e_{v_{l_2}} \end{bmatrix}
$$

and
$$
V_\xi=\frac{c}{\varepsilon}, \quad V_\eta=\frac{u}{h_s}.
$$

Correspondingly, the two-step scheme (3.5), (3.6) becomes

$$
\begin{cases} A^k\dfrac{\widetilde{\boldsymbol{X}}^{k+1}-\boldsymbol{X}^k}{\Delta\xi}+B^k\left[\left(\dfrac{\partial\widetilde{\boldsymbol{X}}}{\partial\eta}\right)^{k+1}+\left(\dfrac{\partial\boldsymbol{X}}{\partial\eta}\right)^k\right]\Big/2=\boldsymbol{H}^k, & (4.6) \\ \dfrac{\widetilde{A}^{k+1}+A^k}{2}\dfrac{\boldsymbol{X}^{k+1}-\boldsymbol{X}^k}{\Delta\xi}+\dfrac{1}{2}(\widetilde{B}^{k+1}+B^k) & \\ \quad\times\left[\left(\dfrac{\partial\boldsymbol{X}}{\partial\eta}\right)^{k+1}+\left(\dfrac{\partial\boldsymbol{X}}{\partial\eta}\right)^k\right]\Big/2=\dfrac{1}{2}[\widetilde{\boldsymbol{H}}^{k+1}+\boldsymbol{H}^k]. & (4.7) \end{cases}
$$

The usual method, e.g., the Gauss elimination method, can be used to solve (4.6) and (4.7).

2. *Equations on the Axis*

For the equations on the axis, some special treatments must be undertaken. It is easy to see that in the explicit scheme the third equation of (4.2) becomes an identity $0\equiv0$ on the axis. In order to increase the accuracy of the approximation for u, the interpolation polynomial with u_0' is used, where u_0' is the value of the derivative u' on the axis. To obtain the equations satisfied by u_0', the third equation of (4.2) is differentiated with respect to η. Note that because $u_0'=\varepsilon_0'=p_0'=f_{20}'=0$ at $\eta=0$, where the quantities with the prime and with the subscript 0 denote the derivatives with respect to η on the axis, therefore the following equation can be obtained:

$$
\frac{\partial u_0'}{\partial\xi}=\frac{1}{c}\left[F_1'-\frac{\sigma'}{\rho}\frac{\partial p}{\partial\xi}\right]_0, \tag{4.8}
$$

where
$$
\begin{cases} F_{1,0}'=-\varepsilon\left[\dfrac{u_0'^2}{h_s}+\dfrac{1}{h_s\rho}p_0''+a_{s,0}'\right], \\ a_{s,0}'=\dfrac{vu_0'}{h_sR}, \\ \sigma_0'=-\dfrac{(\xi\varepsilon''+f_2'')}{h_s}. \end{cases}
$$

In addition, by using the l'Hospitale rule for the terms 0/0 in b, and noting that $\frac{\partial}{\partial s}=\frac{\partial}{\partial \eta}$, we have

$$b_0=\frac{\rho}{h_s}(u_0'+2v/R). \tag{4.9}$$

Similarly, for the implicit scheme the first equation of (4.4) is differentiated with respect to η and the l'Hospitale rule is used for the terms 0/0 in the remaining equations of (4.4). With slight manipulation we obtain the required equation

$$\bar{A}\frac{\partial \bar{\boldsymbol{X}}}{\partial \xi}+\bar{\boldsymbol{D}}\frac{\partial^2 p}{\partial \eta^2}=\bar{\boldsymbol{H}}, \tag{4.10}$$

where

$$\left\{\begin{aligned}
\bar{A}&=\begin{bmatrix} V_\xi & 0 & \frac{\sigma'}{\varepsilon\rho} & 0 & & & \\ 0 & V_\xi & \frac{1}{\varepsilon\rho} & 0 & & \mathbf{0} & \\ 0 & \frac{\rho}{\varepsilon} & 0 & V_\xi & & & \\ 0 & 0 & 1 & -a^2 & & & \\ & & & & 1 & & \\ & \mathbf{0} & & & & \ddots & \\ & & & & & & 1 \end{bmatrix},\\
\bar{\boldsymbol{D}}&=\begin{bmatrix} \frac{1}{\rho h_s} \\ 0 \\ \vdots \\ 0 \end{bmatrix},\\
\bar{\boldsymbol{H}}&=-\begin{bmatrix} \frac{1}{h_s}\left(\frac{u_0'v}{R}+u_0'^2\right) \\ 0 \\ \frac{2\rho}{h_s}\left(u_0'+\frac{v}{R}\right) \\ g/V_\xi \\ -\omega_1/\rho V_\xi \\ \vdots \\ -\omega_{l_1}/\rho V_\xi \\ -E_1/V_\xi \\ \vdots \\ -E_{l_2}/V_\xi \end{bmatrix}, \quad \bar{\boldsymbol{X}}=\begin{bmatrix} u_0' \\ v \\ p \\ \rho \\ c_1 \\ \vdots \\ c_{l_1} \\ e_{v_1} \\ \vdots \\ e_{v_{l_2}} \end{bmatrix},
\end{aligned}\right. \tag{4.11}$$

Due to the fact that $V_\xi=0$ at the stagnation point, it is necessary to make a further treatment for (4.10). From the second equation of (4.10) we know that $\frac{\partial p}{\partial \xi}=0$. Therefore the first equation of (4.10) becomes

$$u_0'^2+\frac{1}{\rho}\frac{\partial^2 p}{\partial \eta^2}=-\frac{v u_0'}{R}. \tag{4.12}$$

For this equation we may adopt the following two step scheme: in the first step we take

$$u_0'^k \tilde{u}_0'^{k+1}+\frac{1}{2\rho^k}\left[\left(\frac{\partial^2 \tilde{p}}{\partial \eta^2}\right)^{k+1}+\left(\frac{\partial^2 p}{\partial \eta^2}\right)^k\right]=0, \tag{4.13}$$

and in the second step

$$\left(\frac{\tilde{u}_0'^{k+1}+u_0'^k}{2}\right)u_0'^{k+1}+\frac{1}{4}\left(\frac{1}{\tilde{\rho}^{k+1}}+\frac{1}{\rho^k}\right)$$
$$\times\left[\left(\frac{\partial^2 p}{\partial \eta^2}\right)^{k+1}+\left(\frac{\rho^2 p}{\partial \eta^2}\right)^k\right]=0. \tag{4.14}$$

In addition, the chemical reaction equations and vibration relaxation equations must be replaced by the corresponding equilibrium equations.

3. *Boundary Conditions*

The boundary conditions on the shock and body are given by (2.2) and (2.3). Moreover, the condition for u_0' is required. The terms related to u in the Rankine–Hugoniot relations are differentiated with respect to η. Noting that on the axis

$$n_1=\frac{-m_1}{\sqrt{1+m_1^2}}=0, \quad n_1'=-m_1',$$

we can get

$$u_0'=u_{\infty,0}'-\left(1-\frac{\rho_\infty}{\rho}\right)V_{\infty,N}n_1'$$
$$=u_{\infty,0}'+\left(1-\frac{\rho_\infty}{\rho}\right)m_1'V_{\infty,N}, \tag{4.15}$$

where $m_1=\frac{h_n}{h_s}f_1'$, and the quantities with a prime denote the derivatives with respect to η.

4. *Geometric Quantities and Free Stream Conditions*

Suppose that the equation of the basic line of the boundary layer coordinate system is $f(r, z)=0$ for the cylindrical coordinates. Obviously, the following relationships between geometric quantities exist:

$$\begin{cases} s=\int_0^r \sqrt{1+(\dot{r}_0^{-1})^2}\,dr, \\ \dot{r}_0^{-1}=-\dfrac{f_r}{f_z}, \\ R^{-1}=\dfrac{|f_r^2 f_{zz}+f_z^2 f_{rr}-2f_r f_z f_{rz}|}{(f_r^2+f_z^2)^{3/2}}, \end{cases} \tag{4.16}$$

where $\dot{r}$ denotes the derivative of r with respect to z, and where f_r, f_{rr} etc. are the first derivative or the second derivative of f with respect to the subscripts. Suppose that the equation of the basic coordinate line is $r^m+z^m=l^m(m\geqslant 2)$. This becomes $r^m+z^m=1$ after it is nondimensionalized by l. Then we have

$$\begin{cases} \dot{r}_0^{-1}=\dfrac{-r^{m-1}}{z^{m-1}}, \\ R^{-1}=(m-1)\dfrac{r^{m-2}z^{m-2}}{[r^{2(m-1)}+z^{2(m-1)}]^{3/2}}. \end{cases} \tag{4.17}$$

In this case, the expressions for the free stream quantities u_∞, v_∞ are

$$\begin{cases} u_\infty=\dfrac{|\dot{r}_0^{-1}|}{\sqrt{1+(\dot{r}_0^{-1})^2}}, \\ v_\infty=\dfrac{-\operatorname{sign}(\dot{r}_0^{-1})}{\sqrt{1+(\dot{r}_0^{-1})^2}}, \end{cases} \tag{4.18}$$

and owing to the fact that

$$u'_\infty=\frac{1}{R}\frac{\operatorname{sign}(\dot{r}_0^{-1})}{\sqrt{1+(\dot{r}_0^{-1})^2}},$$

$$\dot{r}_0^{-1}=0,$$

the expression for $u'_{\infty,0}$ is

$$u'_{\infty,0}=\frac{1}{R}. \tag{4.19}$$

5. *Interpolation Polynomials*

Let us take $(k+1)$ lines in $[0, \eta_{\max}]$, i.e., $\eta=\eta_j=\text{constant}$ $(j=0, 1, \cdots, k, \eta_0=0, \eta_k=\eta_{\max})$. Because of symmetry of flow, we can construct interpolation polynomials of order $2k$ for even functions and of order $2k+1$ for odd functions, that is, we adopt the following expressions:

$$\begin{cases} g=\sum\limits_{i=0}^{k} g_i^0\eta^{2i}, \quad g=v,\ p,\ \rho,\ c_i,\ e_{v_i},\ f_1, \\ u=\sum\limits_{i=0}^{k} u_i^0\eta^{2i+1} \quad \text{or} \quad \tilde{u}=\dfrac{u}{\eta}=\sum\limits_{i=0}^{k} u_i^0\eta^{2i}, \end{cases} \tag{4.20}$$

where f_1 denotes the shock position. Since f_1 is independent of ξ, $\dfrac{\partial f_1}{\partial \eta}=\dfrac{\partial f_1}{\partial s}$. The derivatives with respect to η are determined by (4.20). As done in § 4 of Chapter 3, the derivatives can be written in the following form:

$$\begin{cases} g'(\eta)=(\boldsymbol{M}^{*-1}\boldsymbol{d}_1)^*\boldsymbol{g}\equiv\boldsymbol{b}_1^*\boldsymbol{g}, \\ g''(\eta)=(\boldsymbol{M}^{*-1}\boldsymbol{d}_2)^*\boldsymbol{g}\equiv\boldsymbol{b}_2^*\boldsymbol{g}, \\ u'=\eta\dfrac{\partial\tilde{u}}{\partial\eta}+\tilde{u}=[\eta(\boldsymbol{M}^{*-1}\boldsymbol{d}_1)+\boldsymbol{M}^{*-1}\boldsymbol{d}_0]\tilde{\boldsymbol{u}}\equiv\boldsymbol{b}_3^*\tilde{\boldsymbol{u}}, \end{cases} \tag{4.21}$$

where

$$\begin{cases} \boldsymbol{M}^*=\begin{bmatrix} 1 & 1 & \cdots & 1 \\ \eta_0^2 & \eta_1^2 & & \eta_k^2 \\ \vdots & \vdots & & \vdots \\ \eta_0^{2k} & \eta_1^{2k} & \cdots & \eta_k^{2k} \end{bmatrix}, \\ \boldsymbol{d}_0^*=(1,\ \eta^2,\ \cdots,\ \eta^{2k}), \\ \boldsymbol{d}_1^*=(0,\ 2\eta,\ \cdots,\ 2k\eta^{2k-1}), \\ \boldsymbol{d}_2^*=(0,\ 2,\ \cdots,\ 2k(2k-1)\eta^{2k-2}), \\ \boldsymbol{g}^*=(g_0,\ g_1,\ \cdots,\ g_k), \\ \tilde{\boldsymbol{u}}^*=\left(u_0',\ \dfrac{u_1}{\eta_1},\ \cdots,\ \dfrac{u_k}{\eta_k}\right), \end{cases} \tag{4.22}$$

and where $\boldsymbol{b}_1$, $\boldsymbol{b}_2$, $\boldsymbol{b}_3$ are calculated first and then stored for further application.

§ 5 Calculation of the Three-dimensional Flow

1. *Coordinates and Governing Equations*

Suppose that we take spherical coordinates $(\theta,\ r,\ \varphi)$, and that the flow is symmetric with respect to planes $\varphi=0,\ \pi$. Hence we need only to solve the problem for $0\leqslant\varphi\leqslant\pi$. In the following, only the explicit scheme is listed.

The governing equations are still (2.1) or (3.1). Because of the fact that

$$\begin{cases} h_s=r,\ h_n=1,\ h_\varphi=r\sin\theta, \\ a_s=\dfrac{1}{r}(uv-w^2\operatorname{ctg}\theta), \\ a_n=-\dfrac{1}{r}(u^2+w^2), \\ a_\varphi=\dfrac{1}{r}(vw+uw\operatorname{ctg}\theta), \\ b=\dfrac{\rho}{r}(2v+u\operatorname{ctg}\theta), \end{cases}$$

therefore F_i in (3.2) and σ, λ, β in (2.1) can be written as

$$\left\{\begin{aligned}
&F_1=-\frac{\varepsilon}{r}\left[u\frac{\partial u}{\partial\eta}+\frac{w}{\sin\theta}\left(\frac{\partial u}{\partial\psi}-\cos\theta w\right)+\frac{1}{\rho}\frac{\partial p}{\partial\eta}+uv\right],\\
&F_2=-\frac{\varepsilon}{r}\left[u\frac{\partial v}{\partial\eta}+\frac{w}{\sin\theta}\frac{\partial v}{\partial\psi}-(u^2+w^2)\right],\\
&F_3=-\frac{\varepsilon}{r}\left[u\frac{\partial w}{\partial\eta}+\frac{1}{\sin\theta}\left[w\left(\frac{\partial w}{\partial\psi}+\cos\theta u\right)+\frac{1}{\rho}\frac{\partial p}{\partial\psi}\right]+vw\right],\\
&F_4=-\frac{\varepsilon}{r}\left[u\frac{\partial\rho}{\partial\eta}+\frac{w}{\sin\theta}\frac{\partial\rho}{\partial\psi}+\rho\left(2v+\frac{\partial u}{\partial\eta}\right)\right.\\
&\qquad\left.+\frac{\rho}{\sin\theta}\left(\frac{\partial w}{\partial\psi}+u\cos\theta\right)\right],\\
&F_5=-\frac{\varepsilon}{r}\left[u\left(\frac{\partial p}{\partial\eta}-a^2\frac{\partial\rho}{\partial\eta}\right)+\frac{w}{\sin\theta}\left(\frac{\partial p}{\partial\psi}-a^2\frac{\partial\rho}{\partial\psi}\right)\right],\\
&\sigma=-\frac{1}{r}\left(\xi\frac{\partial\varepsilon}{\partial\theta}+\frac{\partial f_2}{\partial\theta}\right)=-\frac{1}{r}\left(\xi\frac{\partial\varepsilon}{\partial\eta}+\frac{\partial f_2}{\partial\eta}\right),\\
&\lambda=1,\\
&\beta=-\frac{1}{r\sin\theta}\left(\xi\frac{\partial\varepsilon}{\partial\varphi}+\frac{\partial f_2}{\partial\varphi}\right)=-\frac{1}{r\sin\theta}\left(\xi\frac{\partial\varepsilon}{\partial\psi}+\frac{\partial f_2}{\partial\psi}\right),
\end{aligned}\right. \tag{5.1}$$

where u, v, w are the components of velocity in the θ, r, φ directions, respectively (see (1.13) of Chapter 4). For simplicity we do not consider the nonequilibrium flow.

2. *Equations on the Axis*

Suppose that the axis $\theta=0$ lies on the symmetric plane. Clearly, equations (2.1) can be rewritten as

$$\left\{\begin{aligned}
&c\frac{\partial u}{\partial\xi}+\frac{\sigma}{\rho}\frac{\partial p}{\partial\xi}=F_1,\\
&c\frac{\partial v}{\partial\xi}+\frac{\lambda}{\rho}\frac{\partial p}{\partial\xi}=F_2,\\
&c\frac{\partial w}{\partial\xi}+\frac{\beta}{\rho}\frac{\partial p}{\partial\xi}=F_3,\\
&c\frac{\partial\rho}{\partial\xi}+\rho\left(\sigma\frac{\partial u}{\partial\xi}+\lambda\frac{\partial v}{\partial\xi}+\beta\frac{\partial w}{\partial\xi}\right)=F_4,\\
&c\left(\frac{\partial p}{\partial\xi}-a^2\frac{\partial\rho}{\partial\xi}\right)=F_5.
\end{aligned}\right. \tag{5.2}$$

Since $\sin\theta$ occurs in the denominator in (5.1), the equations (5.2) on the axis $\theta=0$ must be dealt with. Let u^* denote $u(\xi,\ 0,\ 0)$ and the axis $\theta=0$ be located on the symmetric plane. Obviously, we have the following relations

$$\begin{cases} u(\xi,\ 0,\ \psi)=u^*(\xi)\cos\psi, \\ v(\xi,\ 0,\ \psi)=v(\xi,\ 0,\ 0), \\ w(\xi,\ 0,\ \psi)=-u^*(\xi)\sin\psi, \\ p(\xi,\ 0,\ \psi)=p(\xi,\ 0,\ 0), \\ \rho(\xi,\ 0,\ \psi)=\rho(\xi,\ 0,\ 0) \end{cases}$$

and

$$\frac{\partial p}{\partial\psi}=\frac{\partial\rho}{\partial\psi}=\frac{\partial v}{\partial\psi}=\frac{\partial u}{\partial\psi}-w\cos\theta=\frac{\partial w}{\partial\psi}+u\cos\theta=0$$

on the axis $\theta=0$. Therefore, by using the I'Hospitale rule for terms 0/0 in (5.1), we get

$$\begin{cases} \beta=\dfrac{\partial\sigma}{\partial\psi},\ \dfrac{1}{\sin\theta}\dfrac{\partial p}{\partial\psi}=\dfrac{\partial^2 p}{\partial\theta\,\partial\psi}=\dfrac{\partial^2 p}{\partial\eta\,\partial\psi}, \\ \dfrac{1}{\sin\theta}\dfrac{\partial\rho}{\partial\psi}=\dfrac{\partial^2\rho}{\partial\eta\,\partial\psi},\quad \dfrac{1}{\sin\theta}\dfrac{\partial v}{\partial\psi}=\dfrac{\partial^2 v}{\partial\eta\,\partial\psi}, \\ \dfrac{1}{\sin\theta}\left(\dfrac{\partial u}{\partial\psi}-w\cos\theta\right)=\dfrac{\partial^2 u}{\partial\psi\,\partial\eta}-\dfrac{\partial w}{\partial\eta}, \\ \dfrac{1}{\sin\theta}\left(\dfrac{\partial w}{\partial\psi}+u\cos\theta\right)=\dfrac{\partial^2 w}{\partial\eta\,\partial\psi}+\dfrac{\partial u}{\partial\eta}. \end{cases}$$

Substituting these expressions into (5.2), we get the equations on the axis $\theta=0$. In principle, the equations on each ψ coordinate plane might be taken as the equations desired, but because of the numerical error, the results obtained for different planes may be discordant, which may be avoided by integrating the equations for ψ from 0 to π. In fact, if we integrate the equation

$$\left(c\frac{\partial u}{\partial\xi}+\frac{\sigma}{\rho}\frac{\partial p}{\partial\xi}\right)\cos\psi-\left(c\frac{\partial w}{\partial\xi}+\frac{\beta}{\rho}\frac{\partial p}{\partial\xi}\right)\sin\psi=F_1\cos\psi-F_3\sin\psi$$

for ψ from 0 to π, and note the relations

$$\begin{cases} \dfrac{1}{\pi}\displaystyle\int_0^\pi c\left(\frac{\partial u}{\partial\xi}\cos\psi-\frac{\partial w}{\partial\xi}\sin\psi\right)d\psi=\frac{1}{\pi}\int_0^\pi\left(\sigma u+\lambda v+w\frac{\partial\sigma}{\partial\psi}\right)\frac{\partial u^*}{\partial\xi}\,d\psi \\ \quad=\dfrac{1}{\pi}\displaystyle\int_0^\pi\left(\sigma u^*\cos\psi+\lambda v-u^*\sin\psi\frac{\partial\sigma}{\partial\psi}\right)\frac{\partial u^*}{\partial\xi}\,d\psi \\ \quad=(\lambda v+u^*\sigma^*)\dfrac{\partial u^*}{\partial\xi}\equiv c^*\dfrac{\partial u^*}{\partial\xi}, \\ \dfrac{1}{\rho}\displaystyle\int_0^\pi\left(\sigma\cos\psi-\frac{\partial\sigma}{\partial\psi}\sin\psi\right)\frac{\partial p}{\partial\xi}\,d\psi\equiv\frac{\sigma^*}{\rho}\frac{\partial p}{\partial\xi}, \\ \dfrac{1}{\pi}\displaystyle\int_0^\pi(F_1\cos\psi-F_3\sin\psi)d\psi=-\frac{2\varepsilon}{\pi r}\left\{\int_0^\pi\left[u^*\left(\frac{\partial u}{\partial\eta}\cos^2\psi-\frac{\partial w}{\partial\eta}\sin\psi\cos\psi\right)\right.\right. \\ \quad\left.\left.+\dfrac{1}{\rho}\dfrac{\partial p}{\partial\eta}\cos\psi\right]d\psi+\dfrac{\pi}{2}vu^*\right\}=F_1^*, \end{cases}$$

we obtain

$$c^* \frac{\partial u^*}{\partial \xi} + \sigma^* \frac{\partial p}{\partial \xi} = F_1^*,$$

where

$$\begin{cases} c^* = \lambda v + u^* \sigma^*, \\ \sigma^* = \dfrac{1}{\pi} \displaystyle\int_0^\pi \left(\sigma \cos \psi - \frac{\partial \sigma}{\partial \psi} \sin \psi \right) d\psi = \frac{2}{\pi} \int_0^\pi \sigma \cos \psi \, d\psi. \end{cases}$$

Similarly, we can get

$$\begin{cases} c^* \dfrac{\partial v}{\partial \xi} + \dfrac{\lambda}{\rho} \dfrac{\partial p}{\partial \xi} = F_2^*, \\ c^* \dfrac{\partial \rho}{\partial \xi} + \rho \left(\lambda \dfrac{\partial v}{\partial \xi} + \sigma^* \dfrac{\partial u^*}{\partial \xi} \right) = F_4^*, \\ c^* \left(\dfrac{\partial p}{\partial \xi} - a^2 \dfrac{\partial \rho}{\partial \xi} \right) = F_5^*, \end{cases}$$

where

$$\begin{cases} F_2^* = -\dfrac{2\varepsilon}{\pi r} \left\{ u^* \displaystyle\int_0^\pi \frac{\partial v}{\partial \eta} \cos \psi \, d\psi - \frac{\pi}{2} u^{*2} \right\}, \\ F_4^* = -\dfrac{2\varepsilon}{\pi r} \left\{ u^* \displaystyle\int_0^\pi \frac{\partial \rho}{\partial \eta} \cos \psi \, d\psi + \rho \int_0^\pi \frac{\partial u}{\partial \eta} \, d\psi + \rho v \pi \right\}, \\ F_5^* = -\dfrac{2\varepsilon}{\pi r} u^* \displaystyle\int_0^\pi \left(\frac{\partial p}{\partial \eta} - a^2 \frac{\partial \rho}{\partial \eta} \right) \cos \psi \, d\psi. \end{cases}$$

Rewriting these equations in the explicit form for $\dfrac{\partial p}{\partial \xi}, \cdots, \dfrac{\partial w}{\partial \xi}$, we obtain the following desired equations:

$$\begin{cases} \dfrac{\partial p}{\partial \xi} = \dfrac{a^2 [F_4^* c - \rho(\sigma^* F_1^* + \lambda F_2^*)] + c^* F_5^*}{c^{*2} - \tau^{*2} a^2}, \\ \dfrac{\partial \rho}{\partial \xi} = \dfrac{1}{a^2} \left[-\dfrac{F_5^*}{c^*} + \dfrac{\partial p}{\partial \xi} \right], \\ \dfrac{\partial u^*}{\partial \xi} = \dfrac{1}{c^*} \left[F_1^* - \dfrac{\sigma^*}{\rho} \dfrac{\partial p}{\partial \xi} \right], \\ \dfrac{\partial v}{\partial \xi} = \dfrac{1}{c^*} \left[F_2^* - \dfrac{\lambda}{\rho} \dfrac{\partial p}{\partial \xi} \right], \end{cases} \tag{5.3}$$

where $\tau^{*2} = \lambda^2 + \sigma^{*2}$.

3. *Boundary Conditions*

The boundary conditions on the shock and body are given by (2.2) and (2.3). But some treatment for $\theta = 0$ is required. The unit normal to the surface $r = f(\theta, \varphi) \equiv \xi [f_1(\theta, \varphi) - f_2(\theta, \varphi)] + f_2(\theta, \varphi)$ can be expressed by

$$\boldsymbol{N}=\begin{pmatrix} n_1 \\ n_2 \\ n_3 \end{pmatrix}=\frac{1}{\sqrt{\left(\frac{f'_\theta}{f}\right)^2+1+\left(\frac{f''_{\theta\varphi}}{f}\right)^2}}\begin{pmatrix} -f'_\theta/f \\ 1 \\ -f''_{\theta\varphi}/f \end{pmatrix}$$

$$=\frac{1}{\sqrt{\sigma^2+1+\sigma_\varphi^2}}\begin{pmatrix} \sigma \\ 1 \\ \sigma_\varphi \end{pmatrix}.$$

Since the normal should be unique in the physical space, it can also be expressed by

$$\boldsymbol{N}=\frac{1}{\sqrt{\sigma^{*2}+1}}\begin{pmatrix} \sigma^*\cos\psi \\ 1 \\ -\sigma^*\sin\psi \end{pmatrix}, \tag{5.4}$$

where σ^* is the value of σ in the plane $\psi=0$. Because of numerical error the σ^* obtained for different ψ will be different. By an integration procedure similar to that described above, we get

$$\sigma^*=\frac{\int_0^\pi \sigma\cos\psi\, d\psi}{\int_0^\pi \cos^2\psi\, d\psi}=\frac{2}{\pi}\int_0^\pi \sigma\cos\psi\, d\psi,$$

which will be used when we calculate $\boldsymbol{N}$ by (5.4). Obviously, as long as (2.2) and (2.3) hold in a certain ψ–plane, they also hold in the other ψ–planes. Therefore, the plane can be arbitrarily taken, e.g., $\psi=0$.

4. *Geometric Quantities and Free Stream Conditions*

The coordinates (θ, r, φ) are shown in Fig. 5.1. The axis $\theta=0$ lies on the symmetric plane and coincides with z. Suppose that the angle between z and the horizontal axis z_1 is $\bar{\delta}$ and the angle between z_1 and the free stream direction is α. Then the free stream conditions can be expressed by

$$\begin{cases} u_\infty=\cos(\alpha-\bar{\delta})\sin\theta-\sin(\alpha-\bar{\delta})\cos\theta\cos\varphi, \\ v_\infty=-(\cos(\alpha-\bar{\delta})\cos\theta+\sin(\alpha-\bar{\delta})\sin\theta\cos\varphi), \\ w_\infty=\sin(\alpha-\bar{\delta})\sin\varphi, \\ u_\infty^*=-\sin(\alpha-\bar{\delta}). \end{cases} \tag{5.5}$$

Some formulas of geometric quantities for a certain body are given below. Assume that the shape of the body is expressed by

$$(x_0^2+y_0^2)^{\frac{m}{2}}+z_0^m=1 \tag{5.6}$$

for the Cartesian coordinates (x_0, y_0, z_0), and there are the following relations between two systems (x_1, y_1, z_1) and (x_0, y_0, z_0):

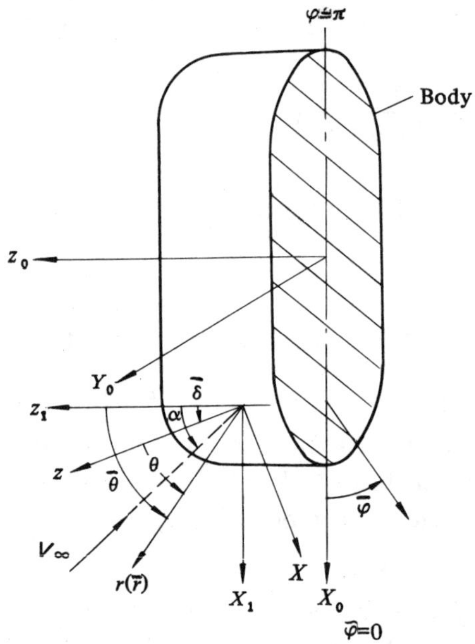

Figure 5.1

$$\begin{cases} x_0 = x_1 + a_1, \\ y_0 = y_1, \\ z_0 = z_1 + b_1. \end{cases}$$

Then the shape of the body is expressed by

$$[(x_1+a_1)^2 + y_1^2]^{\frac{m}{2}} + (z_1+b_1)^m = 1$$

or

$$G = [(x_1+a_1)^2 + y_1^2]^{\frac{m}{2}} + (z_1+b_1)^m - 1 = 0$$

for the coordinates (x_1, y_1, z_1).

Moreover, for the spherical coordinates $(\bar{r}, \bar{\theta}, \bar{\varphi})$ whose axis $\bar{\theta}=0$ is coincident with z_1, its expression is

$$\bar{G} = [(\bar{r}\sin\bar{\theta}\cos\bar{\varphi} + a_1)^2 + (\bar{r}\sin\bar{\theta}\sin\bar{\varphi})^2]^{\frac{m}{2}} + (\bar{r}\cos\bar{\theta} + b_1)^m - 1 = 0.$$

Assume that it can be written as $\bar{r} = f_2^*(\bar{\theta}, \bar{\varphi})$. Then the equation of the body surface for the coordinates (r, θ, φ) is $r = f_2(\theta, \varphi) = f_2^*(\bar{\theta}(\theta, \varphi), \bar{\varphi}(\theta, \varphi))$, where between $\bar{\theta}$, $\bar{\varphi}$ and θ, φ there exist the following relations:

$$\begin{cases} \cos\bar{\theta} = \cos\theta\cos\bar{\delta} - \sin\theta\sin\bar{\delta}\cos\varphi, \\ \sin\bar{\varphi} = \dfrac{\sin\theta\sin\varphi}{\sin\bar{\theta}}. \end{cases}$$

Therefore we have

$$\begin{cases} \dfrac{\partial f_2}{\partial \eta} = \dfrac{\partial f_2}{\partial \theta} = \dfrac{\partial f_2^*}{\partial \bar\theta}\dfrac{\partial \bar\theta}{\partial \theta} + \dfrac{\partial f_2^*}{\partial \bar\varphi}\dfrac{\partial \bar\varphi}{\partial \theta}, \\ \dfrac{\partial f_2}{\partial \psi} = \dfrac{\partial f_2}{\partial \varphi} = \dfrac{\partial f_2^*}{\partial \bar\theta}\dfrac{\partial \bar\theta}{\partial \varphi} + \dfrac{\partial f_2^*}{\partial \bar\varphi}\dfrac{\partial \bar\varphi}{\partial \varphi}, \end{cases} \tag{5.7}$$

where

$$\begin{cases} \dfrac{\partial f_2^*}{\partial \bar\theta} = \dfrac{-\dfrac{\partial \bar G}{\partial \bar\theta}}{\dfrac{\partial \bar G}{\partial \bar r}} = \dfrac{e}{d}, \\ \dfrac{\partial f_2^*}{\partial \bar\varphi} = \dfrac{-\dfrac{\partial \bar G}{\partial \bar\varphi}}{\dfrac{\partial \bar G}{\partial \bar r}} = \dfrac{h}{d}, \\ e = r\sin\bar\theta(r\cos\bar\theta + b_1)^{m-1} \\ \qquad - [(r\sin\bar\theta\cos\bar\varphi + a_1)^2 + (r\sin\bar\theta\sin\bar\varphi)^2]^{\frac{m-2}{2}} \\ \qquad \times [(r\sin\bar\theta\cos\bar\varphi + a_1) r\cos\bar\theta\cos\bar\varphi + r^2\sin\bar\theta\cos\bar\theta\sin^2\bar\varphi], \\ h = [(r\sin\bar\theta\cos\bar\varphi + a_1)^2 + (r\sin\bar\theta\sin\bar\varphi)^2]^{\frac{m-2}{2}} \\ \qquad \times [(r\sin\bar\theta\cos\bar\varphi + a_1) r\sin\bar\theta\sin\bar\varphi - r^2\sin^2\bar\theta\sin\bar\varphi\cos\bar\varphi], \\ d = [(r\sin\bar\theta\cos\bar\varphi + a_1)^2 + (r\sin\bar\theta\sin\bar\varphi)^2]^{\frac{m-2}{2}} \\ \qquad \times [(r\sin\bar\theta\cos\bar\varphi + a_1)\sin\bar\theta\cos\bar\varphi + r\sin^2\bar\theta\sin^2\bar\varphi] \\ \qquad + (r\cos\bar\theta + b_1)^{m-1}\cos\bar\theta, \\ \dfrac{\partial \bar\theta}{\partial \theta} = \dfrac{\sin\theta\cos\delta + \cos\theta\cos\varphi\sin\delta}{\sin\bar\theta} \\ \qquad = \dfrac{-\sin\theta + \sin\bar\theta\sin\delta\sin\bar\varphi\dfrac{\partial \bar\varphi}{\partial \theta}}{-\sin\bar\theta\cos\delta + \cos\bar\theta\sin\delta\cos\bar\varphi}, \\ \dfrac{\partial \bar\theta}{\partial \varphi} = -\dfrac{\sin\theta\sin\varphi\sin\delta}{\sin\bar\theta} = -\sin\bar\varphi\sin\delta, \\ \dfrac{\partial \bar\varphi}{\partial \theta} = \dfrac{\cos\theta\sin\varphi - \cos\bar\theta\sin\bar\varphi\dfrac{\partial \bar\theta}{\partial \theta}}{\sin\bar\theta\cos\bar\varphi}, \\ \dfrac{\partial \bar\varphi}{\partial \varphi} = \dfrac{\sin\theta\cos\varphi - \cos\bar\theta\sin\bar\varphi\dfrac{\partial \bar\theta}{\partial \varphi}}{\sin\bar\theta\cos\bar\varphi}. \end{cases}$$

On the axis $\bar\theta = 0$, some treatments are needed. Suppose that $\delta > 0$. Then $\varphi = \pi$, $\theta = \delta$. From the above expressions, it is seen that

$$\begin{cases} \dfrac{\partial\bar{\theta}}{\partial\theta} = -\cos\bar{\varphi}, & \dfrac{\partial\bar{\theta}}{\partial\varphi} = -\sin\bar{\varphi}\sin\delta, \\ \sin\bar{\theta}\,\dfrac{\partial\bar{\varphi}}{\partial\theta} = \sin\bar{\varphi}, & \sin\bar{\theta}\,\dfrac{\partial\bar{\varphi}}{\partial\varphi} = -\cos\bar{\varphi}\sin\delta, \\ \dfrac{\partial f_2^*}{\partial\bar{\theta}} = \left(\dfrac{\partial f_2^*}{\partial\bar{\theta}}\right)_{\bar{\theta}=0,\,\bar{\varphi}=\pi} (-\cos\bar{\varphi}), \\ \dfrac{1}{\sin\bar{\theta}}\,\dfrac{\partial f_2^*}{\partial\bar{\varphi}} = \left(\dfrac{\partial f_2^*}{\partial\bar{\theta}}\right)_{\bar{\theta}=0,\,\bar{\varphi}=\pi} \sin\bar{\varphi}, \end{cases}$$

and

$$\frac{\partial f_2}{\partial\theta} = \left(\frac{\partial f_2^*}{\partial\bar{\theta}}\right)_{\bar{\theta}=0,\,\bar{\varphi}=0}, \qquad \frac{\partial f_2}{\partial\varphi} = 0$$

on the axis $\bar{\theta}=0$.

5. *Interpolation Polynomials*

We make $k+1$ planes, $\psi=j\dfrac{\pi}{k}\equiv\psi_j=\text{constant}$, $j=0, \cdots, k$, and $n+1$ cones, $\eta=\eta_i=\text{constant}$, $i=0, \cdots, n$, $\eta_0=0$, $\eta_n=\eta_{\max}$. These cones intersect the planes $\psi=\psi_j$, $\psi=\psi_j+\pi$ at $2n+1$ lines. Because of symmetry, we construct an interpolation polynomial of order $2n$ in η by using the values on the planes $\psi=\psi_j$ and $\psi=\pi-\psi_j$, i.e., we construct the following polynomial:

$$g=\sum_{i=0}^{2n} g_i^0 \eta^i.$$

The method described in § 4 of Chapter 3 is used to calculate the combination coefficients, and the derivative may be determined by

$$g'_\eta = (M^{*-1}\boldsymbol{d}_1)^*\boldsymbol{g},$$

where

$$\begin{cases} M^* = \begin{bmatrix} 1 & 1 & \cdots & 1 \\ \eta_0 & \eta_1 & & \eta_{2n} \\ \eta_0^2 & \eta_1^2 & & \vdots \\ \vdots & \vdots & & \vdots \\ \eta_0^{2n} & \eta_1^{2n} & \cdots & \eta_{2n}^{2n} \end{bmatrix}, \\ \boldsymbol{d}_1^* = (0,\ 1,\ 2\eta,\ \cdots,\ 2n\eta^{2n-1}), \\ \boldsymbol{g}^* = (g_0,\ g_1,\ \cdots,\ g_{2n}), \quad g_i = g\,|_{\eta=\eta_i}. \end{cases} \tag{5.8}$$

For a fixed η and a fixed ξ, we construct the trigonometric interpolation polynomial in ψ as follows. For even functions we take the following form:

$$g=\sum_{i=0}^{k} g_i^0 \cos^i \psi.$$

Then we have

$$g'_\psi=(M^{*-1}d_1)^* g,$$

where

$$\left\{\begin{aligned} &M^*=\begin{bmatrix} 1 & 1 & \cdots & 1 \\ \cos\psi_0 & \cos\psi_1 & & \cos\psi_k \\ \vdots & \vdots & & \vdots \\ \cos^k\psi_0 & \cos^k\psi_1 & \cdots & \cos^k\psi_k \end{bmatrix}, \\ &d_1^*=(0,\ -\sin\psi,\ \cdots,\ -k\sin\psi\cos^{k-1}\psi), \\ &g^*=(g_0,\ g_1,\ \cdots,\ g_k). \end{aligned}\right. \tag{5.9}$$

For $\frac{1}{\pi}\int_0^\pi g\,d\psi$, $\frac{1}{\pi}\int_0^\pi g\cos\psi\,d\psi$ we have

$$\left\{\begin{aligned} &\frac{1}{\pi}\int_0^\pi g\,d\psi=(M^{*-1}d_3)^* g, \\ &\frac{1}{\pi}\int_0^\pi g\cos\psi\,d\psi=(M^{*-1}d_4)^* g, \end{aligned}\right.$$

where

$$\left\{\begin{aligned} &d_3^*=\left(1,\ 0,\ \frac{1}{2},\ 0,\cdots,\ \frac{1}{2}\cdot\frac{3}{4}\cdot\cdots\cdot\frac{k-1}{k}\right), \\ &d_4^*=\left(0,\ \frac{1}{2},\ 0,\ \cdots,\ \frac{1}{2}\cdot\frac{3}{4}\cdot\cdots\cdot\frac{k-1}{k},\ 0\right), \\ &g^*=(g_0,\ g_1,\ \cdots,\ g_k). \end{aligned}\right.$$

Here, we assume that k is even and the relations

$$\frac{1}{\pi}\int_0^\pi \cos^i\psi\,d\psi=\begin{cases} 1 & (i=0), \\ 0 & (\text{odd } i), \\ \frac{1}{2}\cdot\frac{3}{4}\cdots\frac{i-1}{i} & (\text{even } i) \end{cases}$$

are used to derive d_3^*, d_4^*.

For odd functions we take

$$g=\sum_{i=0}^{k-2} g_i^0 \cos^i\psi\sin\psi.$$

Then we have

$$g'_\psi=(M^{*-1}d_1)^* g,$$

where

$$\left\{\begin{aligned}
M^* &= \begin{bmatrix} \sin\psi_1 & \sin\psi_2 & \cdots & \sin\psi_{k-1} \\ \sin\psi_1\cos\psi_1 & \sin\psi_2\cos\psi_2 & \cdots & \sin\psi_{k-1}\cos\psi_{k-1} \\ \vdots & \vdots & & \vdots \\ \sin\psi_1\cos^{k-2}\psi_1 & \sin\psi_2\cos^{k-2}\psi_2 & \cdots & \sin\psi_{k-1}\cos^{k-2}\psi_{k-1} \end{bmatrix}, \\
d_1^* &= (\cos\psi,\ -\sin^2\psi+\cos^2\psi,\ \cdots, \\
&\qquad -(k-2)\cos^{k-3}\psi\sin^2\psi+\cos^{k-1}\psi), \\
g^* &= (g_1,\ g_2,\ \cdots,\ g_{k-1}).
\end{aligned}\right. \tag{5.10}$$

In addition to the above, we employ the following method to construct interpolation polynomials. For even functions, we take

$$g=\sum_{i=0}^{n}\sum_{j=0}^{k} g_{i,j}\eta^i\cos^j\psi,$$

and for odd functions, we take

$$w=\left(\sum_{i=0}^{n}\sum_{j=0}^{k-2} w_{i,j}\eta^i\cos^j\psi\right)\sin\psi.$$

In order to guarantee that the functions and their derivatives with respect to η at $\eta=0$ have unique values, it is necessary to set some restrictions on the polynomials. Now let us describe those restrictions.

For example, for the shock $r=f_1(\eta,\ \psi)$ we demand that it be independent of ψ at $\eta=0$. This means that $f_{0,j}=0$ $(j\neq 0)$. At the same time, in order to guarantee that the normal has a definite direction, we require that it should be of the form $(b\cos\psi,\ a,\ -b\sin\psi)$, where a, b are arbitrary constants. This means that $f_{1,j}=0$ $(j\neq 1)$. Therefore, the polynomial for f_1 should possess the following form:

$$f_1(\eta,\ \psi)=f_{0,0}+f_{1,1}\eta\cos\psi+\sum_{i=2}^{n}\sum_{j=0}^{k} f_{i,j}\eta^i\cos^j\psi. \tag{5.11}$$

For the other quantities, the concrete forms of the polynomials can be derived as follows. Set

$$\left\{\begin{aligned}
g &= g_{0,0}+\sum_{i=1}^{n}\sum_{j=0}^{k} g_{i,j}\eta^i\cos^j\psi \quad (g=p,\ \rho,\ v), \\
u &= u_{0,1}\cos\psi+\sum_{i=1}^{n}\sum_{j=0}^{k} u_{i,j}\eta^i\cos^j\psi, \\
w &= -u_{0,1}\sin\psi+\sum_{i=1}^{n}\sum_{j=0}^{k-2} w_{i,j}\eta^i\cos^j\psi\sin\psi,
\end{aligned}\right.$$

where $p|_{\eta=0}$, $\rho|_{\eta=0}$, $v|_{\eta=0}$ are single-valued and where $u_{0,j}=0$ $(j\neq 1)$, $w_{0,j}=0(j\neq 0)$, $u_{0,1}=-w_{0,0}$ because $u(\xi,\ 0,\ \psi)=A\cos\psi$, $w(\xi,\ 0,\ \psi)=-A\sin\psi$. Substitute the above expressions into (5.2). Noting that

$$\left.\frac{\partial f_2}{\partial\eta}\right|_{\eta=0}=f'_{2\eta}(0,\ 0)\cos\psi,$$

$$\frac{1}{\sin\theta}\left.\frac{\partial f_2}{\partial\psi}\right|_{\eta=0}=-f'_{2\eta}(0,\ 0)\sin\psi,$$

and letting $\eta \to 0$, we get

$$\begin{cases} \left(c_0 \dfrac{\partial u_{0,1}}{\partial \xi} - \dfrac{f'_{2\eta,0}+\xi(f_{1,1}-f'_{2\eta,0})}{r_0\rho_{0,0}} \dfrac{\partial p_{0,0}}{\partial \xi}\right)\cos\psi = \lim\limits_{\eta\to 0} F_1, \\ c_0 \dfrac{\partial v_{0,0}}{\partial \xi} + \dfrac{\lambda}{\rho_{0,0}} \dfrac{\partial p_{0,0}}{\partial \xi} = \lim\limits_{\eta\to 0} F_2, \\ -\left(c_0 \dfrac{\partial u_{0,1}}{\partial \xi} - \dfrac{f'_{2\eta,0}+\xi(f_{1,1}-f'_{2\eta,0})}{r_0\rho_{0,0}} \dfrac{\partial p_{0,0}}{\partial \xi}\right)\sin\psi = \lim\limits_{\eta\to 0} F_3, \\ c_0 \dfrac{\partial \rho_{0,0}}{\partial \xi} + \rho_{0,0}\lambda \dfrac{\partial v_{0,0}}{\partial \xi} - \rho_{0,0}\left[\dfrac{f'_{2\eta,0}+\xi(f_{1,1}-f'_{2\eta,0})}{r_0}\right]\dfrac{\partial u_{0,1}}{\partial \xi} \\ \quad = \lim\limits_{\eta\to 0} F_4, \\ c_0\left(\dfrac{\partial p_{0,0}}{\partial \xi} - a_0^2 \dfrac{\partial \rho_{0,0}}{\partial \xi}\right) = \lim\limits_{\eta\to 0} F_5, \end{cases} \tag{5.12}$$

where

$$\begin{cases} c_0 = \lambda v_{0,0} - u_{0,1} \dfrac{f'_{2\eta,0}+\xi(f_{1,1}-f'_{2\eta,0})}{r_0}, \\ r_0 = f_{2,0}+\xi(f_{1,0}-f_{2,0}), \\ f'_{2\eta,0} = \dfrac{\partial}{\partial \eta}(f_2(0,0)), \\ a_0^2 = \dfrac{\gamma p_{0,0}}{\rho_{0,0}}. \end{cases}$$

Expanding the even functions on the right side of the above equations for polynomials in $\cos\psi$ and using the independence of $1, \cos\psi, \cdots, \cos^k\psi$, we can conclude that p, ρ, v should take the form (5.11) and u, w should take the following form:

$$\begin{cases} u = u_{1,0}\cos\psi + u_{1,0}\eta + u_{1,2}\eta\cos^2\psi + \sum\limits_{i=2}^{n}\sum\limits_{j=0}^{k} u_{i,j}\eta^i\cos^j\psi, \\ w = -u_{0,1}\sin\psi - u_{1,2}\eta\cos\psi\sin\psi + \left(\sum\limits_{i=2}^{n}\sum\limits_{j=0}^{k-2} w_{i,j}\eta^i\cos^j\psi\right)\sin\psi. \end{cases} \tag{5.13}$$

Simultaneously. we obtain the following equations at $\eta=0$:

$$\begin{cases} c_0 \dfrac{\partial u_{0,1}}{\partial \xi} - \dfrac{f'_{2\eta,0}+\xi(f_{1,1}-f'_{2\eta,0})}{r_0\rho_{0,0}} \dfrac{\partial p_{0,0}}{\partial \xi} \\ \quad = -\dfrac{f_{1,0}-f_{2,0}}{r_0}\left[u_{0,1}(u_{1,0}+u_{1,2}+v_{0,0}) + \dfrac{p_{1,1}}{\rho_{0,0}}\right], \\ c_0 \dfrac{\partial v_{0,0}}{\partial \xi} + \dfrac{\lambda}{\rho_{0,0}} \dfrac{\partial p_{0,0}}{\partial \xi} = -\dfrac{f_{1,0}-f_{2,0}}{r_0} u_{0,1}(v_{1,1}-u_{0,1}), \\ c_0 \dfrac{\partial \rho_{0,0}}{\partial \xi} + \rho_{0,0}\lambda \dfrac{\partial v_{0,0}}{\partial \xi} - \dfrac{\rho_{0,0}}{r_0}[f'_{2\eta,0}+\xi(f_{1,1}-f'_{2\eta,0})] \dfrac{\partial u_{0,1}}{\partial \xi} \\ \quad = -\dfrac{f_{1,0}-f_{2,0}}{r_0}[u_{0,1}\rho_{1,1}+\rho_{0,0}(u_{1,2}+2u_{1,0}+2v_{0,0})], \\ c_0\left(\dfrac{\partial p_{0,0}}{\partial \xi} - a_0^2 \dfrac{\partial \rho_{0,0}}{\partial \xi}\right) = -\dfrac{f_{1,0}-f_{2,0}}{r_0} u_{0,1}(p_{1,1}-a_0^2\rho_{1,1}). \end{cases} \tag{5.14}$$

§ 6 Results

We have applied the above method for calculating axisymmetric and three-dimensional supersonic inviscid flows around blunt bodies. The bodies are in the form of ellipsoids or in the shape of discs (for which the equation of bodies is $(x^2+y^2)^{m/2}+z^m=l^m$, $m \geqslant 2$. The free stream Mach number ranges from 1.5 to ∞. The flows of air in frozen, equilibrium and nonequilibrium states are calculated. The method described in § 4 of Chapter 4 is used for calculating thermodynamic functions of equilibrium air, and the model of nonequilibrium air given in § 5 of Chapter 4 is adopted. The method presented in the appendix of this chapter is used for calculating the supersonic flow around the afterbody of a sphere.

The accuracy of results is determined in two different ways. One way is to determine the accuracy of results, through comparing those results by using different numbers of lines or different integration steps, or through verifying the validity of certain conditions deduced from gas dynamic equations, for example, the conservation law of total energy. The other way is to check our results by comparing them with those obtained by other methods and with experimental data. All the comparisons show that the present method gives satisfactory results.

In the following, we shall give part of the results and make some remarks. The results for frozen air flow around a sphere are obtained by explicit and implicit schemes. Both results are in good agreement with each other. The other results are obtained by the explicit scheme.

Fig. 6.1 shows the shocks and sonic lines for frozen air flow around a sphere at different M_∞. It is seen that the shocks get closer to the body as M_∞ increases. When M_∞ is large enough, the position of sonic lines on the body almost remains unchanged. The numerical results are given in Table 6.1.

Table 6.1 Positions η_a (degree) of sonic lines on a body

M_∞	4	10	15	20	∞
η_a	42.70	40.97	40.77	40.71	40.63

Fig. 6.2 gives the relations between stand-off distances ε_0 at the stagnation point and M_∞. For frozen air flow ε_0 almost remains unchanged when $M_\infty > 10$.

Fig 6.3 gives the pressure curves along the body for frozen air flow around a sphere at different M_∞.

The pressure curves along the stagnation line for frozen air flow around a sphere are given in Fig. 6.4. From this figure we see that the pressure profiles almost remain unchanged when $M_\infty > 10$.

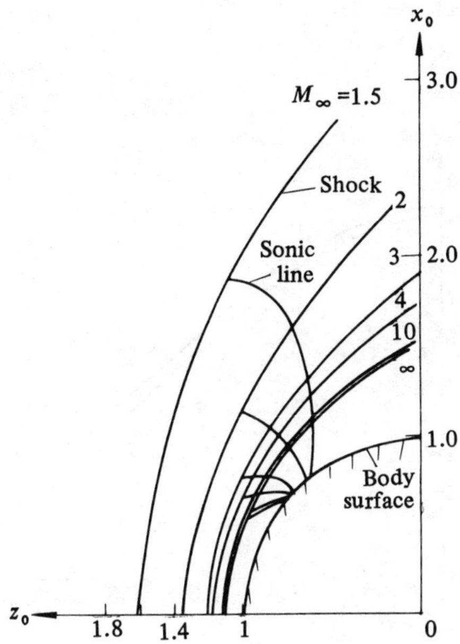

Figure 6.1 Shapes of shocks and sonic lines of frozen air flow around spheres

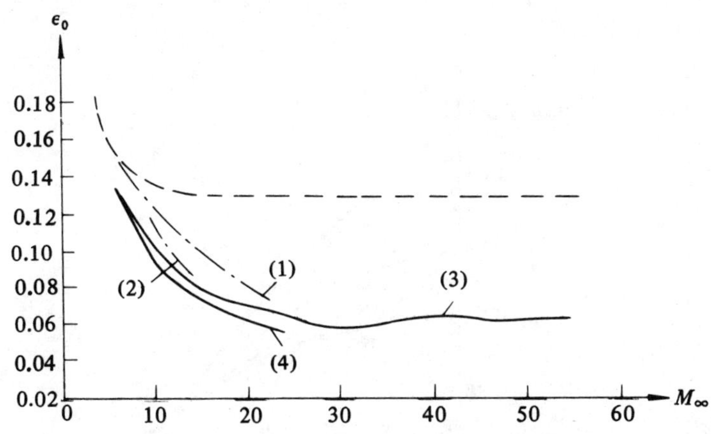

Figure 6.2 Variation of the stand-off distance ε_0 of shocks at the stagnation point with M_∞

—·— Nonequilibrium air, ---- Frozen air, —— Equilibrium air.

(1) $\begin{cases} p_\infty = 0.94\times 10^3 \text{ dyn/cm}^2 \\ \rho_\infty = 0.12\times 10^{-5} \text{ g/cm}^3 \\ R = 5 \text{ cm} \end{cases}$ (2) $\begin{cases} p_\infty = 0.803\times 10^4 \text{ dyn/cm}^2 \\ \rho_\infty = 0.118\times 10^{-4} \text{ g/cm}^3 \\ R = 5 \text{ cm} \end{cases}$

(3) $\begin{cases} p_\infty = 0.122\times 10^6 \text{ dyn/cm}^2 \\ \rho_\infty = 0.192\times 10^{-3} \text{ g/cm}^3 \end{cases}$ (4) $\begin{cases} p_\infty = 0.803\times 10^4 \text{ dyn/cm}^2 \\ \rho_\infty = 0.118\times 10^{-4} \text{ g/cm}^3 \end{cases}$

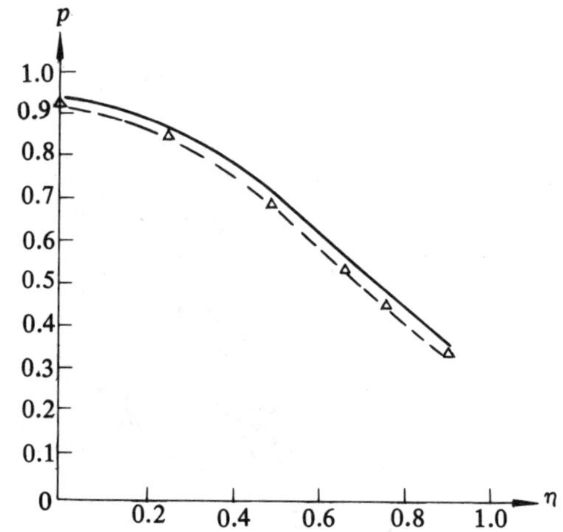

Figure 6.3 Pressure distributions on spheres ($\gamma=1.4$)
—— $M_\infty=4$, △ $M_\infty=8$, ---- $M_\infty=20$.

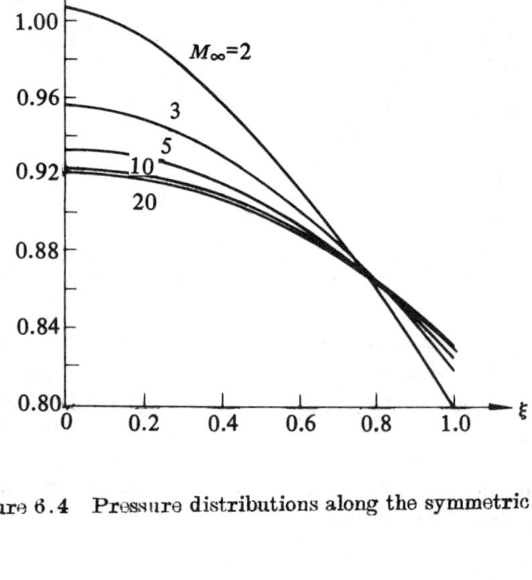

Figure 6.4 Pressure distributions along the symmetric axis

Fig. 6.5 shows the shocks and the sonic lines for various bodies. For the body whose shape is close to a sphere the radius of curvature of the shock on the symmetric axis is greater than that of the body, and the stand-off distance of the shock increases as the distance from the axis increases. When the bluntness of bodies increases the results will inverse, i.e., the radius of curvature of the shock on the symmetric axis is less than that of the body, and the stand-off distance of the shock decreases as the distance from the axis increases and reaches the minimum near the sonic line. For a very blunt body, e.g., $m>20$, the shock will almost remain unchanged.

In order to show the influence of the curvature of bodies near the sonic point on the stand-off distance of the shock, Fig. 6.6 gives the variation of ε_0 with m for the bodies like discs at $M_\infty=3$. It is easy to see that ε_0. almost remains unchanged when $m>20$. Table 6.2 demonstrates the pressure along the symmetric axis for various bodies at $M_\infty=3$.

Table 6.2 Pressure p on the symmetric axis ($M_\infty=3$)

Parameters of bodies		ξ					
		1.0	0.8	0.6	0.4	0.2	0.0
δ	1	0.8195	0.8664	0.9036	0.9317	0.9501	0.9573
	1.5	0.8201	0.8674	0.9047	0.9325	0.9505	0.9576
m	10	0.8201	0.8710	0.9098	0.9370	0.9527	0.9567
	40	0.8201	0.8712	0.9099	0.9369	0.9524	0.9569

Table 6.2 shows that the influence of bodies on the pressure on the axis is very weak.

Fig. 6.7 indicates the surface pressure for various bodies at $M_\infty=3$ and the surface pressure on a body with $m=40$ for different M_∞. It shows that the pressure remains unchanged when $m>10$, and for very blunt bodies the pressure changes slowly in a large region near the stagnation point and drops down rapidly in the region near the sonic point. Moreover, the influence of M_∞ on the surface pressure is very weak.

Figs. 6.2, 6.8—6.11 give some results for the equilibrium air flow around a sphere.

Fig. 6.8 shows the shocks and sonic lines at different M_∞. The free stream conditions are $M_\infty=4$, $p_\infty=0.87\times10^3\,\mathrm{dyn/cm^2}$, $\rho_\infty=0.95\times10^{-6}\,\mathrm{g/cm^3}$ and $M_\infty=20$, $p_\infty=0.122\times10^6\,\mathrm{dyn/cm^2}$, $\rho_\infty=0.192\times10^{-3}\,\mathrm{g/cm^3}$. This figure shows that the dissociation causes obvious difference between the frozen and equilibrium air flows for shocks and sonic lines. The thickness of shock layers for equilibrium air flow is smaller.

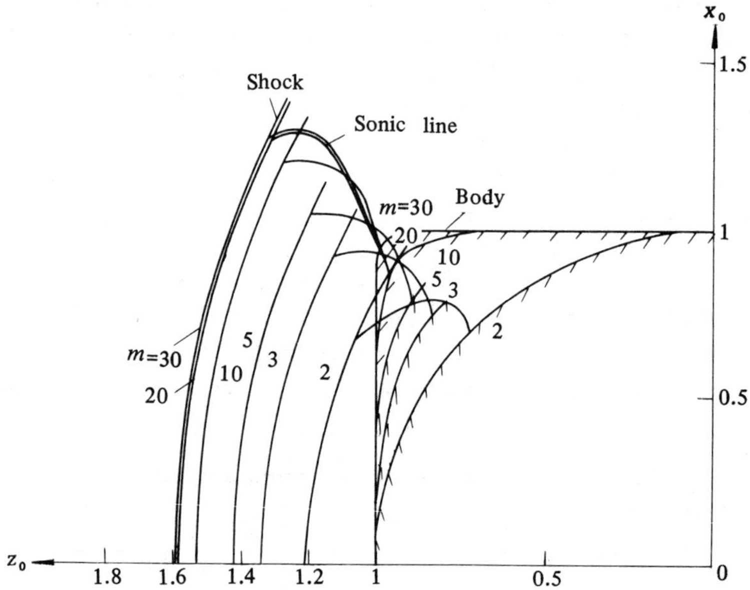

Figure 6.5 Shapes of shocks and locations of sonic lines ($M_\infty = 3$)

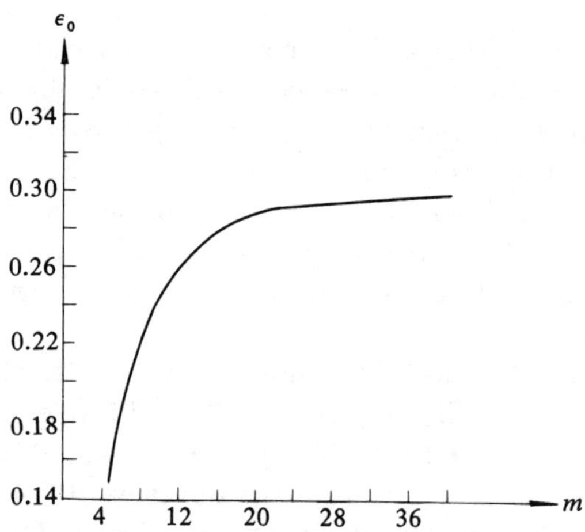

Figure 6.6 Variation of the stand-off distance of shocks at the stagnation point with m ($M_\infty = 3$)

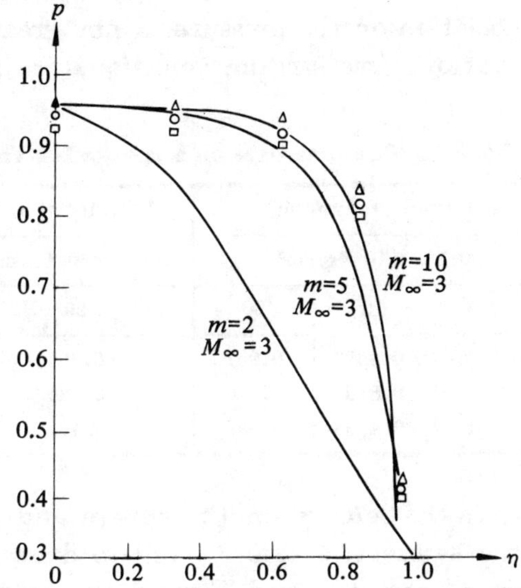

Figure 6.7 Pressure distributions on various bodies

○ $m=40$, $M_\infty=4$; △ $m=40$, $M_\infty=3$; □ $m=40$, $M_\infty=10$.

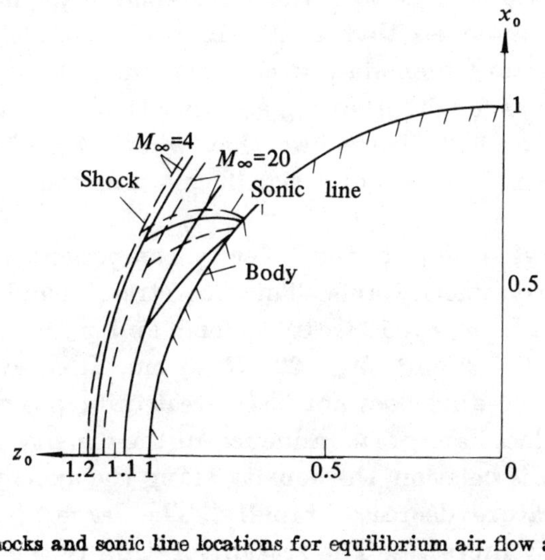

Figure 6.8 Shocks and sonic line locations for equilibrium air flow around spheres

---- Frozen air, —— Equilibrium air.

Fig. 6.2 demonstrates that as M_∞ increases the change of ε_0 is not monotonic for equilibrium air flow, and depends on the free stream conditions (p_∞, ρ_∞).

Fig. 6.9 gives the surface pressure on the sphere. It shows that the influence of dissociation on the pressure is not great. The pressures on the sphere for various free stream conditions are illustrated in Table 6.3.

Table 6.3 The pressures on a sphere for various η

p_∞		0.122×10^6 dyn/cm^2			0.103×10^5 dyn/cm^2	0.310×10^3 dyn/cm^2
ρ_∞		0.192×10^{-3} g/cm^3			0.117×10^{-4} g/cm^3	0.324×10^{-6} g/cm^3
M_∞		4	15	20	20	20
η_4	0	0.944	0.953	0.957	0.965	0.971
	0.355	0.818	0.815	0.818	0.824	0.823
	0.710	0.531	0.499	0.498	0.498	0.496

Fig. 6.10 gives the density on the sphere and Fig. 6.11 gives the density along the symmetric axis. Owing to dissociation, the density obviously increases. Fig. 6.2 and Figs. 6.12—6.17 show some results for nonequilibrium air flow around a sphere. For the model of nonequilibrium air flow, see § 5 of Chapter 4. Fig. 6.12 shows the shocks and sonic lines for nonequilibrium flow around a sphere. The free stream conditions are $p_\infty=0.947\times10^3$ dyn/cm^3, $\rho_\infty=0.123\times10^{-5}$ g/cm^3, $R=5$ cm, where R is the radius of the sphere. When $M_\infty>10$ the shock changes markedly, which is different from the case for frozen air flow. In addition, it is worthwhile to notice the special shape of sonic line at $M_\infty=20$. Fig. 6.2 indicates that as M_∞ increases, ε_0 changes markedly and a transition from frozen state to equilibrium state appears. Fig. 6.2 also gives the curves for different p_∞, showing that ε_0 changes markedly as p_∞ changes. Fig. 6.2 also shows that when $M_\infty>7$ the effect of nonequilibrium air flow becomes significant and must be taken into account.

Fig. 6.13 shows profiles of the velocity component v, pressure and density along the symmetric axis. The free stream conditions are $p_\infty=0.947\times10^3$ dyn/cm^2, $\rho_\infty=0.123\times10^{-5}$ g/cm^3 and $p_\infty=0.239\times10^3$ dyn/cm^2, $\rho_\infty=0.340\times10^{-6}$ g/cm^3, $M_\infty=22$, $R=5$ cm. It is evident that the nonequilibrium procedure does not have great influence on the pressure and velocity, but does have great influence on the density and the temperature. Owing to dissociation, the density along the axis increases rapidly while the temperature decreases rapidly. The larger p_∞ is, the more rapidly the density increases. The reason for this is the following: the rate of dissociation depends on the collision probability of particles, and

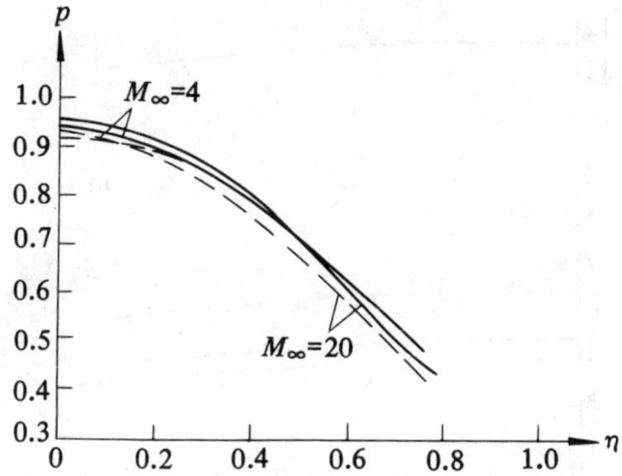

Figure 6.9 Pressure distributions on spheres (the equilibrium air flow)

---- Frozen air, —— Equilibrium air $\begin{cases} p_\infty = 0.122 \times 10^6 \text{ dyn/cm}^2, \\ \rho_\infty = 0.192 \times 10^{-3} \text{ g/cm}^3. \end{cases}$

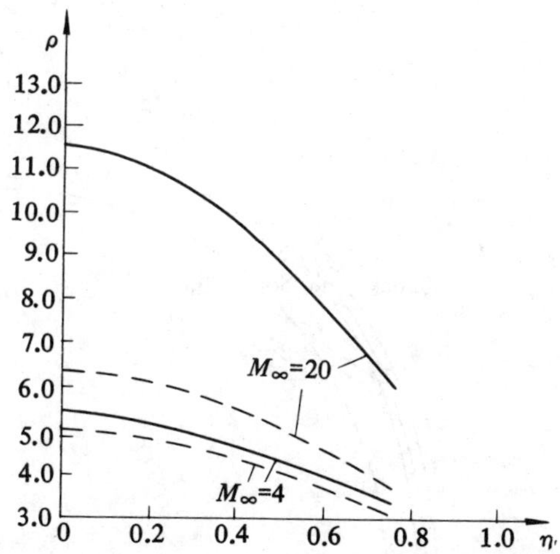

Figure 6.10 Density distributions on spheres (the equilibrium air flow)

---- Frozen air, —— Equilibrium air $\begin{cases} p_\infty = 0.122 \times 10^6 \text{ dyn/cm}^2, \\ \rho_\infty = 0.192 \times 10^{-3} \text{ g/cm}^3. \end{cases}$

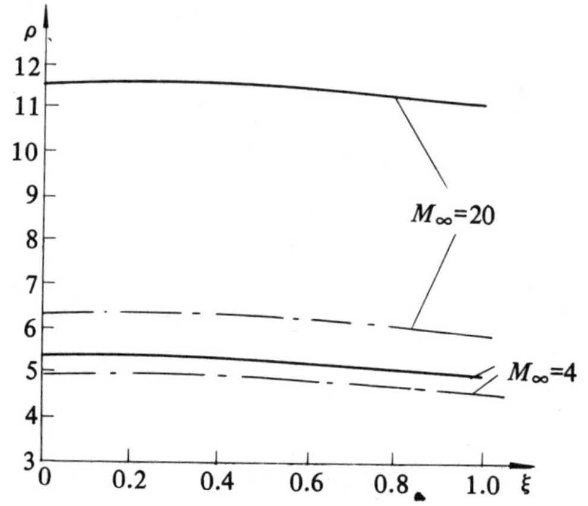

Figure 6.11 Density distributions along the symmetric axis (the equilibrium air flow around spheres)

—·— Frozen air, —— Equilibrium air $\begin{cases} p_\infty = 0.122 \times 10^6 \text{ dyn/cm}^2, \\ \rho_\infty = 0.192 \times 10^{-3} \text{ g/cm}^3. \end{cases}$

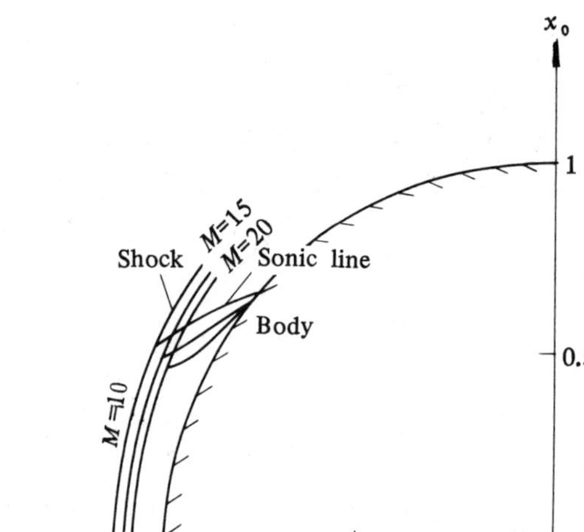

Figure 6.12 Shapes of shocks and sonic lines of nonequilibrium air flow around spheres

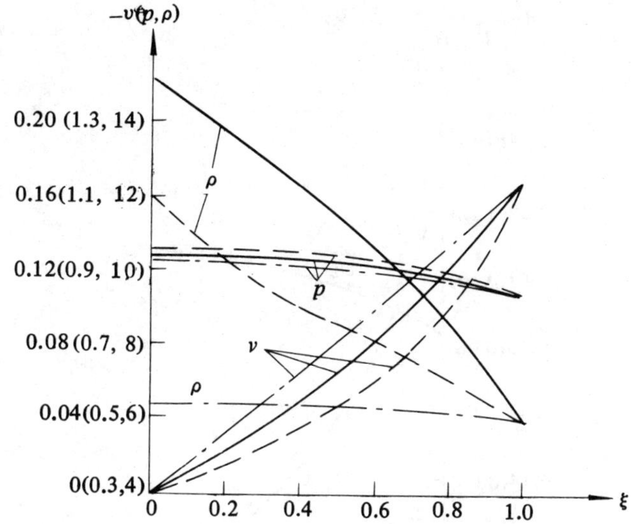

Figure 6.13 Distributions of velocity component v, pressure and density along the symmetric axis ($M_\infty=22$, $R=5\,\mathrm{cm}$)

—— $p_\infty=0.947\times10^3\ \mathrm{dyn/cm^2}$, $\rho_\infty=0.123\times10^{-5}\ \mathrm{g/cm^3}$,

- -·- $p_\infty=0.239\times10^3\ \mathrm{dyn/cm^2}$, $\rho_\infty=0.34\times10^{-6}\ \mathrm{g/cm^3}$,

—·— Frozen air.

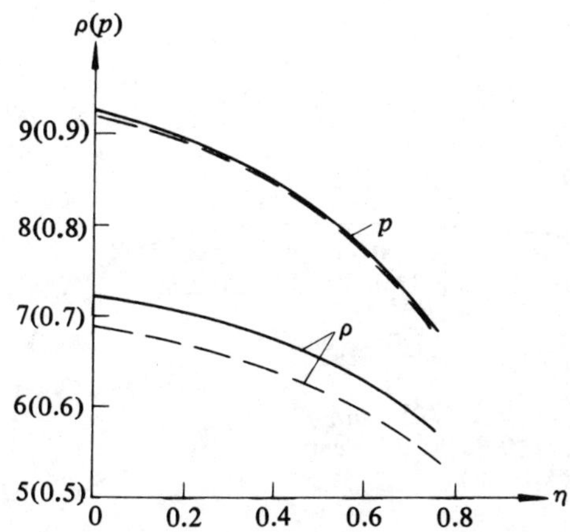

Figure 6.14 Pressure and density distributions on bodies ($M_\infty=10$, $p_\infty=10^{-3}\,\mathrm{atm}$, $T_\infty=250\,°\mathrm{K}$.)

—— Without coupling, ---- Coupling.

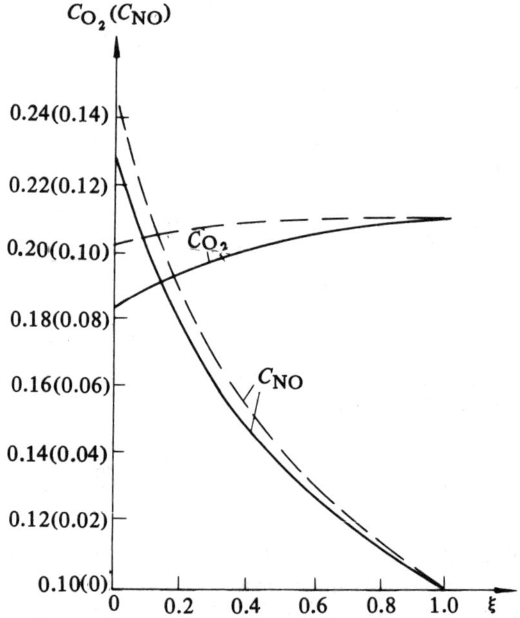

Figure 6.15 Distribution of the concentration of species along the symmetric axis ($M_\infty=10$)
—— Without coupling, ---- Coupling.

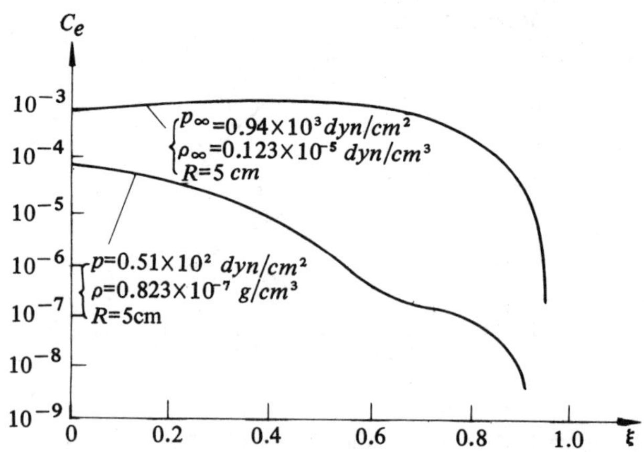

Figure 6.16 Electron density distribution along the symmetric axis ($M_\infty=22$)

the increase of p_∞ causes the increase of density behind shocks, resulting in an increase of collision probability of particles.

The positions of sonic points on the sphere are listed in Table 6.4 for different free stream conditions.

Table 6.4 Positions η_a of sonic points on a sphere

p_∞	0.947×10^3 dyn/cm²		0.8033×10^4 dyn/cm²
ρ_∞	0.123×10^{-5} g/cm³		0.1187×10^{-4} g/cm³
M_∞	10	22	22
η_a	40.98°	40.14°	41.37°

Figs. 6.14, 6.15 show profiles of the pressure and density, and concentrations of the molecular species O_2 and NO along the symmetric axis for the flow around a sphere with and without the couple of dissociation and vibration relaxation. The free stream conditions are $p_\infty=10^{-3}$ atm, $T_\infty=250\,°K$, $R=1.5$ cm, $M_\infty=10$. It is evident that the couple does not have great influence on the pressure, but does have great influence on the density and concentration of the species.

Fig. 6.16 shows the concentration of electrons along the axis for $M_\infty=22$ and under different p_∞ and ρ_∞.

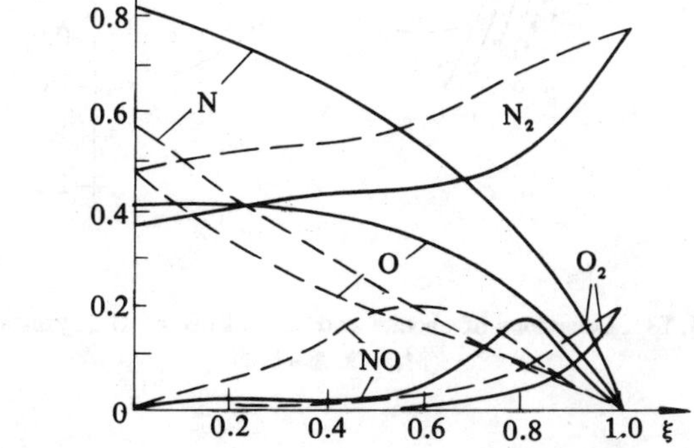

Figure 6.17 Distribution of the concentration of species along the symmetric axis
—— $p_\infty=0.947\times10^3$ dyn/cm², $\rho_\infty=0.123\times10^{-5}$ g/cm³, $R=5$cm,
---- $p_\infty=0.239\times10^3$ dyn/cm², $\rho_\infty=0.34\times10^{-6}$ g/cm³.

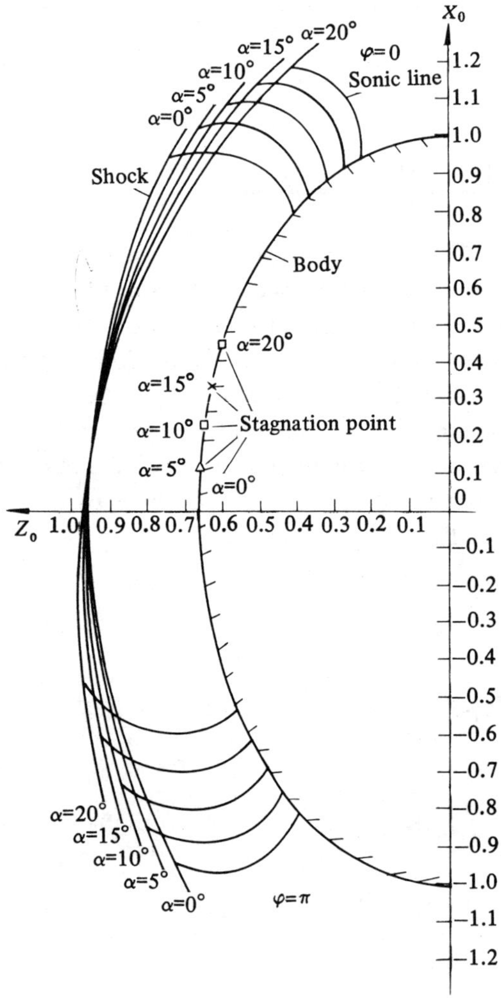

Figure 6.18 Locations of shocks and sonic lines on the symmetric plane (M_∞=3, δ=1.5)

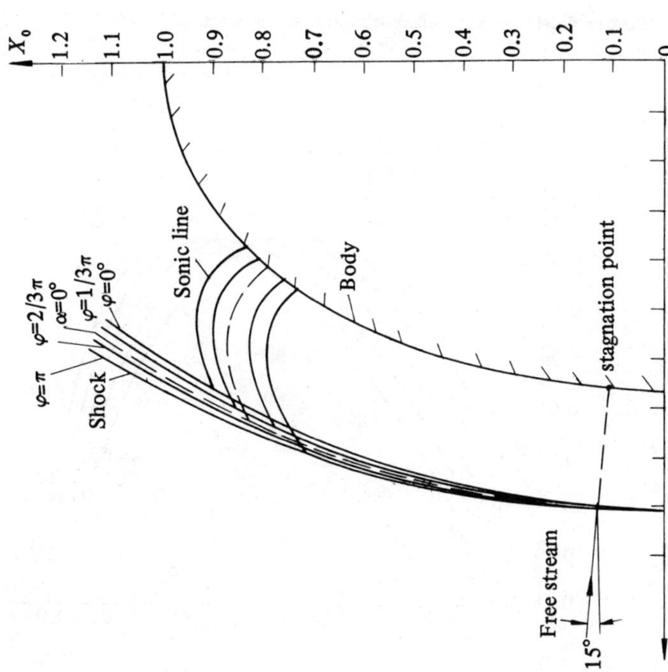

Figure 6.19 Shapes of shocks and sonic lines ($M_\infty=4$, $\delta=1.5$, $\alpha=5°$)

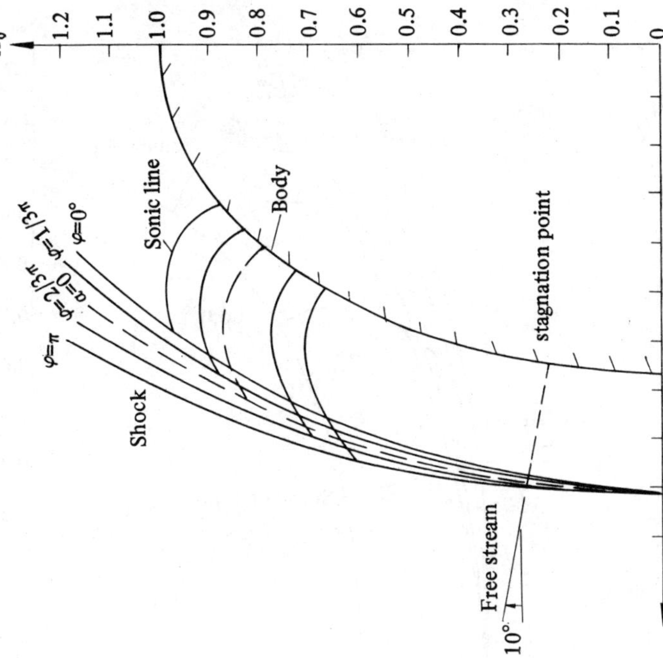

Figure 6.20 Shapes of shocks and sonic lines ($M_\infty=4$, $\delta=1.5$, $\alpha=10°$)

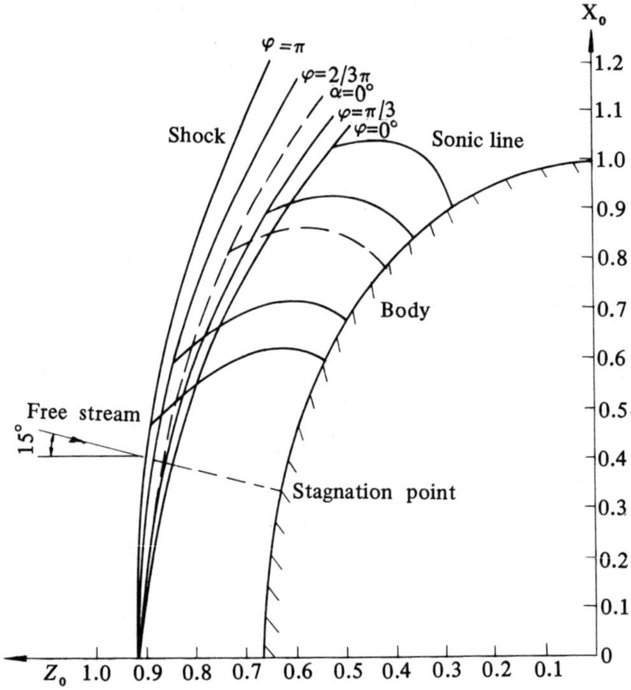

Figure 6.21 **Shapes of shocks and sonic lines** ($M_\infty=4$, $\delta=1.5$, $\alpha=15°$)

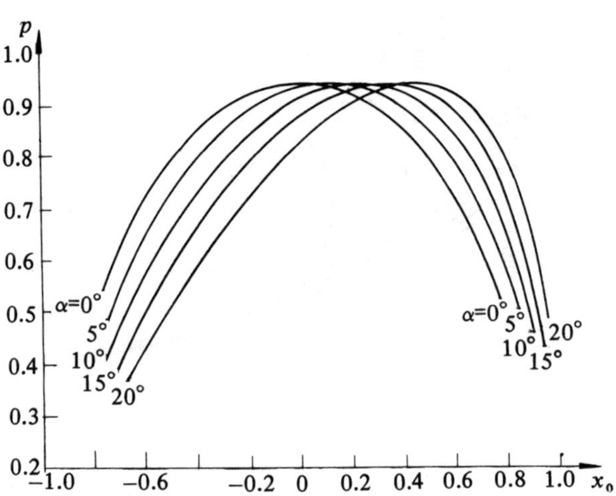

Figure 6.22 **Surface pressure distributions on the symmetric plane** ($M_\infty=4$, $\delta=1.5$)

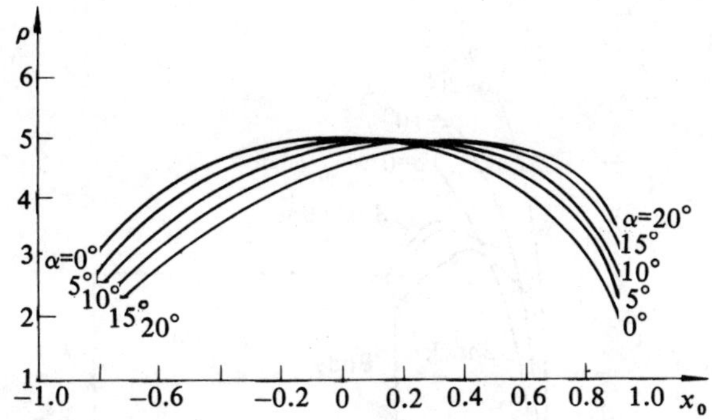

Figure 6.23 Variations of density along an ellipsoid ($M_\infty=4$, $\delta=1.5$)

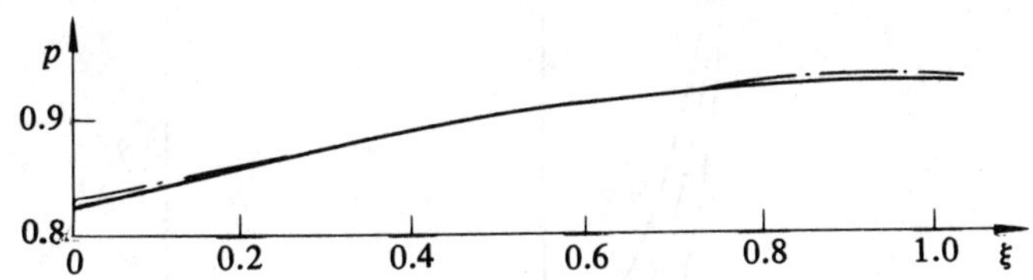

Figure 6.24 Pressure distributions along the axis $\theta=0$ ($M_\infty=4$, $\delta=1.5$)
—·— $\alpha=0°$, —— $\alpha=15°$.

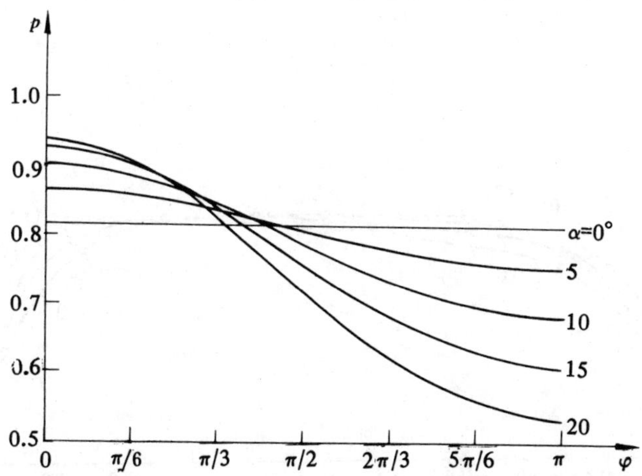

Figure 6.25 Variations of the surface pressure on bodies with φ
($M_\infty=4$, $\delta=1.5$, $\theta=0.3125$)

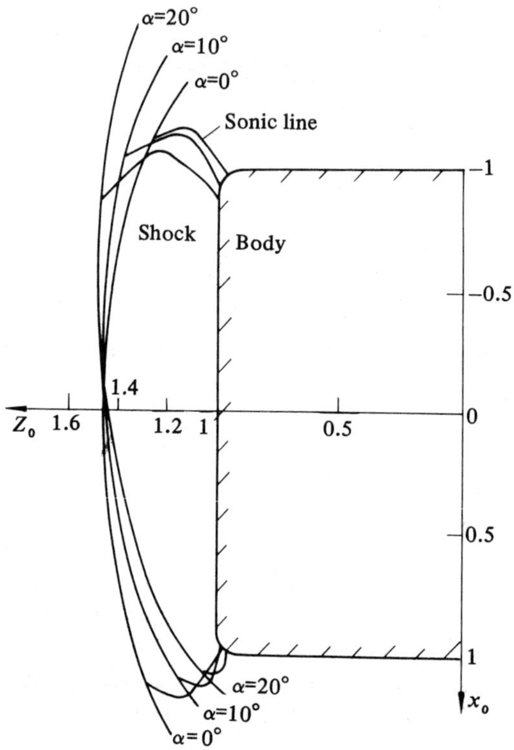

Figure 6.26 Shapes of shocks and sonic lines (M_∞=5.8, m=20)

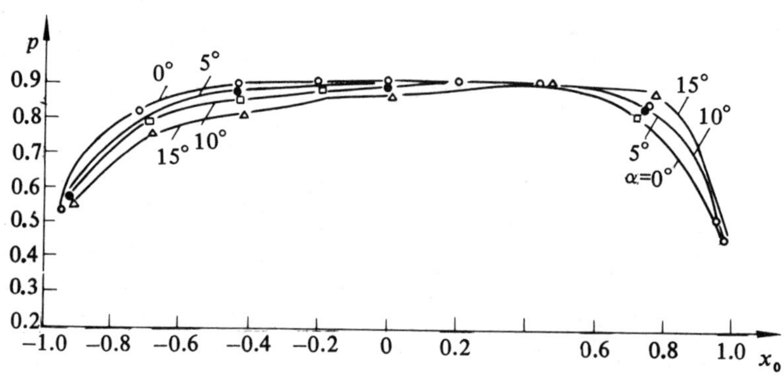

Figure 6.27 Pressure distribu ions along bodies (M_∞=5.8, m=20)

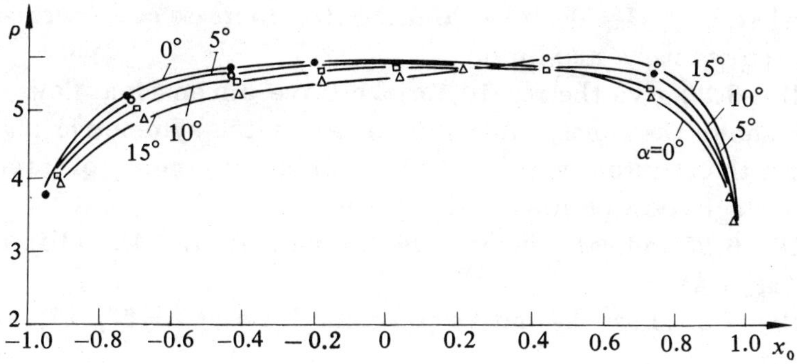

Figure 6.28 Density distributions along bodies (M_∞=5.8, m=20)

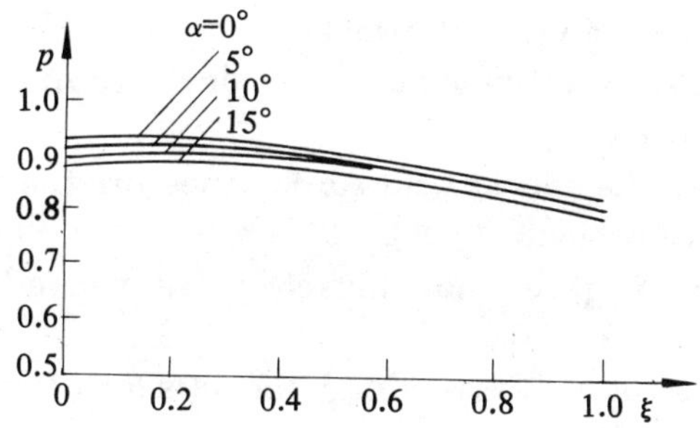

Figure 6.29 Variations of pressure along the axis $\theta=0$ (M_∞=5.8, m=20)

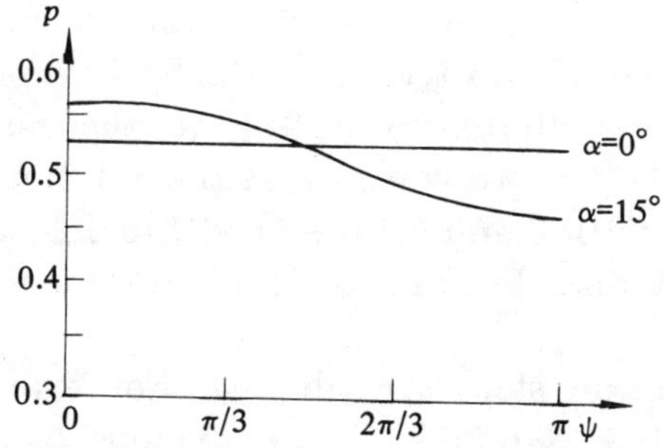

Figure 6.30 Variations of pressure with ψ (M_∞=5.8, m=20, θ=1.099)

Fig. 6.17 gives the concentrations of the species O_2, N_2, NO, O, N along the axis for $M_\infty=22$, $p_\infty=0.947\times10^3$ dyn/cm², $\rho_\infty=0.123\times10^{-5}$ g/cm³ and $p_\infty=0.239\times10^3$ dyn/cm², $\rho_\infty=0.34\times10^{-6}$ g/cm³, $R=5$ cm. It is easy to see that C_{O_2}, C_{N_2} decrease and C_O, C_N increase as ξ increases, while the C_{NO}–curve has a maximum.

Figs. 6.18—6.30 give the results for the three–dimensional flow.

Fig. 6.18 shows the shocks and sonic lines on the symmetric plane for flow around the ellipsoid with $\delta=1.5$ at different angles of attack ($M_\infty=3$). The stagnation points are also given.

Figs. 6.19—6.25 indicate the results for flow around the ellipsoid with $\delta=1.5$ ($M_\infty=4$).

Figs. 6.19—6.21 show the shocks and sonic lines at $\alpha=5°$, $10°$ and $15°$ respectively.

Figs. 6.22, 6.23 show the surface pressure and density on the symmetric plane at different angles of attack.

Fig. 6.24 gives the pressure along the axis $\theta=0$ at different angles of attack.

Fig. 6.25 shows the change of surface pressure on bodies with φ at $\theta=0.3125$ for different angles of attack.

Figs. 6.26—6.30 give the results for flow around bodies like disks with $m=20$ ($M_\infty=5.8$).

Fig. 6.26 shows the shocks and sonic lines on the symmetric plane at different angles of attack.

Figs. 6.27, 6.28 give the surface pressure and density on the symmetric plane.

Fig. 6.29 gives the pressure along the axis $\theta=0°$ at different angles of attack.

Fig. 6.30 shows the change of the surface pressure with ψ at $\theta=1.099$.

The results are checked by many ways, some of which are given below.

In our calculations, we have used different numbers of lines and different distributions of lines. Table 6.5 lists the results for the axisymmetric flow around a sphere at $M_\infty=6$, $\gamma=1.4$, $\eta_{max}=0.7505$. s represents the stand–off distance of shocks. "Equidistant" means that the lines are distributed equidistantly, and "Chebyshev zero points" means that the lines are distributed according to the zero points of the Chebyshev polynomial[A62]. It is seen that the relative difference of the two results obtained with 3 and 5 lines is within 1%, and the two results obtained by different distributions of lines are in agreement to three significant figures.

Different integration steps are adopted. For example, for the flow around a sphere with $M_\infty=6$ and $\gamma=1.4$, 10 steps, 20 steps and 80 steps are taken from the shock to the body. The relative difference of results obtained by using 10 steps and 20 steps is less than 0.3%. Moreover, the

Table 6.5

$N+1$ (Number of lines)			3 (Equidistant)	5 (Equidistant)	5 (Chebyshev zero points)
$\eta=0$	$\xi=1$	e	0.1486	0.1488	0.1488
		p	0.8300	0.8300	0.8300
		ρ	5.268	5.268	5.268
		u	0	0	0
		v	−0.1898	−0.1898	−0.1898
	$\xi=0$	p	0.9288	0.9288	0.9288
		ρ	5.709	5.709	5.709
		u	0	0	0
		v	0	0	0
$\eta=0.7505$	$\xi=1$	e	0.2093	0.2104	0.2104
		p	0.5710	0.5732	0.5732
		ρ	4.993	4.996	4.996
		u	0.5768	0.5747	0.5747
		v	−0.07579	−0.07473	−0.07475
	$\xi=0$	p	0.4673	0.4679	0.4678
		ρ	3.506	3.505	3.507
		u	0.4503	0.4519	0.4525
		v	0	0	0

Table 6.6

M_∞			4		20		∞	
			Explicit	Implicit	Explicit	Implicit	Explicit	Implicit
$\eta=0$	$\xi=1$	e	0.1754	0.1751	0.1304	0.1303	0.1286	0.1285
		p	0.8259	0.8259	0.8330	0.8330	0.8333	0.8333
		ρ	4.571	4.571	5.926	5.926	6.0	6.0
	$\xi=0$	p	0.9406	0.9405	0.9204	0.9205	0.9196	0.9196
		ρ	5.016	5.016	6.364	6.364	6.437	6.437
$\eta=0.7505$	ξ	e	0.2503	0.2501	0.1822	0.1821	0.1794	0.1793
		p	0.5885	0.5885	0.5604	0.5604	0.5591	0.5590
		u	0.5621	0.5621	0.5858	0.5859	0.5870	0.5871
	$\xi=0$	p	0.4913	0.4912	0.4497	0.4494	0.4478	0.4476
		u	0.4720	0.4705	0.4415	0.4321	0.4411	0.4302

results of 20 steps and 80 steps are in agreement to at least three significant figures. This also means that there is no need to worry about serious growth of rounding error.

Table 6.6 gives results obtained by using explicit and implicit schemes for frozen–air flow around a sphere with $\gamma=1.4$. We can see that both results are in good agreement with each other.

We have also checked the total energy on all nodes. For axisymmetric flow, the relative error of the total energy is within 1%. For three–dimensional flow, Table 6.7 gives the results for $M_\infty=4$, $\delta=1.5$ at different angles of attack. The relative error of the total energy is found to be less than 1% if $\alpha\leqslant 15°$. For bodies like disks, the error increases a little, e.g., for $M_\infty=5.8$, $m=20$, $\alpha=10°$, the relative error of the total energy reaches 3%.

Table 6.7 Error of the total energy ($M_\infty=4$, total energy=0.65625)

α	Maximal error of the total energy	Relative error
0°	0.0010	0.15%
5°	0.0019	0.30%
10°	0.0034	0.51%
15°	0.0051	0.78%
20°	0.0087	1.33%

Also, we have made use of the integral check, i.e., the check for accuracy of the integrals

$$\iint_\sigma \rho\boldsymbol{u}\cdot d\boldsymbol{\sigma}=0,$$

$$\iint_\sigma u_i\rho\boldsymbol{u}\cdot d\boldsymbol{\sigma}+\iint_\sigma p\boldsymbol{x}_i\, d\sigma=0 \quad (i=1,\ 2,\ 3),$$

$$\iint_\sigma p\rho^{-\gamma}\rho\boldsymbol{u}\cdot d\boldsymbol{\sigma}=0,$$

where $\boldsymbol{x}_i$ denotes the axial unit vectors in the Cartesian coordinates, u_i is the projection of vector $\boldsymbol{u}$ on $\boldsymbol{x}_i$, and $\boldsymbol{\sigma}$ is an arbitrary surface in the computational region. For $M_\infty=4$, $\delta=1.5$, $\alpha=10°$, $20°$, the results of the integrals are given in Table 6.8. This table shows that the results are correct.

The entropy on the nodes of bodies has also been calculated. Table 6.9 gives the results for $M_\infty=4$, $\delta=1.5$, $\alpha=0°$, $10°$, $20°$. It is seen that the error of the entropy on the body is only one on the third significant figure, if $\alpha\leqslant 15°$, and only three on the third significant figure if $\alpha=20°$.

For the explicit scheme, the quantities on a body are obtained by extrapolations. The three–point and two–point–extrapolations are used, and the extrapolations using different ξ points are also tested. For example, Table 6.10 gives results obtained by using $\xi=0.3$, 0.2, 0.1 and $\xi=0.2$, 0.1 and $\xi=0.3$, 0.2, 0.07 for equilibrium air flow around a sphere. The free stream conditions are $p_\infty=0.87\times10^3$ dyn/cm², $\rho_\infty=0.95\times10^{-6}$ g/cm³, $M_\infty=20$, $\eta_{max}=0.7505$. The method of extrapolation has

Table 6.8 Integral checks ($M_\infty=4,\ \delta=1.5,\ \bar{\delta}=0.6\alpha$)

α	Equations	Flux on the shock	Flux on the body	Flux on the cone with $\eta=\eta_{max}$	Sum
10°	Mass	-0.64327×2	0.00066×2	0.64517×2	0.00256×2
	Momentum in the X direction	0.04595×2	-0.00731×2	-0.03849×2	0.00015×2
	Momentum in the Y direction	0	0	0	0
	Momentum in the Z direction	0.67039×2	-0.32282×2	-0.34862×2	-0.00105×2
	Entropy	-0.05789×2	0.00006×2	0.05809×2	0.00026×2
20°	Mass	-0.63088×2	0.00084×2	0.63298×2	0.00294×2
	Momentum in the X direction	0.09011×2	-0.01579×2	-0.07398×2	0.00035×2
	Momentum in the Y direction	0	0	0	0
	Momentum in the Z direction	0.65271×2	-0.30987×2	-0.34423×2	-0.00139×2
	Entropy	-0.056379×2	0.00008×2	0.05663×2	0.00033×2

Table 6.9 Entropy on the nodes of bodies ($\ln p-\gamma\ln\rho$)
($M_\infty=4,\ \delta=1.5,\ \bar{\delta}=0.6\alpha$, entropy behind the normal shock $=-2.31904$)

θ	φ	α = 0°	10°	20°
0°	0	-2.31903	-2.31537	-2.30611
11°	0	-2.31853	-2.31346	-2.30865
	$\frac{\pi}{3}$	-2.31853	-2.31474	-2.30962
	$\frac{2\pi}{3}$	-2.31853	-2.31833	-2.31361
	π	-2.31853	-2.32091	-2.31858
22°	0	-2.31951	-2.32040	-2.32554
	$\frac{\pi}{3}$	-2.31951	-2.31942	-2.32117
	$\frac{2\pi}{3}$	-2.31951	-2.31939	-2.31759
	π	-2.31951	-2.32077	-2.32111
33°	0	-2.31723	-2.31276	-2.30283
	$\frac{\pi}{3}$	-2.31723	-2.31543	-2.31361
	$\frac{2\pi}{3}$	-2.31723	-2.31854	-2.31903
	π	-2.31723	-2.32045	-2.32421

Table 6.10 The surface pressure on a sphere by extrapolation

η	The ways of extrapolation		
	Three points (0.1)	Two points (0.1)	Three points (0.07)
0	0.9686	0.9691	0.9686
0.7505	0.4585	0.4589	0.4600

Table 6.11
($\delta=1.5$, $\gamma=1.4$, $\eta=0$, $M_\infty=4$)

	ε	ρ ($\xi=0$)	ρ ($\xi=1$)
M. I. R.*	0.158	5.0160	4.571
Method of lines	0.1589	5.0167	4.5714

* "M. I. R." stands for the method of integral relations.

Table 6.12
($\delta=1$, $\gamma=1.4$, $\eta=0$)

M_∞		ε	ρ ($\xi=0$)	ρ ($\xi=1$)
4	M. I. R. Method of lines	0.175 0.1754	5.016 5.016	4.572 4.571
10	M. I. R. Method of lines	0.136 0.136	6.153 6.153	5.714 5.714
∞	M. I. R. Method of lines	0.128 0.1286	6.438 6.437	6.0 6.0

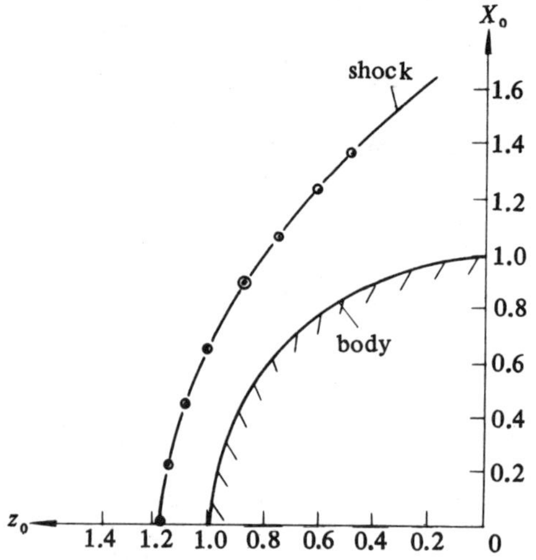

Figure 6.31 Shock for the frozen air flow around spheres ($M_\infty=4$, $\gamma=1.4$)
—— Calculation, ⊙ Experiment.

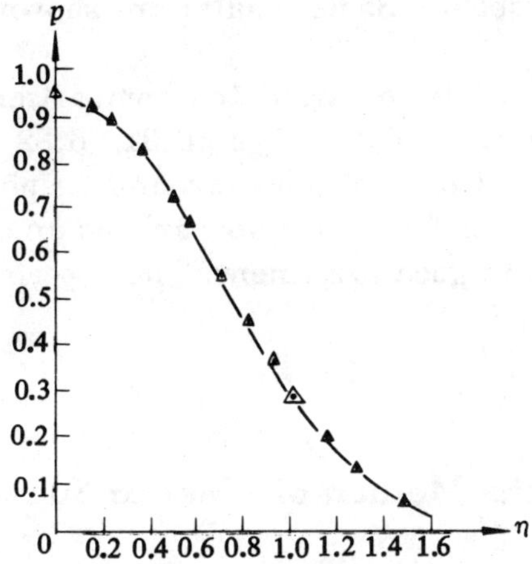

Figure 6.32 Pressure distribution along a sphere ($M_\infty=4$, $\gamma=1.4$)
—— Calculation, △ Experiment.

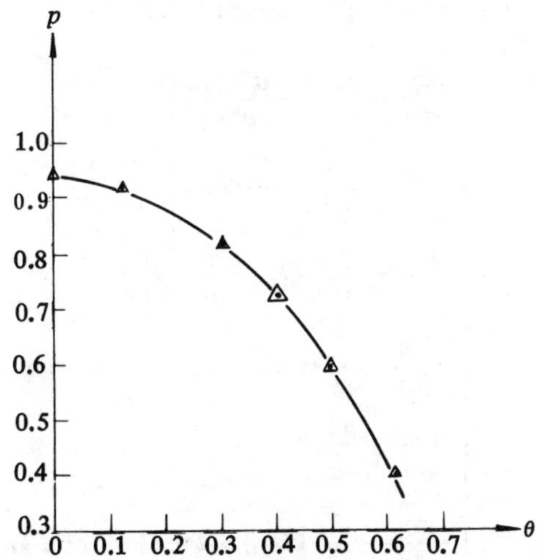

Figure 6.33 Pressure distribution along an ellipsoid
($M_\infty=4$, $\gamma=1.4$, $\alpha=0°$, $\delta=1.5$)
—— Calculation, △ Experiment.

no significant effect on the pressure.

The results have been compared with those obtained by Belotserkovskii[A6] who used the method of integral relations. For example, both results are in agreement to three significant figures for the ellipsoid with $\delta=1.5$ and for the sphere. Some results are shown in Tables 6.11 and 6.12.

Figs. 6.31—6.33 demonstrate the comparison between calculated results and experimental data. Figs. 6.31, 6.32 give the shock and surface pressures for frozen-air flow around a sphere. Fig. 6.33 shows the surface pressure for frozen-air flow around an ellipsoid. It is evident that the results are in good agreement. The experimental data are taken from [A20].

Appendix

Application of the Method of Lines to Supersonic Regions of Flow

Suppose that Π is a space-like surface, as shown in Fig. 2.1, and that the solution of the transonic flow before Π has been obtained. Then the problem of obtaining the solution for the supersonic flow behind Π is reduced to an initial-boundary-value problem of hyperbolic equations. In the following, we describe the method of lines for solving this problem.

Writing (2.1) in the explicit form for $\frac{\partial p}{\partial \eta}, \frac{\partial \rho}{\partial \eta}, \cdots, \frac{\partial e_{v_i}}{\partial \eta}$, we get

$$\left\{\begin{aligned} &\frac{\partial p}{\partial \eta}=\frac{G_5 u+a^2(G_4 u-\rho G_1)}{u^2-a^2}, \\ &\frac{\partial \rho}{\partial \eta}=\frac{1}{a^2 u}\left(-G_5+u\frac{\partial p}{\partial \eta}\right), \\ &\frac{\partial u}{\partial \eta}=\frac{1}{u}\left(G_1-\frac{1}{\rho}\frac{\partial p}{\partial \eta}\right), \\ &\frac{\partial v}{\partial \eta}=\frac{1}{u}G_2, \\ &\frac{\partial w}{\partial \eta}=\frac{1}{u}G_3, \\ &\frac{\partial c_i}{\partial \eta}=\frac{1}{u}G_{6,i}, \quad i=1,2,\cdots,l_1, \\ &\frac{\partial e_{v_i}}{\partial \eta}=\frac{1}{u}G_{7,i}, \quad i=1,2,\cdots,l_2, \end{aligned}\right. \tag{A-1}$$

where

$$
\begin{cases}
G_1=-h_s\left[\dfrac{c}{\varepsilon}\dfrac{\partial u}{\partial\xi}+\dfrac{\sigma}{\rho\varepsilon}\dfrac{\partial p}{\partial\xi}+\dfrac{w}{h_\varphi}\dfrac{\partial u}{\partial\psi}+a_s\right],\\
G_2=-h_s\left[\dfrac{c}{\varepsilon}\dfrac{\partial v}{\partial\xi}+\dfrac{\lambda}{\rho\varepsilon}\dfrac{\partial p}{\partial\xi}+\dfrac{w}{h_\varphi}\dfrac{\partial v}{\partial\psi}+a_n\right],\\
G_3=-h_s\left[\dfrac{c}{\varepsilon}\dfrac{\partial w}{\partial\xi}+\dfrac{\beta}{\rho\varepsilon}\dfrac{\partial p}{\partial\xi}+\dfrac{w}{h_\varphi}\dfrac{\partial w}{\partial\psi}+\dfrac{1}{\rho h_\varphi}\dfrac{\partial p}{\partial\psi}+a_\varphi\right],\\
G_4=-h_s\left[\dfrac{\rho}{\varepsilon}\left(\sigma\dfrac{\partial u}{\partial\xi}+\lambda\dfrac{\partial v}{\partial\xi}+\beta\dfrac{\partial w}{\partial\xi}\right)+\dfrac{c}{\varepsilon}\dfrac{\partial\rho}{\partial\xi}\right.\\
\qquad\left.+\dfrac{\rho}{h_\varphi}\dfrac{\partial w}{\partial\psi}+\dfrac{w}{h_\varphi}\dfrac{\partial\rho}{\partial\psi}+b\right],\\
G_5=-h_s\left[\dfrac{c}{\varepsilon}\dfrac{\partial p}{\partial\xi}+\dfrac{w}{h_\varphi}\dfrac{\partial p}{\partial\psi}-a^2\left(\dfrac{c}{\varepsilon}\dfrac{\partial\rho}{\partial\xi}+\dfrac{w}{h_\varphi}\dfrac{\partial\rho}{\partial\psi}\right)-g\right],\\
G_{6,i}=-h_s\left[\dfrac{c}{\varepsilon}\dfrac{\partial c_i}{\partial\xi}+\dfrac{w}{h_\varphi}\dfrac{\partial c_i}{\partial\psi}-\dfrac{w_i}{\rho}\right],\\
G_{7,i}=-h_s\left[\dfrac{c}{\varepsilon}\dfrac{\partial e_{v_i}}{\partial\xi}+\dfrac{w}{h_\varphi}\dfrac{\partial e_{v_i}}{\partial\psi}-E_i\right].
\end{cases}
$$

Suppose that the flow has a symmetric plane. As we do for the transonic region, we take k_1+1 planes in the ξ direction and k_2+1 planes in the ψ direction. The intersection of these planes forms $(k_1+1)\times(k_2+1)$ lines. The unknowns are the flow quantities on each line and the locations of the shock on each ψ plane. The total number is equal to $l(k_1+1)(k_2+1)+k_2+1$ $(l=5+l_1+l_2)$. In order to get the equations which the unknowns satisfy, we do the following.

Letting equations (2.1) hold for each inner line (i.e., $\xi\neq 0, 1$), we get $l(k_1-1)(k_2+1)$ equations. For the body, there exist the boundary conditions (2.3). In addition, we take $l-2$ compatibility relations on the stream surface and a compatibility relation on the second kind of wave–characteristic surface. The total number of equations is $l(k_2+1)$. For the shock, there exist the Rankine–Hugoniot relations (2.2). We take a compatibility relation on the first kind of wave–characteristic surface. Hence we get $(l+1)(k_2+1)$ equations. All together, we obtain $l(k_1+1)\times(k_2+1)+k_2+1$ equations. As we have done for the transonic flow, we use the interpolation polynomials to calculate the partial derivatives with respect to ξ and ψ. Then the problem under consideration is reduced to solving a system of ordinary differential and transcendental equations.

We now derive the equations on the boundaries. For simplicity, the chemical reaction equations and vibration relaxation equations are omitted.

For the body, (2.3) can be rewritten in the differential form

$$
\begin{cases}
\dfrac{\partial u}{\partial\eta}n_1+\dfrac{\partial v}{\partial\eta}n_2+\dfrac{\partial w}{\partial\eta}n_3=G,\\
G=-\left(u\dfrac{\partial n_1}{\partial\eta}+v\dfrac{\partial n_2}{\partial\eta}+w\dfrac{\partial n_3}{\partial\eta}\right),
\end{cases}
\tag{A-2}
$$

where

$$\boldsymbol{n}_b=\begin{pmatrix} n_1 \\ n_2 \\ n_3 \end{pmatrix}$$

is the outward unit normal vector to the body. In addition, we take three independent compatibility relations on the stream surface:

$$\begin{cases} u\cdot①+v\cdot②+w\cdot③=0, \\ t_1\cdot①+t_2\cdot②+t_3\cdot③=0, \\ ⑤=0, \end{cases} \tag{A-3}$$

where $\boldsymbol{t}=\begin{pmatrix} t_1 \\ t_2 \\ t_3 \end{pmatrix}=\frac{1}{V}\boldsymbol{V}\times\boldsymbol{n}_b$ and Ⓘ[1] represents the left-hand side of the i-th equation in (2.1). (The deduction of the compatibility relation on the stream surface can be found in (1.30), (1.37), (1.38) and (1.39) of Chapter 4.) Finally, we take a compatibility relation on the second kind of wave-characteristic surface. According to (1.31) and (1.40) of Chapter 4, the compatibility relation on the wave-characteristic surface is

$$\rho a(①\cdot N_1+②\cdot N_2+③\cdot N_3)-a^2\cdot④-⑤=0,$$

i.e.

$$\begin{aligned} \rho a\Big[\Big(u\frac{\partial u}{\partial\eta}+\frac{1}{\rho}\frac{\partial p}{\partial\eta}\Big)N_1+u\frac{\partial v}{\partial\eta}N_2+u\frac{\partial w}{\partial\eta}N_3\Big] \\ -a^2\Big(u\frac{\partial\rho}{\partial\eta}+\rho\frac{\partial u}{\partial\eta}\Big)-u\frac{\partial p}{\partial\eta}+a^2u\frac{\partial\rho}{\partial\eta} \\ =\rho a(G_1N_1+G_2N_2+G_3N_3)-a^2G_4-G_5=-F, \end{aligned} \tag{A-4}$$

where

$$\boldsymbol{N}=\begin{pmatrix} N_1 \\ N_2 \\ N_3 \end{pmatrix}$$

satisfies

$$\boldsymbol{V}\cdot\boldsymbol{N}=a,\quad |\boldsymbol{N}|=1. \tag{A-5}$$

The number of $\boldsymbol{N}$ satisfying the above conditions is infinite. In the following, let $\boldsymbol{N}$ be in the plane determined by some vector $\boldsymbol{m}$ and $\boldsymbol{V}$. In view of (A-5), $\boldsymbol{N}$ can be expressed as

$$\boldsymbol{N}=b_1\boldsymbol{m}+b_2\boldsymbol{V}, \tag{A-6}$$

where

1) Ⓘ=①, ②, ③, ④ or ⑤.

$$\begin{cases} b_1 = \pm\sqrt{\dfrac{V^2-a^2}{(\boldsymbol{m}\cdot\boldsymbol{m})V^2-(\boldsymbol{m}\cdot\boldsymbol{V})^2}}, \\ b_2 = \dfrac{a-b_1(\boldsymbol{m}\cdot\boldsymbol{V})}{V^2}, \\ V = |\boldsymbol{V}|. \end{cases}$$

For the body, we take $\boldsymbol{n}_b$ as $\boldsymbol{m}$. Therefore, noticing that $\boldsymbol{n}_b\cdot\boldsymbol{V}=0$, we get

$$\boldsymbol{N} = \sqrt{1-\frac{1}{M^2}}\,\boldsymbol{n}_b + \frac{1}{MV}\,\boldsymbol{V}. \tag{A-7}$$

With slight manipulation, we obtain the equations for the body

$$\begin{cases} u\dfrac{u}{V}\dfrac{\partial u}{\partial\eta}+u\dfrac{v}{V}\dfrac{\partial v}{\partial\eta}+u\dfrac{w}{V}\dfrac{\partial w}{\partial\eta}+\dfrac{u}{V}\dfrac{1}{\rho}\dfrac{\partial p}{\partial\eta} \\ \quad =\dfrac{1}{V}(G_1u+G_2v+G_3w)\equiv -H_1, \\ ut_1\dfrac{\partial u}{\partial\eta}+ut_2\dfrac{\partial v}{\partial\eta}+ut_3\dfrac{\partial w}{\partial\eta}+\dfrac{t_1}{\rho}\dfrac{\partial p}{\partial\eta} \\ \quad =G_1t_1+G_2t_2+G_3t_3\equiv -H_2, \\ un_1\dfrac{\partial u}{\partial\eta}+un_2\dfrac{\partial v}{\partial\eta}+un_3\dfrac{\partial w}{\partial\eta}=Gu\equiv -H_3, \\ -\rho a\sqrt{1-\dfrac{1}{M^2}}\dfrac{n_1}{\rho}\dfrac{\partial p}{\partial\eta}+a^2\left(u\dfrac{\partial\rho}{\partial\eta}+\rho\dfrac{\partial u}{\partial\eta}\right)+\left(u\dfrac{\partial p}{\partial\eta}-a^2u\dfrac{\partial\rho}{\partial\eta}\right) \\ \quad =-\left\{\rho a\sqrt{1-\dfrac{1}{M^2}}\left[H_3+(n_1G_1+n_2G_2+n_3G_3)\right]-a^2G_4\right\}+G_5 \\ \quad \equiv -(J_4+J_5), \\ u\dfrac{\partial p}{\partial\eta}-a^2u\dfrac{\partial\rho}{\partial\eta}=G_5\equiv -J_5. \end{cases} \tag{A-8}$$

Notice that the matrix

$$U=\begin{pmatrix} u/V & n_1 & t_1 \\ v/V & n_2 & t_2 \\ w/V & n_3 & t_3 \end{pmatrix}$$

is an orthogonal one. Thus multiplying the first three equations in (A-8) by u/V, t_1, n_1, respectively, and then adding them up, we obtain

$$u\frac{\partial u}{\partial\eta}+\frac{1}{\rho}(u^2/V^2+t_1^2)\frac{\partial p}{\partial\eta}=-[H_1u/V+H_2t_1+H_3n_1]\equiv -J_1.$$

Similarly, we get

$$\begin{cases} u\dfrac{\partial v}{\partial\eta}+\dfrac{1}{\rho}\left(\dfrac{uv}{V^2}+t_1t_2\right)\dfrac{\partial p}{\partial\eta} \\ \quad =-[H_1v/V+H_2t_2+H_3n_2]\equiv -J_2, \\ u\dfrac{\partial w}{\partial\eta}+\dfrac{1}{\rho}\left(\dfrac{uw}{V^2}+t_1t_3\right)\dfrac{\partial p}{\partial\eta} \\ \quad =-[H_1w/V+H_2t_3+H_3n_3]\equiv -J_3. \end{cases} \tag{A-9}$$

By using these relations, (A–8) can further be rewritten in the following form, which is convenient for practical calculation:

$$\begin{cases} \dfrac{\partial p}{\partial \eta}=\dfrac{-(J_4+J_5)u+a^2\rho J_1}{u\left(u-n_1 a\sqrt{1-\dfrac{1}{M^2}}\right)-a^2W_1}, \\ \dfrac{\partial \rho}{\partial \eta}=\dfrac{J_5+u\dfrac{\partial p}{\partial \eta}}{a^2 u}, \\ \dfrac{\partial u}{\partial \eta}=\dfrac{-J_1-\dfrac{W_1}{\rho}\dfrac{\partial p}{\partial \eta}}{u}, \\ \dfrac{\partial v}{\partial \eta}=\dfrac{-J_2-\dfrac{W_2}{\rho}\dfrac{\partial p}{\partial \eta}}{u}, \\ \dfrac{\partial w}{\partial \eta}=\dfrac{-J_3-\dfrac{W_3}{\rho}\dfrac{\partial p}{\partial \eta}}{u}, \end{cases} \tag{A-10}$$

where

$$\begin{cases} W_1=\dfrac{u^2}{V^2}+t_1^2, \\ W_2=\dfrac{uv}{V^2}+t_1t_2, \\ W_3=\dfrac{uw}{V^2}+t_1t_3. \end{cases} \tag{A-11}$$

On the shock, besides (2.2) we have the compatibility relation on the first kind of wave–characteristic surface. Now we take $\boldsymbol{n}_s$ as $\boldsymbol{m}$, where $\boldsymbol{n}_s$ is the outward unit normal vector to the shock. The compatibility relation is still (A–4), but $\boldsymbol{N}$ is as follows:

$$\begin{cases} \boldsymbol{N}=b_1\boldsymbol{n}_s+b_2\boldsymbol{V}, \\ b_1=-\sqrt{\dfrac{V^2-a^2}{(\boldsymbol{n}_s\cdot\boldsymbol{n}_s)V^2-(\boldsymbol{n}_s\cdot\boldsymbol{V})^2}}, \\ b_2=\dfrac{a-b_1(\boldsymbol{n}_s\cdot\boldsymbol{V})}{V^2}. \end{cases} \tag{A-12}$$

The flow quantities on the shock are determined by the normal to the shock and the direction of the free stream (suppose that $V_\infty=1$). The direction of the uniform free stream depends only on two parameters, i.e., among u_∞, v_∞, w_∞ only two are independent. Suppose that $m_3=-\dfrac{v_\infty}{u_\infty}$, $m_4=-\dfrac{w_\infty}{u_\infty}$ and $m_1=\dfrac{h_n}{h_s}f'_{1s}$, $m_2=\dfrac{h_n}{h_\varphi}f'_{1\varphi}$, where

$n=f_1(s, \varphi)$ is the equation of shock and the quantity with a prime denotes the derivative with respect to the subscript. Then every flow quantity q on the shock can be expressed as a function of m_1, m_2, m_3 and m_4, i.e.,

$$q=q(m_1, m_2, m_3, m_4). \tag{A-13}$$

Thus on the shock $\xi=1$,

$$\frac{\partial q}{\partial \eta}=\sum_{i=1}^{4} \frac{\partial q}{\partial m_i} \frac{\partial m_i}{\partial \eta}. \tag{A-14}$$

Correspondingly, (A-4) can be rewritten as

$$L_1 \frac{\partial m_1}{\partial \eta}+L_2 \frac{\partial m_2}{\partial \eta}+L_3 \frac{\partial m_3}{\partial \eta}+L_4 \frac{\partial m_4}{\partial \eta}=-F, \tag{A-15}$$

where

$$L_i=\rho a\left[\left(u \frac{\partial u}{\partial m_i}+\frac{1}{\rho} \frac{\partial p}{\partial m_i}\right)N_1+u \frac{\partial v}{\partial m_i} N_2+u \frac{\partial w}{\partial m_i} N_3\right]$$
$$-\rho a^2 \frac{\partial u}{\partial m_i}-u \frac{\partial p}{\partial m_i}, \quad i=1, 2, 3, 4. \tag{A-16}$$

Since m_3 and m_4 denote the direction of the free stream for the coordinates (s, n, φ), they are known. Therefore $\frac{\partial m_3}{\partial \eta}$, $\frac{\partial m_4}{\partial \eta}$ are also known. Moreover, we have

$$\frac{\partial m_2}{\partial \eta}=\frac{h_n}{h_\varphi} \frac{\partial (f'_{1\varphi})}{\partial \eta}+f'_{1\varphi} \frac{\partial\left(\frac{h_n}{h_\varphi}\right)}{\partial \eta}.$$

Since $f_1(s, \varphi)$ is independent of ξ, we further have

$$\frac{\partial (f'_{1\varphi})}{\partial \eta}=\frac{\partial (f'_{1\varphi})}{\partial s} \frac{\partial s}{\partial \eta}+\frac{\partial (f'_{1\varphi})}{\partial \varphi} \frac{\partial \varphi}{\partial \eta}=\frac{\partial (f'_{1\varphi})}{\partial s}=\frac{\partial (f'_{1s})}{\partial \varphi}.$$

If f_1, f'_{1s} for each φ are known, then $f'_{1\varphi}$, $\frac{\partial (f'_{1s})}{\partial \varphi}$ can be calculated by the interpolation polynomials and therefore $\frac{\partial m_2}{\partial \eta}$ is obtained. Clearly, the equation (A-15) can be rewritten as

$$\frac{\partial m_1}{\partial \eta}=\frac{-F-\sum_{i=2}^{4} L_i \frac{\partial m_i}{\partial \eta}}{L_1}. \tag{A-17}$$

In (A-17) only $\frac{\partial m_1}{\partial \eta}$ now is unknown. Hence (A-17) is convenient for practical use. In addition, we should point out that (A-17) is a second order differential equation for f_1. Therefore two initial conditions are required, i.e., f_1, f'_{1s} need to be given.

The interpolation polynomial is quite similar to that described for the transonic flow, and is omitted here.

The standard methods, such as the Runge–Kutta method, are used for integrating the ordinary equations (A–1), (A–10) and (A–17).

As pointed out above, (A–17) is a second order equation. In calculation it is usually reduced to two first order equations. If we now take m_1, f_1 as unknowns, then the two first order equations are (A–17) and

$$\frac{\partial f_1}{\partial s}=\frac{\partial f_1}{\partial \eta}=m_1\,\frac{h_s}{h_n}. \tag{A–18}$$

Chapter 6

Calculation of Supersonic Conical Flow

§ 1 Introduction

In this chapter we deal with the numerical calculation of supersonic conical flow. Since the late forties, many methods have been developed for this problem. They can be divided into two kinds, namely, the method of stabilization and the steady state method. The steady state method covers the finite difference method, the method of integral relations, the series expansion method, the method of lines and so on.

From § 1 of Chapter 4 we know that when the "spherical velocity" (i.e., the projection of velocity vector on the sphere surface centered at the vertex of cone) is less than the local sound speed in all the computational domain, the problem is reduced to the boundary-value problem of elliptical equations, and when the "spherical velocity" is greater than the local sound speed in a part of the domain, the problem is reduced to the boundary-value problem of mixed-type equations. Here we mainly discuss the application of the method of lines described in Chapter 3 to the conical flow and present some results.

We have carried out extensive calculations for supersonic conical flow around elliptical cones at angle of attack, where the frozen and equilibrium air flows are considered. When the "spherical velocity" is greater than the local sound speed, the calculation is limited to the windward region.

The accuracy of the results is determined in two different ways. One way is to determine the accuracy of the results through comparing those results obtained by using different numbers of lines or different integration steps or through verifying the validity of certain conditions, for example, the conservation law of total energy. The other way is to check the computed results through comparing our results with those obtained by other methods or with experimental data. The comparisons show that the method of lines gives us good results.

As to the improperly posed problem, our calculations show, as pointed out in § 5 of Chapter 3, that the growth of the rounding error is not a serious problem as long as a proper number of lines is taken.

§ 2 Formulation of Problems

1. *Governing Equations and Coordinates*

We consider the inviscid, non–heat conducting, steady state flow with chemical reactions. In orthogonal curvilinear coordinates the fluid dynamic equations have the form (1.11) of Chapter 4. Suppose we take cylindrical coordinates, as shown in Fig. 2.1.

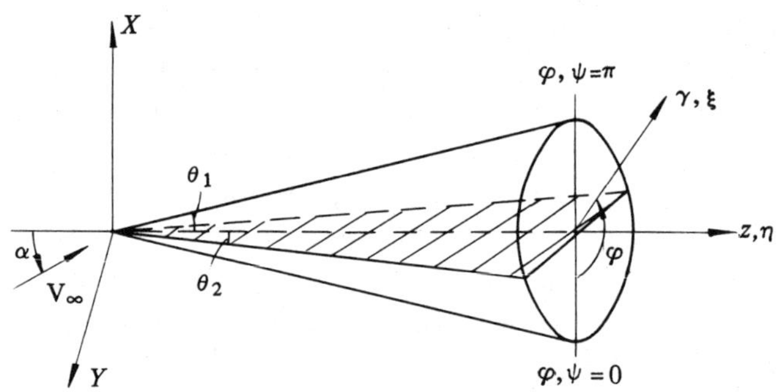

Figure 2.1

Thus the Lamé coefficients are $h_1=1$, $h_2=1$, $h_3=r$. For maintaining the conical property of flow, we shall not consider the nonequilibrium flow. From (1.11) and (1.12) of Chapter 4, the equations are as follows

$$\begin{cases} u\dfrac{\partial u}{\partial z}+v\dfrac{\partial u}{\partial r}+\dfrac{w}{r}\dfrac{\partial u}{\partial \varphi}+\dfrac{1}{\rho}\dfrac{\partial p}{\partial z}=0, \\ u\dfrac{\partial v}{\partial z}+v\dfrac{\partial v}{\partial r}+\dfrac{w}{r}\dfrac{\partial v}{\partial \varphi}+\dfrac{1}{\rho}\dfrac{\partial p}{\partial r}-\dfrac{w^2}{r}=0, \\ u\dfrac{\partial w}{\partial z}+v\dfrac{\partial w}{\partial r}+\dfrac{w}{r}\dfrac{\partial w}{\partial \varphi}+\dfrac{1}{\rho r}\dfrac{\partial p}{\partial \varphi}+\dfrac{vw}{r}=0, \\ \rho\left(\dfrac{\partial u}{\partial z}+\dfrac{\partial v}{\partial r}+\dfrac{1}{r}\dfrac{\partial w}{\partial \varphi}\right)+u\dfrac{\partial \rho}{\partial z}+v\dfrac{\partial \rho}{\partial r}+\dfrac{w}{r}\dfrac{\partial \rho}{\partial \varphi}+\dfrac{\rho v}{r}=0, \\ u\left(\dfrac{\partial p}{\partial z}-a^2\dfrac{\partial \rho}{\partial z}\right)+v\left(\dfrac{\partial p}{\partial r}-a^2\dfrac{\partial \rho}{\partial r}\right)+\dfrac{w}{r}\left(\dfrac{\partial p}{\partial \varphi}-a^2\dfrac{\partial \rho}{\partial \varphi}\right)=0, \end{cases} \tag{2.1}$$

where u, v, w stand for the components of velocity $\boldsymbol{V}$ in z, r, φ directions, respectively, p stands for pressure, ρ density, and a local sound speed.

Suppose that the vertex of the cone is coincident with the origin of the coordinate system. Owing to the conical property of flow, the body and shock can be expressed by $r=zf_2(\varphi)$ and $r=zf_1(\varphi)$, respectively, where f_1 and f_2 depend only on φ. For convenience, we introduce the following coordinate transformation:

$$\begin{cases} \eta = z, \\ \xi = \dfrac{r - zf_2(\varphi)}{z(f_1(\varphi) - f_2(\varphi))}, \\ \psi = \varphi. \end{cases}$$

Under the transformation, the shock and body lie on planes $\xi = 1$ and $\xi = 0$, respectively. As pointed out in Chapter 4, the equations can now be written in the form of (1.16). Since

$$\begin{cases} \dfrac{\partial \eta}{\partial z} = 1, & \dfrac{\partial \xi}{\partial z} = -\dfrac{(\xi(f_1 - f_2) + f_2)}{z(f_1 - f_2)}, & \dfrac{\partial \psi}{\partial z} = 0, \\ \dfrac{\partial \eta}{\partial r} = 0, & \dfrac{\partial \xi}{\partial r} = \dfrac{1}{z(f_1 - f_2)}, & \dfrac{\partial \psi}{\partial r} = 0, \\ \dfrac{\partial \eta}{\partial \varphi} = 0, & \dfrac{\partial \xi}{\partial \varphi} = -\dfrac{z(\xi(f_1' - f_2') + f_2')}{z(f_1 - f_2)}, & \dfrac{\partial \psi}{\partial \varphi} = 1, \end{cases}$$

and

$$\frac{\partial u}{\partial \eta} = \frac{\partial v}{\partial \eta} = \frac{\partial w}{\partial \eta} = \frac{\partial p}{\partial \eta} = \frac{\partial \rho}{\partial \eta} = 0,$$

the conical flow equations in coordinates (ξ, η, ψ) should be

$$\begin{cases} \dfrac{c}{\varepsilon}\dfrac{\partial u}{\partial \xi} + \dfrac{\sigma}{\rho v}\dfrac{\partial p}{\partial \xi} + \dfrac{w}{r}\dfrac{\partial u}{\partial \psi} = 0, \\ \dfrac{c}{\varepsilon}\dfrac{\partial v}{\partial \xi} + \dfrac{1}{\varepsilon\rho}\dfrac{\partial p}{\partial \xi} + \dfrac{w}{r}\dfrac{\partial v}{\partial \psi} - \dfrac{w^2}{r} = 0, \\ \dfrac{c}{\varepsilon}\dfrac{\partial w}{\partial \xi} + \dfrac{\beta}{\varepsilon\rho}\dfrac{\partial p}{\partial \xi} + \dfrac{w}{r}\dfrac{\partial w}{\partial \psi} + \dfrac{1}{\rho r}\dfrac{\partial p}{\partial \psi} + \dfrac{vw}{r} = 0, \\ \dfrac{\rho}{\varepsilon}\left(\sigma\dfrac{\partial u}{\partial \xi} + \dfrac{\partial v}{\partial \xi} + \beta\dfrac{\partial w}{\partial \xi}\right) + \dfrac{c}{\varepsilon}\dfrac{\partial \rho}{\partial \xi} + \dfrac{\rho}{r}\dfrac{\partial w}{\partial \psi} + \dfrac{w}{r}\dfrac{\partial \rho}{\partial \psi} + \dfrac{\rho v}{r} = 0, \\ \dfrac{c}{\varepsilon}\left(\dfrac{\partial p}{\partial \xi} - a^2\dfrac{\partial \rho}{\partial \xi}\right) + \dfrac{w}{r}\left(\dfrac{\partial p}{\partial \psi} - a^2\dfrac{\partial \rho}{\partial \psi}\right) = 0, \end{cases} \tag{2.2}$$

where

$$\begin{cases} \sigma = -[\xi(f_1 - f_2) + f_2] \\ \beta = -\dfrac{z}{r}[\xi(f_1' - f_2') + f_2'], \\ \varepsilon = z(f_1 - f_2), \\ c = \sigma u + v + \beta w, \\ r = z[\xi(f_1 - f_2) + f_2]. \end{cases} \tag{2.3}$$

The quantities with a prime denote the derivatives with respect to φ.

2. *Boundary Conditions*

From (2.4) of Chapter 4, it is known that at the shock $r = zf_1(\varphi)$ there are the Rankine–Hugoniot conditions

$$\begin{cases} u=u_\infty-\left(1-\dfrac{\rho_\infty}{\rho}\right)n_1V_{\infty,N}, \\ v=v_\infty-\left(1-\dfrac{\rho_\infty}{\rho}\right)n_2V_{\infty,N}, \\ w=w_\infty-\left(1-\dfrac{\rho_\infty}{\rho}\right)n_3V_{\infty,N}, \\ p+\rho V_N^2=p_\infty+\rho_\infty V_{\infty,N}^2, \\ h+\dfrac{1}{2}V_N^2=h_\infty+\dfrac{1}{2}V_{\infty,N}^2, \end{cases} \tag{2.4}$$

where n_1, n_2, n_3 are components of the unit normal to the shock in z, r, φ directions, i.e.,

$$N=\begin{pmatrix} n_1 \\ n_2 \\ n_3 \end{pmatrix}=\frac{1}{\sqrt{1+f_1^2+\left(\dfrac{f_1'}{f_1}\right)^2}}\begin{pmatrix} -f_1 \\ 1 \\ -\dfrac{f_1'}{f_1} \end{pmatrix},$$

and $V_{\infty,N}=\boldsymbol{V}_\infty\cdot\boldsymbol{N}$, $V_N=\boldsymbol{V}\cdot\boldsymbol{N}$. The quantities with subscript ∞ represent those before the shock and the quantities without subscript represent those behind the shock. h is the specific enthalpy.

According to (3.1) of Chapter 4, at the body $r=zf_2(\varphi)$ the zero flux condition must be satisfied, i.e.,

$$q=-uf_2+v-w\frac{f_2'}{f_2}=0. \tag{2.5}$$

All the equations described above are nondimensionized. The dimensionless manner is V, u, v, $w\sim V_\infty$, $p\sim\rho_\infty V_\infty^2$, $\rho\sim\rho_\infty$, $h\sim V_\infty^2$, $r,z\sim R_0$. Without loss of generality, we assume that $z=1$ in the following.

3. *Formulation of Problems*

To solve supersonic conical flow is to solve equations (2.2) with boundary conditions (2.4), (2.5) in the domain L: $0\leqslant\xi\leqslant1$, $-\pi\leqslant\psi\leqslant\pi$. In the following, we consider only two cases:

Firstly, the solution is smooth in the domain L except for the singularity of entropy. If the "spherical velocity" is less than the local sound speed in the whole domain, the solution usually belongs to this case. In this case, equations (2.2) are elliptic. Therefore the problem to be solved is reduced to a boundary-value problem of elliptical equations. Of course, here is also included the case where the solution is still smooth although the equations have become mixed-type.

Secondly, the solution is no longer smooth in the whole domain L, but there is a subdomain L_1: $0\leqslant\xi\leqslant1$, $\psi_1\leqslant\psi\leqslant\psi_2$, in which the solution is smooth and can be solved alone. For example, it is the case that the "spherical velocity" is greater than the local sound speed in a part of domain L_1, the lines $\psi=\psi_1$ and $\psi=\psi_2$ are space-like curves (i.e., w is greater than local sound speed at these curves), and the fluid flows out of L_1. In fact, this case is similar to that for two-dimensional supersonic flow around blunt bodies. The problem in this case is reduced to the boundary-value problem of mixed-type equations.

In short, the problems to be solved are the boundary-value problems of elliptical or mixed-type equations with a smooth solution. Clearly, the formulation of the problem is coincident with that described in Chapter 3. Therefore, the method discussed in Chapter 3 can be used here.

§ 3 Methods of Solution

Here only the explicit scheme of the method of lines described in Chapter 3 is discussed.

Suppose that the flow is symmetrical with the planes $\psi=0$ and $\psi=\pi$. Therefore, what we need to do is to solve the problem for $0\leqslant\psi\leqslant\psi_{max}\leqslant\pi$. For the first case stated above $\psi_{max}=\pi$, and for the second $\psi_{max}=\psi_2$.

The procedure of calculation can be outlined as follows: We introduce some lines $\psi=\psi_i$=constant into the domain. For a fixed ξ, we construct interpolation polynomials in terms of functional values on the lines, and hence the corresponding derivatives with respect to ψ can be determined by the polynomials. Let equations. (2.2) hold on the lines. Then we obtain a boundary-value problem of ordinary differential equations. For this problem the iteration method can be used. If a shock is given the flow quantities behind the shock can be obtained by (2.4). Then we start integrating the ordinary differential equations. After getting flow quantities at the body, we check whether the quantities satisfy the boundary condition (2.5). If it is not satisfied, then we adjust the shock and repeat the above procedure until (2.5) is satisfied within a we tolerant error.

1. *Scheme*

Rewriting equations. (2.2) in the explicit form for $\frac{\partial u}{\partial \xi}, \frac{\partial v}{\partial \xi}, \cdots, \frac{\partial \rho}{\partial \xi}$, obtain the following equations

$$
\left\{
\begin{aligned}
&\frac{\partial p}{\partial \xi}=\frac{a^2[F_4c-\rho(\sigma F_1+F_2+\beta F_3)]+F_5c}{c^2-\tau^2a^2}\\
&\frac{\partial \rho}{\partial \xi}=\frac{1}{a^2}\left[-\frac{F_5}{c}+\frac{\partial p}{\partial \xi}\right],\\
&\frac{\partial u}{\partial \xi}=\frac{1}{c}\left[F_1-\frac{\sigma}{\rho}\frac{\partial p}{\partial \xi}\right],
\end{aligned}
\right.
\tag{3.1}
$$

$$\begin{cases} \dfrac{\partial v}{\partial \xi}=\dfrac{1}{c}\left[F_2-\dfrac{1}{\rho}\dfrac{\partial p}{\partial \xi}\right], \\ \dfrac{\partial w}{\partial \xi}=\dfrac{1}{c}\left[F_3-\dfrac{\beta}{\rho}\dfrac{\partial p}{\partial \xi}\right], \end{cases}$$

where

$$\begin{cases} \tau^2=1+\sigma^2+\beta^2, \\ F_1=-\dfrac{\varepsilon}{\tau}w\dfrac{\partial u}{\partial \psi}, \\ F_2=-\dfrac{\varepsilon}{\tau}\left(w\dfrac{\partial v}{\partial \psi}-w^2\right), \\ F_3=-\dfrac{\varepsilon}{\tau}\left(w\dfrac{\partial w}{\partial \psi}+\dfrac{1}{\rho}\dfrac{\partial p}{\partial \psi}+vw\right), \\ F_4=-\dfrac{\varepsilon}{\tau}\left(\rho\dfrac{\partial w}{\partial \psi}+w\dfrac{\partial \rho}{\partial \psi}+\rho v\right), \\ F_5=-\dfrac{\varepsilon w}{\tau}\left(\dfrac{\partial p}{\partial \psi}-a^2\dfrac{\partial \rho}{\partial \psi}\right). \end{cases} \tag{3.2}$$

For the integration of equations (3.1), the standard explicit method, e.g., the Runge–Kutta method, can be used.

2. *Equations on the Symmetrical Planes*

In order to increase the accuracy of the approximation for w, the interpolation polynomial with derivatives $(w')_{0,\pi}$ is used. $(w')_{0,\pi}$ denote the values of derivative w' on the planes $\psi=0$ and $\psi=\pi$. Obviously, in order to realize this we have to deduce the equations for $\left(\dfrac{\partial w'}{\partial \xi}\right)_{0,\pi}$. Differentiating the fifth equation of (3.1) with respect to ψ, setting $\psi=0$, π, and noticing $w_{0,\pi}=\varepsilon'_{0,\pi}=(f'_2)_{0,\pi}=p'_{0,\pi}=0$, we get the following equations

$$\left(\frac{\partial w'}{\partial \xi}\right)_{0,\pi}=\frac{1}{c}\left[\left(\frac{\partial F_3}{\partial \psi}\right)_{0,\pi}-\frac{\left(\dfrac{\partial \beta}{\partial \psi}\right)_{0,\pi}}{\rho}\frac{\partial p}{\partial \xi}\right], \tag{3.3}$$

where

$$\begin{cases} \left(\dfrac{\partial F_3}{\partial \psi}\right)_{0,\pi}=-\dfrac{\varepsilon}{\tau}\left[(w'_{0,\pi})^2+\dfrac{1}{\rho}p''+vw'_{0,\pi}\right], \\ \left(\dfrac{\partial \beta}{\partial \psi}\right)_{0,\pi}=-\dfrac{1}{\tau}[\xi(f''_1-f''_2)+f''_2], \end{cases} \tag{3.4}$$

and the quantities with a single prime denote the first–order derivatives with respect to ψ and those with a double prime the second–order derivatives.

3. *Interpolation Polynomials*

As pointed out at the beginning of this section. For a fixed ξ the interpolation polynomials are constructed in terms of functional values on the lines and the derivatives with respect to ψ are determined by polynomials. However, when concrete polynomials are constructed two ways are used according to the variation of functions with ψ.

When the variation of functions with ψ is smooth, fewer lines are taken and the global interpolation polynomials are used in the whole domain. Owing to periodicity of functions with ψ, the trigonometric interpolation polynomials are constructed. Since u, v, $w/\sin\psi$, p, ρ, f_1 belong to even functions, we let their polynomials have the following form

$$g=\sum_{i=0}^{n} a_i \cos^i \psi.$$

We need to point out that since

$$\frac{w}{\sin\psi}\bigg|_{\psi=0}=\left(\frac{\partial w}{\partial\psi}\right)_0=w'_0,$$

$$\frac{w}{\sin\psi}\bigg|_{\psi=\pi}=-\left(\frac{\partial w}{\partial\psi}\right)_\pi=-w'_\pi,$$

the polynomials with derivative are actually used for w. In order to save running time the combination coefficients are computed first, as described in Chapter 3. For the present, we have

$$\begin{cases} g(\psi)=(\boldsymbol{M}^{*-1}\boldsymbol{d}_0)^*\boldsymbol{g}, \\ g'(\psi)=(\boldsymbol{M}^{*-1}\boldsymbol{d}_1)^*\boldsymbol{g}, \\ g''(\psi)=(\boldsymbol{M}^{*-1}\boldsymbol{d}_2)^*\boldsymbol{g}, \end{cases} \tag{3.5}$$

where

$$\begin{cases} \boldsymbol{M}^*=\begin{bmatrix} 1 & 1 & \cdots & 1 \\ \cos\psi_0 & \cos\psi_1 & \cdots & \cos\psi_n \\ \cos^2\psi_0 & \cos^2\psi_1 & \cdots & \cos^2\psi_n \\ \vdots & \vdots & & \vdots \\ \cos^n\psi_0 & \cos^n\psi_1 & \cdots & \cos^n\psi_n \end{bmatrix}, \\ \boldsymbol{d}_0^*=(1,\ \cos\psi,\ \cos^2\psi,\ \cdots,\ \cos^n\psi), \\ \boldsymbol{d}_1^*=\dfrac{d\boldsymbol{d}_0^*}{d\psi}=(0,\ -\sin\psi,\ -2\cos\psi\sin\psi,\ \cdots, \\ \qquad -n\cos^{n-1}\psi\sin\psi), \\ \boldsymbol{d}_2^*=\dfrac{d^2\boldsymbol{d}_0^*}{d\psi^2}=(0,\ -\cos\psi,\ 2(\sin^2\psi-\cos^2\psi),\ \cdots, \\ \qquad \cdots n(n-1)\cos^{n-2}\psi\sin^2\psi-n\cos^n\psi), \\ \boldsymbol{g}^*=(g_0,\ g_1,\cdots,\ g_n). \end{cases} \tag{3.6}$$

It is therefore straightforward to calculate

$$\frac{\partial u}{\partial\psi},\ \frac{\partial v}{\partial\psi},\ \frac{\partial p}{\partial\psi},\ \frac{\partial^2 p}{\partial\psi^2},\ \frac{\partial\rho}{\partial\psi},\ \frac{\partial f_1}{\partial\psi},\ \frac{\partial^2 f_1}{\partial\psi^2}$$

by these formulas. As to $\frac{\partial w}{\partial \psi}$ it can be calculated in the following way. Since

$$\frac{\partial w}{\partial \psi}=\sin \psi \frac{\partial\left(\frac{w}{\sin \psi}\right)}{\partial \psi}+\frac{w}{\sin \psi} \cos \psi,$$

then

$$\frac{\partial w}{\partial \psi}=[M^{*-1}(\cos \psi \boldsymbol{d}_0+\sin \psi \boldsymbol{d}_1)]^* \tilde{\boldsymbol{w}},$$

where

$$\tilde{\boldsymbol{w}}^*=\left(\frac{w_0}{\sin \psi_0}, \frac{w_1}{\sin \psi_1}, \dots, \frac{w_n}{\sin \psi_n}\right),$$

$\left(\frac{\partial w / \partial \psi}{\cos \psi}\right)_i$ being used instead of $\frac{w_i}{\sin \psi_i}$ if $\sin \psi_i=0$.

When functions vary rapidly with ψ, it is not desirable to construct polynomials in the above way. In this case more lines are needed and local interpolation polynomials are constructed. For example, a five-point central difference can be used to calculate the derivative with respect to ψ, that is, the derivative at the i-th line can be expressed by

$$g_i'(\psi)=\frac{4}{3}\left[\frac{g_{i+1}-g_{i-1}}{2 \Delta \psi}\right]-\frac{1}{3}\left[\frac{g_{i+2}-g_{i-2}}{4 \Delta \psi}\right]. \tag{3.7}$$

Of course, when $i=0, 1, n-1, n$, it is necessary to use the symmetry of functions with ψ, i.e., for even functions to use the relations $g_{-1}=g_1$, $g_{-2}=g_2$, $g_{n+2}=g_{n-2}$, $g_{n+1}=g_{n-1}$, and for odd functions, $w_{-1}=-w_1$, $w_{-2}=-w_2$, $w_{n+2}=-w_{n-2}$, $w_{n+1}=-w_{n-1}$.

4. *Free Stream Conditions and Geometric Quantities*

Assume that the velocity vector of free stream lies on the symmetric plane, and its angle with the axis z is α (see Fig. 2.1). Then the formulas for the free stream velocity are

$$\begin{cases} u_\infty=\cos \alpha, \\ v_\infty=-\sin \alpha \cos \psi, \\ w_\infty=\sin \alpha \sin \psi. \end{cases} \tag{3.8}$$

We now derive the expression for $w'_{0,\pi}$. In fact, differentiating the third equation of (2.4) with respect to ψ and letting $\psi=0, \pi$, we immediately obtain

$$w'_{0,\pi}=(w'_\infty)_{0,\pi}-\left[\left(1-\frac{\rho_\infty}{\rho}\right) \frac{\partial n_3}{\partial \psi} V_{\infty,N}\right]_{0,\pi}, \tag{3.9}$$

where

$$\begin{cases} (w'_\infty)_{0,\pi} = \sin\alpha(\cos\psi)_{0,\pi}, \\ \left(\dfrac{\partial n_3}{\partial\psi}\right)_{0,\pi} = \left(\dfrac{-f_1''}{f_1\sqrt{1+f_1^2+\left(\dfrac{f_1'}{f_1}\right)^2}}\right)_{0,\pi}. \end{cases}$$

The quantity with a double prime denotes the second-order derivative with respect to ψ and the quantities with the subscripts 0 and π the values on the symmetrical planes $\psi=0$, π respectively.

In the following, some formulas for geometric quantities of elliptical cones are given.

Suppose that in the Cartesian coordinates (x, y, z) the shape of the body is expressed by

$$\frac{x^2}{a^2}+\frac{y^2}{b^2}=z^2,$$

the x–z plane is coincident with the 0–π plane, the semiangle on the x–z plane is θ_1, and that on the y–z plane is θ_2. Obviously, when $\theta_1=\theta_2$ it is a cone (see Fig. 2.1).

Since

$$\begin{cases} x=r\cos\varphi, \\ y=r\sin\varphi, \end{cases}$$

therefore

$$r^2\frac{\cos^2\varphi}{a^2}+r^2\frac{\sin^2\varphi}{b^2}=z^2.$$

Because $a=\operatorname{tg}\theta_1$ and $b=\operatorname{tg}\theta_2$, we have

$$\begin{cases} r=z\left(\sqrt{\dfrac{\cos^2\varphi}{\operatorname{tg}^2\theta_1}+\dfrac{\sin^2\varphi}{\operatorname{tg}^2\theta_2}}\right)^{-1}. \\ f_2=\dfrac{1}{\sqrt{\dfrac{\cos^2\varphi}{\operatorname{tg}^2\theta_1}+\dfrac{\sin^2\varphi}{\operatorname{tg}^2\theta_2}}}, \\ f_2'=\dfrac{\left(\dfrac{1}{\operatorname{tg}^2\theta_1}-\dfrac{1}{\operatorname{tg}^2\theta_2}\right)\sin\varphi\cos\varphi}{\left(\dfrac{\cos^2\varphi}{\operatorname{tg}^2\theta_1}+\dfrac{\sin^2\varphi}{\operatorname{tg}^2\theta_2}\right)^{\frac{3}{2}}}, \\ (f_2'')_{0,\pi}=\left[\dfrac{\left(\dfrac{1}{\operatorname{tg}^2\theta_1}-\dfrac{1}{\operatorname{tg}^2\theta_2}\right)\cos^2\varphi}{\left(\dfrac{\cos^2\varphi}{\operatorname{tg}^2\theta_1}+\dfrac{\sin^2\varphi}{\operatorname{tg}^2\theta_2}\right)^{\frac{3}{2}}}\right]_{0,\pi} \\ \qquad =\operatorname{tg}^3\theta_1\left(\dfrac{1}{\operatorname{tg}^2\theta_1}-\dfrac{1}{\operatorname{tg}^2\theta_2}\right). \end{cases}$$

5. *Extrapolation at the Body*

(2.3) and (2.5) show that $c=0$ at the body. Therefore equations (3.1) are singular at the body, and the integration cannot be marched to the body by using explicit methods. In order to obtain the flow quantities at the body, extrapolation is used. That is, when integrating equations (3.1) to some ξ near the body (e.g., $\xi=0.1$), we use the values at several points (e.g., $\xi=0.3$, 0.2, 0.1) to extrapolate the values at $\xi=0$.

As pointed out by Ferri[F34], in conical flow, a thin vortex layer exists near the body, where the entropy, density and velocity change drastically. Strictly speaking, except for p and $\left(u\frac{\partial\xi}{\partial z}+v\frac{\partial\xi}{\partial r}+\frac{w}{r}\frac{\partial\xi}{\partial\varphi}\right)$, the values on the body obtained by extrapolation are not accurate. They can be considered as the values at an outer-edge of the vortex layer. The real values ρ and $|\boldsymbol{V}|$ at the body can be obtained by using the extrapolated values p and $\left(u\frac{\partial\xi}{\partial z}+v\frac{\partial\xi}{\partial r}+\frac{w}{r}\frac{\partial\xi}{\partial\varphi}\right)$, the Bernoulli equation, the iso-entropy condition and a stream-characteristic compatibility relation.

6. *Calculation Procedure*

Let us now outline the procedure for calculation using the formulas above. The problem to be solved is to determine the shock and flow quantities corresponding to the given free stream condition and the body. Suppose that a shock is given. Thus, using (3.5), (3.8), (2.4) and (3.9), we can obtain u, v, w, p, ρ and w'_0 (if necessary, w'_π) at $\xi=1$. Taking these values as initial ones and using (3.5), we can integrate equations (3.1) and (3.3) from $\xi=1$ to $\xi=\xi^*$ (e.g., $\xi^*=0.1$). After obtaining the values at the body by extrapolation, we can check whether condition (2.5) is satisfied. If it is not satisfied, the shock is adjusted and we repeat the above procedure until (2.5) is satisfied within a tolerant error. The iteration formula (3.11) in § 3 of Chapter 3 is used to adjust the shock. At present, the adjusted parameters are the shock positions $f_{1,i}$. In order to guarantee the convergence of iteration, a good initial shock is required. Therefore, as we did in Chapter 5, the way of successive transition for a certain parameter such as α, M_∞, θ_1 or θ_2 is employed.

§ 4 Results

Using the above method, we carry out extensive calculations for supersonic flow around elliptical cones at angles of attack. The free stream Mach number ranges from 1.5 to 40, the semiangle of the circular cone ranges from 4.5° to 40°, and the ratio b/a of the long axis to the short one for elliptical cones ranges from 0.5 to 3.68. For the flow around the cones, the maximum ratio of angles of attack to semiangles of cones is 1.3, when $\psi_{\max}=\pi$. When the "spherical velocity" is greater than the local sound speed, calculations are performed only in the windward region, and the maximal angle of attack is 40°. In addition, we also calculate the conical flow around the body described by the equation

$$r = z\,\text{tg}\,\theta_1\sqrt{\cos^2\varphi + \left(\frac{\text{tg}\,\theta_2}{\text{tg}\,\theta_1}\right)^2 \sin^2\varphi}\,.$$

The frozen and equilibrium air flows are considered. The thermodynamic functions of equilibrium air are calculated by using the approximate formulas presented in Chapter 4, § 4.

The accuracy of the results is determined through two different ways. One way is to determine the accuracy of results through comparing those results obtained by using different numbers of lines or different integration steps or through verifying the validity of certain conditions, for example, the conservation law of total energy. The other way is to check the results through comparing the calculated results with those obtained by other methods (e.g., the method of stabilization) or with experimental data. All the comparisons show that the method of lines give us satisfactory results.

In the following, we give some results and make some corresponding remarks. For $N \leqslant 7$, the derivatives with respect to ψ are obtained by global trigonometric interpolation polynomials, and for $N \geqslant 9$, by the five-point central difference formula.

Figs. 4.1—4.3 show the shocks at $M_\infty = 7$ for the flow around the cones with semiangles 10°, 40°, 25°, respectively. We find that for the small semiangle cone the stand-off distance of the shock on the windward symmetric plane is less than that on the leeward symmetric plane, while for the big semiangle cone the situation is inverse. Fig. 4.1 indicates that the difference between stand-off distances on the windward and

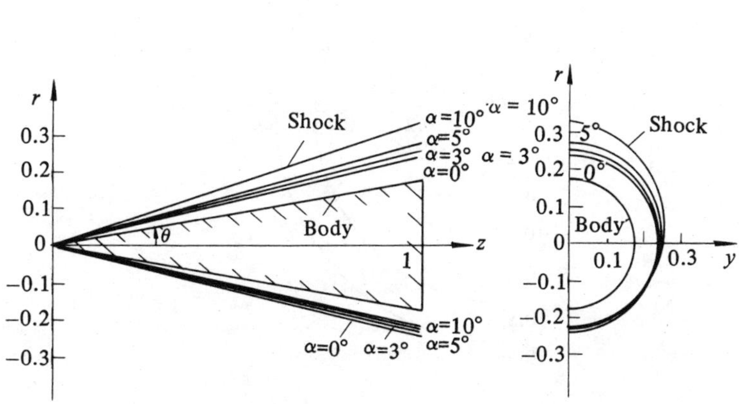

Figure 4.1 Shapes of shocks of the flow around a cone ($M_\infty = 7$, $\theta = 10°$)

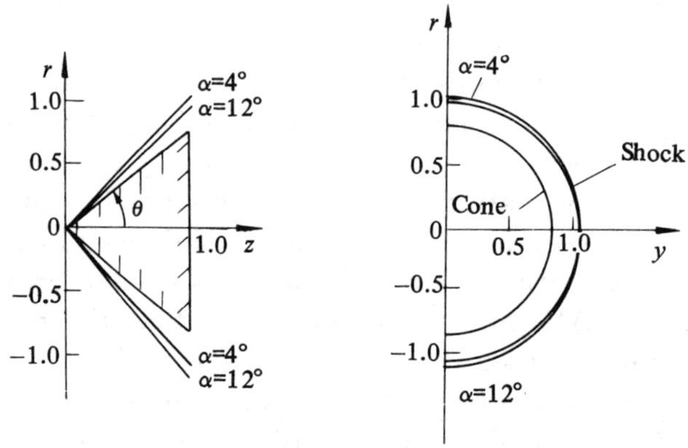

Figure 4.2 Shapes of shocks of the flow around a cone ($M_\infty=7$, $\theta=40°$)

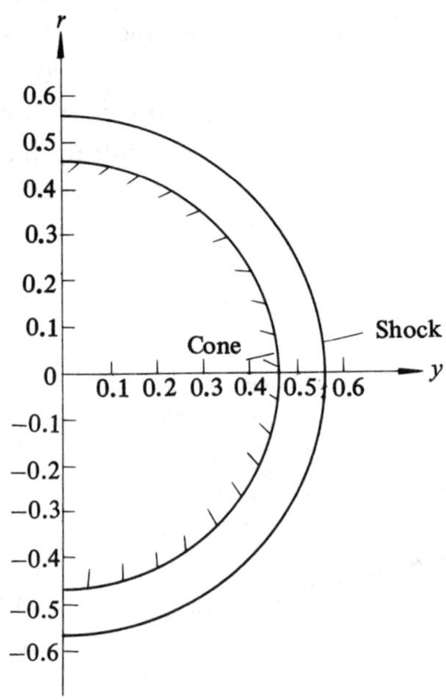

Figure 4.3 Shape of a shock of the flow around a cone ($M_\infty=7$, $\theta=25°$, $\alpha=10°$)

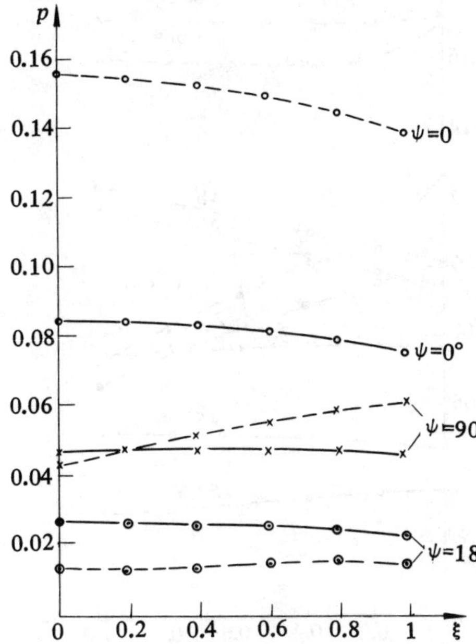

Figure 4.4 Variations of pressure with ξ ($M_\infty=7$, $\theta=10°$) —— $\alpha=5°$, —— $\alpha=12°$.

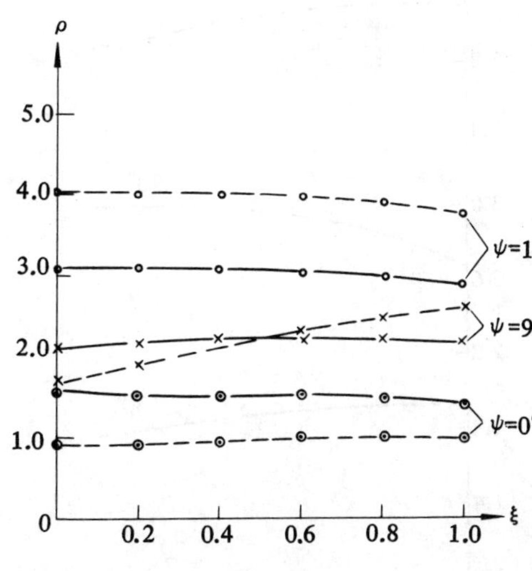

Figure 4.5 Variations of density with ξ ($M_\infty=7$, $\theta=10°$) - - - $\alpha=12°$, —— $\alpha=5°$.

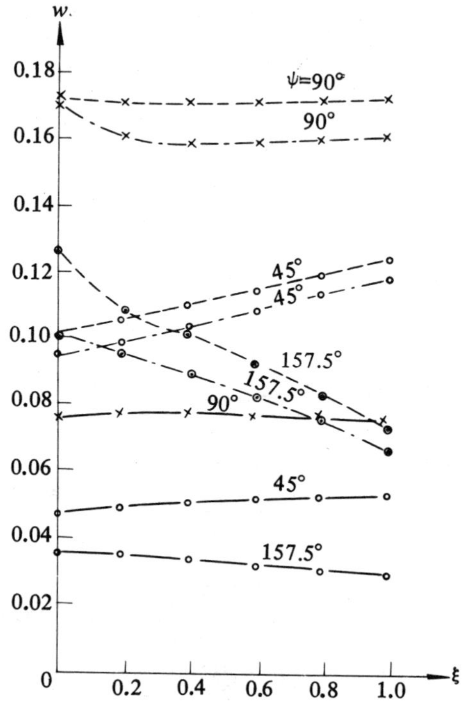

Figure 4.6 Variations of velocity component w with ξ ($M_\infty=7$, $\theta=10°$)
—— $\alpha=12°$, —·— $\alpha=11°$, —— $\alpha=5°$t

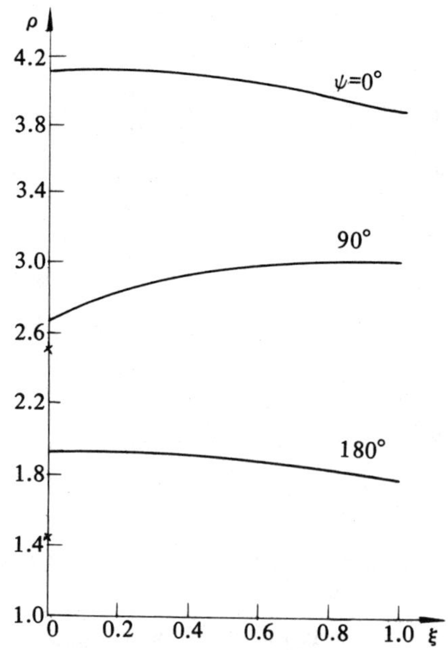

Figure 4.7 Variations of density with ξ ($M_\infty=6$, $\theta_1=\theta_2=17.5°$, $\alpha=9°$)

leeward symmetrical planes could be quite large, and for the thin cone the stand–off distance at some ψ is almost independent of α. Fig. 4.3 shows that the maximum stand–off distance of the shock may not be on the symmetric plane.

Figs. 4.4—4.6 give the profiles of p, ρ and w along the ξ for the flow around the cone with $\theta=10°$ ($M_\infty=7$).

Fig. 4.7 shows the density curves along ξ for the flow around the cone with $\theta=17.5°$ ($M_\infty=6$). In this figure, the cross points "×" denote the real values of ρ at the body, and the points on the lines the values at the outer–edge of vortex layer. From Figs. 4.4—4.7 it is seen that the variations of flow quantities along ξ are not very great.

Figs. 4.8—4.10 show the variations of p, ρ, w with ψ for the flow around the cone with $\theta=10°$ ($\alpha=10°$, $M_\infty=7$). Clearly, the variation of flow quantities with ψ is nonlinear and may be quite large. There is a maximum value for w.

The variation curves of p, w with θ for the flow around the cones for $M_\infty=7$, $\alpha=5°$ are given in Fig. 4.11. This figure shows the variation of flow quantities with θ is also nonlinear, and its value may change for several–fold.

Fig. 4.12 shows the variations of p, w with α for the flow around the cone with $\theta_1=\theta_2=10°$ ($M_\infty=7$). Fig. 4.13 gives the variation of p with α for $M_\infty=20$, $\theta_1=\theta_2=40°$. We find also in these figures that only for small angles of attack the variation is approximately linear, while for a large α it is nonlinear.

Fig. 4.14 shows the variations of surface pressure at $\psi=0$, $180°$ with M_∞. It shows that p decreases monotonously with M_∞ and when $M_\infty>15$, it is independent of M_∞.

Figs. 4.15—4.20 give the flow around the body whose equation is

$$r=z\ \mathrm{tg}\ \theta_1\sqrt{\cos^2\varphi+\frac{\mathrm{tg}^2\theta_2}{\mathrm{tg}^2\theta_1}\sin^2\varphi}.$$

The shocks for $M_\infty=3.09$, $\theta_1=12.7°$, $\theta_2=22°$ are given in Fig. 4.15, and for $M_\infty=3.09$, $\theta_1=8.9°$, $\theta_2=30°$ in Fig. 4.16.

Figs. 4.17—4.19 give the variation curves of surface pressure in the ψ–direction for $M_\infty=3.09$, $\theta_1=8.9°$, $\theta_2=30°$, $M_\infty=5.8$, $\theta_1=6°$, $\theta_2=11.8°$, and $M_\infty=3.09$, $\theta_1=12.7°$, $\theta_2=22°$, respectively.

Fig. 4.20 shows the variation of surface pressure in the ψ–direction for $M_\infty=3.09$, $\theta_2=30°$ and various θ_1.

Fig. 4.21 gives the shocks for the flow around the elliptic cone with $\theta_1=5.95°$, $\theta_2=11.8°$ ($M_\infty=5.8$, $b/a\approx2$).

Fig. 4.22 gives the shocks and sonic lines (i.e., the lines on which the "spherical velocity" is equal to the local sound speed) for $M_\infty=7$, $\theta_1=\theta_2=10°$ and $\theta_1=\theta_2=20°$.

Figs. 4.23—4.25 give the equilibrium air flow around the cone with $\theta_1=\theta_2=10°$ for $M_\infty=20$, $\alpha=20°$ and $p_\infty=0.1206\times10^5$ dyn/cm², $\rho_\infty=0.1894\times10^{-4}$ g/cm³.

The shocks are given in Fig. 4.23, and the profiles of $p(\xi, \psi)$ and $\rho(\xi, \psi)$ on plane $\psi=0$ are given in Figs. 4.24, 4.25. For comparison, the values for frozen air flow under the same conditions are also given in these figures. Obviously, the equation of state has significant influence on the flow quantities.

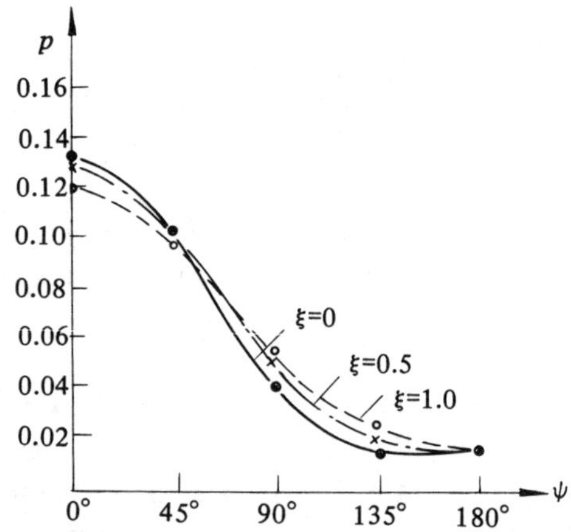

Figure 4.8 Variations of pressure with ψ ($M_\infty=7$, $\theta_1=\theta_2=10°$, $\alpha=10°$)

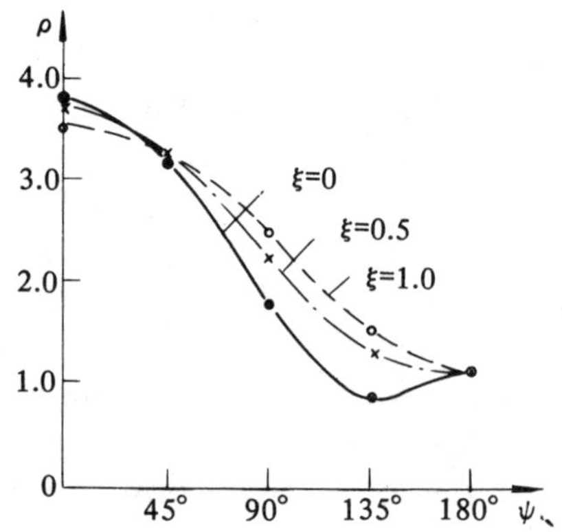

Figure 4.9 Variations of density with ψ ($M_\infty=7$, $\theta_1=\theta_2=10°$, $\alpha=10°$)

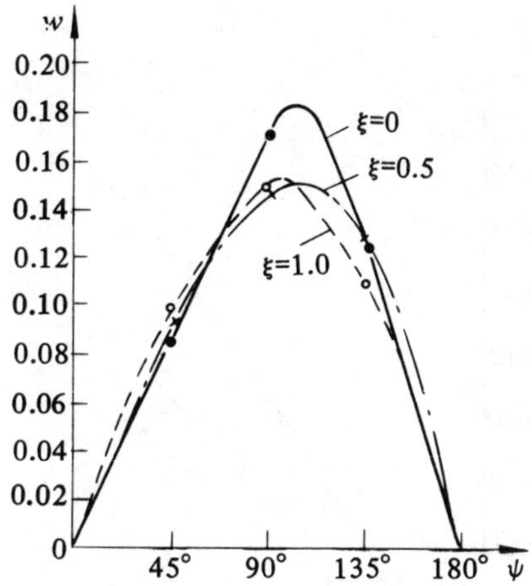

Figure 4.10 Variations of velocity component w with ψ ($M_\infty=7$, $\theta_1=\theta_2=10°$, $\alpha=10°$)

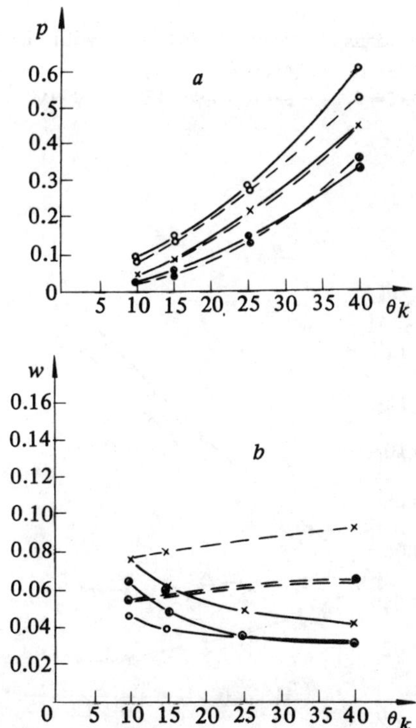

Figure 4.11 Variations of pressure and velocity component w with the semivertex angle of cones ($M_\infty=7$, $\alpha=5°$)
—— $\xi=0$, ---- $\xi=1$; ○ $\psi=45°$, × $\psi=90°$, ⊙ $\psi=135°$.

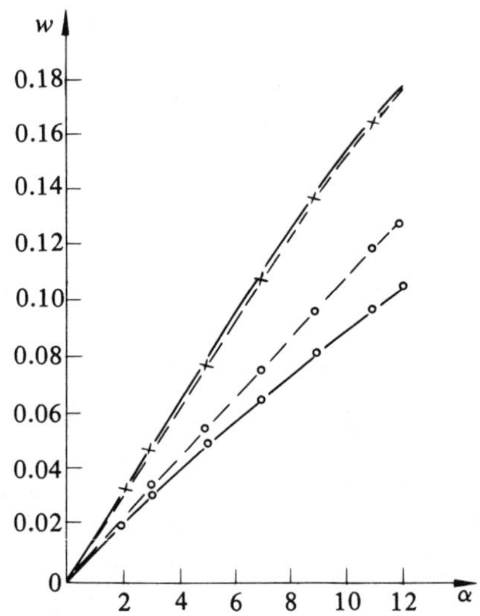

Figure 4.12a Variations of velocity component w with the angle of attack ($M_\infty=7, \theta=10°$)
—— $\xi=0$, — — $\xi=1$; ○ $\psi=45°$, × $\psi=90°$.

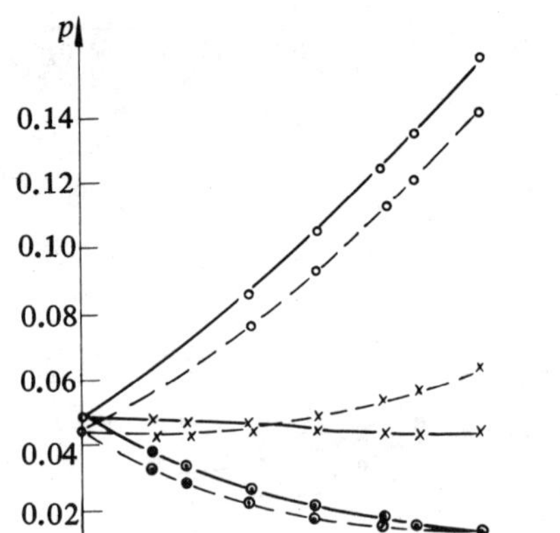

Figure 4.12b Variations of pressure with the angle of attack ($M_\infty=7, \theta=10°$)
○ $\psi=0°$, × $\psi=90°$, ⊙ $\psi=180°$; —— $\xi=0$, — — $\xi=1$.

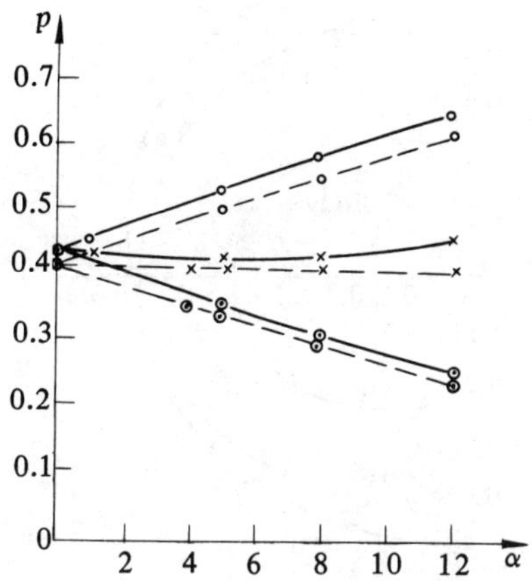

Figure 4.13 Variations of pressure with the angle of attack ($M_\infty=20$, $\theta=40°$)
○ $\psi=0°$, × $\psi=90°$, ⊙ $\psi=180°$; —— $\xi=0$, — $\xi=1$.

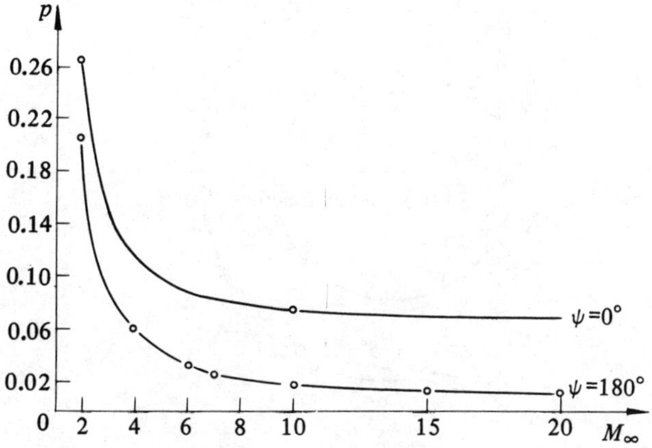

Figure 4.14 Variations of pressure with M_∞ ($\theta=10°$, $\alpha=5°$, $\xi=0$)

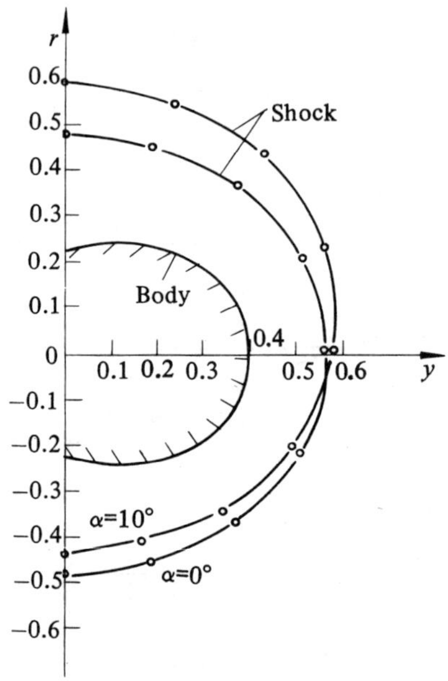

Figure 4.15 Shapes of shocks of the flow around a noncircular cone ($M_\infty=3.09, \theta_1=12.7°, \theta_2=22°$)

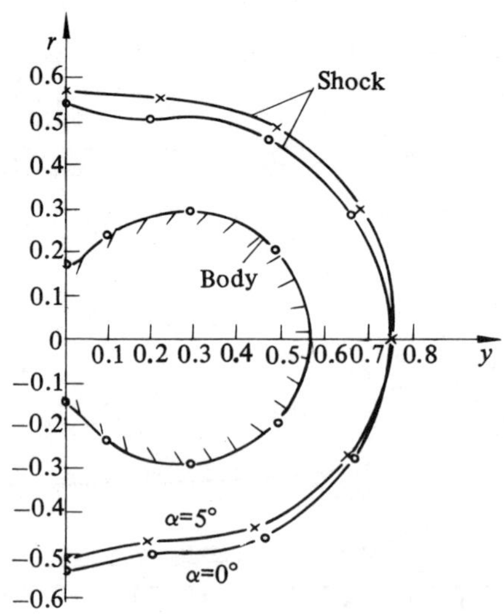

Figure 4.16 Shapes of shocks of the flow around a noncircular cone ($M_\infty=3.09, \theta_1=8.9°, \theta_2=30°$)

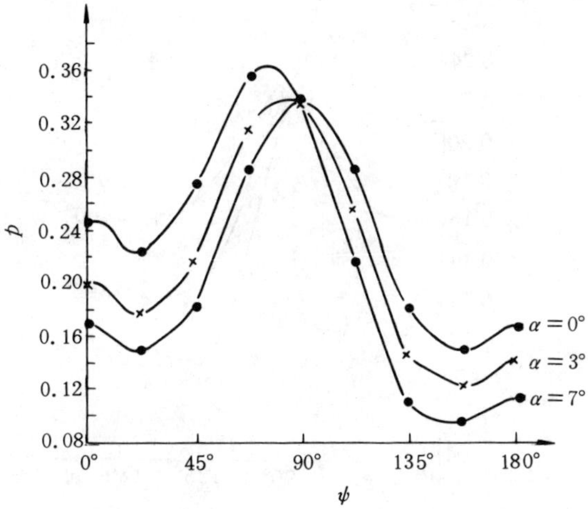

Figure 4.17 Variations of pressure with ψ ($M_\infty = 3.09$, $\theta_1 = 8.9°$, $\theta_2 = 30°$)

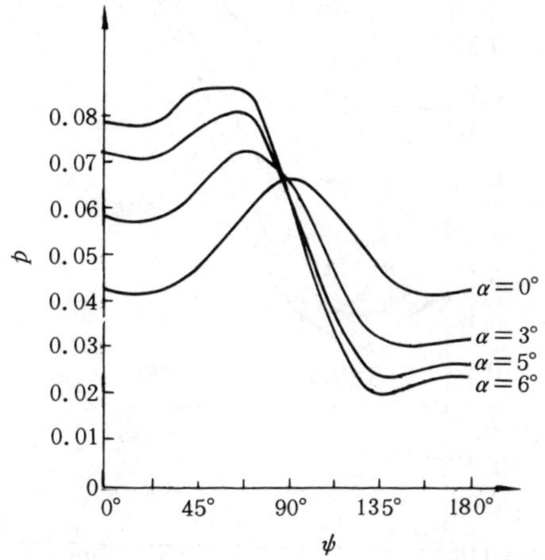

Figure 4.18 Variations of pressure with ψ ($M_\infty = 5.8$, $\theta_1 = 6°$, $\theta_2 = 11.8°$)

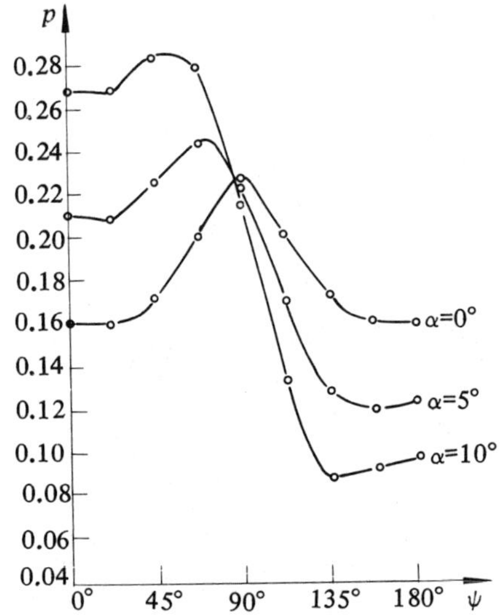

Figure 4.19 Variations of pressure with ψ ($M_\infty=3.09$, $\theta_1=12.7°$, $\theta_2=22°$)

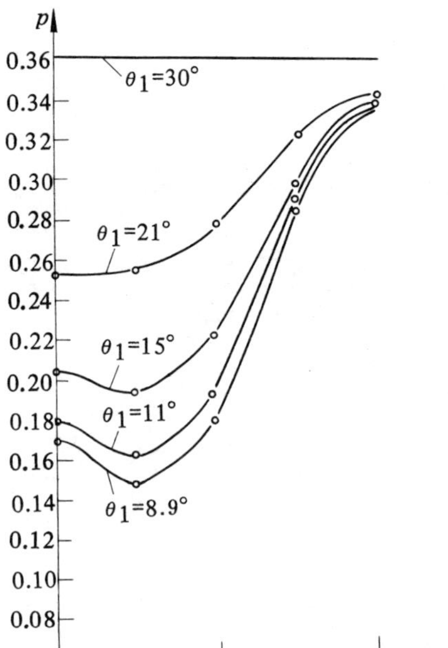

Figure 4.20 Surface pressure distributions for various θ_1 ($M_\infty=3.09$, $\theta_2=30°$, $\alpha=0°$)

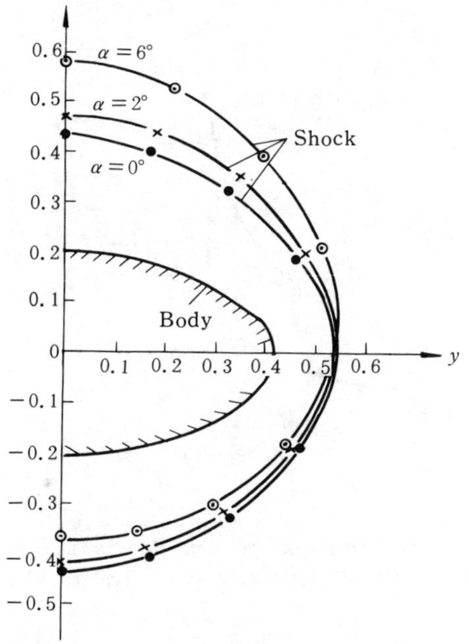

Figure 4.21 Shapes of shocks of the flow around an elliptic cone ($M_\infty=5.8$, $\theta_1=5.95°$, $\theta_2=11.8°$, $b/a\approx2$)

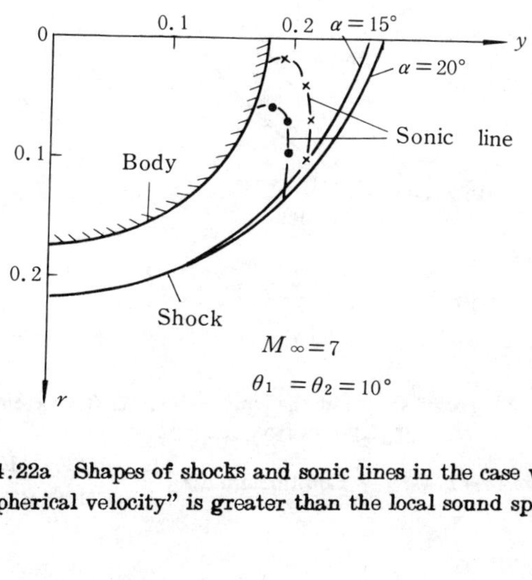

Figure 4.22a Shapes of shocks and sonic lines in the case where the "spherical velocity" is greater than the local sound speed

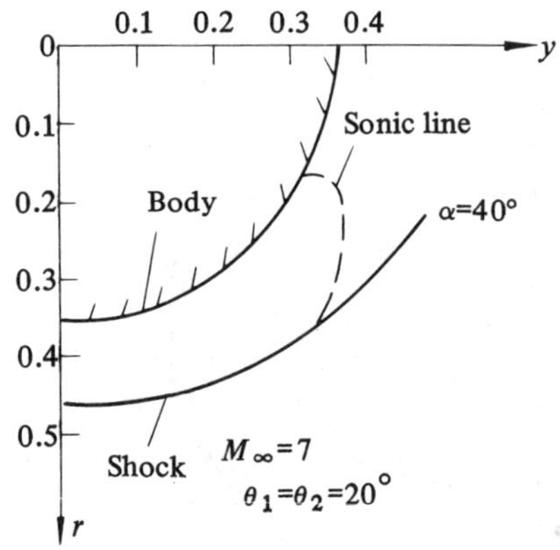

Figure 4.22b Shapes of shocks and sonic lines in the case where the "spherical velocity" is greater than the local sound speed

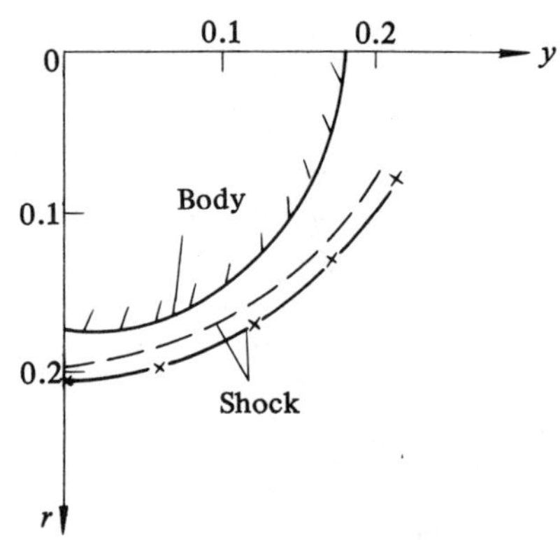

Figure 4.23 Shapes of shocks of the equilibrium air flow around a cone ($M_\infty=20$, $\theta=10°$, $\alpha=20°$)

—— Frozen air ($\gamma=1.4$), —— Equilibrium air $\begin{cases} p_\infty=0.1206\times10^5\ \text{dyn/cm}^2, \\ \rho_\infty=0.1894\times10^{-4}\ \text{g/cm}^3. \end{cases}$

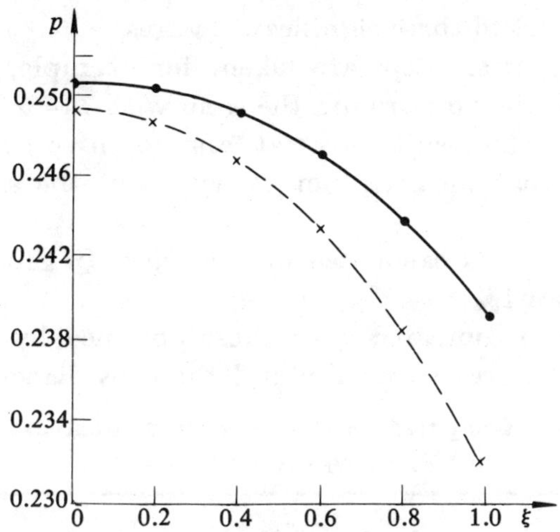

Figure 4.24 Variations of pressure with ξ for equilibrium air flow around cones ($M_\infty=20$, $\theta_1=\theta_2=10°$, $\alpha=20°$, $\psi=0°$)

—— Frozen air ($\gamma=1.4$), ---- Equilibrium air $\begin{cases} p_\infty=0.1206\times10^5\ \text{dyn/cm}^2, \\ \rho_\infty=0.1894\times10^{-4}\ \text{g/cm}^3. \end{cases}$

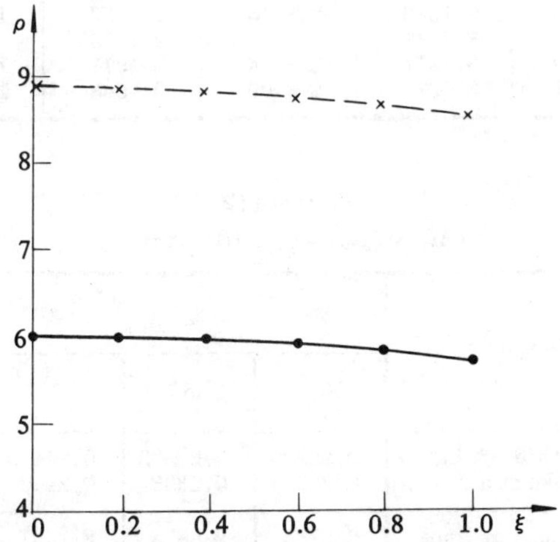

Figure 4.25 Variations of density with ξ for equilibrium air flow around cones ($M_\infty=20$, $\theta_1=\theta_2=10°$, $\alpha=20°$, $\psi=0°$)

—— Frozen air ($\gamma=1.4$), —— Equilibrium air $\begin{cases} p_\infty=0.1206\times10^5\ \text{dyn/cm}^2, \\ \rho_\infty=0.1894\times10^{-4}\ \text{g/cm}^3. \end{cases}$

The results are checked in many ways. We list some of them below:

Different numbers of lines and different approximate formulas are used in the calculation. Table 4.1 gives the shock f_1, the surface pressure p and the density ρ for $M_\infty=7$, $\theta_1=\theta_2=10°$, $\alpha=5°$. It is seen that all results agree at least to three significant figures.

Various integration steps are taken, for example. $\Delta\xi=0.1$, 0.025, 0.01 are taken for the flow around the cone with $\theta_1=\theta_2=10°$ for $\alpha=5°$, $M_\infty=7$ and $N=5$. The results agree at least to three significant figures. This fact implies that we need not worry about the serious growth of rounding error.

The total energy is calculated on all nodes. In general, the relative errors are less than 1%.

The results are compared with those obtained by other methods. Table 4.2 gives the present results and those by Babenko et al.[B47] for

Table 4.1 Computed results for various numbers of lines
($M_\infty=7$, $\theta_1=\theta_2=10°$, $\alpha=5°$)

Flow parameters	N	ψ				
		0°	45°	90°	135°	180°
f_1	5	0.22836	0.23222	0.24415	0.26144	0.27162
	7	0.22838		0.24408		0.27158
	9	0.22839	0.23222	0.24406	0.26152	0.27161
	17	0.22838	0.23222	0.24408	0.26151	0.27150
p	5	0.08535	0.07264	0.04762	0.03106	0.02720
	7	0.08535		0.04767		0.02719
	9	0.08536	0.07261	0.04767	0.03103	0.02720
	17	0.08535	0.07260	0.04766	0.03104	0.02719
ρ	5	3.1270	2.8115	2.1350	1.6494	1.5560
	7	3.1277		2.1375		1.5557
	9	3.1277	2.8114	2.1371	1.6466	1.5560
	17	3.1277	2.8109	2.1368	1.6476	1.5557

Table 4.2
($M_\infty=7$, $\theta_1=\theta_2=10°$, $\alpha=5°$)

Flow parameters	Methods	ψ				
		0°	45°	90°	135°	180°
f_1	The method of lines	0.22836	0.23222	0.24415	0.26143	0.27162
	Babenko et al.	0.2285	0.2323	0.2440	0.2614	0.2719
ρ_{shock}	The method of lines	2.8785	2.6875	2.1921	1.6494	1.4096
	Babenko et al.	2.8791	2.6871	2.1915	1.6503	1.4059
ρ_{wall}	The method of lines	3.1275	2.7873	2.0614	1.5194	1.3817
	Babenko et al.	3.1275	2.7858	2.0620	1.5179	1.3804

Table 4.3

($M_\infty=6$, $\theta_1=\theta_2=17.5°$, $\alpha=5°$)

Flow parameters	Methods	ψ 0°	45°	90°	135°	180°
p_{shock}	The method of lines	0.1597	0.14445	0.11049	0.08032	0.06894
	The difference method	0.1595	0.1443	0.1105	0.08055	0.06925
ρ_{shock}	The method of lines	3.5087	3.3645	2.9746	2.5168	2.3059
	The difference method	3.507	3.363	2.975	2.521	2.312
p_{wall}	The method of lines	0.17504	0.15639	0.11729	0.08664	0.07661
	The difference method	0.1751	0.1565	0.1176	0.08696	0.07688
ρ_{wall}	The method of lines	3.7462	3.5095	2.9928	2.5927	2.4864
	The difference method	3.784	3.539	3.001	2.576	2.460

$M_\infty=7$, $\theta_1=\theta_2=10°$, $\alpha=5°$. Table 4.3 shows the present results and those obtained by the difference method similar to that described in Chapter 7. p and ρ at the shock and body are listed in this table for $M_\infty=6$, $\theta_1=\theta_2=17.5°$, $\alpha=5°$. It is seen that the relative errors are less than 1%.

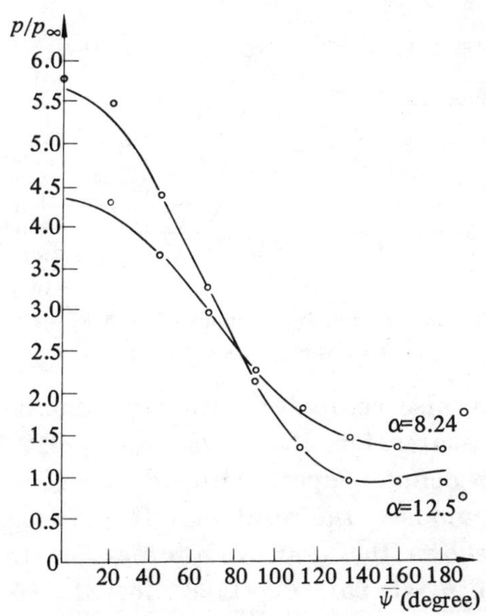

Figure 4.26 Surface pressure distributions ($M_\infty=4.25$, $\theta_1=\theta_2=12.5°$)
—— Calculation, ● Experiment.

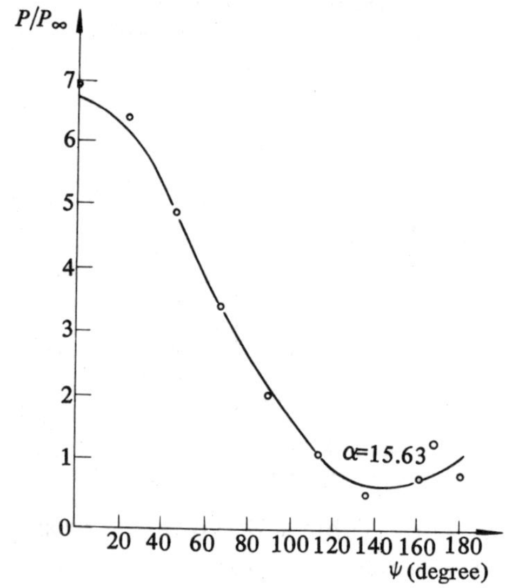

Figure 4.27 Surface pressure distributions ($M_\infty=4.25$, $\theta_1=\theta_2=12.5°$)
—— Calculation, ○ Experiment.

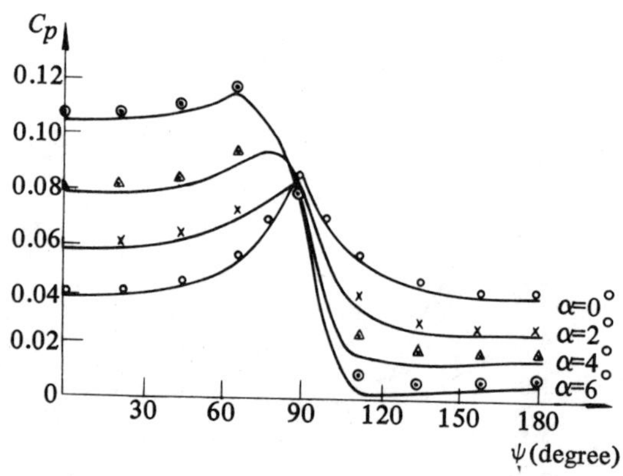

Figure 4.28 Surface pressure distributions ($M_\infty=5.8$, $\theta_1=5.95°$, $\theta_2=11.8°$)
—— Calculation, ○×△⊙ Experiment.

The results are also compared with experimental data. Figs. 4.26, 4.27 give the pressures for $M_\infty=4.25$, $\theta_1=\theta_2=12.5°$, and $\alpha=8.24°$, 12.5°, 15.63°. Dots denote experimental data and the lines denote the present results. Obviously, the results are in good agreement with each other. Fig. 4.28 shows the pressure profiles for the flow around the elliptical cone with $\theta_1=5.95°$, $\theta_2=11.8°$ for $M_\infty=5.8$, and $\alpha=0°$, 2°, 4°, 6°. The lines denote the present results and the symbols ⊙, △, •, × are experimental data. This figure also shows that the two kinds of results agree.

Chapter 7

Solution of Supersonic Regions of Flow around Combined Bodies

§ 1 Introduction

In this chapter, we discuss a numerical method for supersonic regions of three–dimensional flow around combined bodies. The flow under consideration is the steady inviscid, non–heat–conducting flow of a perfect gas or air in chemical equilibrium. If the initial data are given on a space–like surface near the nose of a body, and if the flow in the downstream region of the initial surface is supersonic, then the flow in the supersonic region can be determined by solving an initial–boundary–value problem for hyperbolic systems.

In Chapters 5 and 6, we have discussed the numerical methods for the region near the nose. To calculate the flow field in the supersonic region the initial data needed can usually be obtained by using the methods developed there. This is true, for example, for usual axisymmetric bodies whose noses are cones, spheres, and other smooth bodies. In fact, if the nose of a combined body is a cone, then the flow near the nose under certain conditions belongs to conical flow. In this case, the initial data needed to solve the supersonic region can be obtained by using the method described in Chapter 6. If the combined body has a smooth and blunt nose, the initial data needed can be provided by the method described in Chapter 5 for the subsonic–transonic region. If the nose is a sphere, the desired data can be obtained from the table of flow around a sphere (we have computed a complete table of flow around a sphere by using the method discussed in Chapter 5) by using the interpolation.

In the supersonic region of a flow field, besides the main shock, there may appear some other discontinuity surfaces, such as the boundary surfaces of centered expansion waves starting from edges on the convex parts of the body, the secondary shocks starting from edges on the concave parts of the body, and the embedded shocks near smooth concave surfaces of the body. Moreover, these discontinuity surfaces in the flow field might intersect, so some new discontinuities could be produced. For example, a new main shock, a contact discontinuity and a reflected expansion wave are produced when the main shock intersects a secondary shock (see Fig. 1.1). Hence, if a "through" difference method is adopted, then the results are usually not satisfactory because the truncation error is greater.

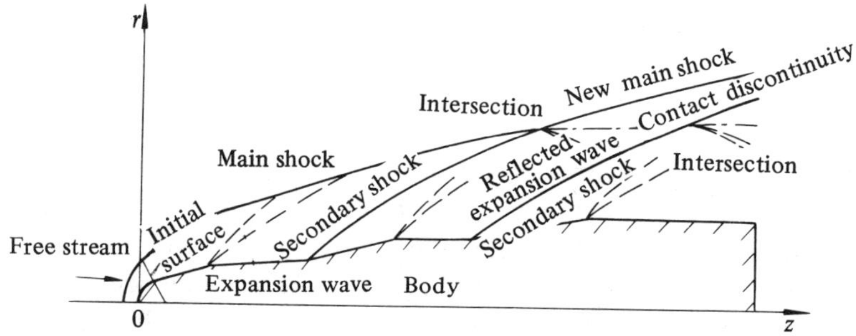

Figure 1.1 Construction of the flow field of a gas around a combined body on a meridian plane

In order to raise the accuracy of results, we take these discontinuity surfaces as boundaries in our process of calculation. Moreover, in the flow field around a slender combined body, a thin layer called the entropy layer appears in the region near the body far from the nose, where certain flow variables, such as the entropy, and the density, vary rapidly. In order that the computed results may reflect the real change of flow variables in that layer, it is necessary to take a smaller mesh–size there. For this, we take a stream surface near the body as an internal boundary, and several mesh points are put in the region between the stream surface and the body. When the entropy layer gets thinner and thinner, the mesh–size there gets smaller and smaller automatically. Thus, this net adapts to the rapid change of functions very well, and accurate results can be obtained.

When those surfaces mentioned above are taken as internal boundaries, the solution of flow field in the supersonic region is reduced to the solution of an initial–boundary–value problem for first order quasi–linear hyperbolic systems with several internal boundaries.

In § 2 of this chapter, we shall give the concrete expressions of the boundary conditions or the internal boundary conditions on those boundary surfaces mentioned above, and shall point out that the problem here is of the form of that considered in Chapter 2. This means that the problem here can be solved by the numerical method presented there. In

§ 3, we shall discuss in detail certain problems which appear in the application of the numerical method of Chapters 1 and 2 to the problem of supersonic flow around combined bodies. In § 4, a number of results about the flow around several typical combined bodies will be given.

§ 2 Formulation of Problems

1. *Partial Differential Equations and Coordinate Transformations*

(1) *Basic Partial Differential Equations*

According to (1.11) and (1.12) of Chapter 4, the equations of three-dimensional steady flow of inviscid, non-heat-conducting perfect gas or air in chemical equilibrium in the cylindrical coordinate system $\{z, r, \varphi\}$ are

$$A\frac{\partial U}{\partial z}+B\frac{\partial U}{\partial r}+C\frac{\partial U}{\partial \varphi}+D=0, \tag{2.1}$$

where

$$\begin{cases} A=\begin{pmatrix} u & 0 & 0 & 1/\rho & 0 \\ 0 & u & 0 & 0 & 0 \\ 0 & 0 & u & 0 & 0 \\ \rho & 0 & 0 & 0 & u \\ 0 & 0 & 0 & u & -a^2u \end{pmatrix}, \quad B=\begin{pmatrix} v & 0 & 0 & 0 & 0 \\ 0 & v & 0 & 1/\rho & 0 \\ 0 & 0 & v & 0 & 0 \\ 0 & \rho & 0 & 0 & v \\ 0 & 0 & 0 & v & -a^2v \end{pmatrix}, \\ C=\frac{1}{r}\begin{pmatrix} w & 0 & 0 & 0 & 0 \\ 0 & w & 0 & 0 & 0 \\ 0 & 0 & w & 1/\rho & 0 \\ 0 & 0 & \rho & 0 & w \\ 0 & 0 & 0 & w & -a^2w \end{pmatrix}, \quad D\equiv\begin{pmatrix} d_1 \\ d_2 \\ d_3 \\ d_4 \\ d_5 \end{pmatrix}=\begin{pmatrix} 0 \\ -w^2/r \\ vw/r \\ \rho v/r \\ 0 \end{pmatrix}, \quad U\equiv\begin{pmatrix} u \\ v \\ w \\ p \\ \rho \end{pmatrix}, \end{cases}$$

u, v, w being respectively the axial, radial and circumferential velocities in the cylindrical coordinate system, p the pressure, ρ the density, and a the sonic speed. For frozen air, the expression of the sonic speed is $a^2=\gamma p/\rho$, $\gamma=1.4$; for air in chemical equilibrium, an approximating expression for a^2, as a function of p and ρ, is given in Chapter 4, § 4.

(2) *Coordinate Transformations*

In the process of calculation, we introduce three new independent variables η, ξ, ψ through the relations

$$\begin{cases} z=E(r, \eta, \psi), \\ r=G_{l-1}(\eta, \psi)+(\xi-l+1)\left[G_l(\eta, \psi)-G_{l-1}(\eta, \psi)\right]\equiv G(\eta, \xi, \psi), \\ \qquad \text{if } l-1\leqslant\xi\leqslant l,\ l=1, 2, \cdots, L, \\ \varphi=H_{m-1}(\eta)+(\psi-m+1)\left[H_m(\eta)-H_{m-1}(\eta)\right], \\ \qquad \text{if } m-1\leqslant\psi\leqslant m,\ m=1, 2, \cdots, M, \end{cases} \tag{2.2}$$

where we suppose that

$$G_{l-1}(\eta, \psi) \leqslant G_l(\eta, \psi), \ (l=1, 2, \cdots, L), \ H_{m-1}(\eta) < H_m(\eta),$$

and $E(r, \eta, \psi)$, $G_l(\eta, \psi)$, $H_m(\eta)$ are given functions or functions that will be determined together with the solution of (2.1) in the process of calculation.

The purpose of introducing the coordinate transformation is that we may solve the problem here by using the difference method described in Chapter 2 with rectangular meshes. That is, it aims at transforming a series of specified surfaces in the physical space (discontinuity surfaces, space-like surfaces, $\cdots$) into a series of planes in the $\{\eta, \xi, \psi\}$-space. Let $A(\eta^*)$, $B(\xi^*)$ and $C(\psi^*)$ denote the surfaces in the $\{z, r, \varphi\}$-space determined by (2.2) with $\eta=\eta^*$, $\xi=\xi^*$ and $\psi=\psi^*$ respectively, i.e., we suppose that the relations (2.2) establish a one-to-one correspondence between the planes $\eta=\eta^*$, $\xi=\xi^*$, $\psi=\psi^*$ in the $\{\eta, \xi, \psi\}$-space and the surfaces $A(\eta^*)$, $B(\xi^*)$, $C(\psi^*)$ in the $\{z, r, \varphi\}$-space. Then the realization of the above procedure can be described as follows: E, G_l, H_m are chosen such that each specified surface belongs to the family of surfaces $A(\eta^*)$, $B(\xi^*)$ or $C(\psi^*)$.

In what follows, we shall give some explanation about the choice of G_l, E, H_m.

A proper choice of G_l can make certain specified surfaces in the $\{z, r, \varphi\}$-space correspond to some planes $\xi=$constant in the $\{\eta, \xi, \psi\}$-space. If we require that a series of surfaces $r=F_l(z, \varphi)$ $(l=0, 1, \cdots, L)$, for example, the body, the main shock and certain other surfaces in the region between the body and the main shock, be treated as boundaries in the ξ-direction, then we may choose $G_l(\eta, \psi) \equiv F_l(z(\eta, \xi, \psi), \varphi(\eta, \xi, \psi))|_{\xi=l}$. In this case, the surface $B(l)$ is just the surface $r=F_l(z,\varphi)$.

A proper choice of E can make certain specified surfaces in the $\{z, r, \varphi\}$-space correspond with certain planes $\eta=$constant in the $\{\eta, \xi, \psi\}$-space. Its concrete form should be chosen according to the problem we need to solve. Now, we shall just describe a general principle, and give several expressions used in our practical computation. When choosing $E(r, \eta, \psi)$, one should make the initial surface near the nose be one of $A(\eta)$, for example $A(\eta_0)$. Also, it must be guaranteed that every surface of $A(\eta)$ in the computational region is space-like. Moreover, the form of E should be simple in order to decrease the amount of computation. We now list several forms of $E(r, \eta, \psi)$ used in practical computation.

(i) If initial values are given at a plane $z=z_0$, and if the planes $z=$ constant are space-like when $z \geqslant z_0$, then we choose

$$E(r, \eta, \psi)=\eta. \tag{2.3}$$

(ii) In some problems of flow around bodies, the planes $z=$constant are not space-like in some regions. In order to compute these problems, we have to adopt the technique of curved-surface-march. In certain cases, when we take

$$E(r, \eta, \psi)=\eta-\frac{r \operatorname{ctg} \theta_0(\eta-\eta_1^*)}{\eta_0^*-\eta_1^*}, \tag{2.4}$$

for

$$\eta_0^* \leqslant \eta \leqslant \eta_1^*,$$

where η_0^*, η_1^*, θ_0 are constants, every "marching" surface will be space-like.

(iii) Another form of E considered is

$$E(r, \eta, \psi)=\eta_1^*+\frac{\eta-\eta_1^*}{\eta_0^*\left(\frac{\pi}{2}\right)-\eta_1^*}(\eta_0^*(\varphi)-\eta_1^*)$$

$$+\frac{\eta-\eta_1^*}{\eta_0^*\left(\frac{\pi}{2}\right)-\eta_1^*} \operatorname{ctg} \theta_0 r, \tag{2.5a}$$

for

$$\eta_0^*\left(\frac{\pi}{2}\right) \leqslant \eta \leqslant \eta_1^*,$$

where η_1^* is a constant and η_0^*, θ_0 are functions of φ. For example, the following polynomials of degree three in $\cos \varphi$

$$\left\{\begin{array}{l}\eta_0^*=\eta_{0,0}+\eta_{0,1} \cos \varphi+\eta_{0,2} \cos^2 \varphi+\eta_{0,3} \cos^3 \varphi, \\ \theta_0=\theta_{0,0}+\theta_{0,1} \cos \varphi+\theta_{0,2} \cos^2 \varphi+\theta_{0,3} \cos^3 \varphi\end{array}\right. \tag{2.5b}$$

have been taken as η_0^* and θ_0, where the coefficients are constants.

In practical computation, we take (2.3) to be E as often as possible. Only in the regions where the planes $z=$constant are not space-like is (2.4) or (2.5) used.

(iv) When the intersection and reflection of discontinuities appear in the flow field, in order to obtain accurate results, we choose $E(r, \eta, \psi)$ such that the intersection line of two discontinuities is located at some surface $\eta=$constant. In this way, the structures of flow field in all φ-planes are made to be the same for every η both in the upstream and in the downstream regions of that surface $A(\eta)$. Therefore, we can quite easily carry out an accurate computation of the interaction of discontinuities.

Now we come to discuss the choice of the function $H_m(\eta)$. If the flow field has a symmetric plane, we can take $H_0(\eta)=0$ and $H_M(\eta)=\pi$. If a section-wise equidistant net is adopted in the φ-direction, we take $H_m(\eta)=\varphi_m(\eta)$, $0<\varphi_m(\eta)<\pi$, $\varphi_m(\eta)$ being a given function. In other cases, $H_m(\eta)$ should be determined in the light of specific conditions. For example, if two discontinuities intersect in the flow field, and if the flow field is divided into several parts in the φ–direction such that the structure of flow field on all φ–planes of every part is the same, then $H_m(\eta)$ must be determined together with the solution of the partial differential equations. If the flow field does not have a symmetric plane, we can choose $H_m(\eta)$ in a similar way.

Before going on in our discussion, we shall derive several expressions of derivatives of the functions $z(\eta, \xi, \psi)$, $r(\eta, \xi, \psi)$, $\varphi(\eta, \xi, \psi)$, defined by (2.2), and of the inverse functions $\eta(z, r, \varphi)$, $\xi(z, r, \varphi)$, $\psi(z, r, \varphi)$. In what follows, we shall use the common symbol $\dfrac{\partial f}{\partial g} \equiv f_g$. Obviously, we can derive the following relations from (2.2):

$$\begin{cases} z_\eta = E_r r_\eta + E_\eta, \\ z_\xi = E_r r_\xi, \\ z_\psi = E_r r_\psi + E_\psi, \end{cases} \tag{2.6a}$$

$$\begin{cases} r_\eta = G_{\eta,l-1} + (\xi - l + 1)(G_{\eta,l} - G_{\eta,l-1}) = G_\eta, \\ r_\xi = G_l - G_{l-1} = G_\xi, \\ r_\psi = G_{\psi,l-1} + (\xi - l + 1)(G_{\psi,l} - G_{\psi,l-1}) = G_\psi, \\ \qquad \text{if } l-1 \leqslant \xi \leqslant l, \ l = 1, 2, \cdots, L, \end{cases} \tag{2.6b}$$

$$\begin{cases} \varphi_\eta = H_{\eta,m-1} + (\psi - m + 1)(H_{\eta,m} - H_{\eta,m-1}), \\ \varphi_\xi = 0, \\ \varphi_\psi = H_m - H_{m-1}, \\ \qquad \text{if } m-1 \leqslant \psi \leqslant m, \ m = 1, 2, \cdots, M. \end{cases} \tag{2.6c}$$

On the other hand, at a point where the transformation (2.2) is invertible, there exists the relation

$$\begin{pmatrix} z_\eta & z_\xi & z_\psi \\ r_\eta & r_\xi & r_\psi \\ \varphi_\eta & \varphi_\xi & \varphi_\psi \end{pmatrix} \begin{pmatrix} \eta_z & \eta_r & \eta_\varphi \\ \xi_z & \xi_r & \xi_\varphi \\ \psi_z & \psi_r & \psi_\varphi \end{pmatrix} = \begin{pmatrix} 1 & 0 & 0 \\ 0 & 1 & 0 \\ 0 & 0 & 1 \end{pmatrix},$$

which can be rewritten as

$$\begin{pmatrix} \eta_z & \eta_r & \eta_\varphi \\ \xi_z & \xi_r & \xi_\varphi \\ \psi_z & \psi_r & \psi_\varphi \end{pmatrix} = \begin{pmatrix} z_\eta & z_\xi & z_\psi \\ r_\eta & r_\xi & r_\psi \\ \varphi_\eta & \varphi_\xi & \varphi_\psi \end{pmatrix}^{-1}.$$

Therefore, using the expressions (2.6a) of z_η, z_ξ, z_ψ, we obtain

$$\left\{\begin{aligned}
\boldsymbol{\eta}&=\begin{pmatrix}\eta_z\\ \eta_r\\ \dfrac{\eta_\varphi}{r}\end{pmatrix}=\begin{pmatrix}\varphi_\psi/(E_\eta\varphi_\psi-E_\psi\varphi_\eta)\\ -E_r\varphi_\psi/(E_\eta\varphi_\psi-E_\psi\varphi_\eta)\\ -E_\psi/r(E_\eta\varphi_\psi-E_\psi\varphi_\eta)\end{pmatrix}=\frac{1}{E_\eta\varphi_\psi-E_\psi\varphi_\eta}\begin{pmatrix}\varphi_\psi\\ -E_r\varphi_\psi\\ \dfrac{-E_\psi}{r}\end{pmatrix},\\
\boldsymbol{\xi}&=\begin{pmatrix}\xi_z\\ \xi_r\\ \dfrac{\xi_\varphi}{r}\end{pmatrix}=\begin{pmatrix}-(r_\eta\eta_z+r_\psi\psi_z)/r_\xi\\ (1-r_\eta\eta_r-r_\psi\psi_r)/r_\xi\\ -(r_\eta\eta_\varphi+r_\psi\psi_\varphi)/rr_\xi\end{pmatrix}=\frac{1}{r_\xi}\left[\begin{pmatrix}0\\1\\0\end{pmatrix}-r_\eta\boldsymbol{\eta}-r_\psi\boldsymbol{\psi}\right]^{1)},\\
\boldsymbol{\psi}&=\begin{pmatrix}\psi_z\\ \psi_r\\ \dfrac{\psi_\varphi}{r}\end{pmatrix}=\begin{pmatrix}-\varphi_\eta/(E_\eta\varphi_\psi-E_\psi\varphi_\eta)\\ E_r\varphi_\eta/(E_\eta\varphi_\psi-E_\psi\varphi_\eta)\\ E_\eta/r(E_\eta\varphi_\psi-E_\psi\varphi_\eta)\end{pmatrix}=\frac{1}{E_\eta\varphi_\psi-E_\psi\varphi_\eta}\begin{pmatrix}-\varphi_\eta\\ E_r\varphi_\eta\\ \dfrac{E_\eta}{r}\end{pmatrix}.
\end{aligned}\right. \tag{2.7}$$

Clearly, $\boldsymbol{\eta}$, $\boldsymbol{\xi}$, $\boldsymbol{\psi}$ are the respective gradients of the surfaces $\eta(z, r, \varphi)=$ constant, $\xi(z, r, \varphi)=$constant, and $\psi(z, r, \varphi)=$constant in the cylindric coordinate system $\{z, r, \varphi\}$.

The surface $\xi=$constant defined by (2.2) can be expressed as $r=F(z, \varphi)$ or $r=G(\eta, \psi)$ in the physical space. In the cylindric coordinate system $\{z, r, \varphi\}$, the unit normal vector of this surface is

$$\boldsymbol{n}=\frac{1}{\sqrt{F_z^2+1+(F_\varphi/F)^2}}\begin{pmatrix}F_z\\ -1\\ F_\varphi/F\end{pmatrix}=\frac{\boldsymbol{\xi}}{|\boldsymbol{\xi}|}, \tag{2.8}$$

from which, noticing the expression for $\boldsymbol{\xi}$ in (2.7), we have

$$\begin{pmatrix}F_z\\ -1\\ F_\varphi/F\end{pmatrix}=K_1\left[\begin{pmatrix}0\\1\\0\end{pmatrix}-G_\eta\boldsymbol{\eta}-G_\psi\boldsymbol{\psi}\right]\equiv K_1[\boldsymbol{s}-G_\eta\boldsymbol{\eta}], \tag{2.9}$$

K_1 being a scalar, and

$$\boldsymbol{s}=\begin{pmatrix}0\\1\\0\end{pmatrix}-G_\psi\boldsymbol{\psi}.$$

By using the expressions for $\boldsymbol{\eta}$ and $\boldsymbol{\psi}$ in (2.7), (2.9) can be further rewritten as

$$\begin{pmatrix}F_z\\ -1\\ F_\varphi/F\end{pmatrix}=K_2\begin{pmatrix}(-G_\eta\varphi_\psi+G_\psi\varphi_\eta)/(E_\eta\varphi_\psi-E_\psi\varphi_\eta)\\ 1+E_r(G_\eta\varphi_\psi-G_\psi\varphi_\eta)/(E_\eta\varphi_\psi-E_\psi\varphi_\eta)\\ (G_\eta E_\psi-G_\psi E_\eta)/F(E_\eta\varphi_\psi-E_\psi\varphi_\eta)\end{pmatrix},$$

where K_2 is a scalar. From the above relations, it is easy to get the expressions for F, F_z, F_φ as functions of G, G_η, G_ψ:

1) When $r_\xi=0$, the treatment described in Chapter 2, § 2 is adopted. The relation $\gamma_\xi=0$ here corresponds to the relation $X_l-X_{l-1}=0$ there.

$$\begin{cases} F=G, \\ F_z=\dfrac{G_\eta\varphi_\psi-G_\psi\varphi_\eta}{E_\eta\varphi_\psi-E_\psi\varphi_\eta+E_r(G_\eta\varphi_\psi-G_\psi\varphi_\eta)}, \\ F_\varphi=\dfrac{-G_\eta E_\psi+G_\psi E_\eta}{E_\eta\varphi_\psi-E_\psi\varphi_\eta+E_r(G_\eta\varphi_\psi-G_\psi\varphi_\eta)}, \end{cases} \tag{2.10}$$

and the expressions for G, G_η, G_ψ as functions of F, F_z, E_φ:

$$\begin{cases} G=F, \\ G_\eta=(F_zE_\eta+F_\varphi\varphi_\eta)/(1-F_zE_r), \\ G_\psi=(F_zE_\psi+F_\varphi\varphi_\psi)/(1-F_zE_r). \end{cases} \tag{2.11}$$

(3) *Basic Equations and Compatibility Relations in the New Coordinate System*

The basic equations in a curvilinear coordinate system $\{\xi_1, \xi_2, \xi_3\}$ were derived in Chapter 4, § 1. Substituting the independent variables η, ξ, ψ for the independent variables ξ_1, ξ_2, ξ_3 there, we can immediately obtain the basic equations in the new coordinate system:

$$A_1\frac{\partial \boldsymbol{U}}{\partial\eta}+A_2\frac{\partial \boldsymbol{U}}{\partial\xi}+A_3\frac{\partial \boldsymbol{U}}{\partial\psi}+\boldsymbol{D}=0, \tag{2.12}$$

where

$$\begin{cases} A_1=\begin{pmatrix} V_\eta & 0 & 0 & \eta_z/\rho & 0 \\ 0 & V_\eta & 0 & \eta_r/\rho & 0 \\ 0 & 0 & V_\eta & \eta_\varphi/r\rho & 0 \\ \rho\eta_z & \rho\eta_r & \rho\eta_\varphi/r & 0 & V_\eta \\ 0 & 0 & 0 & V_\eta & -a^2V_\eta \end{pmatrix}, \\ A_2=\begin{pmatrix} V_\xi & 0 & 0 & \xi_z/\rho & 0 \\ 0 & V_\xi & 0 & \xi_r/\rho & 0 \\ 0 & 0 & V_\xi & \xi_\varphi/r\rho & 0 \\ \rho\xi_z & \rho\xi_r & \rho\xi_\varphi/r & 0 & V_\xi \\ 0 & 0 & 0 & V_\xi & -a^2V_\xi \end{pmatrix}, \\ A_3=\begin{pmatrix} V_\psi & 0 & 0 & \psi_z/\rho & 0 \\ 0 & V_\psi & 0 & \psi_r/\rho & 0 \\ 0 & 0 & V_\psi & \psi_\varphi/r\rho & 0 \\ \rho\psi_z & \rho\psi_r & \rho\psi_\varphi/r & 0 & V_\psi \\ 0 & 0 & 0 & V_\psi & -a^2V_\psi \end{pmatrix}, \\ V_\eta=\boldsymbol{V}\cdot\boldsymbol{\eta},\ V_\xi=\boldsymbol{V}\cdot\boldsymbol{\xi},\ V_\psi=\boldsymbol{V}\cdot\boldsymbol{\psi}, \\ \boldsymbol{V}=\begin{pmatrix} u \\ v \\ w \end{pmatrix}. \end{cases}$$

Since $\det A_1 = -V_\eta^3(V_\eta^2 - a^2\boldsymbol{\eta}\cdot\boldsymbol{\eta})$, the matrix A_1 is invertible if the surface $\eta(z, r, \varphi) = \text{constant}$ is space–like. In this case, (2.12) can be rewritten in the form similar to (2.3) in Chapter 2:

$$\frac{\partial \boldsymbol{U}}{\partial \eta} + B_1 \frac{\partial \boldsymbol{U}}{\partial \xi} + C_1 \frac{\partial \boldsymbol{U}}{\partial \psi} + \boldsymbol{D}_1 = 0, \tag{2.13}$$

where

$$B_1 = A_1^{-1}A_2, \quad C_1 = A_1^{-1}A_3, \quad \boldsymbol{D}_1 = A_1^{-1}\boldsymbol{D}.$$

In order to use the numerical method of Chapter 2, the equations (2.13) should be rewritten in the form (2.11) of Chapter 2. The equations of that form are, in reality, the compatibility relations on the characteristic surfaces through the line

$$\begin{cases} \eta(z, r, \varphi) = \text{constant}, \\ \xi(z, r, \varphi) = \text{constant}. \end{cases} \tag{2.14}$$

According to the results of Chapter 4, § 1, we know that the compatibility relations of (2.13) on characteristic surfaces (two wave–characteristic surfaces and one stream–characteristic surface) through the line (2.14) are as follows (see (1.58) of Chapter 4):

$$\begin{cases} \boldsymbol{G}^+ \cdot \left(\dfrac{\partial \boldsymbol{V}}{\partial \eta} + \sigma_+ \dfrac{\partial \boldsymbol{V}}{\partial \xi}\right) + g_4^+\left(\dfrac{\partial p}{\partial \eta} + \sigma_+ \dfrac{\partial p}{\partial \xi}\right) \\ \quad + \boldsymbol{R}^+ \cdot \dfrac{\partial \boldsymbol{V}}{\partial \psi} + R_4^+ \dfrac{\partial p}{\partial \psi} = -f_1, \\ V_\eta \boldsymbol{V} \cdot \left(\dfrac{\partial \boldsymbol{V}}{\partial \eta} + \sigma \dfrac{\partial \boldsymbol{V}}{\partial \xi}\right) + \dfrac{V_\eta}{\rho}\left(\dfrac{\partial p}{\partial \eta} + \sigma \dfrac{\partial p}{\partial \xi}\right) \\ \quad + V_\psi \left(\boldsymbol{V} \cdot \dfrac{\partial \boldsymbol{V}}{\partial \psi} + \dfrac{1}{\rho} \dfrac{\partial p}{\partial \psi}\right) = -f_2, \\ V_\eta \boldsymbol{T} \cdot \left(\dfrac{\partial \boldsymbol{V}}{\partial \eta} + \sigma \dfrac{\partial \boldsymbol{V}}{\partial \xi}\right) + \dfrac{V_\eta}{\rho}(\boldsymbol{V} \times \boldsymbol{\eta}) \cdot \boldsymbol{\xi}\left(\dfrac{\partial p}{\partial \eta} + \sigma \dfrac{\partial p}{\partial \xi}\right) \\ \quad + V_\psi \boldsymbol{T} \cdot \dfrac{\partial \boldsymbol{V}}{\partial \psi} + \dfrac{1}{\rho} \boldsymbol{T} \cdot \boldsymbol{\psi} \dfrac{\partial p}{\partial \psi} = -f_3, \\ V_\eta \left(\dfrac{\partial p}{\partial \eta} + \sigma \dfrac{\partial p}{\partial \xi}\right) - a^2 V_\eta \left(\dfrac{\partial \rho}{\partial \eta} + \sigma \dfrac{\partial \rho}{\partial \xi}\right) \\ \quad + V_\psi \left(\dfrac{\partial p}{\partial \psi} - a^2 \dfrac{\partial \rho}{\partial \psi}\right) = -f_4, \\ \boldsymbol{G}^- \cdot \left(\dfrac{\partial \boldsymbol{V}}{\partial \eta} + \sigma_- \dfrac{\partial \boldsymbol{V}}{\partial \xi}\right) + g_4^-\left(\dfrac{\partial p}{\partial \eta} + \sigma_- \dfrac{\partial p}{\partial \xi}\right) \\ \quad + \boldsymbol{R}^- \cdot \dfrac{\partial \boldsymbol{V}}{\partial \psi} + R_4^- \dfrac{\partial p}{\partial \psi} = -f_5, \end{cases} \tag{2.15}$$

where

$$
\left\{
\begin{aligned}
&\sigma_\pm=\frac{\beta_2\pm\sqrt{\beta_2^2-\beta_1\beta_3}}{\beta_1}, \quad \beta_1=V_\eta^2-a^2\boldsymbol{\eta}\cdot\boldsymbol{\eta},\\
&\beta_2=V_\eta V_\xi-a^2\boldsymbol{\eta}\cdot\boldsymbol{\xi},\\
&\beta_3=V_\xi^2-a^2\boldsymbol{\xi}\cdot\boldsymbol{\xi},\\
&\boldsymbol{N}_W^\pm=a(\sigma_\pm\boldsymbol{\eta}-\boldsymbol{\xi})/(\sigma_\pm V_\eta-V_\xi),\\
&\boldsymbol{G}^\pm=\rho a(V_\eta\boldsymbol{N}_W^\pm-a\boldsymbol{\eta})=\frac{\rho a^2}{\sigma_\pm V_\eta-V_\xi}\boldsymbol{N}_s,\\
&g_4^\pm=a\boldsymbol{N}_W^\pm\cdot\boldsymbol{\eta}-V_\eta=\pm\sqrt{\beta_2^2-\beta_1\beta_3}/(V_\xi-\sigma_\pm V_\eta),\\
&\boldsymbol{R}^\pm=\rho a(V_\psi\boldsymbol{N}_W^\pm-a\boldsymbol{\psi}),\\
&R_4^\pm=a\boldsymbol{N}_W^\pm\cdot\boldsymbol{\psi}-V_\psi,\\
&\sigma=V_\xi/V_\eta, \quad \boldsymbol{N}_s=V_\xi\boldsymbol{\eta}-V_\eta\boldsymbol{\xi},\\
&\boldsymbol{T}=\begin{pmatrix}t_1\\t_2\\t_3\end{pmatrix}\equiv\boldsymbol{V}\times\boldsymbol{N}_s,\\
&\boldsymbol{D}_{1-3}=\begin{pmatrix}0\\-w^2/r\\vw/r\end{pmatrix},\\
&f_1=\rho a(\boldsymbol{N}_W^+\cdot\boldsymbol{D}_{1-3}-av/r),\\
&f_2=\boldsymbol{V}\cdot\boldsymbol{D}_{1-3}=0,\\
&f_3=\boldsymbol{T}\cdot\boldsymbol{D}_{1-3},\\
&f_4=0,\\
&f_5=\rho a(\boldsymbol{N}_W^-\cdot\boldsymbol{D}_{1-3}-av/r).
\end{aligned}
\right.
$$

If we introduce the following symbols,

$$
\widetilde{G}\equiv\begin{pmatrix}
g_{1,1} & g_{1,2} & g_{1,3} & g_{1,4} & g_{1,5}\\
g_{2,1} & g_{2,2} & g_{2,3} & g_{2,4} & g_{2,5}\\
g_{3,1} & g_{3,2} & g_{3,3} & g_{3,4} & g_{3,5}\\
g_{4,1} & g_{4,2} & g_{4,3} & g_{4,4} & g_{4,5}\\
g_{5,1} & g_{5,2} & g_{5,3} & g_{5,4} & g_{5,5}
\end{pmatrix}
$$

$$
=\begin{pmatrix}
g_1^+ & g_2^+ & g_3^+ & g_4^+ & 0\\
V_\eta u & V_\eta v & V_\eta w & V_\eta/\rho & 0\\
V_\eta t_1 & V_\eta t_2 & V_\eta t_3 & \dfrac{V_\eta(\boldsymbol{V}\times\boldsymbol{\eta})\cdot\boldsymbol{\xi}}{\rho} & 0\\
0 & 0 & 0 & V_\eta & -a^2V_\eta\\
g_1^- & g_2^- & g_3^- & g_4^- & 0
\end{pmatrix},
$$

$$\Lambda \equiv \begin{pmatrix} \lambda_1 & 0 & 0 & 0 & 0 \\ 0 & \lambda_2 & 0 & 0 & 0 \\ 0 & 0 & \lambda_3 & 0 & 0 \\ 0 & 0 & 0 & \lambda_4 & 0 \\ 0 & 0 & 0 & 0 & \lambda_5 \end{pmatrix} = \begin{pmatrix} \sigma_+ & 0 & 0 & 0 & 0 \\ 0 & \sigma & 0 & 0 & 0 \\ 0 & 0 & \sigma & 0 & 0 \\ 0 & 0 & 0 & \sigma & 0 \\ 0 & 0 & 0 & 0 & \sigma_- \end{pmatrix},$$

$$\boldsymbol{W} = -\begin{pmatrix} f_1 + \boldsymbol{R}^+ \cdot \dfrac{\partial \boldsymbol{V}}{\partial \psi} + R_4^+ \dfrac{\partial p}{\partial \psi} \\ f_2 + V_\psi \left(\boldsymbol{V} \cdot \dfrac{\partial \boldsymbol{V}}{\partial \psi} + \dfrac{1}{\rho} \dfrac{\partial p}{\partial \psi} \right) \\ f_3 + V_\psi \boldsymbol{T} \cdot \dfrac{\partial \boldsymbol{V}}{\partial \psi} + \dfrac{1}{\rho} \boldsymbol{T} \cdot \boldsymbol{\psi} \dfrac{\partial p}{\partial \psi} \\ f_4 + V_\psi \dfrac{\partial p}{\partial \psi} - a^2 V_\psi \dfrac{\partial \rho}{\partial \psi} \\ f_5 + \boldsymbol{R}^- \cdot \dfrac{\partial \boldsymbol{V}}{\partial \psi} + R_4^- \dfrac{\partial p}{\partial \psi} \end{pmatrix},$$

then (2.15) can be rewritten in the matrix form

$$\tilde{G} \frac{\partial \boldsymbol{U}}{\partial \eta} + \Lambda \tilde{G} \frac{\partial \boldsymbol{U}}{\partial \xi} = \boldsymbol{W}. \tag{2.16}$$

According to the results of Chapter 4, § 1, we also know that if the surface $\eta(z, r, \varphi) =$constant is space-like, then the equations (2.15), i.e., (2.16) are equivalent to (2.12). Moreover, $\sigma_\pm$ and σ are eigenvalues of $B_1 = A_1^{-1} A_2$, and there is the following relation

$$\sigma_+ > \sigma > \sigma_- \tag{2.17}$$

if $V_\eta > 0$ holds.

2. *Boundary Conditions and Internal Boundary Conditions*

In § 1, we pointed out that we try to "separate" all the shocks, the centered expansion waves, and the entropy layers from the flow field in our process of calculation, and that we want to get accurate results of the intersection and reflection of discontinuities. Thus, the types of boundaries we need to consider in the problem of flow around bodies are surfaces of bodies, shocks, contact discontinuities, wave–characteristics, and stream–characteristics. Therefore, in order to solve the problem of flow around bodies by using the singularity–separating method, we must have the boundary conditions on these types of surfaces. In what follows, we will describe concrete expressions of the boundary conditions and the internal boundary conditions on these surfaces, which are taken as the surfaces $\xi(z, r, \varphi) =$constant in the process of our calculation, and we will show that those conditions satisfy the compatibility conditions for boundary conditions in Chapter 2. Because those conditions are compatible with (2.13), the method of Chapter 2 can be used for the case here.

(1) *Bodies*

Let the equation of the body be $r=F(z, \varphi)$ in the cylindric coordinate system $\{z, r, \varphi\}$. According to (3.1) of Chapter 4, the boundary conditions on bodies are

$$\begin{cases} r=F(z, \varphi), \\ uF_z-v+wF_\varphi/F=0. \end{cases} \tag{2.18}$$

In our process of calculation, we need to introduce the concepts of the lower–body and the upper–body. If gas flows in the region $r>F(z, \varphi)$, it is called the lower–body; if gas flows in the region $r<F(z, \varphi)$, it is called the upper–body.

In our process of computation, the surface of the body is the surface $\xi=0$ if it is a lower–body or $\xi=L$ if it is an upper–body. Thus, its normal vector is (2.8). Accordingly, using the second expression of (2.18), we obtain the relation

$$V_\xi=\boldsymbol{V}\cdot\boldsymbol{\xi}=\frac{1}{r_\xi}(v-V_\eta G_\eta-V_\psi G_\psi)=0. \tag{2.19}$$

Therefore, on bodies where

$$\sigma=\frac{V_\xi}{V_\eta}=0,$$

and from the inequality (2.17), we further obtain the following relation

$$\sigma_+>\sigma=0>\sigma_-.$$

This means that the characteristic number q_L on an upper–body is equal to 4. In this case, the number of boundary conditions γ_L is equal to 2, and the number of the unknowns N is equal to 5. Hence, the condition (1.7) of Chapter 2, i.e., the equality $\gamma_L+q_L=N+1$, holds. Similarly, the characteristic number p_0'[1] on a lower–body is equal to 4, so the condition $\gamma_0+p_0'=N+1$ also holds.

(2) *Shocks*

The quantities on the two sides of a shock should satisfy the Rankine–Hugoniot conditions (see (2.4) of Chapter 4)

$$\begin{cases} h(p_1, \rho_1)+\dfrac{1}{2}V_{1,n}^2=h(p_0, \rho_0)+\dfrac{1}{2}V_{0,n}^2, \\ p_1+\rho_1V_{1,n}^2=p_0+\rho_0V_{0,n}^2, \\ \boldsymbol{V}_1=\boldsymbol{V}_0-\left(1-\dfrac{\rho_0}{\rho_1}\right)V_{0,n}\boldsymbol{n}, \end{cases} \tag{2.20}$$

where $\boldsymbol{n}$ is the unit normal vector to the shock, a quantity with the subscript 0 (or 1) denotes a quantity in front of (or behind) the shock,

1) In order to distinguish the characteristic number p in Chapter 2 from the pressure p, the characteristic number p in Chapter 2 will be substituted by p' in this chapter.

$\boldsymbol{V}_{0,n}=\boldsymbol{V}_0\cdot\boldsymbol{n}$, $V_{1,n}=\boldsymbol{V}_1\cdot\boldsymbol{n}$, and $h(p, \rho)$ is the specific enthalpy of gas. For the perfect gas, i.e.,

$$h=\frac{\gamma p}{(\gamma-1)\rho};$$

for air in chemical equilibrium, an approximating expression for $h(p, \rho)$ is given in Chapter 4, § 4.

The quantities in front of the main shock——the quantities of the free stream——are known. Suppose that the free stream is uniform, and that the velocity, pressure and density are V_∞, p_∞ and ρ_∞ respectively. Furthermore, suppose that the velocity of the free stream is parallel to the plane $\varphi=0$ and $\varphi=\pi$ ($\varphi=0$ being the windward side) and that the angle between the velocity and the z-axis, the so-called angle of attack, is α. Then the quantities in front of the main shock are

$$\begin{cases} \dfrac{u_0}{V_\infty}=\cos\alpha, \\ \dfrac{v_0}{V_\infty}=-\sin\alpha\cos\varphi, \\ \dfrac{w_0}{V_\infty}=\sin\alpha\sin\varphi, \\ \dfrac{p_0}{\rho_\infty V_\infty^2}=\dfrac{1}{\gamma M_\infty^2}, \\ \dfrac{\rho_0}{\rho_\infty}=1. \end{cases} \tag{2.21}$$

When introducing a coordinate transformation, we always choose a function $E(r, \eta, \psi)$ such that the surfaces $\eta(z, r, \varphi)=$constant are space-like, and such that η increases along the direction of flow. Moreover, as is pointed out, $\boldsymbol{\eta}$ is the gradient of the surface $\eta(z, r, \varphi)=$ constant. Thus we have the relations

$$\begin{cases} V_{0,\eta}=\boldsymbol{V}_0\cdot\boldsymbol{\eta}>a_0|\boldsymbol{\eta}|, \\ V_{1,\eta}=\boldsymbol{V}_1\cdot\boldsymbol{\eta}>a_1|\boldsymbol{\eta}|. \end{cases}$$

On the other hand, ξ increases from the body to the main shock, and $\boldsymbol{\xi}$ is the gradient of the surface $\xi(z, r, \varphi)=$constant. Thus, for the first family of shocks, $V_\xi=\boldsymbol{V}\cdot\boldsymbol{\xi}<0$. Moreover, the normal velocities in front of and behind the shock should satisfy the relation (2.16) of Chapter 4. Therefore, we have

$$V_{0,\xi}=\boldsymbol{V}_0\cdot\boldsymbol{\xi}\leqslant -a_0|\boldsymbol{\xi}|, \quad 0>V_{1,\xi}=\boldsymbol{V}_1\cdot\boldsymbol{\xi}\geqslant -a_1|\boldsymbol{\xi}|.$$

From these relations, we can see that, in front of the first family of shocks, there are the following inequalities:

$$\begin{cases} \beta_{0,1}=V_{0,\eta}^2-a_0^2\boldsymbol{\eta}\cdot\boldsymbol{\eta}>0, \\ \beta_{0,2}=V_{0,\eta}V_{0,\xi}-a_0^2\boldsymbol{\eta}\cdot\boldsymbol{\xi}<0, \\ \beta_{0,3}=V_{0,\xi}^2-a_0^2\boldsymbol{\xi}\cdot\boldsymbol{\xi}\geqslant 0. \end{cases}$$

Furthermore, we can also obtain

$$\sigma_{0+}=\frac{\beta_{0,2}+\sqrt{\beta_{0,2}^2-\beta_{0,1}\beta_{0,3}}}{\beta_{0,1}}\leqslant 0.$$

Similarly, for the quantities behind the first family of shocks, we have

$$\begin{cases}\beta_{1,1}=V_{1,\eta}^2-a_1^2\boldsymbol{\eta}\cdot\boldsymbol{\eta}>0,\\ \beta_{1,3}=V_{1,\xi}^2-a_1^2\boldsymbol{\xi}\cdot\boldsymbol{\xi}\leqslant 0,\end{cases}$$

and
$$\sigma_{1+}=\frac{\beta_{1,2}+\sqrt{\beta_{1,2}^2-\beta_{1,1}\beta_{1,3}}}{\beta_{1,1}}\geqslant 0.$$

From these relations and (2.17), the following inequalities are immediately obtained:

$$\begin{cases}0\geqslant\sigma_{0+}>\sigma_0>\sigma_{0-},\\ \sigma_{1+}\geqslant 0>\sigma_1>\sigma_{1-}.\end{cases}$$

This means that for the first family of shocks, $p_l'=5$ and $q_l=1$ if the positive ξ–direction (ξ increases along this direction) is "opposite" to the direction of flow, i.e., if $\boldsymbol{V}\cdot\boldsymbol{\xi}<0$. In the case here, $\gamma_l=5$. Thus, if it is an internal boundary, we have the relation $\gamma_l+p_l'+q_l=2N+1$; if it is a boundary, we have the relation $\gamma_L+q_L=N+1$. That is, (1.7) of Chapter 2 holds in this case.

For the second family of shocks, if the positive ξ–direction agrees with the direction of flow, i.e., if $\boldsymbol{V}\cdot\boldsymbol{\xi}>0$, then we can obtain

$$\begin{cases}\sigma_{0+}>\sigma_0>\sigma_{0-}\geqslant 0,\\ \sigma_{1+}>\sigma_1>0\geqslant\sigma_{1-}.\end{cases}$$

Therefore, $p_l'=1$ and $q_l=5$. We can easily see from these two equations that (1.7) of Chapter 2 also holds in this case.

(3) *Contact Discontinuities*

Under our selection of independent variables, the normal direction to a contact discontinuity is just the direction $\boldsymbol{\xi}$. Thus, the jump conditions, which the quantities on the two sides of a contact discontinuity should satisfy, i.e., the conditions (2.3)of Chapter 4, can be rewritten as

$$\begin{cases}p_0=p_1,\\ V_{0,\xi}=\boldsymbol{V}_0\cdot\boldsymbol{\xi}=0,\\ V_{1,\xi}=\boldsymbol{V}_1\cdot\boldsymbol{\xi}=0.\end{cases}\tag{2.22}$$

From (2.22), (2.17) and the relation $\sigma=\dfrac{V_\xi}{V_\eta}$, we know that on every side of a contact discontinuity, there is the relation

$$\sigma_+>\sigma=0>\sigma_-.$$

This means that $p_l'=q_l=4$ on a contact discontinuity. Here $\gamma_l=3$. Thus the condition $\gamma_l+p_l'+q_l=2N+1$ is satisfied.

(4) *Wave–Characteristic Surfaces*

According to the definition of a wave–characteristic surface (see

Chapter 4, § 1), we have the relation

$$|\boldsymbol{V}\cdot\boldsymbol{n}|=a,$$

where $\boldsymbol{n}$ is a unit normal vector to a wave–characteristic surface. On the other hand, we can obtain the relation $\boldsymbol{n}=(\boldsymbol{s}-G_\eta\boldsymbol{\eta})/|\boldsymbol{s}-G_\eta\boldsymbol{\eta}|$ from (2.9). Substituting this expression of $\boldsymbol{n}$ into the above relation and squaring the terms on both sides of that relation, we derive a quadratic equation for G_η

$$\alpha_1 G_\eta^2-2\alpha_2 G_\eta+\alpha_3=0,$$

where

$$\begin{cases}\alpha_1=V_\eta^2-a^2\boldsymbol{\eta}\cdot\boldsymbol{\eta},\\ \alpha_2=V_\eta(v-V_\psi G_\psi)-a^2\boldsymbol{\eta}\cdot\boldsymbol{s},\\ \alpha_3=(v-V_\psi G_\psi)^2-a^2\boldsymbol{s}\cdot\boldsymbol{s}.\end{cases}$$

Thus, we have

$$G_\eta=(\alpha_2\pm\sqrt{\alpha_2^2-\alpha_1\alpha_3})/\alpha_1. \tag{2.23}$$

Here α_1 is always greater than zero since we choose independent variables such that the surfaces $\eta(z, r, \varphi)=$constant are space–like. Therefore, the positive sign corresponds to the first family of wave–characteristic surfaces, and the negative sign to the second family of wave–characteristic surfaces.

On a wave–characteristic surface, in addition to the fact that the relation (2.23) should be satisfied, the flow parameters u, v, w, p, ρ on both sides should be the same. Because the flow parameters on each side of a wave–characteristic surface should satisfy a wave–compatibility relation, there can be only four independent parameters. This means that, in order to guarantee that the flow variables are continuous, only four conditions are required. For example, as connection conditions we can require that the other four "characteristic–combined quantities" on one side be equal to those on the other side. For the first family of wave–characteristic surfaces, the four conditions are

$$\begin{cases}(V_\eta\boldsymbol{V}\cdot\boldsymbol{V}+V_\eta p/\rho)_0=(V_\eta\boldsymbol{V}\cdot\boldsymbol{V}+V_\eta p/\rho)_1,\\ (V_\eta\boldsymbol{T}\cdot\boldsymbol{V}+V_\eta p(\boldsymbol{V}\times\boldsymbol{\eta})\cdot\boldsymbol{\xi}/\rho)_0\\ \quad=(V_\eta\boldsymbol{T}\cdot\boldsymbol{V}+V_\eta p(\boldsymbol{V}\times\boldsymbol{\eta})\cdot\boldsymbol{\xi}/\rho)_1,\\ (V_\eta p-a^2V_\eta\rho)_0=(V_\eta p-a^2V_\eta\rho)_1,\\ (\boldsymbol{G}^-\cdot\boldsymbol{V}+g_4^-p)_0=(\boldsymbol{G}^-\cdot\boldsymbol{V}+g_4^-p)_1.\end{cases} \tag{2.24a}$$

For the second family of wave–characteristic surfaces, the four conditions are

$$\begin{cases}(\boldsymbol{G}^+\cdot\boldsymbol{V}+g_4^+p)_0=(\boldsymbol{G}^+\cdot\boldsymbol{V}+g_4^+p)_1,\\ (V_\eta\boldsymbol{V}\cdot\boldsymbol{V}+V_\eta p/\rho)_0=(V_\eta\boldsymbol{V}\cdot\boldsymbol{V}+V_\eta p/\rho)_1,\\ (V_\eta\boldsymbol{T}\cdot\boldsymbol{V}+V_\eta p(\boldsymbol{V}\times\boldsymbol{\eta})\cdot\boldsymbol{\xi}/\rho)_0=(V_\eta\boldsymbol{T}\cdot\boldsymbol{V}+V_\eta p(\boldsymbol{V}\times\boldsymbol{\eta})\cdot\boldsymbol{\xi}/\rho)_1,\\ (V_\eta p-a^2V_\eta\rho)_0=(V_\eta p-a^2V_\eta\rho)_1.\end{cases} \tag{2.24b}$$

According to the definition of independent variables aforementioned and noticing that $\boldsymbol{\xi}/|\boldsymbol{\xi}|$ is the unit normal vector to the surface, we know that $V_\xi=\boldsymbol{V}\cdot\boldsymbol{\xi}=-a|\boldsymbol{\xi}|<0$ on the first family of wave-characteristic surfaces. Hence, $\beta_1>0$, $\beta_2<0$ and $\beta_3=0$, and we further have

$$\sigma_+=\frac{\beta_2+\sqrt{\beta_2^2-\beta_1\beta_3}}{\beta_1}=0.$$

Therefore, from the inequality (2.17) we have $p_l'=5$, $q_l=1$. On the second family of wave-characteristic surfaces, we have $V_\xi=a|\boldsymbol{\xi}|$, so that $\sigma_-=0$ and $p_l'=1$, $q_l=5$.

In this case, the total number of the given boundary or internal boundary conditions is 5. Therefore, the condition $\gamma_l+p_l'+q_l=2N+1$ is always satisfied on the wave-characteristic surfaces.

We should point out that the internal boundary conditions (2.24a) or (2.24b) are not convenient for practical application. The two compatibility relations on the two sides of a wave-characteristic surface, the condition (2.23) and the connection conditions (2.24a) or (2.24b), are equivalent to the compatibility relation on one side of the wave-characteristic surface, the condition (2.23), and the continuity conditions

$$\boldsymbol{U}_0=\boldsymbol{U}_1. \tag{2.24c}$$

Clearly, the conditions (2.23) and (2.24c) can be rewritten in the form (3.13) of Chapter 4.

(5) *Stream-Characteristic Surfaces*

If $\boldsymbol{n}$ stands for a normal direction to a stream-characteristic surface, then the velocity on the surface should satisfy (see § 1 of Chapter 4)

$$\boldsymbol{V}\cdot\boldsymbol{n}=0,$$

which can be rewritten as

$$G_\eta=\frac{v-V_\psi G_\psi}{V_\eta} \tag{2.25}$$

by using (2.9).

Besides (2.25), the flow variables $\boldsymbol{U}$ on two sides should satisfy the continuity condition. Since the flow variables on each side of a stream-characteristic surface should satisfy three compatibility relations on the stream-characteristic surface, there are only two independent parameters on each side. Hence, in order to guarantee that the flow variables are continuous, two more conditions are needed. For example, as connection conditions we can require that the two "wave-characteristic-combined quantities" on one side be equal to those on the other side:

$$\begin{cases}(\boldsymbol{G}^+\cdot\boldsymbol{V}+g_4^+p)_0=(\boldsymbol{G}^+\cdot\boldsymbol{V}+g_4^+p)_1,\\ (\boldsymbol{G}^-\cdot\boldsymbol{V}+g_4^-p)_0=(\boldsymbol{G}^-\cdot\boldsymbol{V}+g_4^-p)_1.\end{cases} \tag{2.26}$$

Because $\sigma=\dfrac{V_\xi}{V_\eta}=0$ on a stream-characteristic surface, from (2.17) we have

$$\sigma_+ > \sigma = 0 > \sigma_-.$$

Thus, $p_l' = q_l = 4$ on a stream–characteristic surface. On the other hand, there are three internal boundary conditions. Therefore, the condition $\gamma_l + p_l' + q_l = 2N + 1$ holds.

The connection conditions (2.26) are not convenient for practical application. The stream–compatibility relations on two sides of a stream–characteristic surface, the condition (2.25) and the connection condition (2.26), are equivalent to the three stream–compatibility relations on one side, the condition (2.25), and the continuity conditions (2.24c). In practical computation, the three stream–compatibility relations on one side of a stream–characteristic surface, the condition (2.25), and the continuity conditions (2.24c) are used. Clearly, (2.25) and (2.24c) can be rewritten as (3.12) in Chapter 4.

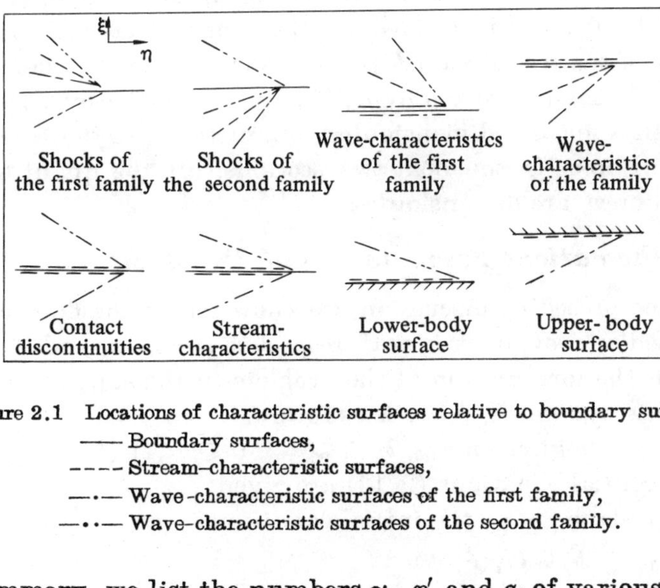

Figure 2.1 Locations of characteristic surfaces relative to boundary surfaces
—— Boundary surfaces,
---- Stream–characteristic surfaces,
—·— Wave-characteristic surfaces of the first family,
—··— Wave-characteristic surfaces of the second family.

In summary, we list the numbers γ_l, p_l' and q_l of various boundaries for the problem of flow around bodies in Table 2.1. As mentioned previously, those boundaries are planes $\xi = \text{constant}$ in the space $\{\eta, \xi, \psi\}$. Hence, p_l' and q_l are numbers of backward characteristic surfaces passing through the intersection line of a boundary plane $\xi = l$ with a plane $\eta = \text{constant}$ respectively in the upper and the lower regions (see Fig. 2.1). We have given the concrete expressions of boundary or internal boundary conditions on various boundaries, and have proved

Table 2.1 The values of γ_l, p'_l and q_l on various boundaries

Types of boundaries	γ_l	p'_l	q_l
Main shock	5	—	1
Shock of the first family	5	5	1
Shock of the second family	5	1	5
Wave–characteristic of the first family	5	5	1
Wave–characteristic of the second family	5	1	5
Contact discontinuity	3	4	4
Stream–characteristic	3	4	4
Upper–body surface	2	—	4
Lower–body surface	2	4	—

that the number γ_l satisfies the condition (1.7) of Chapter 2. As shown in practical computation, for those boundary conditions and internal boundary conditions listed previously, the systems corresponding to (1.8) or (2.10) of Chapter 2 usually have reasonable solutions. This means that those boundary conditions and internal boundary conditions are compatible with the equations (2.13).

For convenience, we use the following dimensionless quantities in the computation: z and r, divided by the length factor L^*; u, v, w, a by the velocity of the free stream V_∞^*; the density ρ by the density of the free stream ρ_∞^*; the pressure p by $\rho_\infty^* V_\infty^{*2}$; and the enthalpy h by V_∞^{*2}. It is easy to see that for the dimensionless quantities, all the equations and relations still hold. In the next two sections, all the quantities, except some special ones, are dimensionless.

3. *Mathematical Formulation of Problems*

From the preceding discussion, we know that if the flow field around combined bodies is symmetric with respect to $\varphi=0\sim\pi$, and if $H_0=0$ and $H_1=\pi$, then the formulation of the problem in the supersonic region in the coordinate system $\{\eta, \xi, \psi\}$ is as follows:

I. In every region $\{\eta>\eta_0, l-1\leqslant\xi\leqslant l, 0\leqslant\psi\leqslant1\}$ $(l=1, 2,\cdots, L)$, the partial differential equations (2.13) are given.

II. At a plane $\eta=\eta_0$, the initial conditions

1. $\boldsymbol{U}(\eta_0, \xi, \psi)=\boldsymbol{U}_l^0(\xi, \psi)$,

$$(l-1\leqslant\xi\leqslant l,\ l=1,\ 2,\ \cdots,\ L;\ 0\leqslant\psi\leqslant1);$$

2. $$\begin{cases} G_l(\eta_0, \psi)=G_l^0(\psi), \\ G_{\eta,l}(\eta_0, \psi)=G_{\eta,l}^0(\psi), \end{cases} \quad (l=0,\ 1,\ 2,\ \cdots,\ L;\ 0\leqslant\psi\leqslant1)$$

are given

III. On the boundaries $\{\eta>\eta_0, \xi=l, 0\leqslant\psi\leqslant1\}$ $(l=0, 1, \cdots, L)$, the conditions corresponding to types of boundary surfaces, such as (2.18) to (2.26), are given.

IV. On the plane $\psi=0$ (i.e., $\varphi=0$), for u, v, p, ρ, the symmetric condition

$$f(\Delta\psi)=f(-\Delta\psi) \tag{2.27a}$$

is given; for w, the anti-symmetric condition

$$f(\Delta\psi)=-f(-\Delta\psi) \tag{2.27b}$$

is given. On the plane $\psi=1$ (i.e., $\varphi=\pi$), for u, v, p, ρ, the symmetric condition

$$f(1+\Delta\psi)=f(1-\Delta\psi) \tag{2.28a}$$

is given; for w, the anti-symmetric condition

$$f(1+\Delta\psi)=-f(1-\Delta\psi) \tag{2.28b}$$

is given.

We need to determine $G_l(\eta, \psi)$ (the shapes of boundaries) and the flow parameters in all the regions which satisfy the above equations and conditions.

It is easy to see that the formulation of problem here is the same as that for the problems (2.3)—(2.5) and (1.10a—b) of Chapter 2. Thus, the method of Chapter 2, § 2, can be applied here. If a section-wise equidistant net is adopted in the φ-direction, the difference scheme can be applied only with a little change, because the treatment in the φ-direction of the difference scheme of Chapter 2, § 2, is explicit. The only change is that on the planes $\psi=m$ ($m=1, 2, \cdots, M-1$), a certain special treatment is needed, which will be described in § 3.

If the flow field has no symmetry, the formulation of this problem is in the form of I—IV in Chapter 2, § 1. Thus, the scheme of Chapter 2, § 2, can also be used.

§ 3 Numerical Methods

1. *Construction of Difference Equations*

We see from Chapter 2, § 2, that the difference equations in the l-th region[1] are constructed in the following way. We first divide the characteristic form equations (2.15) into three groups according to the characteristic number p'_{l-1} on the upper side of the lower boundary of that region and to the characteristic number q_l on the lower side of the upper boundary. Then, every group of partial differential equations is approximated by a certain difference scheme. Thus, in constructing difference equations, we can merge all the various boundaries with the same characteristic numbers p'_l and q_l into a type of boundary. Therefore, according to Table 2.1, all the boundaries considered can be divided into three types, namely,

1) It denotes the region $l-1\leqslant\xi\leqslant l$, $m-1\leqslant\psi\leqslant m$.

(1) the first type of wave–characteristic boundaries with $p'_l=5$ and $q_l=1$, which contains the first family of shocks[1] and the first family of wave–characteristic surfaces;

(2) the type of stream–characteristic boundaries with $p'_l=q_l=4$, which contains the stream–characteristic boundaries, contact discontinuities and surfaces of bodies[2];

(3) the second type of wave–characteristic boundaries with $p'_l=1$ and $q_l=5$, which contains the second family of shocks and the second family of wave–characteristic surfaces.

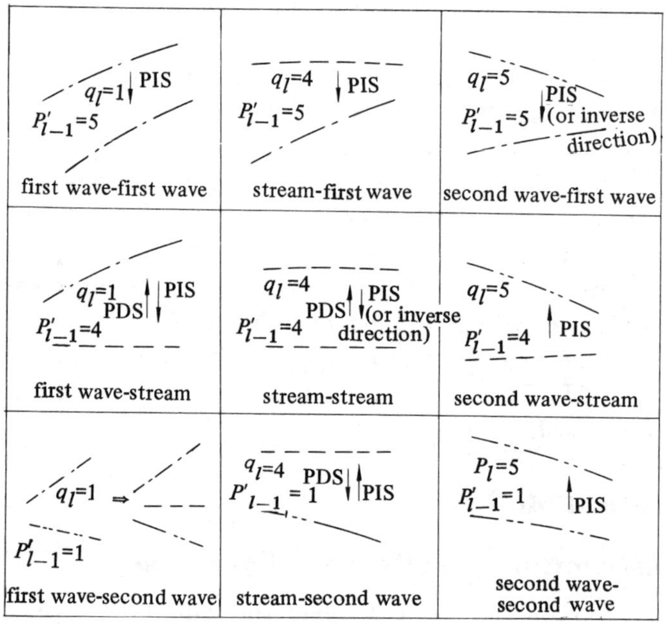

Figure 3.1 The types of regions and the directions of the processes of the "direct sweep" and the "inverse sweep" for each type

—·— The first type of wave–characteristic boundaries,

---- The type of stream–characteristic boundaries,

—··— The second type of wave–characteristic boundaries,

PDS——The process of "direct sweep", PIS——The process of "inverse sweep".

---

1) In this chapter, the main shock belongs to the first family of shocks. It is always an upper boundary, and only its characteristic number q_l is used.

2) The lower–body is always a lower boundary of regions, and only its characteristic number p'_l is used in this case; the upper–body is always an upper boundary of regions, and only its characteristic number q_l is used in this case.

Therefore, there are nine types of regions in the problem of flow around bodies. We show the types of the upper and lower boundaries and the characteristic numbers p'_{l-1}, q_l for the nine types of regions in Fig. 3.1. As pointed out in Chapter 2, the scheme of computation there is applicable only to the regions with $p'_{l-1}+q_l \geqslant N$ (now $N=5$). Eight types among the nine types of regions satisfy the condition $p'_{l-1}+q_l \geqslant N$, only the "first wave–second wave" region[1] where $p'_{l-1}+q_l=2$, not satisfying that condition. In order to be able to use the method of Chapter 2, we introduce a stream-surface as an internal boundary in that type of region. Then the original region is divided into two. One is a "first wave–stream" region[2], the other is a "stream–second wave" region[3]. Hence, it is necessary for us to consider only the eight types of regions with $p'_{l-1}+q_l \geqslant 5$, but not the "first wave–second wave" region.

In any region with $p'_{l-1}+q_l \geqslant 5$, the equations (2.15) are divided into three groups, depending on the characteristic numbers p'_{l-1} and q_l. The first through $(5-p'_{l-1})$-th equations are called the first group of equations. The $(5-p'_{l-1}+1)$-th through q_l-th equations are called the second group. The (q_l+1)-th through the fifth equations are called the third group. It is possible that one or two groups involve no equations. Table 3.1 shows which equations each group contains in each region.

Using Table 3.1 and the difference scheme of Chapter 2, § 2, we can obtain the desired difference equations in each region. Since we allow ourselves to use a "section-wise" equidistant net in the φ-direction, a certain treatment on the planes $\psi=m$ $(m=0, 1, \cdots, M)$ is needed. We shall now describe the process of constructing difference equations.

Suppose that in the region $\{\eta>\eta_0, l-1\leqslant\xi\leqslant l, m-1\leqslant\psi\leqslant m\}$ $(l=1, 2, \cdots, L; m=1, 2, \cdots, M)$, the three mesh-steps are

$$\Delta\eta, \quad \Delta\xi_{l,m}=\frac{1}{I_{l,m}} \ (l=1, 2, \cdots, L), \quad \Delta\psi_m=\frac{1}{J_m} \ (m=1, 2, \cdots, M).$$

Here $f^k_{i,l,j,m}$ stands for the value of $f(\eta, \xi, \psi)$ at a mesh-point $\{\eta_k=\eta_0+k\Delta\eta,\ \xi_{i,l,m}=l-1+i\Delta\xi_{l,m},\ \psi_{j,m}=m-1+j\Delta\psi_m\}$. Moreover, on a discontinuity boundary, its value in the l-th region is given if it has a subscript l. For simplicity, if no confusion results when the subscript m or the subscript l is omitted, then we omit it. In what follows, we shall describe how to obtain the difference equations for a fixed m, l and j. (We take $j=0, 1, \cdots, J$ when the quantities at a normal level are computed, and take $j=\frac{1}{2}, \cdots, J-\frac{1}{2}$ when the quantities at an interim level are computed.) The three groups of partial differential equations

1) The "first wave–second wave" region means that its upper boundary belongs to the first type of wave-characteristic boundaries, and the lower boundary belongs to the second type of wave-characteristic boundaries.

2), 3) Their meanings are similar to that of the "first wave–second wave" region.

Table 3.1 The grouping about equations (2.15) in various regions

The type of the lower boundary	The type of the upper boundary: "First wave"	"Stream"	"Second wave"
"First wave"	① No equation* ② A compatibility relation on the first family of wave–characteristic surfaces ③ Three compatibility relations on the stream characteristic surfaces and a compatibility relation on the second family of wave–characteristic surfaces	① No equation ② A compatibility relation on the first family of wave–characteristic surfaces and three compatibility relations on the stream characteristic surfaces ③ A compatibility relation on the second family of wave–characteristic surfaces	① No equation ② All equations ③ No equation
"Stream"	① A compatibility relation on the first family of wave–characteristic surfaces ② No equation ③ Three compatibility relations on the stream characteristic surfaces and a compatibility relation on the second family of wave–characteristic surfaces	① A compatibility relation on the first family of wave–characteristic surfaces ② Three compatibility relations on the stream characteristic surfaces ③ A compatibility relation on the second family of wave–characteristic surfaces	① A compatibility relation on the first family of wave–characteristic surfaces ② Three compatibility relations on the stream characteristic surfaces and a compatibility relation on the second family of wave–characteristic surfaces ③ No equation
"Second wave"	Not to be considered	① A compatibility relation on the first family of wave–characteristic surfaces and three compatibility relations on the stream characteristic surfaces ② No equation ③ A compatibility relation on the second family of wave–characteristic surfaces	① A compatibility relation on the first family of wave–characteristic surfaces and three compatibility relations on the stream characteristic surfaces ② A compatibility relation on the second family of wave–characteristic surfaces ③ No equation

* "① No equation" means that the first group contains no equations.

are approximated in the following way.

(1) Every equation of the first group is approximated at $i=1, 2, \cdots, I$. If $|\lambda|>\varepsilon$ at a point, Scheme I is adopted; otherwise, Scheme II is adopted, where λ stands for σ_+, σ, or σ_-. Adopting Scheme I means using (2.13) of Chapter 2 for an interim level and using (2.14) of Chapter 2 for a normal level; adopting Scheme II means using (2.19) of Chapter 2 for an interim level and using (2.20) or (2.21) of Chapter 2 for a normal level, The subscript i here corresponds to the subscript m in Chapter 2.

(2) Every equation of the second group is approximated by Scheme II at $i=0, 1, \cdots, I$.

(3) Every equation of the third group is approximated at $i=0, 1, \cdots, I-1$. If $|\lambda|>\varepsilon$ at a point, Scheme I is adopted; otherwise Scheme

We need to point out that when a partial differential equation is approximated on a plane $j=0$ or J in the manner described in (1)—(3), certain quantities at certain points outside the region appear in difference equations as given quantities. Those quantities are obtained in the following way. (We suppose that the flow field has a symmetric plane $\varphi=0, \pi$. If not, those quantities can be obtained in a similar way.)

a. If $j=0$ (or $j=J$) corresponds to $\varphi=0$ (or $\varphi=\pi$), then those needed quantities outside the region can be obtained by the conditions (2.27) and (2.28) as follows. In this case, the quantities on the line $\eta=\eta_k+\alpha\Delta\eta$, $\psi=-\frac{\Delta\psi}{2}$ $\left(\text{i.e., } \varphi=-\frac{\Delta\varphi}{2}\right)$ or $\psi=M+\frac{\Delta\psi}{2}$ $\left(\text{i.e., } \varphi=\pi+\frac{\Delta\varphi}{2}\right)$, which does not belong to the computational region, appear in difference equations, and they can be obtained from the quantities on the line $\eta=\eta_k+\alpha\Delta\eta$, $\psi=\frac{\Delta\psi}{2}$ or $\psi=M-\frac{\Delta\psi}{2}$, which belongs to the computational region, by using (2.27) and (2.28) (see Fig. 3.2).

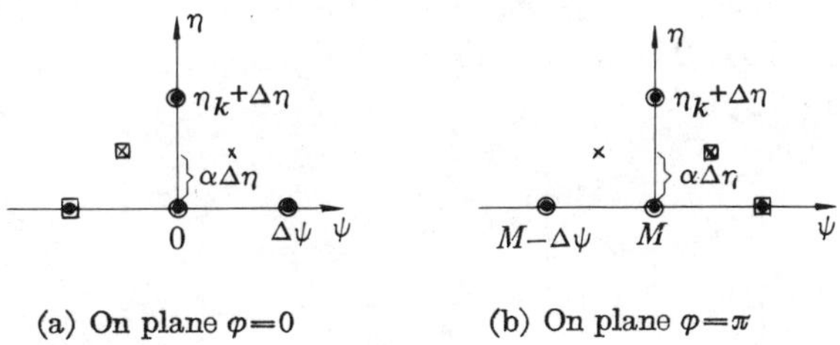

Figure 3.2 Schematic diagram of computational process on a symmetric plane
"⊙", "×"——Mesh lines on a normal level and on an interim level, respectively,
"⊡" "⊠"——Mesh lines on a normal level and on an interim level outside the computational region, respectively.

b. If the plane $j=0$ (or $j=J$) is a border between two regions with different mesh steps in the φ-direction, then the quantities at points outside the computational region are obtained from the quantities in the other region by interpolation. For example, suppose that the interval $0\leqslant\varphi\leqslant\pi$ is divided into two intervals $0\leqslant\varphi\leqslant\varphi^*$ and $\varphi^*\leqslant\varphi\leqslant\pi$, and that the mesh step in the φ-direction in $0\leqslant\varphi\leqslant\varphi^*$ is $\Delta\varphi_1$, and the mesh step in $\varphi^*\leqslant\varphi\leqslant\pi$ is $\Delta\varphi_2$, i.e., the mesh step is $\Delta\psi_1$ if $0\leqslant\psi\leqslant1$, and $\Delta\psi_2$ if $1\leqslant\psi\leqslant2$ (see Fig. 3.3). In this case, we cannot directly construct the difference equations on $\psi=1$ (i.e., $\varphi=\varphi^*$) in the manner described in (1)—(3) because quantities in different regions are involved in the difference equations. In order to use the method described in (1)—(3), we calculate the quantities on an auxiliary line at the known level $\eta=\eta_k$ before the "normal" computation is carried out. For example, the quantities on the line $\eta=\eta_k$, $\varphi=\varphi^*-\Delta\varphi_2$ (this line is denoted by ⊡ with a number 4 in Fig. 3.3) can be obtained from the quantities in the region $0\leqslant\varphi\leqslant\varphi^*$ by interpolation. In order to match the interpolation with a second order difference scheme, the quantities on that line are obtained from the three lines with numbers 7, 8 and 5 by the quadratic interpolation. After that, we can obtain the difference equations on the plane $\psi=1-\frac{1}{2}\Delta\psi_2$ $\left(\text{i.e., } \varphi=\varphi^*-\frac{\Delta\varphi_2}{2}\right)$ in the manner of (1)—(3) which we need in order to compute the quantities on the auxiliary line "2" at an interim level. Furthermore, we can also get the difference equations on $\psi=1$ which we need in order to compute the quantities on the line "1". That is, after we introduce two auxiliary lines "4" and "2", the quantities on the line "1" can be obtained from the quantities on the three lines "4", "5" and "6" in the same manner as the quantities on the other lines in the region $1\leqslant\psi\leqslant2$. However, we should notice that the definitions of G_ψ are different in different regions. Hence, we have to transform certain values of G_ψ into the values to be used in the computation. The transformation of values of G_ψ can be done in the following way. First, F, F_z, F_φ are determined from the given G, G_η, G_ψ, by using (2.10). Then, the desired G, G_η, G_ψ are obtained from F, F_z, F_φ by using (2.11).

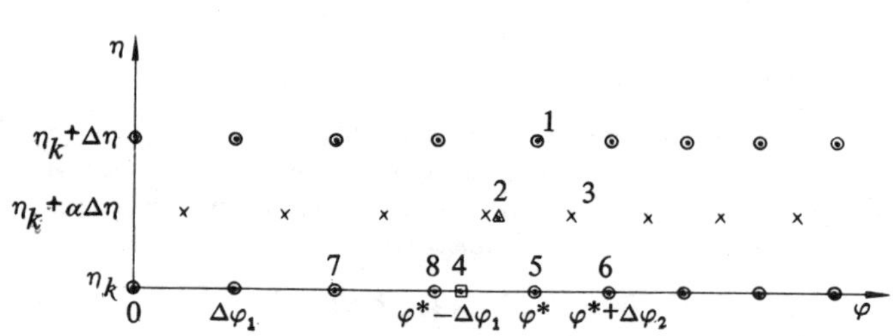

Figure 3.3 Schematic diagram of computational process on the plane which is a border between two regions with different mesh steps in the φ-direction "⊙", "×", "⊡", "⊠"——For their meanings, see Fig. 3.2.

Besides those difference equations mentioned above, two equations to determine the geometric quantities $G_{l,j}$, $G_{\psi,l,j}$ of the l-th boundary are needed for every couple of subscripts (l, j). (2.22) or (2.23) of Chapter 2 provides the two desirable equations.

Clearly, the total number of the difference equations obtained and the boundary and internal boundary conditions on boundaries $\xi=l\,(l=0, 1, \cdots, L)$ is always equal to the total number of the unknowns u, v, w, p, ρ at all points and the geometric quantities G, G_η, G_ψ on all boundaries.

If the partial differential equations are approximated by Scheme II everywhere, i.e., if the treatment in the ξ-direction is explicit everywhere, then the quantities at each point can be determined independently. If the partial differential equations are approximated in a mixed manner by Schemes I and II, the system of equations obtained will be solved by using the block-double-sweep method of Chapter 1, § 5.

2. *Solution of Systems of Difference Equations*

In Chapter 2, we pointed out that since the treatment in the φ-direction (i.e., in the ψ-direction) is explicit, we can solve the problem one φ-plane by one φ-plane. That is, in the coordinate system $\{\eta, \xi, \psi\}$, the quantities on a coordinate line in the ξ-direction can be obtained independently. Corresponding with the fact that the flow field is divided into several flow regions, a coordinate line in the ξ-direction is also divided into several intervals. We have pointed out that there are only eight types of flow regions we need to consider. Thus we have only eight types of intervals which need to be considered in our process of computation. In the following, we first describe how to apply the method of Chapter 1, § 5, to different intervals and then we describe the process of computing the quantities on the whole line.

(1) *Solution of Difference Systems in Different Types of Intervals*

As mentioned above, equations (2.15) are approximated in the following way. They are first divided into three groups, in terms of Table 3.1, and are then approximated in different manners for different groups. Thus, the system of difference equations in the l-th interval can be written in the following form

$$\begin{cases} \widetilde{G}_1^{(i+1)}\boldsymbol{U}_{i+1}+\Omega_1^{(i+1)}\widetilde{G}_1^{(i+1)}\boldsymbol{U}_i=\widetilde{\boldsymbol{F}}_1^{(i+1)}, \quad i=0, 1, \cdots, I-1, & (3.1\text{a}) \\ \widetilde{G}_2^{(i)}\boldsymbol{U}_i=\widetilde{\boldsymbol{F}}_2^{(i)}, \quad i=0, 1, \cdots, I, & (3.1\text{b}) \\ \Omega_3^{(i)}\widetilde{G}_3^{(i)}\boldsymbol{U}_{i+1}+\widetilde{G}_3^{(i)}\boldsymbol{U}_i=\widetilde{\boldsymbol{F}}_3^{(i)}, \quad i=0, 1, \cdots, I-1, & (3.1\text{c}) \end{cases}$$

where $\widetilde{G}_1^{(i+1)}$, $\widetilde{G}_2^{(i)}$, $\widetilde{G}_3^{(i)}$ are $s_1\times 5$, $s_2\times 5$, $s_3\times 5$-matrices respectively; $\Omega_1^{(i+1)}$, $\Omega_3^{(i)}$ are $s_1\times s_1$, $s_3\times s$'-diagonal matrices respectively; $\boldsymbol{U}_i$, $\widetilde{\boldsymbol{F}}_1^{(i)}$, $\widetilde{\boldsymbol{F}}_2^{(i)}$, $\widetilde{\boldsymbol{F}}_3^{(i)}$ are 5, s_1, s_2, s_3-dimensional vectors respectively; $s_1=5-p'_{l-1}$, $s_2=p'_{l-1}+q_l-5$, $s_3=5-q_l$.

Clearly, this system is of the same form as (5.1a—c) of Chapter 1, § 5. Thus, it can be solved by the block-double-sweep method for "incomplete" systems there. Moreover, we point out in Chapter 1, § 5, that if $s_1=0$ or $s_3=0$, then the procedure of elimination (the process of "direct sweep") can be cancelled, i.e., only the procedure of calculation of the unknowns (the process of the "inverse sweep") is necessary. Before turning to the description of the concrete algorithm, we should show which types of intervals among the eight types of intervals require both the procedure of elimination and the procedure of calculation of the unknowns, which types of intervals require only the procedure of calculation of the unknowns, and from which boundary we should start to eliminate the unknowns in order to minimize the amount of computation. (The proper directions of the procedure of elimination and of the procedure of calculation of the unknowns for each type of intervals are shown in Fig. 3.1.)

In the "first wave–first wave", "stream–first wave" and "second wave–first wave" intervals, there is the relation $s_1=5-p'_{l-1}=0$, i.e., (3.1a) is empty. Hence, when the group-dividing method is used, the procedure of elimination in these intervals is cancelled. In this case, we first calculate the quantities $\boldsymbol{U}_I$ at the upper boundary point. Starting from that point, we can then calculate the quantities $\boldsymbol{U}_i$ at all the other points in that interval in the order $i=I-1, I-2, \cdots, 0$, by using the equations (3.1b) and (3.1c).

In the "second wave–first wave", "second wave–stream" and "second wave–second wave" intervals, there is the relation $s_3=5-q_l=0$, i.e., (3.1c) is empty. Hence when the group-dividing method is adopted, only the procedure of calculation of the unknowns is needed, i.e., after the quantities at the lower boundary point are obtained, then $\boldsymbol{U}_i$ $(i=1, 2, \cdots, I)$ can be obtained by using (3.1a) and (3.1b). We note that in the "second wave–first wave" region, the procedure of calculation of the unknowns can be carried out in any direction. The reason for this is that the quantities $\boldsymbol{U}_i$ at any point in that interval can be obtained independently by using (3.1b) alone, since both (3.1a) and (3.1c) are empty.

In the "first wave–stream", "stream–stream" and "stream–second wave" intervals, neither s_1 nor s_3 is equal to zero, i.e., both (3.1a) and (3.1c) are not empty. Thus both the procedure of elimination and the procedure of calculation of the unknowns are necessary. On a stream boundary, $\sigma=0$ and the three compatibility relations on it are approximated by using Scheme II. Hence three rows of the matrix $\Omega_1^{(I)}$ (if the stream surface is an upper boundary) or $\Omega_3^{(0)}$ (if the stream surface is a lower boundary) are equal to zero. As is pointed out in Chapter 1, § 5, the three difference equations corresponding to the three stream-compatibility relations on the stream boundary can be separated from the procedure of elimination when the calculation is started with the stream boundary. Therefore the amount of calculation and the memory capacity needed are less if the calculation of elimination is started with the stream boundary for the "first wave–stream" or "stream–second wave" interval. That is why we choose the method of solution where the calculation of elimination begins with a stream boundary. For a "stream–stream" interval, the calculation of elimination may be initiated with either boundary. In our program, the calculation of elimination begins with the lower boundary in this case.

From the above analysis we see the following: when the block-double-sweep method of Chapter 1, § 5, is used for the intervals in the problem of flow around bodies, where both the "direct sweep" and the "inverse sweep" are necessary, then the matrices μ_0, μ_1 and μ have only one row, and only one row of the matrices ν_0, ν_1 and ν varies in the "direct sweep". Moreover, because the derivative of ρ appears only in the entropy equation (the fourth equation of (2.15)), and because it is always approximated by Scheme II on a stream boundary, therefore the difference equations involving ρ may be separated from the "direct sweep", i.e., we can take $N=4$ in the "direct sweep". Thus, we are able to use an algorithm that involves less computation for the problem of flow around bodies.

For a "first wave-stream" interval, the equations (3.1 a—c) can be rewritten as

$$\begin{cases} \overline{G}_1^{(i+1)}\overline{\boldsymbol{U}}_{i+1}+\Omega_1^{(i+1)}\overline{G}_1^{(i+1)}\overline{\boldsymbol{U}}_i=\widetilde{\boldsymbol{F}}_1^{(i+1)}, & i=1, 2, \cdots, I-1, \quad (3.2\text{a}) \\ \overline{\boldsymbol{\Omega}}_3^{(i)}\overline{G}_3^{(i)}\overline{\boldsymbol{U}}_{i+1}+\overline{G}_3^{(i)}\overline{\boldsymbol{U}}_i=\overline{\boldsymbol{F}}_3^{(i)}, & i=1, 2, \cdots, I-1, \quad (3.2\text{b}) \\ \overline{G}_1^{(1)}\overline{\boldsymbol{U}}_1+\Omega_1^{(1)}\overline{G}_1^{(1)}\overline{\boldsymbol{U}}_0=\widetilde{\boldsymbol{F}}_1^{(1)}, & (3.2\text{c}) \\ (0,\ 0,\ 1)\,(\overline{\boldsymbol{\Omega}}_3^{(0)}\overline{G}_3^{(0)}\overline{\boldsymbol{U}}_1+\overline{G}_3^{(0)}\overline{\boldsymbol{U}}_0)=(0,\ 0,\ 1)\,\overline{\boldsymbol{F}}_3^{(0)}, & (3.2\text{d}) \end{cases}$$

$$\Omega_4^{(i)}(g_{4,4}^{(i)}p_{i+1}+g_{4,5}^{(i)}\rho_{i+1})+g_{4,4}^{(i)}p_i+g_{4,5}^{(i)}\rho_i=\bar{f}_4^{(i)}, \quad (i=1, 2, \cdots, I-1), \tag{3.3}$$

$$\begin{cases} g_{2,1}^{(0)}u_0+g_{2,2}^{(0)}v_0+g_{2,3}^{(0)}w_0+g_{2,4}^{(0)}p_0=\bar{f}_2^{(0)}, \\ g_{3,1}^{(0)}u_0+g_{3,2}^{(0)}v_0+g_{3,3}^{(0)}w_0+g_{3,4}^{(0)}p_0=\bar{f}_3^{(0)}, \\ g_{4,4}^{(0)}p_0+g_{4,5}^{(0)}\rho_0=\bar{f}_4^{(0)}, \end{cases} \tag{3.4}$$

where $\overline{G}_1$, $\overline{G}_3$, Ω_1, $\overline{\boldsymbol{\Omega}}_3$ belong to 1×4, 3×4, 1×1, 3×3-matrices respectively; $\widetilde{\boldsymbol{F}}_1$, $\overline{\boldsymbol{F}}_3$ belong to 1, 3-dimensional vectors respectively;

$$\overline{\boldsymbol{U}}=\begin{pmatrix} u \\ v \\ w \\ p \end{pmatrix};$$

and the others are scalars. Here, (3.2a) and (3.2c) are the difference equations corresponding to the compatibility relation on the first family of wave-characteristic surfaces (3.1a), $\overline{G}_1$ being composed of the first four elements of $\widetilde{G}_1$. And equations (3.2b) are the difference equations corresponding to the compatibility relations on the stream-characteristic surfaces other than the entropy equation, and to the compatibility relations on the second family of wave-characteristic surfaces. That is, (3.2b) are the first, second, and fourth equations of (3.1c) with $i=1$, $2, \cdots I-1$, $\overline{G}_3$ being composed of the elements of $\widetilde{G}_3$ other than the third row and the fifth column; $\overline{\boldsymbol{\Omega}}_3$ being composed of the elements of Ω_3 other than the third row and the third column; and $\overline{\boldsymbol{F}}_3$ being composed of the elements of $\widetilde{\boldsymbol{F}}_3$ other than the third element. Equation (3.2d) is the difference equation with $i=0$ corresponding to the compatibility relation on the second family of wave-characteristic surfaces, i.e., (3.2d) is the fourth equation of (3.1c) with $i=0$. The difference equations corresponding to the entropy equation, i.e., the third equations of (3.1c), with $i=1, 2, \cdots, I-1$, are separately written as (3.3), since they can be separated from the "direct sweep". Equations (3.4) are the difference equations corresponding to the three stream-compatibility relations on the stream boundary, i.e., (3.4) are the first, second, and third equations of (3.1c), with $i=0$.

That is to say, at this time we only need to consider the following steps. First, the "direct sweep" for (3.2a—d) is completed, i.e., $\overline{\boldsymbol{U}}_1, \overline{\boldsymbol{U}}_2, \cdots, \overline{\boldsymbol{U}}_{I-1}$ are successively eliminated and two relations between $\overline{\boldsymbol{U}}_0$ and $\overline{\boldsymbol{U}}_I$ are obtained. Secondly, by using these two relations, the boundary conditions and (3.4), we obtain $\boldsymbol{U}_0$ and $\boldsymbol{U}_I$. Lastly, $\overline{\boldsymbol{U}}_i$ $(i=I-1, I-2, \cdots, 1)$ are successively calculated by using certain equations obtained in the process of the "direct sweep", and ρ_i are computed by using (3.3) at the same time. Moreover, as is pointed out in Chapter 1, § 5, in the case where s_1 is less (here $s_1=1$), the following scheme may be adopted in order to minimize the amount of computation and the memory capacity needed. Instead of the coefficients of $\overline{\boldsymbol{U}}_0$, the coefficients of $\mu_1^{(i)}\overline{\boldsymbol{U}}_0$ are introduced into the calculation in the process of the "direct sweep", and they are multiplied by $\mu_1^{(i)}\overline{\boldsymbol{U}}_0$ in the process of the "inverse sweep". Therefore, the concrete processes of the "direct and inverse sweeps" can be described as follows:

(i) The Process of the "Direct Sweep" of (3.2a—d)

For $i=1, 2, \cdots, I-1$, we change the matrix of coefficients from the form

$$H=\begin{bmatrix} \times & \times & \times & \times & \times & \times & \times & \times & \times & 0 & 0 \\ \times & \times & \times & \times & \times & \times & \times & \times & \times & 0 & 0 \\ \times & \times & \times & \times & \times & \times & \times & \times & \times & 0 & 0 \\ \times & \times & \times & \times & \times & \times & \times & \times & \times & 0 & 0 \\ 0 & 0 & 0 & 0 & \times & \times & \times & \times & \times & 1 & 0 \\ 0 & 0 & 0 & 0 & \times & \times & \times & \times & \times & 0 & 1 \end{bmatrix}$$

coefficients of $\overline{\boldsymbol{U}}_{i+1}$ (columns 1–4); coefficients of $\overline{\boldsymbol{U}}_i$ (columns 5–8); the term on the right-hand side (column 9); coefficients of $\mu_1^{(i)}\overline{\boldsymbol{U}}_0$ (column 10); coefficients of $\tilde{\nu}_1^{(i)}\overline{\boldsymbol{U}}_0$ (column 11)

to the form

$$H^* = \begin{bmatrix} \times & \times & \times & \times & 0 & 0 & 0 & 0 & \times & a_i & 0 \\ \otimes & \otimes & \otimes & \otimes & 1 & 0 & 0 & 0 & \otimes & \otimes & 0 \\ \otimes & \otimes & \otimes & \otimes & 0 & 1 & 0 & 0 & \otimes & \otimes & 0 \\ \otimes & \otimes & \otimes & \otimes & 0 & 0 & 1 & 0 & \otimes & \otimes & 0 \\ \otimes & \otimes & \otimes & \otimes & 0 & 0 & 0 & 1 & \otimes & \otimes & 0 \\ \times & \times & \times & \times & 0 & 0 & 0 & 0 & \times & b_i & 1 \end{bmatrix},$$

coefficients of $\overline{U}_{i+1}$ (columns 1–4); coefficients of $\overline{U}_i$ (columns 5–8); the term on the right-hand side (column 9); coefficients of $\mu_1^{(i)}\overline{U}_0$ (column 10); coefficients of $\tilde{\nu}_1^{(i)}\overline{U}_0$ (column 11)

by using the pivoting–elimination method. Here the pivot element is chosen from among the elements within the square in the middle of H, and the transposition of the second and the fifth rows usually is needed. In the matrix H, the first row corresponds to (3.2a); the second, third and fourth rows correspond to (3.2b); the fifth row corresponds to the relation $\mu_0^{(i)}\overline{U}_i+\mu_1^{(i)}\overline{U}_0=\mu^{(i)}$ (it is (3.2c) with $i=1$); and the sixth row corresponds to the relation $\tilde{\nu}_0^{(i)}\overline{U}_i+\tilde{\nu}_1^{(i)}\overline{U}_0=\tilde{\nu}^{(i)}$ (it is (3.2d) with $i=1$). Indeed, since the last column of H is not involved in the calculation of the "direct sweep", it is not necessary to list this column in H. The reason for listing this column is to explain the process of the "direct sweep" completely.

After obtaining a_i and b_i of the matrix H^* for each i, we compute $\mu_1^{(i+1)}$, $\tilde{\nu}_1^{(i+1)}$ by using the formulas

$$\begin{cases} \mu_1^{(i+1)}=a_i\mu_1^{(i)}, \\ \tilde{\nu}_1^{(i+1)}=\tilde{\nu}_1^{(i)}+b_i\mu_1^{(i)}, \end{cases} \quad (i=1,\ 2,\ \cdots,\ I-1). \tag{3.5}$$

In the meantime, we obtain all the coefficients of the two relations

$$\begin{cases} \mu_0^{(i+1)}\overline{U}_{i+1}+\mu_1^{(i+1)}\overline{U}_0=\mu^{(i+1)}, \\ \tilde{\nu}_0^{(i+1)}\overline{U}_{i+1}+\tilde{\nu}_1^{(i+1)}\overline{U}_0=\tilde{\nu}^{(i+1)}, \end{cases} \quad (i=1,\ 2,\ \cdots,\ I-1), \tag{3.6}$$

where $\mu_0^{(i+1)}$, $\mu^{(i+1)}$ and $\tilde{\nu}_0^{(i+1)}$, $\tilde{\nu}^{(i+1)}$ are the elements of H^* denoted by $\times$ in the first and the sixth rows respectively. Equations (3.6) with $i=I-1$ are the equations which we need in order to get the quantities at boundary points.

In H^*, there are also the coefficients of the following equations to be used in the process of the "inverse sweep":

$$Q_1^{(i+1)}\overline{U}_{i+1}+\overline{U}_i+\boldsymbol{Q}_2^{(i+1)}\mu_1^{(i)}\overline{U}_0=\boldsymbol{Q}^{(i+1)}, \quad i=1,\ 2,\ \cdots,\ I-1, \tag{3.7}$$

where $Q_1^{(i+1)}$ is a 4×4–matrix and $\boldsymbol{Q}_2^{(i+1)}$ and $\boldsymbol{Q}^{(i+1)}$ are four-dimensional vectors composed of the elements of H^* denoted by $\otimes$. The coefficients of (3.7), the terms on the right–hand side of (3.7), $\mu_1^{(i)}$, the coefficients of (3.3), and the terms on the right–hand side of (3.3), need to be kept since they will be used in the process of the "inverse sweep".

(ii) The Process of the "Inverse Sweep"

$\boldsymbol{U}_i$ are computed in the order of $i=I-1,\ I-2,\ \cdots,\ 1$, by using the relations

$$\begin{cases} y_i=\mu_1^{(i)}\overline{\boldsymbol{U}}_0, \\ \overline{\boldsymbol{U}}_i=\boldsymbol{Q}^{(i+1)}-Q_1^{(i+1)}\overline{\boldsymbol{U}}_{i+1}-\boldsymbol{Q}_2^{(i+1)}y_i, \\ \rho_i=\dfrac{1}{g_{4,5}^{(i)}}[\bar{f}_4^{(i)}-\Omega_4^{(i)}(g_{4,4}^{(i)}p_{i+1}+g_{4,5}^{(i)}\rho_{i+1})-g_{4,4}^{(i)}p_i], \\ \qquad (i=I-1,\ I-2,\ \cdots,\ 1). \end{cases} \tag{3.8}$$

For the "stream–second wave" interval, the situation is similar. The only difference is that the directions of the "direct and inverse sweeps" are opposite to those for the "first wave–stream" interval. That is, the difference equations of (3.2 a—d), corresponding to the compatibility relations on the first and second families of wave–characteristic surfaces, should be substituted by those on the second and first families of wave–characteristic surfaces, respectively. Moreover, the superscript i, in (3.2)—(3.8), should be substituted by $I-i$, the equations with $i=0$ by those with $i=I$, and $\overline{\boldsymbol{U}}_0$ by $\overline{\boldsymbol{U}}_I$. After all these substitutions, the calculation in both the processes of the "direct sweep" and of the "inverse sweep" can be carried out by using the method for the "first wave–stream" interval.

The difference between the equations (3.1a—c) in a "stream–stream" interval and those in a "first wave–stream" interval is as follows: for the former, the compatibility relations on the stream–characteristic surfaces belong to (3.1b) and $i=0,\ 1,\ \cdots,\ I$; for the latter, these relations belong to (3.1c) and $i=0,\ 1,\ \cdots,\ I-1$. When the double–sweep method of Chapter 1, § 5, is applied to (3.1a—c) and the "direct sweep" starts with $i=1$, then the treatment for (3.1b) is actually the same as that for (3.1c). Hence the calculation in the processes of "the direct and the inverse sweeps" for a "stream–stream" interval can be done by using the method for a "first wave–stream" interval. However, the calculation for boundary points is slightly different. This is because (3.1a—c) in a "stream–stream" interval involve not only (3.2a—d), (3.3), (3.4), but also the three difference equations on the stream characteristic boundary with $i=I$:

$$\begin{cases} g_{2,1}^{(I)}u_I+g_{2,2}^{(I)}v_I+g_{2,3}^{(I)}w_I+g_{2,4}^{(I)}p_I=\bar{f}_2^{(I)}, \\ g_{3,1}^{(I)}u_I+g_{3,2}^{(I)}v_I+g_{3,3}^{(I)}w_I+g_{3,4}^{(I)}p_I=\bar{f}_3^{(I)}, \\ g_{4,4}^{(I)}p_I+g_{4,5}^{(I)}\rho_I=\bar{f}_4^{(I)}. \end{cases} \tag{3.9}$$

Therefore, when the quantities at boundary points are calculated, (3.9) are needed in addition to two relations obtained in the "direct sweep", the boundary conditions, and (3.4).

We can see from the above discussion that, for a region where both the "direct sweep" and the "inverse sweep" are needed, the processes of "the direct and the inverse sweeps" are actually the same as those with $N=4$ described in Chapter 1, § 5. However, since the matrix $\overline{\boldsymbol{\Omega}}_3^{(0)}$ in (3.2) has two zero-rows (suppose that the "direct sweep" starts with a stream surface), therefore among (5.2e) of Chapter 1, i.e., among $\boldsymbol{\nu}_0^{(4)}\overline{\boldsymbol{U}}_i+\boldsymbol{\nu}_1^{(4)}\overline{\boldsymbol{U}}_0=\boldsymbol{\nu}^{(4)}$, there are two equations which are not needed in the calculation of the "direct sweep". Hence they are cancelled when we write down the formulas. For an interval where only the "inverse sweep" is needed, the process of calculation is the same as that of Chapter 1, § 5. However, since the derivative of ρ appears only in the entropy equation, the difference equations involving ρ_i are separated in order to decrease the number of operations. That is to say, the problem of solving the original system of five algebraic equations for $\boldsymbol{U}_i$ is reduced to one of solving a system of four equations for $\overline{\boldsymbol{U}}_i$ first and then calculating ρ_i from p_i by using the entropy difference equation. Therefore, according to the results of Chapter 1, § 6, the procedures of elimination and of calculation of the unknowns $\overline{\boldsymbol{U}}_i$ are stable. Moreover, since $|\Omega_4^{(4)}|<1-\varepsilon$, ε being a small positive constant (see Chapter 1, § 5), it is easy to see that the process of calculating ρ_i is also stable.

(2) *The Procedure for Solution of a System on a Line*

A system on a coordinate line in the ξ-direction will be solved by using the group-dividing method suggested in Chapter 1, § 5. Hence we first need to determine that every type of boundary here corresponds to which type of boundary for a group of intervals. According to the definitions of the upper boundary of a group of intervals (B_u), the lower boundary of a group of intervals (B_l), and a non-boundary of a group of intervals (B_n), an external boundary is, of course, always an upper boundary of a group if it is at the top, or a lower boundary of a group if it is at the bottom. For internal boundaries, a first-wave boundary is a lower boundary of a group, since $p_l'=5$ and $s_1=5-p_l'=0$; a second-wave boundary is an upper boundary of a group, since $q_l=5$ and $s_3=5-q_l=0$; a stream boundary is a non-boundary of a group of intervals, since $p_l'=q_l=4$ and $s_1=s_3=1\neq0$.

After determining the type of every boundary, we can solve the system on a line in the following way. First, we find out all the groups of intervals. Then, a system in each group of intervals is solved in the following way: (i) In every interval where the "direct sweep" is necessary, the calculation of the "direct sweep" is carried out; (ii) By using certain relations obtained in the process of the "direct sweep", the difference equations at the boundary points, the boundary conditions, and the internal boundary conditions, then the unknowns at the boundary points can be calculated; (iii) The calculation of the "inverse sweep" is done, i.e., the unknowns at interior points in this group of intervals are obtained. After solving a system in a group of intervals several times, we obtain the values of all the unknowns on the line. Now, let us illustrate this procedure with two examples.

Example 1. In the downstream region of an expansion edge of a combined body, there are usually four boundaries, namely, the main shock (a shock of the first family), two wave–characteristic surfaces of the first family, and the body, in the order from the top to the bottom (see Fig. 3.4a).

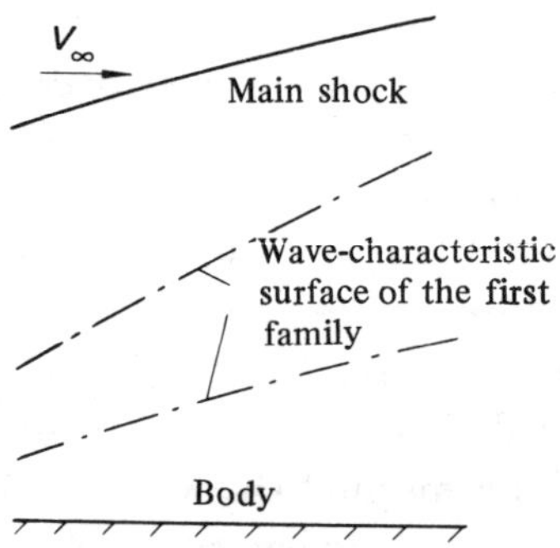

Figure 3.4a Construction of a flow field

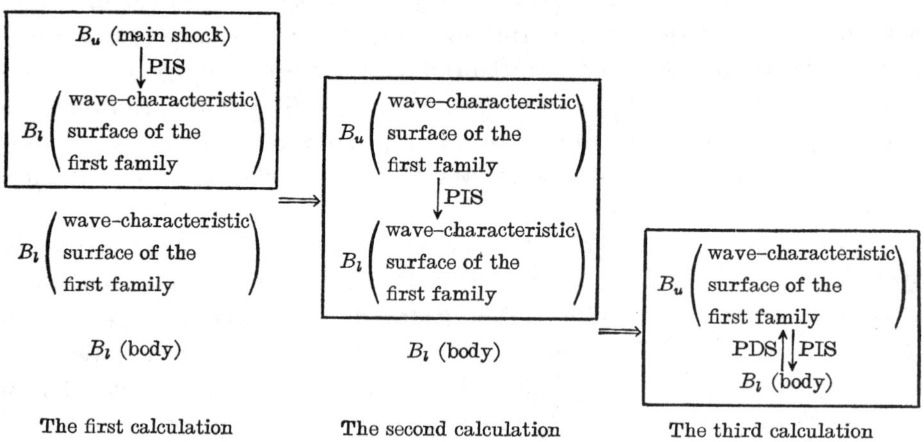

Figure 3.4b Schematic diagram of the procedure of calculation of the group–dividing method for the flow field whose construction is shown in Fig. 3.4a
PDS——The process of "direct sweep", PIS——The process of "inverse sweep".

In this case, the procedure of calculation by the group–dividing method is as follows (see Fig. 3.4b). At the beginning, the interval between the main shock and the upper wave–characteristic surface of the first family forms a group of intervals, where only the "inverse sweep" is needed. After the unknowns in that interval are obtained, the lower boundary of the group of intervals——the wave–characteristic surface—— becomes an upper boundary of a group. Hence the interval between two wave–characteristic surfaces makes up a group of intervals, where only the "inverse sweep" is again needed. Finally, the interval between the lower wave–characteristic surface and the body constitutes a group of intervals, where both the "direct sweep" and the "inverse sweep" are needed in order to obtain the values of all the unknowns in that interval.

Example 2. Fig. 1.1 shows the flow field around a multistage combined body. After the main shock intersects a secondary shock, there are seven boundaries in the flow field, namely the main shock (a shock of the first family), a contact discontinuity, two wave–characteristic surfaces of the second family, a secondary shock of the first family, a stream surface, and the body, in the order from the top to the bottom (see Fig. 3.5a) (supposing that a stream surface near the body is introduced in order to consider the entropy layer).

In this case, the process of calculation by the group–dividing method is as follows (see Fig. 3.5b). At the beginning, there is only one group of intervals which is composed of the interval between the lower wave–characteristic surface of the second family and the secondary shock. In this interval only the "inverse sweep" is needed (the "inverse sweep" here may be carried out from the top to the bottom, or from the bottom to the top). After the unknowns in this interval are obtained, the wave–characteristic surface of the second family becomes a lower boundary of a group, and the secondary shock becomes an upper boundary of a group. Thus the second "round" of calculation can proceed. There are now two groups of intervals. One is composed of the interval between two wave–characteristic surfaces, where only the "inverse sweep" is needed (the "inverse sweep" is carried out from the lower boundary to the upper boundary). The other is composed of the intervals between the secondary shock and the body, where there is one non–boundary of a group——the stream surface between the secondary shock and the body. In that case, both the "direct sweep" and the "inverse sweep" are needed. The "direct sweep" is carried out from the stream surface to the other two boundaries, and the "inverse sweep" is carried out in the opposite directions. After the quantities in those intervals are obtained, the interval between the main shock and the upper wave–characteristic surface of the second family constitutes a group of intervals, which also has one non–boundary of a group——the contact discontinuity between the main shock and the

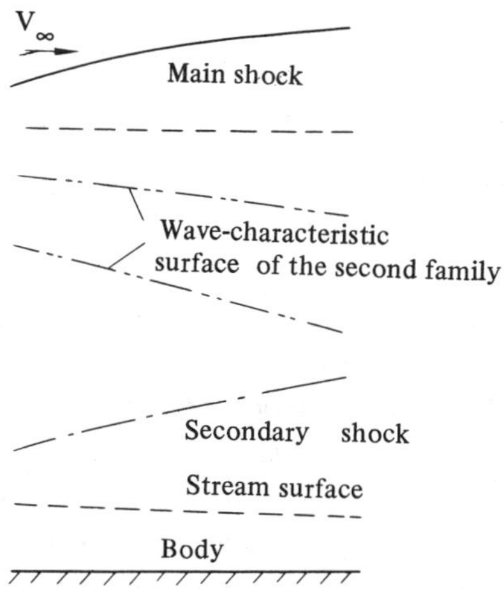

Figure 3.5a Construction of a flow field with seven boundaries

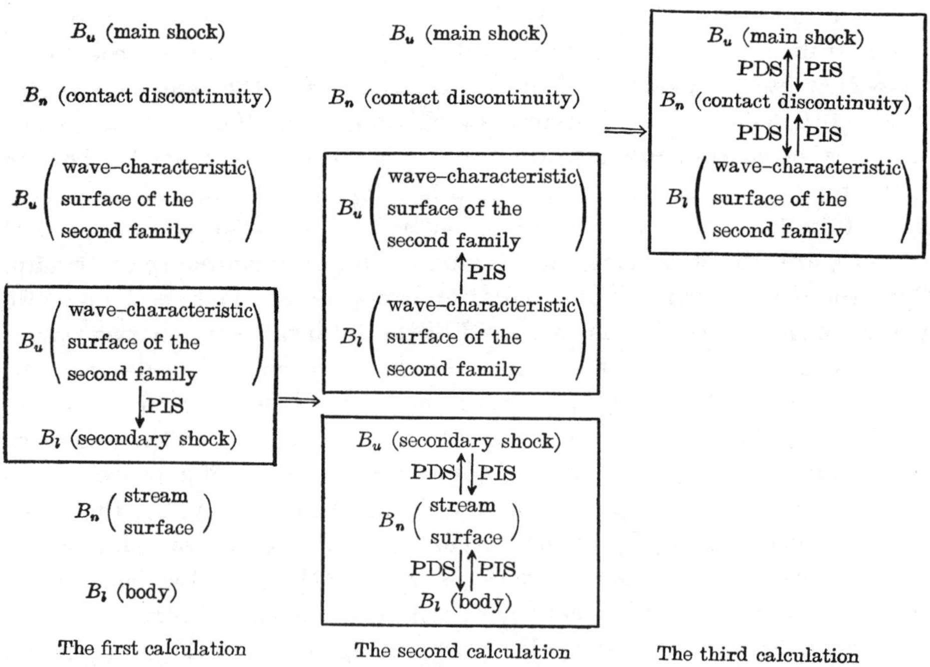

Figure 3.5b Schematic diagram of the procedure of calculation of the group-dividing method for the flow field whose construction is shown in Fig. 3.5a
PDS——The process of "direct sweep", PIS——The process of "inverse sweep".

wave-characteristic surface and where both the "direct sweep" and the "inverse sweep" are again needed. Thus the third "round" of calculation can be conducted. After the third "round" of calculation is done, the values of all the unknowns on a line are obtained, i.e., the whole system is solved.

(3) *Calculation of the Quantities at Boundary Points*

In order to determine the unknowns at the boundary points of a group of intervals, a simultaneous system of equations involving all the unknowns at boundary points needs to be solved. In accordance with Chapter 1, § 5, this problem is reduced to solving several systems, every one of which involves merely the unknowns at one boundary. Especially where the "direct sweep" is not necessary in all the intervals, the unknowns at a boundary point can be obtained by using the difference equations and the boundary conditions or the internal boundary conditions at the point. Thus the remaining problem is to find a method for each type of boundary. In what follows, all the quantities in equations are understood as given except the quantities considered at that boundary point. Moreover, among G, G_η, G_ψ of an unknown boundary, only G_η is understood as the unknown because G and G_ψ can be obtained in advance by using (2.22) and (2.23) of Chapter 2.

(i) Calculation of the Quantities at Body Points

We first consider the case where the body is a lower boundary. In this case, there are three types of intervals, namely, the "first wave-stream" interval, the "stream-stream" interval and the "second wave-stream" interval. As for the first two types of intervals, we obtain the relation $\tilde{\boldsymbol{\nu}}_0^{(I)}\overline{\boldsymbol{U}}_I+\tilde{\boldsymbol{\nu}}_1^{(I)}\overline{\boldsymbol{U}}_0=\tilde{\boldsymbol{\nu}}^{(I)}$ in the process of the "direct sweep". As for the third type of interval, we obtain a difference equation only involving $\overline{\boldsymbol{U}}_0$ and approximating a second-wave-compatibility relation:

$$g_{5,1}^{(0)}u_0+g_{5,2}^{(0)}v_0+g_{5,3}^{(0)}w_0+g_{5,4}^{(0)}p_0=\bar{f}_5^{(0)}.$$

Because $\sigma=0$ at a body point, we always have three difference equations at a body point which are obtained by using Scheme II and which approximate the three stream-compatibility relations respectively.

Moreover, the quantities at a body point should satisfy the condition (2.19).

Therefore, in order to get the quantities at a body point, we need to solve a system of equations for $\boldsymbol{U}_0$ in the following form:

$$\begin{cases} \tilde{\boldsymbol{\nu}}_1^{(I)}\overline{\boldsymbol{U}}_0=\tilde{\boldsymbol{\nu}}^{(I)}-\tilde{\boldsymbol{\nu}}_0^{(I)}\overline{\boldsymbol{U}}_I, \\ g_{2,1}^{(0)}u_0+g_{2,2}^{(0)}v_0+g_{2,3}^{(0)}w_0+g_{2,4}^{(0)}p_0=\bar{f}_2^{(0)}, \\ g_{3,1}^{(0)}u_0+g_{3,2}^{(0)}v_0+g_{3,3}^{(0)}w_0+g_{3,4}^{(0)}p_0=\bar{f}_3^{(0)}, \\ \boldsymbol{V}_0\cdot\boldsymbol{\xi}=0, \\ g_{4,4}^{(0)}p_0+g_{4,5}^{(0)}\rho_0=\bar{f}_4^{(0)}. \end{cases} \tag{3.10}$$

The first four equations of (3.10) are a system of linear algebraic equations for

$$\overline{U}_0=\begin{pmatrix} u_0 \\ v_0 \\ w_0 \\ p_0 \end{pmatrix},$$

which can be solved by using the pivoting–elimination method. After p_0 has been determined, ρ_0 can easily be obtained from the fifth equation of (3.10).

If the body is an upper boundary, the process of calculation is similar.

(ii) Calculation of the Quantities at Shock Points

We first consider the first family of shocks. If a shock is an external boundary, the quantities in front of the shock are given by the free stream conditions (2.21). If it is an internal boundary, the quantities in front of the shock can also be obtained before the other quantities at the shock point are determined. In fact, it is a lower boundary of a group of intervals in the group–dividing method. Thus the quantities in the interval where the shock is a lower boundary are determined before the quantities in the interval where the shock is an upper boundary are obtained. That is, the quantities in front of a shock can be determined independently of the boundary conditions. Hence in any case the quantities in front of a shock can be understood as given in the Rankine–Hugoniot conditions. Therefore, in order to determine all the quantities at a shock point, the remaining problem is to find a method for solving the following nonlinear equations for G_η and $\boldsymbol{U}_I$ (the quantities behind the shock):

$$\begin{cases} h(p_I,\ \rho_I)+V_{I,n}^2/2=h(p_0^*,\ \rho_0^*)+V_{0,n}^{*2}/2, \\ p_I+\rho_I V_{I,n}^2=p_0^*+\rho_0^* V_{0,n}^{*2}, \\ \boldsymbol{V}_I=\boldsymbol{V}_0^*-(1-\rho_0^*/\rho_I)V_{0,n}^*\boldsymbol{n}, \\ \mu_0^{(I)}\overline{\boldsymbol{U}}_I=\mu^{(I)}-\mu_1^{(I)}\overline{\boldsymbol{U}}_0. \end{cases} \tag{3.11}$$

For the last equation of (3.11), there are two possibilities. If the downstream region of the shock is a "first wave–first wave" interval, it is the difference equation $g_{1,1}^{(I)}u_I+g_{1,2}^{(I)}v_I+g_{1,3}^{(I)}w_I+g_{1,4}^{(I)}p_I=\bar{f}_1^{(I)}$, approximating the first–wave compatibility relation by Scheme II; if the downstream region of the shock is a "first wave–stream" interval, it is the equation $\mu_0^{(I)}\overline{\boldsymbol{U}}_I+\mu_1^{(I)}\overline{\boldsymbol{U}}_0=\mu^{(I)}$, obtained in the process of the "direct sweep". Clearly, both relations can be rewritten in that form. The other equations of (3.11) are the jump conditions at a shock (2.20), where $\boldsymbol{n}$ is a unit normal vector to the shock whose components are functions of G_η (see (2.8)). In (3.11), we put a superscript "*" on every quantity in front of the shock, in order to distinguish these quantities from the quantities $\boldsymbol{U}_0$ in the downstream region of the shock.

The nonlinear equations (3.11) can usually be solved by iteration methods. In our calculation, the following "two-stage method of false position" is adopted. In fact, if G_η is given, then $\boldsymbol{U}_I$ can be determined by using the first five[1] relations of (3.11). That is, the first five relations of (3.11) determine the functions $\boldsymbol{U}_I(G_\eta)$. Substituting the relations $\boldsymbol{U}_I(G_\eta)$ into the last equation of (3.11), we have

$$f_1(G_\eta) \equiv \mu_0^{(I)} \overline{\boldsymbol{U}}_I(G_\eta) + \mu_1^{(I)} \overline{\boldsymbol{U}}_0 - \mu^{(I)} = 0.$$

Thus if the functions $\boldsymbol{U}_I(G_\eta)$ have explicit expressions, then solving (3.11) may be reduced to solving the nonlinear equation for G_η, $f_1(G_\eta) = 0$. In this case, when a value of G_η is given, the corresponding value of $f_1(G_\eta)$ can be obtained straightforwardly. Hence if two initial values $G_\eta^{(i)}$ $(i=0, 1)$ are given, we can solve this equation by using the method of false position:

$$G_\eta^{(i+1)} = G_\eta^{(i)} - f_1(G_\eta^{(i)}) (G_\eta^{(i)} - G_\eta^{(i-1)}) / [f_1(G_\eta^{(i)}) - f_1(G_\eta^{(i-1)})],$$
$$i = 1, 2, \cdots.$$

When $|f_1(G_\eta^{(i)})| < \varepsilon_1$, ε_1 being an error-controlling constant, we consider that $\boldsymbol{U}_I^{(i+1)}$ and $G_\eta^{(i+1)}$ are approximate solutions desired.

However, $h(p, \rho)$ of air in chemical equilibrium has a complicated expression. Hence when G_η is given, iteration is needed in order to determine $\boldsymbol{U}_I$ by the jump conditions at a shock. We also adopt the method of false position for this problem. The concrete algorithm is as follows. From the second to the fifth relations of (3.11), we can easily get the explicit expressions for $\overline{\boldsymbol{U}}_I$ as functions of ρ_I and G_η. Substituting these expressions into the first equation of (3.11) and regarding G_η as a given quantity, we obtain an equation for ρ_I,

$$f_2(\rho_I) \equiv h(p_I(\rho_I), \rho_I) - h(p_0^*, \rho_0^*) + \frac{1}{2}[V_{I,n}^2(\rho_I) - V_{0,n}^{*2}] = 0.$$

Therefore, if $G_\eta^{(i)}$ is given, solving $\boldsymbol{U}_I$ is reduced to solving the nonlinear equation $f_2(\rho_I) = 0$. The equation can be solved by using the following method of false position:

$$\rho_I^{(i,j+1)} = \rho_I^{(i,j)} - f_2(\rho_I^{(i,j)}) (\rho_I^{(i,j)} - \rho_I^{(i,j-1)}) / [f_2(\rho_I^{(i,j)}) - f_2(\rho_I^{(i,j-1)})],$$
$$j = 1, 2, \cdots.$$

In order to begin carrying out this iteration, two initial values $\rho_I^{(i,j)}$ $(j = 0, 1)$ of the quantity ρ_I and two corresponding values $f_2(\rho_I^{(i,j)})$ $(j=0, 1)$ of the function $f_2(\rho_I)$ are needed. When $|f_2(\rho_I^{(i,j)})| < \varepsilon_2$, ε_2 being an error-controlling constant, the approximate values of $\boldsymbol{U}_I$ corresponding to $G_\eta^{(i)}$ are obtained.

---

1) The vector system of equations contains three relations.

In our calculation, we take the corresponding values at the "given" level as initial values, i.e., we take the corresponding values of G_η and ρ_I at $\eta=\eta^*$ as the initial values $G_\eta^{(0)}$ and $\rho_I^{(i,0)}$ at $\eta=\eta^*+\Delta\eta$. Furthermore, let $G_\eta^{(1)}=G_\eta^{(0)}-\varepsilon_3$ and $\rho_I^{(i,1)}=\rho_I^{(i,0)}-\varepsilon_4$, ε_3 and ε_4 being small constants.

In our practical calculation, the iteration described above is usually convergent.

The calculation for the second family of shocks is similar. After changing the subscripts "0", "I", into "I", "0", and the last equation of (3.11) into an approximate second wave-compatibility relation or its equivalent, we can solve this problem in the same way.

(iii) Calculation of the Quantities at Contact Discontinuity Points

If the contact discontinuity under consideration has a subscript l, then the interval in which it is an upper boundary has the subscript l and the other interval in which it is a lower boundary has the subscript $l+1$. Clearly, $\sigma=0$ on both its sides. Thus, in the $(l+1)$-th interval, we have three approximate stream-compatibility relations obtained by using Scheme II and one "approximate second wave-compatibility relation":

$$\begin{cases} g_{2,1}^{(0,l+1)}u_{0,l+1}+g_{2,2}^{(0,l+1)}v_{0,l+1}+g_{2,3}^{(0,l+1)}w_{0,l+1} \\ \qquad +g_{2,4}^{(0,l+1)}p_{0,l+1}=\bar{f}_2^{(0,l+1)}, \\ g_{3,1}^{(0,l+1)}u_{0,l+1}+g_{3,2}^{(0,l+1)}v_{0,l+1}+g_{3,3}^{(0,l+1)}w_{0,l+1} \\ \qquad +g_{3,4}^{(0,l+1)}p_{0,l+1}=\bar{f}_3^{(0,l+1)}, \\ g_{4,4}^{(0,l+1)}p_{0,l+1}+g_{4,5}^{(0,l+1)}\rho_{0,l+1}=\bar{f}_4^{(0,l+1)}, \\ \tilde{\nu}_0^{(I',l+1)}\overline{U}_{I',l+1}+\tilde{\nu}_1^{(I',l+1)}\overline{U}_{0,l+1}=\tilde{\nu}^{(I',l+1)}. \end{cases} \tag{3.12}$$

In the l-th interval, we have three approximate stream-compatibility relations obtained by using Scheme II, and one "approximate first wave-compatibility relation":

$$\begin{cases} \mu_0^{(I,l)}\overline{U}_{I,l}+\mu_1^{(I,l)}\overline{U}_{0,l}=\mu^{(I,l)}, \\ g_{2,1}^{(I,l)}u_{I,l}+g_{2,2}^{(I,l)}v_{I,l}+g_{2,3}^{(I,l)}w_{I,l}+g_{2,4}^{(I,l)}p_{I,l}=\bar{f}_2^{(I,l)}, \\ g_{3,1}^{(I,l)}u_{I,l}+g_{3,2}^{(I,l)}v_{I,l}+g_{3,3}^{(I,l)}w_{I,l}+g_{3,4}^{(I,l)}p_{I,l}=\bar{f}_3^{(I,l)}, \\ g_{4,4}^{(I,l)}p_{I,l}+g_{4,5}^{(I,l)}\rho_{I,l}=\bar{f}_4^{(I,l)}. \end{cases} \tag{3.13}$$

The "approximate second wave-compatibility relation" in (3.12) is either a certain relation obtained in the process of the "direct sweep" or an approximate wave-compatibility relation obtained by using Scheme II. In the latter case $\tilde{\nu}_0^{(I',l+1)}\overline{U}_{I',l+1}$ is equal to zero. For the "approximate first wave-compatibility relation" in (3.13), the situation is similar.

Besides (3.12) and (3.13), the jump conditions (2.22) should be satisfied.

Therefore, in order to determine the quantities at a contact discontinuity, the following equations for $U_{0,l+1}$, G_η and $U_{I,l}$ should be solved:

$$\left\{\begin{aligned}&\boldsymbol{V}_{0,l+1}\cdot\boldsymbol{\xi}=0,\\&g_{2,1}^{(0,l+1)}u_{0,l+1}+g_{2,2}^{(0,l+1)}v_{0,l+1}+g_{2,3}^{(0,l+1)}w_{0,l+1}\\&\qquad+g_{2,4}^{(0,l+1)}p_{0,l+1}=\bar{f}_2^{(0,l+1)},\\&g_{3,1}^{(0,l+1)}u_{0,l+1}+g_{3,2}^{(0,l+1)}v_{0,l+1}+g_{3,3}^{(0,l+1)}w_{0,l+1}\\&\qquad+g_{3,4}^{(0,l+1)}p_{0,l+1}=\bar{f}_3^{(0,l+1)},\\&\tilde{\nu}_1^{(I',l+1)}\overline{\boldsymbol{U}}_{0,l+1}=\tilde{\nu}^{(I',l+1)}-\tilde{\nu}_0^{(I',l+1)}\overline{\boldsymbol{U}}_{I',l+1},\\&g_{4,4}^{(0,l+1)}p_{0,l+1}+g_{4,5}^{(0,l+1)}\rho_{0,l+1}=\bar{f}_4^{(0,l+1)};\end{aligned}\right.\tag{3.14a}$$

$$\left\{\begin{aligned}&\mu_0^{(I,l)}\overline{\boldsymbol{U}}_{I,l}=\mu^{(I,l)}-\mu_1^{(I,l)}\overline{\boldsymbol{U}}_{0,l},\\&g_{2,1}^{(I,l)}u_{I,l}+g_{2,2}^{(I,l)}v_{I,l}+g_{2,3}^{(I,l)}w_{I,l}+g_{2,4}^{(I,l)}p_{I,l}=\bar{f}_2^{(I,l)},\\&g_{3,1}^{(I,l)}u_{I,l}+g_{3,2}^{(I,l)}v_{I,l}+g_{3,3}^{(I,l)}w_{I,l}+g_{3,4}^{(I,l)}p_{I,l}=\bar{f}_3^{(I,l)},\\&\boldsymbol{V}_{I,l}\cdot\boldsymbol{\xi}=0,\\&g_{4,4}^{(I,l)}p_{I,l}+g_{4,5}^{(I,l)}\rho_{I,l}=\bar{f}_4^{(I,l)};\end{aligned}\right.\tag{3.14b}$$

$$p_{0,l+1}=p_{I,l},\tag{3.14c}$$

where $\boldsymbol{\xi}$ are functions of G_η.

When G_η, i.e. $\boldsymbol{\xi}$, is given, both (3.14a) and (3.14b) are systems of linear algebraic equations involving five unknowns. Thus they determine implicitly the functions $\boldsymbol{U}_{0,l+1}(G_\eta)$ and $\boldsymbol{U}_{I,l}(G_\eta)$. (When G_η is given, $\boldsymbol{U}$ can be calculated by using the method for solving the quantities at a body point.) Substituting the relations into (3.14c), we have the following nonlinear equation for G_η:

$$f_3(G_\eta)=p_{0,l+1}(G_\eta)-p_{I,l}(G_\eta)=0.$$

Therefore, solving (3.14a—c) can be reduced to solving this nonlinear equation. It may also be solved by using the method of false position. The choice of the initial value of G_η can be done in a way similar to that for the case of shock points.

(iv) Calculation of the Quantities on Wave-Characteristic Boundaries

We first consider a first wave–characteristic boundary. Suppose that it has a subscript l. We know from the preceding discussion that only the "inverse sweep" is necessary in the $(l+1)$-th interval, where the first wave–characteristic boundary is the lower boundary. Thus, when the calculation is done by using the group–dividing method, both $\boldsymbol{U}_{0,l+1}$ and the quantities at the interior points in the $(l+1)$-th interval are first obtained in the process of the "inverse sweep". Then the quantities $\boldsymbol{U}_{I,l}$ on the lower side are obtained by the continuity conditions (2.24c). G_η can be determined by (2.23) with the positive sign.

In the case of a second wave–characteristic boundary, the situation is similar, i.e., only the "inverse sweep" is necessary in the l–th interval where the second wave–characteristic surface is the upper boundary. Thus, if the group–dividing method is used, both $\boldsymbol{U}_{I,l}$ and the quantities at interior points in the l–th interval are obtained in the process of the "inverse sweep". After that, $\boldsymbol{U}_{0,l+1}$ and G_η can be determined immediately by using the continuity conditions (2.24c) and the relation (2.23) with the negative sign.

(v) Calculation of the Quantities on Stream – Characteristic Boundaries

Suppose that a stream boundary has a subscript l. As we have pointed out, it is a non–boundary of a group. When the group–dividing method is used, we always obtain the relation

$$\tilde{\nu}_0^{(I',l+1)}\overline{\boldsymbol{U}}_{I',l+1}+\tilde{\nu}_1^{(I',l+1)}\overline{\boldsymbol{U}}_{0,l+1}=\tilde{\nu}^{(I',l+1)}$$

in the $(l+1)$–th interval whose lower boundary is the stream surface. This relation is either a certain relation obtained in the process of the "direct sweep" or an approximate second wave–compatibility relation obtained by using Scheme II. For simplicity, we write these two relations in the same form. Moreover, in the l–th interval, we obtain the relation

$$\mu_0^{(I,l)}\overline{\boldsymbol{U}}_{I,l}+\mu_1^{(I,l)}\overline{\boldsymbol{U}}_{0,l}=\mu^{(I,l)},$$

which is either a certain relation obtained in the process of the "direct sweep" or an approximate first wave–compatibility relation obtained by using Scheme II. In addition, on either of its sides, for example, on its upper side, there are three approximate stream–compatibility relations obtained by Scheme II:

$$\begin{cases} g_{2,1}^{(0,l+1)}u_{0,l+1}+g_{2,2}^{(0,l+1)}v_{0,l+1}+g_{2,3}^{(0,l+1)}w_{0,l+1}+g_{2,4}^{(0,l+1)}p_{0,l+1}=\bar{f}_2^{(0,l+1)}, \\ g_{3,1}^{(0,l+1)}u_{0,l+1}+g_{3,2}^{(0,l+1)}v_{0,l+1}+g_{3,3}^{(0,l+1)}w_{0,l+1}+g_{3,4}^{(0,l+1)}p_{0,l+1}=\bar{f}_3^{(0,l+1)}, \\ g_{4,4}^{(0,l+1)}p_{0,l+1}+g_{4,5}^{(0,l+1)}\rho_{0,l+1}=\bar{f}_4^{(0,l+1)}. \end{cases}$$

Obviously, the quantities on a stream–characteristic boundary should also satisfy the continuity conditions (2.24c) and the boundary condition (2.25). Therefore, the quantities on a stream–characteristic boundary can be determined in the following way. First, from

$$\begin{cases} \mu_0^{(I,l)}\overline{\boldsymbol{U}}_{0,l+1}=\mu^{(I,l)}-\mu_1^{(I,l)}\overline{\boldsymbol{U}}_{0,l}, \\ g_{2,1}^{(0,l+1)}u_{0,l+1}+g_{2,2}^{(0,l+1)}v_{0,l+1}+g_{2,3}^{(0,l+1)}w_{0,l+1} \\ \qquad +g_{2,4}^{(0,l+1)}p_{0,l+1}=\bar{f}_2^{(0,l+1)}, \\ g_{3,1}^{(0,l+1)}u_{0,l+1}+g_{3,2}^{(0,l+1)}v_{0,l+1}+g_{3,3}^{(0,l+1)}w_{0,l+1} \\ \qquad +g_{3,4}^{(0,l+1)}p_{0,l+1}=\bar{f}_3^{(0,l+1)}, \\ \tilde{\nu}_1^{(I',l+1)}\overline{\boldsymbol{U}}_{0,l+1}=\tilde{\nu}^{(I',l+1)}-\tilde{\nu}_0^{(I',l+1)}\overline{\boldsymbol{U}}_{I',l+1}, \\ g_{4,4}^{(0,l+1)}p_{0,l+1}+g_{4,5}^{(0,l+1)}\rho_{0,l+1}=\bar{f}_4^{(0,l+1)}, \end{cases} \tag{3.15}$$

$\boldsymbol{U}_{0,l+1}$ are calculated by using the method for solving the quantities at a body point. Then the quantities $\boldsymbol{U}_{I,l}$ are obtained from (2.24c). Finally, G_η is determined by

$$G_\eta=(v_{0,l+1}-G_\psi \boldsymbol{V}_{0,l+1}\cdot\boldsymbol{\psi})/(\boldsymbol{V}_{0,l+1}\cdot\boldsymbol{\eta}).$$

3. *Certain Treated Problems*

(1) *Calculation of the Quantities on Body Singular Lines*

On a combined body, there are usually some expansion edges and compression edges, where the flow variables are multivalued. We suppose that these edges are space–like singular lines, and that we can find such a coordinate system in which every edge is located on a space–like march surface, η=constant. (If the body is axisymmetrical and the axis of symmetry of the body is taken as the z–axis, any edge of the body is a circle on a plane z=constant. Thus we may take $\eta=z$ in the process of calculation if u is supersonic.) When the march–calculation in the upstream region of an edge has been completed, we have to obtain all the quantities on the edge before proceeding with the calculation. In Chapter 4, § 2, we have given the method of computing the flow parameters on these two types of edges. Let us now givean an explanation in detail.

(i) Expansion Edges on Bodies

When a body has an expansion edge, there appears a centered expansion wave starting from the expansion edge in the flow field. (see Fig. 3.6) As is pointed out in Chapter 4, § 2, the flow field near the edge can be divided into three parts by the two boundaries of the expansion wave: the region in front of the expansion wave (the region between the front body and the front boundary of the expansion wave), the region of the expansion wave, and the region behind the wave (the region between the back boundary of the wave and the back body). On the edge, the flow parameters in the regions both in front of the wave and behind the wave are constants. However, the flow variables in the region of the expansion wave are multivalued. They satisfy the ordinary differential equations (2.48) of Chapter 4. The initial shape of the front

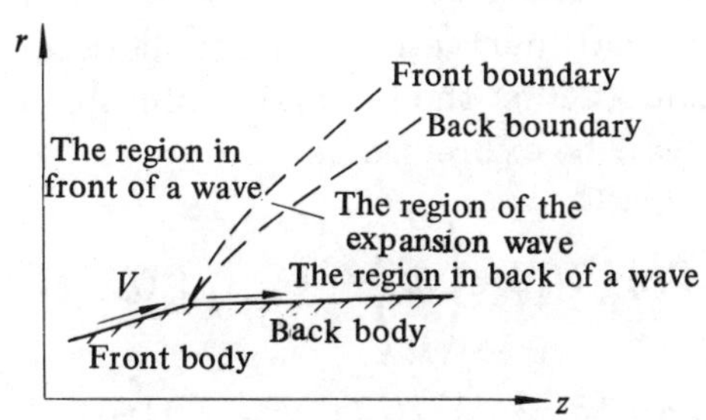

Figure 3.6 Construction of the flow field near an expansion edge

boundary of the expansion wave is determined by the quantities in front of the wave. However, the initial shape of the back boundary of the expansion wave is determined by the quantities in front of the wave, by the shape of the back body, and by the ordinary differential equations which the flow variables in the region of the expansion wave should satisfy.

When an expansion edge of bodies appears in the process of calculation, we cancel the body in front of the expansion wave, and add three new boundaries, namely, the front and the back boundaries of the expansion wave, and the body behind the wave, into the flow field. Thus two new regions, the region of the expansion wave and the region behind the wave, appear in the flow field between the main shock and the body. In those two new regions, the quantities on the edge may be determined in the following way: we take the quantities in front of the wave as initial values, then the ordinary differential equations (2.48) of Chapter 4 are integrated until the flow parameters satisfy the condition on the body behind the wave. That is, calculation is carried out until the angle β is equal to the value corresponding to the body behind the wave. The definition of β is given in Fig. 2.6 of Chapter 4. The values of β corresponding to the bodies in front of and behind the wave are known. Thus, through integrating (2.48) of Chapter 4, we can obtain the values of flow variables as functions of β in the region of the wave. In the region behind the wave, the values of flow variables can be easily determined because they are "single–valued". From (2.46) in Chapter 4, we can also determine the value of the angle θ corresponding to a value of β. (The definition of θ and the relation between θ and β are given in Figs. 2.5 and 2.6 of Chapter 4.) Hence F, F_z, F_φ of the front and back boundaries of the wave on the edge can be determined from (2.67) in Chapter 4. F, F_z, F_φ of the body are known. Therefore, G, G_η, G_ψ of the front and the back boundaries of the wave and of the body behind the wave can be obtained by using the transformation relation (2.11). However, in order to continue with the computation, we need the values of flow variables on the edge as functions of ξ. Thus we need to know the relation between ξ and β (or θ) in the region of the expansion wave. In what follows we shall derive the relation between ξ and θ.

From (2.9) and (2.6b) the normal vector to a surface ξ=constant in the l–th region can be expressed as

$$\begin{aligned}\boldsymbol{n}(\xi) &= \begin{pmatrix}0\\1\\0\end{pmatrix} - G_\eta \boldsymbol{\eta} - G_\psi \boldsymbol{\psi} = \begin{pmatrix}0\\1\\0\end{pmatrix} - [G_{\eta,l-1} + (\xi - l + 1)(G_{\eta,l} - G_{\eta,l-1})]\boldsymbol{\eta} \\ &\quad - [G_{\psi,l-1} + (\xi - l + 1)(G_{\psi,l} - G_{\psi,l-1})]\boldsymbol{\psi} \\ &= (l - \xi)\boldsymbol{n}(l-1) + (1 - l + \xi)\boldsymbol{n}(l), \quad l-1 \leqslant \xi \leqslant l,\end{aligned}$$

where

$$\boldsymbol{n}(i)=\begin{pmatrix}0\\1\\0\end{pmatrix}-G_{\eta,i}\eta-G_{\psi,i}\psi,\quad i=l-1,\ l.$$

We know from the discussion in Chapter 4, § 2, that $\boldsymbol{n}(i)$ of the front and the back boundaries of the expansion wave have the following expressions:

$$\boldsymbol{n}(i)=|\boldsymbol{n}(i)|(-\sin\theta_i\boldsymbol{t}_1(0)+\cos\theta_i\boldsymbol{t}_2(0)),\quad i=l-1,\ l,\tag{3.16}$$

where

$$|\boldsymbol{n}(i)|=\left|\begin{pmatrix}0\\1\\0\end{pmatrix}-G_{\eta,i}\eta-G_{\psi,i}\psi\right|,$$

and where the definitions of θ, $\boldsymbol{t}_1(0)$ and $\boldsymbol{t}_2(0)$ are given in Figs. 2.5 and 2.6 of Chapter 4. Therefore, for a surface $\xi=$constant in the region of the expansion wave, there exists the relation

$$\boldsymbol{n}(\xi)=|\boldsymbol{n}(\xi)|(-\sin\theta(\xi)\boldsymbol{t}_1(0)+\cos\theta(\xi)\boldsymbol{t}_2(0)),\tag{3.17}$$

where

$$\begin{cases}|\boldsymbol{n}(\xi)|=\{[(l-\xi)|\boldsymbol{n}(l-1)|\sin\theta_{l-1}+(1-l+\xi)|\boldsymbol{n}(l)|\sin\theta_l]^2\\ \qquad\qquad+[(l-\xi)|\boldsymbol{n}(l-1)|\cos\theta_{l-1}+(1-l+\xi)|\boldsymbol{n}(l)|\cos\theta_l]^2\}^{\frac{1}{2}},\\ \sin\theta(\xi)=\{(l-\xi)|\boldsymbol{n}(l-1)|\sin\theta_{l-1}+(1-l+\xi)|\boldsymbol{n}(l)|\sin\theta_l\}/|\boldsymbol{n}(\xi)|,\\ \cos\theta(\xi)=\{(l-\xi)|\boldsymbol{n}(l-1)|\cos\theta_{l-1}+(1-l+\xi)|\boldsymbol{n}(l)|\cos\theta_l\}/|\boldsymbol{n}(\xi)|.\end{cases}$$

The value of θ corresponding to a value of ξ can be determined from the above relations, and thus the values of flow variables on the edge corresponding to a value of ξ can also be obtained.

(ii) Compression Edges on Bodies

If there is a compression edge on a body, a secondary shock starting from the edge usually appears in the flow field (see Fig. 3.7). We know, from the discussion in Chapter 4, § 2, that the flow field near the edge is divided into two parts by the secondary shock, namely, into the region in front of the shock and the region behind the shock. The flow variables on the edge in every region are single-valued. The initial shape of the secondary shock and the flow parameters behind it are determined by the flow parameters in front of the shock and by the shape of the body behind it.

When a compression edge appears, we cancel the body in front of the shock and add the secondary shock and the body behind it into the flow field. Thus a new region——the region behind the shock——appears in the flow field. The problem of determining the initial shape of the secondary shock and the flow parameters on the edge in the region behind

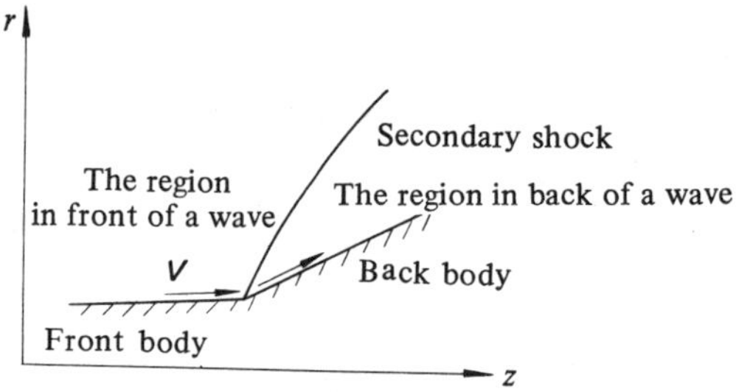

Figure 3.7 Construction of the flow field near a compression edge

the shock is reduced to solving the system which consists of the jump conditions (2.20) on shocks and the condition (2.19) on bodies. We have described the concrete algorithm in Chapter 4, § 2. The equation (2.63) in Chapter 4 can be solved by using the method of false position aforementioned. After F, F_θ, F_φ are obtained from (2.64) of Chapter 4, then G, G_η, G_ψ can be computed from (2.11). The flow variables on the edge in the region behind the shock are single-valued, so their values for any ξ are equal to the values on the body behind the shock. G, G_η, G_ψ of the body behind the secondary shock can easily be obtained from the equation of the body. Using all these quantities, we can proceed with calculation along the η direction.

(2) *Treatment of Intersection of Boundaries*

We assume discontinuities and some weak discontinuities to be boundaries in our calculation, so that solving the supersonic flow field is reduced to solving an initial–boundary–value problem for a quasilinear hyperbolic system with several boundaries. Thus the intersection of boundaries appears in the flow field around a multistage combined body. Flow variables or their derivatives on the lines of intersection are usually not unique, and consequently those lines of intersection are either singular or weak singular. In principle, if boundaries intersect, the structure of the reflected waves should be accurately determined. However, we need to do a great deal of work if we wish to treat all the intersections accurately. In practical calculation, we treat them in the following simplified way.

(i) If at least one of the intersecting boundaries is a weak discontinuity, then we shall not determine the structure of the reflected waves accurately. We cancel a boundary just before the intersection of two boundaries. If a discontinuity intersects a weak discontinuity, the weak discontinuity is cancelled; if two weak discontinuities intersect, one of them is cancelled. The flow variables on weak discontinuities are themselves continuous. Only their derivatives are discontinuous. The error made by cancelling a weak discontinuity is therefore acceptable in the case when the cancelled weak discontinuity is not very strong. That is, this treatment is reasonable under certain conditions.

(ii) If both the intersecting boundaries are strong discontinuities, for example, if the main shock intersects a secondary shock, if a shock intersects a contact discontinuity, or if a shock is reflected by a body, then the structure of the reflected wave needs to be accurately determined. In Chapter 4, § 2, the determination of the structures of the reflected waves is described in detail. (The problems discussed in Chapter 4 concern the interaction between two shocks, the interaction between a shock and a contact discontinuity, the reflection of a shock on a body and other phenomena which appear in flow around bodies.) By using the method there and from the quantities given on the singular line in those old regions, we can determine the quantities in the new regions.

Roughly speaking, an accurate calculation of interaction between discontinuities can be done in the following way. In the process of calculation, we choose a function E such that the intersection line of two discontinuities is involved in a surface $\eta^*(z, r, \varphi)=$constant. In that case, as the calculation comes to $\eta=\eta^*$, where two discontinuities intersect, the two intersecting boundaries are cancelled from the flow field, and the new boundaries produced, starting from the intersection line, are added into the flow field. G, G_η, G_ψ of those new boundaries on the intersection line and the flow parameters as functions of ξ in those regions between the new boundaries can be determined by the method described in Chapter 4, § 2, and from (2.11), (3.16), (3.17). After obtaining those quantities, we can proceed with calculation along η.

(3) *Treatment of Entropy Layers*

As is well known, if the blunt body is slender, then there is the so-called "entropy layer" in the region near the body far from the nose, where certain flow variables, such as entropy, density, temperature, vary rapidly in the r-direction. In order to obtain an accurate result in the entropy layer, a smaller mesh size in the physical space should be used in that region. However, because flow variables in other parts of the flow field vary slowly, a larger mesh size should be adopted if the same accuracy is required. Moreover, the difference between these two mesh sizes should get larger as the entropy layer gets thinner. How do we realize this requirement? We will adopt the following procedure. In the region near the body, we choose a stream surface as an internal boundary properly, such that the region between the stream surface and the body involves the whole entropy layer and is nearly the same as the entropy layer. Moreover, a certain number of mesh points are put in the region between the stream surface and the body. Clearly, the mesh size in that region in the physical space gets smaller and smaller automatically as the entropy layer gets thinner and thinner. In this way, we make the results in the entropy layer quite accurate even though the entropy layer is very thin.

§ 4. Computed Results

We have made a program for the scheme described in the preceding sections, and done a series of calculations of supersonic flow fields around the bodies. The bodies considered are various types of combined bodies with circular or elliptic cross sections, whose noses are spheres or cones. For example, sphere–cones, multistage cone–cylinder–flare combined bodies and combined bodies with elliptic cross sections are under our consideration. In our calculation we take the vertex of the body as the origin of the coordinate system, and the axis of the body as the z–axis. For slender combined bodies, the range of angle of attack of our computed results is $0°\leqslant\alpha<10°$. For short combined bodies, the range of angle of attack of our computed results is $0°\leqslant\alpha\leqslant30°$. The gases we consider in our calculation are the perfect gas with $\gamma=1.4$ and air in chemical equilibrium. For the sonic speed, enthalpy, temperature, ··· of air in chemical equilibrium, we use the approximate expressions in Chapter 4, § 4. Our computed results have been examined in several different ways. We have made comparisons among the results with different mesh sizes, examined how accurately the conservation law of the total energy is satisfied at every point, and compared the computed results with the experimental data. We know from these examinations that our numerical method is successful and satisfactory if the functions are smooth in the φ–direction. In what follows, we shall give several results and briefly analyse some of the results. For the sake of convenience, in several figures and tables we shall use the following abbreviations:

MS——a main shock;
S_1——a shock of the first family;
S_2——a shock of the second family;
W_1——a wave–characteristic of the first family;
W_2——a wave–characteristic of the second family;
CD——a contact discontinuity;
SC——a stream–characteristic;
BS——a body surface.

1. *Results of the Flow around Sphere–cones*

In Figs. 4.1 and 4.2, we show several curves of pressure distribution and density distribution on sphere–cones. Here $z\leqslant20$, the angle of

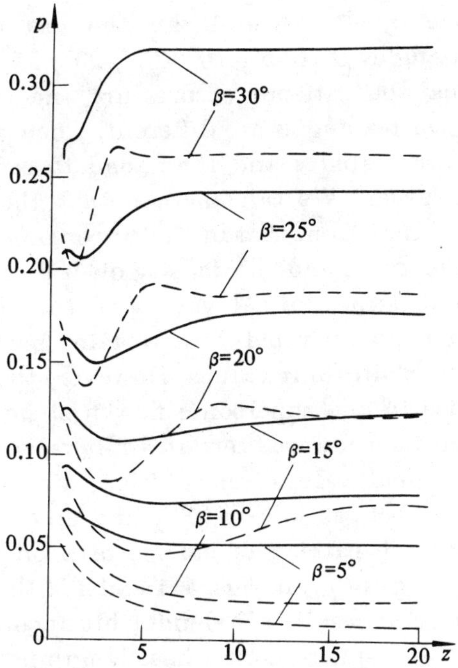

Figure 4.1 Pressure distributions on the bodies for perfect gas flows around a sphere-cone, where β is the semi-vertex angle, the radius of the sphere $R_0=1$ and the angle of attack $\alpha=0°$
—— $M_\infty=4$, - - - $M_\infty=20$.

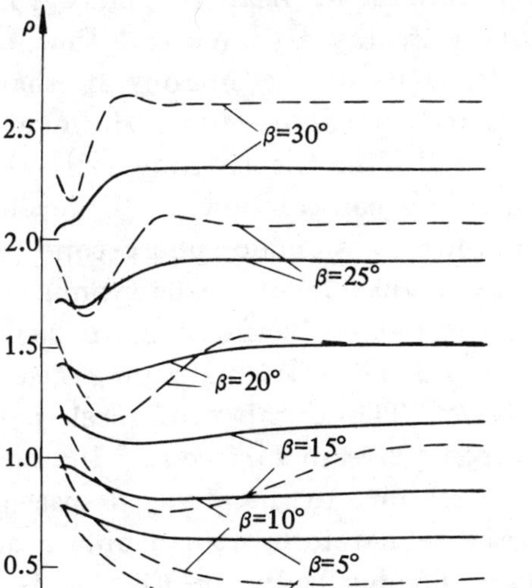

Figure 4.2 Density distributions on the bodies for perfect gas flows around a sphere-cone, where β is the semi-vertex angle, the radius of the sphere $R_0=1$ and the angle of attack $\alpha=0°$
—— $M_\infty=4$, - - - $M_\infty=20$.

attack is equal to zero, $M_\infty=4$ and 20, the gas is a perfect gas, and the semi-vertex angles β are 5°, 10°, 15°, 20°, 25°, 30°. We can see from those figures that the patterns of pressure and density curves are different if the semi-vertex angles are different, They vary solwly if β is smaller. They first vary rapidly and then tend to certain limit values immediately if β is larger. We can also see that the character of the supersonic flow is different from that of the hypersonic flow. When $\beta=$ 20°, 25° or 30°, the pressure and the density on bodies in the hypersonic flow first go down to certain minimal values behind the tangent points on sphere-cones, then go up rapidly to certain maximal values, and finally tend towards certain limit values. However, the pressure and the density on the bodies in the supersonic flow first go down to certain minimal values, then tend towards certain limit values directly, without the appearance of maximal values. In addition, since the expression of the entropy for the perfect gas is $s=p/\rho^{1.4}$, and since the entropy on a body does not change, the pattern of change of p on the bodies is the same as that of the change of ρ. In Figs. 4.1 and 4.2, this fact can be seen.

We now show several results of "slender blunt cones". The so-called "slender blunt cone" is a sphere-cone whose length is much greater than the radius of the sphere. Intuitively, we recognize that the flow field around a long sphere-cone in the region far from the nose should be close to the flow field around the corresponding cone. However, because of the sphere-nose, there is a thin layer near the body where the entropy, density, ⋯, and others vary rapidly. We call that layer the entropy layer. In most cases, the value of the entropy in that layer is greater than that of the corresponding conical flow. However, there is often a small region where the value of the entropy is less than that of the entropy of the corresponding conical flow. In order to get an accurate result of the flow field around a slender sphere-cone, we "separate" the entropy layer. This means that in our calculations we will introduce several stream surfaces as internal boundaries, in such a way, that the region between the body and the farthest stream surface from the body is almost the entropy layer. The number of mesh-points between two stream surfaces or between a stream surface and the body is fixed, so that the mesh sizes in the r direction, in the physical space get smaller as the entropy layer gets thinner. Therefore, we can obtain accurate results of the flow field around a slender body. In Figs. 4.3—4.14 and Tables 4.1—4.6, we give several results about sphere-cones at zero-angle of attack, where the gas is a perfect gas with $\gamma=1.4$. Several stream surfaces are introduced when these results are calculated.

In Figs. 4.3—4.6 and Table 4.1 we show pressure distributions behind the shock and on the body in the flow field around sphere-cones

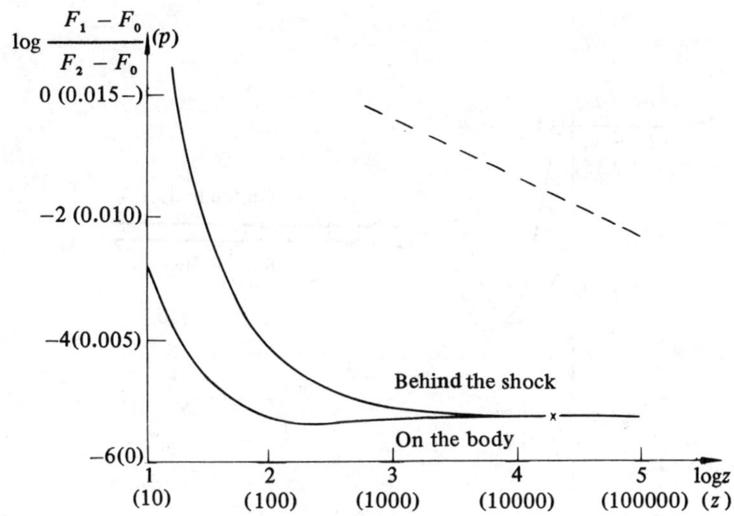

Figure 4.3 Pressure distributions behind the shock and on the body (M_∞=20, perfect gas, β=0°, R_0=1)

—— Pressure, × Pressure of free stream, - - - $\log \frac{F_1 - F_0}{F_2 - F_0}$.

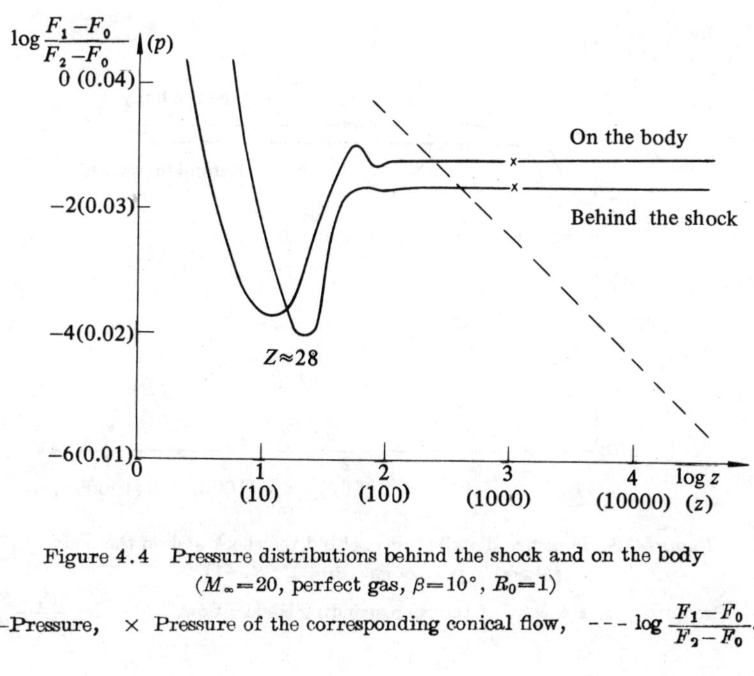

Figure 4.4 Pressure distributions behind the shock and on the body (M_∞=20, perfect gas, β=10°, R_0=1)

——Pressure, × Pressure of the corresponding conical flow, - - - $\log \frac{F_1 - F_0}{F_2 - F_0}$.

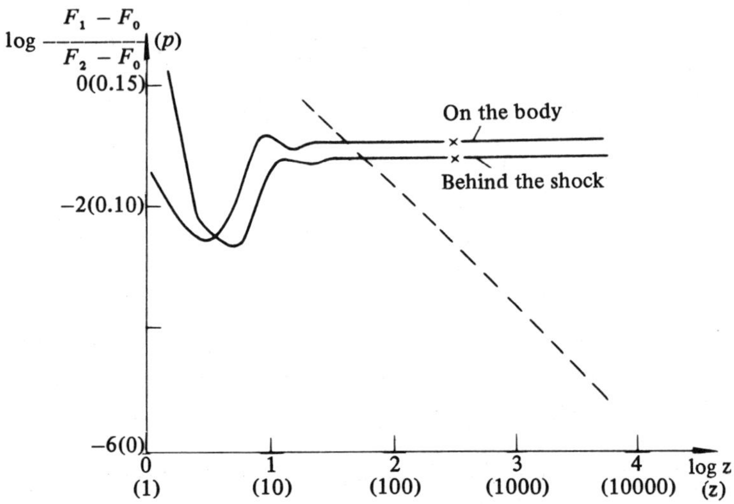

Figure 4.5 Pressure distributions behind the shock and on the body ($M_\infty = 20$, perfect gas, $\beta = 20°$, $R_0 = 1$)

—— Pressure, × Pressure of the corresponding conical flow, --- $\log \frac{F_1 - F_0}{F_2 - F_0}$.

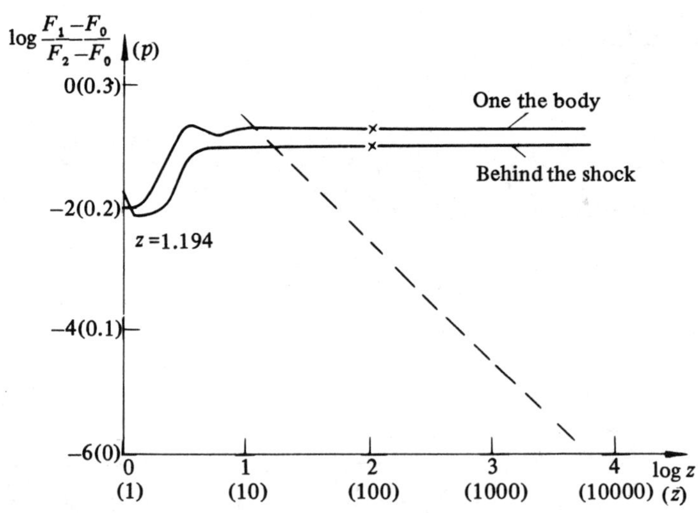

Figure 4.6 Pressure distributions behind the shock and on the body ($M_\infty = 20$, perfect gas, $\beta = 30°$, $R_0 = 1$)

—— Pressure, × Pressure of the corresponding conical flow, --- $\log \frac{F_1 - F_0}{F_2 - F_0}$.

Table 4.1 Comparisons of surface pressure between the flow around sphere–cones and the corresponding conical flow
(perfect gas, $\gamma=1.4$, $\alpha=0°$)

	z	Pressures behind shocks	Pressures on bodies
	10	0.02575	0.00849
	20	0.01370	0.00482
	50	0.00722	0.00256
	100	0.00478	0.00180
$\beta=0°$	200	0.00345	0.00154
$M_\infty=20$	500	0.00256	0.00166
	1000	0.00222	0.00176
	5000	0.00191	0.00178
	50000	0.00179	0.00179
	Conical flow	0.00179	0.00179
	10	0.02959	0.02175
	20	0.02040	0.02339
	50	0.03100	0.03481
	100	0.03138	0.03368
$\beta=10°$	200	0.03167	0.03378
$M_\infty=20$	500	0.03170	0.03380
	1000	0.03170	0.03380
	5000	0.03170	0.03380
	50000	0.03170	0.03380
	Conical flow	0.03170	0.03380
	10	0.1131	0.1282
	20	0.1169	0.1240
	50	0.1182	0.1246
	100	0.1183	0.1247
$\beta=20°$	200	0.1183	0.1247
$M_\infty=20$	500	0.1183	0.1247
	1000	0.1183	0.1247
	5000	0.1183	0.1247
	50000	—	—
	Conical flow	0.1183	0.1247
	10	0.2503	0.2641
	20	0.2512	0.2643
	50	0.2514	0.2644
	100	0.2514	0.2644
$\beta=30°$	200	0.2514	0.2644
$M_\infty=20$	500	0.2514	0.2644
	1000	0.2514	0.2644
	5000	0.2514	0.2644
	50000	—	—
	Conical flow	0.2514	0.2644

with $M_\infty = 20$, and compare the results with the values of the corresponding conical flow. We see from these results that the process of the pressure tending to the values of the conical flow from the values at the nose is usually non-monotone. For the sphere-cones with the semi-vertex angle $\beta=0°, 10°, 20°, 30°$, the values of pressure are equal to the limit values with an accuracy of three significant digits respectively at $z\approx 50000$, 200, 40 and 15, where the radius of the nose is equal to 1 and the vertex of the body is at $z=0$. In these figures, the curves of $\log\frac{F_1-F_0}{F_2-F_0}$ are also given, where F_0, F_1, F_2 are the values of r on the body, a certain stream surface and the shock respectively. Those curves are approximately straight lines, whose slopes are equal to -1 if $\beta=0°$ or -2 if $\beta\neq 0°$. Hence, for larger z, we have the approximate expression:

$$\log\frac{F_1-F_0}{F_2-F_0}\approx a_1+b_1\log z,$$

where a_1 is a function of β, and

$$b_1=\begin{cases}-1, & \text{if } \beta=0°,\\ -2, & \text{if } \beta\neq 0°.\end{cases}$$

Therefore, the relative thickness $\frac{F_1-F_0}{F_2-F_0}$ of the entropy layer tends to zero in the pattern of $O(z^{-2})$ if $\beta\neq 0°$ or $O(z^{-1})$ if $\beta=0°$ as z tends towards infinity.

In Figs. 4.7—4.10, several pressure curves and density curves in the ξ-direction in the flow field around sphere-cones are given. Here $M_\infty=20$, and $\beta=0°, 10°, 20°$ and $30°$. We see from these figures that for $\beta=0°, 10°, 20°, 30°$, the curves in the interval $\xi=1$—2, are almost the same as those of the corresponding conical flow when $z\approx 20000$, 400, 100, 50 respectively. (The results of conical flow are obtained by using the method of Chapter 6.) From Figs. 4.3—4.6 we know $\frac{F_1-F_0}{F_2-F_0}\leqslant 0.02$ in these cases. This means that the thickness of the entropy layer is less than 2% of the thickness of the region between the shock and the body. Hence the flow is almost a conical flow.

We can also see from Figs. 4.7—4.10 and Table 4.2 that in the z, ξ-coordinate system the flow parameters almost keep constant as z varies, if z is very large. Therefore a certain similarity law exists. Hence, if we have all flow parameters on a cross section $z=$constant (very large), then we can determine the flow parameters on any cross section in the downstream region of the cross section. (In Tables 4.3—4.6, we provide these flow parameters in the cases of $\beta=0°, 10°, 20°, 30°$ and $M_\infty=20$.) In fact, the surface of shock of flow around a sphere-cone is close to the surface of shock of the corresponding conical flow, if z is very large. Hence we

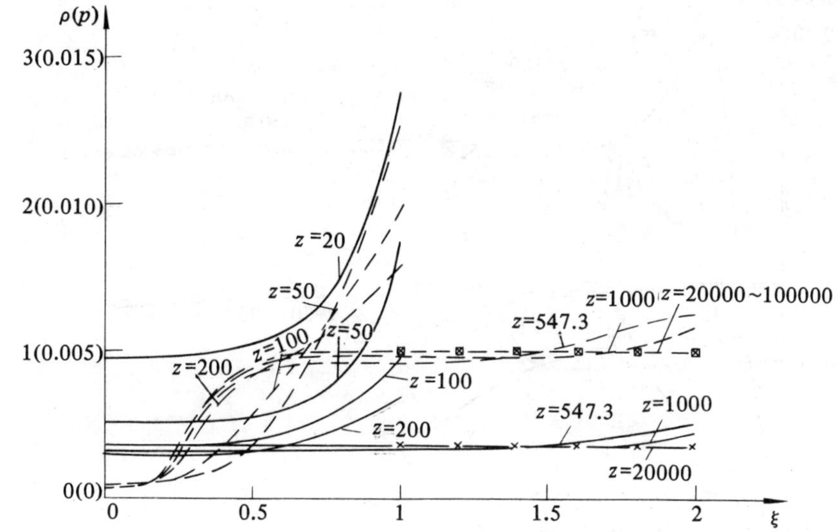

Figure 4.7 Variations of pressure and density with ξ
($M_\infty=20$, perfect gas, $\beta=0°$, $R_0=1$)
—— Pressure, × Pressure of free stream,
---- Density, ⊠ Density of free stream.

The construction of the flow field $\begin{cases} \text{MS–BS, } z=20\text{—}200. \\ \text{MS–SC–BS, } z=547.3\text{—}100000. \end{cases}$

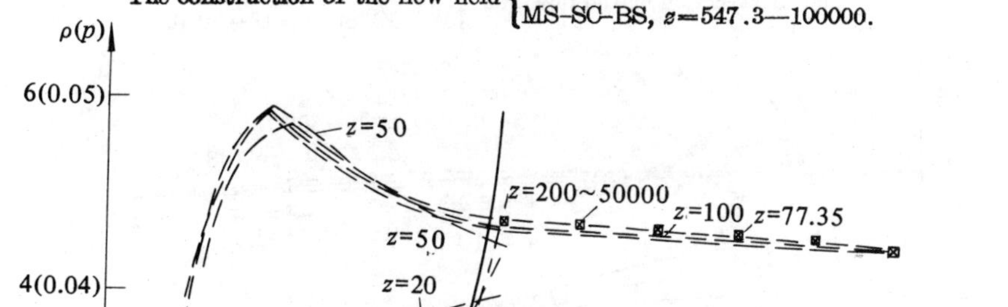

Figure 4.8 Variations of pressure and density with ξ
($M_\infty=20$, perfect gas, $\beta=10°$, $R_0=1$)
—— Pressure, × Pressure of the corresponding conical flow.
---- Density, ⊠ Density of the corresponding conical flow.

The construction of the flow field $\begin{cases} \text{MS–BS, } z=5\text{—}50. \\ \text{MS–SC–BS, } z=77.35\text{—}50000. \end{cases}$

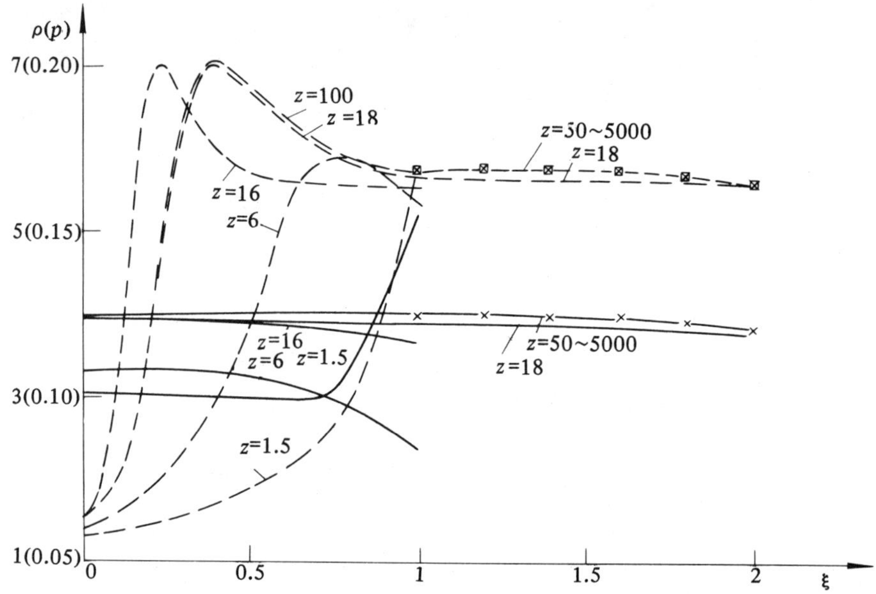

Figure 4.9 Variations of pressure and density with ξ

($M_\infty=20$, perfect gas, $\beta=20°$, $R_0=1$)

—— Pressure, × Pressure of the corresponding conical flow,

---- Density, ⊠ Density of the corresponding conical flow.

The construction of the flow field $\begin{cases}\text{MS-BS}, z=1.5\text{—}16.\\ \text{MS-SC-BS}, z=18\text{—}5000.\end{cases}$

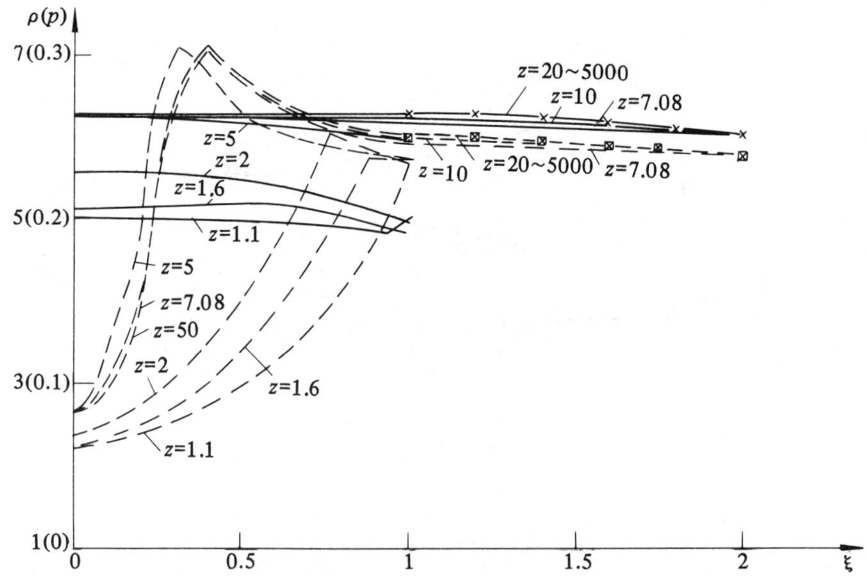

Figure 4.10 Variations of pressure and density with ξ

($M_\infty=20$, perfect gas, $\beta=30°$, $R_0=1$)

—— Pressure, × Pressure of the corresponding conical flow,

---- Density, ⊠ Donsity of the corresponding conical flow.

The construction of the flow field $\begin{cases}\text{MS-BS}, z=1.1\text{—}5.\\ \text{MS-SC-BS}, z=7.08\text{—}5000.\end{cases}$

Table 4.2

Body: sphere–cone, $\beta=30°$, $R_0=1$
Free stream: $M_\infty=20$ (perfect gas), $\alpha=0°$

Construction of flow field: MS–SC–SC–BS

$z_3=20$, $F_3=13.818$, $F_{3,z}=0.65742$

ξ	u	v	p	ρ
3.000	0.75060	0.37935	0.25118	5.76136
2.900	0.74835	0.38282	0.25337	5.79753
2.800	0.74613	0.38628	0.25536	5.83048
2.700	0.74393	0.38975	0.25715	5.86035
2.600	0.74175	0.39324	0.25875	5.88719
2.500	0.73958	0.39677	0.26016	5.91107
2.400	0.73743	0.40035	0.26138	5.93233
2.300	0.73532	0.40400	0.26239	5.95199
2.200	0.73330	0.40777	0.26320	5.97238
2.100	0.73152	0.41177	0.26380	5.99870
2.000	0.72928	0.41560	0.26417	6.00528

$z_2=20$, $F_2=12.284$, $F_{2,z}=0.56988$

ξ	u	v	p	ρ
2.000	0.72928	0.41560	0.26417	6.00528
1.900	0.72911	0.41590	0.26419	6.00577
1.800	0.72913	0.41631	0.26421	6.01364
1.700	0.72927	0.41679	0.26423	6.02587
1.600	0.73004	0.41762	0.26424	6.06224
1.500	0.73137	0.41878	0.26425	6.12070
1.400	0.73352	0.42041	0.26427	6.20293
1.300	0.73803	0.42340	0.26428	6.40865
1.200	0.74604	0.42841	0.26429	6.79316
1.100	0.75077	0.43154	0.26430	7.03676
1.000	0.75298	0.43322	0.26430	7.17530

$z_1=20$, $F_1=12.167$, $F_{1,z}=0.57535$

Table 4.2 (continued)

ξ	u	v	p	ρ
1.000	0.75298	0.43322	0.26430	7.17530
0.975	0.74899	0.43097	0.26431	6.96187
0.950	0.74383	0.42803	0.26431	6.70102
0.925	0.73846	0.42498	0.26431	6.44797
0.900	0.73349	0.42216	0.26431	6.23584
0.875	0.72794	0.41900	0.26431	6.01354
0.850	0.72225	0.41576	0.26431	5.80629
0.825	0.71651	0.41249	0.26431	5.61722
0.800	0.71041	0.40902	0.26431	5.42659
0.775	0.70436	0.40557	0.26431	5.25204
0.750	0.69821	0.40206	0.26431	5.08748
0.725	0.69170	0.39835	0.26431	4.92161
0.700	0.68499	0.39452	0.26431	4.76474
0.675	0.67811	0.39059	0.26431	4.61737
0.650	0.67125	0.38667	0.26431	4.48475
0.625	0.66437	0.38274	0.26431	4.36251
0.600	0.65718	0.37863	0.26431	4.23901
0.575	0.64979	0.37440	0.26431	4.11807
0.550	0.64231	0.37013	0.26431	4.00016
0.525	0.63494	0.36591	0.26431	3.88846
0.500	0.62769	0.36177	0.26431	3.78425
0.475	0.62033	0.35755	0.26431	3.68254
0.450	0.61294	0.35333	0.26431	3.58603
0.425	0.60561	0.34913	0.26431	3.49341
0.400	0.59831	0.34495	0.26431	3.40529
0.375	0.59100	0.34077	0.26431	3.32342
0.350	0.58368	0.33658	0.26431	3.24540
0.325	0.57643	0.33243	0.26431	3.17221
0.300	0.56908	0.32821	0.26431	3.10064
0.275	0.56172	0.32400	0.26431	3.03455
0.250	0.55449	0.31985	0.26431	2.97595

Table 4.2 (continued)

ξ	u	v	p	ρ
0.225	0.54725	0.31571	0.26431	2.92046
0.200	0.53998	0.31154	0.26431	2.86924
0.175	0.53263	0.30733	0.26431	2.82075
0.150	0.52545	0.30321	0.26431	2.77917
0.125	0.51817	0.29903	0.26431	2.74211
0.100	0.51095	0.29490	0.26431	2.71104
0.075	0.50375	0.29077	0.26431	2.68276
0.050	0.49647	0.28659	0.26431	2.65955
0.025	0.48938	0.28252	0.26431	2.64813
0.000	0.48222	0.27841	0.26431	2.63389

$z_0=20,\ F_0=12.124,\ F_{0,z}=0.57735$

Construction of flow field: MS–SC–SC–BS

$z_3=50,\ F_3=33.548,\ F_{3,z}=0.65775$

ξ	u	v	p	ρ
3.000	0.75043	0.37943	0.25136	5.76152
2.900	0.74798	0.38317	0.25370	5.79993
2.800	0.74557	0.38691	0.25582	5.83450
2.700	0.74318	0.39066	0.25771	5.86534
2.600	0.74080	0.39444	0.25938	5.89249
2.500	0.73843	0.39826	0.26081	5.91597
2.400	0.73606	0.40214	0.26202	5.93577
2.300	0.73369	0.40608	0.26299	5.95193
2.200	0.73131	0.41012	0.26372	5.96463
2.100	0.72896	0.41428	0.26418	5.97512
2.000	0.72730	0.41897	0.26437	6.00854

$z_2=50,\ F_2=29.511,\ F_{2,z}=0.57606$

ξ	u	v	p	ρ
2.000	0.72730	0.41897	0.26437	6.00854

Table 4.2 (continued)

ξ	u	v	p	ρ
1.900	0.72728	0.41902	0.26438	6.00873
1.800	0.72744	0.41919	0.26438	6.01625
1.700	0.72772	0.41942	0.26438	6.02821
1.600	0.72862	0.42000	0.26438	6.06396
1.500	0.73008	0.42092	0.26438	6.12200
1.400	0.73237	0.42230	0.26438	6.20382
1.300	0.73699	0.42504	0.26438	6.40797
1.200	0.74512	0.42980	0.26438	6.79130
1.100	0.75005	0.43272	0.26438	7.03758
1.000	0.75242	0.43416	0.26438	7.17675

$z_1=50$, $F_1=29.4623$, $F_{1,z}=0.57701$

ξ	u	v	p	ρ
1.000	0.75242	0.43416	0.26438	7.17675
0.975	0.74846	0.43187	0.26438	6.96360
0.950	0.74332	0.42892	0.26438	6.70307
0.925	0.73798	0.42584	0.26438	6.45010
0.900	0.73303	0.42299	0.26438	6.23792
0.875	0.72750	0.41980	0.26438	6.01591
0.850	0.72183	0.41654	0.26438	5.80861
0.825	0.71610	0.41324	0.26438	5.61950
0.800	0.71003	0.40975	0.26438	5.42898
0.775	0.70400	0.40627	0.26438	5.25429
0.750	0.69787	0.40274	0.26438	5.08970
0.725	0.69138	0.39900	0.26438	4.92388
0.700	0.68468	0.39514	0.26438	4.76687
0.675	0.67782	0.39118	0.26438	4.61941
0.650	0.67097	0.38724	0.26438	4.48661
0.625	0.66410	0.38328	0.26438	4.36428
0.600	0.65694	0.37915	0.26438	4.24079
0.575	0.64955	0.37489	0.26438	4.11975

Table 4.2 (continued)

ξ	u	v	p	ρ
0.550	0.64209	0.37059	0.26438	4.00176
0.525	0.63473	0.36635	0.26438	3.88995
0.500	0.62749	0.36217	0.26438	3.78562
0.475	0.62014	0.35794	0.26438	3.68386
0.450	0.61277	0.35369	0.26438	3.58726
0.425	0.60544	0.34946	0.26438	3.49457
0.400	0.59815	0.34526	0.26438	3.40637
0.375	0.59085	0.34105	0.26438	3.32442
0.350	0.58354	0.33684	0.26438	3.24634
0.325	0.57630	0.33266	0.26438	3.17309
0.300	0.56895	0.32843	0.26438	3.10148
0.275	0.56160	0.32419	0.26438	3.03533
0.250	0.55438	0.32002	0.26438	2.97666
0.225	0.54715	0.31586	0.26438	2.92113
0.200	0.53988	0.31167	0.26438	2.86987
0.175	0.53254	0.30743	0.26438	2.82136
0.150	0.52537	0.30329	0.26438	2.77974
0.125	0.51809	0.29910	0.26438	2.74265
0.100	0.51088	0.29494	0.26438	2.71156
0.075	0.50369	0.29079	0.26438	2.68327
0.050	0.49642	0.28660	0.26438	2.66005
0.025	0.48934	0.28251	0.26438	2.64861
0.000	0.48218	0.27839	0.26438	2.63437

$z_0=50,\ F_0=29.4449,\ F_{0,z}=0.57735$

Construction of flow field: MS–SC–SC–BS

$z_3=500,\ F_3=329.5503,\ F_{3,z}=0.65780$

ξ	u	v	p	ρ
3.000	0.75040	0.37944	0.25138	5.76154
2.900	0.74794	0.38323	0.25375	5.80045
2.800	0.74547	0.38704	0.25591	5.83542

Table 4.2 (continued)

ξ	u	v	p	ρ
2.700	0.74305	0.39084	0.25780	5.86648
2.600	0.74062	0.39469	0.25950	5.89374
2.500	0.73822	0.39856	0.26092	5.91713
2.400	0.73579	0.40251	0.26215	5.93666
2.300	0.73338	0.40651	0.26309	5.95217
2.200	0.73092	0.41061	0.26381	5.96355
2.100	0.72846	0.41480	0.26423	5.97058
2.000	0.72689	0.41967	0.26439	6.00886

$z_2=500,\ F_2=289.2592,\ F_{2,z}=0.57735$

ξ	u	v	p	ρ
2.000	0.72689	0.41967	0.26439	6.00886
1.900	0.72690	0.41967	0.26440	6.00912
1.800	0.72710	0.41979	0.26440	6.01671
1.700	0.72742	0.41997	0.26440	6.02903
1.600	0.72834	0.42051	0.26440	6.06484
1.500	0.72984	0.42138	0.26440	6.12305
1.400	0.73218	0.42272	0.26440	6.20619
1.300	0.73684	0.42541	0.26440	6.41044
1.200	0.74495	0.43010	0.26440	6.79204
1.100	0.74991	0.43296	0.26440	7.03819
1.000	0.75231	0.43435	0.26439	7.17703

$z_1=500,\ F_1=289.2543,\ F_{1,z}=0.57735$

ξ	u	v	p	ρ
1.000	0.75231	0.43435	0.26439	7.17703
0.975	0.74930	0.43261	0.26440	7.01382
0.950	0.74462	0.42991	0.26440	6.77459
0.925	0.73925	0.42681	0.26440	6.51539
0.900	0.73394	0.42374	0.26440	6.27954
0.875	0.72857	0.42065	0.26440	6.06186
0.850	0.72296	0.41741	0.26440	5.85300

Table 4.2 (continued)

ξ	u	v	p	ρ
0.825	0.71716	0.41406	0.26440	5.65507
0.800	0.71118	0.41061	0.26440	5.46683
0.775	0.70506	0.40707	0.26440	5.28707
0.750	0.69881	0.40346	0.26440	5.11599
0.725	0.69239	0.39975	0.26440	4.95201
0.700	0.68572	0.39590	0.26440	4.79347
0.675	0.67884	0.39193	0.26440	4.64284
0.650	0.67187	0.38791	0.26440	4.50384
0.625	0.66486	0.38386	0.26440	4.37590
0.600	0.65774	0.37975	0.26440	4.25450
0.575	0.65043	0.37553	0.26440	4.13491
0.550	0.64295	0.37121	0.26440	4.01651
0.525	0.63547	0.36689	0.26440	3.90234
0.500	0.62808	0.36262	0.26440	3.79465
0.475	0.62076	0.35839	0.26440	3.69307
0.450	0.61343	0.35416	0.26440	3.59617
0.425	0.60607	0.34991	0.26440	3.50308
0.400	0.59870	0.34566	0.26440	3.41411
0.375	0.59137	0.34143	0.26440	3.33050
0.350	0.58405	0.33720	0.26440	3.25196
0.325	0.57672	0.33297	0.26440	3.17757
0.300	0.56933	0.32870	0.26440	3.10599
0.275	0.56192	0.32442	0.26440	3.03871
0.250	0.55459	0.32019	0.26440	2.97828
0.225	0.54731	0.31599	0.26440	2.92276
0.200	0.54001	0.31177	0.26440	2.87099
0.175	0.53266	0.30753	0.26440	2.82264
0.150	0.52541	0.30334	0.26440	2.78020
0.125	0.51814	0.29915	0.26440	2.74315
0.100	0.51090	0.29497	0.26440	2.71164
0.075	0.50369	0.29080	0.26440	2.68361
0.050	0.49642	0.28661	0.26440	2.66051
0.025	0.48932	0.28251	0.26440	2.64849
0.000	0.48217	0.27838	0.26440	2.63453

$z_0=500,\ F_0=289.2525,\ F_{0,z}=0.57735$

Table 4.3

Body: sphere–cone, $\beta=0°$, $R_0=1$

Free stream: $M_\infty=20$ (perfect gas), $\alpha=0°$

Construction of flow field: MS–SC–BS

$z_2=20000$, $F_2=1037.5$, $F_{2,z}=0.0503$

ξ	u	v	p	ρ
2.000	0.99998	0.00033	0.00180	1.00662
1.937	1.00000	−0.00009	0.00178	0.99804
1.875	1.00000	−0.00002	0.00178	0.99955
1.812	1.00000	0.00000	0.00179	0.99981
1.750	1.00000	0.00000	0.00179	0.99988
1.687	1.00000	0.00000	0.00179	0.99991
1.625	1.00000	0.00000	0.00179	0.99993
1.562	1.00000	0.00000	0.00179	0.99994
1.500	1.00000	0.00000	0.00179	0.99994
1.437	1.00000	0.00000	0.00179	0.99995
1.375	1.00000	0.00000	0.00179	0.99995
1.312	1.00000	0.00000	0.00179	0.99996
1.250	1.00000	0.00000	0.00179	0.99996
1.187	1.00000	0.00000	0.00179	0.99996
1.125	1.00000	0.00000	0.00179	0.99993
1.062	1.00000	0.00000	0.00179	0.99979
1.000	0.99994	0.00000	0.00179	0.99008

$z_1=20000$, $F_1=20.5$, $F_{1,z}=0.000$

ξ	u	v	p	ρ
1.000	0.99994	0.00000	0.00179	0.99008
0.937	0.99992	0.00000	0.00179	0.98793
0.875	0.99991	0.00000	0.00179	0.98506
0.812	0.99988	0.00000	0.00179	0.98115
0.750	0.99985	0.00000	0.00179	0.97569
0.687	0.99979	0.00000	0.00179	0.96779
0.625	0.99971	0.00000	0.00179	0.95593
0.562	0.99958	0.00000	0.00179	0.93730
0.500	0.99934	0.00000	0.00179	0.90640
0.437	0.99884	0.00000	0.00179	0.85201
0.375	0.99764	0.00000	0.00179	0.75240
0.312	0.99457	0.00000	0.00179	0.57892
0.250	0.98703	0.00000	0.00179	0.33878
0.187	0.97132	0.00000	0.00179	0.14181
0.125	0.94840	0.00000	0.00179	0.09695
0.062	0.92785	0.00000	0.00179	0.08874
0.000	0.91901	0.00000	0.00179	0.07412

$z_0=20000$, $F_0=1.00$, $F_{0,z}=0.000$

Table 4.4

Body: sphere–cone, $\beta=10°$, $R_0=1$

Free stream: $M_\infty=20$ (perfect gas), $\alpha=0°$

Construction of flow field: MS–SC–BS

$z_2=500$, $F_2=100.88$, $F_{2,z}=0.1998$

ξ	u	v	p	ρ
2.000	0.97009	0.14969	0.03170	4.52631
1.937	0.96983	0.15098	0.03195	4.55175
1.875	0.96958	0.15225	0.03218	4.57540
1.812	0.96934	0.15350	0.03239	4.59733
1.750	0.96909	0.15476	0.03259	4.61762
1.687	0.96885	0.15600	0.03278	4.63633
1.625	0.96861	0.15725	0.03295	4.65348
1.562	0.96838	0.15849	0.03310	4.66912
1.500	0.96814	0.15974	0.03324	4.68327
1.437	0.96791	0.16099	0.03336	4.69593
1.375	0.96768	0.16225	0.03347	4.70712
1.312	0.96744	0.16352	0.03357	4.71683
1.250	0.96721	0.16480	0.03365	4.72506
1.187	0.96698	0.16609	0.03371	4.73182
1.125	0.96675	0.16740	0.03376	4.73758
1.062	0.96654	0.16872	0.03379	4.74594
1.000	0.96640	0.17008	0.03380	4.76615

$z_1=500$, $F_1=89.1709$, $F_{1,z}=0.1760$

ξ	u	v	p	ρ
1.000	0.96640	0.17008	0.03380	4.76615
0.937	0.96655	0.17013	0.03380	4.79665
0.875	0.96677	0.17019	0.03380	4.83889
0.812	0.96705	0.17026	0.03380	4.89729
0.750	0.96744	0.17035	0.03380	4.97824
0.687	0.96795	0.17046	0.03380	5.08841
0.625	0.96859	0.17059	0.03380	5.23196
0.562	0.96933	0.17074	0.03380	5.40776
0.500	0.97008	0.17089	0.03380	5.59929
0.437	0.97063	0.17101	0.03380	5.74853
0.375	0.97055	0.17101	0.03380	5.73918
0.312	0.96921	0.17080	0.03380	5.42765
0.250	0.96535	0.17014	0.03380	4.69963
0.187	0.95484	0.16831	0.03380	3.48408
0.125	0.92683	0.16339	0.03380	1.84659
0.062	0.87051	0.15348	0.03380	0.70073
0.000	0.77684	0.13698	0.03380	0.60607

$z_0=500$, $F_0=89.0026$, $F_{0,z}=0.1763$

Table 4.5

Body: sphere–cone, $\beta=20°$, $R_0=1$

Free stream: $M_\infty=20$ (perfect gas), $\alpha=0°$

Construction of flow field: MS–SC–BS

$z_2=100$, $F_2=41.5204$, $F_{2,z}=0.40728$

ξ	u	v	p	ρ
2.000	0.88352	0.28600	0.11827	5.51544
1.937	0.88275	0.28789	0.11902	5.54040
1.875	0.88199	0.28977	0.11972	5.56373
1.812	0.88125	0.29164	0.12037	5.58545
1.750	0.88050	0.29351	0.12098	5.60566
1.687	0.87977	0.29537	0.12154	5.62430
1.625	0.87904	0.29724	0.12205	5.64151
1.562	0.87831	0.29911	0.12252	5.65723
1.500	0.87759	0.30099	0.12295	5.67155
1.437	0.87687	0.30287	0.12333	5.68444
1.375	0.87615	0.30477	0.12367	5.69597
1.312	0.87543	0.30669	0.12396	5.70611
1.250	0.87472	0.30862	0.12420	5.71495
1.187	0.87400	0.31057	0.12439	5.72246
1.125	0.87330	0.31255	0.12454	5.72947
1.062	0.87270	0.31459	0.12464	5.74270
1.000	0.87210	0.31666	0.12468	5.75450

$z_1=100$, $F_1=37.1860$, $F_{1,z}=0.36311$

ξ	u	v	p	ρ
1.000	0.87210	0.31666	0.12468	5.75450
0.937	0.87242	0.31683	0.12468	5.78042
0.875	0.87295	0.31707	0.12468	5.82173
0.812	0.87377	0.31741	0.12468	5.88724
0.750	0.87498	0.31790	0.12468	5.98575
0.687	0.87659	0.31853	0.12468	6.12123
0.625	0.87858	0.31930	0.12468	6.29680
0.562	0.88096	0.32021	0.12468	6.52088
0.500	0.88361	0.32123	0.12468	6.78687
0.437	0.88578	0.32207	0.12468	7.02487
0.375	0.88556	0.32203	0.12468	7.03654
0.312	0.87875	0.31960	0.12468	6.47462
0.250	0.85822	0.31218	0.12468	5.10102
0.187	0.81617	0.29693	0.12468	3.25882
0.125	0.73723	0.26623	0.12468	2.24053
0.062	0.68044	0.24668	0.12468	1.70452
0.000	0.62756	0.22841	0.12468	1.53909

$z_0=100$, $F_0=37.0972$, $F_{0,z}=0.36397$

Table 4.6

Body: sphere–cone, $\beta=30°$, $R_0=1$

Free stream: $M_\infty=20$ (perfect gas), $\alpha=0°$

Construction of flow field: MS–SC–BS

$z_2=50$, $F_2=33.5486$, $F_{2,z}=0.6577$

ξ	u	v	p	ρ
2.000	0.75043	0.37943	0.25135	5.76151
1.937	0.74890	0.38178	0.25285	5.78607
1.875	0.74737	0.38412	0.25426	5.80910
1.812	0.74586	0.38647	0.25558	5.83062
1.750	0.74436	0.38882	0.25681	5.85068
1.687	0.74286	0.39117	0.25795	5.86928
1.625	0.74137	0.39354	0.25900	5.88643
1.562	0.73988	0.39593	0.25996	5.90214
1.500	0.73840	0.39833	0.26083	5.91641
1.437	0.73691	0.40075	0.26162	5.92923
1.375	0.73542	0.40320	0.26231	5.94061
1.312	0.73393	0.40568	0.26290	5.95056
1.250	0.73244	0.40819	0.26340	5.95911
1.187	0.73095	0.41075	0.26380	5.96644
1.125	0.72947	0.41335	0.26410	5.97292
1.062	0.72806	0.41604	0.26429	5.98077
1.000	0.72759	0.41933	0.26437	6.02281

$z_1=50$, $F_1=29.4972$, $F_{1,z}=0.5763$

ξ	u	v	p	ρ
1.000	0.72759	0.41933	0.26437	6.02281
0.937	0.72798	0.41960	0.26437	6.03842
0.875	0.72873	0.42008	0.26437	6.06811
0.812	0.72995	0.42084	0.26437	6.11964
0.750	0.73166	0.42187	0.26437	6.18584
0.687	0.73387	0.42318	0.26437	6.27754
0.625	0.73680	0.42492	0.26437	6.40376
0.562	0.74063	0.42718	0.26437	6.57636
0.500	0.74520	0.42986	0.26437	6.79406
0.437	0.74943	0.43235	0.26437	7.00872
0.375	0.74925	0.43230	0.26437	7.03112
0.312	0.73641	0.42493	0.26437	6.52836
0.250	0.70047	0.40424	0.26437	5.27708
0.187	0.64138	0.37018	0.26437	3.76307
0.125	0.58511	0.33774	0.26437	3.13670
0.062	0.53366	0.30807	0.26437	2.74583
0.000	0.48219	0.27839	0.26437	2.63436

$z_0=50$, $F_0=29.4449$, $F_{0,z}=0.5774$

have $F_2 \approx a_2 + b_2 z$, where a_2 and b_2 are functions of β. Moreover, we have pointed out that

$$\log \frac{F_1 - F_0}{F_2 - F_0} \approx a_1 + b_1 \log z.$$

We can then determine a_1, a_2 and b_2 from the values of F_2, $F_{2,z}$, F_1, F_0 on a plane z=constant. Therefore, if the coordinates z^*, r^* of a point in the region between the shock and the body are given, then the flow parameters at the point can be obtained in the following way. Because the locations of the shock and the stream surface corresponding to z^* can be approximately determined by these formulae:

$$\begin{cases} F_2^* = a_2 + b_2 z^*, \\ F_1^* = F_0^* + (F_2^* - F_0^*) 10^{a_1} (z^*)^{b_1}. \end{cases}$$

The value of ξ corresponding to r^* should be

$$\xi^* = \begin{cases} 1 + \dfrac{r^* - F_1^*}{F_2^* - F_1^*}, & \text{if } F_1^* \leqslant r^* \leqslant F_2^*, \\ \dfrac{r^* - F_0^*}{F_1^* - F_0^*}, & \text{if } F_0^* \leqslant r^* \leqslant F_1^*. \end{cases}$$

Therefore, the flow parameters at the point (z^*, r^*) can be evaluated from the given data by using the similarity law. These formulae can be used only when z^* is greater than or close to a special value of z, which is 20000, 500, 100, 50 for $\beta = 0°$, $10°$, $20°$, $30°$ respectively (see Tables 4.3—4.6). If z^* is much less than the special values of z, the accuracy will decrease.

Illustrated in Fig. 4.11 are the pressure distributions behind the shock and on the body in the flow field around a sphere–cone with $\beta = 10°$, where $M_\infty = 4$. Fig. 4.4 and 4.11 show that for the pressure both behind the shock and on the body, the difference of the patterns of pressure distribution between the hypersonic flow and the supersonic flow around sphere–cones, with large semi–vertex angles, still appears for $\beta = 10°$. However, the process of change slows down with the decrease of β.

The pressure curves behind the shock in Figs. 4.4—4.6 and 4.11 always have a minimal value. The free stream is uniform, so the slope of the shock and the entropy also take minimal values at the place where the pressure is equal to its minimal value. Moreover, the entropy keeps constant on any stream line. Therefore, the entropy curve in the ξ–direction has a minimal value. (If we take $p/\rho^{1.4}$ as the entropy, the minimal value is 20—30% less than the value of the conical flow.) When z is very large, the entropy layer is very thin. Moreover, the pressure in the entropy layer keeps almost constant in the ξ–direction. Thus, there is a maximal value of the density in the entropy layer in correspondence with the minimal value of the entropy. This is the reason for the

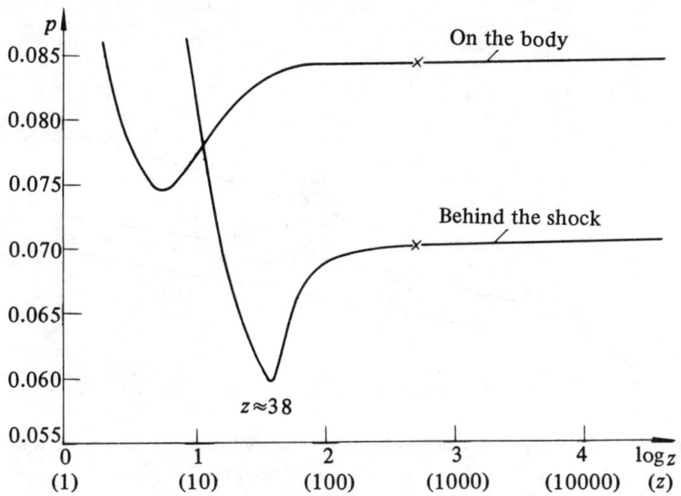

Figure 4.11 Pressure distributions behind the shock and on the body ($M_\infty=4$, perfect gas, $\beta=10°$, $R_0=1$)
—— Pressure, × Pressure of the corresponding conical flow.

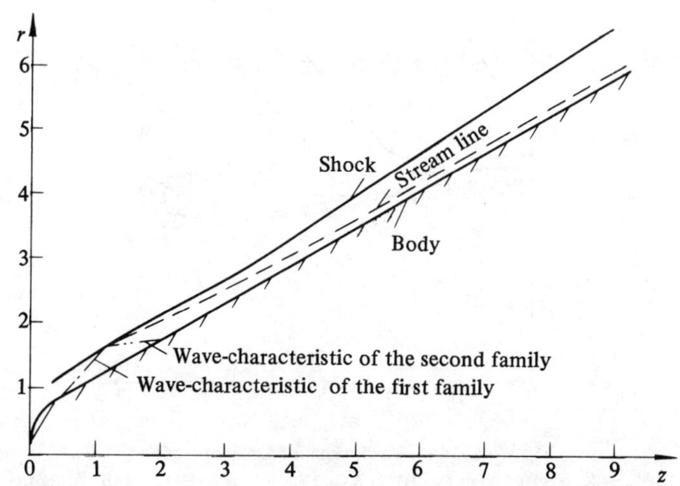

Figure 4.12 Shapes of the shock, wave-characteristics, stream lines and body surfaces ($M_\infty=20$, perfect gas, $\beta=30°$, $R_0=1$)

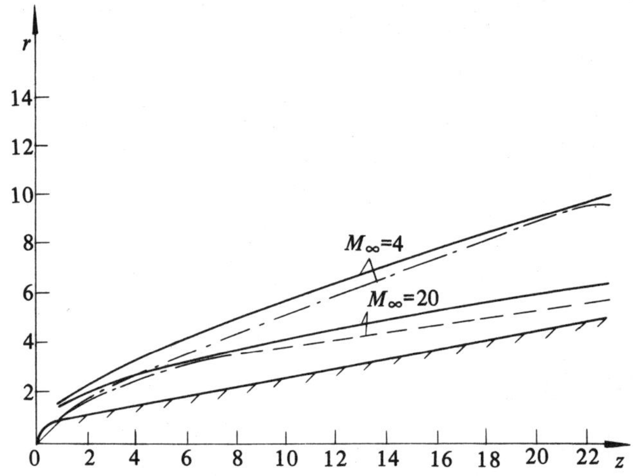

Figure 4.13 Shapes of the shocks, wave-characteristics, stream lines and body surfaces ($M_\infty=4$, 20, perfect gas, $\beta=10°$, $R_0=1$)
—— Shock, —·— Wave-characteristic of the first family, ---- Stream line.

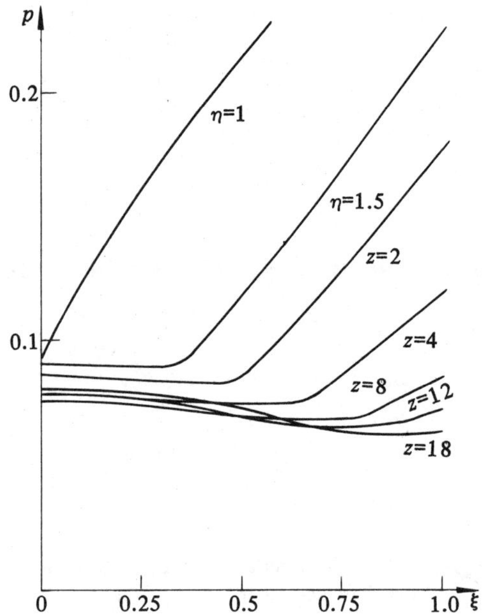

Figure 4.14a Variations of pressure with ξ
($M_\infty=4$, perfect gas, $\beta=10°$, $R_0=1$), where $\eta=1$, 1.5 denote respectively the lines through the points $z=1$, 1.5 on the symmetric axis and with $\frac{dz}{dr}=-\text{tg}\,10°$, $-\frac{1}{2}\text{tg}\,10°$.

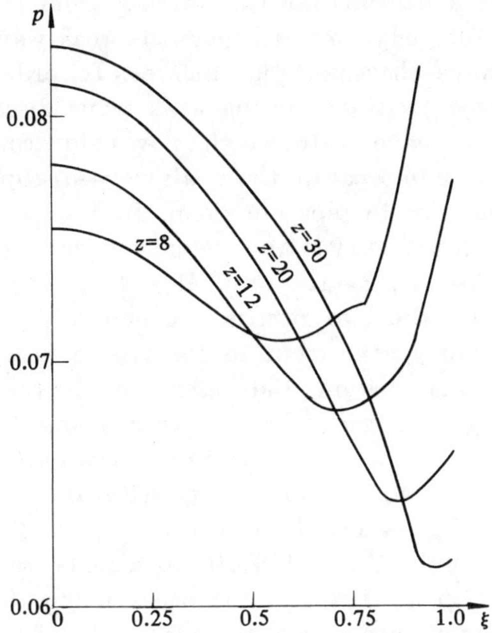

Figure 4.14b Variations of pressure with ξ ($M_\infty=4$, perfect gas, $\beta=10°$, $R_0=1$)

appearance of the "peaks" in the density curves in Figs. 4.8—4.10. It is worth while noticing that the minimal value of the pressure curves behind the shock is sometimes related to the discontinuity of the curvature of the body, and sometimes not. Figs. 4.6 and 4.12 show that for the flow around a sphere–cone with $\beta=30°$ and $M_\infty=20$, the minimal value of the pressure curve behind the shock appears at the point of intersection of the shock with the weak discontinuity starting from the point of the discontinuity of the curvature of the body. The slope of the pressure curve behind the shock at this point is discontinuous, and the stream line starting from that point is also a weak discontinuity, i.e., the "peak" of the density curve in the ξ-direction is a weak discontinuity (see Fig. 4.10). However, the situation is different for the flow around a sphere–cone with $\beta=10°$ and $M_\infty=4$. Figs. 4.4, 4.11 and 4.13 show that the minimal point of pressure behind the shock is located far from the point of intersection of the shock with the weak discontinuity starting from the point of the curvature–discontinuity of the body. After observing Figs. 4.11 and 4.14 simultaneously, we can find that the minimal point of pressure behind the shock appears, roughly speaking, when the minimal point of the pressure curve in the ξ-direction reaches the shock.

If the curvature of a body is discontinuous at some points, then there appear several weak discontinuities, starting from those points in the flow field around the body. We call them the weak wave–discontinuities, since they are wave–characteristic surfaces. In order to estimate the effect of different algorithms on the weak wave–discontinuities in the computed results, we have evaluated the flow field around sphere–cones, first in such a way as to separate the weak wave–discontinuities and then in such a way as not to separate them. In Figs. 4.15 and 4.16, we present several pressure curves and density curves in the ξ–direction, where $M_\infty=20$, $\beta=30°$, $z=1.1$ and $M_\infty=4$, $\beta=10°$, $z=5$. Making a comparison between the two results, we can find that the difference between them is not great, except in the region near the weak wave–discontinuity. In that region, the difference of pressure is, relatively speaking, rather great, the relative error being 1/80. As the weak discontinuity is not separated, the pressure curves are smeared.

Several pressure distributions and density distributions on sphere–cones are given in Figs. 4.17—4.20, where $M_\infty=4$, 20, $\alpha=0°$, $\beta=10°$, 20°, 30°, R_0 (the nose radius) $=1$. Here we consider two gases, one being the perfect gas with $\gamma=1.4$ and the other being the air in chemical equilibrium. (The pressure p_∞ and the density ρ_∞ of the free stream are given in these figures.) We see from these figures that the pressure distribution of the perfect gas is almost the same as that of the air in chemical equilibrium if $M_\infty=4$. The difference between the two density distributions is great only when the semi–vertex angle is quite large in the case of $M_\infty=4$. However, if $M_\infty=20$, the difference between the pressure distributions of the two gases is sensible for any semi–vertex angle, and the difference between two density distributions is very great. This means that if the free stream is supersonic, the effect of the physical–chemical process may not be considered. However, if the free stream is hypersonic, the effect of the physical–chemical process must be considered, since the difference of flow parameters between two gases is very great.

We have illustrated the results regarding the flow around sphere–cones at the zero–angle of attack. In what follows, we shall present a series of results regarding sphere–cones at the angle of attack. In Fig. 4.21 we show several curves of pressure p and circumferential velocity w on the body in the φ–direction, where $M_\infty=20$, $\alpha=15°$ and 20° and $\beta=10°$. In Fig. 4.22 we present several curves of pressure and density on the body in the φ–direction, where $M_\infty=20$, $\alpha=30°$ and $\beta=10°$. We can see from these figures that the pressure and density on the leeward side are very low, since the angle of attack is large. The ratio of the value on the windward side to the value on the leeward side is more than ten. We can also find that there is a minimal point on the curves of pressure

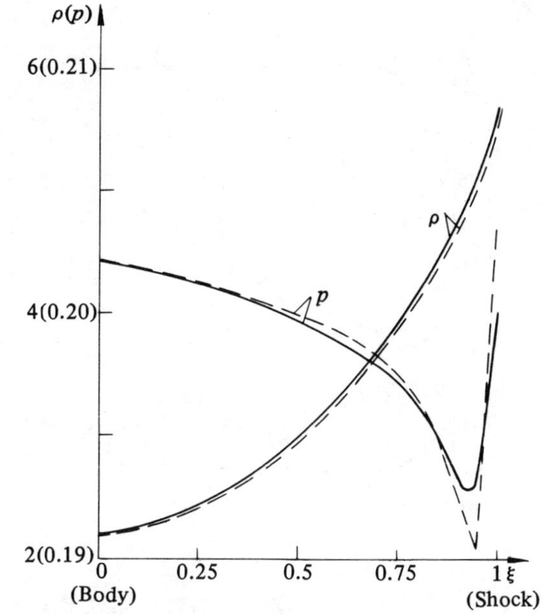

Figure 4.15 Variations of pressure and density with ξ ($M_\infty=20$, perfect gas, $\beta=30°$, $R_0=1$, $z=1.1$)
—— The weak discontinuity is not separated,
- - - - The weak discontinuity is separated.

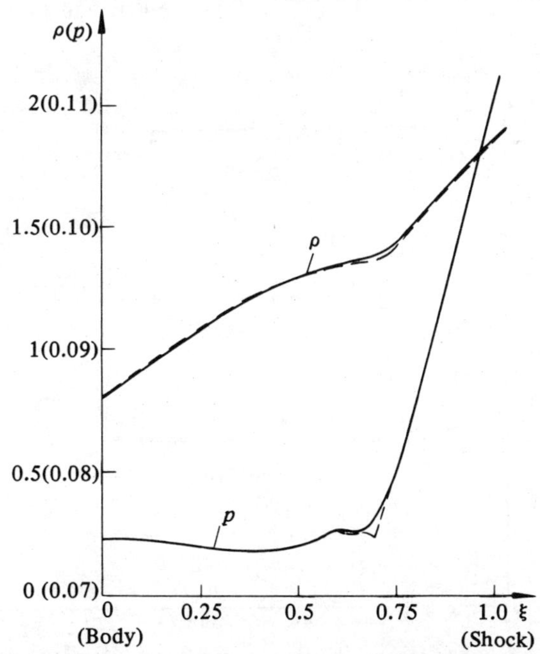

Figure 4.16 Variations of pressure and density with ξ ($M_\infty=4$, perfect gas, $\beta=10°$, $R_0=1$, $z=5$)
—— The weak discontinuity is not separated,
----- The weak discontinuity is separated.

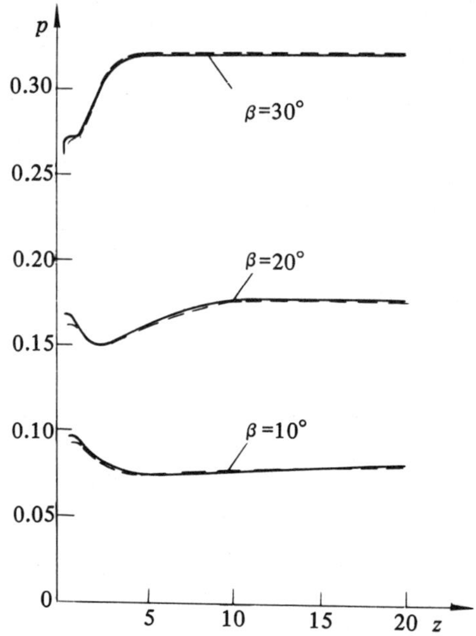

Figure 4.17 Pressure distributions on sphere-cones for perfect gas and equilibrium air ($M_\infty=4$, $\beta=10°, 20°, 30°$, $R_0=1$, $\alpha=0°$)

---- Perfect gas, —— Equilibrium air $\begin{cases} p_\infty=0.8874\times 10\,\text{kgf/m}^2, \\ \rho_\infty=0.9717\times 10^{-4}\,\text{kgf}\cdot\text{sec}^2/\text{m}^4. \end{cases}$

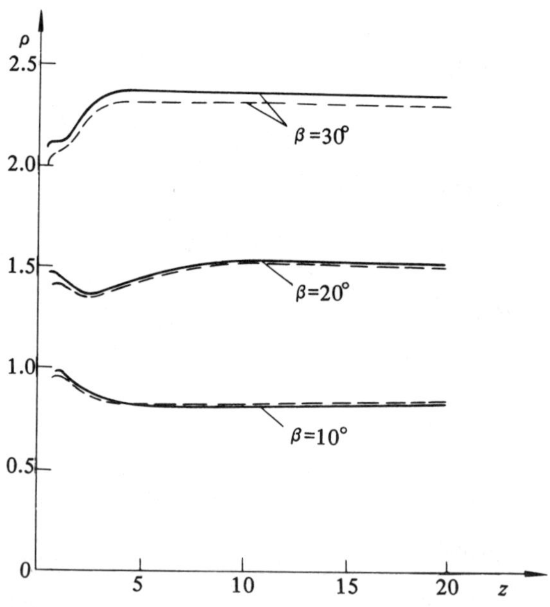

Figure 4.18 Density distributions on sphere-cones for perfect gas and equilibrium air ($M_\infty=4$, $\beta=10°, 20°, 30°$, $R_0=1$, $\alpha=0°$)

---- Perfect gas, —— Equilibrium air $\begin{cases} p_\infty=0.8874\times 10\,\text{kgf/m}^2, \\ \rho_\infty=0.9717\times 10^{-4}\,\text{kgf}\cdot\text{sec}^2/\text{m}^4. \end{cases}$

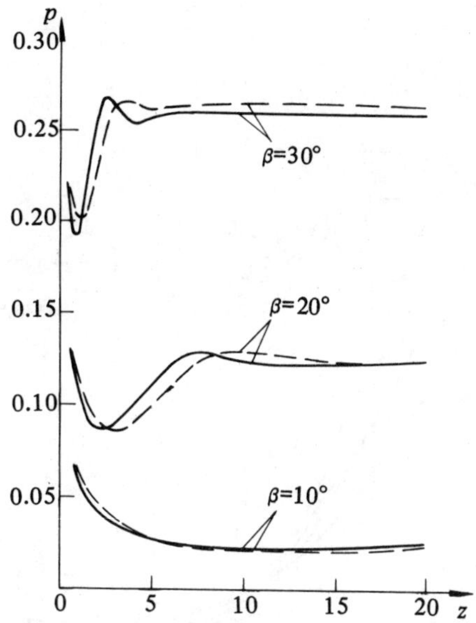

Figure 4.19 Pressure distributions on sphere-cones for perfect gas and equilibrium air (M∞=20, β=10°, 20°, 30°, R_0=1, α=0°)

--- Perfect gas, —— Equilibrium air $\begin{cases} p_\infty = 0.1244\times10^4\,\mathrm{kgf/m^2}, \\ \rho_\infty = 0.1958\times10^{-1}\,\mathrm{kgf\cdot sec^2/m^4}. \end{cases}$

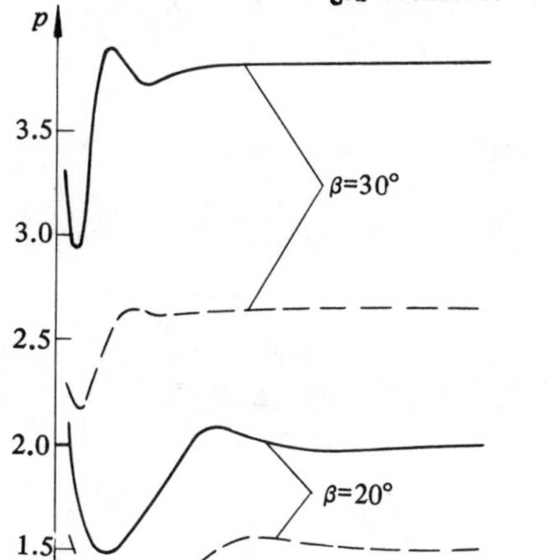

Figure 4.20 Density distributions on sphere-cones for perfect gas and equilibrium air (M∞=20, β=10°, 20°, 30°, R_0=1, α=0°)

--- Perfect gas; —— Equilibrium air $\begin{cases} p_\infty = 0.1244\times10^4\,\mathrm{kgf/m^2}, \\ \rho_\infty = 0.1958\times10^{-1}\,\mathrm{kgf\cdot sec^2/m^4}. \end{cases}$

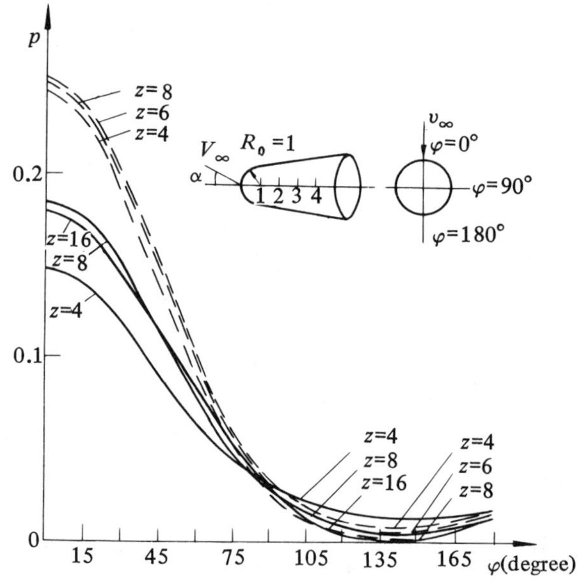

Figure 4.21a Pressure distributions on sphere-cones in the φ-direction ($M_\infty=20$, $\beta=10°$)
—— $\alpha=15°$, ---- $\alpha=20°$.

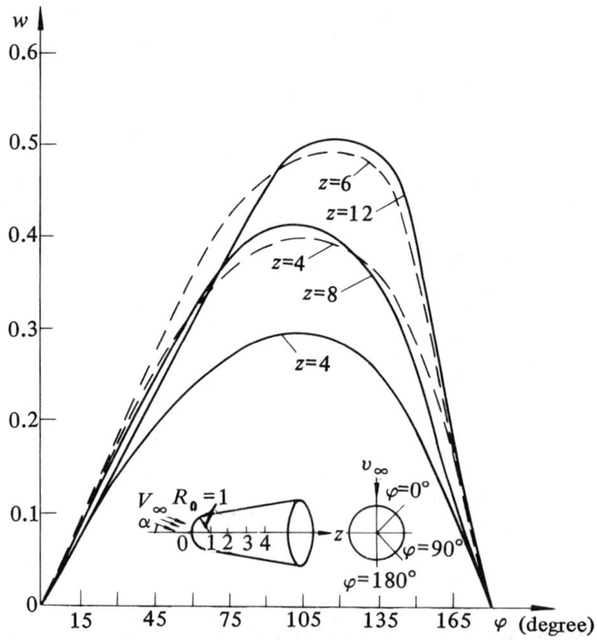

Figure 4.21b Circumferential velocity distributions on sphere-cones in the φ-direction ($M_\infty=20$, $\beta=10°$)
—— $\alpha=15°$, —— $\alpha=20°$.

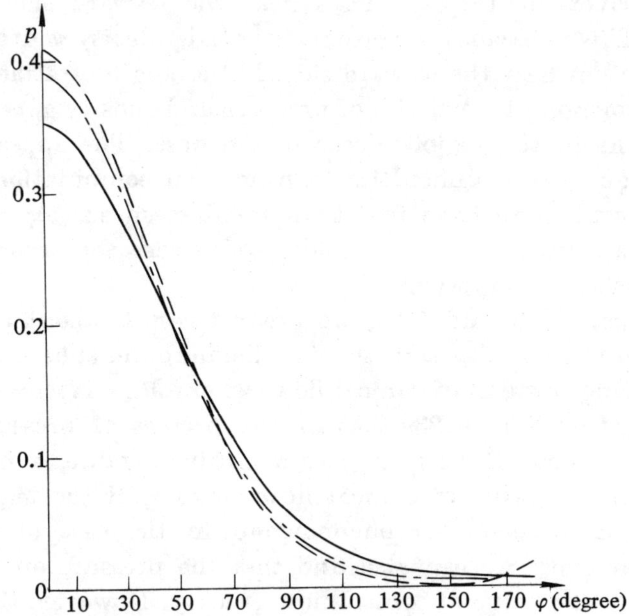

Figure 4.22a Pressure distributions on sphere-cones in the φ-direction ($M_\infty=20$, $\alpha=30°$, $\beta=10°$)
—— $z=1.02$, —·— $z=2.039$, ---- $z=3.251$.

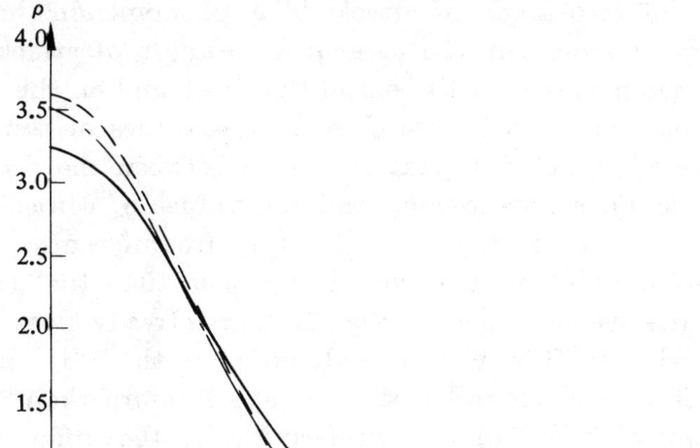

Figure 4.22b Density distributions on sphere-cones in the φ-direction ($M_\infty=20$, $\alpha=30°$, $\beta=10°$)
—— $z=1.02$, —·— $z=2.039$, ----- $z=3.251$.

and density on the body in the φ–direction, in the region far from the nose. Moreover, the pressure rises near the leeward side with a large gradient. This is because the circumferential velocity w drops rapidly in the φ–direction near the leeward side. If the angle of attack gets larger, this phenomenon of rapid rise of pressure and density gets more serious, and happens in the region closer to the nose. The appearance of this phenomenon creates difficulties in numerical computation. In order to get good results, we have to take a small mesh size $\Delta\varphi$ in the region where certain functions vary rapidly, otherwise the error will be large and oscillation will appear.

In Figs. 4.23 and 4.24, we present results about pressure p and circumferential velocity w in the flow field around sphere–cones and the corresponding results of conical flow, where $M_\infty=17$, $\alpha=5°$, $\beta=12.5°$. We can see from Fig. 4.23a that in the process of pressure tending to the value of conical flow, besides a minimal value, a maximal value larger than the value of conical flow appears, if the Mach number is large. We have seen this phenomenon in the case of zero–angle of attack. Moreover, we can also find that the pressure on the windward side tends to the value of conical flow quickly. However, the pressure on the leeward side tends to the value of conical flow slowly, there being sensible difference between the two values even when $z=80$. We have pointed out that the process of pressure tending to the value of conical flow gets faster and faster as the semi–vertex angle gets larger and larger in the case of zero–angle of attack. The phenomenon here coincides with the phenomenon in the case of zero–angle of attack. Fig. 4.23b shows that the pressures both behind the shock and on the body, on the windward side, at $z=58.51$, are close to the values of conical flow, and there is a sensible, but not great difference between the pressures of the flow field around a sphere–cone and the values of conical flow on the leeward side. However, the status for the circumferential velocity w is much different. We can see from Fig. 4.24a that the values of the circumferential velocity behind the shock are already close to the values of the conical flow. However, the value of w on the body, on the leeward side in the flow field around a sphere–cone is more than ten times the value of conical flow. On the windward side, the difference is not so great, the former difference being several times the latter. Moreover, we can find from Fig. 4.24b that w near the body goes down rapidly with the increase of ξ at $z=58.51$. Hence a great difference between the value of a sphere–cone and the value of a cone appears only in a certain small region near the body. Therefore, except in a certain small region near the body, the flow field at $z=60$—80 is close to the flow field of a cone. However, in that small region near the body, especially on the leeward side, there is still a great difference between the two values for certain

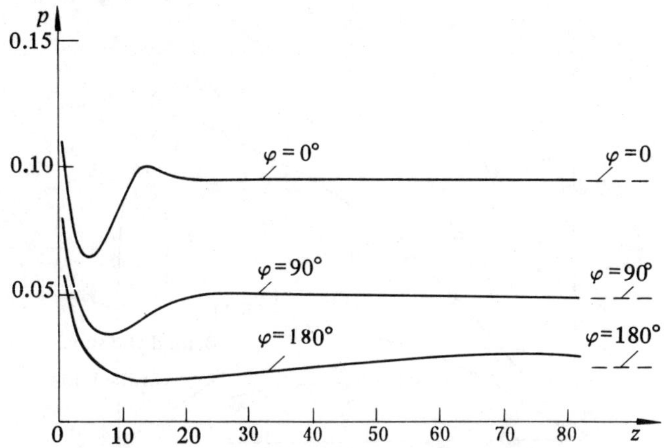

Figure 4.23a Pressure distribution on a body
(M_∞=17, perfect gas, α=5°, β=12.5°, R_0=1)
—— Sphere-cone, ---- Pointed cone.

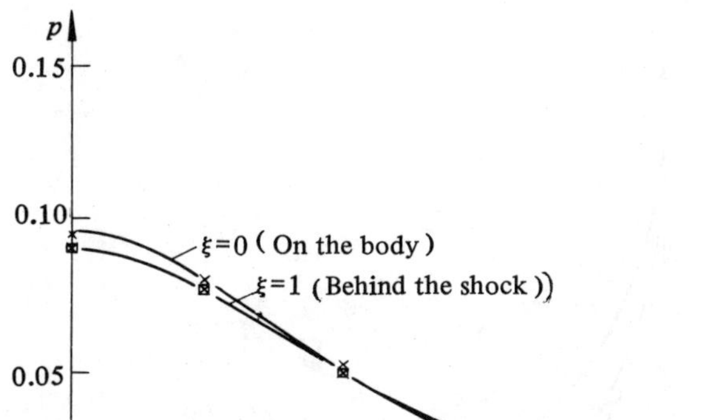

Figure 4.23b Variations of pressure with φ
(M_∞=17, perfect gas, α=5°, β=12.5°, R_0=1, z=58.51)
× Pressure on cone, ⊠ Pressure behind the shock for conical flow.

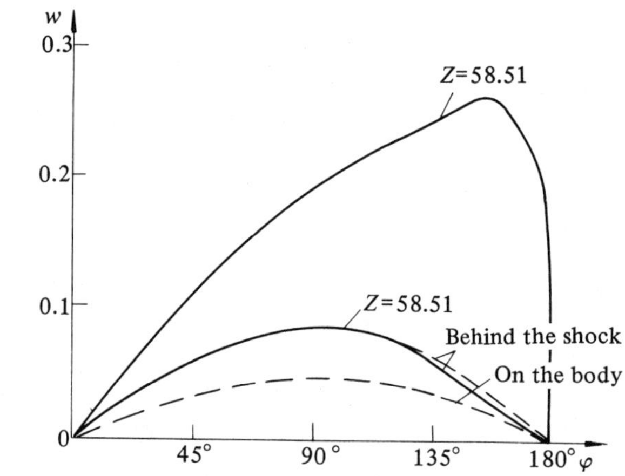

Figure 4.24a Variations of circumferential velocity with φ ($M_\infty=17$, perfect gas, $\alpha=5°$, $\beta=12.5°$, $R_0=1$)
—— Sphere-cone, ---- Pointed cone.

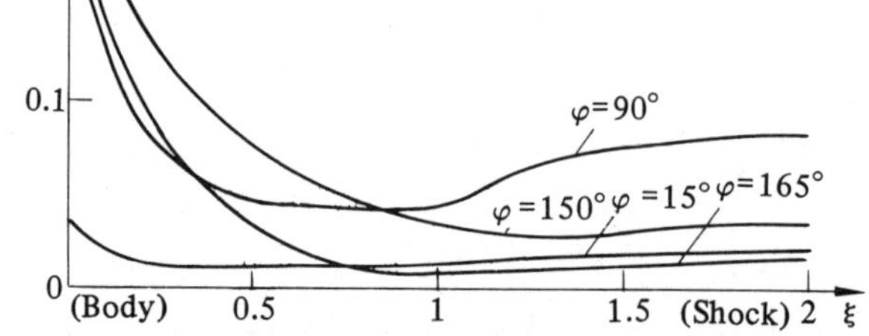

Figure 4.24b Variations of circumferential velocity with ξ ($M_\infty=17$, perfect gas, $\alpha=5°$, $\beta=12.5°$, $R_0=1$, $z=58.51$)

flow parameters.

We can now see the results taken from the entropy layer in the flow field around the sphere–cones at an angle of attack. In Fig. 4.25, the shapes of the shock and a certain stream surface on the windward and leeward plane are drawn, where $M_\infty=17$, $\alpha=5°$, $\beta=12.5°$. The thickness of the entropy layer on the windward plane is approximately one fifth of the distance between the stream surface and the body. Its thickness on the leeward plane is approximately nine tenths of the distance between that stream surface and the body. (We have not shown these facts in Fig. 4.25.) According to these facts and those in Fig. 4.25, we find that the entropy layer on the windward plane is much thinner than that on the leeward plane. This phenomenon can be intuitively understood in the following way. The gas through the forward shock on the windward side approaches the body quickly because of the angle of attack. Thus, the entropy layer on the windward plane thins out very quickly. That is, it is very thin when z is rather large. On the leeward side, because of the "negative" angle of attack and the large circumferential velocity, the gas through the forward shock approaches the body slowly. This means that the entropy layer on the leeward plane thins out very slowly. Therefore, the phenomenon described above occurs.

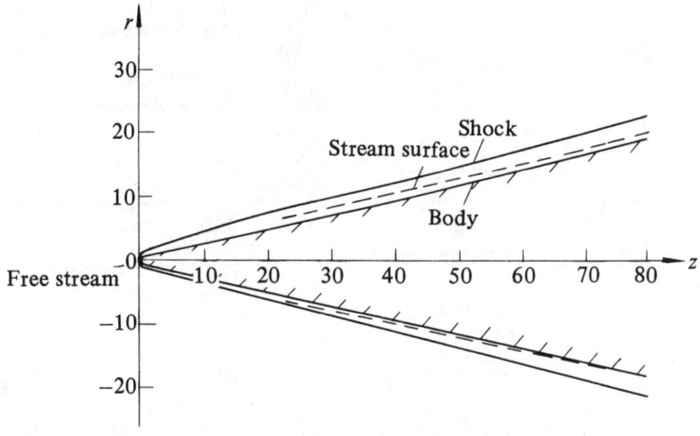

Figure 4.25 Shapes of the shock, stream surface and body ($M_\infty=17$, perfect gas, $\alpha=5°$, $\beta=12.5°$, $R_0=1$)

The strength of the forward shock is close to that of a normal shock, so the temperature in the entropy layer is very high. In the region outside the entropy layer, the temperature is closer to that of a conical flow, so it is not so high. Therefore, the phenomenon mentioned above means that the region of high temperature on the windward plane is much thinner than that on the leeward plane, when z is large (see Fig. 4.26a).

The appearance of this phenomenon sometimes causes the temperature in a certain region near the body on the windward plane to become lower than the temperature in the "corresponding" region on the leeward plane. In fact, if z is very large, then the high temperature on the body goes down rapidly to the temperature of the conical flow on the windward plane, because the entropy layer is thin. However, this process

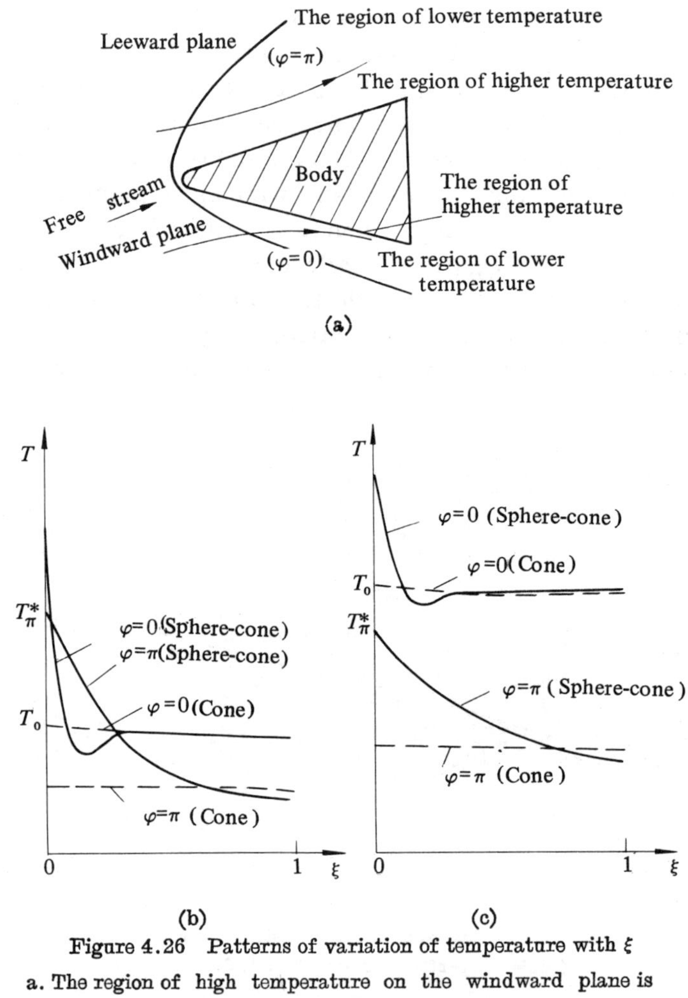

Figure 4.26 Patterns of variation of temperature with ξ

a. The region of high temperature on the windward plane is thinner than that on the leeward plane.

b. If $T_\pi^* \gg T_0$, then it appears that the temperature in a certain region on the leeward plane is higher than the temperature in a corresponding region on the windward plane.

c. If $T_\pi^* \ll T_0$, then it does not appear that the temperature in a certain region on the leeward plane is higher than the temperature in a corresponding region on the windward plane.

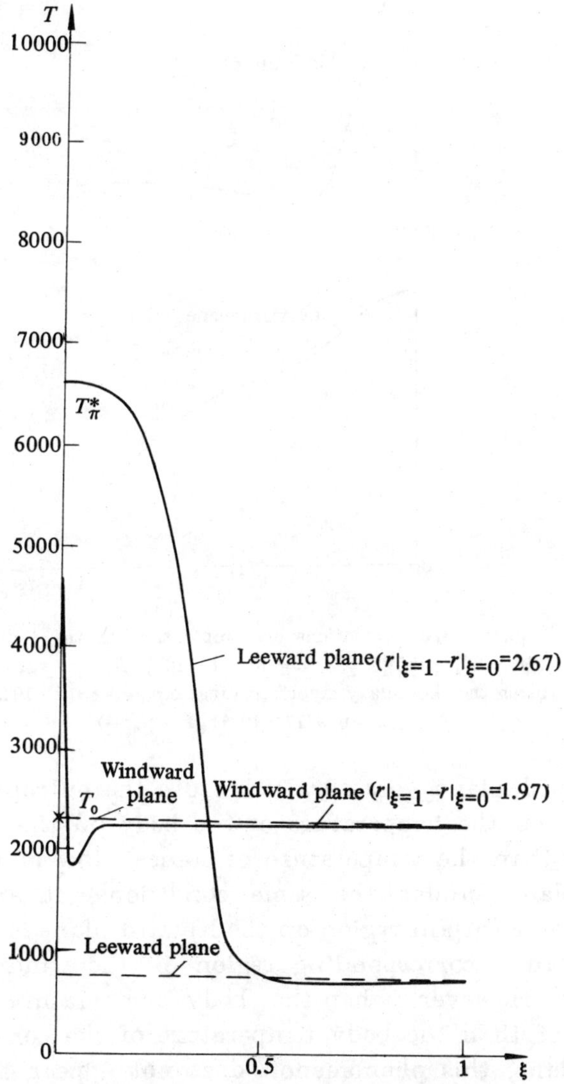

Figure 4.27a Variations of temperature in the ξ-direction ($M_\infty=17$, perfect gas, $\alpha=5°$, $\beta=12.5°$, $R_0=1$, $z=58.51$, $p_\infty=0.8874\times10\,\mathrm{kgf/m^2}$, $\rho_\infty=0.9717\times10^{-4}\,\mathrm{kgf\cdot sec^2/m^4}$) —— Sphere-cone, - - - Cone.

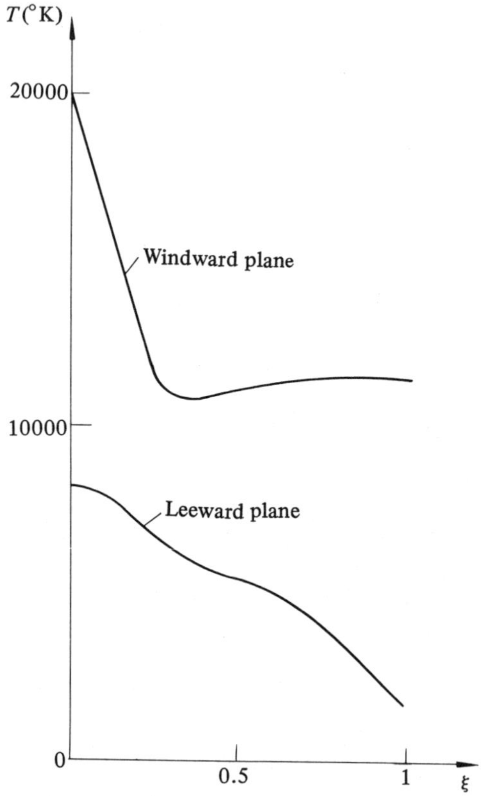

Figure 4.27b Variations of temperature in the ξ-direction ($M_\infty=20$, perfect gas, $\alpha=30°$, $\beta=10°$, $R_0=1$, $z_{wall}=3.25$, using the "boundary layer" coordinates, $p_\infty=0.8874\times10$ kgf/m^2, $\rho_\infty=0.9717\times10^{-4}$ kgf·sec^2/m^4)

on the leeward plane goes slowly because the entropy layer is thick. Therefore, when the temperature of the body on the leeward plane is much higher than the temperature of conical flow of the body on the windward plane, under the same conditions, it appears that the temperature in a certain region on the leeward plane is higher than the temperature in a corresponding region on the windward plane (see Fig. 4.26b). However, when the body temperature on the leeward plane is lower than the body temperature of the conical flow on the windward plane, this phenomenon does not appear even though the difference between thicknesses of the entropy layers on the two planes still exists (see Fig. 4.26c). We know from Figs. 4.27a—b that if $M_\infty=17$, $\alpha=5°$, $\beta=12.5°$, the flow field around a sphere–cone belongs to the first case and if $M_\infty=20$, $\alpha=30°$, $\beta=10°$, the flow field around a sphere–cone belongs to the second case.

We want to point out that by using the resultson conical flow, we can roughly judge whether or not this phenomenon appears using the following criteria. If z is quite large, the body pressure of the flow field around the sphere–cone on the leeward plane is close to the body pressure of the conical flow. Moreover, the entropy there is equal to the entropy behind a normal shock. Hence, we can roughly evaluate the temperature T_π^* on the sphere–cone on the leeward plane. Then, let the temperature of conical flow on the body on the windward plane be T_0. Clearly, if $T_\pi^* \gg T_0$, this phenomenon will appear and if $T_\pi^* \ll T_0$, this phenomenon will not appear. If $T_\pi^* \approx T_0$, both possibilities exist. However if this occurs, then the phenomenon will happen only in a small region, and the temperature on the leeward plane will be just a little higher than that on the windward plane. For example, for the conical flow of the perfect gas with $M_\infty=17$, $\alpha=5°$, $\beta=12.5°$, we have $p|_{\varphi=0}=0.09585$, $\rho|_{\varphi=0}=5.417$, $p|_{\varphi=\pi}=0.02185$ on the body. For $M_\infty=17$, at the point behind the normal shock, we have $p=0.8329$, $\rho=5.8980$, and the entropy

$$s=\frac{p}{\rho^{1.4}}=\frac{0.8329}{(5.8980)^{1.4}}=0.0696.$$

Therefore, for the flow field around the corresponding sphere–cone, we have

$$p_\pi^* \approx 0.02185, \quad \rho_\pi^*=\left(\frac{p_\pi^*}{s}\right)^{\frac{1}{1.4}}=\left(\frac{0.02185}{0.0696}\right)^{\frac{1}{1.4}}$$

on the body on the leeward plane if z is very large. Furthermore, we have

$$\frac{T_\pi^*}{T_0}=\frac{0.02185}{\left(\frac{0.02185}{0.0696}\right)^{\frac{1}{1.4}}} \div \frac{0.09585}{5.417} \approx 2.8.$$

Therefore, this phenomenon should appear in this case and we can see from Fig. 4.27a that this phenomenon does happen in this case.

We also want to point out that because this phenomenon appears in the flow around sphere–cones, and because the thin entropy layer is usually smeared in the computed results, the computed temperature on the body on the windward plane is sometimes lower than that on the leeward plane. Though this result is not reasonable for the temperature on the body, it is reasonable for the temperature near the body.

In order to estimate the accuracy of the computed results, we compared the results of the flow around sphere–cones, obtained by taking different mesh sizes. In Tables 4.7 and 4.8, we give several parameters of the flow around sphere–cones at $z=5$, 10, 15, obtained by taking $\Delta\xi=\frac{1}{20}, \frac{1}{40}, \frac{1}{60}$, where $M_\infty=4$, 20, $\alpha=0°$, $\beta=10°$, $R_0=1$. Table 4.9 gives

Table 4.7 Comparisons of computed results of the flow around sphere-cones obtained by taking different $\Delta\xi$ ($M_\infty=4$, $\alpha=0°$, $\beta=10°$, $R_0=1$)

z	ξ	u			v			p			ρ		
		$\Delta\xi=1/20$	$\Delta\xi=1/40$	$\Delta\xi=1/60$	$\Delta\xi=1/20$	$\Delta\xi=1/40$	$\Delta\xi=1/60$	$\Delta\xi=1/20$	$\Delta\xi=1/40$	$\Delta\xi=1/60$	$\Delta\xi=1/20$	$\Delta\xi=1/40$	$\Delta\xi=1/60$
5	1.0	0.93249	0.93263	0.93260	0.16492	0.16470	0.16474	0.11215	0.11202	0.11204	1.88825	1.88678	1.88707
	0.8	0.95119	0.95072	0.95039	0.12563	0.12660	0.12733	0.08584	0.08629	0.08664	1.53289	1.53854	1.54298
	0.6	0.95374	0.95369	0.95381	0.11029	0.11034	0.11003	0.07527	0.07529	0.07517	1.34824	1.34859	1.34695
	0.4	0.93712	0.93693	0.93695	0.11671	0.11674	0.11675	0.07389	0.07391	0.07391	1.22949	1.22869	1.22886
	0.2	0.89169	0.89148	0.89135	0.12879	0.12878	0.12877	0.07432	0.07432	0.07432	1.03656	1.03718	1.03688
	0	0.80993	0.80992	0.80991	0.14281	0.14281	0.14281	0.07474	0.07475	0.07475	0.82212	0.82218	0.82223
10	1.0	0.96217	0.96196	0.96190	0.10878	0.10923	0.10937	0.08247	0.08268	0.08274	1.53994	1.54254	1.54332
	0.8	0.97205	0.97213	0.97227	0.08682	0.08655	0.08622	0.07054	0.07043	0.07030	1.37119	1.36962	1.36793
	0.6	0.97085	0.97086	0.97084	0.09093	0.09099	0.09100	0.06939	0.06941	0.06941	1.34278	1.34331	1.34326
	0.4	0.96198	0.96198	0.96199	0.10868	0.10865	0.10865	0.07243	0.07242	0.07241	1.35077	1.35066	1.35070
	0.2	0.93500	0.93485	0.93479	0.13084	0.13077	0.13076	0.07559	0.07558	0.07557	1.25686	1.25541	1.25490
	0	0.80708	0.80709	0.80709	0.14231	0.14231	0.14231	0.07671	0.07670	0.07670	0.83755	0.83749	0.83747
15	1.0	0.97332	0.97295	0.97291	0.08258	0.08350	0.08362	0.07132	0.07169	0.07174	1.39329	1.39827	1.39890
	0.8	0.97751	0.97743	0.97742	0.07418	0.07433	0.07435	0.06626	0.06632	0.06633	1.31961	1.32029	1.32038
	0.6	0.97317	0.97321	0.97321	0.08875	0.08863	0.08861	0.06945	0.06941	0.06941	1.35963	1.35909	1.35900
	0.4	0.96515	0.96521	0.96522	0.11122	0.11111	0.11110	0.07460	0.07457	0.07456	1.41641	1.41631	1.41625
	0.2	0.94971	0.94982	0.94985	0.13491	0.13481	0.13480	0.07830	0.07827	0.07827	1.39714	1.39718	1.39724
	0	0.80299	0.80300	0.80300	0.14159	0.14159	0.14159	0.07959	0.07958	0.07958	0.85991	0.85984	0.85984

Table 4.8 Comparisons of computed results of the flow around sphere-cones obtained by taking different $\Delta\xi$ ($M_\infty=20$, $\alpha=0°$, $\beta=10°$, $R_0=1$)

z	ξ	u			v			p			ρ		
		$\Delta\xi=1/20$	$\Delta\xi=1/40$	$\Delta\xi=1/60$	$\Delta\xi=1/20$	$\Delta\xi=1/40$	$\Delta\xi=1/60$	$\Delta\xi=1/20$	$\Delta\xi=1/40$	$\Delta\xi=1/60$	$\Delta\xi=1/20$	$\Delta\xi=1/40$	$\Delta\xi=1/60$
5	1.0	0.95275	0.95213	0.95196	0.18837	0.18957	0.18990	0.04904	0.04966	0.04983	4.95405	4.96473	4.96766
	0.8	0.93923	0.93925	0.93918	0.16495	0.16458	0.16467	0.03384	0.03377	0.03382	2.29714	2.29013	2.29235
	0.6	0.90947	0.90916	0.90911	0.15398	0.15383	0.15382	0.03003	0.03010	0.03010	1.29398	1.29661	1.29780
	0.4	0.87108	0.87151	0.87145	0.14585	0.14594	0.14591	0.02864	0.02870	0.02871	0.85669	0.84722	0.85137
	0.2	0.82554	0.82623	0.82613	0.14022	0.14038	0.14028	0.02827	0.02832	0.02833	0.63737	0.63519	0.63779
	0	0.78890	0.78871	0.78869	0.13910	0.13907	0.13907	0.02826	0.02834	0.02835	0.53328	0.53441	0.53455
10	1.0	0.97221	0.97217	0.97214	0.14412	0.14424	0.14430	0.02957	0.02962	0.02964	4.44869	4.45041	4.45130
	0.8	0.96710	0.96674	0.96669	0.13683	0.13682	0.13680	0.02393	0.02396	0.02397	2.88423	2.83538	2.82928
	0.6	0.94732	0.94701	0.94700	0.13521	0.13521	0.13518	0.02212	0.02216	0.02217	1.60359	1.59105	1.59396
	0.4	0.90913	0.90850	0.90837	0.13636	0.13653	0.13659	0.02177	0.02178	0.02177	0.90352	0.90520	0.90637
	0.2	0.85804	0.85585	0.85522	0.13844	0.13839	0.13838	0.02174	0.02173	0.02173	0.57301	0.58495	0.58734
	0	0.80511	0.80515	0.80510	0.14196	0.14197	0.14196	0.02177	0.02175	0.02177	0.44264	0.44242	0.44266
15	1.0	0.97852	0.97849	0.97849	0.12593	0.12602	0.12601	0.02327	0.02330	0.02329	4.16089	4.16238	4.16219
	0.8	0.97613	0.97615	0.97615	0.12491	0.12500	0.12502	0.02086	0.02088	0.02089	3.31858	3.32191	3.32374
	0.6	0.96489	0.96458	0.96467	0.13104	0.13091	0.13098	0.02090	0.02086	0.02087	2.29386	2.24656	2.26026
	0.4	0.93306	0.93263	0.93260	0.13798	0.13790	0.13784	0.02139	0.02136	0.02135	1.21377	1.20525	1.20567
	0.2	0.87501	0.87566	0.87573	0.14104	0.14123	0.14128	0.02155	0.02157	0.02158	0.65111	0.64724	0.64961
	0	0.80561	0.80554	0.80554	0.14205	0.14204	0.14204	0.02159	0.02161	0.02161	0.43996	0.44033	0.44032

Table 4.9 Comparisons of the aerodynamic force coefficients of the flow around sphere–cones obtained by taking different $\Delta\varphi$ ($M_\infty=20$, $\beta=10°$, $R_0=1$)

Angles of attack α	z	C_R		C_N		C_M		C_D	
		$\Delta\varphi=15°$	$\Delta\varphi=10°$	$\Delta\varphi=15°$	$\Delta\varphi=10°$	$\Delta\varphi=15°$	$\Delta\varphi=10°$	$\Delta\varphi=15°$	$\Delta\varphi=10°$
20°	5	0.05920	0.05930	0.07058	0.07079	0.01059	0.01063	0.1500	0.1502
15°	4 8 12	0.05319 0.06960 0.09264	0.05322 0.06976 0.09287	0.03437 0.09508 0.1864	0.03436 0.09539 0.1865	0.003824 0.02374 0.07193	0.003827 0.02384 0.07192	0.1113 0.2497 0.3859	0.1114 0.2499 0.3857

Table 4.10 Comparisons of parameters of the flow around sphere–cones at $z=6$ obtained by taking different $\Delta\varphi$ ($M_\infty=20$, $\alpha=15°$, $\beta=10°$, $R_0=1$)

		$\varphi=0°$			$\varphi=90°$			$\varphi=180°$		
		$\xi=1$	$\xi=0.5$	$\xi=0$	$\xi=1$	$\xi=0.5$	$\xi=0$	$\xi=1$	$\xi=0.5$	$\xi=0$
u	$\Delta\varphi=15°$ $\Delta\varphi=10°$	0.90649 0.90556	0.90368 0.90410	0.60830 0.60747	0.93340 0.93323	0.87053 0.87222	0.71155 0.70996	0.89303 0.89262	0.87026 0.86980	0.83209 0.83272
v	$\Delta\varphi=15°$ $\Delta\varphi=10°$	0.08289 0.08326	0.11921 0.12031	0.10726 0.10711	0.18453 0.18503	0.16634 0.16747	0.12546 0.12518	0.40269 0.40323	0.28073 0.28176	0.14672 0.14683
w	$\Delta\varphi=15°$ $\Delta\varphi=10°$	0 0	0 0	0 0	0.18307 0.18223	0.19514 0.19384	0.34799 0.34887	0 0	0 0	0 0
p	$\Delta\varphi=15°$ $\Delta\varphi=10°$	0.14738 0.14868	0.17440 0.17558	0.18124 0.18254	0.05281 0.05319	0.03776 0.03811	0.02800 0.02837	0.03496 0.03522	0.01118 0.01148	0.01339 0.01316
ρ	$\Delta\varphi=15°$ $\Delta p=10°$	5.6046 5.6078	6.7512 6.8223	2.0126 2.0219	5.0161 5.0220	1.4037 1.4234	0.52948 0.53465	4.6316 4.6394	0.41405 0.42082	0.31186 0.30844

Table 4.11 Comparisons of parameters of the flow around sphere–cones at $z=12$ obtained by taking different $\Delta\varphi$
($M_\infty=20$, $\alpha=15°$, $\beta=10°$, $R_0=1$)

		$\varphi=0°$			$\varphi=90°$			$\varphi=180°$		
		$\xi=1$	$\xi=0.5$	$\xi=0$	$\xi=1$	$\xi=0.5$	$\xi=0$	$\xi=1$	$\xi=0.5$	$\xi=0$
u	$\Delta\varphi=15°$	0.89073	0.88530	0.60894	0.93416	0.93648	0.65160	0.92080	0.87432	0.84826
	$\Delta\varphi=10°$	0.89101	0.88536	0.60913	0.93301	0.93635	0.65433	0.92038	0.87951	0.84908
v	$\Delta\varphi=15°$	0.10356	0.13011	0.10737	0.18073	0.15538	0.11490	0.36163	0.29084	0.14957
	$\Delta\varphi=10°$	0.10320	0.12992	0.10741	0.18282	0.15627	0.11538	0.36233	0.28369	0.14972
w	$\Delta\varphi=15°$	0	0	0	0.19414	0.16485	0.45013	0	0	0
	$\Delta\varphi=10°$	0	0	0	0.19519	0.16488	0.44808	0	0	0
p	$\Delta\varphi=15°$	0.16821	0.17719	0.18044	0.04921	0.03697	0.02578	0.01877	0.00810	0.00972
	$\Delta\varphi=10°$	0.16785	0.17713	0.18046	0.05005	0.03751	0.02676	0.01899	0.00758	0.00930
ρ	$\Delta\varphi=15°$	5.6507	5.8607	2.0062	4.9570	3.0390	0.50534	3.8801	0.31661	0.24638
	$\Delta\varphi=10°$	5.6600	5.8578	2.0054	4.9715	3.1082	0.51611	3.8960	0.30508	0.24058

the aerodynamic force coefficients of the flow around sphere–cones obtained by taking different $\Delta\varphi$ ($\Delta\varphi=15°$, $10°$) and the same $\Delta\xi$, where $M_\infty=20$, $\alpha=15°$, $20°$, $\beta=10°$. When we evaluate these aerodynamic force coefficients, we let the nose radius R_0 be equal to 1, the reference lenght L_r be equal to 20, the reference area S_r be equal to the area of the cross section of the body at $z=20$. Tables 4.10 and 4.11 show several parameters of the flow around sphere–cones at $z=6$ and 12, obtained by taking $\Delta\varphi=15°$, $10°$, where $M_\infty=20$, $\alpha=15°$, $\beta=10°$ and $R_0=1$.

We can find from Tables 4.7 and 4.8 that for a short sphere–cone

$$\Delta\xi=\frac{1}{20}-\frac{1}{40},$$

the computed quantities on the body already have four significant digits. However, in order to make the relative error of the computed results in the whole flow field, less than 1%, we have to take $\Delta\xi=\frac{1}{40}$ or less. In passing, we point out that there is a rather great difference among the results obtained by taking different $\Delta\xi$ at $z=10$, $\xi=0.2$, 0.8 (see Table 4.8). The reason for this is that the weak discontinuity has not been separated, so the error is rather large if $\Delta\xi$ is not small enough.

Table 4.9 tells us that for short sphere–cones at small angle of attack, the difference between the aerodynamic force coefficients obtained by taking $\Delta\varphi=15°$ and $10°$ is less than 0.5%. Therefore, we have already obtained satisfactory results for the aerodynamic force coefficients when taking 10—20 φ-planes in the φ-direction (suppose $0°\leqslant\varphi\leqslant180°$). However, we have to take more than 20 φ-planes if we want to get satisfactory results for flow parameters in the whole flow field. In fact, we can see from Tables 4.10 and 4.11 that the difference between the two results obtained by taking $\Delta\varphi=15°$ and $10°$, for flow parameters, on the windward side, is already less than 1%, and that the difference in the other region is about 5%. However, the pressure in the other region is quite small compared with the pressure on the windward side, so the contribution of the pressure, in that region, to the aerodynamic force coefficients is much less. Therefore, the aerodynamic force coefficients still have a higher degree of accuracy.

In Figs. 4.28a—c, we show several C_N^*/α, C_M^*/α and C_D^*–curves of sphere–cones with $\beta=12.5°$, and we compare the computed results with the experimental data[52] where $M_\infty=17$, where the gas is a perfect gas and where the independent variable is the ratio of bluntness r_n/r_b. Here the following symbols are adopted: r_n——the radius of the sphere; r_b——the radius of the base; C_N^*——the normal–force coefficient (taking the area of the base as the reference area); C_M^*——the moment coefficient. It is relative to the center of gravity in the body, i.e., $C_M^*=C_N^*(C_D^*-C_g^*)$, C_D^* being the ratio of the distance between the center of pressure and the base to the diameter of the base, C_g^* being the ratio of the distance between the center of gravity of the body and the base to the diameter of the base ($C_g^*=0.79$ in [58]). Clearly, C_N^* and C_M^* are equal to zero if $\alpha=0$. Hence, the ratios C_N^*/α and C_M^*/α (the unit of α is the radian) are respectively the approximate values of

$$\left.\frac{\partial C_N^*}{\partial\alpha}\right|_{\alpha=0}, \quad \left.\frac{\partial C_M^*}{\partial\alpha}\right|_{\alpha=0}.$$

By using the extrapolation formula, we can obtain the results of $\alpha=0°$ from the results of $\alpha=0.5°$, $2°$, $5°$. In these figures, the experimental data is $\frac{\partial C_N^*}{\partial\alpha}$, $\frac{\partial C_M^*}{\partial\alpha}$, C_D^* of $\alpha=0$. We can see from these figures that the computed results agree with the experimental data. We can also find that C_N^*/α and C_M^*/α change with the angle of attack. This means that C_N^* and C_M^* are not linear functions of α. Moreover, C_N^*/α, C_M^*/α, C_D^* are not monotone functions of r_n/r_b and are very close to the values of the conical flow when $r_n/r_b\leqslant 0.06$.

In Figs. 4.29 and 4.30, we give the C_D'–curve and the C_N^*–curve of sphere–cones with $\beta=12.5°$, where $M_\infty=17$, $\alpha=2°$. C_D' is the ratio of the distance between the center of pressure and the vertex of the sphere–cone to the length of the body, and C_N^* is the normal–force coefficient (taking the area of the base as the reference area). In these figures, we give both the results of the perfect gas ($\gamma=1.4$) and the results of the air in chemical equilibrium ($p_\infty=0.1244\times10^4$ kgf/m^2, $\rho_\infty=0.1958\times10^{-1}$ kgf·sec^2/m^4), and take the length of the body as the independent variable. We can see from those figures that the values of C_D' and C_N^* for

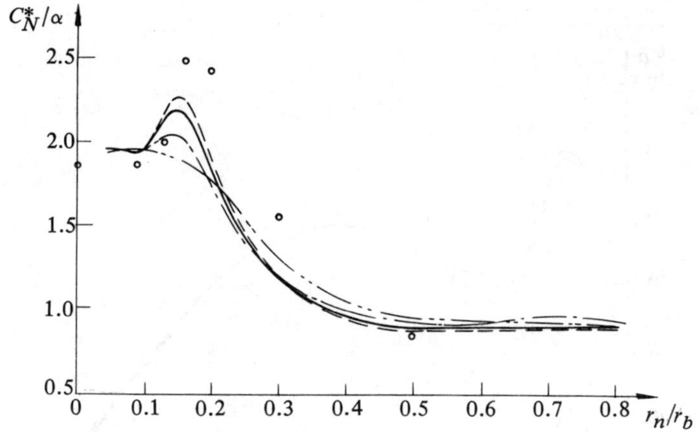

Figure 4.28a Variations of C_N^*/α of the flow around sphere–cones with the ratio of bluntness r_n/r_b ($M_\infty=17$, $\beta=12.5°$)

- - - $\alpha=0°$
—— $\alpha=0.5°$
—·— $\alpha=2°$
—··— $\alpha=5°$
} Computed results; ○ Experimental data[58], △ Conical flow ($\alpha=2°$).

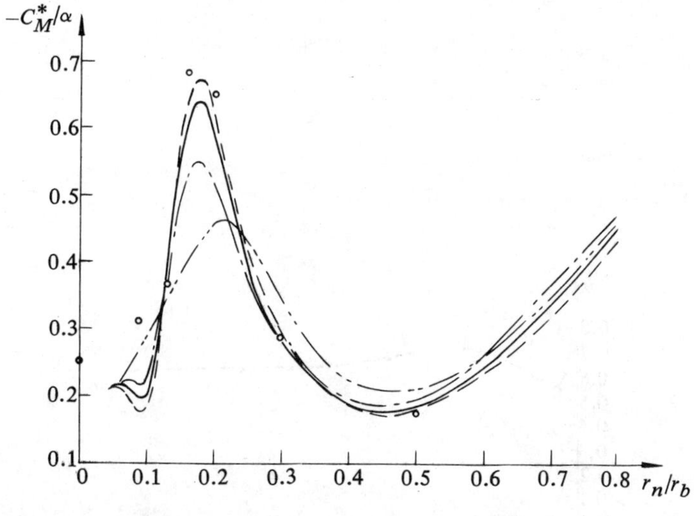

Figure 4.28b Variations of C_M^*/α of the flow around sphere–cones with the ratio of bluntness r_n/r_b ($M_\infty=17$, $\beta=12.5°$)

- - - $\alpha=0°$
—— $\alpha=0.5°$
—·— $\alpha=2°$
—··— $\alpha=5°$
} Computed results; ○ Experimental data[58], △ Conical flow ($\alpha=2°$).

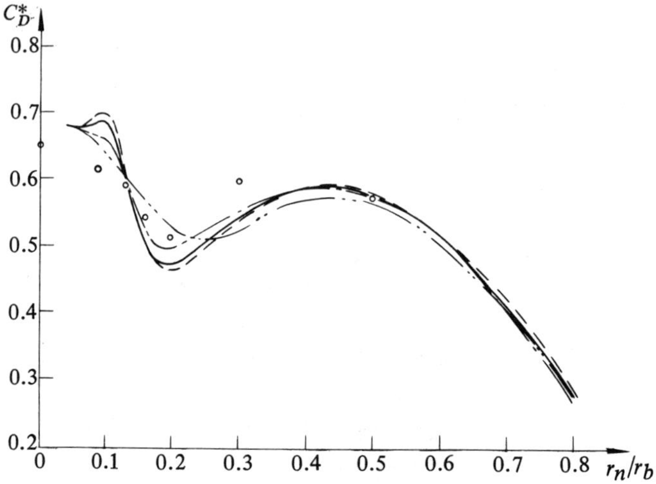

Figure 4.28c Variations of C_D^* of the flow around sphere-cones with the ratio of bluntness r_n/r_b ($M_\infty=17$, $\beta=12.5°$)

- - - $\alpha=0°$; —— $\alpha=0.5°$; —·— $\alpha=2°$; —··— $\alpha=5°$ } Computed results, ○ Experimental data [58], △ Conical flow.

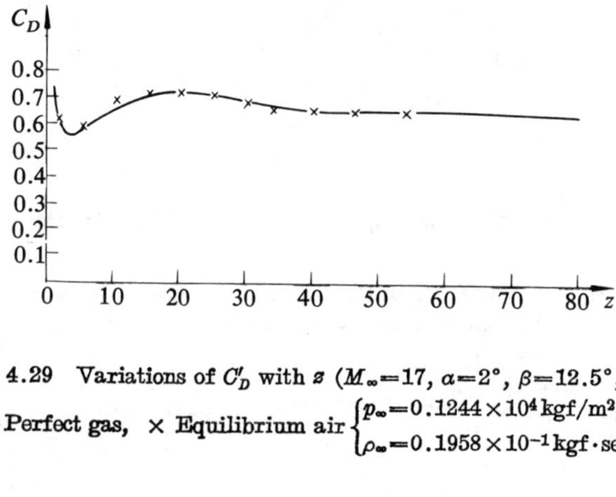

Figure 4.29 Variations of C'_D with z ($M_\infty=17$, $\alpha=2°$, $\beta=12.5°$, $R_0=1$)

—— Perfect gas, × Equilibrium air $\begin{cases} p_\infty=0.1244\times10^4\,\text{kgf/m}^2, \\ \rho_\infty=0.1958\times10^{-1}\,\text{kgf}\cdot\text{sec}^2/\text{m}^4. \end{cases}$

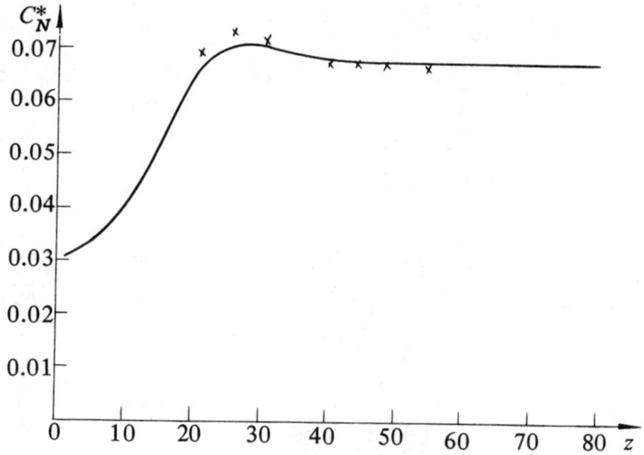

Figure 4.30 Variations of the normal-force coefficient with z ($M_\infty=17$, $\alpha=2°$, $\beta=12.5°$, $R_0=1$)
—— Perfect gas, × Equilibrium air $\begin{cases} p_\infty=0.1244\times10^4\,\text{kgf/m}^2, \\ \rho_\infty=0.1958\times10^{-1}\,\text{kgf}\cdot\text{sec}^2/\text{m}^4. \end{cases}$

different models of gases are different, and that the difference is quite large somewhere.

2. *Results of the Flow around Multi-stage Combined Bodies*

Fig. 4.31 represents the structure of the perfect gas flow field around the combined body I which is a slender cone–cylinder–flare combined body with three expansion edges and two compression edges, where $M_\infty=6$, and $\alpha=0°$. The figure shows that the first secondary shock intersects the main shock at $z\approx4.33$, and a new main shock, a contact discontinuity and a reflected expansion wave are produced. The slope of the new main shock is greater than that of the old main shock, and the slope at the point of intersection is discontinuous. This means that the strength of the main shock increases suddenly at this point. The contact discontinuity intersects the second secondary shock at $z\approx7.56$, and a new secondary shock, a new contact discontinuity and a reflected expansion wave are produced. The slope of the new contact discontinuity is greater than that of the old one. This means that the airstream bends towards the left in passing through the shock. Obviously, both the slopes of the secondary shock and of the contact discontinuity are discontinuous. In both cases, the distances between the front boundary and the back boundary of the reflected expansion wave are small, which means that both the reflected expansion waves are weak.

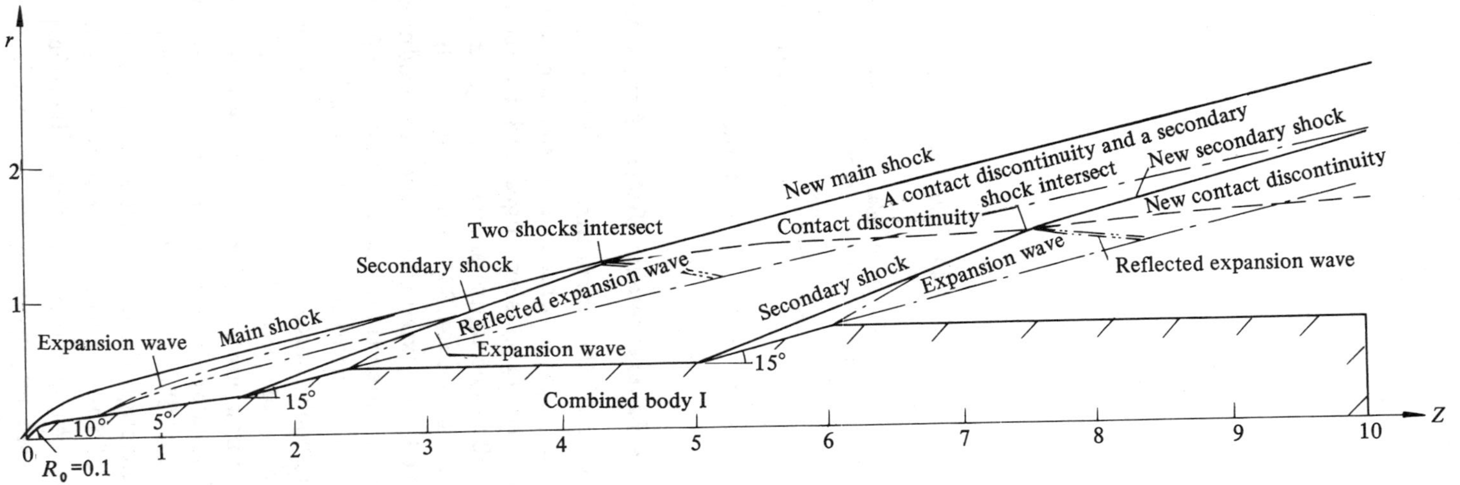

Figure 4.31 Construction of the perfect gas flow field around the combined body I ($M_\infty=6$, $\alpha=0°$)

Generator of the combined body I: circular arc ($R_0=0.1$) $\xrightarrow[\text{(tangent)}]{}$ straight line ($\theta_1=10°$) $\xrightarrow[z_2=0.6]{}$ straight line ($\theta_2=5°$) $\xrightarrow[z_3=1.6]{}$ straight line ($\theta_3=15°$) $\xrightarrow[z_4=2.4]{}$ straight line ($\theta_4=0°$) $\xrightarrow[z_5=5]{}$ straight line ($\theta_5=15°$) $\xrightarrow[z_6=6]{}$ straight line ($\theta_6=0°$). Where $z_i=\times\times\times\times$ represents the location of a connection point between different parts of the body, for example, "straight line ($\theta_1=10°$) $\xrightarrow[z_2=0.6]{}$ straight line ($\theta_2=5°$)" represents that at $z_2=0.6$ the generator changes from a straight line with $\theta_1=10°$ to a straight line with $\theta_2=5°$, and $R_0=0.1$ represents that the radius of circular arc is equal to 0.1.

In Figs. 4.32 and 4.33, we present several pressure curves and density curves in the ξ-direction in the perfect gas flow field around the combined body I, where $M_\infty=6$, $\alpha=0°$, $z=8.4$ and 10. Those figures show that the pressure and density in the shock layer (the layer between the main shock and the body) have a complicated variation. However, the curves in each interval are smooth, there being no oscillation.

Fig. 4.34 represents the shapes of the shocks and contact discontinuities in the perfect gas flow fields around the combined body II (a two-stage body) and the combined body III (a five-stage body), where $M_\infty=3$, $\alpha=0°$. In the flow field around the combined body II, the main shock intersects the secondary shock at $z\approx1.55$, and a shock, a contact discontinuity and a very weak reflected expansion wave are produced. (The weak reflected expansion wave is not drawn in Fig. 4.34.) In the flow field around the combined body III, the secondary shocks starting from the compression corners first converge to a strengthened secondary shock. In this process, several contact discontinuities with weak intensity and weak expansion waves are produced, and there exist intersections of contact discontinuities with secondary shocks. For clarity, these are not drawn in Fig. 4.34. The strengthened secondary shock then intersects the main shock at $z\approx1.64$, and a new main shock, a contact discontinuity and an expansion wave are produced. The parameters of the flow fields around these two combined bodies on several cross sections are given in Tables 4.12 and 4.13. We can find from Table 4.12 that the flow field around the combined body II after $z\approx20$ is almost conical.

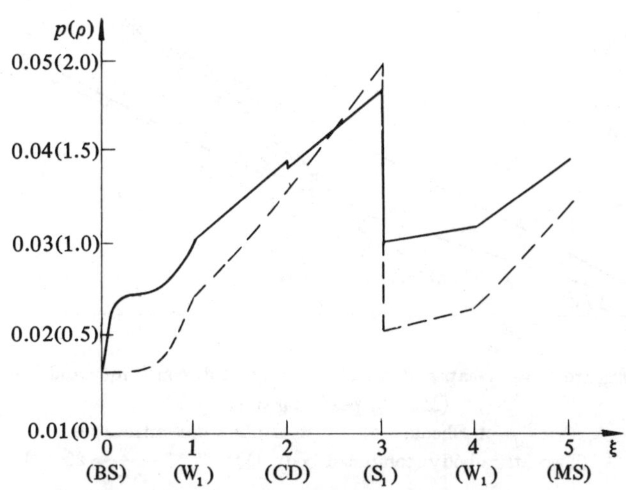

Figure 4.32 Variations of pressure and density with ξ at $z=8.4$ ($M_\infty=6$, $\alpha=0°$, combined body I)
---- Pressure, —— Density.

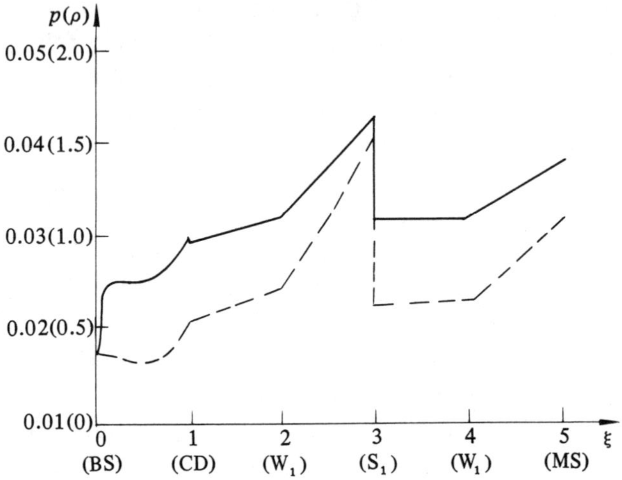

Figure 4.33 Variations of pressure and density with ξ at $z=10$ ($M_\infty=6$, $\alpha=0°$, combined body I)
---- Pressure, —— Density.

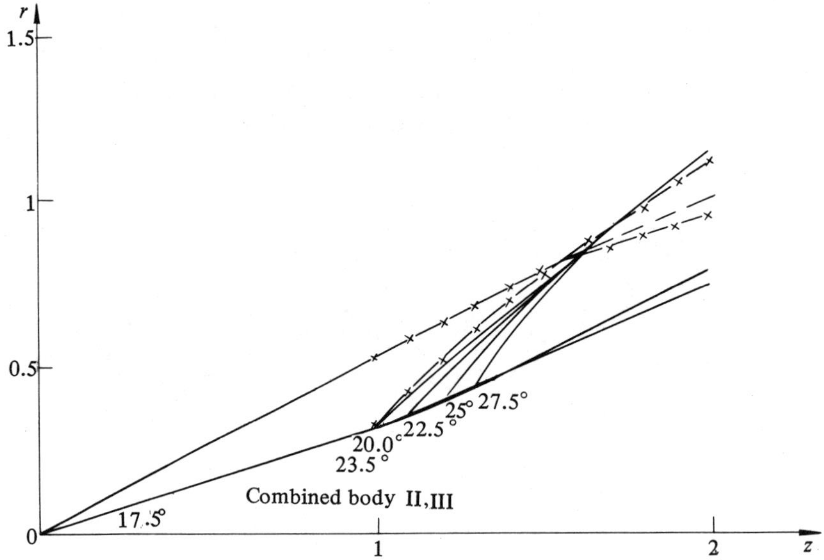

Figure 4.34 Shapes of the shock, contact discontinuity and body ($M_\infty=3$, perfect gas, $\alpha=0°$)
×——× Shock, ×----× Contact discontinuities.
(Two-stage body (combined body II): 17.5° $\xrightarrow[z_1=1]{}$ 23.5°)
—— Shock, ---- Contact discontinuities, ////// Wall surface.
(Five-stage body (combined body III): 17.5° $\xrightarrow[z_1=1]{}$ 20.0° $\xrightarrow[z_2=1.1]{}$ 22.5° $\xrightarrow[z_3=1.2]{}$ 25.0° $\xrightarrow[z_4=1.3]{}$ 27.5°)

Table 4.12

Body: two-stage body, $17.5° \xrightarrow[z=1]{} 23.5°$

Free stream: $M_\infty=3$ (perfect gas), $\alpha=0°$

Construction of flow field: MS–CD–W_2–W_2–BS

$z_4=1.5471,\ F_4=0.8010,\ F_{4,z}=0.66856$

ξ	u	v	p	ρ
4.000	0.83518	0.24654	0.24419	2.14401
3.500	0.83518	0.24654	0.24419	2.14401
3.000	0.83518	0.24654	0.24419	2.14401

$z_3=1.5471,\ F_3=0.8010,\ F_{3,z}=0.29519$

ξ	u	v	p	ρ
3.000	0.84696	0.25002	0.24419	2.20357
2.500	0.84696	0.25002	0.24419	2.20357
2.000	0.84696	0.25002	0.24419	2.20357

$z_2=1.5471,\ F_2=0.8010,\ F_{2,z}=-0.17710$

ξ	u	v	p	ρ
2.000	0.84696	0.25002	0.24419	2.20357
1.750	0.84691	0.24974	0.24443	2.20513
1.500	0.84686	0.24947	0.24467	2.20669
1.250	0.84681	0.24919	0.24492	2.20825
1.000	0.84676	0.24892	0.24516	2.20982

$z_1=1.5471,\ F_1=0.8010,\ F_{1,z}=-0.17885$

Table 4.12 (continued)

ξ	u	v	p	ρ
1.000	0.84676	0.24892	0.24516	2.20982
0.955	0.84440	0.25279	0.24749	2.22492
0.909	0.84210	0.25662	0.24969	2.23914
0.864	0.83980	0.26047	0.25182	2.25286
0.818	0.83752	0.26434	0.25388	2.26609
0.773	0.83525	0.26822	0.25586	2.27879
0.727	0.83302	0.27213	0.25775	2.29085
0.682	0.83081	0.27606	0.25954	2.30234
0.636	0.82861	0.28004	0.26124	2.31324
0.591	0.82642	0.28407	0.26286	2.32359
0.545	0.82422	0.28816	0.26440	2.33338
0.500	0.82200	0.29232	0.26586	2.34258
0.455	0.81978	0.29656	0.26722	2.35119
0.409	0.81754	0.30090	0.26850	2.35920
0.364	0.81529	0.30534	0.26967	2.36659
0.318	0.81302	0.30989	0.27075	2.37340
0.273	0.81073	0.31457	0.27171	2.37956
0.227	0.80841	0.31940	0.27256	2.38491
0.182	0.80605	0.32437	0.27329	2.38940
0.136	0.80366	0.32953	0.27388	2.39307
0.091	0.80122	0.33487	0.27432	2.39587
0.045	0.79873	0.34042	0.27460	2.39756
0.000	0.79621	0.34620	0.27470	2.39808

$z_0=1.5471,\ F_0=0.5532,\ F_{0,z}=0.43481$

Table 4.12 (continued)

Construction of flow field: MS–CD–BS

$z_2=2.0000,\ F_2=1.1020,\ F_{2,z}=0.66200$

ξ	u	v	p	ρ
2.000	0.83867	0.24370	0.24070	2.12522
1.909	0.83639	0.24721	0.24287	2.13877
1.818	0.83411	0.25071	0.24500	2.15190
1.727	0.83191	0.25416	0.24699	2.16423
1.636	0.82971	0.25762	0.24894	2.17618
1.545	0.82756	0.26106	0.25079	2.18743
1.455	0.82540	0.26453	0.25259	2.19835
1.364	0.82328	0.26800	0.25430	2.20867
1.273	0.82115	0.27150	0.25595	2.21858
1.182	0.81903	0.27502	0.25753	2.22805
1.091	0.81691	0.27857	0.25904	2.23689
1.000	0.81474	0.28218	0.26051	2.24539

$z_1=2.0000,\ F_1=0.9475,\ F_{1,z}=0.34635$

ξ	u	v	p	ρ
1.000	0.82668	0.28632	0.26051	2.30776
0.923	0.82452	0.29036	0.26206	2.31779
0.846	0.82225	0.29452	0.26360	2.32768
0.769	0.82006	0.29869	0.26496	2.33634
0.692	0.81779	0.30303	0.26635	2.34532
0.615	0.81559	0.30738	0.26752	2.35286
0.538	0.81327	0.31191	0.26870	2.36035
0.462	0.81098	0.31651	0.26970	2.36671
0.385	0.80866	0.32124	0.27061	2.37248
0.308	0.80623	0.32620	0.27149	2.37810
0.231	0.80395	0.33115	0.27202	2.38148
0.154	0.80134	0.33650	0.27275	2.38606
0.077	0.79891	0.34187	0.27304	2.38785
0.000	0.79693	0.34652	0.27308	2.38793

$z_0=2.000,\ F_0=0.7501,\ F_{0,z}=0.43481$

Table 4.12 (continued)

Construction of flow field: MS–CD–BS

$z_2=15.000$, $F_2=9.5554$, $F_{2,z}=0.64897$

ξ	u	v	p	ρ
2.000	0.84563	0.23787	0.23373	2.08722
1.955	0.84283	0.24221	0.23650	2.10483
1.909	0.84012	0.24648	0.23911	2.12141
1.864	0.83749	0.25070	0.24159	2.13706
1.818	0.83491	0.25489	0.24394	2.15188
1.773	0.83238	0.25907	0.24617	2.16594
1.727	0.82989	0.26325	0.24830	2.17929
1.682	0.82742	0.26745	0.25032	2.19197
1.636	0.82498	0.27169	0.25225	2.20398
1.591	0.82255	0.27597	0.25407	2.21537
1.545	0.82013	0.28031	0.25580	2.22614
1.500	0.81772	0.28472	0.25744	2.23628
1.455	0.81530	0.28921	0.25897	2.24577
1.409	0.81287	0.29380	0.26040	2.25462
1.364	0.81043	0.29850	0.26173	2.26279
1.318	0.80797	0.30332	0.26294	2.27025
1.273	0.80549	0.30828	0.26403	2.27697
1.227	0.80297	0.31339	0.26500	2.28288
1.182	0.80042	0.31868	0.26583	2.28792
1.136	0.79781	0.32415	0.26652	2.29197
1.091	0.79512	0.32981	0.26704	2.29481
1.045	0.79221	0.33565	0.26738	2.29572
1.000	0.78792	0.34118	0.26753	2.28847

$z_1=15.000$, $F_1=6.4293$, $F_{1,z}=0.43302$

ξ	u	v	ρ	p
1.000	0.79966	0.34626	0.26753	2.35203
0.500	0.79956	0.34694	0.26753	2.35311
0.000	0.79940	0.34759	0.26753	2.35320

$z_0=15.000$, $F_0=6.4027$, $F_{0,z}=0.43481$

Table 4.12 (continued)

Construction of flow field: MS–CD–BS

$z_2=20.000$, $F_2=12.8002$, $F_{2,z}=0.64895$

ξ	u	v	p	ρ
2.000	0.84564	0.23786	0.23372	2.08716
1.955	0.84283	0.24222	0.23650	2.10484
1.909	0.84011	0.24651	0.23912	2.12148
1.864	0.83747	0.25074	0.24160	2.13719
1.818	0.83488	0.25495	0.24396	2.15207
1.773	0.83234	0.25914	0.24620	2.16617
1.727	0.82984	0.26334	0.24834	2.17956
1.682	0.82737	0.26756	0.25037	2.19227
1.636	0.82492	0.27181	0.25230	2.20433
1.591	0.82248	0.27611	0.25413	2.21574
1.545	0.82005	0.28047	0.25586	2.22653
1.500	0.81763	0.28490	0.25750	2.23668
1.455	0.81520	0.28942	0.25903	2.24619
1.409	0.81276	0.29403	0.26046	2.25505
1.364	0.81031	0.29875	0.26179	2.26322
1.318	0.80785	0.30360	0.26300	2.27069
1.273	0.80536	0.30859	0.26409	2.27741
1.227	0.80284	0.31373	0.26505	2.28332
1.182	0.80028	0.31905	0.26588	2.28836
1.136	0.79767	0.32456	0.26656	2.29245
1.091	0.79501	0.33029	0.26707	2.29544
1.045	0.79220	0.33622	0.26740	2.29688
1.000	0.78770	0.34171	0.26753	2.28847

$z_1=20.000$, $F_1=8.5966$, $F_{1,z}=0.43381$

ξ	u	v	p	ρ
1.000	0.79942	0.34680	0.26753	2.35204
0.500	0.79944	0.34721	0.26753	2.35310
0.000	0.79940	0.34759	0.26753	2.35318

$z_0=20.000$, $F_0=8.5767$, $F_{0,z}=0.43481$

Table 4.13

Body: five-stage body, 17.5° $\xrightarrow[z=1]{}$ 20.0° $\xrightarrow[z=1.1]{}$ 22.5° $\xrightarrow[z=1.2]{}$ 25.0° $\xrightarrow[z=1.3]{}$ 27.5°

Free stream: $M_\infty=3$ (perfect gas), $\alpha=0°$

Construction of flow field: MS–CD–CD–CD–CD–BS

$z_5=1.6443,\ F_5=0.8512,\ F_{5,z}=0.80195$

ξ	u	v	p	ρ
5.000	0.76642	0.29126	0.31294	2.47997
4.500	0.76642	0.29126	0.31294	2.47997
4.000	0.76642	0.29126	0.31294	2.47997

$z_4=1.6443,\ F_4=0.8512,\ F_{4,z}=0.38003$

ξ	u	v	p	ρ
4.000	0.79183	0.30092	0.31294	2.61403
3.750	0.79047	0.29967	0.31683	2.63738
3.500	0.78951	0.30109	0.31779	2.64336
3.250	0.78856	0.30252	0.31874	2.64921
3.000	0.78761	0.30394	0.31968	2.65494

$z_3=1.6443,\ F_3=0.8335,\ F_{3,z}=0.38590$

ξ	u	v	p	ρ
3.000	0.78963	0.30471	0.31968	2.66651
2.800	0.78835	0.30658	0.32089	2.67388
2.600	0.78709	0.30844	0.32207	2.68104
2.400	0.78583	0.31030	0.32324	2.68808
2.200	0.78457	0.31216	0.32439	2.69502
2.000	0.78332	0.31402	0.32551	2.70174

$z_2=1\ 6443,\ F_2=0.8047,\ F_{2,z}=0.40088$

Table 4.13 (continued)

ξ	u	v	p	ρ
2.000	0.78434	0.31443	0.32551	2.70772
1.750	0.78298	0.31646	0.32667	2.71468
1.500	0.78163	0.31851	0.32778	2.72130
1.250	0.78027	0.32056	0.32890	2.72797
1.000	0.77892	0.32264	0.32998	2.73434

$z_1=1.6443,\ F_1=0.7799,\ F_{1,z}=0.41421$

ξ	u	v	p	ρ
1.000	0.77926	0.32278	0.32998	2.73638
0.944	0.77731	0.32580	0.33147	2.74525
0.889	0.77535	0.32884	0.33294	2.75399
0.833	0.77340	0.33191	0.33434	2.76232
0.778	0.77143	0.33503	0.33568	2.77031
0.722	0.76946	0.33823	0.33693	2.77768
0.667	0.76748	0.34149	0.33809	2.78453
0.611	0.76547	0.34478	0.33922	2.79113
0.556	0.76346	0.34814	0.34026	2.79719
0.500	0.76143	0.35155	0.34123	2.80289
0.444	0.75939	0.35504	0.34214	2.80820
0.389	0.75733	0.35862	0.34293	2.81295
0.333	0.75525	0.36229	0.34365	2.81722
0.278	0.75315	0.36603	0.34429	2.82092
0.222	0.75102	0.36989	0.34480	2.82387
0.167	0.74887	0.37382	0.34524	2.82642
0.111	0.74669	0.37788	0.34555	2.82834
0.056	0.74448	0.38209	0.34575	2.82947
0.000	0.74225	0.38639	0.34582	2.82981

$z_0=1.6443,\ F_0=0.6196,\ F_{0,z}=0.52057$

By using our method, we have accurately evaluated the complicated flow field around the combined bodies I and III, where two types of interactions between discontinuities appear. This means that our singularity-separating method treats this type of problems with great success.

Figs. 4.35—4.43 give several results about the perfect gas flow around the combined body IV, where $M_\infty=4$ and $\alpha=2.05°$. Fig. 4.35 represents the shapes of the main shock, the expansion wave, the secondary shock, and a certain stream surface which encloses the entropy layer between itself and the body. We can see from this figure that the expansion wave has not intersected the main shock at $z=15.48$. Hence, the shape of the main shock there should be the same as that in the flow field around a sphere-cone with $\beta=5.15°$, and the flow field near the main shock is also the same as that around a sphere-cone. Fig. 4.36 gives the distribution of the pressure just behind the main shock. We know from the figure that the pressure decreases rapidly in this entire region. This means that the pressure is far from that of conical flow. This phenomenon agrees with the results of the flow around

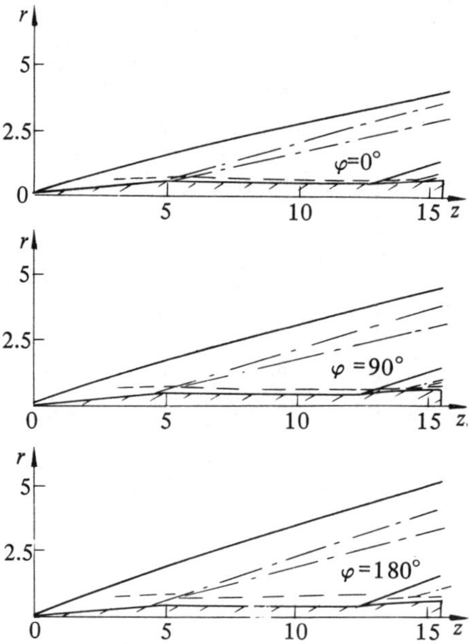

Figure 4.35 Shapes of boundaries ($M_\infty=4$, $\alpha=2.05°$)

Generator of the combined body IV: circular arc ($R_0=0.08995$) $\xrightarrow{\text{(tangent)}}$ straight line ($\theta_1=5.15°$) $\xrightarrow{z_2=4.68}$ straight line ($\theta_2=0°$) $\xrightarrow{z_3=12.417}$ straight line ($\theta_3=6°$) $\xrightarrow{z_4=14.472}$ straight line ($\theta_4=0°$).

—— Shock, —·— Expansion wave of the first family, ----- Stream surface, ⫽⫽ Wall surface.

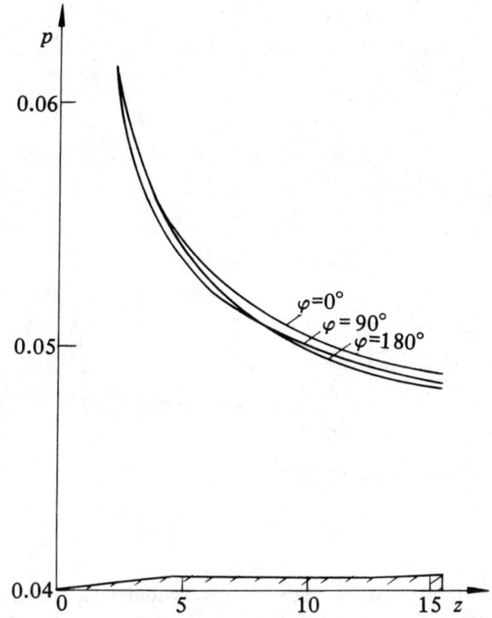

Figure 4.36 Variations of pressure behind shock with z ($M_\infty=4$, $\alpha=2.05°$, combined body IV)

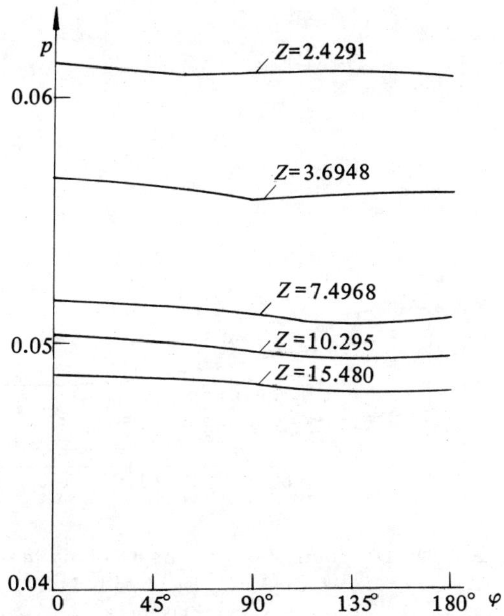

Figure 4.37 Variations of pressure behind shock with φ ($M_\infty=4$, $\alpha=2.05°$, combined body IV)

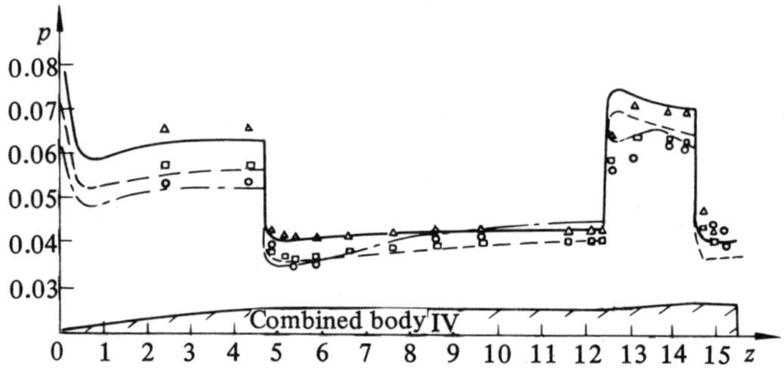

Figure 4.38 Distributions of pressure on a body and comparison between computed results and experimental data ($M_\infty=4$, $\alpha=2.05°$, combined body IV)

—— $\varphi=0$, ----- $\varphi=\pi/2$, —·— $\varphi=\pi$ } Computed result, △ $\varphi=0$, □ $\varphi=\pi/2$, ○ $\varphi=\pi$ } Experimental data[60].

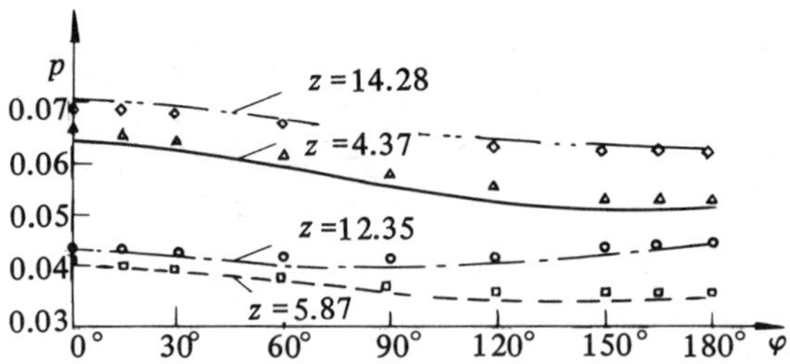

Figure 4.39 Distributions of pressure on a body in the φ-direction ($M_\infty=4$, $\alpha=2.05°$, combined body IV)

——, -----, —·—, —··— } Computed result, ◇, △, ○, □ } Experimental data[60].

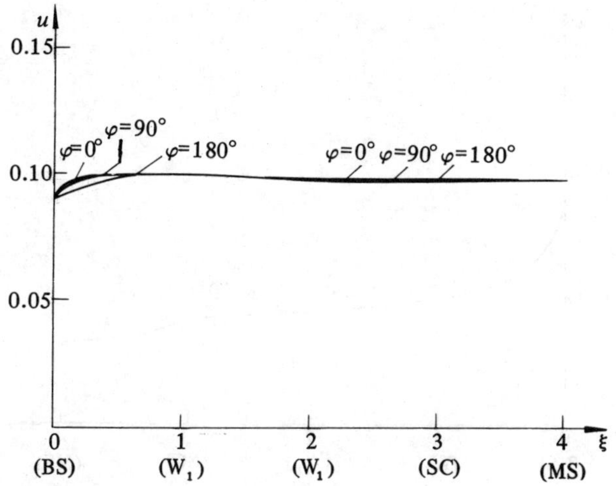

Figure 4.40a Variations of u with ξ ($M_\infty=4$, $\alpha=2.05°$, combined body IV, $z=5.2278$)

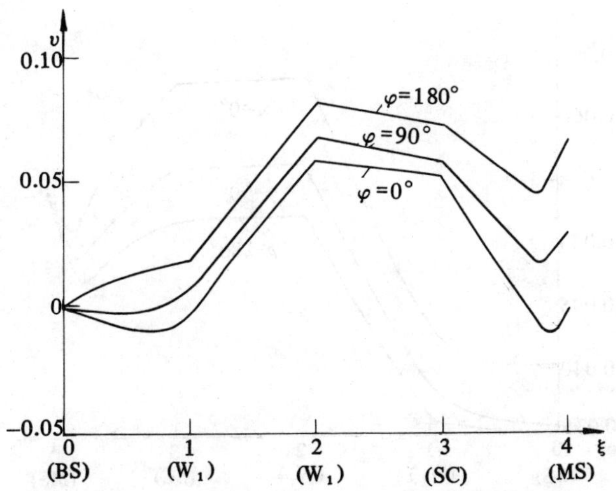

Figure 4.40b Variations of v with ξ ($M_\infty=4$, $\alpha=2.05°$, combined body IV, $z=5.2278$)

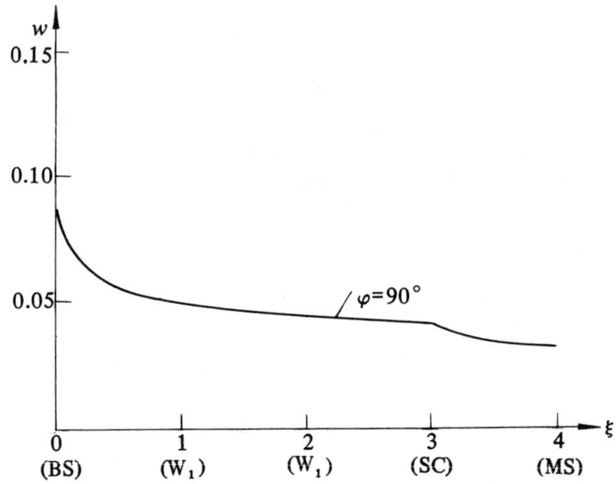

Figure 4.40c Variation of circumferential velocity with ξ ($M_\infty=4$, $\alpha=2.05°$, combined body IV, $z=5.2278$)

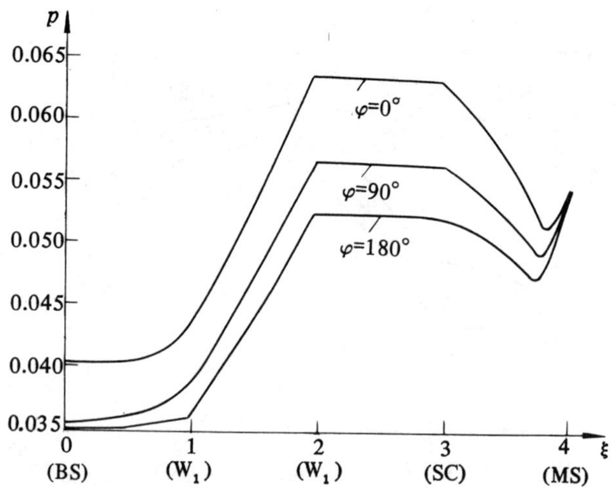

Figure 4.40d Variations of pressure with ξ ($M_\infty=4$, $\alpha=2.05°$, combined body IV, $z=5.2278$)

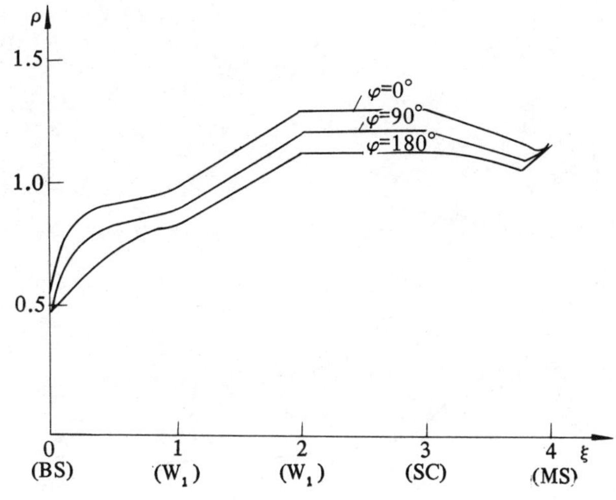

Figure 4.40e Variations of density with ξ ($M_\infty=4$, $\alpha=2.05°$, combined body IV, $z=5.2278$)

sphere–cones at zero–angle of attack. In fact, we know from Fig. 4.11 that, for the flow around the sphere–cone with $\beta=10°$ ($M_\infty=4$, $\alpha=0°$), the pressure is not close to the asymptotic value until the length of the body is close to 500 radii of the sphere. Moreover, Figs. 4.3—4.6 also show that the process of tending to the asymptotic value gets slower with decrease of β. In the case of Fig.4.36, $\beta=5.15°$, $\alpha=2.05°$, $\beta+\alpha<10°$, $R_0=0.0899$, and $z=15.48\approx172R_0$. Therefore, by using the results of sphere–cones at zero–angle of atta ck, we also obtain the conclusion that the flow near the shock at $z=15.48\approx172R_0$ is not yet close to the conical flow. Fig. 4.38 represents the distribution of pressure on the body. This figure shows that the gas is "under–expansive" when flowing around the first expansion edge, that is, the pressure on the body continues to decrease after the gas flows around the edge. When flowing around the compression edge, the gas is "under–compressive" on $\varphi=0°$, $\varphi=90°$, but "over–compressive" on $\varphi=180°$. There are several factors which affect the pattern of change of pressure behind edges. Besides the three–dimensional effect of an edge, the distributions of pressure, Mach–number and velocity in the ξ–direction before the edge also influence the variation of pressure. It seems that the reason why the pressure behind the first expansion edge continues to decrease is the three–dimensional effect of the edge. The reason why the pressure behind the compression edge on $\varphi=0°$ and $\varphi=90°$ continues to increase is mainly due to the fact that the Mach number in the region near the body boefre the compression edge increases rapidly with increase of ξ. (Since there is an

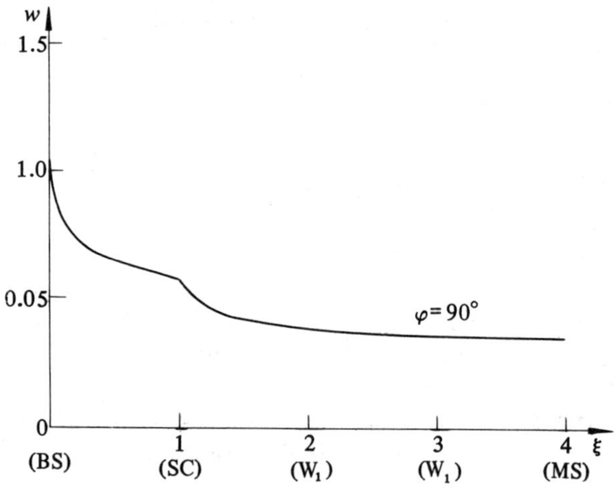

Figure 4.41a Variation of circumferential velocity with ξ ($M_\infty=4$, $\alpha=2.05°$, combined body IV, $z=11.657$)

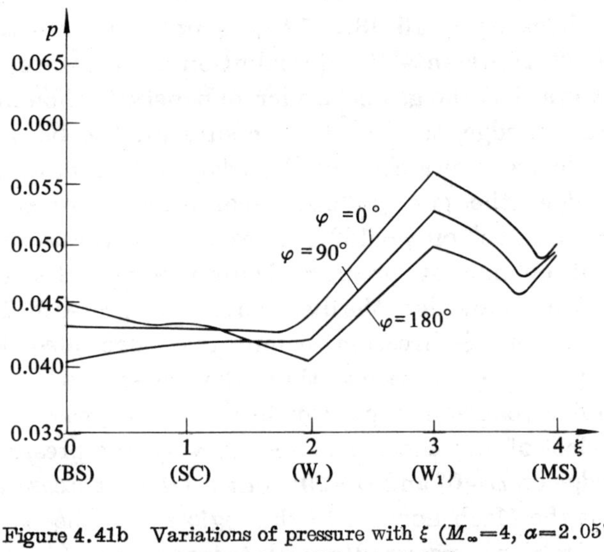

Figure 4.41b Variations of pressure with ξ ($M_\infty=4$, $\alpha=2.05°$, combined body IV, $z=11.657$)

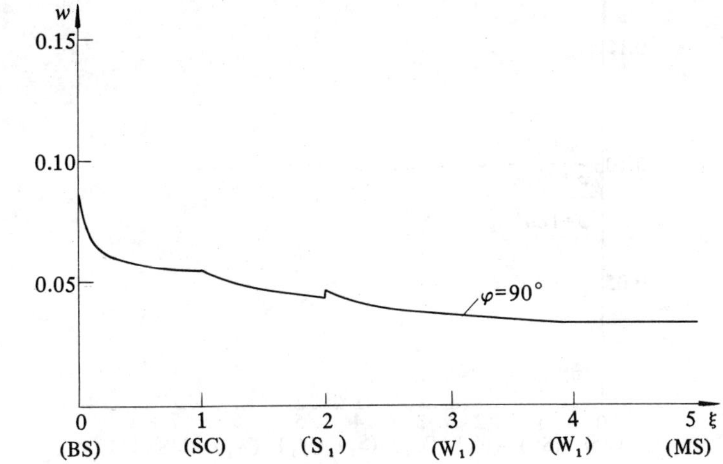

Figure 4.42a Variation of circumferential velocity with ξ ($M_\infty=4$, $\alpha=2.05°$, combined body IV, $z=13.986$)

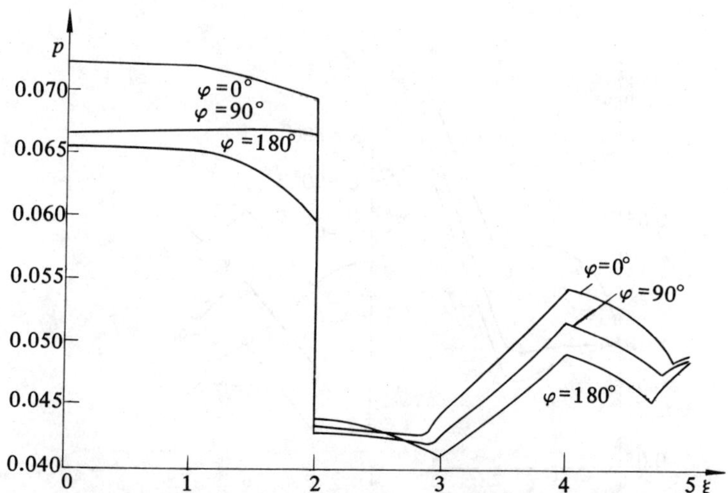

Figure 4.42b Variations of pressure with ξ ($M_\infty=4$, $\alpha=2.05°$, combined body IV, $z=13.986$)

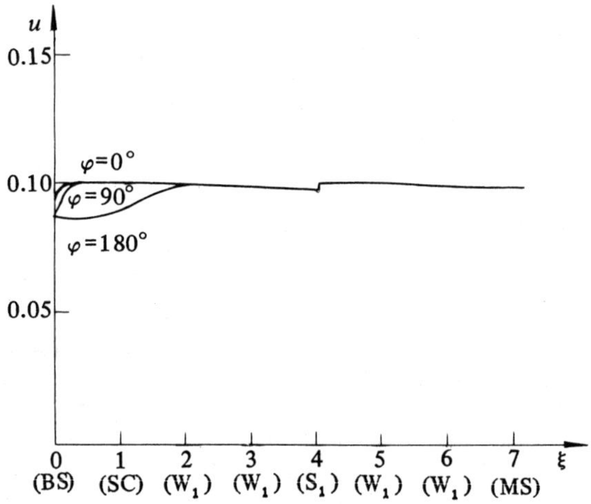

Figure 4.43a Variations of u with ξ ($M_\infty=4$, $\alpha=2.05°$, combined body IV, $z=15.480$)

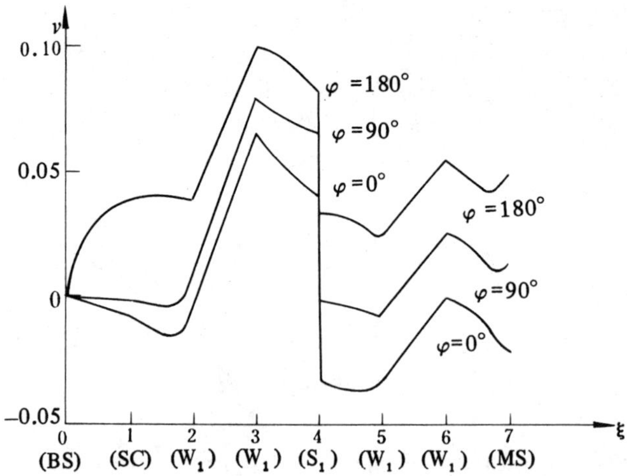

Figure 4.43b Variations of v with ξ ($M_\infty=4$, $\alpha=2.05°$, combined body IV, $z=15.480$)

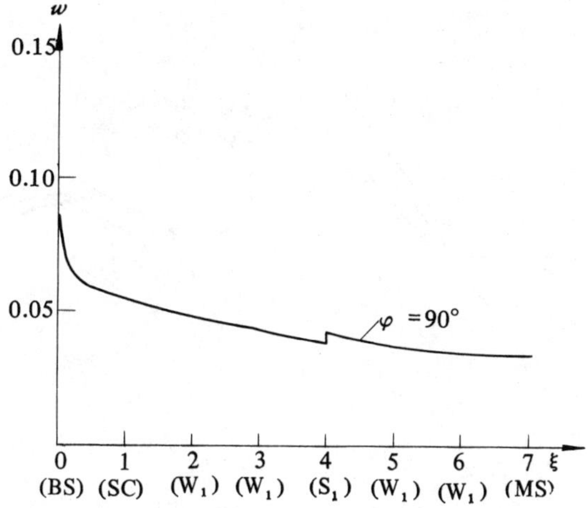

Figure 4.43c Variation of circumferential velocity with ξ ($M_\infty=4$, $\alpha=2.05°$, combined body IV, $z=15.480$)

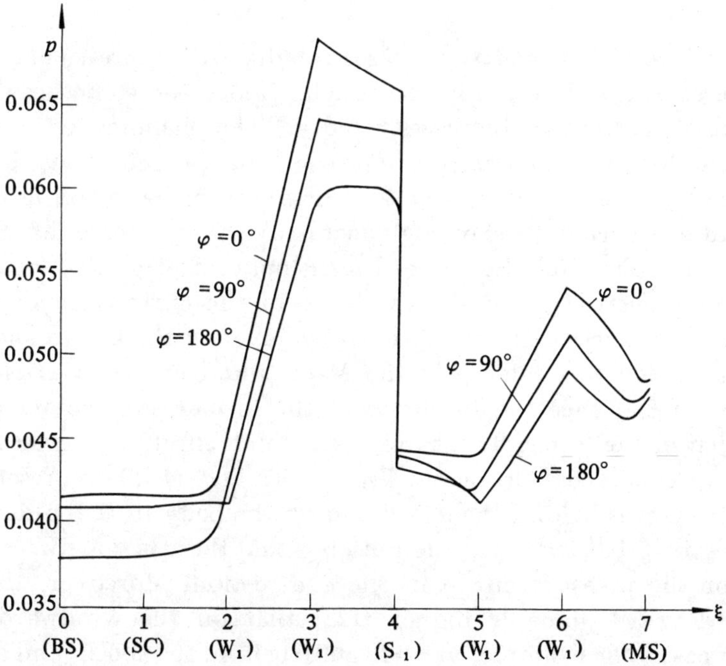

Figure 4.43d Variations of pressure with ξ ($M_\infty=4$, $\alpha=2.05°$, combined body IV, $z=15.480$)

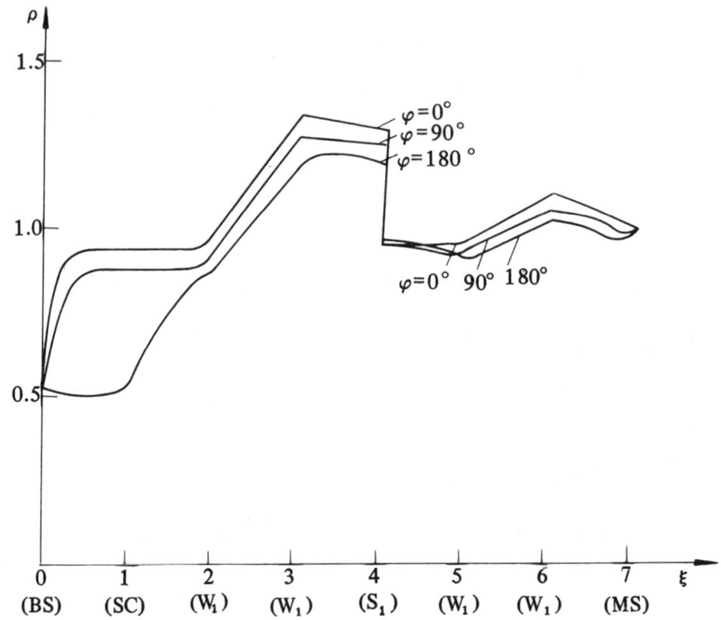

Figure 4.43e Variations of density with ξ ($M_\infty=4$, $\alpha=2.05°$, combined body IV, $z=15.480$)

entropy layer, the density increases rapidly with increase of ξ. However, the pressure does not change so much. Hence the sonic speed decreases and the Mach number increases rapidly.) The phenomenon of decrease of pressure behind the compression edge on $\varphi=180°$ can be roughly explained in the following way. The pressure in the region near the body in front of the compression edge decreases with increase of ξ (see Fig. 4.41b). This fact and the three-dimensional effect of the edge cause the pressure to decrease, and the existence of the entropy layer causes the pressure to increase. Here, the entropy layer is thick (we shall explain the reason for this later), so the Mach number increases slowly with increase of ξ. Hence the influence of the former two factors is stronger than that of the latter. Therefore, as a total effect, the pressure behind the compression edge decreases. Figs. 4.37 and 4.39 represent several pressure curves behind the shock and on the body in the φ-direction. In those figures, it is worth while noticing that there is sometimes a minimal point on the pressure curve in the φ-direction. Moreover, the pressure on the leeward plane is higher than that on the windward plane in certain cases, for example, at $z=2.4291$ behind the shock, and at $z=12.35$ on the body. This phenomenon appears not only in the computed results but also in the experimental data[60]. Hence we think that this phenomenon truely exists. Roughly speaking, the following is the reason for this phenomenon. Because of the angle of attack, the gas is compressed more violently on the windward side. If we observe the problem from this point, then the pressure should decrease with increase of φ (suppose that the plane $\varphi=0°$ is the windward plane). On the other hand, as the body is at angle of attack, the circumferential velocity w near the leeward plane decreases with increase of φ. If we observe the problem from this viewpoint, then the pressure on the leeward side should increase in the φ-direction. These two factors, plus several other factors, influence the way in which the pressure on the leeward side varies. In the case here, the second factor listed is strong, so the pressure on the leeward side increases with increase of φ. Furthermore, the increment of pressure on the leeward side is greater than the decrement of pressure on the windward side. Hence it appears that the pressure on the leeward plane is higher than that on the windward plane.

In Figs. 4.40—4.43, we give several flow parameter curves in the ξ-direction in four cross sections $z=$constant, where $\varphi=0°$, $90°$, $180°$. Those figures show that, since there are several expansion edges and compression edges, the change of certain quantities, for example, the pressure, the density, and the radial velocity v, is complicated. For example, the pressure curves at $z=15.48$ have two minimal points, two maximal points and a discontinuous point (see Fig. 4.43d). Moreover, since the body is quite long, there is a thin entropy layer near the body where the density varies rapidly (see Figs. 4.40e and 4.43e). We have seen this phenomenon in the flow field around a sphere-cone at zero-angle of attack. We can find from Figs. 4.40c, 4.41a, 4.42a and 4.43c, that the value of w near the body is two or three times the value of w behind the shock. This phenomenon is similar to that in the flow around a sphere-cone at angle of attack. The v-curves in Figs. 4.40b and 4.43b show that the value of v near the body on the windward plane is negative, and the value of v on the leeward plane is positive at $z=5.2278$ and 15.48. This means that the gas near the body on the windward plane gets closer and closer to the body and that the gas on the leeward plane gets farther and farther from the body. This is why the entropy layer on the windward plane is thinner than that on the leeward plane. We can see, from Figs. 4.40e and 4.43e, that the density near the body varies rapidly on the plane $\varphi=0°$, but slowly on the plane $\varphi=180°$. From this, we can also see that the entropy layer on the plane $\varphi=0°$ is thinner than that on the plane $\varphi=180°$. This conclusion is the same as what we obtain from the v-curves and from the shape of the stream surface given in Fig. 4.35. Figs. 4.40a and 4.43a represent several u-curves, which show that u does not change very much with φ and ξ, except in the region near the body.

Figs. 4.44—4.50 show several results about the perfect gas flow around the combined body V, where $M_\infty=10$, $\alpha=20°$. Fig. 4.44 represents the shapes of the shock and the expansion wave. We can find that the "shock layer" on the windward plane is much thinner than that on the leeward plane and that the slope of the shock gets smaller very quickly on the windward plane after the expansion wave intersects the shock at $z\approx5$. Figs. 4.45 and 4.46 represent the distributions of pressure just behind the shock in the z-direction and in the φ-direction. Before $z\approx5$, the shape of the pressure curves on $\varphi=0°$ is similar to that of the flow field around the sphere-cone in Fig. 4.6, where $M_\infty=20$, $\beta=30°$, $\alpha=0°$. This is reasonable. Because $\beta+\alpha=35°$, the shape of the pressure curve on the windward plane should be similar to that of the flow field around a sphere-cone with the semi-vertex angle $\beta=35°$ at zero-angle of attack. Behind $z\approx5$, the pressure on the plane $\varphi=0°$ goes down rapidly This is because the expansion wave has reached the shock.

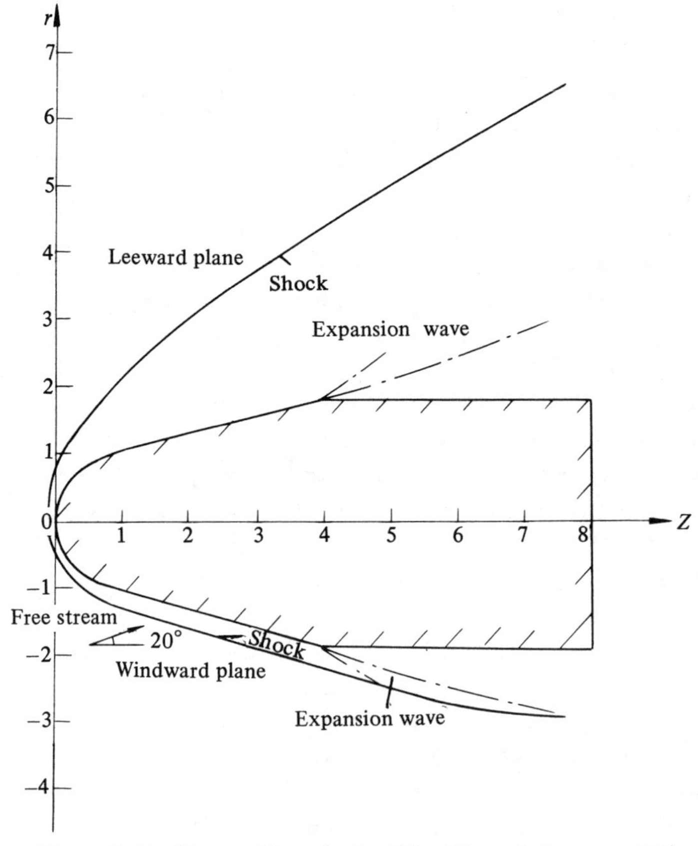

Figure 4.44 Shapes of boundaries ($M_\infty=10$, perfect gas, $\alpha=20°$)
Generator of the combined body V: circular arc ($R_0=1$) $\xrightarrow[\text{(tangent)}]{}$ straight line ($\theta_1=15°$) $\xrightarrow[z_2=4]{}$ straight line ($\theta_2=0°$).

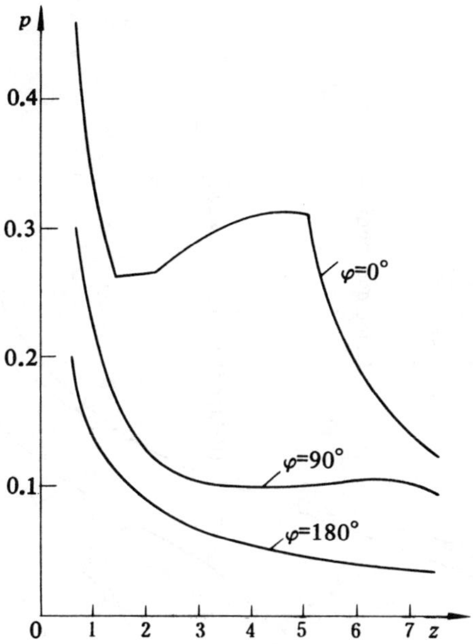

Figure 4.45 Pressure distributions behind the shock ($M_\infty=10$, $\alpha=20°$, combined body V)

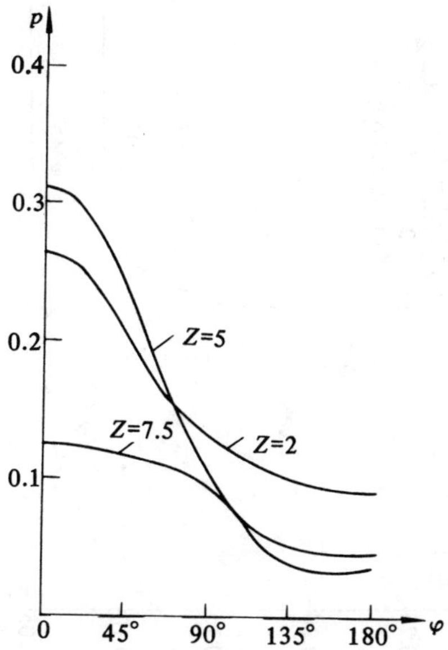

Figure 4.46 Variations of pressure behind the shock in the φ-direction ($M_\infty=10$, $\alpha=20°$, combined body V)

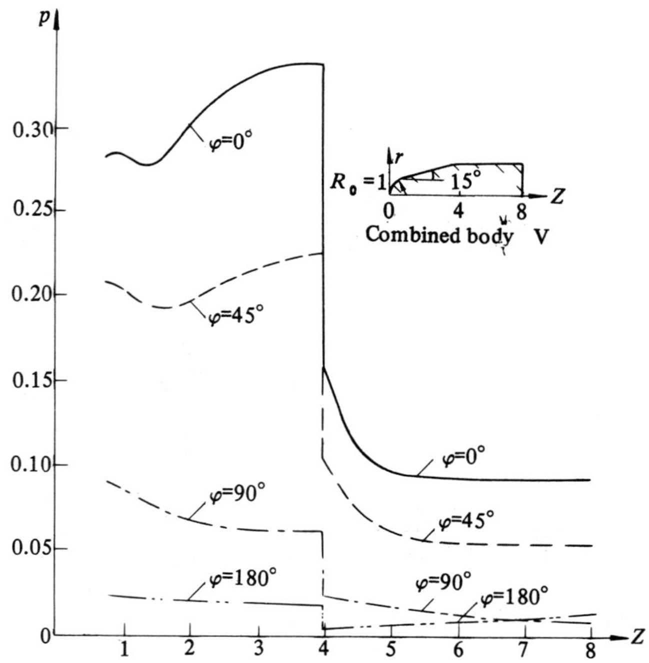

Figure 4.47 Pressure distributions on the combined body V in the z-direction (M_∞=10, perfect gas, α=20°)

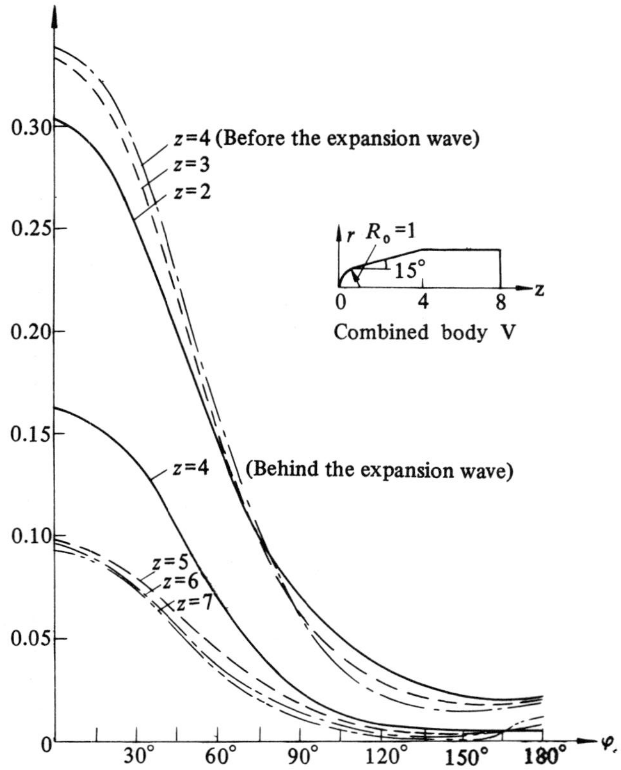

Figure 4.48 Pressure distributions on th combined body V in the φ-direction (M_∞=10, perfect gas, α=20°)

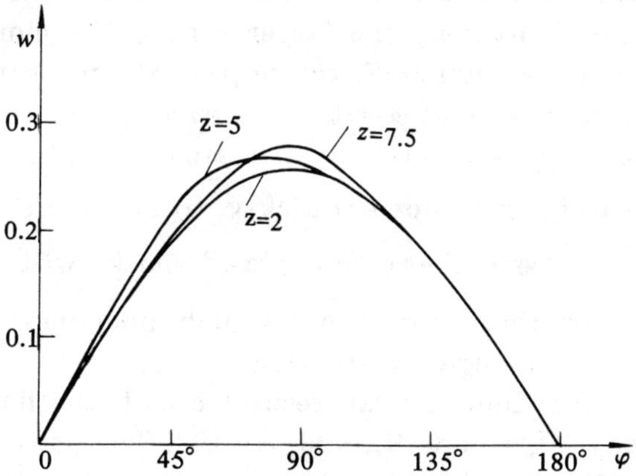

Figure 4.49 **Variations of w behind the shock in the φ-direction ($M_\infty=10$, $\alpha=20°$, combined body V)**

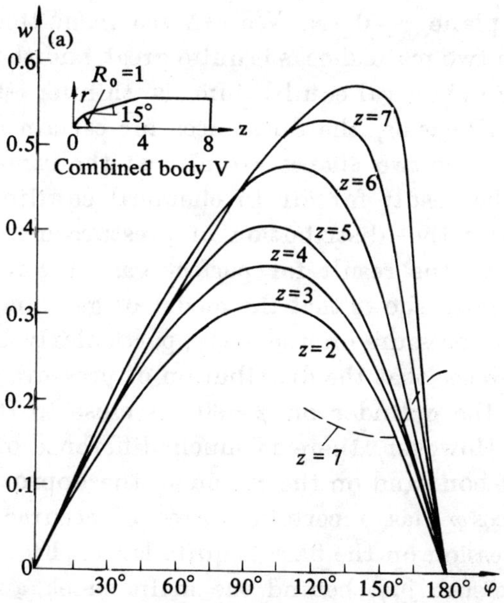

Figure 4.50 Distributions of w on the combined body V in the φ-direction ($M_\infty=10$, $\alpha=20°$)
—— Circumferential velocity w, ------ Sonic velocity a.

Figs. 4.47 and 4.48 represent the distributions of pressure on the body in the z–direction and in the φ–direction. Figs. 4.49 and 4.50 show several curves of w on the shock and on the body in the φ–direction. We can see from Fig. 4.47 that the pressure on $\varphi=0°$ and $90°$ continues to go down behind the expansion edge. However, the pressure on the leeward plane goes up behind the expansion edge. We can find from Fig. 4.48 that the pressure on the cylinder on the leeward side goes up in the φ–direction. Moreover, the larger z gets, the more the pressure goes up. On the cross section $z=7$, the slope of the pressure curve in the φ–direction gets quite large at $\varphi=160°$. In correspondence with this, we can also find from Fig. 4.50 that, at the same place, the circumferential speed w on the body goes down quickly from supersonic $\left(\frac{w}{a}>1\right)$ to subsonic $\left(\frac{w}{a}<1\right)$. It seems that a "crossflow" shock will appear. Figs. 4.46 and 4.49 show that there are no such phenomena for p and w behind the shock in the region $z=0$—7.5.

Figs. 4.51—4.61 show several results about the flow field around the combined body VI, where $M_\infty=20$, $\alpha=10°$. The gases considered are the perfect gas with $\gamma=1.4$ and air in chemical equilibrium ($p_\infty=0.1244 \times 10^4$ kgf/m^2, $\rho_\infty=0.1958\times 10^{-1}$ kgf·sec^2/m^4). The curves with small circles represent results for the perfect gas, and the curves without small circles results for air in chemical equilibrium. Fig. 4.51 shows the shapes of the main shock, the expansion wave, the secondary shock and the body on the plane $\varphi=0$—π. We can see from this figure that the difference between two main shocks is quite great, and that the shock layer of the flow of air in chemical equilibrium is thinner than that of the perfect gas flow. However, the differences for certain other boundaries are little, and even the two shapes are almost the same. (If this is the case, then only the result for air in chemical equilibrium is drawn.) Fig. 4.52 indicates the distribution of pressure on the body. In the figure, we also give the result for perfect gas at zero–angle of attack ($M_\infty=20$). The figure shows that the model of gas obviously influences the distribution of pressure on the body, particularly on the windward plane. We can also see that the distribution of pressure of the perfect gas on the cone and the cylinder on $\varphi=90°$ is close to the corresponding result at $\alpha=0°$. However, there is much difference on the flare. This means that on the cone and on the cylinder, the approximate expression $p(\alpha, \varphi)\approx p_0+\alpha p_1\cos\varphi$ has a certain degree of accuracy. However, the error of this expression on the flare is quite large. Fig. 4.53 represents the curves of pressure just behind the main shock and the secondary shock in the z–direction. Moreovere, observing Figs. 4.51 and 4.53, we can see that the reason why the pressure behind the main shock on $\varphi=$ $°$ goes down rapidly behind $z\approx5.25$ for the flow of air in chemical equilibrium (or behind $z \approx 5.6$ for the flow of the perfect gas) is due to the influence of the expansion wave on the main shock. Fig. 4.54 shows the curves of density just behind the main shock and the secondary shock in the z-direction, and shows that the difference between the results of two models of gases might be 50%. Figs. 4.55a—c and 4.56 a—b give the pressure curves, the density curves and other curves in the ξ-direction on $z=8$ and 10, and indicate that the results, even the patterns of flow quantity variation, are very different if the models of gases are different. Moreover, there is a "peak" of the density curve on the windward plane near the body, which also appears in the flow field around sphere-cones. The "peaks" of the density curve and the Mach

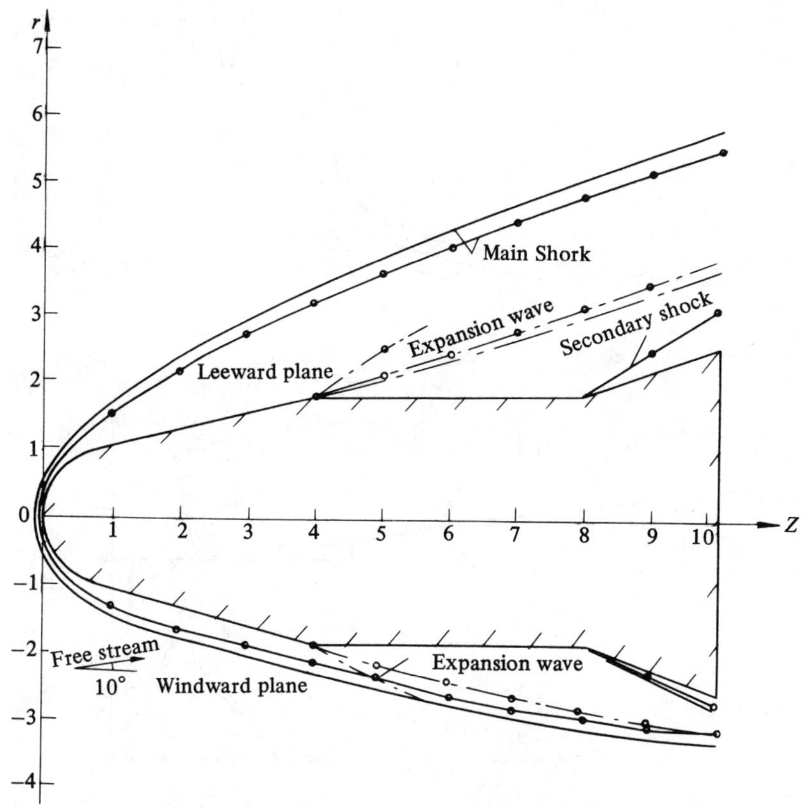

Figure 4.51 Shapes of boundaries ($M_\infty=20$, $\alpha=10°$)
Generator of the combined body VI: circular arc ($R_0=1$) $\xrightarrow[\text{(tangent)}]{}$ straight line ($\theta_1=15°$) $\xrightarrow[z_2=4]{}$ straight line ($\theta_2=0°$) $\xrightarrow[z_3=8]{}$ straight line ($\theta_3=20°$).

—— Shock (perfect gas),

—·— Wave-characteristic of the first family (perfect gas),

∘——∘ Shock $\left(\text{equilibrium air} \begin{cases} p_\infty=0.1244\times10^4\ \text{kgf/m}^2 \\ \rho_\infty=0.1958\times10^{-1}\ \text{kgf}\cdot\text{sec}^2/\text{m}^4 \end{cases}\right)$,

•—·—∘ Wave-characteristic of the first family (equilibrium air).

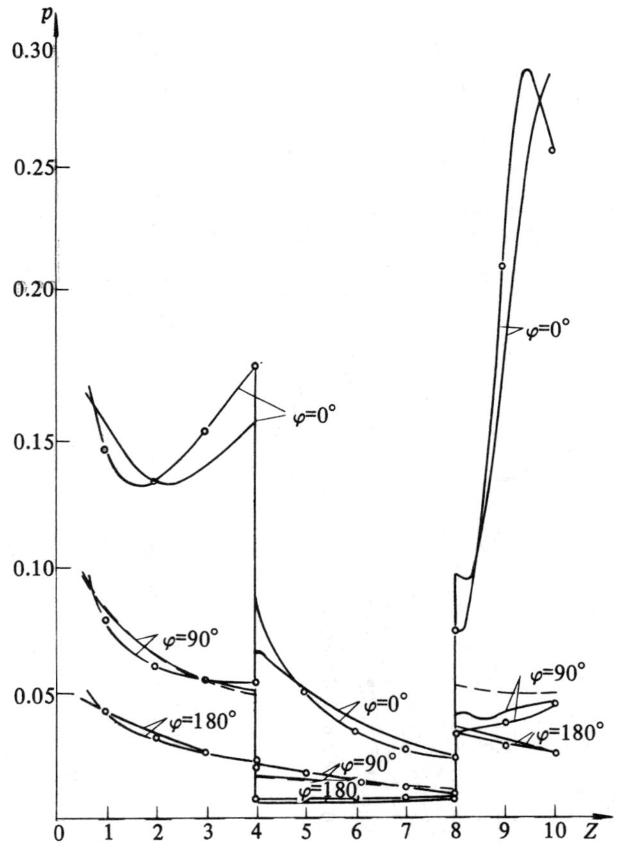

Figure 4.52 Pressure distributions on the combined body VI ($M_\infty=20$, $\alpha=10°$)

—— Perfect gas, ---- Perfect gas ($\alpha=0°$),

∘——∘ Equilibrium air $\begin{cases} p_\infty=0.1244\times10^4 \text{ kgf/m}^2, \\ \rho_\infty=0.1958\times10^{-1} \text{ kgf}\cdot\text{sec}^2/\text{m}^4. \end{cases}$

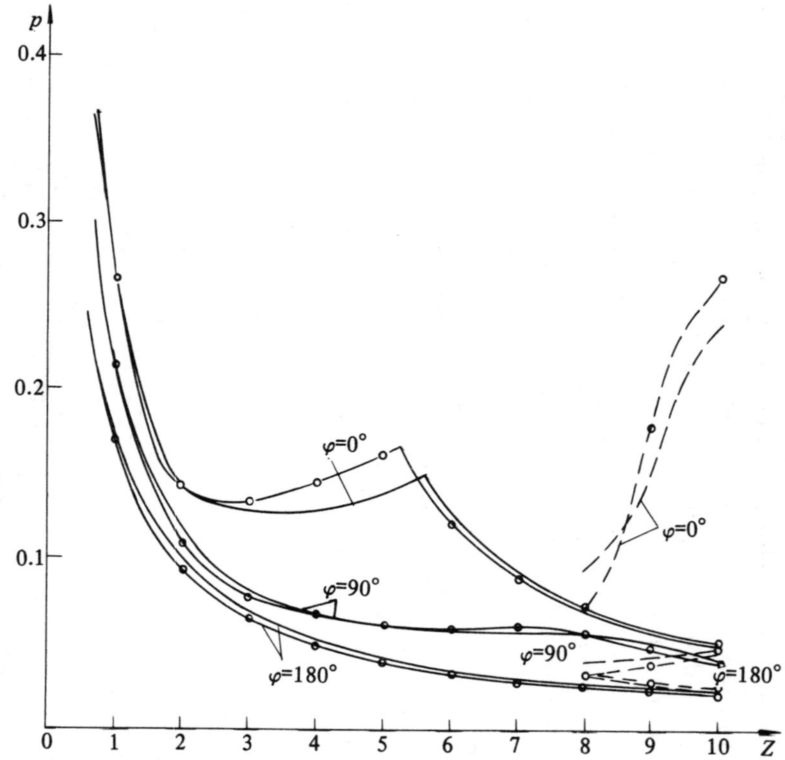

Figure 4.53 Pressure distributions behind the main shock and the secondary shock in the z-direction ($M_\infty=20$, $\alpha=10°$, combined body VI)

—— Perfect gas, ∘——∘ Equilibrium air: Main shock,
---- Perfect gas, ∘——∘ Equilibrium air: Secondary shock.

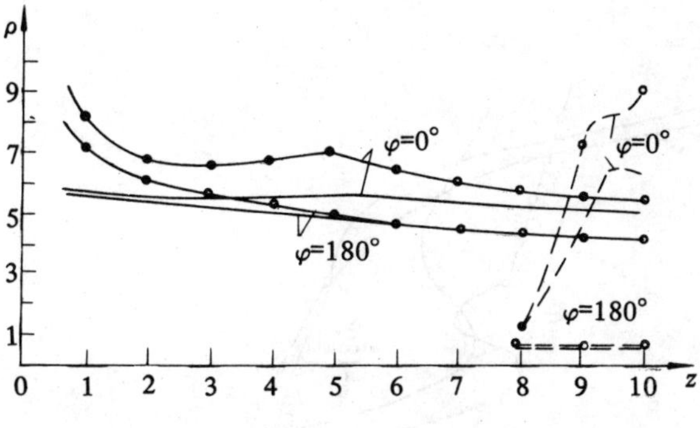

Figure 4.54 Variations of density behind the shock ($M_\infty=20$, $\alpha=10°$, combined body VI)

—— Perfect gas, ∘——∘ Equilibrium air } Main shock, ---- Perfect gas, ∘----∘ Equilibrium air } Secondary shock.

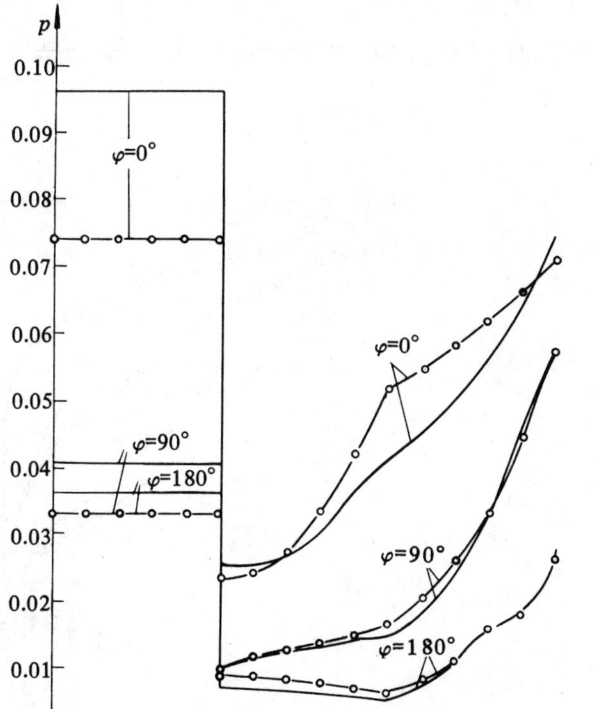

Figure 4.55a Variations of pressure in the ξ-direction ($M_\infty=20$, $\alpha=10°$, combined body VI, $z=8$)

—— Perfect gas, ∘——∘ Equilibrium air.

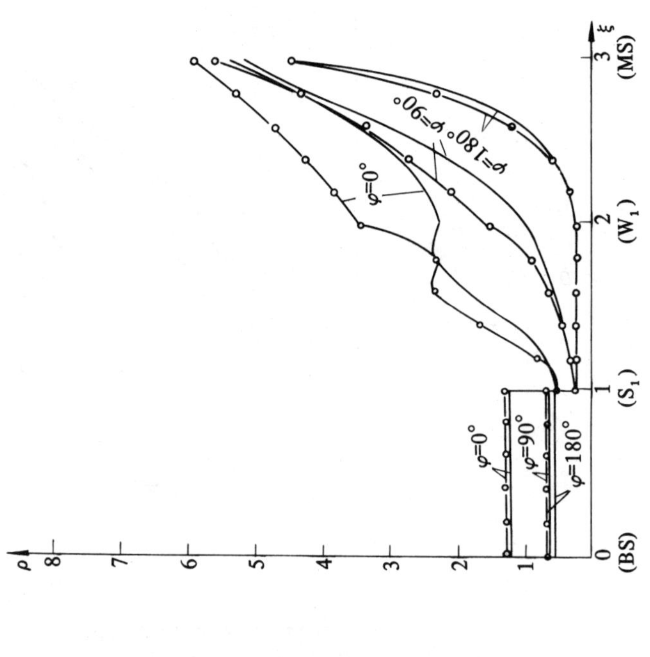

Figure 4.55b Variations of density in the ξ-direction (M_∞=20, α=10°, combined body VI, z=8)
—— Perfect gas, ∘——∘ Equilibrium air.

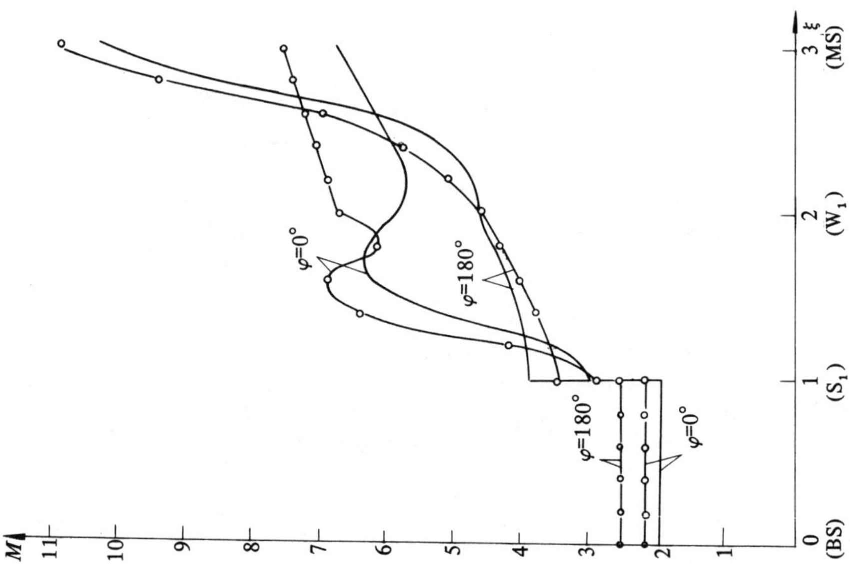

Figure 4.55c Variations of the Mach number in the ξ-direction (M_∞=20, α=10°, combined body VI, z=8)
—— Perfect gas, ∘——∘ Equilibrium air.

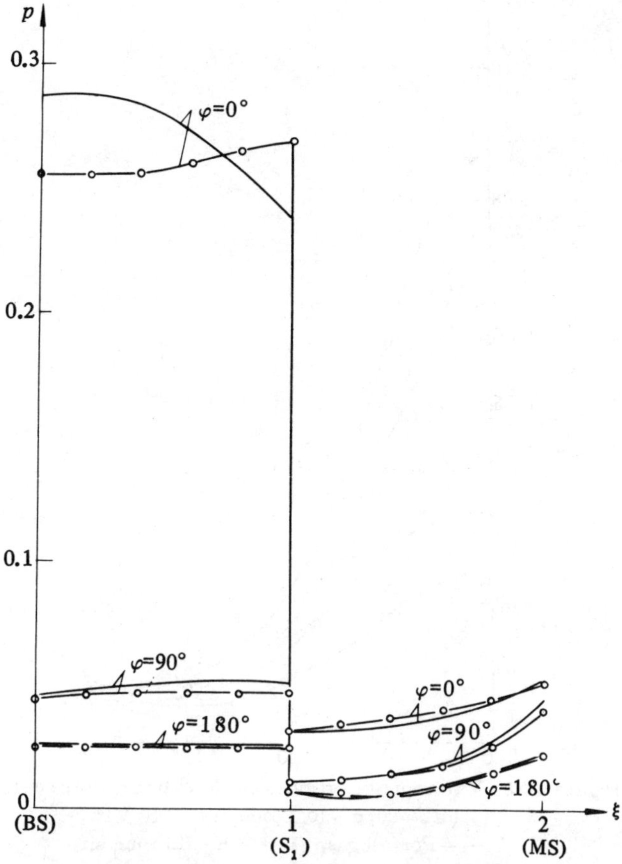

Figure 4.56a Variations of pressure in the ξ-direction
($M_\infty=20$, $\alpha=10°$, combined body VI, $z=10$)
—— Perfect gas, ∘——∘ Equilibrium air.

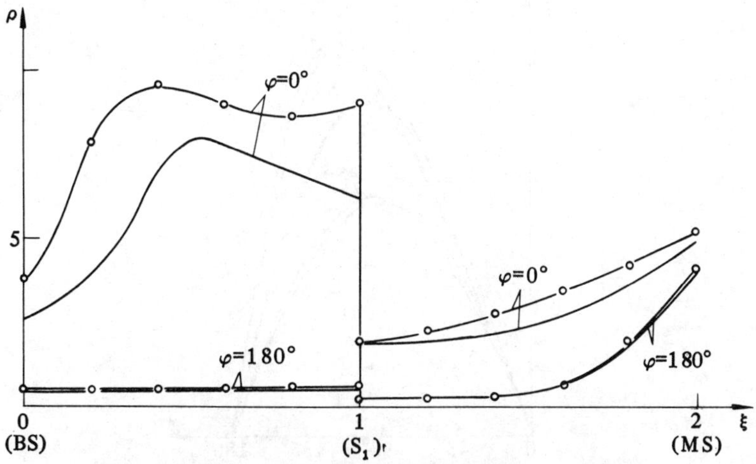

Figure 4.56b Variations of density in the ξ-direction
($M_\infty=20$, $\alpha=10°$, combined body VI, $z=10$)
—— Perfect gas, ∘——∘ Equilibrium air.

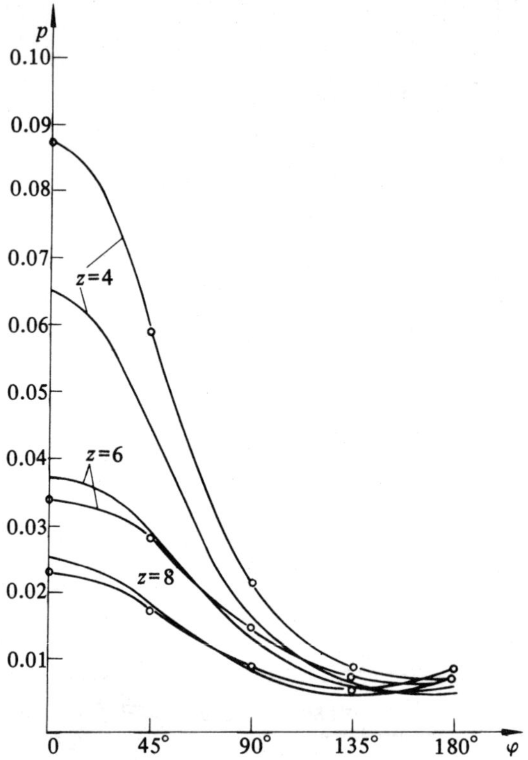

Figure 4.57 Variations of pressure on the cylinder in the φ-direction ($M_\infty=20$, $\alpha=10°$, combined body VI)
—— Perfect gas, ∘——∘ Equilibrium air,

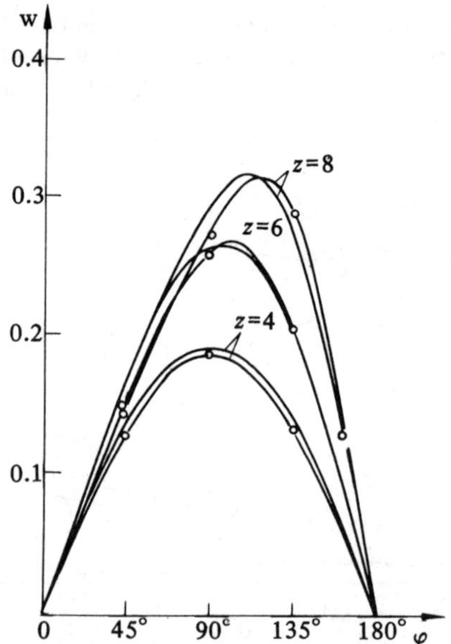

Figure 4.58 Variations of circumferential velocity w on the cylinder in the φ-direction ($M_\infty=20$, $\alpha=10°$, combined body VI)
—— Perfect gas, ∘——∘ Equilibrium air.

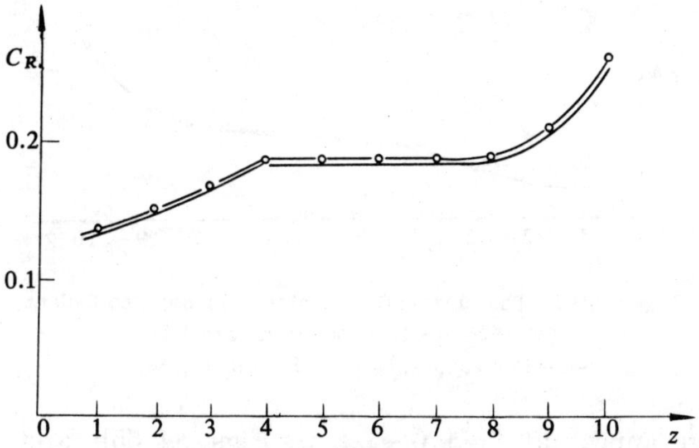

Figure 4.59 The curves of the axial force coefficient
($M_\infty=20$, $\alpha=10°$, combined body VI)
—— Perfect gas, ∘——∘ Equilibrium air.

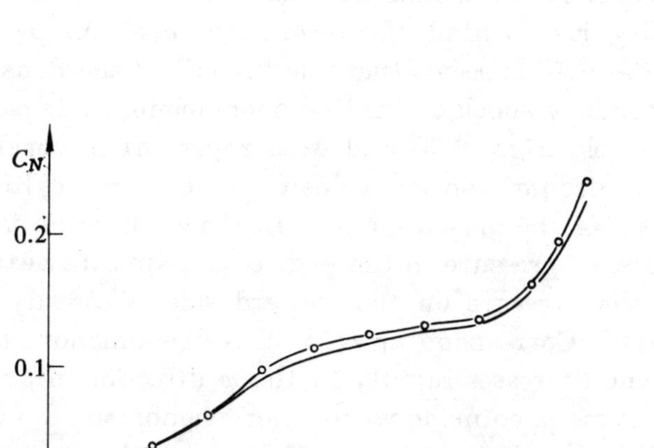

Figure 4.60 The curves of the normal force coefficient
($M_\infty=20$, $\alpha=10°$, combined body VI)
—— Perfect gas, ∘——∘ Equilibrium air.

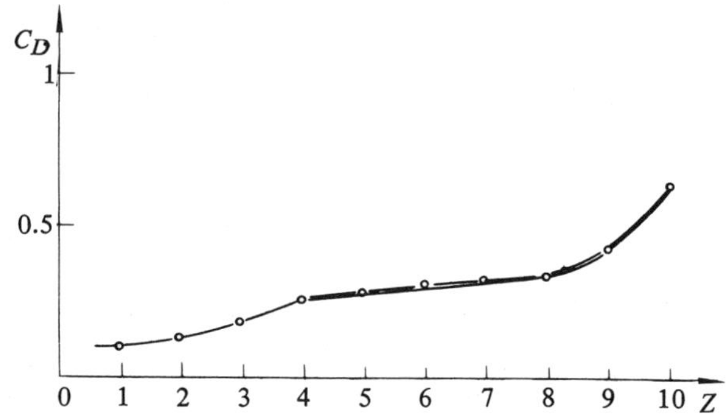

Figure 4.61 The curves of the center of pressure coefficient ($M_\infty=20$, $\alpha=10°$, combined body VI) —— Perfect gas, ∘——∘ Equilibrium air.

number curve appear at $\xi=1.6$—1.8 in Figs. 4.55b and 4.55c, and the "peak" of the density curve appears at $\xi=0.4$—0.6 in Fig. 4.56b. The reason for this phenomenon is as follows. Fig. 4.53 shows that the curve of pressure just behind the main shock on the plane $\varphi=0°$ has a minimal point at $z\approx$, i.e., there is a minimal point of slope of the main shock. Hence there is a minimal point of the entropy, which usually appears on the density curve as a maximal point. We find from Fig. 4.56b that the "peak" of the density curve at $z=10$ is located between the secondary shock and the body. Fig. 4.54 shows that the curve of density just behind the secondary shock on $\varphi=0°$ suddenly bends at $z=9.25$—9.5. It seems that the "peak" of the density curve in front of the secondary shock causes this phenomenon as it passes through the secondary shock. Figs. 4.57 and 4.58 represent several curves of pressure p and circumferential velocity w on the cylinder in the φ-direction. At $z=4$, the pressure curve in the φ-direction is monotonic. Then, an increase of pressure in the φ-direction appears near the leeward side. At $z=8$, the pressure on the leeward side is already close to the pressure at $\varphi=90°$. Corresponding with this phenomenon, w gets large near the body and decreases rapidly in the φ-direction near the leeward side. These phenomena coincide with other phenomena we have seen in the flow around bodies. Figs. 4.59—4.61 represent the curves of the axial force coefficient C_R, the normal force coefficient C_N and the center of pressure coefficient C_D, where we take the area of the base of the body as the reference area and the ratio of the distance between the center of pressure and the vertex of the body to the whole length of the body as the definition of the center of pressure coefficient. These figures show that C_R and

C_N for air in chemical equilibrium are larger than those for the perfect gas. The two values of C_D are almost the same. It is only on the flare that C_D for air in chemical equilibrium is slightly larger than C_D for the perfect gas.

In Figs. 4.62 and 4.63, we give several results of the perfect gas flow around the combined body VII, where $M_\infty=2$ and $\alpha=2°$. The shapes of the main shock, the secondary shock and the body are given in Fig. 4.62, and the pressure curves and the density curves in the ξ–direction in Fig. 4.63. We can see from Fig. 4.63 that, even though the functions are discontinuous on the secondary shock, they are smooth in the region between the main shock and the secondary shock, and in the region between the secondary shock and the body.

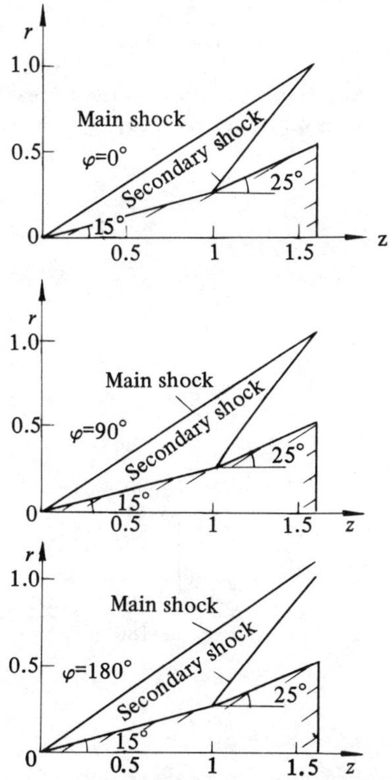

Figure 4.62 Shapes of boundaries ($M_\infty=2$, perfect gas, $\alpha=2°$)
Generator of the combined body VII: straight line $(\theta_0=15°) \xrightarrow[z_1=1]{}$ straight line $(\theta_1=25°)$.

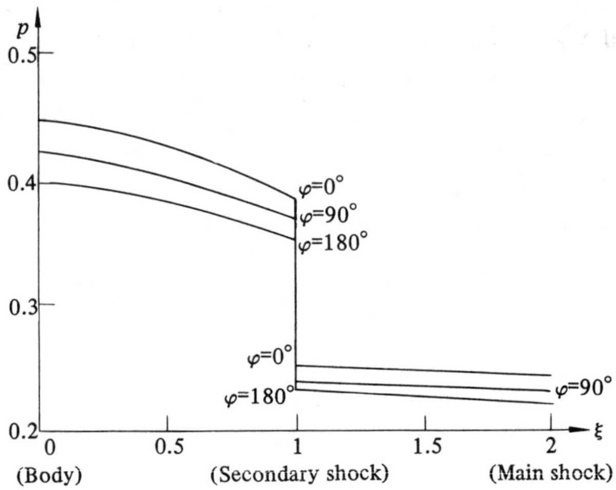

Figure 4.63a Variations of pressure in the ξ-direction ($M_\infty=2$, perfect gas, $\alpha=2°$, combined body VII, $z=1.57$)

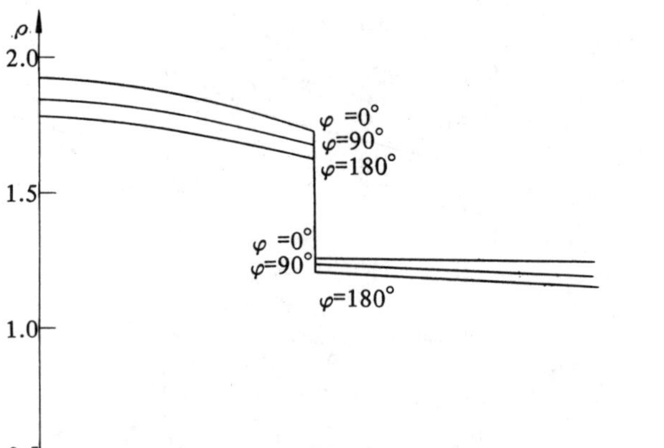

Figure 4.63b Variations of density in the ξ-direction ($M_\infty=2$, perfect gas, $\alpha=2°$, combined body VII, $z=1.57$)

Figs. 4.64—4.67 show several results of the perfect gas flow around the combined body VIII, where $M_\infty=5$. This combined body is different from the preceding bodies. It does not have any edge. However, its generator has two arcs of circles, namely, the compression arc and the expansion arc. In this case, the structure of the flow field is also different from the preceding. There is no centered expansion wave. Moreover, the secondary shock does not start from the body. It is embedded in the flow field. Figs. 4.64 and 4.66 represent the shapes of the main shock and the secondary shock at $\alpha=0°$ and $\alpha=5°$ respectively. In Figs. 4.65 and 4.67, we give the pressure curves in the ξ-direction at $z=4.9$ for $\alpha=0°$ and $5°$ respectively. The location of the embedded secondary shock is automatically determined, and it is taken as an internal boundary in the process of calculation. There is no oscillation on our computed curves (see Figs. 4.65 and 4.67). This means that we have evaluated the embedded shocks with great success.

Besides the flow around axisymmetric bodies, we have also evaluated the flow around bodies with elliptic cross sections. Let the axis of the body be the z-axis of a cylindrical coordinate system $\{z, r, \varphi\}$, one of the semi-axes of the elliptic cross section always be on the plane $\varphi=\beta_0$, and the other always be on the plane $\varphi=\beta_0+90°$. Furthermore, let the function of the generator on the plane $\varphi=\beta_0$ be $r=f_1(z)$, and that on the plane $\varphi=\beta_0+90°$ be $r=f_2(z)$. Then the equation for shape of the body with elliptic cross sections is

$$\frac{r^2\cos^2(\varphi-\beta_0)}{f_1^2(z)}+\frac{r^2\sin^2(\varphi-\beta_0)}{f_2^2(z)}=1.$$

Figs. 4.68 and 4.69 represent several results for perfect gas flow around the combined body IX with elliptic cross sections, whose generator on the plane $\varphi=0°$ is $r=f_1(z)$ and on the plane $\varphi=90°$ is $r=f_2(z)$. The generator on the plane $\varphi=0°$, composed of several sections of straight lines and circular arcs, is the same as that of the combined body VI. The generator on the plane $\varphi=90°$, composed of several sections of straight lines, circular arcs and elliptic arcs, is different from that of the combined body VI. We take $f_1(z)\geqslant f_2(z)$, so the cross section of this combined body is flatter than the cross section of the combined body VI at $\varphi=90°$. In Figs. 4.68a and 4.68b, we draw the shapes of the main shock and the secondary shock on the planes $\varphi=0°$ and $\varphi=90°$, where $M_\infty=20$, $\alpha=0°$. We also give the corresponding results of the combined body VI in those figures. We can find from those figures that the two shocks are very close on the plane $\varphi=0°$, and that the distance from the shock corresponding to the combined body IX to the z-axis is shorter than the distance from the shock corresponding to the combined body VI to the z-axis. Fig. 4.69 represents the pressure distribution on the

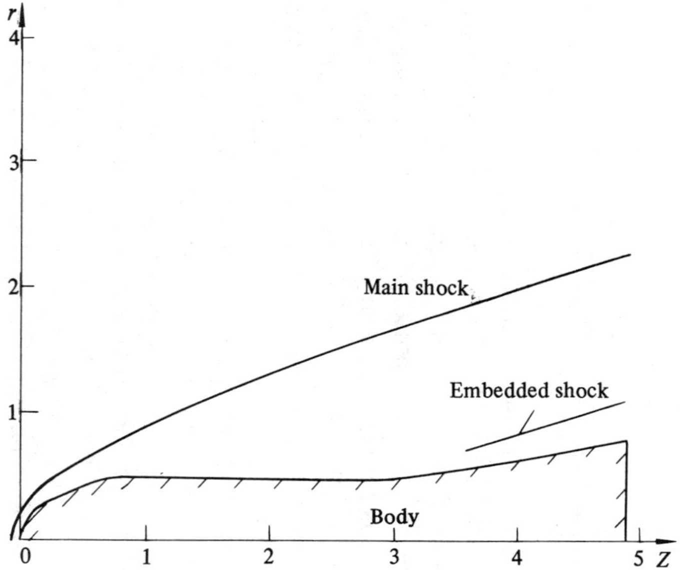

Figure 4.64 Shapes of the main shock and the embedded shock ($M_\infty=5$, $\alpha=0°$, perfect gas)

Generator of the combined body VIII: circular arc ($R_0=0.3$) $\xrightarrow[z_1=0.173]{}$ straight line ($\theta_1=25°$) $\xrightarrow[z_2=0.522]{}$ circular arc ($R_2=0.7$) $\xrightarrow[z_3=0.818]{}$ straight line ($\theta_3=0°$) $\xrightarrow[z_4=2.85]{}$ circular arc ($R_4=4$) $\xrightarrow[z_5=3.54]{}$ straight line ($\theta_5=10°$).

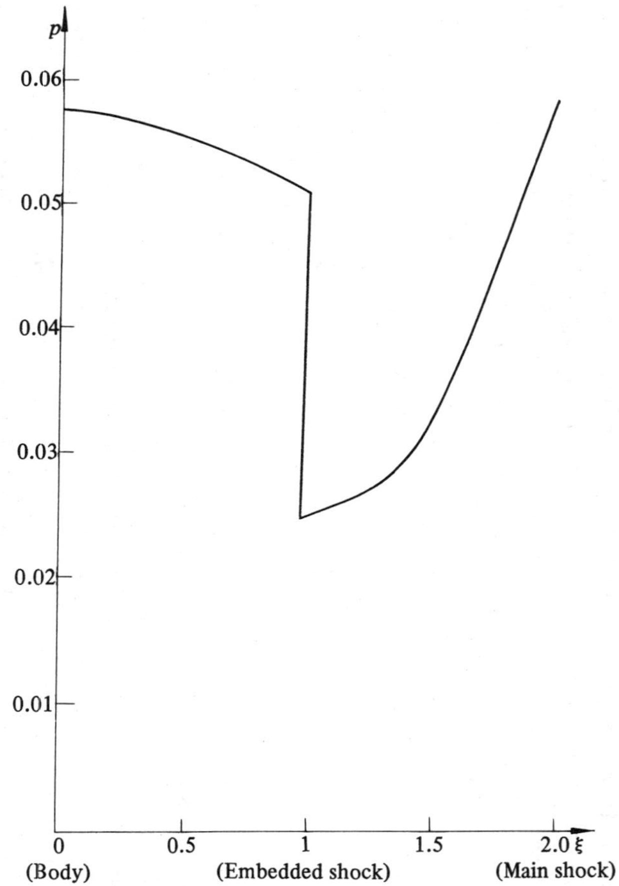

Figure 4.65 Variations of pressure in the ξ-direction ($M_\infty=5$, perfect gas, $\alpha=0°$, combined body VIII, $z=4.9$)

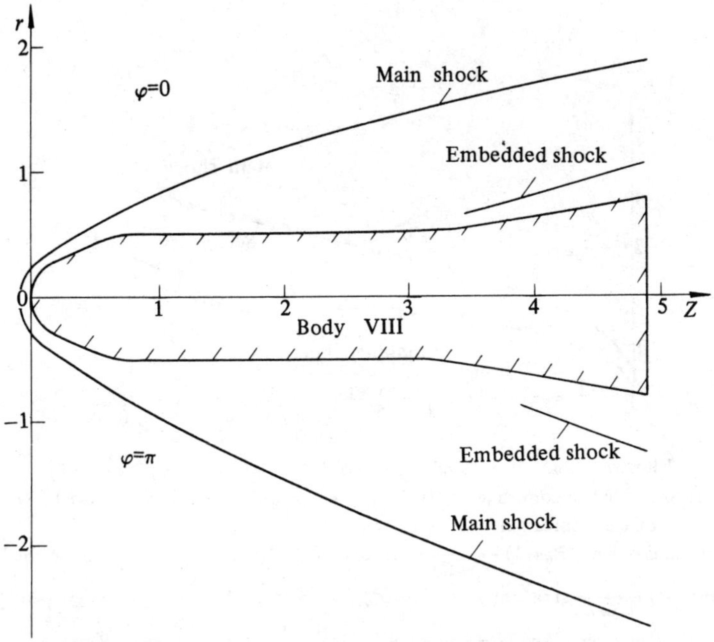

Figure 4.66 Shapes of the main shock and the embedded shock (M_∞=5, perfect gas, α=5°, combined body VIII)

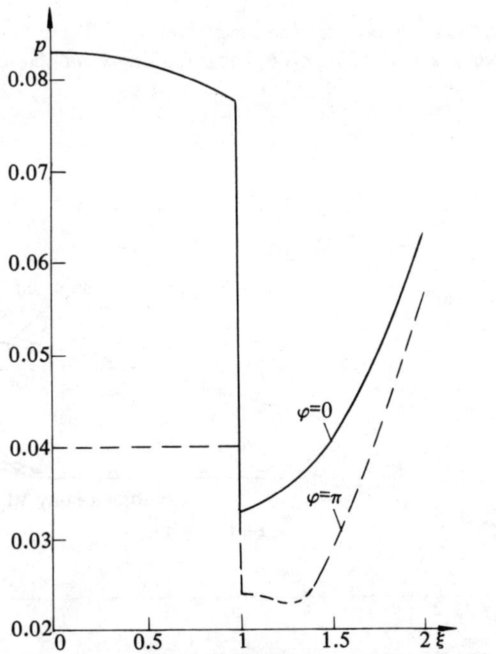

Figure 4.67 Variations of pressure in the ξ-direction (M_∞=5, perfect gas, α=5°, combined body VIII, z=4.9)

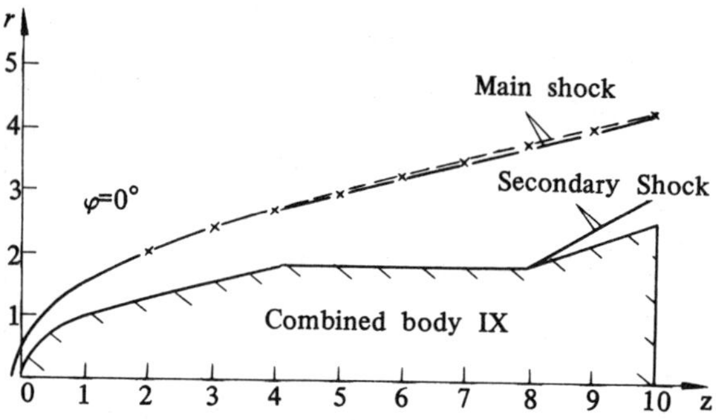

Figure 4.68a Shapes of the shock ($M_\infty=20$, perfect gas, $\alpha=0°$)
—— Result for the combined body IX, ×—× Result for the combined body VI.
Generator of the combined body IX:

1. Circular arc ($R_0=1$) $\xrightarrow[\theta_1=30°]{}$ circular arc($R_1=1$) $\xrightarrow[\text{(tangent)}]{}$ straight line ($\theta_2=15°$) $\xrightarrow[z_3=4]{}$ straight line ($\theta_3=0°$) $\xrightarrow[z_4=8]{}$ straight line($\theta_4=20°$) [$\beta_0=0°$].
2. Circular arc ($R_0=1$) $\xrightarrow[\theta_1=30°]{}$ elliptical arc ($a_1=1$, $b_1=1.12$) $\xrightarrow[\text{(tangent)}]{}$ straight line ($\theta_2=14°$) $\xrightarrow[z_3=4]{}$ straight line ($\theta_3=0°$) $\xrightarrow[z_4=8]{}$ straight line ($\theta_4=18°$), where "circular arc ($R_0=1$) $\xrightarrow[\theta_1=30°]{}$ elliptical arc ($a_1=1$, $b_1=1.12$)" represents that the angle between the common tangent line of the two arcs and the z—axis is equal to 30°, a_1, b_1 being the lengths of the semi-axes.

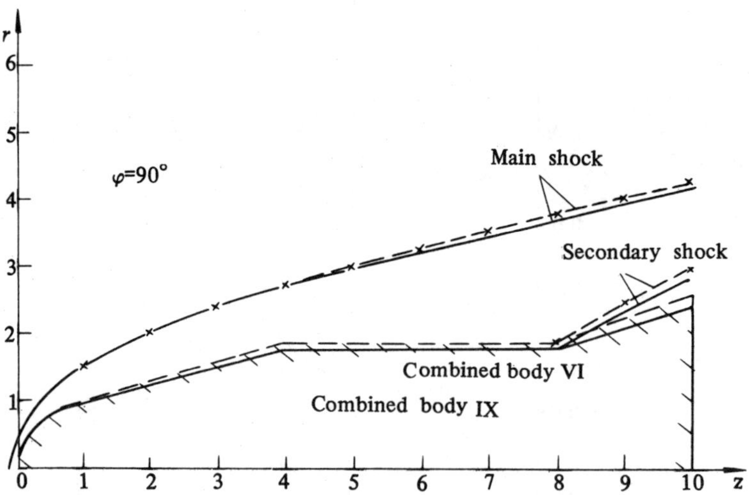

Figure 4.68b Shapes of the shock ($M_\infty=20$, perfect gas, $\alpha=0°$)
—— Result of the combined body IX,
×—× Result of the combined body VI.

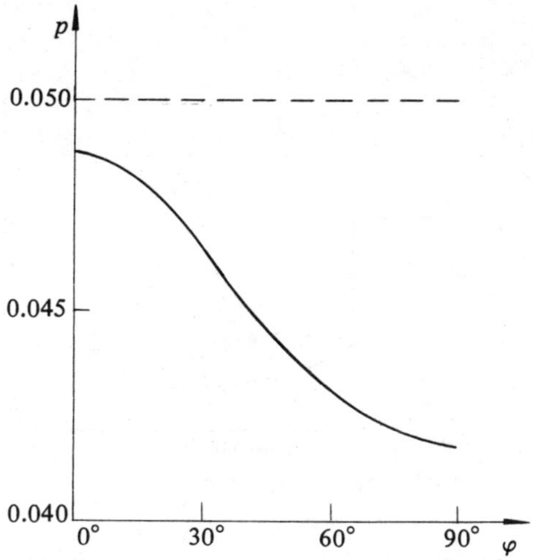

Figure 4.69 Variations of pressure on two bodies
($M_\infty=20$, perfect gas, $\alpha=0°$, $z=10$)
—— Result of the combined body IX,
---- Result of the combined body VI.

body in the φ–direction at $z=10$. The results in that figure show that the shape of the body greatly influences the pressure.

Figs. 4.70 and 4.71 represent the pressure distribution on the body along the axis–direction in the perfect gas flow around the combined body X, where $M_\infty=6.8$, $\alpha=0°$, $3°$. For comparison, the experimental data[59] are also given. Figs. 4.38, 4.39 and 4.72, 4.73 demonstrate the pressure distributions on the body in the axis–direction and in the φ–direction in the perfect gas flow around the combined body IV. The experimental data[60] are also given in order to make a comparison. We can see from these figures that the computed results agree with the experimental data if the angle of attack is small. However. the difference between the computed results and the experimental data in the region near edges is greater than that in other regions. If the angle of attack is larger, obvious difference also appears on the leeward side, for example, in the region $z\geqslant7$, $90°\leqslant\varphi\leqslant180°$ in the flow around the combined body IV ($M_\infty=4$, $\alpha=6.19°$). This is because we do not consider the effect of viscosity in our computation, even though the influence of viscosity in those regions is rather great. In the computed results, pressure goes up rapidly with increase of φ in the region near the leeward plane. In the experimental data, pressure goes up slowly in that region because of the influence of viscosity.

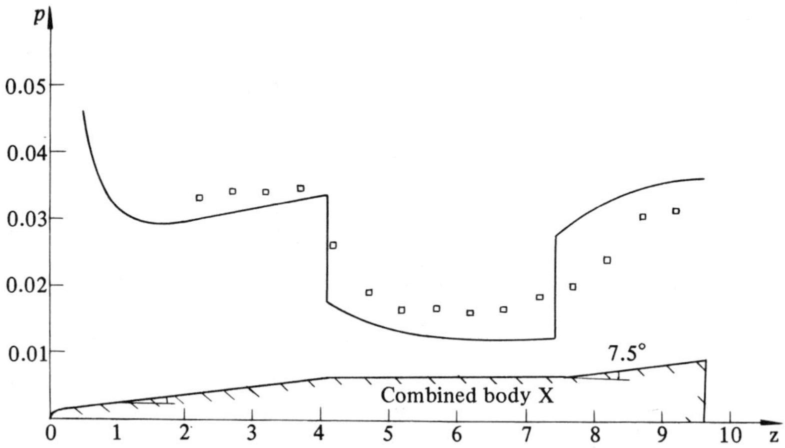

Figure 4.70 Pressure distributions on the combined body X ($M_\infty=6.8$, perfect gas, $\alpha=0°$)
—— Computed result, ⊡ Experimental data[59].
Generator of the combined body X: circular arc ($R_0=0.11428$) $\xrightarrow[\text{(tangent)}]{}$ straight line ($\theta_1=7.5°$) $\xrightarrow[z_2=4.1]{}$ straight line ($\theta_2=0°$) $\xrightarrow[z_3=7.45]{}$ straight line ($\theta_3=7.5$).

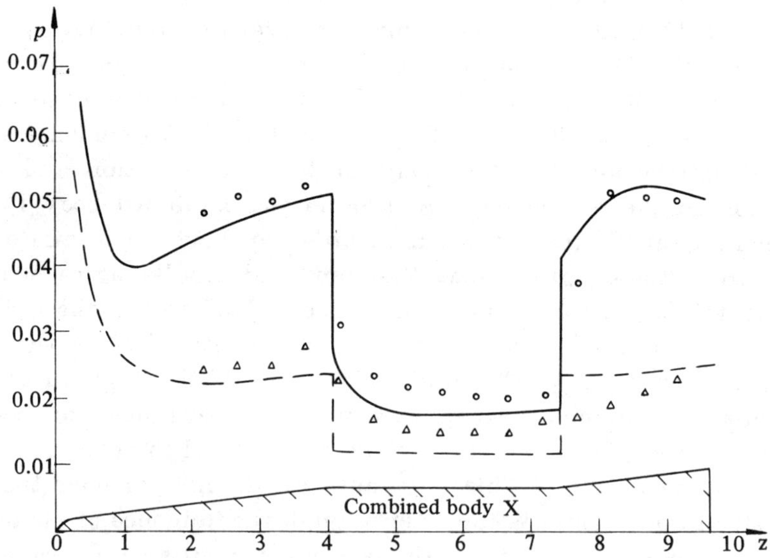

Figure 4.71 Pressure distributions on the combined body X ($M_\infty=6.8$, perfect gas, $\alpha=3°$)
—— $\varphi=0$, - - - - $\varphi=\pi$ Computed result, ○ $\varphi=0$, △ $\varphi=\pi$ Experimental data[59].

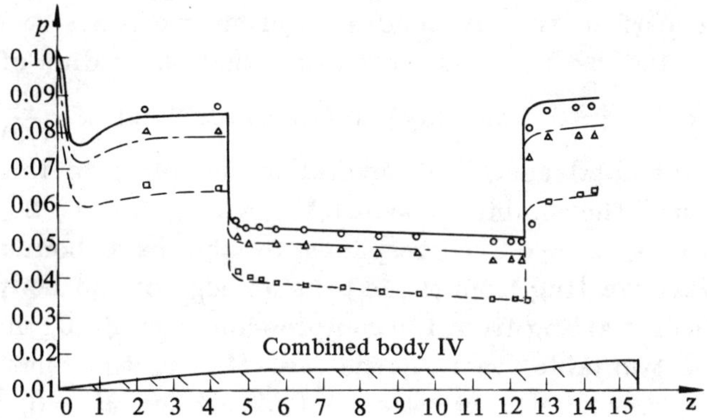

Figure 4.72 Pressure distributions on the combined body IV ($M_\infty=4$, $\alpha=6.19°$)

—— $\varphi=0°$; – · – $\varphi=30°$; – – – $\varphi=60°$ } Computed result, ○ $\varphi=0°$; △ $\varphi=30°$; □ $\varphi=60°$ } Experimental data[60].

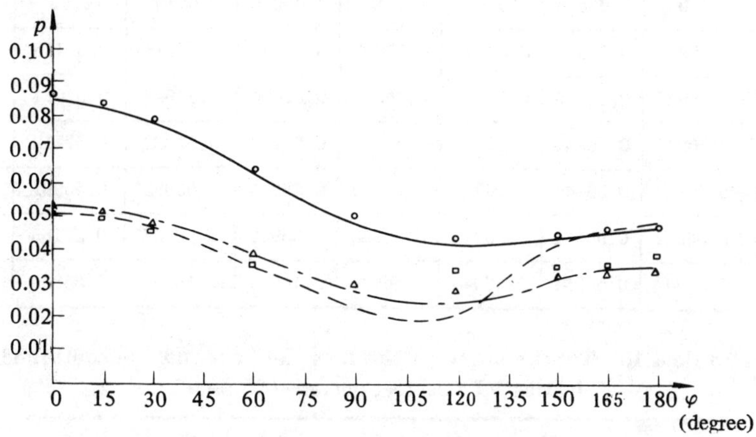

Figure 4.73 Pressure distributions on the combined body IV in the φ–direction ($M_\infty=4$, $\alpha=6.19°$)

—— $z=4.368$; – · – $z=5.868$; – – – $z=9.618$ } Computed result, ○ $z=4.368$; △ $z=5.868$; □ $z=9.618$ } Experimental data[60].

Because of the uniformity of free stream and the conservation of energy, the total energy per unit mass of gas $H=h(p, \rho)+\frac{1}{2}V^2$ should be equal to the value of free stream everywhere. Hence we can estimate the accuracy of the computed results by calculating the total energy at mesh-points. In Table 4.14 we list the total energy at a series of mesh-points in the perfect gas flow field around sphere-cones, where $M_\infty=6$, $\alpha=0°$, $\beta=0°$, $10°$, $z=5$, 20, 40 (supposing that the radius of the sphere R_0 is equal to 1), $\xi=0$ (the body), 1 (the shock) and $\Delta\xi=\frac{1}{40}$. In Table 4.15 we list the total energy at several mesh-points in the perfect gas flow field around the combined bodies VI, where $M_\infty=6$, 20, $\alpha=0°$, $z=10$, and $\xi=0$, 1, 2. $\xi=2$ corresponds to the back boundary of the expansion wave starting from the expansion edge of the body; $\xi=1$, the secondary shock starting from the compression edge; $\xi=0$, the body. We see from these two tables that in the case $M_\infty=6$, the relative error of the total energy, whose exact value is 0.56944 for $M_\infty=6$, is less than 0.5% everywhere and that, in the case $M_\infty=20$, the maximal relative error of the total energy, whose exact value is 0.50625 for $M_\infty=20$, is 1%. We can also see from Table 4.14 that the accuracy of the total energy on the cone in the flow field around sphere-cones is high, and that the error

Table 4.14 Total energy in the flow field around sphere-cones
($M_\infty=6$, perfect gas, $\alpha=0°$)

β	z	ξ							
		1.0	0.8	0.6	0.4	0.2	0.1	0.05	0
0°	5	0.56944	0.56945	0.56949	0.57005	0.56949	0.56946	0.57178	0.57015
	20	0.56944	0.56945	0.56946	0.56943	0.57004	0.57048	0.57024	0.57015
	40	0.56944	0.56945	0.56945	0.56945	0.56935	0.57032	0.57112	0.57015
10°	5	0.56944	0.56945	0.56944	0.56952	0.56975	0.56992	0.56991	0.57019
	20	0.56944	0.56944	0.56944	0.56945	0.56939	0.56957	0.57163	0.57019
	40	0.56944	0.56944	0.56944	0.56944	0.56944	0.56932	0.56889	0.57019

Table 4.15 Total energy in the flow field around the combined bodies VI (perfect gas, $\alpha=0°$, $z=10$)

M_∞	ξ				
	2.0	1.5	1.0	0.5	0
6	0.56932	0.56919	0.56909	0.57104	0.56997
20	0.50671	0.50612	0.50720	0.50862	0.50652

there comes from the initial values. For general combined bodies if all the discontinuities are accurately treated, and if a proper small mesh size is taken in the region where the flow parameters vary rapidly, then the error of total energy is quite small.

Lastly, we shall give several results about the interaction between discontinuities in the three-dimensional flow around bodies. Suppose that two discontinuities $r=F_0(z, \varphi)$ and $r=F_1(z, \varphi)$ intersect in the computational region, and that the equations of the intersection curve are $z=z^*(\varphi)$ and $r=r^*(\varphi)$, $0\leqslant\varphi\leqslant 2\pi$. Clearly, if we take

$$E(r, \eta, \varphi)=z_0(\varphi)+\frac{\eta-z_0(0)}{z_1(0)-z_0(0)}(z_1(\varphi)-z_0(\varphi))$$
$$+\left[z_{r,0}(\varphi)+\frac{\eta-z_0(0)}{z_1(0)-z_0(0)}(z_{r,1}(\varphi)-z_{r,0}(\varphi))\right]r,$$

where the functions $z_0(\varphi)$, $z_{r,0}(\varphi)$, $z_{r,1}(\varphi)$ are properly chosen by certain requirements, and $z_1(\varphi)=z^*(\varphi)-z_{r,1}(\varphi)r^*(\varphi)$, then the marching surface $\eta=z_1(0)$ always passes through the intersection curve. Hence two discontinuities intersect wholly on a marching surface, and we are able to evaluate this problem by directly using the method of § 3. It is apparent that the functions $z^*(\varphi)$ and $r^*(\varphi)$ are unknown. Therefore, it is impossible to select an $E(r, \eta, \varphi)$ in advance such that the curve of intersection is located on the surface with $\eta=z_1(0)$. However, this can be realized by a "track" method, i.e., in our process we often evaluate the approximate curve of intersection and select $E(r, \eta, \varphi)$ so as to locate the curve of intersection on the surface with $\eta=z_1(0)$. When two discontinuities tend towards their curve of intersection, the approximate curve of intersection tends towards the accurate one. Hence by the "track" method the curve of intersection will be accurately located on the surface with $\eta=z_1(0)$ when the two discontinuities intersect. In passing, we should point out that as E has the above form, the marching surfaces $\eta=$constant are ruled surfaces, which intersect the planes $\varphi=$constant always on a straight line.

Figs. 4.74—4.76 show several results about the perfect gas flow around the combined body XI. Fig. 4.74 represents the structure of the flow field on the windward plane and the leeward plane. On the windward plane, the first secondary shock intersects the main shock at $z=8.87$, and the second secondary shock intersects the contact discontinuity at $z=30.69$; on the leeward plane the first secondary shock and the main shock intersect at $z=12.75$, and the second secondary shock and the contact discontinuity intersect at $z=31.95$. That is, the intersection on the leeward plane happens much later than that on the windward plane. Fig. 4.75 indicates the shapes of boundaries on three cross sections $z=8$, 10, 13, located, respectively, on the left-hand side, on the inside

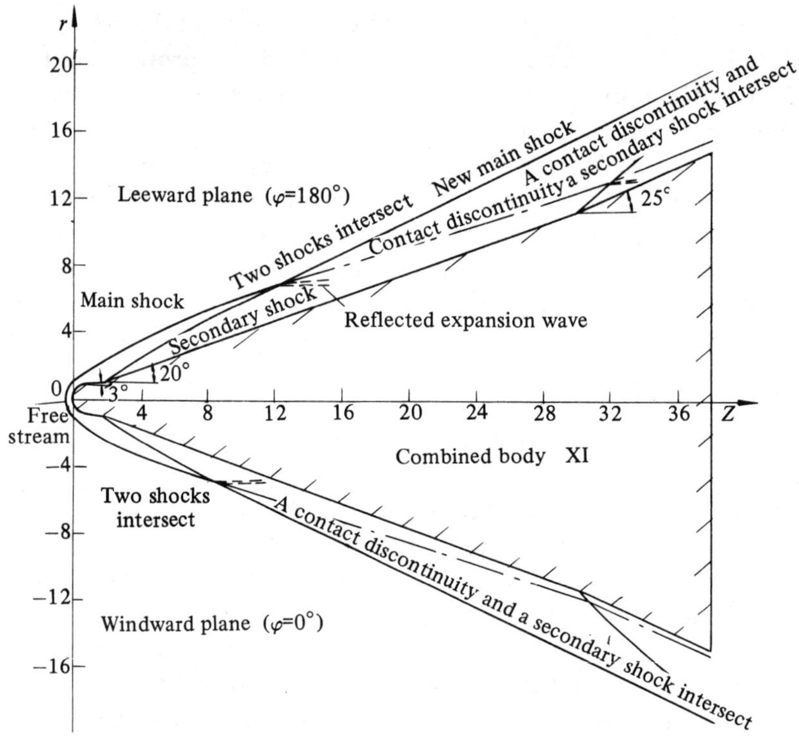

Figure 4.74 Construction of the flow field on the windward and leeward planes ($M_\infty=4$, $\alpha=2°$, perfect gas, $\gamma=1.4$)

Generator of the combined body XI: circular arc ($R_0=1$) $\xrightarrow[\text{(tangent)}]{}$ straight line ($\theta_1=3°$) $\xrightarrow[z_2=2]{}$ straight line ($\theta_2=20°$) $\xrightarrow[z_3=30]{}$ straight line ($\theta_3=25°$).

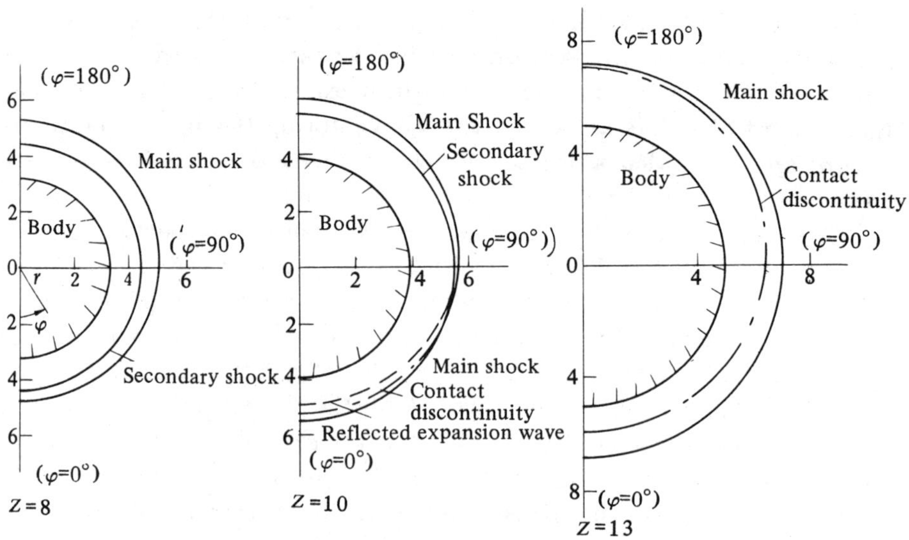

Figure 4.75 Shapes of boundaries on the cross sections $z=8$, 10, 13 ($M_\infty=4$, $\alpha=2°$, perfect gas, $\gamma=1.4$, combined body XI)

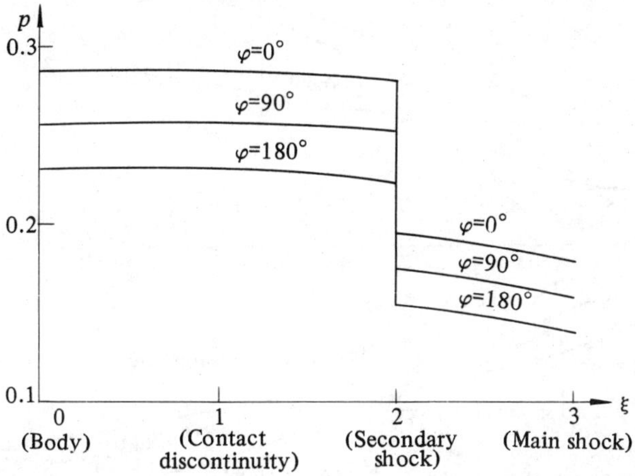

Figure 4.76a Pressure curves in the ξ-direction ($M_\infty=4$, $\alpha=2°$, perfect gas, $\gamma=1.4$, combined body XI, $z=34$)

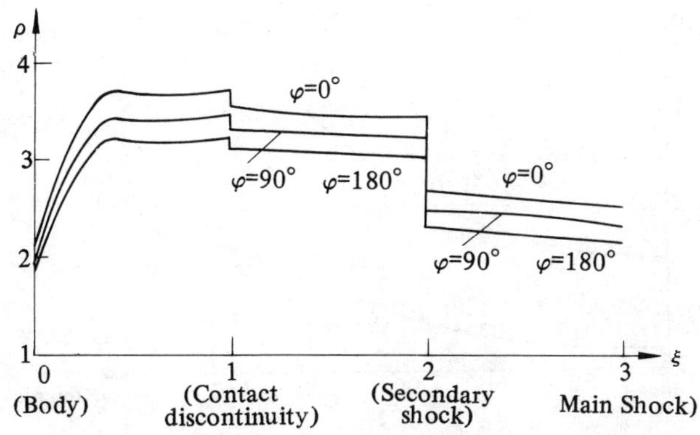

Figure 4.76b Density curves in the ξ-direction ($M_\infty=4$, $\alpha=2°$, perfect gas, $\gamma=1.4$, combined body XI, $z=34$)

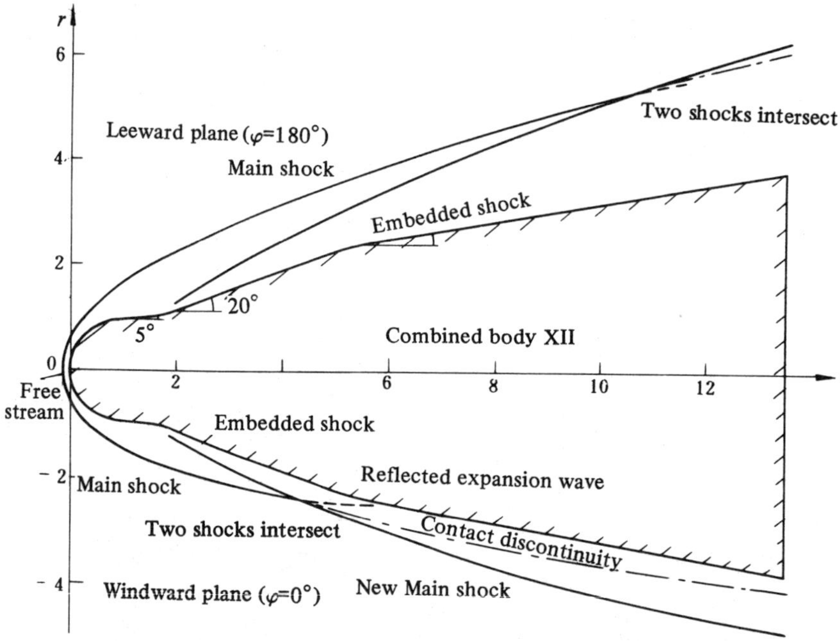

Figure 4.77 Construction of the flow field on the windward and leeward planes ($M_\infty=15$, $\alpha=6°$, perfect gas, $\gamma=1.4$)

Generator of the combined body XII: circular arc ($R_0=1$) $\xrightarrow[\text{(tangent)}]{}$ straight line ($\theta_1=5°$) $\xrightarrow[z_2=1.5]{}$ circular arc ($R_2=2$) $\xrightarrow[\text{(tangent)}]{}$ straight line ($\theta_3=20°$) $\xrightarrow[z_4=5]{}$ circular arc ($R_4=4$) $\xrightarrow[\text{(tangent)}]{}$ straight line ($\theta_5=10°$).

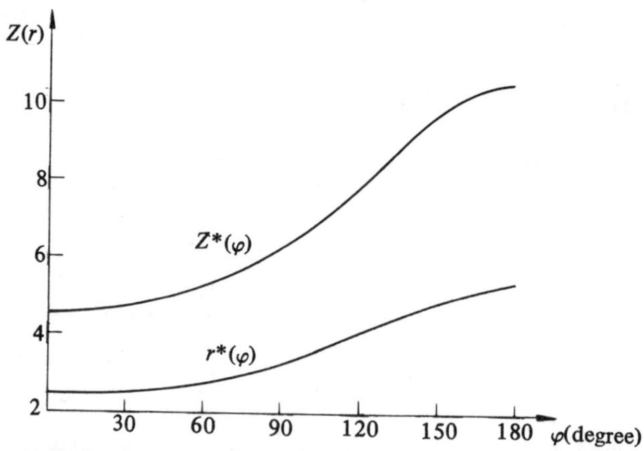

Figure 4.78 Curves of the equations $z=z^*(\varphi)$ and $r=r^*(\varphi)$ of the spatial curve at which the main shock intersects the embedded shock ($M_\infty=15$, $\alpha=6°$, perfect gas, $\gamma=1.4$, combined body XII)

and on the right-hand side of the region where the first secondary shock intersects the main shock. On the cross section $z=10$, near the windward side the first secondary shock has intersected the main shock, and the new main shock, the contact discontinuity and the expansion wave have been reflected. But near the leeward side, the first secondary shock has not intersected the main shock, so the structures of flow in these two regions are different. Fig. 4.76 shows several pressure curves and density curves in the ξ-direction on the plane $z=34$, on which there is no oscillation.

Figs. 4.77, 4.78 give several results about the perfect gas flow around the combined body XII for $M_\infty=15$, $\alpha=6°$. This combined body does not have any edges, but has a surface which presses continuously on the flow to cause an embedded shock formed in the flow near the surface. Fig. 4.77 shows the structure of the flow field on the windward and the leeward planes. As we can see from Fig. 4.77, the embedded shock intersects the main shock on a spatial curve. Fig. 4.78 demonstrates curves for the equations $z=z^*(\varphi)$ and $r=r^*(\varphi)$ which give a parametric representation of the spatial curve.

We see from the above results that the method employed here has solved the interaction between three-dimensional discontinuities with great success.

Appendix

A Numerical Method with High Accuracy for Calculating the Interactions between Discontinuities in Three Independent Variables

Abstract

The main purpose of this paper is to present a complete numerical method with high accuracy for calculating the interactions between discontinuities of hyperbolic systems in three independent variables. The method has been applied to the calculation of the interactions between two shocks and between a shock and a contact discontinuity in three-dimensional steady flows. This paper gives some results of the calculations.

The secondary purpose is to present a method of calculating embedded shocks accurately and automatically. The method is based on the characteristic theory. We have applied the method to calculating the embedded shocks in three-dimensional steady flows. This paper gives one numerical result of embedded shocks.

Introduction

In [1] we have accurately calculated the interactions between discontinuities and the embedded shocks in two independent variables. In three independent variables, these problems are more complicated. This paper presents a complete numerical method with high accuracy for calculating the interactions between discontinuities and the embedded shocks in three independent variables.

The contents contain the following:

(1) We briefly discuss a method of solving a class of initial–boundary–value problems for hyperbolic systems in three independent variables.

(2) We discuss how to use a curved–surface–march method of making two discontinuities intersect simultaneously in our calculating procedures so that our difference scheme can be used to solve the interactions between discontinuities completely and accurately.

(3) We briefly describe a numerical method of solving "a Riemann problem in three independent variables".

(4) We point out the procedure of forming shocks in three independent variables and present an accurate method of calculating embedded shocks.

(5) We give some results about three–dimensional steady flows.

We did this work on the basis of [1—3]. Therefore, for details of this method, the reader is referred to [1—3] and [4].

1. *A Numerical Method for Initial–Boundary–Value Problems*

We consider the following quasilinear hyperbolic system with three independent variables

$$\begin{cases} \overline{A}\dfrac{\partial \overline{U}}{\partial z}+\overline{B}\dfrac{\partial \overline{U}}{\partial r}+\overline{C}\dfrac{\partial \overline{U}}{\partial \varphi}=\overline{F}, \\ \bar{r}_{l-1}(z, \varphi)\leqslant r\leqslant \bar{r}_l(z, \varphi), \ l=1, 2, \cdots, L, \\ 0\leqslant\varphi\leqslant 2\pi, \ z\geqslant E_0(r, \varphi), \end{cases} \tag{1.1}$$

where z, r and φ are independent variables in the physical space; $\overline{U}$ and $\overline{F}$ are N–dimensional vectors; $\overline{A}$, $\overline{B}$ and $\overline{C}$ are matrices of order N; $r=\bar{r}_l(z, \varphi)$, $l=0, 1, \cdots, L$ are the boundary surfaces; $z=E_0(r, \varphi)$ is the initial surface.

In order to calculate boundary points conveniently, it is commonly required that the boundary surfaces be coordinate ones. Moreover several special curves must be located on several coordinate surfaces. To fulfill these requirements, we introduce three new independent variables η, ξ, ϕ in the following way:

$$\begin{cases} z=E(\eta,\ r,\ \varphi)^{1)}, \\ \xi=\dfrac{r-\bar{r}_{l-1}(z,\ \varphi)}{\bar{r}_l(z,\ \varphi)-\bar{r}_{l-1}(z,\ \varphi)}+l-1, \\ \quad \text{when } \bar{r}_{l-1}(z,\ \varphi)\leqslant r\leqslant \bar{r}_l(z,\ \varphi),\ 0\leqslant\varphi\leqslant 2\pi,\ l=1,\ 2,\ \cdots,\ L, \\ \phi=\varphi, \end{cases} \tag{1.2}$$

where every surface $z=E(\eta,\ r,\ \varphi)$ is a space-like surface, and $E(0,\ r,\ \varphi)=E_0(r,\ \varphi)$.

Let

$$\boldsymbol{U}(\eta,\ \xi,\ \phi)\equiv\overline{\boldsymbol{U}}(z(\eta,\ \xi,\ \phi),\ r(\eta,\ \xi,\ \phi),\ \phi),$$

and $r=r_l(\eta,\ \phi)$ denote the relation which is determined by $r=\bar{r}_l(E(\eta,\ r,\ \phi),\ \phi)$. Then it is easy to rewrite (1.1) as the following equation with the independent variables $\eta,\ \xi,\ \phi$:

$$\begin{cases} A\dfrac{\partial \boldsymbol{U}}{\partial\eta}+B\dfrac{\partial \boldsymbol{U}}{\partial\xi}+C\dfrac{\partial \boldsymbol{U}}{\partial\phi}=\boldsymbol{F}, \\ \eta\geqslant 0,\ l-1\leqslant\xi\leqslant l,\ l=1,\ 2,\ \cdots,\ L,\ 0\leqslant\phi\leqslant 2\pi, \end{cases} \tag{1.3}$$

where

$$\begin{cases} A(\boldsymbol{U},\ \boldsymbol{R},\ \boldsymbol{R}_\eta,\ \boldsymbol{R}_\phi,\ \eta,\ \xi,\ \phi)=\dfrac{\partial\eta}{\partial z}\bar{A}+\dfrac{\partial\eta}{\partial r}\bar{B}+\dfrac{\partial\eta}{\partial\phi}\bar{C}, \\ B(\boldsymbol{U},\ \boldsymbol{R},\ \boldsymbol{R}_\eta,\ \boldsymbol{R}_\phi,\ \eta,\ \xi,\ \phi)=\dfrac{\partial\xi}{\partial z}\bar{A}+\dfrac{\partial\xi}{\partial r}\bar{B}+\dfrac{\partial\xi}{\partial\phi}\bar{C}, \\ C(\boldsymbol{U},\ \boldsymbol{R},\ \eta,\ \xi,\ \phi)=\bar{C}(\overline{\boldsymbol{U}},\ z,\ r,\ \phi), \\ \boldsymbol{F}(\boldsymbol{U},\ \boldsymbol{R},\ \eta,\ \xi,\ \phi)=\overline{\boldsymbol{F}}(\overline{\boldsymbol{U}},\ z,\ r,\ \phi), \\ \boldsymbol{R}=\begin{bmatrix} r_0 \\ r_1 \\ \vdots \\ r_L \end{bmatrix},\ \boldsymbol{R}_\eta=\begin{bmatrix} r_{0\eta} \\ r_{1\eta} \\ \vdots \\ r_{L\eta} \end{bmatrix},\ \boldsymbol{R}_\phi=\begin{bmatrix} r_{0\phi} \\ r_{1\phi} \\ \vdots \\ r_{L\phi} \end{bmatrix}, \\ r_{l\eta}=\dfrac{\partial r_l}{\partial\eta},\quad r_{l\phi}=\dfrac{\partial r_l}{\partial\phi}. \end{cases} \tag{1.4}$$

Now, we may consider the following initial-boundary-value problem in the new independent variables:

(1) In the L regions $\eta\geqslant 0,\ l-1\leqslant\xi\leqslant l,\ 0\leqslant\phi\leqslant 2\pi,\ l=1,\ 2,\ \cdots,\ L$, equation (1.3) is given.

(2) On the initial surface $\eta=0$, we give the initial conditions

$$\begin{cases} \boldsymbol{U}(0,\ \xi,\ \phi)=\boldsymbol{D}_l(\xi,\ \phi),\ l-1\leqslant\xi\leqslant l,\ 0\leqslant\phi\leqslant 2\pi,\ l=1,\ 2,\ \cdots,\ L, \\ \boldsymbol{R}(0,\ \phi)=\boldsymbol{C}_0(\phi),\ \boldsymbol{R}_\eta(0,\ \phi)=\boldsymbol{C}_1(\phi),\ 0\leqslant\phi\leqslant 2\pi, \end{cases} \tag{1.5}$$

where $\boldsymbol{C}_0$ and $\boldsymbol{C}_1$ are $(L+1)$-dimensional vectors.

(3) On the boundaries and the internal boundaries, the boundary and internal boundary conditions are given:

1) $E(\eta,\ r,\ \varphi)$ will be given according to our special needs.

$$\begin{cases} \boldsymbol{H}_0(\boldsymbol{U}_{0^+}, r_0, r_{0\eta}, r_{0\phi}, \eta, \phi)=0, \\ \boldsymbol{H}_l(\boldsymbol{U}_{l^-}, \boldsymbol{U}_{l^+}, r_l, r_{l\eta}, r_{l\phi}, \eta, \phi)=0, \quad l=1, 2, \cdots, L-1, \\ \boldsymbol{H}_L(\boldsymbol{U}_{L^-}, r_L, r_{L\eta}, r_{L\phi}, \eta, \phi)=0, \end{cases} \tag{1.6}$$

which are compatible with (1.3)[1], and where $\boldsymbol{U}_{l^+}$ and $\boldsymbol{U}_{l^-}$ denote the values of $\boldsymbol{U}$ on the upper side and on the lower side, respectively, of the plane $\xi=l$.

(4) In the ϕ-direction, we have the periodicity condition

$$f(\phi)=f(\phi+2\pi), \tag{1.7}$$

where f denotes any one of the above functions.

We must solve $\boldsymbol{R}(\eta, \phi)$ in the region $\eta\geqslant 0$, $0\leqslant\phi\leqslant 2\pi$ and $\boldsymbol{U}(\eta, \xi, \phi)$ in the region $\eta\geqslant 0$, $0\leqslant\xi\leqslant L$, $0\leqslant\phi\leqslant 2\pi$.

The description of this problem is similar to that in paper [1], only there is a ϕ-direction periodicity condition in this problem, so that the method in [1] can be expanded to apply in this case.

In the following, we briefly describe the numerical method of calculating this class of problems.

In the region $l-1\leqslant\xi\leqslant l$, $0\leqslant\phi\leqslant 2\pi$, $\eta\geqslant 0$ we make a net with mesh size $\Delta\xi_l=\dfrac{1}{M_l}$, $\Delta\phi=\dfrac{\pi}{J}$, $\Delta\eta$, $l=1, 2, \cdots, L$. Let $f^k_{l,m,j}$ denote the value of f at the point $\xi=m\,\Delta\xi_l+l-1$, $\phi=j\Delta\phi$, $\eta=k\,\Delta\eta$. (Hereafter, some indexes of f will be omitted if no confusion occurs.) We adopt the scheme with an interim step. The calculation procedure is roughly as follows:

First, equation (1.3) is rewritten as[2]

$$\boldsymbol{G}^*_n\left[\frac{\partial \boldsymbol{U}}{\partial\eta}+\lambda_n\frac{\partial \boldsymbol{U}}{\partial\xi}\right]+\boldsymbol{C}^*_n\frac{\partial \boldsymbol{U}}{\partial\phi}=f_n, \quad n=1, 2, \cdots, N, \tag{1.8}$$

then the difference equations of (1.8) are obtained by utilization of Schemes I and II given below, and the difference equations of the equations

$$\begin{cases} r_l=\displaystyle\int r_{l\eta}\,d\eta, \\ r_{l\phi}=\displaystyle\int\frac{\partial^2 r_l}{\partial\phi\partial\eta}\,d\eta=\int\frac{\partial r_{l\eta}}{\partial\phi}\,d\eta \end{cases} \tag{1.9}$$

are obtained by utilization of schemes (1.16) and (1.17) given below.

---

1) For the definition of compatibility, please refer to [1] or [4].

2) It is easy to rewrite (1.3) as (1.8) because we can find a matrix P such that

$$PA=\overline{G}, \quad PB=\Lambda\overline{G},$$

where

$$\overline{G}=\begin{bmatrix} \boldsymbol{G}^*_1 \\ \boldsymbol{G}^*_1 \\ \vdots \\ \boldsymbol{G}^*_N \end{bmatrix}, \quad \Lambda=\begin{bmatrix} \lambda_1 & & & \\ & \lambda_2 & & \\ & & \ddots & \\ & & & \lambda_N \end{bmatrix},$$

$\boldsymbol{G}_n$ being an N-dimensional vector, λ_n being a real number, $n=1, 2, \cdots, N$.

Finally, to obtain the values we need, these difference equations and the boundary conditions are solved simultaneously. That is, using the values with $\eta=k\Delta\eta$

$$\begin{cases} U^k_{l,m,j}, & m=0,1,\cdots,M_l, \quad l=1,2,\cdots,L, \\ r^k_{l,j},\ r^k_{l\eta,j},\ r^k_{l\phi,j}, & l=0,1,\cdots,L, \quad j=0,1,\cdots,2J-1, \end{cases}$$

we can calculate the values with $\eta=\left(k+\frac{1}{2}\right)\Delta\eta$

$$\begin{cases} U^{k+\frac{1}{2}}_{l,m,j+\frac{1}{2}}, & m=0,1,\cdots,M_l, \quad l=1,2,\cdots,L, \\ r^{k+\frac{1}{2}}_{l,j+\frac{1}{2}},\ r^{k+\frac{1}{2}}_{l_\eta,j+\frac{1}{2}},\ r^{k+\frac{1}{2}}_{l\phi,j+\frac{1}{2}}, & l=0,1,\cdots,L, \quad j=0,1,\cdots,2J-1. \end{cases}$$

Moreover, using the values with $\eta=k\Delta\eta$ and $\eta=\left(k+\frac{1}{2}\right)\Delta\eta$, we can calculate the values with $\eta=(k+1)\Delta\eta$.

We do not discuss here how to apply Schemes I and II to the above initial-boundary-value problem, so as to establish difference equations, and how to solve the difference equations. For details, please refer to [1] and [4].

In the following we will give the concrete schemes. First, let us introduce the following notation:

$$\sigma=\lambda\frac{\Delta\eta}{\Delta\xi}, \quad \tau=\frac{\Delta\eta}{\Delta\phi},$$

$$\Delta_m U_{m,j}=U_{m+1/2,j}-U_{m-1/2,j}, \quad \Delta_{m+}U_{m,j}=U_{m+1,j}-U_{m,j},$$

$$\Delta_{m-}U_{m,j}=U_{m,j}-U_{m-1,j}, \quad \Delta_j U_{m,j}=U_{m,j+1/2}-U_{m,j-1/2},$$

$$\mu_m f_{m,j}=\frac{1}{2}(f_{m+1/2,j}+f_{m-1/2,j}), \quad \mu_{m\pm}f_{m,j}=\frac{1}{2}(f_{m,j}+f_{m\pm1,j}),$$

$$\mu_j f_{m,j}=\frac{1}{2}(f_{m,j+1/2}+f_{m,j-1/2}), \quad \mu f_{m,j}=\mu_m\mu_j f_{m,j}.$$

Scheme I. For the first step, equation (1.8) is approximated by the following equation (hereafter the index n will be omitted):

$$\begin{aligned} &\mu G^{*k}_{m\mp1/2,j+1/2}\left[\mu_m U^{k+1/2}_{m\mp1/2,j+1/2}+\frac{1}{2}\mu\sigma^k_{m\mp1/2,j+1/2}\Delta_m U^{k+1/2}_{m\mp1/2,j+1/2}\right] \\ &=\mu G^{*k}_{m\mp1/2,j+1/2}\,\mu U^k_{m\mp1/2,j+1/2}-\frac{1}{2}\tau\mu C^{*k}_{\eta\ \mp1/2,j+1/2}\,\mu_m\Delta_j U^k_{m\mp1/2,j+1/2} \\ &\quad+\frac{1}{2}\Delta\eta\mu f^k_{m\mp1/2,j+1/2}. \end{aligned} \tag{1.10}$$

From now on, when "±" (or "∓") appears, if $\sigma>0$, then "+" (or "−") is chosen; if $\sigma<0$, then "−" (or "+") is selected. We call this scheme the interim step of Scheme I.

For the second step, equation (1.8) is approximated by

$$\mu G^{*k+1/2}_{m\mp1/2,j}\left[\mu_m U^{k+1}_{m\mp1/2,j}+\frac{1}{2}\mu\sigma^{k+1/2}_{m\mp1/2,j}\Delta_m U^{k+1}_{m\mp1/2,j}\right]$$

$$=\mu G^{*k+1/2}_{m\mp1/2,j}\left[\mu_m U^{k}_{m\mp1/2,j}-\frac{1}{2}\mu\sigma^{k+1/2}_{m\mp1/2,j}\Delta_m U^{k}_{m\mp1/2,j}\right]$$

$$-\tau\mu C^{*k+1/2}_{m\mp1/2,j}\mu_m\Delta_j U^{k+1/2}_{m\mp1/2,j}+\Delta\eta\mu f^{k+1/2}_{m\mp1/2,j}. \qquad (1.11)$$

We call the scheme a regular step of Scheme I.

Scheme II. For the first step, equation (1.8) is approximated by

$$\mu_j G^{*k}_{m,j+1/2}U^{k+1/2}_{m,j+1/2}=\mu_j G^{*k}_{m,j+1/2}\left[\mu_j U^{k}_{m,j+1/2}-\frac{1}{2}\mu_j\sigma^{k}_{m,j+1/2}\Delta_{m\mp}\mu_j U^{k}_{m,j+1/2}\right]$$

$$-\frac{1}{2}\tau\mu_j C^{*k}_{m,j+1/2}\mu_{m\mp}\Delta_j U^{k}_{m,j+1/2}+\frac{1}{2}\Delta\eta\mu_j f^{k}_{m,j+1/2}. \qquad (1.12)$$

We call this scheme an interim step of Scheme II. For the second step, the difference scheme is based on the following two schemes:

(i) First we rewrite (1.8) as

$$G^*\left[\frac{\partial U}{\partial\eta}\mp\frac{\Delta\xi}{\Delta\eta}\frac{\partial U}{\partial\xi}+\left(\lambda\pm\frac{\Delta\xi}{\Delta\eta}\right)\frac{\partial U}{\partial\xi}\right]+C^*\frac{\partial U}{\partial\phi}=f,$$

which is then approximated by

$$\mu G^{*k+1/2}_{m\pm1/2,j}U^{k+1}_{m,j}=\mu G^{*k+1/2}_{m\pm1/2,j}$$

$$\times\left[U^{k}_{m\pm1,j}\pm\frac{1}{4}\Delta_j^2\Delta_m U^{k}_{m\pm1/2,j}-(\mu\sigma^{k+1/2}_{m\pm1/2,j}\pm1)\mu_j\Delta_m U^{k+1/2}_{m\pm1/2,j}\right]$$

$$-\tau\mu C^{*k+1/2}_{m\pm1/2,j}\mu_m\Delta_j U^{k+1/2}_{m\pm1/2,j}+\Delta\eta\mu f^{k+1/2}_{m\pm1/2,j}\equiv S_1U^{k}_{m,j}. \qquad (1.13)$$

(ii) First we rewrite (1.8) as

$$G^*\left[\frac{\partial U}{\Delta\eta}\pm\frac{\Delta\xi}{\Delta\eta}\frac{\partial U}{\partial\xi}+\left(\lambda\mp\frac{\Delta\xi}{\Delta\eta}\right)\frac{\partial U}{\partial\xi}\right]+C^*\frac{\partial U}{\partial\phi}=f,$$

which is then approximated by

$$\mu G^{*k+1/2}_{m\mp1/2,j}U^{k+1}_{m,j}=\mu G^{*k+1/2}_{m\mp1/2,j}[U^{k}_{m\mp1,j}-(\mu\sigma^{k+1/2}_{m\mp1/2,j}\mp1)\mu_j\Delta_m U^{k+1/2}_{m\mp1/2,j}]$$

$$-\tau\mu C^{*k+1/2}_{m\mp1/2,j}\mu_m\Delta_j U^{k+1/2}_{m\mp1/2,j}+\Delta\eta\mu f^{k+1/2}_{m\mp1/2,j}$$

$$\equiv S_2U^{k}_{m,j}. \qquad (1.14)$$

Applying schemes (1.13) and (1.14), we make the following scheme for initial–boundary–value problems.

$$\begin{cases}[\beta_{1,m}\mu G^{*k+1/2}_{m\pm1/2,j}+(1-\beta_{1,m})\mu G^{*k+1/2}_{m\mp1/2,j}]\,U^{k+1}_{m,j}\\ \quad=\beta_{1,m}S_1U^{k}_{m,j}+(1-\beta_{1,m})S_2U^{k}_{m,j}, \quad m\in[m_0,m_1],\\ \mu G^{*k+1/2}_{m\pm1/2,j}U^{k+1}_{m,j}=S_1U^{k}_{m,j}, \quad m\in[m_1+1,\ m_2-1],\\ [(1-\beta_{2,m})\mu G^{*k+1/2}_{m\pm1/2,j}+\beta_{2,m}\mu G^{*k+1/2}_{m\mp1/2,j}]\,U^{k+1}_{m,j}\\ \quad=(1-\beta_{2,m})S_1U^{k}_{m,j}+\beta_{2,m}S_2U^{k}_{m,j}, \quad m\in[m_2,m_3],\end{cases} \qquad (1.15)$$

where m_0, m_1, m_2, m_3, $\beta_{1,m}$ and $\beta_{2,m}$ are selected in the same way as in [1].

Besides (1.8), equation (1.9) also needs to be approximated. At the interim step of the scheme for calculating r_l and $r_{l\phi}$, we take

$$\begin{cases} r_{l,j+1/2}^{k+1/2}=\mu_j r_{l,j+1/2}^k+\dfrac{1}{2}\mu_j r_{l\eta,j+1/2}^k\Delta\eta, \\ r_{l\phi,j+1/2}^{k+1/2}=\mu_j r_{l\phi,j+1/2}^k+\dfrac{1}{2}\tau\Delta_j r_{l\eta,j+1/2}^k. \end{cases} \tag{1.16}$$

For the regular step, we have

$$\begin{cases} r_{l,j}^{k+1}=r_{l,j}^k+\mu_j r_{l\eta,j}^{k+1/2}\Delta\eta, \\ r_{l\phi,j}^{k+1}=r_{l\phi,j}^k+\tau\Delta_j r_{l\eta,j}^{k+1/2}. \end{cases} \tag{1.17}$$

2. *A Numerical Method of Accurately Calculating the Interactions between Discontinuities*

In Section 1, we have pointed out that if all discontinuities are treated as internal boundaries, then the flow which has several discontinuities can be accurately calculated by the method in Section 1. In this section we discuss how to apply the method of Section 1 to accurately calculating the interactions between discontinuities.

We suppose that a discontinuity $r=\bar{r}_i(z, \varphi)$ intersects another discontinuity $r=\bar{r}_j(z, \varphi)$ in the region of calculation, and that this discontinuous solution has to be accurately solved. First, the solution in the upstream region of the intersection line of the two discontinuities should be obtained. From the description of the initial-boundary-value problem in Section 1, we know that if we can find a family of space-like surfaces such that one surface contains the curve at which the surface $r=\bar{r}_i(z, \varphi)$ intersects $r=\bar{r}_j(z, \varphi)$, then we can apply the method of Section 1 to calculating the solution prior to the intersection of the two discontinuities. In general, $\bar{r}_i(z, \varphi)$ and $\bar{r}_j(z, \varphi)$ are unknown functions, and thus the curve at which they intersect is also unknown. Therefore, it is impossible to select an $E(\eta, r, \varphi)$ in advance such that the curve of intersection is located on one surface with $\eta=$const. However, this can be realized by a "track" method, i.e., in our procedure we often calculate the approximate curve of intersection and select $E(\eta, r, \varphi)$ so as to locate the intersection line on one-surface with $\eta=$const. When two discontinuities tend to their curve of intersection, the approximate curve of intersection tends to the accurate one, so by the "track" method the curve of intersection will be accurately located on one surface with $\eta=$ const., when the two discontinuities intersect.

When we have calculated the solution in the upstream region of the intersection line of the two discontinuities, we must then ascertain, in order to continue the solution of this problem, what kinds of discontinuities and weak discontinuities are to be reflected, their initial shapes, and the values of functions at the singular curve in every region between two discontinuities or weak discontinuities. (We call the curve of intersection a singular curve because functions have many values at the curve.) The solution of the above problem is essentially that of a Riemann problem in three independent variables. To solve the problem, the differential equations of the central wave must be derived from equation (1.1). After that, to obtain all data we need, we only have to solve the differential equations of the central wave and the jump conditions simultaneously. We should point out that this is a special system of simultaneous equations. In fact, when we solve this, not only do we calculate the values of the functions, but we also determine how to connect the differential equations of the central wave with the jump conditions, i.e., the construction of the solution and the solution are determined simultaneously. Moreover, since the problem here is a three-dimensional one, we need to solve a series of simultaneous equations to obtain all data required at the singular curve. For details of the equations of three-dimensional steady flows, refer to Chapter 4 of monograph [4].

The new values of functions at the singular curve and the initial shapes of these new boundaries having been obtained, we can continue the solution of this problem by the method of Section 1. The only difference is that the numbers, types and shapes of the boundaries have changed.

In short, if we have a method for solving the Riemann problem in three independent variables, then, by the "track" method and the difference method in Section 1, we can accurately solve the interactions between discontinuities in three independent variables.

Obviously, the method described here can also be applied to the accurate calculation of the reflections of discontinuities when they meet external boundaries.

3. *A Numerical Method of Accurate Calculation for the Automatically Formed Shocks*

In the field of numerical methods of discontinuous solutions, another difficulty is how to calculate the automatically formed discontinuities. As is well known, for the solutions of quasilinear hyperbolic systems, even though the initial value is smooth, the discontinuities may appear later. In order to calculate accurately such discontinuous solutions, we must find out how the smooth solutions become discontinuous solutions. From studying the solutions of equation $\frac{\partial u}{\partial \eta}+\lambda(u, \eta, \xi)\frac{\partial u}{\partial \xi}=0$ (for example, suppose $\lambda(u, \eta, \xi)\equiv u$), we know that the procedure of change is as follows: Before the formation of discontinuities, the procedures of concentration of characteristics appear, that is, $\frac{\partial \lambda}{\partial \xi}$ is negative and it decreases when η increases. When $\frac{\partial \lambda}{\partial \xi}$ becomes $-\infty$, a discontinuity appears. In general, at this time the strength of the discontinuity is zero. After that, the strength of the discontinuity gradually increases. It is evident that $T \equiv -\frac{\partial\lambda}{\partial\xi}$ expresses the speed at which the characteristics are concentrated. Thus we call T the speed of concentration of characteristics.

Therefore, in the case of two independent variables, automatically formed shocks can be calculated in the following way: In the procedures of calculation, the values of T_i, $i=1, 2, \cdots, N$ are often monitored. When one value of T_{i_0} is rather great at one point, a characteristic corresponding to index i_0 is added at that point as a new internal boundary and it is treated as a discontinuity when the calculation is continued[1). When this method is used, wherever a discontinuity forms, the strength of the added discontinuity gets greater and greater, untill, finally, an obvious discontinuity appears; whereas wherever no discontinuity forms, the strength is always less. The results in [1] are obtained by this method.

A problem with two independent variables is a special case of the problem with three independent variables. When a problem with two independent variables is generalized to one with three independent variables, characteristic curves are generalized to characteristic surfaces, and the procedure of the concentration of characteristic curves corresponds to the procedure of the concentration of characteristic surfaces. Therefore, we know that in the problems with three independent variables, characteristic surfaces will be concentrated before discontinuities appear. Thus, if the circumstances of the concentration of characteristic surfaces are known, automatically formed discontinuities can be accurately calculated. However, in two-dimensional problems, if a point is given, the speed of concentration of every kind of characteristic may be calculated. In three-dimensional problems, there are infinite characteristic surfaces which pass through a point. Only if the point, the direction, and the kind are given, may the characteristics and the speeds of concentration be calculated. Moreover, in different directions, the speeds of concentration are different. Obviously, the speed related to the formation of discontinuities is the maximal speed of concentration at every point, that is, the maximum among all speeds of concentration at every point.

As a result, the automatically formed shocks in three-dimensional problems may be accurately calculated in the following way: In the procedures of calculation the maximal speed of concentration of characteristics at every point is calculated. When the maximal

---

1) When characteristics are concentrated, sometimes no discontinuity forms. For example, in steady flow, when stream lines are concentrated no contact discontinuity forms, but an entropy layer forms which gets thinner and thinner. Then we suppose that when characteristics corresponding to index i_0 are concentrated, a discontinuity may form.

speeds of concentration of a kind of characteristics at some points are large enough, one surface of characteristics of the kind which passes through a curve close to these points is added as an internal boundary. When the calculation is continued, it is calculated as a discontinuity. Owing to the limited space here, we shall not discuss the details of the method.

4. *Some Results*

In steady flows, there are only two kinds of discontinuities, namely, shocks and contact discontinuities. Moreover, contact discontinuities cannot intersect one another and it is impossible to form automatically a contact discontinuity. Therefore in steady flows there are only two kinds of interactions between discontinuities, namely, the interaction between two shocks and the interaction between a shock and a contact discontinuity, and there is only one kind of discontinuity which can be automatically formed, namely, the shocks. On certain conditions, these problems and those of reflection of shocks, when they meet bodies or isobaric surfaces, can be accurately calculated by using the aforementioned method. We have made a program for calculating three-dimensional supersonic steady flows around bodies by the aforementioned method, and calculated accurately the interactions between two shocks of the same family, the interactions between a shock and a contact discontinuity and the automatically formed shocks.

We have obtained several results for these problems. For the concrete results, see the last four paragraphs of § 4 in Chapter 7 in [4].

References

[1] Zhu, Youlan, et al., *Scientia Sinica* (1979), Special Issue (II), 261—280 (in English).

[2] Zhu, Youlan, et al., *Acta Mathematicae Applicatae Sinica* (1977), No. 3, 12—27 (in Chinese).

[3] Zhu, Youlan, et al., *Acta Mechanica Sinica* (1979), No. 3, 209—218 (in Chinese).

[4] Zhu, Youlan, Zhong, Xichang, Chen, Bingmu and Zhang, Zuomin, Difference Methods for Initial-Boundary-Value Problems and Flow around Bodies, Science Press, Beijing, 1980 (in Chinese) or Springer-Verlag, Heidelberg and Science Press, Beijing, 1986 (in English).

REFERENCES

General References

[1] Keller, H. B. and Thomée, V., Unconditionally stable difference methods for mixed problems for quasilinear hyperbolic systems in two dimesions, *Comm. Pure Appl. Math.*, **15** (1962), 1, 63—73.

[2] Thomée, V., A stable difference scheme for the mixed boundary problem for a hyperbolic first order system in two dimensions, *J. Soc. Indust. Appl. Math.*, **10** (1962), 2, 229—245.

[3] Godunov, S. K. and Ryabenkii, V. S., Spectral criteria of stability of boundary-value problems for non-self-adjoint difference equations, Uspekhi Mat. Nauk., **X, VIII** (1963), 3, 3—14 (in Russian).

[4] Babenko, K. I., Voskresenskii, G. P., Lyubimov, A. N. and Rusanov, V. V., Spacial flow of the perfect gas around smooth bodies, "Nauka", Moscow, 1964 (in Russian).

[5] Strang, W. G., Wiener-Hopf difference equations, *J. Math. Mech.*, **13** (1964), 85—96.

[6] Kreiss, H.-O., On difference approximations, Symposium on numerical solution of partial differential equations, Univ. of Maryland, May, 1965, 51—58.

[7] Kreiss, H.-O., Difference approximations for the initial-boundary value problem for hyperbolic differential equations, Numerical Solutions of Nonlinear Differential Equations, Wiley, New York, 1966, 141—166.

[8] Kreiss, H.-O., Stability theory for difference approximations of mixed initial-boundary value problems I, *Math. Comp.*, **22** (1968), 104, 703—714.

[9] Osher, S., Stability of difference approximations of dissipative type for mixed initial boundary value problems I, *Math. Comp.*, **23** (1969), 106, 335—340.

[10] Kreiss, H.-O., Difference approximations for initial boundary value problems, *Proc. Roy. Soc. Lond.*, A, **323** (1971), 255—261.

[11] Kreiss, H.-O., Initial boundary value problems for partial differential and difference equations in one space dimension, Numerical Solution of PDE-II, Academic Press, N. Y., 1971, 401—416.

[12] Gustafsson, B., Kreiss, H.-O. and Sundström A., Stability theory of difference approximations for mixed initial boundary value problems II, *Math. Comp.*, **26** (1972), 119, 649—686.

[13] Oliger, J., Fourth order difference methods for the initial boundary value problem for hyperbolic equations, *Math. Comp.*, **28** (1974), 125, 15—25.

[14] Richtmyer, R. D. and Morton, K. W., Difference methods for initial-value problems, Second Edition, Wiley, New York, 1967.

[15] Crank, J. and Nicolson, P., A practical method for numerical evaluation of solutions of partial differential equations of the heat conduction type,*Proc. Cambridge Philos. Soc.*, **43** (1947), Part 1, 50—67.

[16] Courant, R., Isaacson, E. and Rees, M., On the solution of nonlinear hyperbolic differential equations by finite differences, *Comm. Pure and Appl. Math.*, **5** (1952), 3, 243—255.

[17] Lax, P. D., Weak solution of non-linear hyperbolic equations and their numerical computation, *Comm. Pure Appl. Math.*, **7** (1954), 1, 159—193.

[18] Lax, P. D. and Wendroff, B., Systems of conservation laws, *Comm. Pure Appl. Math.*, **13** (1960), 2, 217—237.

[19] Wendroff, B., On centered difference equations for hyperbolic systems, *J. Soc. Indust. Appl. Math.*, **8** (1960), 3, 549—555.

[20] Rusanov, V. V., Computation of interaction of nonstationary shocks with obstacles, *Zh. Vychisl. Mat. i Mat. Fiz.*, **1** (1961), 2, 267—279 (in Russian).

[21] Richtmyer, R. D., A survey of difference methods for non-steady fluid dynamics, NCAR TN 63-2, 1962.

[22] Lax, P. D. and Wendroff, B., Difference schemes for hyperbolic equations with high order of accuracy, *Comm. Pure Appl. Math.*, **17** (1964), 3, 381—398.

[23] MacCormack, R. W., The effects of viscosity in hypervelocity impact cratering, AIAA Paper 69—354, 1969.

[24] Ivanov, M. Ya., Kraiko, A. N. and Mikhailov, N. V., Shock-capturing methods for two and three dimensional supersonic flows I, *Zh. Vychisl. Mat. i Mat. Fiz.*, **12** (1972), 2, 441—463 (in Russian).

[25] Ivanov, M. Ya. and Kraiko, A. N., Shock-capturing methods for two and three dimensional supersonic flows II, *Zh. Vychisl. Mat. i Mat. Fiz.*, **12** (1972), 3, 805—813 (in Russian).

[26] Lavrent'ev, M. M., On the Cauchy problem for the Laplace equation, *Izv. AN SSSR Ser. Mat.*, **20** (1956), 6 (in Russian).

[27] Zhang, G.-q., Improperly posed initial-value problems and difference schemes for them, *Applied Mathematics and Mathematics of Computation*, **2** (1965), No. 1 (in Chinese).

[28] Pucci, C., Sui ploblemi di Cauchy non "ben posti", *Acc. Naz., lincei*, 8, **18** (1955), 5.

[29] Lavrent'ev, M. M., On some improperly posed problems in mathematical physics, Press of Sibirsk AN SSSR, Novosibirsk, 1962 (in Russian).

[30] Hadamard, J., Le Problème de Cauchy, Hermann, Paris, 1932.

[31] Miller, K., Three circle theorems in partial differential equation and application to improperly posed problem, *Archive for Rational Mechanics and Analysis*, **16** (1964), 2.

[32] Chudov, L. A., Difference methods for solving the Cauchy problem for the Laplace equation, *Dokl. AN SSSR*, **143** (1962), 4 (in Russian).

[33] Krein, S. G., On classes of correction for nonstationary boundary problems, *Dokl. AN SSSR*, **114**(1957), 6 (in Russian).

[34] Landau, L. D. and Lifshits, E. M., Continuous medium mechanics, Gostekhizdat, Moscow, 1954. (in Russian).

[35] Ferri, A., The method of characteristics, In: General theory of high speed aerodynamics, Princeton Series, **VI**, Editor W. R. Sears, Princeton Uni. Press, Princeton, N. J., 1954, 583—669.

[36] Tsien, H. S., The Equations of Gas Dynamics, in "Fundamentals of Gas Dynamics, Editor: H. W. Emmons, Princeton University Press, 1958."

[37] Courant, R. and Hilbert, D., Methods of mathematical Physics, **II**, Interscience, New York, 1962.

[38] Rusanov, V. V., Characteristics of general equations of gas dynamics, *Zh. Vychisl. Mat. i Mat. Fiz.*, **3** (1963), 3, 508—527 (in Russian).

[39] Courant, R. and Friedrichs, K. O., Supersonic flow and shock waves, Interscience, New York, 1948.

[40] Predvoditelev, A. S., Stupochenko, E. V. et al., Tables of thermodynamic functions of air,
a) Press of AN SSSR, Moscow, 1957 (for temperature from 6000°K to 12000°K and pressure from 0.001 atm. to 1000 atm.),
b) Press of AN SSSR, Moscow, 1959 (for temperature from 12000°K to 20000°K and pressure from 0.001 atm. to 1000 atm.),
c) Vychisl. Tsentr AN SSSR, Moscow, 1962 (for temperature from 200°K to 6000°K and pressure from 0.00001 atm. to 100 atm.) (in Russian).

[41] Hilsenrath, J., Klein, M. et al., Tables of thermodynamic proporties of air including dissociation and ionization from 1500°K to 15000°K, AEDC TR-59-20 (PB 161311), 1959.

[42] Hilsenrath, J. and Klein, M., Tables of thermodynamic properties of air in chemical equilibrium including second virial corrections from 1500°K to 15000°K, AEDC-TR-65-58 (AD 612301), 1965.

[43] Humphrey, R. L. and Neel, C. A., Tables of thermodynamic properties of air from 90°K to 1500°K, AEDC-TN-61-103 (AD 262692), 1961.

[44] Lewis, C. H. and Neel, C. A., Specific heat and speed of sound data for imperfect air, AEDC-TDR-64-36 (AD 600469), 1964.

[45] Grabau, M. and Brahinsky, H. S., Thermodynamic properties of air from 300°K to 6000°K and from 1 to 1000 amagats, AEDC-TR-66-247 (AD 646172), 1967.

[46] Brahinsky, H. S. and Neel, C. A., Tables of equilibrium thermodynamic properties of air, AEDC-TR-69-89, **I.** constant temperature (AD 686409); **II.** constant pressure (AD 686410); **III.** constant entropy (AD 687092); **IV.** constant temperature with specific heat and speed of sound data (AD 686411), 1969.

[47] Moeckel, W. E. and Weston, K. C., Composition and thermodynamic properties of air in

chemical equilibrium, NACA TN-4265, 1958.

[48] Naumova, I. N., Approximation of thermodynamic functions of air, *Zh. Vychisl. Mat. i Mat. Fiz.*, **1** (1961), 2, 295—300 (in Russian).

[49] Vertushkin, V. K., Approximation of thermodynamic functions of air, *Inzh. Zh.*, **II** (1962), 4, 343—344 (in Russian).

[50] Lewis, C. H. and Burgess, E. G., Empirical equations for the thermodynamic properties of air and nitrogen to 15000°K, AEDC-TDR-63-138 (AD 411624), 1963.

[51] Viegas, J. R. and Howe, J. T., Thermodynamic and transport property correlation formulas for equilibrium air from 1000°K to 15000°K, NASA TND-1429, 1962.

[52] Liu, T.-P., The Riemann problem for general systems of conservation laws, *J. of Differential Equations*, **18** (1975), 1, 218—234.

[53] Wendroff, B., The Riemann problem for materials with nonconvex equations of state, part II, general flow, *J. of Math. Analysis and Applic.*, **38** (1972), 3, 640—658.

[54] Lambrakis, K. C. and Thompson, P. A., Existence of real fluids with a negative fundamental derivative, *Phys. Fluids*, **15** (1972), 5, 933—935.

[55] Lin, S. C. and Teare, J. D., Rate of ionization behind shock wave in air, II. theoretical interpretation, *Phys. Fluids*, **6** (1963), 3.

[56] Treanor, C. E. and Merrone, P. V., Effect of dissociation on the rate of vibrational relaxation, *Phys. Fluids*, **5** (1962), 9.

[57] Merrone, P. V. and Treanor, C. E., Chemical relaxation with preferential dissociation from excited vibrational level, *Phys. Fluids*, **6** (1963), 9.

[58] Malcolm, G. N. and Rakich, J. V., Comparison of free-flight experimental results with theory on the nonlinear aerodynamic effects of bluntness for slender cones at Mach number 17, *AIAA J.*, **9** (1971), 3, 473—478.

[59] Woodley, J. G., Pressure measurements on a cone-cylinder-flare configuration at small incidences for $M_\infty=6.8$, ARC CP 632, 1962.

[60] Washington, W. D. and Humphrey, J. A., Pressure measurements on four cone-cylinder-flare configurations at supersonic speeds, Report No. RD-TM-69-11 (AD 699359), 1969.

[61] Kentzer, C. P., Discretization of boundary condition on moving discontinuities, In "Lecture Notes in Physics, Vol. 8, Springer-Verlag, N. Y., 1971", 108—113.

[62] Li, C. z., Xiao, L., Yang, S. q. and Yuan Z. w., Existence and uniqueness of the solution of the Riemann problem of gas dynamics equations with nonconvex equations of state, Report of Science and Technology University of China, Beijing, China, 1963 (in Chinese).

[63] Zhang, T. and Xiao, L., A Riemann problem for system of conservation laws of aerodynamics without convexity, *Acta Mathematica Sinica*, **22** (1979), No. 6, 719—732 (in Chinese).

[64] Xiao, L. and Zhang, T., Overtaking of shocks belonging to same family in steady plane supersonic flow without viscosity, *Acta Mathematicae Applicatae Sinica*, **3** (1980), No. 4, 343—357 (in Chinese).

[65] Gu, C. h., Some development of theory of symmetric positive equations and its applications, collectanea, Institute of Mathematics of Fudan University, Shanghai, China, 1964 (in Chinese).

[66] Zhu, Y. l., Wang, R. q. and Zhong, X. c., Computation of three-dimensional supersonic flow around blunt bodies, Technical Report, Institute of Computing Technology, Academia Sinica, Beijing, China, 1965 or *Acta Mechanica Sinica*, 1977, No. 4, 270—282 (in Chinese).

[67] Zhu, Y. l., Zhong, X. c. et al., Tables of sphere flow fields, Technical Report, Institute of Computing Technology, Academia Sinica, Beijing, China, 1966 (in Chinese).

[68] Zhu, Y. l., et al., Tables of flow fields around 17.5° cone, Technical Report, Institute of Computing Technology, Academia Sinica, Beijing, China, 1966 (in Chinese).

[69] Zhu, Y. l., Chen, B. m. and Zhang, Z. m., A scheme for calculating perfect gas flow and equilibrium air flow around combined bodies, Technical Report, Institute of Computing Technology, Academia Sinica, Beijing, China, 1968 (in Chinese).

[70] Zhu, Y. l. and Zhong, X. C., Application of the method of lines to supersonic regions of flow around blunt bodies, Technical Report, Institute of Computing Technology, Academia Sinica, Beijing, China, 1968 (in Chinese).

[71] Zhu. Y. l. and Zhong, X. c., The Calculations of perfect gas flow and equilibrium air flow around cones by the method of lines, Technical Report, Institute of Computing Technology, Academia Sinica, Beijing, China, 1968 (in Chinese).

[72] Chen, B. m., Approximation of thermodynamic properties of air in chemical equilibrium, Technical Report, Institute of Computing Technology, Academia Sinica, Beijing, China, 1968. or *Acta Mechanica Sinica*, 1977, No. 3, 176—181 (in Chinese).

[73] Zhu, Y. l. and Zhong, X. c., Tables of conical flow fields at angle of attack, Technical Report, Institute of Computing Technology, Academia Sinica, Beijing, China, 1968 (in Chines).

[74] Zhu, Y. l., A numerical method for a certain class of initial-boundary-value problems of first order quasilinear hyperbolic systems with three independent variables, Technical Report, Institute of Computing Technology, Academia Sinica, Beijing, China, 1973 (in Chinese).

[75] Zhu, Y. l., Chen, B. m., Zhang, Z. m., Zhong, X. c., Qin, B. l. and Zhang, G. q., The numerical calculation of three dimensional supersonic flow around combined bodies, Technical Report, Institute of Computing Technology, Academia Sinica, Beijing, China, 1973 (in Chinese).

[76] Zhong, X. c. and Zhu, Y. l., The calculation of blunt-body supersonic flow, Technical Report, Academia Sinica, Beijing, China, 1974 (in Chinese).

[77] Zhu, Y. l., Chen, B. m., Zhang, Z. m., Zhong, X. c., Qin, B. l. and Zhang, G.-q., A numerical method for initial-boundary-value problems for hyperbolic systems and applications, *Acta Mathematicae Applicatae Sinica*, 1977, No. 3, 12—27 (in Chinese).

[78] Zhu, Y. l., Stability of difference schemes for pure-initial-value problems with variable coefficients, *Mathematicae Numericae Sinica*, 1978, No. 1, 33—43 (in Chinese).

[79] Zhu, Y. l., A block-double-sweep method for "incomplete" linear algebraic systems and its stability, *Mathematicae Numericae Sinica*, 1978, No. 3, 1—27. (in Chinese).

[80] Zhu, Y. l., Chen, B. m., Zhang, Z. m., Zhong, X. c., Qin, B. l. and Zhang, G.-q., Difference schemes for initial-boundary-value problems of hyperbolic systems and examples of application, *Scientia Sinica*, 1979, Special Issue (II) on Mathematics, 261—280.

[81] Zhu, Y. l., Difference schemes for initial-boundary-value problems of first order hyperbolic systems and their stability, *Mathematicae Numericae Sinica*, 1979, No. 1, 1—30 (in Chinese).

[82] Zhong, X. c. and Zhu, Y. l., An implicit method of lines for flow around blunt bodies, *Mathematicae Numericae Sinica*, 1979, No. 2, 112—120 (in Chinese).

[83] Zhu, Y. l., Chen, B. m., Zhang, Z. m., Zhong, X. c., Qin, B. l. and Zhang, G. q., The numerical calculation of the supersonic flow around combined bodies, *Acta Mechanica Sinica*, 1979, No. 3, 209—218 (in Chinese).

[84] Zhu, Y. l. and Chen, B. m., A numerical method with high accuracy for calculating the interactions between discontinuities in three independent variables, *Scientia Sinica*, **23** (1980), No. 12, 1491—1501.

[85] Zhu, Y. l., Stability and convergence of difference schemes for linear initial-boundary-value problems, *Mathematicae Numericae Sinica*, 1982, No. 2, 1—11.

[86] Zhn, Y. l. and Chen, B. m., Difference methods for initial-boundary-value problems and computation of flow around bodies, *Computers and Fluids*, **9** (1981), 339—363.

[87] Zhu, Y. l., Chen, B. m., Wu, X. h. and Xu, Q. s., Some new developments of the singularity-separating difference method, In "Lecture Notes in Physics, Vol. 170, Springer-Verlag, Berlin etc., 1982", 553—559.

[88] Wu, X. h., An application of the singularity-separating method to the computation of unsteady shocks, *Journal on Numerical Methods and Computer Applications*, **3** (1982), No. 3, 142—151 (in Chinese).

[89] Xu, Q. s., A new numerical method for the heat conduction problem with phase-change, *Journal on Numerical Methods and Computer Applications*, **3** (1982), No. 4, 216—226 (in Chinese).

[90] Zhu, Y. l., The Singularity separating method, In "Proceedings of the Fourth International Symposium on Finite Element Methods in Flow Problems, Press of Tokyo

Uni., Japan, July, 26—29-th, 1982".

[91] Chen, B. m. and Zhu, Y. l., Numerical calculation of inviscid supersonic flow field around bent nose cones, *Acta Aerodynamica Sinica*, 1983, No. 2, 11—19 (in Chinese).

[92] Wu, X. h., Numerical calculation of the hyperbolic equation with a non-convex equation of state, *Kexue Tongbao*, 1983, No. 20, 1224—1226 (in Chinese).

[93] Wu, X. h., Huang, D. and Zhu, Y. l., Numerical computation of the flow with a shock wave passing through a "strong explosion" center, *Journal of Computational Mathematics*, **1** (1983), No. 3, 247—258.

[94] Wu, X. h. and Zhu, Y. l., Numerical solution of multimedium flow with various discontinuities, *Journal of Computational Mathematics*, **1** (1983), No. 4, 303—316.

[95] Wu, X. h., Wang, Y., Teng, Z. h. and Zhu, Y. l., Numerical computation of flow field with deflagration and detonation, *Journal of Computational Mathematics*, **2** (1984), No. 3, 247—256.

[96] Ni, L. a., Wu, X. h., Wang, Y. and Zhu, Y. l., Quantitative comparison among several difference schemes, *Journal of Computational Mathematics*, **2** (1984), No. 4, 310—327.

[97] Xu, Q. s. and Zhu, Y. l., Solution of two dimensional Stefan problem by the singularity-separating method, *Journal of Computational Mathematics*, **3** (1985), No. 1, 8—18.

Special References A: Numerical Calculation of Flow in Subsonic and Transonic Regions

The Method of Integral Relations

[1] Dorodnitsyn, A. A., On a numerical method for some aerodynamic problems, In "Trudy III Vsesoyuz. Mat. C'ezda, **III** (1956), Press of AN SSSR, Moscow, 1958", 447—453 (in Russian).

[2] Belotserkovskii, O. M., Flow around a circular cylinder with a detached shock, *Dokl. AN SSSR*, **113** (1957), 3, 509—512 (in Russian).

[3] Belotserkovskii, O. M., Calculation of flow around a circular cylinder with a detached shock, In "Vychisl. Mat., **3**, Press of AN SSSR, 1958", 149—185 (in Russian).

[4] Belotserkovskii, O. M., Flow around a symmetric profile with a detached shock, *Prikl. Mat. i Mekh.*, **22** (1958), 2, 206—219 (in Russian).

[5] Belotserkovskii, O. M., On the calculation of flow around axisymmetric bodies with a detached shock on eletronic computing machines, *Prikl. Mat. i Mekh.*, **24** (1960), 3, 511—517 (in Russian).

[6] Belotserkovskii, O. M., Calculation of flow around axisymmetric bodies with a detached shock (formulae of computation and tables of flow fields), Vychisl. Tsentr AN SSSR, Moscow, 1961 (in Russian).

[7] Belotserkovskii, O. M. and Chushkin, P. I., Supersonic flow past blunt bodies, In: "Fluid Dynamics Transactions, Vol. I, Symposium, Jablonna (Poland), 1961, PWN, Warszawa, 1964", 161—187.

[8] Belotserkovskii, O. M. and Chushkin, P. I., The numerical method of integral relations, *Zh. Vychisl. Mat. i Mat. Fiz.*, **2** (1962), 5, 731—759 (in Russian).

[9] Belotserkovskii, O. M., Supersonic symmetric flow of perfect and real gases around blunt bodies, *Zh. Vychisl. Mat. i Mat. Fiz.*, **2** (1962), 6, 1062—1085 (in Russian).

[10] Belotserkovskii, O. M. and Dushin, V. K., Nonequilibrium supersonic flow around blunt bodies, *Zh. Vychisl. Mat. i Mat. Fiz.*, **4** (1964), 1, 61—77 (in Russian).

[11] Belotserkovskii, O. M., Golomazov, M. M. and Shulishnina, N. P., Calculation of flow of a dissociating gas in equilibrium around blunt bodies with a detached shock, *Zh. Vychisl. Mat. i Mat. Fiz.*, **4** (1964), 2, 306—316 (in Russian).

[12] Lunkin, Yu. P. and Popov, F. D., Influence of nonequilibrium dissociation on supersonic flow around blunt bodies, *Zh. Vychisl. Mat. i Mat. Fiz.*, **4** (1964), 5, 896—904 (in Russian).

[13] Dushin, V. K. and Lunkin, Yu. P., Supersonic nonequilibrium flow of dissociating air around blunt bodies, *Zh. Tekhn. Fiz.*, **35** (1965), 8, 1461—1470 (in Russian).

[14] Lunkin, Yu. P., Popov, F. D., Timofeeva, T. R. and Lipnitskii, Yu. M., On the problem of passing singular points in numerical calculation of supersonic flow around bodies, In "Trudy LPI, 1965," 248, 7—13 (in Russian).

[15] Fomin, V. N., Hypersonic flow around blunt bodies with radiation, *Zh. Vychisl. Mat. i Mat. Fiz.*, **6** (1966), 4, 714—726 (in Russian).

[16] Belotserkovskii, O. M., Sedova, E. S. and Shugaev, F. V., Supersonicflow around bodies of revolution with a "generator–cusp", *Zh. Vychisl. Mat. i Mat. Fiz.*, **6** (1966), 5, 930—934 (in Russian).

[17] Belotserkovskii, O. M., Bulekbaev, A. A. and Grudnitskii, V. G., Algorithms of some numerical schemes of the method of integral relations for mixed flows,*Zh. Vychisl. Mat. i Mat. Fiz.*, **6** (1966), 6, 1064—1081 (in Russian).

[18] Popov, F. D., On a scheme of the method of integral relations for supersonic flow around blunt bodies, *Zh. Tekhn. Fiz.*, **36** (1966), 2, 239—245 (in Russian).

[19] Lunkin, Yu P. and Popov, F. D., Influence of vibrating and dissociating relaxations on supersonic flow around blunt bodies, *Zh. Tekhn. Fiz.*, **36** (1966), 4, 661—671 (in Russian).

[20] Belotserkovskii, O. M. et al., Supersonic flow around blunt bodies (theoretical and experimental investigations), Vychisl. Tsentr AN SSSR, Moscow, 1966, and the second edition, correction and supplement, Vychisl. Tsentr AN SSSR, Moscow, 1967 (in Russian).

[21] Belotserkovskii, O. M. and Fomin, V. N., Calculation of flow of radiating gas in a shock layer, *Zh. Vychisl. Mat. i Mat. Fiz.*, **9** (1969), 2, 397—412 (in Russian).

[22] Lebedev, V. I. and Fomin, V. N., Hypersonic flow around blunt bodies with selective radiation and absorption of energy, *Zh. Vychisl. Mat. i Mat. Fiz.*, **9** (1969), 3, 655—663 (in Russian).

[23] Dushin, V. K., On the solution of system of relaxation equations in calculating flow of a reacting gaseous mixture nearly in equilibrium, *Zh. Vychisl. Mat. i Mat. Fiz.*, **9** (1969), 5, 1121—1136 (in Russian).

[24] Belotserkovskii, O. M., On the calculation of gas flows with second floating shocks, In: "Proceedings of 2–th International Conference on Nnmerical Methods in Fluid Dynamics, Univ. Calif., 1970, Springer-Verlag, New York, 1971", 255—263.

[25] Belotserkovskii, O. M., Numerical methods of some transonic aerodynamics problems, *J. Comput. Phys.*, **5** (1970), 3, 587—611.

[26] Bazzhin, A. P., Calculation of flow in front of blunted bodies of revolution with a "generator–cusp", *Inzh. Zh.*, **1** (1961), 1, 154—159 (in Russian).

[27] Chushkin, P. I., Hypersonic flow around circles and spheres in a magnetic field, *Prikl. Mat. i Mekh.*, **27** (1963), 6, 1089—1094 (in Russian).

[28] Grineva, S. N., Computation of flow around blunt bodies of revolution with a "generator–cusp", *Inzh. Zh.*, **4** (1964), 3, 439—445. (in Russian).

[29] Chushkin, P. I., Magnetized blunted bodies in hypersonic flow, *Magnitnaya Gidrodinamika*, 1965, 3, 67—75 (in Russian).

[30] Traugott, S. C., An approximate solution of the direct supersonic blunt-body problem for arbitrary axisymmetric shapes, *J. Aerospace Sci.*, **27** (1960), 5, 361—370.

[31] Vaglio–laurin, R., Supersonic flow about two–dimensional asymmetric blunt bodies, GASL, Inc. TR–178, (I) and (II), 1960 (AD 256183 and AD 256184).

[32] Holt, M., Direct calculation of pressure distribution on blunt hypersonic nose shapes with sharp corners, *J. Aerospace Sci.*, **28** (1961), 11, 872—876.

[33] Holt, M. and Hoffman, G. H., Calculation of supersonic flow past spheres and ellipsoids, IAS Preprint 61–209–1903, 1961.

[34] Xerikos, J. and Anderson, W. A., A critical study of the direct blunt body integral method, Report SM–42603, Douglas Aircraft Co., 1962 (AD 295992).

[35] Kentzer, C. P., Singular line of the method of integral relations, *AIAA J.*, **1** (1963), 4, 928—929.

[36] Shih, W. C. and Baron, J. R., Nonequilibrium blunt–body flow using the method of integral relations, *AIAA J.*, **2** (1964), 6, 1062—1071.

[37] Brong, E. A. and Leigh, D. C., Method of Belotserkovskii for asymmetric blunt–body flows, *AIAA J.*, **2** (1964), 10, 1852—1853.

[38] Prosnak, W. J., The asymmetric blunt-body problem, *Archiwum Mechaniki Stosowanej*, Warszawa, **16** (1964), 3, 689—708.

[39] Springfield, J. F., Steady, inviscid flow of a relaxing gas about a blunt body with supersonic velocity, In: "Proceedings of the 1964 Heat Transfer and Fluid Mechanics Institute, Standford Univ. Press, 1964", 182—197.

[40] Wilson, K. H. and Hoshizaki, H., Inviscid, nonadiabatic flow about blunt bodies, *AIAA J.*, **3** (1965), 1, 67—74 (or AD 600592).

[41] Kao, H. C., A new technique for the direct calculation of blunt-body flow fields, *AIAA J.*, **3** (1965), 1, 161—163.

[42] Archer, R. D., Supersonic and hypersonic flow of an ideal gas around an elliptic nose, *AIAA J.*, **3** (1965), 5, 987—988.

[43] Xerikos, J. and Anderson, W. A., Blunt-body integral method for air in thermodynamic equilibrium, *AIAA J.*, **3** (1965), 8, 1531—1533.

[44] Chenoweth, D. R., Direct blunt-body integral method, *AIAA J.*, **3** (1965), 9, 1788—1789.

[45] Hermann, R. and Thoenes, J., Hypersonic flow of air past a circular cylinder with nonequilibrium oxygen dissociation including dissociation of the free stream, NASA N 66-12861, 1965.

[46] Jischke, M. C., Radiation coupled wedge flow using method of integral relations, AFOS R 65-1727, 1965 (AD 624554).

[47] Kentzer, C. P., The inverse blunt-body problem, In: "Proceedings of the XV-th International Astronautical Congress, Warsaw, III (1964), Polish Scientific publishers, Warsaw, 1965", 203—212.

[48] Kuby, W., Foster, R. M., Byron, S. R. and Holt, M., Symmetrical equilibrium flow past a blunt-body at superorbital reentry speeds, *AIAA J.*, **5** (1967), 4, 610—617.

[49] Gopinath, R., On flow past blunt bodies by the method of Belotserkovskii, *AIAA J.*, **5** (1967), 8, 1498—1499.

[50] Bailey, F. R., Rollstin, L. R. and Ivereen, J. D., Sonic-point convergence with the method of integral relations, *AIAA J.*, **5** (1967), 10, 1894—1895.

[51] South, J. C. Jr., Calculation of axisymmetric supersonic flow past blunt bodies with sonic corners, including a program description and listing, NASA TND-4563, 1968.

[52] Garrett, L. B., Suttles, J. T. and Perkins, J. N., A modified method of integral relations approach to the blunt body equilibrium air flow field, including comparisons with inverse solutions, NASA TND-5434, 1969.

[53] Rao, P. P., Supersonic flow past large-angle pointed cones, *AIAA J.*, **10** (1972), 12, 1713—1716.

[54] Hunt, B. L., An attempted improvement of the method of integral relations for a blunt body in a supersonic flow, *AIAA J.*, **11** (1973), 11, 1575—1576.

[55] Vaglio-Laurin, R., Supersonic flow about general three-dimensional blunt bodies, **I**: Inviscid supersonic flow about general three-dimensional blunt bodies, ASD-TR-61-727, **I** (1962, AD 292247).

[56] Minailos, A. N., On the calculation of supersonic flow around blunt bodies of revolution at angles of attack, *Zh. Vychisl. Mat. i Mat. Fiz.*, **4** (1964), 1, 171—177 (in Russian).

[57] Waldman, G. D., Integral approach to the yawed blunt body problem, AIAA Paper 65-68, 1965.

[58] Kozlova, I. G. and Minailos, A. N., Asymmetric supersonic flow of a perfect or a real gas around the nose part of a body of revolution, *Zh. Vychisl. Mat. i Mat. Fiz.*, **7** (1967), 3, 594—608 (in Russian).

[59] Golomazov, M. M., Supersonic flow of a dissociating gas around blunt bodies at angles of attack, *Zh. Vychisl. Mat. i Mat. Fiz.*, **11** (1971), 4, 1063—1070 (in Russian).

The Method of Lines

[60] Roslyakov, G. S. and Telenin, G. F., Survey of works on computation of steady

axisymmetric flow, completed in the Computing Center of National University of Moscow, In "Chisl. Metody v Gazovoi Dinamike, **II**, Mosk. Gosud. Uni., 1963", 5—19 (in Russian).

[61] Telenin, G. F. and Tinyakov, G. P., Investigation of supersonic flows of air and carbon dioxide in thermodynamic equilibrium around spheres, *Dokl. AN SSSR*, **159** (1964), 1, 39—42 (in Russian).

[62] Gilinskii, S. M. and Telenin, G. F. and Tinyakov, G. P., A method for computing supersonic flow around blunt bodies with a detached shock, *Izv. AN SSSR, Mekh. i Mashinostr.*, 1964, 4, 9—28 (in Russian).

[63] Gilinskii, S. M. and Telenin, G. F., Supersonic flow around various bodies with a detached shock, *Izv. AN SSSR, Mekh. i Mashinostr.*, 1964, 5, 148—156 (in Russian).

[64] Stulov, V. P. and Telenin, G. F., Nonequilibrium supersonic flow of air around a sphere, *Izv. AN SSSR, Mekh.*, 1965, 1, 3—16 (in Russian).

[65] Gilinskii, S. M. and Lebedev, M. T., Investigation of low supersonic flow around plane and axisymmetric bodies with a detached shock, *Izv. AN SSSR, Mekh.*, 1965, 1, 17—23 (in Russian).

[66] Gilinskii, S. M. and Lebedev, M. G., Computation of supersonic flow of a perfect gas around elliptic cylinders, *Izv. AN SSSR, Mekh.*, 1965, 3, 182—186 (in Russian).

[67] Gilinskii, S. M. and Makarova, N. E., Calculation of supersonic flow of air with physical and chemical reactions in equilibrium around blunt bodies, *Izv. AN SSSR, Mekh. Zhidk. i Gaza*, 1966, 1, 16—21 (in Russian).

[68] Stulov, A. P. and Turchak. L. I., Supersonic flow of air with vibrating relaxation around a sphere, *Izv. AN SSSR, Mekh. Zhidk. i Gaza*, 1966, 5, 1—7 (in Russian).

[69] Gilinskii, S. M., Zapryanov, Z. D. and Chernyi, G. G., Supersonic flow of combustible gaseous mixtures around a sphere, *Izv. AN SSSR, Mekh. Zhidk. i Gaza*, 1966, 5, 8—13 (in Russian).

[70] Sayapin, G.N., Investigation of features of hypersonic nonequilibrium flow around blunt bodies, *Izv. AN SSSR, Mekh. Zhidk. i Gaza*, 1966, 6, 115—116 (in Russian).

[71] Stulov, V. P. and Turchak, G. I., On supersonic flow around blunt bodies with fast nonequilibrium processes, *Izv. AN SSSR, Mekh. Zhidk. i Gaza*, 1967, 5, 88—92 (in Russian).

[72] Gilinskii, S. M. and Chernyi, G. G., Supersonic flow of combustible gaseous mixtures around a sphere (considering the lag time of inflammation), *Izv. AN SSSR, Mekh. Zhidk. i Gaza*, 1968, 1, 20—32 (in Russian).

[73] Gilinskii, S. M., Calculation of flow of combustible mixtures in front of blunt bodies, *Izv. AN SSSR, Mekh. Zhidk. i Gaza*, 1968, 2, 75—80 (in Russian).

[74] Bogolopov, V. V., El'kin, Yu. G. and Neiland, V. Ya.,Computation of flow of an inviscid radiating gas around blunt bodies, *Izv. AN SSSR, Mekh. Zhidk. i Gaza*, 1968, 4, 11—14 (in Russian).

[75] Shkodova, V. P., Supersonic flow of air nearly in equilibrium around bodies of revolution, *Izv. AN SSSR, Mekh. Zhidk. i Gaza*, 1969, 1, 73—76 (in Russian).

[76] Gilinskii, S. M., Computation of combustion of hydrogen in air behind the detached shock formed by supersonic motion of a sphere, *Izv. AN SSSR, Mekh. Zhidk. i Gaza*, 1969, 4, 97—106. (in Russian).

[77] Stulov, V. P. and Turchak, L. I., Non-equilibrium chemical reactions in the shock layer in front of a sphere in flow of a CO_2–N_2–Ar gas mixture, *Izv. AN SSSR, Mekh. Zhidk. i Gaza*, 1969, 5, 147—150. (in Russian).

[78] Stulov, V. P. and Shapiro, E. G., Radiation of the shock layer in hypersonic flow of air around blunt bodies, *Izv. AN SSSR, Mekh. Zhidk. i Gaza*, 1970, 1, 154—160 (in Russian).

[79] Turchak, L. I., Supersonic nonequilibrium flow of the mixture modelled on the atmosphere of the Venus around segmental bodies, *Izv. AN SSSR, Mekh. Zhidk. i Gaza*, 1970, 2, 67—72 (in Russian).

[80] Shapiro, E. G., Radiation of the shock layer in hypersonic flow of air around a spherical segment, *Izv. AN SSSR, Mekh. Zhidk. i Gaza*, 1972, 1, 101—106 (in Russian).

[81] Stulov, V. P., A strong current jetting from a blunt body in supersonic flow, *Izv. AN*

SSSR, Mekh. Zhidk. i Gaza, 1972, 2, 89—97 (in Russian).

[82] Stulov, V. P. and Mirsky, V. N., Radiating flows with intensive evaporation over blunt bodies, In "Lecture Notes in Physics, Vol. 35, Springer–Verlag, N. Y., 1975", 379—384.

[83] Telenin, G. F. and Tinyakov, G. P., A method for computing spacial flow around bodies with a detached shock, *Dokl. AN SSSR*, **154** (1964), 5, 1056—1058 (in Russian).

[84] Tinyakov, G. P., Investigation of three dimensional supersonic flow around ellipsoids of revolution, *Izv. AN SSSR, Mekh.*, 1965, 6, 10—19 (in Russian).

[85] Minostsev, V. B., Telenin, G. F. and Tinyakov, G. P., Investigation of the picture of supersonic spacial flow around segmental bodies, *Dokl. AN SSSR*, **179** (1968), 2, 304—307 (in Russian).

[86] Golomazov, M. M. and Zyuzin, A. P., On a numerical method for computing spacial flow around blunt bodies with a detached shock, *Zh. Vychisl. Mat. i Mat. Fiz.*, **15** (1975), 5, 1349—1355 (in Russian).

Other Methods of Directly Solving the Steady Equations

[87] Van–Dyke, M. D., The supersonic blunt body problem——review and extension, *J. Aerospace Sci.*, **25** (1958), 8, 485—496.

[88] Van–Dyke, M. D. and Gordon, H. D., Supersonic flow past a family of blunt axisymmetric bodies, NASA TR R–1, 1959.

[89] Lick, W., Inviscid flow of a reacting mixture of gases around a blunt body, *J. Fluid Mech.*, **7** (1960), 1, 128—144.

[90] Inouye, M. and Lomax, H., Comparison of experimental and numerical results for the flow of a perfect gas about blunt–nosed bodies, NASA TND–1426, 1962.

[91] Hall, J. G., Eschenroeder, A. O. and Marrone, P. V., Blunt–nosed inviscid airflows with coupled nonequilibrium processes, *J. Aerospace Sci.*, **29** (1962), 9, 1038—1051.

[92] Klaimon, J. H., Hypersonic flow field around a hemisphere in a CO_2–N_2–Ar gas mixture, *AIAA J.*, **2** (1964), 5, 953—954.

[93] Lee, R. H. and Chu, S. T., Nonequilibrium inviscid flow about blunt bodies, Aerospace Corp. SSD–TDR–63–369, 1964 (AD 430426).

[94] Inouye, M., Blunt bodies solutions for spheres and ellipsoids in equilibrium gas mixtures, NASA TN D–2780, 1965.

[95] Inouye, M., Rakich, J. V. and Lomax, H., A description of numerical methods and computer programs for two–dimensional and axisymmetric supersonic flow over blunt–nosed and flared bodies, NASA TND–2970, 1965.

[96] Inouye, M., Shock stand–off distance for equilibrium flow around hemisphere obtained from numerical calculations, *AIAA J.*, **3** (1965), 1, 172—173.

[97] Joss, W. W., Application of the inverse technique to the flow over a blunt body at angle of attack, NASA CR–445, 1966.

[98] Webb, H. G. Jr., Dresser, H. S., Adler, B. K. and Waiter, S. A., Inverse solution of blunt–body flow fields at large angle of attack, *AIAA J.*, **5** (1967), 6, 1079—1085 (or AIAA Paper 66–413, 1966).

[99] Fuller, F. B., Numerical solutions for supersonic flow of an ideal gas around blunt two–dimensional bodies, NASA TN D–791, 1961.

[100] Vaglio–Laurin, R. and Ferri, A., Theoretical investigation of the flow field about blunt–nosed bodies in supersonic flight, *J. Aerospace Sci.*, **25** (1958), 12, 761—770.

[101] Lieberman, E., General description of IBM 704 computer programs for flow field about blunt–nosed bodies of revolution in hypersonic flight, GASL, Inc., TR–208, 1961. (AD 256186).

[102] Vaglio–Laurin, R., On the PLK method and the supersonic blunt–body problem, *J. Aerospace Sci.*, **29** (1962), 2, 185—207 (or IAS Paper 61–22, 1961).

[103] Lunev, V. V., Pavlov, V. G. and Sinchenko, S. G., Hypersonic flow of dissociating air in equilibrium around a sphere, *Zh. Vychisl. Mat. i Mat. Fiz.*, **6** (1966), 1, 121—129 (in Russian).

[104] Mangler, K. W., The calculation of flow field between a blunt body and the bow wave,

In: "Hypersonic Flow, Butterworths Scientific Publications, London, 1960", 219—237.

[105] Swigart, R. J., A theory of asymmetric hypersonic blunt-body flows, *AIAA J.*, **1** (1963), 5, 1034—1042.

[106] Swigart, R. J., Real-gas hypersonic blunt-body flows, *AIAA J.*, **1** (1963), 11, 2642—2644.

[107] Swigart, R. J., Hypersonic blunt body flow fields at angle of attack, *AIAA J.*, **2** (1964), 1, 115—117.

[108] Swigart, R. J., The direct asymmetric hypersonic blunt-body problem, AIAA Paper 66-411, 1966.

[109] Garabedian, P. R., Numerical construction of detached shock waves, *J. Math. and Phys.*, **36** (1957), 3, 192—205.

[110] Lin, C. C., Note on Garabedian's paper "Numerical construction of detached shock waves", *J. Math. and Phys.*, **36** (1957), 3, 206—209.

[111] Garabedian, P. R. and Lieberstein, H. M., On the numerical calculation of detached bow shock waves in hypersonic flow, *J. Aeronaut. Sci.*, **25** (1958), 2, 109—118.

[112] Uchida, S. and Yasuhara, M., The rotational field behind a curved shock wave calculated by the method of flux analysis, *J. Aeronaut. Sci.*, **23** (1956), 9, 830—845.

[113] Maslen, S. H. and Moeckel, W. E., Inviscid hypersonic flow past blunt bodies, *J. Aeronaut. Sci.*, **24** (1957), 9, 683—693.

[114] Kennet, H., The inviscid flow of reacting and radiant air past a hypersonic blunt body, Boeing Airplane Co., Document No. D2-9055, 1960 (AD 438261).

[115] Langelo, V. A., The inviscid reacting flow field about hypersonic bodies, General Electric, Missile and Space Division, IIS R63SD 90, 1963 (AD 431919).

[116] Van-Tuyl, A. H., Use of rational approximations in the calculation of flows past blunt bodies, *AIAA J.*, **5** (1967), 2, 218—225.

[117] Jurak, K., Pert, G. J. and Capjack, C. E., Least-squares solution for the blunt body hypersonic flow problem, *J. Comput. Phys.*, **10** (1972), 2, 369—373.

[118] Severinov, L. I., On application of artificial viscosity to numerical solution of an inverse problem of gas dynamics, *Zh. Vychisl. Mat. i Mat. Fiz.*, **5** (1965), 3, 566—571 (in Russian).

[119] Bratos, M., Burnat, M. and Prosnak, W. J., Application of Lax finite difference scheme to transonic flow problems, In: "Fluid Dynamics Transactions, Vol. 5, Part II, PWN, Warszawa, 1971", 57—65.

[120] Perry, J. C. and Pasiuk, L., A comparison of solutions to a blunt body problem, *AIAA J.*, **4** (1966), 8, 1425—1426.

[121] Kentzer, C. P., Instability numerical solutions of the steady supersonic blunt-body problem, *AIAA J.*, **5** (1967), 5, 1035—1037.

The Time-Dependent Method for Two-Dimensional or Axisymmetric Problems

[122] Godunov, S. K., Zabrodin, A. V. and Prokopov, G. P., 'A difference scheme for two dimensional unsteady problem of gas dynamics and computation of flow with a detached shock, *Zh. Vychisl. Mat. i Mat. Fiz.*, **1** (1961), 6, 1020—1050 (in Russian).

[123] Babenko, K. I. and Rusanov, V. V., Difference methods for solution of spacial problems of gas dynamics, In "Trudy II Vsesoyuz. C'ezda po Teoret. i Prikl. Mekh., Vol. 2, Moscow, 1965", 247—262 (in Russian).

[124] Bohachevsky, I. O. and Rubin, E. L., A direct method for computation of nonequilibrium flows with detached shock waves, *AIAA J.*, **4** (1966), 4, 600—606 (or AIAA Paper 65-24).

[125] Moretti, G. and Abbett, M., A time-dependent computational method for blunt body flows, *AIAA J.*, **4** (1966), 12, 2136—2141.

[126] Burstein, S. Z., Finite difference calculations for hydrodynamic flows containing discontinuities, *J. Comput Phys.*, **1** (1966), 2, 198—222.

[127] Lapidus, A., A detached shock calculation by second order finite differences, *J. Comput. Phys.*, **2** (1967), 2, 154—177.

[128] Bastianon, R., Steady and unsteady solution of the flow field over concave bodies in supersonic free stream, AIAA Paper 68-946, 1968.

[129] Masson, B. S., Taylor, T. D. and Foster, R. M., Application of Godunov's method to blunt-body calculations, *AIAA J.*, **7** (1969), 4, 694—698.

[130] Moretti, G., Importance of boundary conditions in the numerical treatment of hyperbolic equations, *The Physics of Fluids*, **12** (1969), 12, Part II, II-13—II-20 (or PIBAL 68-34 (AD 681365)).

[131] Barnwell, R. W., Inviscid radiating shock layers about spheres traveling at hyperbolic speeds in air, NASA TR R-311, 1969.

[132] Taylor, T. D. and Masson, B. S., Application of the unsteady numerical method of Godunov to computation of supersonic flows past bell-shaped bodies, *J. Comput. Phys.*, **5** (1970), 3, 443—454.

[133] Vliegenthart, A. C., The Shuman filtering operator and the numerical computation of shock waves, *J. of Eng. Math.*, **4** (1970), 4, 341—348.

[134] Barnwell, R. W., Time-dependent numerical method for treating complicated blunt body flow fields, In: "Analytic Methods in Aircraft Aerodynamics, NASA SP-228, 1970", 177—195.

[135] Barnwell, R. W., A time-dependent method for calculating supersonic blunt body flow fields with sharp corners and embedded shock waves, NASA, TND-6031, 1970.

[136] Kyriss, C. L., A time-dependent solution for the blunt body flow of a chemically reacting gas mixture, AIAA Paper 70-771, 1970.

[137] Kosorukov, A. L., Flow around the nosed part of a blunt body with nonequilibrium excitation of vibration freedom, *Izv. AN SSSR, Mekh. Zhidk. i Gaza*, 1970, 1, 40—47 (in Russian).

[138] Belotserkovskii, O. M., Popov, F. D., Tolst'kh, A. I., Fomin, V. N. and Kholodov, A. S., Numerical solution of some problems of gas dynamics, *Zh. Vychisl. Mat. i Mat. Fiz.*, **10** (1970), 2, 401—416 (in Russian).

[139] Coakley, J. F. and Porter, R. W., Time-dependent numerical analysis of MHD blunt body problem, *AIAA J.*, **9** (1971), 8, 1624—1626 (or AIAA Paper 70-760).

[140] Belotserkovskii, O. M. and Davydov, Yu. M., The unsteady method of "large particles" for computation of gas dynamics, *Zh. Vychisl. Mat. i Mat. Fiz.*, **11** (1971), 1, 182—207 (in Russian).

[141] Davydov, Yu. M., Computation of flow around bodies of arbitrary shapes by the method of "large particles", *Zh. Vychisl. Mat. i Mat. Fiz.*, **11** (1971), 4, 1056—1063 (in Russian).

[142] D'Souza, N., Molder, S. and Moretti, G., Numerical method for hypersonic internal flow over blunt leading edges and two blunt bodies, *AIAA J.*, **10** (1972), 5, 617—622.

[143] Nichols, J. O., Artificial viscosity methods for blunt body flow field analysis with thermal radiation, *AIAA J.*, **10**, (1972), 6, 836—838.

[144] Bohachevsky, I. O. and Kostoff, R. N., Supersonic flow over convex and concave shapes with radiation and ablation effects, *AIAA J.*, **10** (1972), 8, 1024—1031 (or AIAA Paper 71-55, 1971).

[145] Harten, A. and Zwas, G., Self-adjusting hybrid schemes for shock computations, *J. of Comput. Phys.*, **9** (1972), 3, 568—583.

[146] Harten, A. and Zwas, G., Switched numerical Shuman filters for shock calculations, *J. of Eng. Math.*, **6** (1972), 2, 207—217.

[147] MacCormack, R. W. and Paullay, A. J., Computational efficiency achieved by time splitting of finite difference operators, AIAA Paper 72-154, 1972.

[148] Rao, P. P. and Lefferdo, J. M., Time-asymptotic solution for sphere cones in hypersonic flow, *AIAA J.*, **12** (1974), 3, 386—388.

[149] MacCormack, R. W. and Paullay, A. J., The influence of the computational mesh on accuracy for initial value problems with discontinuous or nonunique solutions, *Computers & Fluids*, **2** (1974), 314, 339—361.

[150] Belotserkovskii, O. M. and Davydov, Yu. M., A new numerical method for investigation of complicated problems of gasdynamics, In "Trudy Simpoziuma po Mekhanike Sploshnoi Sredy i Rodstvennym Probleman Analiza, Tbilici, II (1971), Metsniereba", Thilici,

1974, 58—73 (in Russian).
[151] Hsieh, T., Hemisphere-cylinder in low supersonic flow, *AIAA J.*, **13** (1975), 12, 1551—1552 (or AIAA Paper 75-83).
[152] Rusanov, V. V., Some properties of the axisymmetric gas flow about the powershape bodies, In "Lecture Notes in Physics, Vol.' **35**, Springer-Verlag, New York, 1975", 353—357.
[153] Krasil'nikov, A. V., Nikulin, A. N. and Kholodov, A. S., Some features of flow around sphere-blunted cones with large semi-vertex angles at hypersonic speed, *Izv. AN SSSR, Mekh. Zhidk. i Gaza*, 1975, 2, 179—181 (in Russian).
[154] Ivanova, V. N. and Radvogin, Yu. B., Some features of supersonic flow around sharp bodies with a detached shock, *Izv. AN SSSR, Mekh. Zhidk. i Gaza*, 1975, 4, 111—115. (in Russian).
[155] Radvogin, Yu. B., Parameters of determining the shape of subsonic section of a shock in an axisymmetric flow around a body, *Izv. AN SSSR, Mekh. Zhidk. i Gaza*, 1976, 4, 77—83 (in Russian).
[156] Lin, T. C., Reeves, B. L. and Siegelman, D., Blunt-body problem in nonuniform flow fields, *AIAA J.*, **15** (1977), 8, 1130—1137 (or AIAA Paper 76-354).
[157] Srinivas, K. and Gururaja, J., On the computation of two-dimensional compressible flow fields by the modified local stability scheme, *Computers and Fluids*, **5** (1977), 3, 139—150.
[158] Gilinskii, M. M. and Lebedev, M. G., On computation of a strong current jetting from a blunt body or a blunt profile, *Izv. AN SSSR, Mekh. Zhidk. i Gaza*, 1977, 1, 117—124 (in Russian).

The Time-Dependent Method for Three-Dimensional Problems

[159] Bohachevsky, I. O. and Mates, R. E., A direct method for calculation of the flow about an axisymmetric blunt body at angle of attack, *AIAA J.*, **4** (1966), 5, 776—782.
[160] Rusanov, V. V., Three-dimensional flow about an arbitrary blunt body, In: "Aerospace Proceedings, 1966, Vol. 1, Macmillan Co., London, 1966", 291—301.
[161] Moretti, G. and Bleich, G., Three-dimensional flow around blunt bodies, *AIAA J.*, **5** (1967), 9, 1557—1562 (or GASL, TR-637, 1966 (NASA N 68-17040)).
[162] Moretti, G., Bleich, B., Abbett, M. and Fort, R., Three dimensional inviscid flow about supersonic blunt cones at angle of attack, Vol. 2, USAEC-SC-CR-68-3728, Vol. 2, 1968.
[163] Xerikos, J. and Anderson, W. A., A time-dependent approach to the numerical solution of flow field about an axisymmetric vehicle at angle of attack, NASA N68-36114, 1968.
[164] Rusanov, V. V., Supersonic spacial flow around blunt bodies, *Zh. Vychisl. Mat. i Mat. Fiz.*, **8** (1968), 3, 616—633 (in Russian).
[165] Rusanov, V. V. and Liubimov, A. N., Studies of flow around blunt bodies by numerical methods, In: "Applied Mechanics, Proceedings of 12-th International Congress of Applied Mechanics, Stanford Uni., 1968, Springer-Verlag, New York, 1969", 356—363.
[166] Cohen, G. A., Foster, R. M. and Dowty, J. R., Synthesis of optimum structural designs for conical and tension shell mass entry capsules, NASA CR-1365, 1969.
[167] Lyubimov, A. N. and Rusanov, V. V., Gas flow around blunt bodies, Part I: Numerical methods and analysis of flow; Part II: Tables of gas dynamic functions, "Nauka", Moscow, 1970 (in Russian).
[168] Barnwell, R. W., Three-dimensional flow around blunt bodies with shoulders, AIAA Paper 71-56, 1971.
[169] Barnwell, R. W., A time-dependent method for calculating supersonic blunt bodies with sharp shoulders and smooth nonaxisymmetric blunt bodies, NASA TN D-6283, 1971.
[170] Barnwell, R. W. and Davis, R. M., A computer program for calculating inviscid, adiabatic flow about blunt bodies, traveling at supersonic and hypersonic speeds at angle of attack, NASA TMX-2334, 1971.
[171] Li, C. P., Time-dependent solutions of nonequilibrium dissociating flow past a blunt

body, *J. of Spacecraft and Rockets*, **8** (1971), 7, 812—814.

[172] Li, C. P., Time-dependent solutions of nonequilibrium air flow past a blunt body, *J. of Spacecraft and Rockets*, **9** (1972), 8, 571—572.

[173] Moretti, G., Grossman, B. and Marconi, F. Jr., A complete numerical technique for the calculation of three-dimensional inviscid supersonic flows, AIAA Paper 72—192, 1972.

[174] Porter, R. W. and Coakley, J. F., Use of characteristics for boundaries in time dependent finite difference analysis of multidimensional gas dynamics, *International Journal for Numerical Methods in Engineering*, **5** (1972), 1, 91—101.

[175] Lipnitskii, Yu. M., Mikhailov, Yu. Ya. and Sabinov, K. G., Computation of spacial flows of a perfect gas without a symmetric plane, *Izv. AN SSSR, Mekh. Zhidk. i Gaza*, 1972, 3, 182—186 (in Russian).

[176] Rizzi, A. W. and Inouye, M., Time-split, finite volume method for three-dimensional blunt-body flow, *AIAA J.*, **11** (1973), 11, 1478—1485 (or AIAA Paper 73-133).

[177] Li, C. P., Recent developments on time-dependent calculation of nonequilibrium flows, *J. of Spacecraft and Rockets*, **11** (1974), 2, 123—125.

[178] Belotserkovskii, O. M., Osetrova, S. D., Fomin, V. N. and Kholodov, A. S., Hypersonic flow of radiating gas around blunt bodies, *Zh. Vychisl. Mat. i Mat. Fiz.*, **14** (1974), 4, 992—1003 (in Russian).

[179] Rizzi, A. W. and Bailey, H. E., Reacting nonequilibrium flow around the space shuttle using a time-split method, In "NASA SP-347, Part II, 1975", 1327—1349.

[180] Förster, K., Roesner, K. and Weiland, C., The numerical solution of blunt body flow fields using Rusanov's method, In "Lecture Notes in Physics, Vol. 35, Springer-Verlag, New York, 1975", 167—175.

[181] Ivanov, M. Ya., On solution of two and three dimensional transonic flows around bodies, *Zh. Vychisl. Mat. i Mat. Fiz.*, **15** (1975), 5, 1222—1240 (in Russian).

[182] Savinov, K. G. and Shkadova, V. P., On application of the method of stabilization to non-equilibrium flow around a blunt body, *Izv. AN SSSR. Mekh. Zhidk. i Gaza*, 1976, 2, 140—145 (in Russian).

[183] Kostrykin, V. S., Fomin, V. N. and Kholodov, A. S., Spacial flow of a radiating gas around blunt cones and ellipsoids of revolution, *Zh. Vychisl. Mat. i Mat. Fiz.*, **16** (1976), 2, 451—459 (in Russian).

Special References B: Numerical Calculation of Conical Flow

Direct Methods

[1] Maslen, S. H., Supersonic conical flow, NACA TN-2651, 1952.

[2] Briggs, B. R., The calculation of supersonic flow past bodies supporting shock waves shaped like elliptic cones, NASA TN D-24, 1959.

[3] Briggs, B. R., The numerical calculation of flow past conical bodies supporting elliptic conical shock waves at finite angles of incidence, NASA TN D-340, 1960.

[4] Chushkin, P. I. and Schennikov, V. V., Computation of some non-axisymmetric conic flows, *Inzh. Fiz. zh.*, **3** (1960), 7, 88—94 (in Russian).

[5] Brook, J. W., The method of integral relations for conical flow——theoretical analysis, Grumman Aircraft Engineering Corporation, Research Department, RM-193, 1961 (AD 266501).

[6] Stocker, P. M. and Mauger, F. E., Supersonic flow past cones of general cross section, *J. Fluid Mech.*, **13** (1962), Part 3, 383—399.

[7] Babaev, D. A., Numerical solution of supersonic flow around the upper surface of a triangular wing, *Zh. Vychisl. Mat. i Mat. Fiz.*, **2** (1962), 278—289 (in Russian).

[8] Babaev, D. A., Numerical solution of supersonic flow around the lower surface of a triangular wing, *Zh. Vychisl. Mat. i Mat. Fiz.*, **2** (1962), 6, 1086—1101 (in Russian).

[9] Belotserkovskii, O. M. and Chushkin, P. I., The numerical method of integral relations, *Zh. Vychisl. Mat. i Mat. Fiz.*, **2** (1962), 5, 731—759 (in Russian).

[10] Kennet, H., The inviscid hypersonic flow on the wingward side of a delta wing, IAS Paper 63-55, 1963.

[11] Syagaev, V. F., A numerical method for solving supersonic flow around conic bodies, *Zh. Vychisl. Mat. i Mat. Fiz.*, **3** (1963), 4, 742—754 (in Russian).

[12] Brook, J. W., The calculation of non-linear supersonic conical flows by the method of integral relations, Grumman Aircraft Engineering Corp., FDL-TDR-64-7, 1964 (AD 605284).

[13] Melnik, W. L., Supersonic and hypersonic flow of an ideal gas about elliptic cones by method of integral relations, ph. D. Dissertation, University of Minnesota, 1964.

[14] Eastman, D. W. and Omar, M. E., Flow fields about highly yawed cones by the inverse method, *AIAA J.*, **3** (1965), 9, 1782—1784.

[15] Makhin, N. A. and Syagaev, V. F., Numerical solution of supersonic flow around conic bodies at angles of attack, *Izv. AN SSSR, Mekh. Zhidk. i Gaza*, 1966, 1, 140—142 (in Russian).

[16] Bazzhin, A. P. and Chelysheva, I. F., Applications of the method of lines to computation of flow around conic bodies at large angles of attack, *Izv. AN SSSR, Mekh. Zhidk. i Gaza*, 1967, 3, 119—123 (in Russian).

[17] Bazzhin, A. P., Trusova, O. N. and Chalysheva, I. F., Computation of a perfect gas flow around elliptic cones at large angles of attack, *Izv. AN SSSR, Mekh. Zhidk. i Gaza*, 1968, 4, 45—51 (in Russian).

[18] Jones, D. J., Numerical solutions of the flow field for conical bodies in a supersonic stream, National Research Council of Canada, Aeronautical Report, LR-507, 1968 (AD 686646).

[19] Jones, D. J., Tables of inviscid supersonic flow about circular cones at incidence, $\gamma=1.4$, AGARD-ograph-137, Parts I and II, 1969 (AD 698779).

[20] Ndefo, D. E., A numerical method for calculating steady unsymmetrical supersonic flow past cones, Rep. No. AS-69-11 (AFOSR Grant 268-68), U. S. Air Force, 1969 (AD 691270).

[21] South, J. C. Jr. and Klunker, E. B., Methods for calculating nonlinear conical flows, In: "Analytic Methods in Aircraft Aerodynamics, NASA SP-228, 1970", 131—158.

[22] Akinrelere, E. A., The calculation of inviscid hypersonic flow past the lower surface of a delta wing, *J. Fluid Mech.*, **44** (1970), 1, 113—127.

[23] Bazzhin, A. P., Some results of calculations of flows around conical bodies at large incidence angles, In: Proceedings of the Second International Conference on Numerical Methods in Fluid Dynamics,Univ.Calif.1970,Springer-Verlag,New-York,1971",223—229.

[24] Chushkin, P. I., Supersonic flows about conical bodies, *J. Comput. Phys.*, **5** (1970), 3, 572—586.

[25] Holt, M. and Ndefo, D. E., A numerical method for calculating steady unsymmetrical supersonic flow past cones, *J. Comput. Phys.*, **5** (1970), 3, 463—486.

[26] Klunker, E. B., South, J. C. Jr. and Davis, R. M., Calculation of nonlinear conical flows by the method of lines, NASA TR R-374, 1971.

[27] Jones, D. J., South, J. C. Jr. and Klunker, E. B., On the numerical solution of elliptic partial differential equations by the method of lines, *J. Comput. Phys.*, **9** (1972),3,496—527.

[28] Fletcher, C. A. J., Vortical singularity behind a highly yawed cone, *AIAA J.*, **13** (1975), 8, 1073—1078.

[29] Fletcher, C. A. J., GTT method applied to cones at large angles of attack, In "Lecture Notes in Physics, Vol. 35, Springer-Verlag, N. Y., 1975", 161—166.

[30] Nakao, S., Supersonic flow past conical bodies at large angle of attack, University of Tokyo, ISAS Report No. 534, 1975.

[31] Kopal, Z., Tables of supersonic flow around cones, Mass. Inst. Tech., Center of Analysis TR-1, 1947.

[32] Kopal, Z., Tables of supersonic flow around yawing cones, Mass. Inst. Tech., Center of Analysis, TR-3, 1947.

[33] Kopal, Z., Tables of supersonic flow around cones of large yaw, Mass. Inst. Tech., Center of Analysis, TR-5, 1949.

[34] Ferri, A., Supersonic flow around circular cones at angles of attack, NACA TN-2236,

1950.

[35] Ames Research Staff, Equations, Tables, and Charts for Compressible Flow, NACA R-1135, 1953.

[36] Kvashnina, S. S. and Korobeinikov, V. P., Solution of some flows of air with dissociation and ionization, *Izv. AN SSSR, Mekh. i Mashinostr.*, 1960, 2, 34—41 (in Russian).

[37] Kelly, P. D., Conical flow parameters for air and nitrogen in vibrational equilibrium, BRL, R-No: 1164, 1962 (AD 278178).

[38] Woods, B. A., The supersonic flow past a circular cone at incidence, Ministry of Aviation, Aeronautical Research Council, RM-3413, 1963.

[39] Hudgins, H. E. Jr., Supersonic flow about right circular cones at zero yaw in air at thermodynamic equilibrium, Part II. Tables of Data, Picatinny Arsenal, TM-1493 (AD 628539).

[40] Sims, J. L., Tables for supersonic flow around right circular cones at zero angle of attack, NASA SP-3004, 1964.

[41] Sims, J. L., Tables for supersonic flow around right circular cones at small angle of attack, NASA SP-3007, 1964.

[42] Woods, B. A., The supersonic flow past an elliptic cone, *The Aeronautical Quarterly*, **20** (1969), Part 4, 382—404.

[43] Vasil'ev, M. M., Supersonic flow around cones at angles of attack, *Izv. AN SSSR, Mekh. Zhidk. i Gaza*, 1970, 1, 33—39 (in Russian).

[44] Kurosaki, M., A study of supersonic conical flow, University of Tokyo, ISAS Report No. 498, 1973.

[45] Sims, J. L., Tables for supersonic flow of helium around right circular cones at zero angle of attack, NASA SP-3078, 1973.

The Method of Stabilization

[46] Babenko, K. I. and Voskresenskii, G. P., A numerical method for computing spacial supersonic flows around bodies, *Zh. Vychisl. Mat. i Mat. Fiz.*, **1** (1961), 6, 1051—1060 (in Russian).

[47] Babenko, K. I., Voskresenskii, G. P., Lyubimov, A. N. and Rusanov, V. V., Spacial flow of a perfect gas around smooth bodies, "Nauka", Moscow, 1964 (in Russian).

[48] Babenko, K. I., Investigation of a three-dimensional supersonic gas flow around conic bodies, In: "Applied Mechanics, Proceedings of 11-th Interna. Congress of Applied Mechanics, Munich, 1966", 747—755.

[49] Moretti, G. Inviscid flow field past a pointed cone at an angle of attack, Part I: Analysis, GASL, Inc. TR-577, 1965 (AD 632440).

[50] Gonidou, R., Écoulements supersoniques autour de cônes en incidence, *ONERA, La Recherche Aérospatiale*, No. 120, Sept. -Oct., 1967, 11—19.

[51] Moretti, G., Inviscid flow field about a pointed cone at an angle of attack, *AIAA J.*, **5** (1967), 4, 789—791.

[52] Voskresenskii, G. P., Numerical solution of supersonic flow around arbitrary triangular wings in the compression region, *Izv. AN SSSR, Mekh. Zhidk. i Gaza*, 1968, 4, 134—142 (in Russian).

[53] Beeman, E. R. Jr. and Powers, S. A., A method for determining the complete flow field around conical wings at supersonic/hypersonic speeds, AIAA Paper 69-646, 1969.

[54] Rakich, J. V., Application of the method of characteristics to noncircular bodies at angle of attack, In: "Analytic Methods in Aircraft Aerodynamics, NASA SP-228, 1970", 159—176.

[55] Kutler, P. and Lomax, H., Shock-capturing finite difference approach to supersonic flows, *J. of Spacecraft and Rockets*, **8** (1971), 12, 1175—1182 (or AIAA Paper 71-99).

[56] Power, S. A. and Beeman, E. R. Jr., Flow fields over sharp edged delta wings with attached shocks, NASA CR-1738, 1971.

[57] Lapygin, V. I., Computation of supersonic flow around V-type wings by the method of stabilization, *Izv. AN SSSR, Mekh. Zhidk. i Gaza*, 1971, 3, 180—185 (in Russian).

[58] Bachmanova, N. S., Lapygin, V. I. and Lipnitskii, Yu. M., Investigation of supersonic flow around circular cones at large angles of attack, *Izv. AN SSSR, Mekh. Zhidk. i Gaza*, 1973, 6, 79—84 (in Russian).

[59] Ivanov, M. Ya. and Kraiko, A. N., Computation of supersonic flow around conic bodies, *Zh. Vychisl. Mat. i Mat. Fiz.*, **13** (1973), 6, 1557—1572 (in Russian).

[60] Bazzhin, A. P., Flat slender delta wings in supersonic stream at small angles of attack, In "Lecture Notes in Physics, Vol. 35, Springer-Verlag, N. Y., 1975", 69—78.

[61] Kosykh, A. P. and Minailos, A. N., Computation of inviscid supersonic flow around pyramidic bodies, modelled on delta-type flying vehicles, *Izv. AN SSSR, Mekh. Zhidk. i Gaza*, 1975, 3, 105—111 (in Russian).

[62] Bachmanova, N. S. and Lipnitskii, Yu. M., On a similarity law of supersonic flow around sharp circular cones at large angles of attack, *Izv. AN SSSR, Mekh. Zhidk. i Gaza*, 1976, 3, 78—83 (in Russian).

[63] Daywitt, J. and Anderson, D., Supersonic flow about circular cones at large anagles of attack; a floating discontinuity approach, AIAA Paper 77-86.

Special References C: Numerical Calculation of Flow in Supersonic Regions

Two-Dimensional and Axisymmetric Flows

[1] Ferri, A., Application of the method of characteristics to supersonic rotational flow, NACA TN 1135, 1946.

[2] Casaccio, A., Theoretical pressure distribution on a hemisphere-cylinder combination, *J. Aero-Space Sci.*, **26** (1959), 1, 63—64.

[3] Ehlers, F. E., The method of characteristics for isoenergetic supersonic flows adapted to high-speed digital computers, *J. Soc. Industr. and Appl. Math.*, **7** (1959), 1, 85—100.

[4] Chushkin, P. I., Blunt bodies with simple shapes in supersonic flow, *Prikl. Mat. i Mekh.*, **24** (1960), 5, 927—930 (in Russian).

[5] Capiaux, R. and Karchmar, L., Flow past slender blunt bodies——a review and extension, IAS Paper 61-210-1904, 1961.

[6] Kennedy, E., Fields, A. and Seidman, M., The calculation of the flow field about a blunted 9° cone, GASL TR-256, Section I, 1961 (AD 425652).

[7] Vaglio-Laurin, R. and Trella, M., A study of flow fields about some typical blunt-nosed slender bodies, *Aerospace Eng.*, **20** (1961), 8, 20, 21 and 80—88.

[8] Katskova, O. N., Naumova, I. N., Shmyglevskii, Yu. D. and Shulishnina, N. P., Some experience of calculation of plane and axisymmetric supersonic flow by the method of characteristics, Vychisl. Tsentr AN SSSR, Moscow, 1961 (in Russian).

[9] Chushkin, P. I. and Shulishnina, N. P., Tables of supersonic flow around blunted cones, Vychisl. Tsentr AN SSSR, Moscow, 1961 (in Russian).

[10] Belotserkovskii, O. M. and Chushkin, P. I., Supersonic flow past blunt bodies, In: "Fluid Dynamics Transactions, Vol. I, Symposium, Jablonna (Poland), 1961, PWN, Warszawa, 1964", 161—187.

[11] Inouye, M. and Lomax, H., Comparison of experimental and numerical results for the flow of a perfect gas about blunt-nosed bodies, NASA TND-1426, 1962.

[12] Sedney, R., South, J. C. and Gerber, N., Characteristics calculation of nonequilibrium flows, In: "The High Temperature Aspects of Hypersonic Flow, Proceedings of the AGARD-NATO Specialists Meeting, Belgium, 1962, Pergamon Press, N. Y., 1964", 89—104.

[13] Traugatt, S. C., Some features of supersonic and hypersonic flow about blunted cones, *J. Aero-Space Sci.*, **29** (1962), 4, 389—399.

[14] Arkhipov, V. N. and Khoroshko, K. S., A method of computing "relaxation" in the problem of flow around cones, *Zh. Prikl. Mekh. i Tekhn. Fiz.*, 1962, 6, 121—124 (in Russian).

[15] Chushkin, P. I., Investigation of hypersonic flow around blunt bodies of revolution, *Zh. Vychisl. Mat. i Mat. Fiz.*, **2** (1962), 2, 255—277. (in Russian).

[16] Capiaux, R. and Washington, M., Nonequilibrium flow past a wedge, *AIAA J.*, **1** (1963), 3, 650—660.

[17] Powers, S. A. and O'Neill, J. B., Determination of hypersonic flow fields by the method of characteristics, *AIAA J.*, **1** (1963), 7, 1693—1694.

[18] Eastman, D. W. and Radke, L. P., Effect of nose bluntness on the flow around a typical ballistic shape, *AIAA J.*, **1** (1963), 10, 2401—2402.

[19] Sedney, R. and Geber, N., Nonequilibrium flow over a cone, *AIAA J.*, **1** (1963), 11, 2482—2486.

[20] South, J. C. Jr., Application of Dorodnitsyn's integral method to nonequilibrium flows over pointed bodies, NASA TND–1942, 1963.

[21] D'yakonov, Yu. N. and Zaitseva, N. A., Supersonic flow of a perfect gas around blunt bodies, *Izv. AN SSSR, Mekh. i Mashinostr.*, 1963, 1, 118—123. (in Russian).

[22] Katskova, O. N. and Kraiko, A. N., Computation of plane and axisymmetric supersonic flows with irreversible processes, *Zh. Prikl. Mekh. i Tekhn. Fiz.*, 1963, 4, 116—118 (in Russian).

[23] Naumova, I. N., Calculation of equilibrium supersonic flow of air, *Zh. Vychisl. Mat. i Mat. Fiz.*, **3** (1963), 5, 964—970 (in Russian).

[24] Roslyakov, G. S. and Drozdova, N. V., Numerical calculation of flow around "stepped" cones, In "Chisl. Metody v Gazovoi Dinamike, Vol. II, Mosk. Gosud. Uni., Moscow, 1963, 61—75 (in Russian).

[25] Ehlers, F. E., The method of characteristics for the supersonic flow of an ideal dissociating gas, I. Equilibrium flow, Boeing Sci. Res. Lab., D1–82–0364, 1964 (AD 607353).

[26] Katskova, O. N. and Chushkin, P. I., A scheme of numerical method of characteristics, *Dokl. AN SSSR*, **154** (1964), 1, 26—29 (in Russian).

[27] Katskova, O. N. and Kraiko, A. N., Calculation of plane and axisymmetric supersonic flow with irreversible processes, Vychisl. Tsentr AN SSSR, Moscow, 1964 (in Russion).

[28] Naumova, I. N., The method of characteristics for equilibrium flows of an imperfect gas, Vychisl. Tsentr AN SSSR, Moscow, 1964 (in Russian).

[29] South, J. C. Jr. and Newman, P. A., Supersonic flow past pointed bodies, *AIAA J.*, **3** (1965), 6, 1019—1021.

[30] Ehlers, F. E., Equilibrium flow of an ideal dissociating gas over a cone, *AIAA J.*, **3** (1965), 8, 1529—1530.

[31] South, J. C. Jr. and Newman, P. A., Application of the method of integral relations to real-gas flows past pointed bodies, *AIAA J.*, **3** (1965), 9, 1645—1652 (or AIAA Paper 65-27).

[32] Inouye, M., Rakich, J. V. and Lomax, H., A Description of numerical methods and computer programs for two dimensional and axi-symmetric supersonic flow over blunt-nosed and flared bodies, NASA TN D–2970, 1965.

[33] Newman, P. A., A modified method of integral relations for supersonic nonequilibrium flow over a wedge, NASA TN D–2654, 1965.

[34] Chushkin, P. I. and Li, L. K., Determination of parameters of two dimensional supersonic flows, *Zh. Vychisl. Mat. i Mat. Fiz.*, **5** (1965), 1, 57—66 (in Russian).

[35] Ehlers, F. E., The method of characteristics for supersonic flow of an ideal dissociating gas, II. Nonequilibrium flow, Boeing Sci. Res. Lab., D1–82–D548, 1966 (AD 640381).

[36] Spurk, J. H., Gerber, N. and Sedney, R., Characteristic calculation of flow fields wtih chemical reactions, *AIAA J.*, **4** (1966), 1, 30—37.

[37] Lunev, V. V., Pavlov, V. G. and Sinchenko, S. G., Hypersonic flow of dissociating air in equilibrium around spheres, *Zh. Vychisl. Mat. i Mat. Fiz.*, **6** (1966), 1, 121—129 (in Russian).

[38] Dushin, V. K., Applications of the method of characteristics to computation of supersonic external flow with non-equilibrium processes, In "Numerical methods for solution of problems in mathematical physics, "Nauka", Moscow, 1966", 194—199 (in Russian).

[39] Zapryanov, Z. D., Investigation of supersonic flow around various axisymmetric bodies, *Izv. AN SSSR, Mekh. Zhidk. i Gaza*, 1967, 2, 121—125 (in Russian).

[40] Fomin, V. N. and Shulishnina, N. P., Supersonic flow with radiation around blunt cones, *Zh. Vychisl. Mat. i Mat. Fiz.*, **7** (1967), 4, 933—937 (in Russian).

[41] D'yakonov, Yu. N., Pchelkina, L. V. and Sandomirskaya, I. D., Computation of two dimensional flow of equilibrium and perfect gases by a difference method, In "Vychisl. Metody i Programmir, Vol. XI, Mosk. Gosud. Uni., 1968", 113—122 (in Russian).

[42] Roslyakov, G. S. and Telenin, G. F., Survey of numerical researches on external and internal problems of aerodynamics, completed in Moscow University, In "Vychisl. Metody i Programmir, Vol. XI, Mosk. Gosud. Uni., 1968", 93—112 (in Russian).

[43] Roslyakov, G. S. and Drozdova, N. V., Supersonic flow around stepped cones, In "Vychisl. Metody i Programmira., Vol. XI, Mosk. Gosud. Uni., 1968", 123—138 (in Russian).

[44] Severinov, L. I., Calculation of the supersonic part of disturbed region in supersonic nonequilibrium flow around blunt bodies, *Zh. Vychisl. Mat. i Mat. Fiz.*, **8** (1968), 3, 634—646 (in Russian).

[45] Rusanov, V. V., Numerical investigation of an axisymmetric flow around long bodies, *The Phys. Fluids*, **12** (1969), 12, Part II, II-126—II-129.

[46] Dushin, V. K., On the solution of system of relaxation equations in calculating flow of a reacting gaseous mixture nearly in equilibrium, *Zh. Vychisl. Mat. i Mat. Fiz.*, **9** (1969), 5, 1121—1136 (in Russian).

[47] Chushkin, P. I., Supersonic flow of combustible gases around bodies, *Fiz. Goreniya i Vzryva*, **5** (1969), 2, 230—235 (in Russian).

[48] Thompson, R. H. and Furey, R. J., Numerical determination of the flow field about axisymmetric and two-dimensional bodies in supersonic flow, Naval Ship Research and Development Center, Report-3032, 1970 (AD 713917).

[49] Kutler, P. and Lomax, H., Shock-capturing finite-difference approach to supersonic flows, *J. of Spacecraft and Rockets*, **8** (1971), 12, 1175—1182 (or AIAA Paper 71-99).

[50] Rusanov, V. V. and Liubimov, A. N., On interior shock waves in gas flow about blunted cones, In: "Fluid Dynamics Transactions, Vol. 6, Part II, PWN. Polish Scientific Publishers, Warszawa, 1971", 517—524.

[51] Abbett, M. J., Boundary condition computational procedures for inviscid, supersonic steady flow field calculations, Aerotherm Report 71-41, 1971.

[52] Walkden, F. and Caine, P., Application of a pseudo-viscous method to the calculation of the steady supersonic flow past a waisted body, *International Journal for Numerical Methods in Engineering*, **5** (1972), 2, 151—162.

[53] Moretti, G. and Pandolfi, M., Entropy layers, *Computers & Fluids*, **1** (1973), 1, 19—35.

[54] Abbett, M. J., Boundary condition calculation procedures for inviscid supersonic flow fields, In "Proc. AIAA Computational Fluid Dynamics Conference, California, 1973", 153—172.

[55] Lebedev, M. G., Pchelkina, L. V. and Sandomirskaya, I. D., Numerical investigation of supersonic flow around plane blunt bodies, In "Vychisl. Metody i Programmir., Vol. XXIII, Mosk. Gosud. Uni., 1974", 126—140 (in Russian).

[56] Vishnevetskii, S. L. and Pakhomova, Z. S., Hypersonic flow around sharp and blunt bodies with a concave generator, *Izv. AN SSSR, Mekh. Zhidk. i Gaza*, 1975, 1, 176—180. (in Russian).

[57] Moretti, G., Circumspect exploration of multidimensional imbedded shocks, *AIAA J.*, **14** (1976), 7, 894—899.

[58] Znamenskii, V. V., Calculation of hypersonic flow around bodies with a cuspidal generator, considering two dimensional transfer of radiation, *Izv. AN SSSR, Mekh. Zhidk. i Gaza*, 1976, 2, 114—121 (in Russian).

[59] Rusanov, V. V., Advanced techniques for computation of supersonic flows, AIAA Paper 77-173, 1977.

Three-Dimensional Flow

[60] Moretti, G., A technique for computing three-dimensional steady inviscid supersonic flows, GASL, Inc. TR-172, 1960 (AD 256179).

[61] Fowell, L. R., Flow field analysis for lifting re-entry configurations by the method of

characteristics, IAS Paper 61-208-1902, 1961.

[62] Babenko, K. I. and Voskresenskii, G. P., A numerical method for computing spacial supersonic flow around bodies, *Zh. Vychisl. Mat. i Mat. Fiz.*, **1** (1961), 6, 1051—1060 (in Russian).

[63] Moretti, G., Sanlorenzo, E. A., Magnus, E. E. and Weilerstein, G., Supersonic flow about general three-dimensional bodies, Vol. III, Flow field analysis of re-entry configurations by a general three-dimensional method of characteristics, ASD-TR-61-727, Vol. III, 1962 (AD 292328).

[64] Belotserkovskii, O. M. and Chushkin, P. I., The numerical method of integral relations, *Zh. Vychisl. Mat. i Mat. Fiz.*, **2** (1962), 5, 731—759 (in Russian).

[65] Moretti, G., Three-dimensional supersonic flow computations, *AIAA J.*, **1** (1963), 9, 2192—2193.

[66] Babenko, K. I., Molchanov, A. M., Rusanov, V. V. and Shnol', E. E., Methods for solving some two dimensional problems, In "Some problems in computational mathematics and computational technique, Mashgiz., Moscow, 1963", 99—103 (in Russian).

[67] Mikhailov, V. N, Calculation of spacial rotational supersonic flow in the neighborhood of a curve on which cusps of streamlines occur, *Prikl. Mat. i Mekh.*, **27** (1963), 6, 1083—1089 (in Russian).

[68] Burnat, M., Kielbasinski, A. and Wakulicz, A., The method of characteristics for a multidimensional gas flow, *Archiwum Mechaniki Stosowanej*, Warszawa, **16** (1964), 2, 179—188.

[69] Saverwein, H. and Sussman, M., Numerical stability of the three-dimensional method of characteristics, *AIAA J.*, **2** (1964), 2, 387—389.

[70] Chu, C.-W., A simple derivation of three dimensional characteristic relations, *AIAA J.*, **2** (1964), 7, 1336—1337.

[71] Rusanov V. V., A three-dimensional supersonic gas flow past smooth blunt bodies, In: "Applied Mechanics, Proc. 11 Internat. Congr. Appl. Mech., Munich, 1964, Springer-Verlag, New York, 1966", 774—778.

[72] Babenko, K. I., Voskresenskii, G. P., Lyubimov, A. N. and Rusanov V. V., Spacial flow of a perfect gas around smooth bodies, "Nauka", Moscow, 1964 (in Russian).

[73] D'yakonov, Yu. N., Spacial flow around blunt bodies with physical and chemical reactions in equilibrium, *Dokl. AN SSSR*, **157** (1964), 4, 822—825 (in Russian).

[74] D'yakonov, Yu. N., Spacial supersonic flow of a perfect gas around blunt bodies, *Izv. AN SSSR, Mekh. i Mashinostr.*, 1964, 4, 150—153 (in Russian).

[75] Zapryanov, Z. D. and Minostsev, V. B., A numerical method for spacial supersonic flow around bodies, *Izv. AN SSSR, Mekh. i Mashinostr.*, 1964, 5, 20—24 (in Russian).

[76] Chushkin, P. I., Numerical solution of some flow problems, *Arch. Mech. Stosow.*, Warszawa, **16** (1964), 3, 577—596 (in Russian).

[77] Pridmore Brown, B. N. and Franks, W. J., A method of characteristics solution in three independent variables, Aerospace Res. Lab., USAF, Report ARL 65-124, 1965 (AD 623405).

[78] Katskova, O. N. and Chushkin, P. I., Three dimensional supersonic equilibrium flow around bodies at angles of attack, *Zh. Vychisl. Mat. i Mat. Fiz.*, **5** (1965), 3, 503—518 (in Russian).

[79] Podladchikov, Yu. N., Calculation of spacial supersonic flow by the method of characteristics, *Dokl. AN SSSR*, **163** (1965), 5, 1092—1095 (in Russian).

[80] Podladchikov, Yu. N., A method of characteristics for computing spacial supersonic flow, *Izv. AN SSSR, Mekh.*, 1965, 4, 3—12 (in Russian).

[81] Magomedov, K. M., A method of characteristics for numerical calculation of spacial flow, *Zh. Vychisl. Mat. i Mat. Fiz.*, **6** (1966), 2, 313—325 (in Russian).

[82] Magomedov, K. M., On calculation of unknown surfaces in the spacial method of characteristics, *Dokl. AN SSSR*, **171** (1966), 6, 1297—1300 (in Russian).

[83] Chu, C.-W., Compatibility relations and a generalized finite-difference approximation for three dimensional steady supersonic flow, *AIAA J.*, **5** (1967), 3, 493—501.

[84] Rakich, J. V., Three-dimensional flow calculation by the method of characteristics, *AIAA J.*, **5** (1967), 10, 1906—1908.

[85] Strom, C. R., The method of characteristics for three-dimensional real gas flows, Technical Rep. AFFDL-TR-67-47, 1967 (AD 661342).

[86] Voskresenskii, G. P., Calculation of supersonic flow around sharp non-axisymmetrical bodies, *Zh. Vychisl. Mat. i Mat. Fiz.*, **7** (1967), 2, 389—400 (in Russian).

[87] Minostsev, V. B., A method for computing supersonic three dimensional flow around smooth bodies, *Izv. AN SSSR, Mekh. Zhidk. i Gaza*, 1967, 2, 126—133 (in Russian).

[88] Katzkova, O. N. and Chushkin, P. I., Numerical calculation of supersonic flow around bodies with "tubes" at angles of attack, *Izv. AN SSSR, Mekh. Zhidk. i Gaza*, 1967, 3, 124—130 (in Russian).

[89] Magomedov, K. M., Computation of spacial flow around blunt bodies with physical and chemical reactions in equilibrium by the method of characteristics, *Izv. AN SSSR, Mekh. Zhidk. i Gaza*, 1967, 3, 130—137 (in Russian).

[90] Magomedov, K. M., and Kholodov, A. S., On supersonic spacial flow around a triangular wing with blunt leading edges, *Izv. AN SSSR, Mekh. Zhidk. i Gaza*, 1967, 4, 159—163 (in Russian).

[91] Rusanov, V. V. and Lyubimov, A. N., Studies of flow around blunt bodies by numerical methods, In: "Applied Mechanics, Proceedings of 12-th International Congress of Applied Mechanics, Stanford Univ., 1968, Springer-Verlag, New York, 1969", 356—363.

[92] Lunev, V. V., Magomedov, K. M. and Povlov, V. G., Hypersonic flow around blunt bodies with physical and chemical reactions in equilibrium, Vychisl. Tsentr AN SSSR, Moscow, 1968 (in Russian).

[93] Minostsev, V. B., Telenin, G. F. and Tinyakov, G. P., Investigation of pictures of supersonic spacial flow around a "segmental body", *Dokl. AN SSSR*, **179** (1968), 2, 304—307 (in Russian).

[94] Chushkin, P. I., The method of integral relations for supersonic spacial flows, *Zh. Vychisl. Mat. i Mat. Fiz.*, **8** (1968), 4, 853—864 (in Russian).

[95] Katskova, O. N. and Chushkin, P. I., Spacial supersonic nonequilibrium flow, *Zh. Vychisl. Mat. i Mat. Fiz.*, **8** (1968), 5, 1049—1062 (in Russian).

[96] Rakich, J. V., A Method of characteristics for steady three-dimensional supersonic flow with application to inclined bodies of revolution, NASA, TND-5341, 1969.

[97] Rakich, J. V., Application of the method of characteristics to noncircular bodies at angle of attack, In: "Analytic Method in Aircraft Aerodynamics, NASA SP-228, 1969", 159—176.

[98] Lyubimov, A. N., On some effects in spacial supersonic flows around blunt bodies, *Izv. AN SSSR, Mekh. Zhidk. i Gaza*, 1969, 2, 85—91 (in Russian).

[99] Magomedov, K. M. and Kholodov, A. S., On construction of difference schemes based on characteristic relations for hyperbolic equations, *Zh. Vychisl. Mat. i Mat. Fiz.*, **9** (1969), 2, 371—386 (in Russian).

[100] Grigor'ev, E. I. and Magomedov, K. M., On one direct method of characteristics for computing spacial flow, *Zh. Vychisl, Mat. i Mat. Fiz.*, **9** (1969), 6, 1413—1419 (in Russian).

[101] Chushkin, P. I., Combustion in supersonic flows around various bodies, *Zh. Vychisl. Mat. i Mat. Fiz.*, **9** (1969), 6, 1367—1377 (in Russian).

[102] Chushkin, P. I., Supersonic flows about conical bodies, *J. Comput. Phys.*, **5** (1970), 3, 572—586.

[103] Kutler, P. and Lomax, H., The computation of supersonic flow fields about wing-body combinations by "shock-capturing" finite difference, In: "Proceedings of the Second International Conference on Numerical Methods in Fluid Dynamics, Univ. Calif., 1970, Spinger-Verlag, New York, 1971", 24—29.

[104] Antonets, A. V., Calculation of spacial supersonic equilibrium and frozen flows in shock layers around blunt bodies with cuspidal generators, *Izv. AN SSSR, Mekh. Zhidk. i Gaza*, 1970, 2, 178—181 (in Russian).

[105] Katskova, O. N. and Chushkin, P. I., Influence of nonequilibrium dissociation on supersonic spacial flow around inverse cones, *Izv. AN SSSR, Mekh. Zhidk. i Gaza*, 1970, 2, 182—184 (in Russian).

[106] D'yakonov, Yu. N., Pchelkina, L. V., Sandomirskaya, I. D. and Uskov, V. I.,

Calculation of spacial inviscid flow by a difference method, In: "Vychisl. Metody i Programmir., Vol. XV, Mosk. Gosud. Uni., 1970", 72—84 (in Russian).

[107] D'yakonov, Yu. N., et al., Spacial flow around blunt bodies with complicated shapes, In: "Vychisl. Metody i Programmir., Vol. XV, Mosk. Gosud. Uni., 1970", 85—**91** (in Russian).

[108] Lyubimov, A. N. and Rusanov, V. V., Gas flows around blunt bodies, Pts. 1 and 2, Moscow, "Nauka", 1970 (in Russian).

[109] Kosarev, V. I., On calculation of supersonic steady flow with inner jumps of "condensation", *Zh. Vychisl. Mat. i Mat. Fiz.*, **11** (1971), 5, 1262—1271 (in Russian).

[110] Thomas, P. D., Vinokur, M., Bastianon, R. A. and Conti, R. J., Numerical solution for three-dimensional inviscid supersonic flow, *AIAA J.*, **10** (1972), 7, 887—894 (or AIAA Paper 71-596).

[111] Chu, C.-W., A new algorithm for three-dimensional method of characteristics, *AIAA J.*, **10** (1972), 11, 1548—1550.

[112] Lewis, C. H. and Black, R. R., Blunt cones at angle of attack in supersonic nonuniform freestreams, *J. of Spacecraft and Rockets*, **9** (1972), 8, 561—562.

[113] Schiff, L. B., Computation of supersonic flow fields about bodies in coning motion using a shock-capturing finite difference technique, AIAA Paper 72-27, 1972.

[114] Rakich, J. V. and Kutler, P., Comparison of characteristics and shock capturing methods with application to the space shuttle vehicle, AIAA Paper 72-191, 1972.

[115] Moretti, G., Grossman, B. and Marconi, F. Jr., A complete numerical technique for the calculation of three-dimensional inviscid supersonic flows, AIAA Paper 72-192, 1972.

[116] Kutler, P., Lomax, H. and Warming, R. F., Computation of space shuttle flow fields using non-centered finite-difference schemes, AIAA Paper 72-193, 1972.

[117] Chu, C. W. and Powers, S. A., Determination of space shuttle flow field by the three-dimensional method of characteristics, In: "NASA TMX-2506, 1972", 47—63.

[118] Kutler, P., Rakich, J. V. and Mateer, G. G., Application of shock capturing and characteristics methods to shuttle flow fields. In: "NASA TMX-2506, 1972", 65—92.

[119] Burdel'nyi, A. K. and Minostsev, V. B, Computation of supersonic region of spacial flow of nonequilibrium air around bodies, *Izv. AN SSSR, Mekh. Zhidk. i Gaza*, 1972, 5, 124—129 (in Russian).

[120] D'yakonov, Yu. N., Pchelkina, L. V. and Sandomirskaya, I. D., On computation of supersonic flow around bodies at large angles of attack, In: "Vychisl. Metody i Programmir., Vol. XIX, Mosk. Gosud. Uni., 1972", 64—70 (in Russian).

[121] Warming, R. F., Kutler, P. and Lomax, H., Second-and third-order noncentered difference schemes for nonlinear hyperbolic equations, *AIAA J.*, **11** (1973), 2, 189—196 (or AIAA Paper 72-193, Part II).

[122] Kutler, P., Warming, R. F. and Lomax, H., Computation of space shuttle flow fields using noncentered finite-difference schemes, *AIAA J.*, **11** (1973), 2, 196—204 (or AIAA Paper 73-193, Part III).

[123] Kutler, P., Reinhardt, W. A. and Warming, R. F., Multishocked three-dimensional supersonic flow fields with real gas effects, *AIAA J.*, **11** (1973), 5, 657—669 (or AIAA Paper 72-702).

[124] Marconi F. and Salas, M., Computation of three dimensional flows about aircraft configurations, *Computers & Fluids*, **1** (1973), 2, 185—195.

[125] Black, R. R., Fireders, M. C. and Lewis, C. H., A computerized analysis of supersonic nonuniform flows over sharp and spherically blunted cones at angle of attack, *Computers & Fluids*, **1** (1973), 4, 359—365.

[126] Chu, C.-W., Supersonic flow about slab delta wings and wing-body configurations, *J. of Spacecraft and Rockets*, **10** (1973), 11, 741—742.

[127] MacCormack, R. W. and Warming, R. F. Survey of computational methods for three-dimensional supersonic inviscid flows with shocks, In: "Advances in Numerical Fluid Dynamics, AGARD Lecture Series 64, 1973".

[128] Rakich, J. V. and Park, C., Nonequilibrium three dimensional supersonic flow computations with application to the space shuttle orbiter design, In: "Two-Day Symposium on Application of Computers to Fluid Dynamics Analysis and Design, 1973". 205—212.

[129] Moretti, G., Experiments in multi-dimensional floating shock fitting, PIBAL Report 73-18, 1973 (AD 768841).

[130] Kosarev, V. I. and Magomedov, K. M., A "divergent" difference scheme for computing supersonic steady flows with complicated structure, *Zh. Vychisl. Mat. i Mat. Fiz.*, **13** (1973), 4, 923—937 (in Russian).

[131] Kutler, P., Supersonic flow in the corner formed by two intersecting wedges, *AIAA J.*, **12** (1974), 5, 577—578 (or AIAA Paper 73-675).

[132] Walkden, F., Laws, G. T. and Caine, P., Shock capturing numerical method for calculating supersonic flows, *AIAA J.*, **12** (1974), 5, 642—647.

[133] Moretti, G., Three-dimensional, supersonic, steady flows with any number of imbedded shocks, AIAA Paper 74-10, 1974.

[134] D'attorre, L., Bilyk, M. A. and Sergeant, R. J., Three-dimensional supersonic flow field analysis of the B-1 airplane by a finite difference technique and comparison with experimental data, AIAA Paper 74-189, 1974.

[135] Walkden, F. and Caine, P., A shock capturing method for calculating supersonic flow fields, Aeronautical Res. Council CP-1290, 1974.

[136] Antonets, A. V., Hypersonic flow of nonequilibrium air around blunt bodies, *Izv. AN SSSR, Mekh. Zhidk. i Gaza*, 1974, 2, 114—120 (in Russian).

[137] Shankar, V., Anderson, D. and Kutler, P., Numerical solutions for supersonic corner flow, *J. Comput. Phys.*, **17** (1975), 2, 160—180.

[138] Shankar, V. S. V., Numerical solutions for inviscid supersonic corner flows, AIAA Paper 75-221, 1975.

[139] Rakich, J. V., Bailey, H. E. and Park, C., Computation of nonequilibrium three dimensional inviscid flow over blunt-nosed bodies flying at supersonic speeds, AIAA Paper 75-835, 1975.

[140] Hsieh, T., Low supersonic three-dimensional flow about a hemisphere-cylinder, AIAA Paper 75-836, 1975.

[141] Kutler, P., Shankar, V., Anderson D. A. and Sorenson, R. L., Internal and external axial corner flows, In: "NASA SP-347, Part I, 1975", 643—658.

[142] Davy, W. C. and Reinhardt, W.A., Computation of shuttle nonequilibrium flow fields on a parallel processor, In: "NASA SP-347, Part II, 1975", 1351—1376.

[143] Marconi, F., Yaeger, L. and Hamilton, H. H., Computation of high-speed inviscid flows about real configurations, In: "NASA SP-347, Part II, 1975", 1411—1455.

[144] Rizzi, A. W., Klavins, A. and MacCormack, R. W., A generalized hyperbolic marching technique for three-dimensional supersonic flow with shocks, In: "Lecture Notes in Physics, Vol. 35, Springer-Verlag, New York, 1975", 341—346.

[145] Sanders, B.R. and Dwyer, H. A., Magnus forces on spinning supersonic cones, Part II: inviscid flow, *AIAA J.*, **14** (1976), 5, 576—582.

[146] Rizzi, A. W. and Bailey, H. E., Split space-marching finite-volume method for chemically reacting supersonic flow, *AIAA J.*, **14** (1976), 5, 621—628.

[147] Camarero, R., A reference-plane method for the solution of three-dimensional supersonic flows, *The Aeronautical Quarterly*, **27** (1976), Part 1, 75—86.

[148] Chaussee, D.S., Kutler, P. and Holtz, T., Inviscid supersonic/hypersonic-body flow field and aerodynamics from shock-capturing technique calculations, *J. of Spacecraft and Rockets*, **13** (1976), 6, 325—331 (or AIAA Paper 75-837).

[149] Antonets, A. V. and Lipnitskii, Yu. M., Investigation of supersonic flow around lengthened blunt bodies with an elliptic cross section, *Izv. AN SSSR, Mekh. Zhidk. i Gaza*, 1976, 6, 155—159 (in Russian).

[150] Solomon, J. M., Ciment, M., Ferguson, R. E. and Bell, J. B., Three dimensional supersonic inviscid flow field calculations on re-entry vehicles with control surfaces, AIAA Paper 77-84, 1977.

[151] Coburn, N. and Dolph, C. L., The method of characteristics in the three dimensional stationary supersonic flow of a compressible gas, In: "Proceedings of Symposia in Applied Mathematics, Vol. I, American Math. Soc., N. Y., 1949", 55—66.

[152] Thornhill, C. K., The numerical method of characteristics for hyperbolic problems in

three independent variables, Aeronautical Research Council, R. & M., No. 2615, London, 1952.

[153] Holt, M., The method of characteristics for steady supersonic rotational flow in three dimensions, *J. Fluid Mech.*, **1** (1956), Part 4, 409—423.

[154] Tsung, C. C., Study of three-dimensional supersonic flow problems by a numerical method based on the method of characteristics, Ph. D. Thesis, Univ. Illinois, 1961.

[155] Sauer, R., Differenzenverfahren für hyperbolische anfangswertprobleme bei mehr als zwei unabhängigen veränderlichen mit hilfe von nobencharakteristiken, *Numerische Mathematik*, **5** (1963), 1, 55—67.

Subject Index